Theoretical and Mathematical Physics

The series founded in 1975 and formerly (until 2005) entitled *Texts and Monographs in Physics* (TMP) publishes high-level monographs in theoretical and mathematical physics. The change of title to *Theoretical and Mathematical Physics* (TMP) signals that the series is a suitable publication platform for both the mathematical and the theoretical physicist. The wider scope of the series is reflected by the composition of the editorial board, comprising both physicists and mathematicians.

The books, written in a didactic style and containing a certain amount of elementary background material, bridge the gap between advanced textbooks and research monographs. They can thus serve as basis for advanced studies, not only for lectures and seminars at graduate level, but also for scientists entering a field of research.

Errico Presutti

Scaling Limits in Statistical Mechanics and Microstructures in Continuum Mechanics

With 23 Figures

Springer

Professor Errico Presutti
Dipartimento di Matematica
Università di Roma Tor Vergata
Via della Ricerca Scientifica 1
00133 Roma
Italy
presutti@mat.uniroma2.it

ISBN 978-3-642-09236-7 e-ISBN 978-3-540-73305-8

DOI 10.1007/978-3-540-73305-8

Theoretical and Mathematical Physics ISSN 1864-5879

Mathematics Subject Classification (2000): 82B26, 49Q20, 74A15

Softcover reprint of the hardcover 1st edition 2009

Cover design: eStudio Calamar S.L.

Printed on acid-free paper

9 8 7 6 5 4 3 2 1

springer.com

Preface

I began this book about eight years ago after Antonio De Simone, Stephan Luckhaus and Stefan Müller asked me to give a series of lectures on statistical mechanics at the Max Planck Institute in Leipzig. I wrote some notes and after many attempts to make them more readable a book finally came out.

The way a continuum description emerges from atomistic models is an intriguing and fascinating subject, which is behind most of my scientific life, in particular the analysis of large scale phenomena in statistical mechanics and their mathematical formulation in terms of thermodynamic and hydrodynamic limits. The theory has remarkably progressed in the last decades with contributions from many different areas of mathematics and physics, and when asked to give in Leipzig an overview of the state of the art I accepted with great pleasure.

There is the reverse direction as well, where starting from a continuum description after successive blow-ups we find microstructures and the prodromes of an underlying microscopic world. Even though the two directions from microscopics to macroscopics and vice versa are in principle symmetric, mathematical techniques and ideas have proceeded quite separately. I discovered during my lectures that there was a great interest in the audience, whose background was mostly in analysis and continuum mechanics, to learn methods and procedures of statistical mechanics. At the same time it became clear to me that notions and theories developed in continuum mechanics and PDEs had direct implications on my research in statistical mechanics and, to state it succinctly, I felt excited by the idea of building bridges between the two areas, and this book is certainly part of such efforts. Since the book is meant for both communities, it is written with the presumption that the readers may not be expert on all topics; thus the analysis starts from the beginning with parts which are rather elementary and open to younger researchers entering in the field. Statistical mechanics is exemplified in Part I of the book in the context of the Ising model, to avoid technical problems about unboundedness of the variables. Part II is devoted to the mesoscopic theory, which is presented by studying a non-local version of the classical scalar Ginzburg–Landau functional, namely the L–P free energy functional introduced by Lebowitz and Penrose in their analysis of Kac potentials. Part III is more specialistic; it shows how variational methods characteristic of the mesoscopic theory can be used to implement the Pirogov–Sinai theory of phase transitions in lattice and continuum models with Kac potentials.

The choice of working on particular models (Ising, the L–P functional) reflects the didactic purposes of the presentation as well as the taste of the author. I have used parts of the book for lectures in schools and for theses. The various topics can be easily singled out to be used for courses or lectures as I have tried to make the parts not too strongly correlated. Chapter 1 is an introduction essentially based on some survey lectures I gave in the last years, and it hopefully gives a first idea of flavor and content of the book, without entering into too many details.

Part I is about the statistical mechanics of the Ising model. Besides the basic elements of the theory I have also included a more advanced part on the structure of the DLR (Dobrushin, Lanford and Ruelle) measures as a corollary of the Rohlin theory of conditional probabilities for Lebesgue measures, which is explained in some detail. This is a very beautiful and instructive piece of mathematics; it has been fundamental in my education, and for this reason I am fond of it and felt that I had to insert it in the book. Using the theory of DLR measures as a technical tool, I have then shown how the Boltzmann hypothesis that the entropy is proportional to the log of the number of states allows one to derive the thermodynamic potentials. With the help of DLR theory, it is possible to implement Cramer's large deviation methods to prove the existence of the thermodynamic potentials. This is not the traditional way followed in statistical mechanics, but it has the advantage to underline connections with other fields like probability and information theory. All this is in Chap. 2. With Chap. 3 begins the analysis of phase transitions. Still, in the context of the Ising model I discuss here the basic theorems about existence and non-existence of phase transitions, in particular the "Peierls argument" and the "Dobrushin uniqueness theorem," which have fundamental importance in the whole book. Chapter 4 is about mean field and Kac potentials; here scalings, coarse graining and free energy functionals appear for the first time: this is the beginning of the bridge towards mesoscopic and continuum theories. Chapter 5 is about stochastic dynamics; it is in a sense a detour from the main line, but dynamics enters too much into the microscopic and macroscopic theories that I could not leave it completely out of the book. To give a flavor of the research in this field I have presented a derivation of the macroscopic limit evolution for Glauber dynamics with Kac potentials, but I have also inserted a still elementary part where stochasticity persists in the limit, discussing spinodal decomposition and tunneling for the mean field interaction.

Part II is devoted to the mesoscopic theory. For readers whose main interest is in continuum mechanics this could be where to start the book with a "smooth introduction towards statistical mechanics." The presentation is in fact self-contained; motivations for the choice of the free energy functionals come from statistical mechanics, but if the reader accepts the functionals as primitive notions, then references to Part I are not necessary. Chapter 6 starts with a derivation of the thermodynamic potentials, which, in the context of the mesoscopic theory, amounts to the study of some variational problems with constraints. The analysis parallels the one in Chap. 2 for the Ising model and it is certainly instructive to see the two in perspective. I then consider dynamics studying a non-local version of the Allen–Cahn equation related to the L–P free energy functional. As in reaction–diffusion equations, properties like the "Barrier Lemma" and the Comparison Theorem are proved. They are then used to study "large deviations" which, in the mesoscopic theory, refer to estimates of the free energy cost of excursions away from equilibrium. Here contours are introduced and Peierls estimates are proved, in analogy with the corresponding notions and results in the Ising model. As mentioned before, this chapter could be seen as an introduction to statistical mechanics and Part I could be read right after this. In Chap. 7 a first application of large deviation estimates are presented by studying surface tension and Wulff problems for the L–P functional. The basic notions here are Gamma convergence and geometric measure theory, which in fact play a fundamental role; yet some of their basic theorems are here recalled without proofs. In the original plan of the book I had in mind to insert in Part III the statistical mechanics analogue in the context of the LMP particle model introduced by Lebowitz, Mazel and Presutti to study phase transitions in the continuum, but the book was already too long and I gave up. Chapter 8 concludes the analysis of the surface tension for the L–P free energy functional with the study of the shape of the interface (instanton) in the one dimensional case. The instanton is the minimizer of the free energy over profiles constrained to approach the plus and minus equilibrium values at plus and minus infinity, it is therefore a blow up of the interface between the two equilibrium phases. Existence and properties of the instanton including the dynamical ones are proved in this chapter; in particular, spectral gap estimates in a L^∞ setting using an extension of the Dobrushin uniqueness theory of Chap. 3, which involves Vaserstein distance and couplings. I had also planned a chapter about motion by curvature but dropped it for reasons of space (maybe next time...).

Part III is more specialistic, and it has been written with several aims. One was to show how ideas and methods of the mesoscopic theory can find applications in statistical mechanics. Chapter 9 shows that the large deviation estimates of Chap. 6 for the L–P functional can be used to prove the Peierls estimates in the Ising model with ferromagnetic Kac potentials and hence the occurrence of phase transitions in $d \geq 2$ dimensions. I like the result, because it gives an example of the power of Kac ideas in implementing the van der Waals theory of liquid–vapor phase transitions. The careful analysis of the L–P functional in Part II allows one to carry out the Kac program without actually taking the scaling limit as $\gamma \to 0$ (range of the interaction to infinity) and it shows that the mean field phase diagram is a good approximation of the true phase diagram if γ is small. Coarse graining, block spins, effective hamiltonian, and renormalization group ideas appear naturally at this stage. For brevity I do not discuss the specific structure of DLR measures also because they are examined in Chaps. 11 and 12 in the more complex context of the LMP model. The analysis of the Ising model is made simple by the spin flip symmetry, which maps the minus and plus phases one into the other. The Pirogov–Sinai theory is an important step in statistical mechanics which provides methods for deriving Peierls estimates on contours when the symmetry is absent. The original theory was designed to study perturbations of the ground states at small positive temperatures and my purpose in the book is to discuss how the theory can be adapted to study perturbations of the minimizers of the free energy functional at small γ. Instead of adding terms to the Ising hamiltonian to break the spin flip symmetry I have thought it more informative to consider the LMP particle model. This is a system of point particles in $\mathbb{R}^d$ which interact via Kac potentials and which in its mean field approximation has a phase transition into a plus and a minus state with distinct particle densities. There is no symmetry between the two phases and the analysis of the small γ perturbations of the minimizers requires the use of the Pirogov–Sinai theory, which is presented in all detail in Chaps. 10 and 11. In Chapter 12 I have added a characterization of the DLR measures for the LMP model which is not in the existing literature. For reasons of space I have dropped the derivation of the Gibbs phase rule in LMP, for which

I refer to the literature. At the end of Parts I, II and III there are short sections with references and notes. In the subject index the reader will find a list of the most used symbols.

Even though I appear as the author of the book I am certainly not the only one. All this is the outcome of many discussions with friends and colleagues, and parts are taken from lectures and then modified with the help and the comments of the audience. I have just tried to reorganize all that, and the result therefore is not only mine. In particular, the program to study Kac potentials keeping the scaling parameter γ fixed was originally conceived and then carried out with Marzio Cassandro to whom I am especially indebted. Dynamics and its macroscopic limits are instead mostly related to my works with Anna De Masi who also helped me a lot in writing this book. Giovanni Bellettini explained to me some of the fascinating ideas of De Giorgi and helped me to approach the subject from the side of the macroscopic theory.

The influence of the Moscow school is evident in the book and more generally in the way mathematical physics looks today, which is, I believe, largely due to the contributions of the Soviet school and this obviously reflects in the book where the names of Dobrushin and Sinai are recurrent. I learned a lot from them and not only mathematics. There are topics which are not yet ready, in particular a whole big chapter about elastic bodies where there are some intriguing ideas of Stephan Luckhaus which we are trying to develop. But this is for the future and I would like to conclude this preface by mentioning Joel Lebowitz who has been for me a teacher and a friend and if there is something good in this book the credit is certainly his.

Contents

Part IV Appendices

Chapter 1
Introduction

Movies in modern theaters have astonishingly realistic effects. While sitting in your chair and watching the movie, you enter into the middle of the action with sounds and images which surround and overwhelm you. Yet, after all, the magic of all that comes from a loss of information. A movie is just a sequence of images (and sounds) and by looking at them one by one, we certainly gather more information than when we watch them running at the right speed. Moreover, each single image can be amplified more and more, with new details revealed at each step, till it turns out to be an array of pixels or grains, which is the ultimate content of the image. But if we analyze the images to such an extent we are very likely to miss the story of the movie, which, after all, is what it was all about.

All that means that each event has its own characteristic scale, and that even though, theoretically, we improve by successive blow-ups, in fact we may be losing the true meaning of the phenomenon. Sometimes, however, atomization is useful; you may for instance remember the movie "Call Northside 77A" (starring James Stewart) with the trick of magnifying the newspaper to read its date, which proved the alibi of the main character of the story. Blow-ups and slow motions are very popular in sport events, revealing for instance a possible off-side or a penalty in a soccer game, which, at normal speed, are easily missed.

If we replace the word "movies" by "fluids" we are not too far from catching the spirit of this book. The transition from statistical mechanics to continuum mechanics requires a perspective view of the system which captures its main global features and, unavoidably, neglects the details. Sometimes, however, the missing details are just those necessary for explaining a phenomenon and the backward way to microscopics becomes necessary.

There are several levels on the way from microscopics to macroscopics, each one with its own primitive notions, basic axioms and a theory developed from them. Scaling limits and blow ups interrelate the different levels, some links are well established, others are only tentative or just missing, yet the theory is overall in a satisfactory enough shape to justify my attempt to write a monograph on the subject. My goal in this introduction is to give a feeling of the general picture. I will be purposely vague, requiring from the reader a great deal of cooperation for tolerance of the missing details. The hope is to reach in a few pages an idea of the spirit of what research is doing in this area. A quantitative analysis of some of the topics will then come in the successive chapters; other issues will not be expanded further and the reader should look in the literature for more information.

A good way to start is to pick up a specific, interesting issue—interfaces and their structure is my choice—and see how it is developed in the various levels.

Errico Presutti, *Scaling Limits in Statistical Mechanics and Microstructures in Continuum Mechanics*, © Springer 2009

1.1 The macroscopic theory

In the macroscopic theory systems are continuum bodies, each point of the body is representative of a microscopic system in thermal equilibrium; namely a "macroscopic point" observed with a "magnifying lens" becomes a fluid in thermodynamic equilibrium. The parameters which characterize the equilibrium are called "order parameters"; they are specified by the thermodynamic properties of the body and are usually finite in number. Hence the basic postulate of the macroscopic theory is that macroscopic states are local equilibrium states described by order parameter valued functions on the spatial domain Ω occupied by the body.

Besides the temperature which in the sequel is kept fixed throughout the whole body (and thus not included among the order parameters) there are other parameters which characterize the equilibrium states of the system. In a gas an order parameter could be the particle or mass density, but the same substance at lower temperatures, when it solidifies, may also be characterized among other things by its crystalline structure, namely the microscopic pattern in which its atoms are arranged. Thus the order parameter may change drastically with the substance or in the same body if we vary the external conditions. To simplify the discussion we restrict ourselves in the sequel to the simplest case of a single, scalar order parameter, and for the sake of definiteness we consider a ferromagnetic crystal in thermal contact with a reservoir which by exchanges of energy keeps each point of the body at a fixed temperature T. As said before, T is fixed and it does not appear as an order parameter. We further simplify the picture by supposing that the magnetization m along some characteristic axis of the body is the only thermodynamically relevant quantity, so that the scalar m is the order parameter of our system. We also suppose that the reservoir exchanges magnetization with the body so that the total magnetization is not a conserved quantity and at equilibrium m will only have finitely many values: if T is larger than the "critical Curie temperature" only one phase exists (the non-magnetized phase) and $m = 0$ is the only equilibrium value. Below the critical temperature there exist magnetized phases, and the simplest case, which is the one considered here, is when m has only two values, which in proper units are ± 1 (recall that the temperature is fixed) and correspond to the "plus" and to the "minus phase."

As mentioned previously the macroscopic states are functions $m(r)$ which in the present scheme become ± 1 valued functions. The constant functions $m(r) \equiv 1$ and $m(r) \equiv -1$ are global equilibria; they describe the plus and the minus phases, respectively. All the other states are non-equilibrium and in the macroscopic theory they are states for which both phases coexist; they are characterized by two distinct regions, Ω_+ and Ω_-, where respectively $m = 1$ and $m = -1$; see Fig. 1.1. The macroscopic theory postulates that the excess free energy $F(m)$ of such a state m (i.e. the difference of the free energy of m and the free energy of either one of the two global equilibrium states, $m(r) \equiv 1$ and $m(r) \equiv -1$, which have the same free energy) concentrates on the "interface" Σ which separates Ω_+ and Ω_-. Supposing Σ to be a regular surface,

$$F(m) = \int_{\Sigma} s_\beta(\nu(r))\, dH^{d-1}(r), \tag{1.1.1}$$

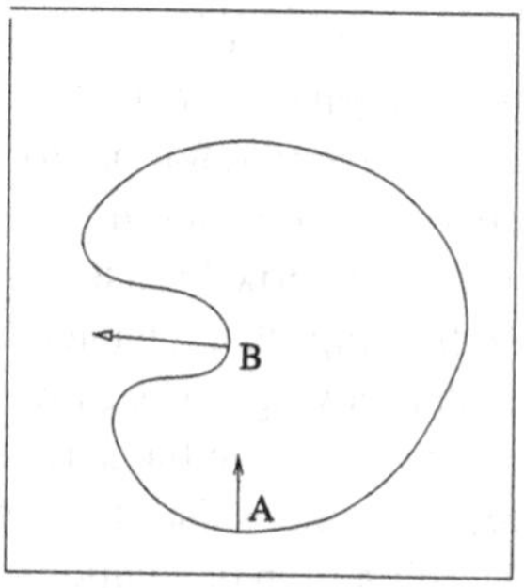

Fig. 1.1 The *thick line* represents the interface between Ω_+ inside and Ω_- outside. The *arrows* are the velocities of A and B under motion by mean curvature: A moves toward the interior, B toward the exterior

where β is "the inverse temperature," more precisely $T = (k\beta)^{-1}$, k the Boltzmann constant; $s_\beta(\nu) > 0$ is the surface tension of a planar interface with unit normal ν, $s_\beta(\nu) = s_\beta(-\nu)$; $\nu(r)$ is the unit normal to Σ at r and dH^{d-1} is the Hausdorff surface measure on Σ.

We can extend (1.1.1) by a limit procedure (for instance in the L^1-sense) to a more general class of states m; the physically acceptable states will be those with bounded free energy, hence the question:

Characterize the states which are limits of regular interface states with bounded free energy. Does the free energy cost of an interface control its regularity?

"Geometric measure theory" gives a full answer to the question. It shows that it is possible to extend the measure of the boundary of a set to "bounded variation," *BV*, sets, and that, measure theoretically, the boundary of a BV set is almost everywhere C^1 [9, 116]. Thus modulo sets of zero measure we may regard the surface as C^1, and (1.1.1) therefore naturally extends to functions in $BV(\Omega, \{-1, 1\})$, namely functions m such that $\{m = 1\}$ and $\{m = -1\}$ are BV sets. Any other state, which is not BV, has an infinite free energy cost: thus the cost of an interface controls its regularity (in the C^1 sense) but only modulo a region of the interface of zero measure.

Let us now turn to dynamics. Thermodynamics states that the free energy must decrease in time till global equilibrium is established. In agreement with this picture the macroscopic theory postulates that time evolution is defined as the gradient flow of the free energy. The gradient flow of a function $f(r)$, $r \in \mathbb{R}^n$, is defined by the evolution equation $dr(t)/dt = -c\,\nabla f(r(t))$ for $r(t)$, $c > 0$ a constant. Namely the velocity is directed opposite to the gradient. As a consequence f decreases along the orbits, $df/dt = -c|\nabla f|^2$, and in fact the gradient dynamics is often used as a tool to search the minimum of a function.

To apply the above to our case we need to define the gradient of the functional (1.1.1), but we shall discuss this issue in the simpler context of the mesoscopic theory. Here we just say that the gradient flow equation associated to (1.1.1) is motion by curvature. In the isotropic case where the surface tension s_β is independent of the orientation, $s_\beta(\nu) = s_\beta$, this is

$$v(r) = c s_\beta\, \kappa(r)\, \nu(r), \quad \kappa(r) = \sum_{i=1}^{d-1} \kappa_i(r), \tag{1.1.2}$$

where $v(r)$ is the velocity of the point r of the interface; $c > 0$ a "mobility coefficient"; $\kappa_i(r)$ the i-th principal curvature, counted negative if directed toward the interior, $\nu(r)$ the unit, external normal to the interface at r; in the anisotropic case [203], the principal curvatures are not weighted equally, the weights being determined by the "stiffness matrix." Existence and uniqueness for the Cauchy problem for (1.1.2), is another important chapter of the PDE theory. Existence of classical solutions (i.e. regular moving surfaces which at all points satisfy (1.1.2)) is proved for finite positive times if the initial surface is regular. It is known that singularities may develop after finite times and the continuation of the solution requires the introduction of a weaker notion of solutions; see for instance [117]. Notice finally that by construction the interface area decreases in time and consequently its free energy does so as well, in agreement with the thermodynamic laws.

The basic postulate of the macroscopic theory that each point is representative of a system in thermodynamic equilibrium looks shaky at the interface, where the excess free energy is concentrated and the whole action takes place. Hence the idea of a blow up and of relaxing the local equilibrium postulate of the macroscopic theory, which leads us into the mesoscopic theory.

1.2 The mesoscopic theory

Also, the mesoscopic theory postulates that each point of the body is representative of a portion of fluid; however, unlike in the macroscopic theory, its smaller size allows for deviations from thermodynamic equilibrium. Thus, referring again to the ferromagnetic crystal in thermal contact with a reservoir at a temperature below the critical one, the order parameter will still be the magnetization density m, but now m may take values different from ± 1, which, as before, we suppose to be the equilibrium values.

Thus, a state is now a function $m \in L^\infty(\Lambda, \mathbb{R})$, Λ the region occupied by the body (there is a reason to use Λ instead of Ω, as we shall see). Again $m \equiv 1$ and $m \equiv -1$ are the global equilibrium states and we set as before their free energies equal to 0. Any other state then has a non-zero free energy. The free energy functional in the mesoscopic theory is again a primitive notion. In the scalar Ginzburg–Landau theory it is supposed to have the form

$$F(m) = \int_\Lambda W(m(r)) + |\nabla m|^2, \tag{1.2.1}$$

m being a continuous function with derivative in L^2, $F(m) = +\infty$ on all the other functions. For simplicity we take Λ to be a torus in $\mathbb{R}^d$ (of side L), $W(\cdot)$ a double well potential with two minima at $m = \pm 1$, and elsewhere $W > 0$; to be definite $W(m) = (m^2 - 1)^2/4$.

In the mesoscopic theory the pure phases are defined as the states which minimize the free energy. In (1.2.1) $W(\cdot)$ penalizes excursions from ± 1 and $|\nabla m|^2$ variations of $m(\cdot)$, hence the minimizers of $F(m)$ are the global equilibria $m \equiv 1$ and $m \equiv -1$.

Interfaces arise in states where the two equilibria coexist, but the simple picture of the macroscopic theory where m takes only the values ± 1 is here inadequate, as the free energy $F(m_\pm)$ of any such state is infinite, due to the term $|\nabla m|^2$ in (1.2.1). Let us then relax the requirement that $m = m_\pm$ by only asking that

$$\fint_\Lambda |m - m_\pm| < \zeta, \qquad \fint_\Lambda f = \frac{1}{|\Lambda|}\int_\Lambda f, \tag{1.2.2}$$

where $\zeta > 0$ is an accuracy parameter which will eventually vanish. Calling L the side of the torus Λ, it has been proved that the inf of $F(m)$ under the constraint (1.2.2) grows proportionally to L^{d-1} as $L \to \infty$; it is also known that minimizing sequences differ appreciably from the state $m_\pm$ only in a neighborhood of order 1 of the interface present in $m_\pm$. To make the statement precise, call Ω the unit torus in $\mathbb{R}^d$, $\psi(r) = L^{-1}r$ the map from Λ to Ω and write for any function u on Ω

$$F_L(u) := L^{-d+1} F(u(\psi^{-1}\cdot)), \quad F \text{ as in (1.2.1)}. \tag{1.2.3}$$

Explicitly

$$F_L(u) := \int_\Omega \epsilon |Du|^2 + \epsilon^{-1} W(u), \quad \epsilon := \frac{1}{L}. \tag{1.2.4}$$

If $v \in BV(\Omega; \{-1, 1\})$ is a macroscopic state with an interface it is then "natural" to define an upper and a lower free energy of the state v as

$$\lim_{\zeta \to 0} \limsup_{L \to \infty} \inf_{|u-v|_{L^1(\Omega)} < \zeta} F_L(u), \qquad \lim_{\zeta \to 0} \liminf_{L \to \infty} \inf_{|u-v|_{L^1(\Omega)} < \zeta} F_\Lambda(u). \tag{1.2.5}$$

This is an example of De Giorgi's Γ-convergence, a notion extensively studied in the last decades; see for instance [49, 87]. In the particular case we are considering, the two limits in (1.2.5) exist and are equal, and they are given by the macroscopic free energy (1.1.1) with $s_\beta(\nu) = s_\beta$ independent of ν and determined by the potential W [173] by the simple formula

$$s_\beta = \int_{-1}^{1} \sqrt{W(s)} ds. \tag{1.2.6}$$

The result (1.2.6) has a nice physical interpretation. Consider the $d = 1$ version of (1.2.1) with $\Lambda = \mathbb{R}$. Then

$$s_\beta = \inf\Big\{F(m), \liminf_{x \to \infty} m(x) > 0; \limsup_{x \to -\infty} m(x) < 0\Big\} = F(\tanh(\cdot)). \tag{1.2.7}$$

The minimizer $x \to \tanh x$ of (1.2.7) is proved to be unique modulo translations and it is called the "instanton." It has also been proved that in the multi-dimensional case if the interface is regular, optimizing functions u are close to the rescaled instanton; i.e. given any point $r_0 \in \Sigma$ consider the line through r_0 along the normal $\nu(r_0)$ to Σ, then on such a line we see (approximately) the one dimensional instanton $\tanh\{L(r - r_0) \cdot \nu(r_0)\}$.

The result gives a positive answer to the questions raised at the end of Sect. 1.1; we have in fact a theory which recovers the macroscopic free energy of any BV state v by taking the macroscopic limit $L \to \infty$, but if more details on the structure of the interface are needed, they can be obtained by keeping L large but fixed. The interface is then called "diffuse" and described by the one dimensional instanton.

Let us now discuss the dynamics. The mesoscopic theory, just like the macroscopic one, postulates that the equations of motion are gradient flows, namely on the torus Λ,

$$\frac{dm}{dt} = -\frac{dF(m)}{dm} = \Delta m - W'(m). \tag{1.2.8}$$

Equation (1.2.8) is a "reaction–diffusion equation" (Δm is the diffusion, $-W'(m)$ the reaction) which is known in the literature as the "Allen–Cahn equation." Existence and uniqueness for the Cauchy problem for (1.2.8) are here well settled questions; the problems present in the motion by curvature arise when we study the macroscopic limit. As seen in the equilibrium theory, when going to macroscopics we shrink space by a factor L^{-1} and since motion by curvature is invariant under parabolic scaling, we scale time as L^2, so that we define on the torus Ω

$$u_L(r,t) := m(Lr, L^2 t), \quad m \text{ a solution of (1.2.8) in } \Lambda. \tag{1.2.9}$$

If $u_L(r,0) \to v$ in $L^1(\Omega)$ as $L \to \infty$ and if $v \in BV(\Omega, \{-1,1\})$ with an interface which is regular, then for t small enough $u_L(\cdot,t) \to v(\cdot,t)$ in $L^1(\Omega)$ and $v(\cdot,t)$ solves the motion by the mean curvature equation (1.1.2) without the coefficient cs_β. Convergence in fact extends to longer times and even past the appearance of singularities via the notion of viscosity solutions, but this is beyond the purposes of the present discussion and we just refer to the literature; see also Sect. 8.7.

The mesoscopic derivation of (1.1.2) fixes the value of the mobility coefficient c as $c = 1/s_\beta$. As we are going to see this is in agreement with the Einstein relation [203], which predicts that the same mobility c which appears in the motion by curvature is also the mobility coefficient which appears in the linear response theory, which describes the response of our magnetic system to the driving force produced by an external magnetic field h. h modifies the free energy by an additive term:

$$F_h(m) = \int_\Lambda |\nabla m|^2 + W(m) - hm, \tag{1.2.10}$$

and therefore the new dynamics is $\frac{dm}{dt} = -\frac{dF_h}{dm}$. In $d = 1$ and with $\Lambda = \mathbb{R}$, for all $h > 0$ small enough there are traveling wave solutions, namely pairs $\tilde{m}$, V, such that $m(x,t) := \tilde{m}(x - Vt)$ solves

$$\frac{dm}{dt} = -\frac{dF_h}{dm} = \frac{d^2 m}{dx^2} - W'(m) + h. \tag{1.2.11}$$

The mobility coefficient in the linear response theory is then defined as

$$c = \lim_{h \to 0} \frac{V}{h}, \tag{1.2.12}$$

namely the ratio of the response, V, to the driving force h in the linear regime, $h \to 0$. By explicit computation the mobility turns out to be equal to $1/s_\beta$, the same as when deriving (1.1.2).

In conclusion, with the mesoscopic theory we can investigate the structure of the interface observing corrections to the [macroscopic theory] postulate of local equilibrium in a region of thickness $\approx \epsilon = 1/L$ around the interface (in macroscopic units) where a special magnetization pattern described by the instanton shape is observed. Deviations from such a behavior are however observed experimentally; they are due to thermal fluctuations which make the interface much less regular than predicted by the macroscopic and mesoscopic theories. The number of ways an interface of minimal energy may be deformed adds in fact a relevant entropic contribution to the surface tension which is totally missed in the mesoscopic description. For instance (here the discussion may be a little cryptic for the non-expert), such a contribution is critical in the Potts model (a variant of the Ising model with spins which take a finite number of values ≥ 3): the energy contribution to the surface tension is the same at the interface between "ordered–ordered and ordered–disordered" states, but the entropic contribution is larger in the latter case. As a consequence [170]–[159], at the interface between two distinct "ordered states" there appears a layer of the disordered phase, a phenomenon known as "wetting."

1.3 The microscopic theory

By a blow up we move from the mesoscopic into the microscopic world: macroscopic points are now fully resolved, and as we shall see they become large regions containing many atoms. The unit microscopic length is defined so that the inter-atomic distance (of near-by atoms) has the order of unity. To simplify the picture we restrict ourselves to the Ising model. Microscopic states are then functions $\sigma_\Lambda : \Lambda \to \{-1, 1\}$, Λ being a finite subset of $\mathbb{Z}^d$ representing the crystal location. σ_Λ is identified with a sequence in $\{-1, 1\}^\Lambda$ and $\sigma_\Lambda(x)$ describes the state of "the spin" of the crystal sitting at x. As before it is convenient to regard Λ as a torus in $\mathbb{Z}^d$ (to avoid to have to discuss the interaction with the "walls" confining the system). The energy of the configuration σ_Λ, thought of as repeated periodically in $\mathbb{Z}^d$, is taken to be of the form

$$H_\Lambda^{\text{per}}(\sigma_\Lambda) = -\frac{1}{2} \sum_{x \in \Lambda, y \in \mathbb{Z}^d, y \neq x} J(x, y) \sigma_\Lambda(x) \sigma_\Lambda(y), \tag{1.3.1}$$

where the coupling constants $J(x, y) \geq 0$ because (1.3.1) is supposed to describe a ferromagnetic crystal. Indeed if $J(x, y) > 0$, the interaction energy $-J(x, y)\sigma_\Lambda(x)\sigma_\Lambda(y)$ of the two spins at x and y is minimal if the two spins are "aligned," $\sigma_\Lambda(x) = \sigma_\Lambda(y)$. Thus the minimizers of the energy $H_\Lambda^{\text{per}}(\sigma_\Lambda)$, i.e. its ground states, are $\sigma_\Lambda \equiv 1$ and $\sigma_\Lambda \equiv -1$, as in such states any single pair interaction is minimized (this is why ferromagnetic interactions are much easier to study).

Unlike in the macroscopic and mesoscopic theories, however, the states $\sigma_\Lambda \equiv 1$ and $\sigma_\Lambda \equiv -1$ are not the equilibrium states, they are only the ground states, and they minimize the energy while the equilibrium states are those which minimize the free energy, i.e. energy minus $T \cdot S$, T being the temperature, S the entropy. According to Boltzmann the entropy is proportional to the log of the number of states, and this is the input of the theory of equilibrium statistical mechanics. It leads in the end to a description of the equilibrium states as probability measures [on the phase space of the system] given by the "Gibbs formula"; the whole issue will be discussed in Chap. 2. In our Ising model the "Gibbs measures" are

$$G^{\mathrm{per}}_{\beta,\Lambda}(\sigma_\Lambda) := \frac{e^{-\beta H^{\mathrm{per}}_\Lambda(\sigma_\Lambda)}}{Z^{\mathrm{per}}_\Lambda}, \tag{1.3.2}$$

where $\beta = 1/kT$, T the absolute temperature, k the Boltzmann constant, $H^{\mathrm{per}}_\Lambda(\sigma_\Lambda)$ is given by (1.3.1), Z^{per}_Λ, "the partition function," is the normalization factor:

$$Z^{\mathrm{per}}_\Lambda := \sum_{\sigma_\Lambda \in \{-1,1\}^\Lambda} e^{-\beta H^{\mathrm{per}}_\Lambda(\sigma_\Lambda)}. \tag{1.3.3}$$

Here the Gibbs measure should be regarded as a primitive notion (in Sect. 2.3 we will discuss the derivation of these measures from the Boltzmann hypothesis); thus we will take the axiomatic view that the equilibrium state at inverse temperature β is $G^{\mathrm{per}}_{\beta,\Lambda}$ (at least when Λ is large enough, as we are going to argue). To connect with the macroscopic theory we must preliminarily identify the "macroscopic points" in the microscopic setting. We regard our microscopic region Λ as a blow up of the macroscopic region Ω. Call L the length of the side of the torus Λ and associate to any point $r \in \Omega$ the empirical magnetization

$$u_R(\sigma_\Lambda; r) := \frac{1}{|B_R(Lr)|} \sum_{x \in B_R(Lr)\cap \mathbb{Z}^d} \sigma_\Lambda(x), \quad r \in \Omega, \tag{1.3.4}$$

where σ_Λ is periodically extended to the whole $\mathbb{Z}^d$; $B_R(r) = \{r' : |r - r'| \le R\}$ is l ball in $\mathbb{R}^d$ of radius R and center r, $|B_R(r)|$ is its volume.

Equation (1.3.4) associates to any macroscopic point r the ball $B_R(rL)$ which quantifies the interpretation of macroscopic points in Sect. 1.1 as representative of portions of the fluid with many molecules, how many being determined by R. The mathematical validity of the statement follows from Theorem 1 below; hereafter for the sake of simplicity $J(x, y) = J > 0$ if $|x - y| = 1$ and $J(x, y) = 0$ otherwise.

Theorem 1 *In $d \ge 2$ there is $\beta_c > 0$ so that if $\beta \le \beta_c$ then for any $\zeta > 0$ (and with L the side of the torus Λ)*

$$\lim_{R\to\infty} \lim_{L\to\infty} G^{\mathrm{per}}_{\beta,\Lambda}\big(\|u_R(\sigma_\Lambda; \cdot)\|_{L^1(\Omega)} \le \zeta\big) = 1. \tag{1.3.5}$$

If $\beta > \beta_c$ there is $m_\beta > 0$ so that on calling $\mathbf{1}_{m_\beta}$ and $\mathbf{1}_{-m_\beta}$ the functions on Ω constantly equal to m_β and to $-m_\beta$

$$\lim_{R\to\infty}\lim_{L\to\infty} G^{\rm per}_{\beta,\Lambda}\big(\|u_R(\sigma_\Lambda;\cdot)-\mathbf{1}_{\pm m_\beta}\|_{L^1(\Omega)}\le\zeta\big)=\frac{1}{2}.$$

Thus the states $\mathbf{1}_{m_\beta}$ and $\mathbf{1}_{-m_\beta}$ when $\beta > \beta_c$ and the state $\mathbf{1}_0$ when $\beta \le \beta_c$ are the equilibrium profiles at the inverse temperature β. We have not normalized the magnetizations and we thus have the equilibrium value m_β instead of 1 as in the previous theories. To proceed with the derivation of the macroscopic theory we will use *the second Gibbs hypothesis*, namely that *the free energy of a macroscopic state is equal to $-\beta^{-1}$ times the log of the Gibbs probability of the macrostate.*

Theorem 2 *For any $d \ge 2$ and $\beta > \beta_c$ there is [a surface tension] $s_\beta(\nu)$, ν ranging among the unit vectors of $\mathbb{R}^d$, so that for any $w \in BV(\Omega,\{-m_\beta,m_\beta\})$,*

$$\lim_{R\to\infty}\lim_{L\to\infty}\frac{1}{L^{d-1}}\log G^{\rm per}_{\beta,\Lambda}\big(\|u_R(\sigma_\Lambda;\cdot)-w\|_{L^1(\Omega)}\le\zeta\big)=-\beta F(w),$$

where $F(w)$ is as in (1.1.1).

Notice that Theorem 2 applied to the equilibrium states $\mathbf{1}_{m_\beta}$ and $\mathbf{1}_{-m_\beta}$ shows that they have zero free energy, and thus $F(w)$ is actually the *excess free energy of the macrostate w*.

While Theorem 1 has been established long ago and has been extended to more general lattice systems, Theorem 2 is much more recent. The first paper in $d = 2$ and for β large is the now famous paper [112] by Dobrushin, Kotecký and Shlosman which started the rigorous statistical mechanics theory of interfaces. Interfaces in [112] are actually studied in a much stronger topology than the L^1 norm used in Theorem 2. The results described in Theorem 2 are more recent; they exploit the L^1 context and are based on the geometric measure theory developed in the macroscopic theory; see for instance [39, 76] and references therein.

Interface fluctuations are hidden in the surface tension in Theorem 2; to make them explicit let us fix an "observation window" $B_R(0)$ in the center, say 0, of the cube $\Lambda \subset \mathbb{Z}^d$. To localize the interface we fix plus boundary conditions outside Λ at $x_d > 0$ (the last coordinate of x) and minus boundary conditions at $x_d < 0$, so that a flat interface (as in the mesoscopic and macroscopic theory) would pass through the window and thus be observed. It has been proved [144] that in $d = 2$ if $R = L^a$, $a < 1/2$, then with probability going to 1 the interface does not appear in $B_R(0)$, see Fig. 1.2 while in $d = 2$ and $R = L^a$, $a > 1/2$, as well as in $d = 3$ [108] with any $R(L) \to \infty$ as $L \to \infty$, the interface appears in the window with probability going to 1 as $L \to \infty$, but only if β is larger than a "critical roughness inverse temperature"; see Fig. 1.3.

The actual statistics of the interface fluctuations when the interface is not macroscopically flat are poorly known and this reflects in the essential absence of results in the dynamics of interfaces. The dynamics in the Ising model is defined as a Markov

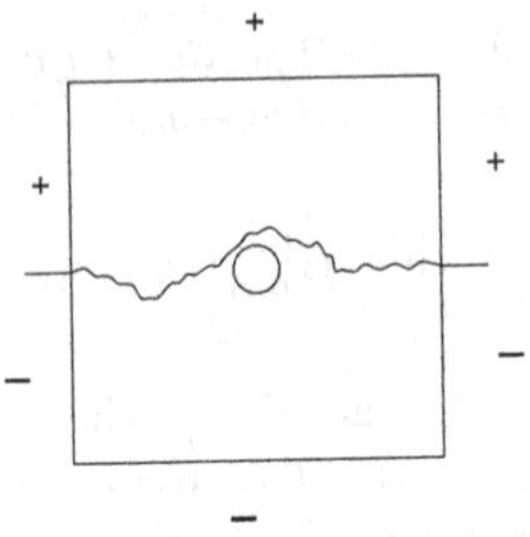

Fig. 1.2 In $d = 2$ and window $B_R(0)$, $R = L^a$, $a < 1/2$

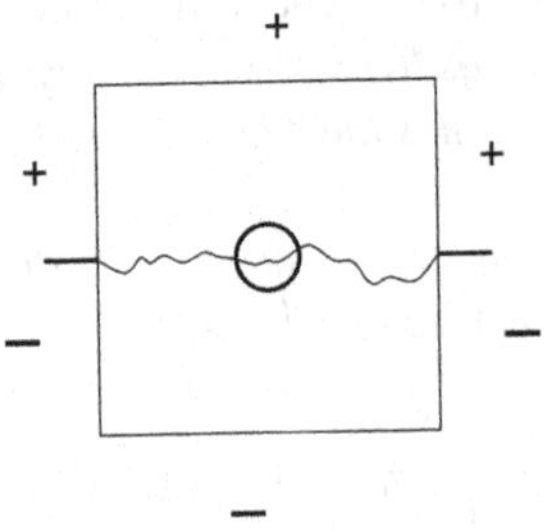

Fig. 1.3 In $d = 2$ and window $B_R(0)$, $R = L^a$, $a > 1/2$ or in $d = 3$ with any $R(L) \to \infty$

process, the Glauber dynamics or the Kawasaki dynamics when the total magnetization is conserved; see Chap. 5. Convergence of the Glauber dynamics to motion by curvature is an open question. Partial results have been obtained in $T = 0$ [202]; for some so called SOS models (where the interface is a graph) [127]; for Kac potentials in some scaling limits [91, 152].

1.4 Kac potentials and the mesoscopic limit

In the previous sections we have seen how and to what extent the macroscopic theory can be derived from an "underlying" mesoscopic or microscopic theory in the scaling limit where the domain Λ invades the whole space. The connection between micro and meso is instead totally missing in the analysis of Sect. 1.3 and indeed it comes from the existence of a third scale which in Sect. 1.3 is not present. There are in principle three main lengths in a system, the inter-atomic distance, which in the Ising model may be taken as the lattice spacing, the interaction range and the size of the spatial domain which defines the macroscopic scale. In Sect. 1.3 we have studied systems where the interaction range is of the same order as the lattice spacing. We have then considered the limit when the region Λ invades the whole space so that the ratio between inter-atomic and macroscopic distances vanishes while the ratio between inter-atomic and effective interaction range stays finite. In this limit the correct thermodynamic behavior emerges and the limit is therefore called "the thermodynamic limit."

There are important examples in physics of long range forces, like the Coulomb and the dipole–dipole interactions, but what we have specifically in mind here are forces à la van der Waals responsible for the liquid–vapor phase transition. Namely a small but rather long range attractive tail in the interaction which (possibly combined with a short range repulsive force) gives rise to condensation phenomena for suitable values of the density and the temperature. With this in mind we are going to consider systems characterized by having the interatomic and the interaction lengths sharply separated. As discussed in Chap. 4 there are then two possible sub-cases: (i) the interaction and the macroscopic lengths have same order—these are mean field models—and (ii) both ratios, lattice spacing/interaction length and interaction length/macroscopic length vanish; this category contains the Kac potentials which will have a central role in the book. It should be mentioned that also the case where the intermolecular distances are much larger than the range of interaction (the opposite of what we are considering) is of great physical relevance, and by a suitable choice of scales it includes the Boltzmann theory of rarefied gases; see for instance the monograph by Cercignani, Illner and Pulvirenti [75].

We will introduce Kac potentials in the version proposed by Alik Mazel, which has then been used by Lebowitz, Mazel and Presutti in the LMP model; see Chap. 10. Let $\gamma^{-1} \gg 1$ be the mesoscopic length in microscopic units. *The basic assumption is that the energy density $e(\cdot)$ is a function of the empirical magnetization density $u_{\gamma^{-1}}(\sigma_\Lambda; \cdot)$*, namely (after a change of notation)

$$H_{\gamma,\Lambda}(\sigma_\Lambda) = \int_{\mathbb{R}^d} e(J_\gamma * \sigma_\Lambda(r))\, dr \tag{1.4.1}$$

where $J_\gamma * \sigma_\Lambda(r) = \sum_{x \in \mathbb{Z}^d} J_\gamma(x, y)\sigma_\Lambda(y)$ and $J_\gamma(r, r') = \frac{1}{|B_{\gamma^{-1}}|} \mathbf{1}_{|r-r'| \le \gamma^{-1}}$ so that $\int J(r, r') dr' = 1$. Observe that for any fixed $\gamma > 0$, $H_{\gamma,\Lambda}(\sigma_\Lambda)$ is a finite range, regular many-body hamiltonian of statistical mechanics. The mesoscopic theory instead requires one to take the limit $\gamma \to 0$, as we are going to see.

Let $m(r)$ be "a mesoscopic magnetization profile" on the torus Λ, $\gamma^{-1}\Lambda$ being its microscopic blow-up (suppose for simplicity that γ^{-1} is an integer). It can then be proved that

$$\lim_{\zeta \to 0} \lim_{\gamma \to 0} -\gamma^d \log G_{\beta,\gamma,\gamma^{-1}\Lambda}\big(\|u_{\gamma^{-1}}(\cdot; \sigma_{\gamma^{-1}\Lambda}) - m(\cdot)\|_{L^1(\Lambda)} \le \zeta\big) =: F(m),$$

where according to "the second Gibbs assumption" $F(m)$ is the free energy of the mesoscopic profile m. It can be shown that

$$\begin{aligned} F(m) &= \int_\Lambda \Big\{ e(J * m(r)) - \frac{1}{\beta} S(m(r)) \Big\}, \\ J(r, r') &= \frac{1}{B_1(0)|} \mathbf{1}_{|r-r'| \le 1}, \end{aligned} \tag{1.4.2}$$

$$S(m) = -\frac{1+m}{2} \log \frac{1+m}{2} - \frac{1-m}{2} \log \frac{1-m}{2}, \tag{1.4.3}$$

and since $e(J * m(r))$ is the energy density, $S(m)$ plays the role of the entropy. In Sect. 4.2 it will be seen that $S(m)$ is related to the log of the number of configurations with magnetization density m, in agreement with the Boltzmann hypothesis.

Suppose now that the energy density is $e(m) = \frac{-m^2}{2}$; then

$$F(m) = \int_\Lambda \left\{ -\frac{1}{2}\, m(r) V * m(r) - \frac{1}{\beta} S(m(r)) \right\}, \quad V = J * J, \tag{1.4.4}$$

which in turns can be rewritten as

$$\begin{aligned} F(m) &= \int_\Lambda \left\{ -\frac{1}{2}\, m(r)^2 - \frac{1}{\beta} S(m(r)) \right\} \\ &\quad + \frac{1}{4} \int_\Lambda \int_\Lambda V(r,r')\big(m(r) - m(r')\big)^2. \end{aligned} \tag{1.4.5}$$

Interpreting the second term as a non-local version of $|\nabla m|^2$, we recover the basic Ginzburg–Landau functional (1.2.1) with $w(m) = -\frac{m^2}{2} - \frac{1}{\beta} S(m)$, which is indeed a double well if $\beta > 1$. On the other hand, when $e(m) = \frac{-m^2}{2}$ the hamiltonian (1.4.1) becomes

$$H_{\gamma,\Lambda}(\sigma_\Lambda) = -\frac{1}{2} \sum_{x,y \in \Lambda} V_\gamma(x,y) \sigma_\Lambda(x) \sigma_\Lambda(y), \tag{1.4.6}$$

where $V_\gamma(x,y) = \gamma^d V(\gamma x, \gamma y)$. This fits with the original formulation of Kac potentials by Kac, Uhlenbeck and Hemmer [147–149], and then widely studied in statistical mechanics starting from the basic paper by Lebowitz and Penrose, [160]. We will study in Chap. 6 the development of the mesoscopic theory starting from (1.4.5), re-deriving the results stated previously for the Ginzburg–Landau functional.

In Part III of this book, which is of a more specialized nature, we will study Kac potentials keeping γ fixed (but sufficiently small). We will see that phase transitions for systems with Kac potentials can be studied as "perturbations" of the mesoscopic states, or, more precisely, that the Pirogov–Sinai theory used to study the small temperature perturbations of the grounds states of lattice models can be extended to study of the small γ perturbations of the minimizers of the limit non-local mesoscopic functionals. In particular, we will apply such considerations first, as a warm up, to the ferromagnetic Ising model with Kac potentials and then to the LMP model [164], for phase transitions in the continuum.

1.5 Notes for the reader

The book is divided into three parts. At the end of each part there is a short section with Notes and References where I collect the references, which in the text are often omitted. In the subject index I have put some of the most used notation. I just mention here the following one which is often used in the book:

$$⨍_\Delta f(r)\,dr := \frac{1}{|\Delta|} \int_\Delta f(r)\,dr, \tag{1.5.1}$$

where $\Delta \subset \mathbb{R}^d$, $|\Delta|$ being its Lebesgue measure, $f \in L^1(\Delta)$. An analogous notation is used for more general measures. When it is clear from the context I may sometimes drop the measure from the integral and just write $\int_\Delta f$.

Part I
Statistical Mechanics of Ising systems

Chapter 2
Thermodynamic limit in the Ising model

Modern statistical mechanics rests on the Gibbs hypothesis that in a system in equilibrium with a reservoir at temperature T, the probability of observing a state is proportional to $e^{-E/kT}$, where E is the energy of the state, k the Boltzmann constant, and T the absolute temperature. Macroscopic equilibrium behavior (in particular thermodynamics) can then be derived via a sharp separation of scales implemented mathematically by requiring that the ratio between inter-atomic distances and macroscopic lengths vanishes. Such a scaling limit procedure is then referred to as "the thermodynamic limit."

We will discuss all that in the simpler context of the Ising model. In Sect. 2.1 we will study the Gibbs measures in bounded domains, and in Sect. 2.2 their infinite volume limits (thermodynamic limit). The limit measures obtained in this way will be characterized by the DLR property (DLR stands for Dobrushin, Lanford and Ruelle); the general structure of the DLR measures will then be examined in detail. Thermodynamic phases will be related to "extremal" DLR measures and consequently the occurrence of a phase transition to the non-uniqueness of DLR measures.

In Sects. 2.3 and 2.4 we will focus on the thermodynamic potentials of the Ising model. Using only the Boltzmann hypothesis to identify the thermodynamic entropy as the log of the number of states with given energy, we will derive the well known formula which relates the pressure to the log of the partition function; see Sect. 2.3. Formulas for all the other thermodynamic potentials will then be obtained by using the thermodynamic relations establishing a bridge between the macroscopic properties of a body and its microscopic interactions. The power of the thermodynamic formalism will become evident in Sect. 2.4, where it is applied in several different contexts. In particular, we shall see that the Gibbs assumption and DLR measures can be actually derived from the thermodynamic potentials, and we shall also show that large deviations naturally fall in the formalism.

2.1 Finite volume Gibbs measures

In this section we will study the finite volume Gibbs measures in the context of the Ising model. Ising configurations are collections of spins on the lattice $\mathbb{Z}^d$. Gibbs measures are then probabilities on the phase space of Ising configurations, they are defined by the Gibbs formula $Ce^{-E/kT}$, C a normalization constant (whose inverse is called the "partition function"), E the energy, k the Boltzmann constant and T the absolute temperature. The energy of an Ising configuration will be specified by the spin–spin interaction, the interaction of the spins with an external magnetic field and with the "external" spins, which act as boundary conditions.

Errico Presutti, *Scaling Limits in Statistical Mechanics and Microstructures in Continuum Mechanics*, © Springer 2009

2.1.1 Spin configurations, phase space

The Ising spin configurations (denoted by σ) are ± 1 valued functions on $\mathbb{Z}^d$; the collection of all spin configurations is the Ising phase space $\mathcal{X} = \{-1, 1\}^{\mathbb{Z}^d}$. A spin configuration is therefore the collection

$$\sigma = \{\sigma(x),\, x \in \mathbb{Z}^d\},$$

$\sigma(x) = \pm 1$ (up or down) being the spin at site x in the configuration σ.

Analogously, spin configurations in $\Lambda \subset \mathbb{Z}^d$ are functions on Λ with values ± 1, namely elements σ_Λ of $\mathcal{X}_\Lambda = \{-1, 1\}^\Lambda$,

$$\sigma_\Lambda = \{\sigma_\Lambda(x),\, x \in \Lambda\}.$$

The restriction map from $\mathcal{X}$ to $\mathcal{X}_\Lambda$, denoted by $\restriction_\Lambda$, is defined as

$$\restriction_\Lambda(\sigma) = \sigma_\Lambda = \{\sigma(x), x \in \Lambda\}, \tag{2.1.1.1}$$

and when there is no room for doubt, we will simply write σ_Λ for $\restriction_\Lambda(\sigma)$. We will often use the expression $(\sigma_\Lambda, \sigma_\Delta)$, $\Lambda \cap \Delta = \emptyset$ to denote the element in $\mathcal{X}_{\Lambda \cup \Delta}$ whose restrictions to Λ and Δ are respectively σ_Λ and σ_Δ.

We regard $\mathcal{X}$ as a topological space with the product topology, namely a sequence $\sigma^{(n)} \to \sigma$ if and only if for any $x \in \mathbb{Z}^d$, $\sigma^{(n)}(x) = \sigma(x)$ for all n large enough. A countable basis of open sets is the collection of cylindrical sets. We define

Cylindrical functions and sets *A function f on $\mathcal{X}$ is cylindrical in Δ if it only depends on the restriction σ_Δ of σ to Δ. A set is cylindrical in Δ if its characteristic function is cylindrical in Δ. A function or a set is cylindrical if it is cylindrical in a bounded region Δ. Elementary cylinders are sets of the form $C_{\sigma_\Lambda} = \{\sigma' \in \mathcal{X} : \sigma'_\Lambda = \sigma_\Lambda\}$ with Λ bounded and $\sigma_\Lambda \in \mathcal{X}_\Lambda$. Their collection is denoted by $\mathcal{C}$.*

Cylindrical sets are both open and closed; cylindrical functions are evidently continuous, and vice versa any continuous function can be approximated in sup norm by cylindrical functions; see Appendix A.

2.1.2 Energy

The energy of an Ising spin system is a family $\{H_\Lambda(\sigma_\Lambda)\}$ with Λ running over all the bounded subsets of $\mathbb{Z}^d$ and $\sigma_\Lambda \in \mathcal{X}_\Lambda$. $H_\Lambda(\sigma_\Lambda)$ is the energy in Λ of the spin configuration σ_Λ, it includes all the interactions of the spins of Λ among themselves and the interaction of the spins of Λ with an external magnetic field, if present. The interaction between two disjoint, bounded regions Λ and Δ is then defined by

$$W_{\Lambda,\Delta}(\sigma_\Lambda, \sigma_\Delta) = H_{\Lambda \cup \Delta}(\sigma_\Lambda, \sigma_\Delta) - H_\Lambda(\sigma_\Lambda) - H_\Delta(\sigma_\Delta), \tag{2.1.2.1}$$

where $(\sigma_\Lambda, \sigma_\Delta) \in \mathcal{X}_{\Lambda \sqcup \Delta}$ is the configuration whose restrictions to Λ and Δ are σ_Λ and σ_Δ. The energy in Λ under the field produced by the spins in Δ, $\Lambda \cap \Delta = \emptyset$ is

$$H_{\Lambda,\Delta}(\sigma_\Lambda | \sigma_\Delta) = H_{\Lambda \sqcup \Delta}(\sigma_\Lambda, \sigma_\Delta) - H_\Delta(\sigma_\Delta), \tag{2.1.2.2}$$

and

$$H_{\Lambda,\Lambda^c}(\sigma_\Lambda | \sigma_{\Lambda^c}) = \lim_{\Delta \nearrow \Lambda^c} \big(H_{\Lambda \sqcup \Delta}(\sigma_\Lambda, \sigma_\Delta) - H_\Delta(\sigma_\Delta)\big), \tag{2.1.2.3}$$

if the limit exists independently of the sequence approximating Λ^c. The above notation is redundant because the region appearing as a suffix in the energy can be read off from the spin configurations, so that when no confusion arises we may drop them from the notation. In particular, $H_\Lambda(\sigma_\Lambda | \sigma_{\Lambda^c})$ will always stand for $H_{\Lambda,\Lambda^c}(\sigma_\Lambda | \sigma_{\Lambda^c})$.

We will restrict in the sequel to energies of the form

$$H_\Lambda(\sigma_\Lambda) = -\frac{1}{2} \sum_{x \neq y \in \Lambda} J(x, y)\sigma_\Lambda(x)\sigma_\Lambda(y) - h \sum_{x \in \Lambda} \sigma_\Lambda(x). \tag{2.1.2.4}$$

If Λ is a singleton, $\Lambda = \{x\}$,

$$H_{\{x\}}(\sigma_{\{x\}}(x)) = -h\sigma_{\{x\}}(x).$$

h is interpreted as an external magnetic field and $-h\sigma_\Lambda(x)$ is the energy of the spin at x under the sole influence of the external magnetic field; indeed such a term is the only one surviving when $\Lambda = \{x\}$. $-J(x, y)\sigma_\Lambda(x)\sigma_\Lambda(y)$ is the interaction energy between the spins at x and y as follows from (2.1.2.1) with $\Lambda = \{x\} \cup \{y\}$. Notice finally that (2.1.2.2) with $\{H_\Lambda(\sigma_\Lambda)\}$ as in (2.1.2.4) becomes

$$H_{\Lambda,\Delta}(\sigma_\Lambda | \sigma_\Delta) = H_\Lambda(\sigma_\Lambda) - \sum_{x \in \Lambda} \sum_{y \in \Delta} J(x, y)\sigma_\Lambda(x)\sigma_\Delta(y). \tag{2.1.2.5}$$

Assumptions on the interaction

- *symmetry*: $J(x, y) = J(y, x)$.
- *translational invariance*: $J(x + z, y + z) = J(x, y)$, *for all* x, y *and* z *in* $\mathbb{Z}^d$.
- *summability*: $\sum_{x \neq 0} |J(0, x)| < \infty$ *(by the summability assumption, the series in* (2.1.2.5) *is convergent)*.

Under the above summability assumption, (2.1.2.3) becomes

$$H_\Lambda(\sigma_\Lambda | \sigma_{\Lambda^c}) = H_\Lambda(\sigma_\Lambda) - \sum_{x \in \Lambda} \sum_{y \in \Lambda^c} J(x, y)\sigma_\Lambda(x)\sigma_{\Lambda^c}(y).$$

Examples

- *Classical Ising model*. The only active bonds are those connecting nearest neighbor (n.n.) sites, with ferromagnetic coupling constants all equal to $J > 0$.

- *Kac potentials*. The coupling constants $J_\gamma(x, y)$ depend on a parameter $\gamma > 0$:

$$J_\gamma(x, y) = \gamma^d J(\gamma x, \gamma y),$$

where $J(r, r') = J(r + a, r' + a) \geq 0$, for all r, r' and a in $\mathbb{R}^d$; $J(0, r)$ is continuous with compact support and normalized as a probability kernel:

$$\int_{\mathbb{R}^d} dr\, J(0, r) = 1.$$

We are interested in small γ, a regime characterized by (i) long range interactions $\approx \gamma^{-1}$, and a large, $\approx \gamma^{-d}$, connectivity of each site (i.e. the number of active bonds starting from that site); (ii) the coupling constants of the bonds are small, $\approx \gamma^d$, and (iii) the total strength of a site (i.e. the sum of all the coupling constants of bonds originating from that site) is ≈ 1.
- *Mean field models*. Here the coupling constants depend on the region Λ where the system is studied. If Λ has N sites $J(x, y) = N^{-1}$, x, y in Λ. The model shares the properties (i), (ii) and (iii) of the previous one, which was indeed conceived of as a refinement of the mean field model to correct its various unphysical features. The mean field model, though, has the great advantage of providing a simple and not too unrealistic mechanism for phase transitions.

2.1.3 Potentials

Often the energy of the system is given indirectly by assigning its potential. The potential is a family $\{U_\Delta(\sigma_\Delta)\}$, Δ running over the bounded sets and $\sigma_\Delta \in \mathcal{X}_\Delta$. Given a potential, its energy is

$$H(\sigma_\Lambda) = \sum_{\Delta \subseteq \Lambda} U_\Delta(\sigma_\Delta), \tag{2.1.3.1}$$

where σ_Δ above is the restriction to Δ of σ_Λ.

As already observed the potential associated to the energy (2.1.2.4) is made by the potential with $U_{\{x\}}(\sigma_{\{x\}}) = -h\sigma_{\{x\}}(x)$, $U_{\{x,y\}}(\sigma_{\{x,y\}}) = -J(x, y)\sigma_{\{x,y\}}(x) \times \sigma_{\{x,y\}}(y)$, while all other $U_\Delta = 0$.

The relation (2.1.3.1) can be inverted, namely, given $\{H_\Lambda(\sigma_\Lambda)\}$, we can recover uniquely $\{U_\Delta(\sigma_\Delta)\}$ in such a way that (2.1.3.1) holds. This is done iteratively starting from sets of cardinality 1 and then increasing progressively the cardinality, noticing that the potential with the set having maximal cardinality can be expressed using (2.1.3.1) in terms of the energy and of the potentials with smaller cardinality.

As stated earlier we will restrict ourselves to one and two body potentials only, but in magnetic systems also quadrupole and multipole interactions may be relevant. Theoretically the many body interactions are quite important, and an example is provided in Sect. 2.4.2 where a relation is established between DLR measures and the thermodynamic pressure as a function of the general many body potential. In general, many body potentials arise after coarse graining transformations and describe effective hamiltonians, as discussed in Chaps. 9 and 10.

Example Potentials are often written as

$$U_{\{x_1,\ldots,x_n\}}(\sigma_{\{x_1,\ldots,x_n\}}) = -J(x_1,\ldots,x_n)\sigma_{\{x_1,\ldots,x_n\}}(x_1)\cdots\sigma_{\{x_1,\ldots,x_n\}}(x_n), \quad (2.1.3.2)$$

where $J(x_1,\ldots,x_n)$ is a symmetric function on $(\mathbb{Z}^d)^n$. Then (2.1.3.1) yields

$$H(\sigma_\Lambda) = -\sum_{n=1}^{|\Lambda|}\frac{1}{n!}\sum_{x_1\neq\cdots\neq x_n\in\Lambda} J(x_1,\ldots,x_n)\sigma_\Lambda(x_1)\cdots\sigma_\Lambda(x_n).$$

One may wonder why one would take only a multi-linear dependence on the spins in (2.1.3.2). This is specific of Ising spins where all functions are necessarily sums of multi-linear terms as shown below. A function f cylindrical in Δ, Δ bounded, can be written as

$$f(\sigma) = \sum_{a_\Delta\in\mathcal{X}_\Delta} f(a_\Delta)\prod_{x\in\Delta}\frac{1+a_\Delta(x)\sigma(x)}{2}, \quad (2.1.3.3)$$

where $f(a_\Delta)$ denotes the values of $f(\sigma)$ when the restriction of σ to Δ is a_Δ. Thus any cylindrical function in Δ is a multi-linear polynomial of the variables $\{\sigma(x), x\in\Delta\}$. In particular, if C is a cylindrical set in Δ, Δ bounded,

$$\mathbf{1}_{\sigma\in C} = \sum_{a_\Delta\in C}\prod_{x\in\Delta}\frac{1+a_\Delta(x)\sigma(x)}{2}.$$

2.1.4 Finite volume Gibbs measures

In the context of the Ising model the Gibbs hypothesis says that the equilibrium state at temperature T of the Ising spin system in a bounded region Λ is described by a probability distribution on $\mathcal{X}_\Lambda$ given by the formula

$$\text{probability of observing } \sigma_\Lambda = \frac{e^{-\beta E_\Lambda(\sigma_\Lambda)}}{Z_\Lambda}, \quad (2.1.4.1)$$

where $\beta = 1/kT$, $E_\Lambda(\sigma_\Lambda)$ is the energy of the configuration σ_Λ and Z_Λ a normalization constant. The energy E_Λ is not necessarily the same as the energy H_Λ of (2.1.2.4), as there may be interactions with the "walls" and/or with "the world" outside Λ.

Remarks

Equation (2.1.4.1) is supposed to describe the equilibrium states of the collection of spins σ_Λ in contact with a thermal reservoir at fixed temperature T which exchanges energy with the spins in Λ at a rate determined by T. The physical assumption is that the interaction is so weak that the energy levels E_Λ are not affected by the reservoir, yet it is at the same time so strong that it eventually drives the system to a final equilibrium, independently of its initial conditions. Equilibrium is eventually established throughout the system due to complex mechanisms which involve various space and time scales, hydrodynamic behaviors and all the other relevant non-equilibrium phenomena. But no matter how complicated the pattern to equilibrium is, according to Gibbs the final equilibrium has the very simple expression (2.1.4.1).

The physical meaning of representing a state as a probability and in particular the equilibrium state as in (2.1.4.1) is that if we (ideally) make repeated observations of the system after equilibrium is established, the frequency with which we observe a configuration σ_Λ is given by the Gibbs formula (2.1.4.1).

The energy $E_\Lambda(\sigma_\Lambda)$

With reference to the system defined in Sect. 2.1.2, the energy $E_\Lambda(\sigma_\Lambda)$ is not necessarily the same as the energy $H_\Lambda(\sigma_\Lambda)$ of (2.1.2.4), as there may be interactions with the "walls," namely interactions with "the world" outside Λ. They are usually quite complex, but they affect significantly only the spins close to the boundaries of Λ. We may thus suppose that

$$E_\Lambda(\sigma_\Lambda) = H_\Lambda(\sigma_\Lambda) - \sum_{x\in\Lambda} h_x^{\Lambda^c}\sigma_\Lambda(x), \qquad (2.1.4.2)$$

where $h_x^{\Lambda^c}$ is an "effective" magnetic field which takes into account the interactions with Λ^c and which decays as $\mathrm{dist}(x,\Lambda^c)$ increases. If the interaction has many body potentials, more general expressions will arise, but for simplicity we will only consider here the case (2.1.4.2). The energy (2.1.4.2) refers to the ideal case where the effective magnetic field $h_x^{\Lambda^c}$ does not fluctuate, i.e. when outside Λ everything is frozen and does not change in time. We will also suppose (most of the times) that $h_x^{\Lambda^c}$ arises from the interaction with a fixed spin configuration σ_{Λ^c} so that

$$h_x^{\Lambda^c} = -\sum_{y\in\Lambda^c} J(x,y)\sigma_{\Lambda^c}(y), \quad \text{and} \quad E_\Lambda(\sigma_\Lambda) = H_\Lambda(\sigma_\Lambda|\sigma_{\Lambda^c}). \qquad (2.1.4.3)$$

This is not a very realistic model of a wall, but it has considerable theoretical importance as we shall see.

The Gibbs measure $G_\Lambda(\sigma_\Lambda|\sigma_{\Lambda^c})$

When we use the choice (2.1.4.3) we will write (2.1.4.1) as

$$G_\Lambda(\sigma_\Lambda|\sigma_{\Lambda^c}) = \frac{e^{-\beta H_\Lambda(\sigma_\Lambda|\sigma_{\Lambda^c})}}{Z_\Lambda(\sigma_{\Lambda^c})}, \quad Z_\Lambda(\sigma_{\Lambda^c}) = \sum_{\sigma'_\Lambda \in \mathcal{X}_\Lambda} e^{-\beta H_\Lambda(\sigma'_\Lambda|\sigma_{\Lambda^c})}. \tag{2.1.4.4}$$

The dependence on β and on the parameters defining the energy are not made explicit; if necessary they will be added as subscripts. The "partition function" $Z_\Lambda(\sigma_{\Lambda^c})$, which in (2.1.4.4) appears just as a normalization factor, has instead a meaning that is important physically, being directly related to the thermodynamic pressures; see Sect. 2.3.

By abuse of notation, we will use the same symbol $G_\Lambda(\cdot|\sigma_{\Lambda^c})$ for the probability on $\mathcal{X}$ which is a finite sum of Dirac deltas, namely calling $\{\sigma'\}$ the set consisting of the singleton σ', $G_\Lambda(\{\sigma'\}|\sigma_{\Lambda^c})$ is equal to 0 for all σ' except those whose restriction to Λ^c is exactly σ_{Λ^c}; thus

$$G_\Lambda(\{\sigma'\}|\sigma_{\Lambda^c}) = \mathbf{1}_{\sigma'_{\Lambda^c}=\sigma_{\Lambda^c}} \frac{e^{-\beta H_\Lambda(\sigma'_\Lambda|\sigma_{\Lambda^c})}}{Z_\Lambda(\sigma_{\Lambda^c})}, \tag{2.1.4.5}$$

where, to make notation easier, we may just write $G_\Lambda(\sigma'|\sigma_{\Lambda^c})$ for $G_\Lambda(\{\sigma'\}|\sigma_{\Lambda^c})$. We shall switch from (2.1.4.4) to (2.1.4.5) freely, and the reader should understand from the context the meaning of $G_\Lambda(\cdot|\sigma_{\Lambda^c})$.

2.1.5 *A consistency property of Gibbs measures*

In this subsection we will prove that the conditional probability of a Gibbs measure is also Gibbs (hence the title of the subsection) and show that this has a nice physical meaning.

Marginal distributions

Let m_Λ be a probability on $\mathcal{X}_\Lambda$, Λ a bounded set in $\mathbb{Z}^d$, and Δ a proper subset of Λ. Then the marginal distribution [or simply the marginal] of m_Λ on $\mathcal{X}_\Delta$ is the probability $(m_\Lambda)_\Delta$ on $\mathcal{X}_\Delta$ defined by

$$(m_\Lambda)_\Delta(\sigma^*_\Delta) = \sum_{\sigma_\Lambda:\sigma_\Lambda(x)=\sigma^*_\Delta(x)\, x\in\Delta} m_\Lambda(\sigma_\Lambda) \equiv \sum_{\sigma_{\Lambda\setminus\Delta}\in\mathcal{X}_{\Lambda\setminus\Delta}} m_\Lambda(\sigma^*_\Delta, \sigma_{\Lambda\setminus\Delta}).$$

The marginal of $G_\Lambda(\sigma_\Lambda|\sigma_{\Lambda^c})$ on $\mathcal{X}_{\Lambda\setminus\Delta}$ is

$$(G_\Lambda(\cdot|\sigma_{\Lambda^c}))_{\Lambda\setminus\Delta}(\sigma^*_{\Lambda\setminus\Delta}) = \sum_{\sigma_\Lambda:\sigma_\Lambda=\sigma^*_{\Lambda\setminus\Delta} \text{ on } \Lambda\setminus\Delta} G_\Lambda(\sigma_\Lambda|\sigma_{\Lambda^c}).$$

Theorem 2.1.5.1 *With the above notation*

$$G_\Delta(\sigma^*_\Delta|\sigma^*_{\Lambda\setminus\Delta},\sigma_{\Lambda^c}) = \frac{G_\Lambda(\sigma^*_\Delta,\sigma^*_{\Lambda\setminus\Delta}|\sigma_{\Lambda^c})}{(G_\Lambda(\cdot|\sigma_{\Lambda^c}))_{\Lambda\setminus\Delta}(\sigma^*_{\Lambda\setminus\Delta})}. \tag{2.1.5.1}$$

Proof

$$\text{r.h.s. of (2.1.5.1)} = \frac{e^{-\beta H_\Lambda(\sigma^*_\Delta,\sigma^*_{\Lambda\setminus\Delta}|\sigma_{\Lambda^c})}}{\sum_{\sigma_\Delta} e^{-\beta H_\Lambda(\sigma_\Delta,\sigma^*_{\Lambda\setminus\Delta}|\sigma_{\Lambda^c})}}. \tag{2.1.5.2}$$

Equation (2.1.5.1) then follows by using the identity

$$H_\Lambda(\sigma'_\Delta,\sigma^*_{\Lambda\setminus\Delta}|\sigma_{\Lambda^c}) = H_\Delta(\sigma'_\Delta|\sigma^*_{\Lambda\setminus\Delta},\sigma_{\Lambda^c}) + H_{\Lambda\setminus\Delta,\Lambda^c}(\sigma^*_{\Lambda\setminus\Delta}|\sigma_{\Lambda^c});$$

see (2.1.2.5) for notation, with $\sigma'_\Delta = \sigma^*_\Delta$ in the numerator of the fraction in (2.1.5.2) and $\sigma'_\Delta = \sigma_\Delta$ in the denominator. □

Remarks

Equation (2.1.5.1) has a nice physical interpretation. Suppose we make an ideal experiment where we measure N times the state of the system in equilibrium. By the Gibbs hypothesis the number of times we find in $\Lambda\setminus\Delta$ the configuration $\sigma^*_{\Lambda\setminus\Delta}$ is $\approx (G_\Lambda(\cdot|\sigma_{\Lambda^c}))_{\Lambda\setminus\Delta}(\sigma^*_{\Lambda\setminus\Delta})N$. We then select those experiments where we have seen $\sigma^*_{\Lambda\setminus\Delta}$ and count the relative frequency of appearance of $\sigma^*_\Delta \in \mathcal{X}_\Delta$, namely the number of times when we see both σ^*_Δ and $\sigma^*_{\Lambda\setminus\Delta}$, that is $\approx N \times G_\Lambda(\sigma^*_\Delta,\sigma^*_{\Lambda\setminus\Delta}|\sigma_{\Lambda^c})$, over the number of times when we see $\sigma^*_{\Lambda\setminus\Delta}$, i.e. $\approx N \times (G_\Lambda(\cdot|\sigma_{\Lambda^c}))_{\Lambda\setminus\Delta}(\sigma^*_{\Lambda\setminus\Delta})$. Thus the conditional frequency in the limit $N\to\infty$ is $\frac{G_\Lambda(\sigma^*_\Delta,\sigma^*_{\Lambda\setminus\Delta}|\sigma_{\Lambda^c})}{(G_\Lambda(\cdot|\sigma_{\Lambda^c}))_{\Lambda\setminus\Delta}(\sigma^*_{\Lambda\setminus\Delta})}$ which is equal to the r.h.s. of (2.1.5.1), hence to $G_\Delta(\sigma^*_\Delta|\sigma^*_{\Lambda\setminus\Delta},\sigma_{\Lambda^c})$, i.e. to the Gibbs probability in Δ as if the spins in $\Lambda\setminus\Delta$ were frozen, just like those in Λ^c. Thus the Gibbs probability $G_\Delta(\sigma^*_\Delta|\sigma^*_{\Lambda\setminus\Delta},\sigma_{\Lambda^c})$ is not only the probability of observing σ^*_Δ when $\sigma^*_{\Lambda\setminus\Delta}$ and σ_{Λ^c} are frozen, but also the conditional probability of observing σ^*_Δ conditioned to having observed $\sigma^*_{\Lambda\setminus\Delta}$ (and with σ_{Λ^c} frozen).

Equation (2.1.5.1) can be read as meaning that *he conditional* $G_\Lambda(\cdot|\sigma_{\Lambda^c})$-*probability of* σ^*_Δ *given* $\sigma^*_{\Lambda\setminus\Delta}$ *is equal to* $G_\Delta(\sigma^*_\Delta|\sigma^*_{\Lambda\setminus\Delta},\sigma_{\Lambda^c})$ because we have the following.

Conditional probabilities

Let m be a probability on a space Ω, $C\subset\Omega$, $m(C)>0$. Then the ratio $\frac{m(A\cap C)}{m(C)} =: m(A|C)$ is the m-conditional probability of A given C. If $\pi = (C_1,\ldots, C_n)$ is a partition of Ω with $m(C_i)>0$, $i=1,\ldots,n$, then

$$m(A) = \sum_{i=1}^n m(A|C_i)m(C_i). \tag{2.1.5.3}$$

Thus (2.1.5.1) is just the statement at the beginning of the subsection that the Gibbs $G_\Lambda(\cdot|\sigma_{\Lambda^c})$ conditional probability of observing σ^*_Δ given $\sigma^*_{\Lambda\setminus\Delta}$ is equal to $G_\Delta(\sigma^*_\Delta|\sigma^*_{\Lambda\setminus\Delta},\sigma_{\Lambda^c})$. Indeed the numerator on the r.h.s. of (2.1.5.1) is the probability of having σ^*_Δ in Δ intersected with the event $\sigma^*_{\Lambda\setminus\Delta}$ in $\Lambda\setminus\Delta$; the denominator, being the marginal probability of $\sigma^*_{\Lambda\setminus\Delta}$, is the probability of $\sigma^*_{\Lambda\setminus\Delta}$. The ratio is by definition the conditional probability of σ^*_Δ given $\sigma^*_{\Lambda\setminus\Delta}$.

Consider (2.1.5.3) with the partition $\pi=\{C_{\sigma^*_{\Lambda\setminus\Delta}},\sigma^*_{\Lambda\setminus\Delta}\in\mathcal{X}_{\Lambda\setminus\Delta}\}$, $C_{\sigma^*_{\Lambda\setminus\Delta}}$ being the set of all σ_Λ whose restriction to $\Lambda\setminus\Delta$ is $\sigma^*_{\Lambda\setminus\Delta}$. Then (2.1.5.3) reads

$$(G_\Lambda(\cdot|\sigma_{\Lambda^c}))_\Delta(\sigma^*_\Delta)$$
$$=\sum_{\sigma^*_{\Lambda\setminus\Delta}\in\mathcal{X}_{\Lambda\setminus\Delta}} G_\Delta(\sigma^*_\Delta|\sigma^*_{\Lambda\setminus\Delta},\sigma_{\Lambda^c})\,(G_\Lambda(\cdot|\sigma_{\Lambda^c}))_{\Lambda\setminus\Delta}(\sigma^*_{\Lambda\setminus\Delta}). \qquad (2.1.5.4)$$

2.1.6 *Random boundary conditions*

The Gibbs measures $G_\Lambda(\cdot|\sigma_{\Lambda^c})$ cannot exhaust the set of all equilibrium measures in Λ as the correct notion should reflect the physical request that if a system is in equilibrium in a region Λ then it is also in equilibrium in all subregions Δ of Λ. However, the equilibrium state $G_\Lambda(\cdot|\sigma_{\Lambda^c})$ observed in the subregion Δ is described by the marginal of $G_\Lambda(\cdot|\sigma_{\Lambda^c})$ on $\mathcal{X}_\Delta$, and it is thus given by (2.1.5.4) which does not have the expression (2.1.4.1) unless $\Delta=\Lambda$.

We thus need to relax the statement that the equilibrium measures have the form $G_\Lambda(\cdot|\sigma_{\Lambda^c})$. Since by (2.1.5.4) the marginal of $G_\Lambda(\cdot|\sigma_{\Lambda^c})$ on $\mathcal{X}_\Delta$ is a convex combination of measures $G_\Delta(\sigma_\Delta|\sigma_{\Delta^c})$, we certainly want to include among the equilibrium measures in Δ probabilities of the form

$$\mu(\sigma^*_\Delta)=\sum_{\sigma^*_{\Lambda\setminus\Delta}\in\mathcal{X}_{\Lambda\setminus\Delta}} G_\Delta(\sigma^*_\Delta|\sigma^*_{\Lambda\setminus\Delta},\sigma_{\Lambda^c})\,m(\sigma^*_{\Lambda\setminus\Delta}), \qquad (2.1.6.1)$$

with m a probability on $\mathcal{X}_{\Lambda\setminus\Delta}$. Referring to Appendix A for definitions and notation relative to measures on Ising spaces we formalize the above considerations as follows.

Definition A probability μ on $\mathcal{X}$ is a "*Gibbs measure in Λ with random boundary conditions*" if ($\mu(f)$ is the integral of f)

$$\mu(f)=\int_{\mathcal{X}}\sum_{\sigma'\in\mathcal{X}} G_\Lambda(\sigma'|\sigma_{\Lambda^c})f(\sigma')\,\nu(d\sigma),$$
$$\text{for any bounded, measurable } f, \qquad (2.1.6.2)$$

where ν is a Borel probability on $\mathcal{X}$ and measurability means Borel measurability. $\mathcal{G}_\Lambda$ denotes the set of Gibbs measures in Λ with random boundary conditions and it is identified with the set of equilibrium measures in Λ.

The integral on the r.h.s. of (2.1.6.2) is well defined because

Lemma 2.1.6.1 *For any bounded, measurable function* f,

$$b_f(\sigma) := G_\Lambda\big(f|\sigma_{\Lambda^c}\big) \equiv \sum_{\sigma'\in\mathcal{X}} G_\Lambda\big(\sigma'|\sigma_{\Lambda^c}\big) f(\sigma') \tag{2.1.6.3}$$

is a bounded, measurable function of σ, *which is continuous if* f *is.*

Proof Obviously $|b_f(\sigma)| \le \|f\|_\infty$. Writing $(\sigma'_\Lambda, \sigma_{\Lambda^c})$ for the configuration whose restrictions to Λ and Λ^c are σ'_Λ and σ_{Λ^c}, we have

$$b_f(\sigma) = \sum_{\sigma'_\Lambda} G_\Lambda\big((\sigma'_\Lambda, \sigma_{\Lambda^c})|\sigma_{\Lambda^c}\big) f\big((\sigma'_\Lambda, \sigma_{\Lambda^c})\big).$$

Since the sum is finite, it is sufficient to prove that each term is measurable. $\sigma \to f((\sigma'_\Lambda, \sigma_{\Lambda^c}))$ is measurable by assumption, while

$$\big|G_\Lambda\big(\sigma'_\Lambda|\sigma_{\Lambda^c}\big) - G_\Lambda\big(\sigma'_\Lambda|\sigma^N_{\Lambda^c}\big)\big| \le \epsilon(N), \tag{2.1.6.4}$$

where σ^N is obtained from σ by replacing $\sigma(x)$ by $+1$ whenever $|x| > N$ and $\epsilon(N)$ is a function of N which vanishes as $N \to \infty$. (2.1.6.4) follows from the summability assumption on the interaction.

Since $\sigma \to G_\Lambda(\sigma'_\Lambda|\sigma^N_{\Lambda^c})$ is continuous (being a cylindrical function), the above shows that $G_\Lambda(\sigma'_\Lambda|\sigma_{\Lambda^c})$ is approximated in sup norm by cylindrical functions and it is therefore itself a continuous function of σ (see Theorem A.1). We have proved that $b_f(\sigma)$ is measurable and continuous as well if f is continuous; hence Lemma 2.1.6.1 is proved. □

We conclude the subsection with another lemma which will play an important role in the sequel.

Lemma 2.1.6.2 $\mu \in \mathcal{G}_\Lambda$ *if and only if for any bounded, measurable* f

$$\mu(f) = \int_{\mathcal{X}} \sum_{\sigma'} G_\Lambda\big(\sigma'|\sigma_{\Lambda^c}\big) f(\sigma')\, \mu(d\sigma). \tag{2.1.6.5}$$

Proof μ is obviously in $\mathcal{G}_\Lambda$ if (2.1.6.5) holds. If μ is given by (2.1.6.2), then for any f $\mu(f) = \nu(b_f)$, b_f as in (2.1.6.3). We denote by $\mathcal{B}_{\Lambda^c}$ the σ-algebra generated by the cylindrical sets in Λ^c. If f is $\mathcal{B}_{\Lambda^c}$ measurable by (2.1.6.3) $b_f(\sigma) = f(\sigma)$, and since in general $\mu(f) = \nu(b_f)$ we have $\mu(f) = \nu(f)$ for all $\mathcal{B}_{\Lambda^c}$ measurable functions f. Take now any (bounded, measurable) f, then $\mu(f) = \nu(b_f)$, by (2.1.6.2), and since b_f is $\mathcal{B}_{\Lambda^c}$ measurable, $\mu(f) = \nu(b_f) = \mu(b_f)$, which is (2.1.6.5). □

Remarks Let ν in (2.1.6.2) be the measure supported by the single configuration σ, then $\mu = G_\Lambda(\cdot|\sigma_{\Lambda^c})$, thus $G_\Lambda(\cdot|\sigma_{\Lambda^c}) \in \mathcal{G}_\Lambda$ as implicit in the whole discussion so far.

2.1.7 Structure of finite volume Gibbs measures

With the identification of the equilibrium measures as Gibbs measures with random boundary conditions, the original Gibbs measures $G_\Lambda(\cdot|\sigma_{\Lambda^c})$ do not only belong to $\mathcal{G}_\Lambda$ as stated in the last remark of Sect. 2.1.6, but, as a consequence of (2.1.5.4), they are also in any $\mathcal{G}_\Delta$, $\Delta \subset \Lambda$. Indeed, we have the following.

Theorem 2.1.7.1 *$\mathcal{G}_\Lambda$ is a convex, weakly compact set and*

$$\mathcal{G}_\Lambda \subset \mathcal{G}_\Delta, \quad \text{for any } \Delta \subset \Lambda.$$

In particular, $G_\Lambda(\cdot|\sigma_{\Lambda^c}) \in \mathcal{G}_\Delta$ for any $\Delta \subset \Lambda$.

Proof It directly follows from the representation (2.1.6.2) that $\mathcal{G}_\Lambda$ is a convex set. We will next prove that $\mathcal{G}_\Lambda$ is closed (and since the space $M(\mathcal{X})$ of all probabilities on $\mathcal{X}$ is compact, see Sect. A.4 of Appendix A, it will also follow that $\mathcal{G}_\Lambda$ is compact).

Let $\mu_n \in \mathcal{G}_\Lambda$, $n \geq 1$, and $\mu_n \to \mu$ weakly, i.e. for all continuous f ($f \in C(\mathcal{X})$), $\mu_n(f) \to \mu(f)$. We need to prove that $\mu \in \mathcal{G}_\Lambda$. Let $f \in C(\mathcal{X})$. By (2.1.6.5), $\mu_n(f) = \mu_n(b_f)$, b_f as in (2.1.6.3). We have proved in Lemma 2.1.6 that $b_f \in C(\mathcal{X})$, i.e. it is continuous, since we are supposing that $f \in C(\mathcal{X})$. By letting $n \to \infty$, $\mu(f) = \mu(b_f)$, which, recalling the expression (2.1.6.3) for b_f proves (2.1.6.2) for continuous f. On the other hand if the integrals of all continuous functions are equal, then the two measures are the same and (2.1.6.2) holds also for all bounded, measurable f, so that $\mu \in \mathcal{G}_\Lambda$ and the latter is weakly closed.

By (2.1.5.1), $G_\Lambda(f|\sigma_{\Lambda^c})$, f a bounded measurable function, is equal to

$$\sum_{\sigma'_{\Lambda\setminus\Delta}} \sum_{\sigma_\Delta} f(\sigma_\Delta, \sigma'_{\Lambda\setminus\Delta}, \sigma_{\Lambda^c}) G_\Delta(\sigma_\Delta|\sigma'_{\Lambda\setminus\Delta}, \sigma_{\Lambda^c}) (G_\Lambda(\cdot|\sigma_{\Lambda^c}))_{\Lambda\setminus\Delta}(\sigma'_{\Lambda\setminus\Delta}),$$

which can be rewritten as

$$\sum_{\sigma' \in \mathcal{X}} \left\{ \sum_{\sigma_\Delta} f(\sigma_\Delta, \sigma'_{\Lambda\setminus\Delta}, \sigma_{\Lambda^c}) G_\Delta(\sigma_\Delta|\sigma'_{\Lambda\setminus\Delta}, \sigma_{\Lambda^c}) \right\} G_\Lambda(\sigma'|\sigma_{\Lambda^c}),$$

so that $G_\Lambda(\cdot|\sigma_{\Lambda^c}) \in \mathcal{G}_\Delta$.

Let now $\mu \in \mathcal{G}_\Lambda$; then $\mu(f) = \int \{\sum_{\sigma'} G_\Lambda(\sigma'|\sigma_{\Lambda^c}) f(\sigma')\}\, \mu(d\sigma)$, hence

$$\mu(f) = \int \left\{ \sum_{\sigma'} G_\Lambda(\sigma'|\sigma_{\Lambda^c}) \sum_{\sigma''} G_\Delta(\sigma''|\sigma'_{\Delta^c}) f(\sigma'') \right\} \mu(d\sigma).$$

Then

$$\mu(f) = \int \sum_{\sigma''} f(\sigma'') G_\Delta(\sigma''|\sigma'_{\Delta^c})\, \nu(d\sigma'), \tag{2.1.7.1}$$

where ν is such that $\nu(g) = \int \sum_{\sigma'} G_\Lambda(\sigma'|\sigma_{\Lambda^c}) g(\sigma')\, \mu(d\sigma)$. Since ν is a probability, the last expression in (2.1.7.1) is by (2.1.6.2) in $\mathcal{G}_\Delta$ and the theorem is proved. □

2.2 Thermodynamic limit and DLR measures

We will derive a macroscopic theory from the Gibbs hypothesis by separating bulk from surface effects. Bulk properties are those which refer to the behavior of the spins [in our Ising model] which are far from the boundaries; if the region occupied by the system is not too weird, indeed most spins will be far from the boundaries. To make this quantitative, we must specify the meaning of "most" and "far from the boundaries." There is here an evident degree of arbitrariness, which has to be lifted if we want a mathematical theory with precise statements. Modern statistical mechanics defines the bulk properties as those which emerge in an infinite volume limit, which is usually referred to as "the thermodynamic limit." In this limit in fact the bulk thermodynamics of the system is singled out.

As we will see in the next two sections the notion is well defined for intensive thermodynamic potentials like pressure and free energy density, for which, under quite general assumptions on the interaction, the infinite volume limit exists and is essentially independent of the sequence of regions and boundary conditions used in the limit procedure. The situation is different when we look at the full equilibrium states, the object of our present analysis, where, as we will argue below, we cannot expect in general the existence of a limit independent of the sequence of regions and boundary conditions.

The modern rigorous theory of equilibrium states is founded on an hypothesis which avoids [or better, it seems to avoid] the thermodynamic limit procedure by extending the original Gibbs hypothesis to one formulated directly in infinite systems, the so called DLR condition. DLR stands for Dobrushin, Lanford and Ruelle, who are the founders of the theory. The condition translates the physically obvious notion that if a system is globally in equilibrium, then it is also locally in equilibrium. According to what we have argued so far, a state is in equilibrium in a bounded region Λ if it is in $\mathcal{G}_\Lambda$; thus DLR define the set of all equilibrium measures $\mathcal{G}$ as

$$\mathcal{G} := \bigcap_{\Lambda \text{ bounded in } \mathbb{Z}^d} \mathcal{G}_\Lambda. \tag{2.2.0.1}$$

The definition (2.2.0.1) immediately raises three questions: the existence of DLR measures, the meaning of possible non-uniqueness, and the way the notion is related to the thermodynamic limit procedure discussed earlier.

We shall prove that $\mathcal{G}$ is non-empty, weakly compact and convex. Convexity allows one to distinguish in $\mathcal{G}$ extremal elements and mixtures of extremal elements.

We will see that the extremal measures in $\mathcal{G}$ are obtained as infinite volume limits, $\Delta_n \to \mathbb{Z}^d$, of Gibbs measures $G_{\Delta_n}(\cdot|\sigma_{\Delta_n^c})$, where σ is a fixed configuration and $\sigma_{\Delta_n^c}$ is the restriction of σ to Δ_n^c. They will be interpreted as pure phases. The non-uniqueness of the limit then means that $\mathcal{G}$ has several pure phases (selected by the appropriate boundary conditions) and we have a phase transition. Thus phase transitions are related to a persistent diversity among Gibbs states in large domains Λ when the boundary conditions are varied: from such a perspective, phase transitions means "sensitive dependence" on the boundary conditions. As we vary the spins at the boundary, we cause a chain reaction which propagates inside Λ, affecting eventually all the spins; thus a volume effect is produced by a comparatively small surface change. The context when phase transitions occur must therefore be critical and phase transitions rare. Indeed, thermodynamics tells us that they occur on surfaces [of the phase diagram] with positive codimension; in this sense they are "exceptional." In the next section we will see that a formulation of this property (that unfortunately is very weak) is true in general.

Outline of the main results As already mentioned, the set $\mathcal{G}$ of DLR measures (at fixed inverse temperature β) is a non-empty, convex, weakly compact set, just as $\mathcal{G}_\Lambda$ and indeed $\mathcal{G}$ is structurally similar to $\mathcal{G}_\Lambda$, and we may in fact think of $\mathcal{G}$ as $\mathcal{G}_\Lambda$ with $\Lambda = \mathbb{Z}^d$, as we are going to argue. Recall that for finite Λ the extremal elements of $\mathcal{G}_\Lambda$ are obtained by taking any $\sigma \in \mathcal{X}$ and constructing the measure $G_\Lambda(\cdot|\sigma_{\Lambda^c})$. Any element in $\mathcal{G}_\Lambda$ can then be written as $\int G_\Lambda(\cdot|\sigma_{\Lambda^c})p(d\sigma)$, p a probability on $\mathcal{X}$; it is namely a convex combination of extremal states, and any such integral defines an element of $\mathcal{G}_\Lambda$. The extremal elements of $\mathcal{G}$ are obtained in a similar way. Fix arbitrarily an increasing sequence $\{\Delta_n\}$ of regions invading $\mathbb{Z}^d$; take any configuration σ, not as before in the whole $\mathcal{X}$, but only in a suitable set $\mathcal{X}_{gg}$ (which depends on $\{\Delta_n\}$, "gg" for very good); take as before the measure $G_{\Delta_n}(\cdot|\sigma_{\Delta_n^c})$ and let $\Delta_n \to \mathbb{Z}^d$. The weak limit, which is proved to exist in $\mathcal{X}_{gg}$, defines an extremal measure G_σ in $\mathcal{G}$. It is also true that any element in $\mathcal{G}$ is an integral $\int G_\sigma(\cdot)p(d\sigma)$ with p a probability with support on $\mathcal{X}_{gg}$.

To continue with the analogy between $\mathcal{G}_\Lambda$ and $\mathcal{G}$, observe that the measures $\{G_\Lambda(\cdot|\sigma_{\Lambda^c}), \sigma \in \mathcal{X}\}$ define a natural partition π_Λ of $\mathcal{X}$: σ' and σ'' are in the same atom of π_Λ if and only if $G_\Lambda(\sigma_\Lambda|\sigma'_{\Lambda^c}) = G_\Lambda(\sigma_\Lambda|\sigma''_{\Lambda^c})$ for all $\sigma_\Lambda \in \mathcal{X}_\Lambda$. If we change σ' only inside Λ, the new configuration σ'' is trivially in the same atom as σ', and the partition is thus called measurable on Λ^c. Analogously the measures $\{G_\sigma, \sigma \in \mathcal{X}_{gg}\}$ are extremal in $\mathcal{G}$ and define a partition π_∞ of $\mathcal{X}_{gg}$ by the equivalence relation $\sigma' \sim \sigma''$ if and only if $G_{\sigma'} = G_{\sigma''}$. The atoms Ω_σ are such that if $\sigma' \in \Omega_\sigma$ then any modification of σ in a bounded set gives rise to a new configuration which however is in the same Ω_σ, for this reason the partition π_∞ is said to be "measurable at infinity." Moreover, let Ω_σ, $\sigma \in \mathcal{X}_{gg}$ the atom of π_∞ containing σ, then $G_\sigma(\Omega_\sigma) = 1$, and distinct extremal measures have therefore disjoint support and the decomposition of an element of $\mathcal{G}$ as $\int G_\sigma(\cdot)p(d\sigma)$ is unique.

We will conclude the section by discussing the group of space translations and its action on $\mathcal{G}$; see Sects. 2.2.6 and 2.2.7. We shall introduce the notion of "ergodic DLR measures" and prove that any translation invariant DLR measure is an integral over ergodic DLR measures, an ergodic decomposition.

2.2.1 DLR measures

We fix hereafter $\beta > 0$ and drop it from the notation when no ambiguity may arise; all measures in the sequel are meant as Gibbs measures at the inverse temperature β. With such an understanding we define:

Definition 1 A probability measure μ on $\mathcal{X}$ is an *equilibrium measure* if μ belongs to $\mathcal{G}_\Lambda$ for any bounded Λ in $\mathbb{Z}^d$. The set of all equilibrium measures is denoted by $\mathcal{G}$ with $\mathcal{G}$ given in (2.2.0.1).

Theorem 2.2.1.1 *The set $\mathcal{G}$ is non-empty, convex and weakly compact. Moreover, if Δ_n is any sequence of increasing regions which invades $\mathbb{Z}^d$, $\mathcal{G}_{\Delta_n}$ is non-increasing and*

$$\mathcal{G} = \bigcap_{\Delta_n} \mathcal{G}_{\Delta_n}. \tag{2.2.1.1}$$

Proof (2.2.1.1) follows from (2.2.0.1) because $\mathcal{G}_\Lambda \subset \mathcal{G}_\Delta$ if $\Delta \subset \Lambda$ (by Theorem 2.1.7.1). The l.h.s. of (2.2.1.1) is a non-empty, convex, weakly compact set, because all $\mathcal{G}_{\Delta_n}$ are convex and weakly compact sets (again by Theorem 2.1.7.1), each one containing the successive one. □

Thus Definition 1 is non-empty and equilibrium measures do indeed exist. The elements of $\mathcal{G}$ have a nice interpretation in terms of conditional probabilities which, as we will see, is the key to their analysis. The definition and properties of conditional probabilities are given in Appendix A, see Sect. A.6; all proofs hereafter strongly depend on Appendix A. Call $\mathcal{B}_{\Lambda^c}$, Λ a finite subset of $\mathbb{Z}^d$, the minimal σ algebra which contains all the cylinders in Λ^c.

Definition 2 [DLR measures] A probability measure μ on $\mathcal{X}$ is called *DLR* if for any bounded $\Lambda \subset \mathbb{Z}^d$, $\{\mathcal{X}, G_\Lambda(\cdot|\sigma_{\Lambda^c})\}$ is a version of the conditional probability of μ given $\mathcal{B}_{\Lambda^c}$; see Sect. A.6.

Theorem 2.2.1.2 *μ is DLR if and only if $\mu \in \mathcal{G}$.*

Proof Suppose $\mu \in \mathcal{G}$ and let Λ be a bounded set. Then $\mu \in \mathcal{G}_\Lambda$ and by (2.1.6.5) with $f = \mathbf{1}_{A\cap B}$, $A \in \mathcal{B}$, $B \in \mathcal{B}_{\Lambda^c}$,

$$\mu(A\cap B) = \int_{\mathcal{X}} \sum_{\sigma'} G_\Lambda\big(\sigma'|\sigma_{\Lambda^c}\big)\mathbf{1}_{\sigma'\in A}\mathbf{1}_{\sigma'\in B}\, \mu(d\sigma).$$

Since $B \in \mathcal{B}_{\Lambda^c}$ and $G_\Lambda(\sigma'|\sigma_{\Lambda^c}) = 0$ unless $\sigma' = \sigma$ on Λ^c, $\mathbf{1}_{\sigma'\in B} = \mathbf{1}_{\sigma\in B}$,

$$\mu(A\cap B) = \int_B \Big\{\sum_{\sigma'} G_\Lambda\big(\sigma'|\sigma_{\Lambda^c}\big)\mathbf{1}_{\sigma'\in A}\Big\}\, \mu(d\sigma) = \int_B G_\Lambda\big(A|\sigma_{\Lambda^c}\big)\, \mu(d\sigma),$$

which, by (A.6.1), proves that $\{\mathcal{X}, G_\Lambda(\cdot|\sigma_{\Lambda^c})\}$ is a version of the conditional probability of μ given $\mathcal{B}_{\Lambda^c}$. By the arbitrariness of Λ, μ is DLR.

Vice versa, suppose that μ is DLR. Then by (A.6.1),

$$\mu(A) = \int_{\mathcal{X}} \sum_{\sigma'} G_\Lambda\big(\sigma'|\sigma_{\Lambda^c}\big)\mathbf{1}_{\sigma' \in A}\, \mu(d\sigma),$$

which proves (2.1.6.3) for functions f which are characteristic functions of Borel sets. By a density argument (details are omitted) the equality extends to all bounded, Borel measurable functions. Thus $\mu \in \mathcal{G}_\Lambda$ and by the arbitrariness of Λ, $\mu \in \mathcal{G}$. □

2.2.2 Thermodynamic limits of Gibbs measures

In this subsection we will establish a first relation between DLR measures and thermodynamic limits of finite volume Gibbs measures. Let $\{\Delta_n\}$ be an increasing sequence of finite regions which invades the whole space; the construction below will depend on the choice of $\{\Delta_n\}$, but, as we shall see, the final conclusions about the elements of $\mathcal{G}$ are structural and independent of $\{\Delta_n\}$. Recalling from Sect. 2.1.3 that $\mathcal{C}$ denotes the family of all elementary cylindrical sets, we introduce "the good set"

$$\mathcal{X}_g = \Big\{\sigma \in \mathcal{X} : \lim_{n\to\infty} G_{\Delta_n}\big(C|\sigma_{\Delta_n^c}\big) \text{ exists for all } C \in \mathcal{C}\Big\}$$

(we shall later introduce a very good set $\mathcal{X}_{gg}$). A priori $\mathcal{X}_g$ may be empty, but the following theorem excludes such a possibility:

Theorem 2.2.2.1 *$\mathcal{X}_g$ is a non-empty Borel set and $\mu(\mathcal{X}_g) = 1$ for any $\mu \in \mathcal{G}$. Moreover, for any $\sigma \in \mathcal{X}_g$ there is a unique measure $G_\sigma(\cdot)$ such that*

$$G_\sigma(f) = \lim_{n\to\infty} G_{\Delta_n}\big(f|\sigma_{\Delta_n^c}\big) \quad \text{for all continuous functions } f. \qquad (2.2.2.1)$$

All G_σ, $\sigma \in \mathcal{X}_g$, are in $\mathcal{G}$.

Proof Let $\mu \in \mathcal{G}$; recall that we have proved in Theorem 2.2.1.1 that $\mathcal{G} \neq \emptyset$. Then μ is DLR and the pair $\{\mathcal{X}, G_{\Delta_n}(\cdot|\sigma_{\Delta_n^c})\}$ is a version of the conditional probability of μ given the σ-algebra $\{\mathcal{B}_{\Delta_n^c}\}$. We can then apply Theorem A.11 with $\Sigma_n = \mathcal{B}_{\Delta_n^c}$, $\mathcal{X}_n = \mathcal{X}$ and $\mu(\cdot|\Sigma_n)(\sigma) = G_{\Delta_n}(\cdot|\sigma_{\Delta_n^c})$. Then the set $\mathcal{X}'$ in (A.10.1) is our set $\mathcal{X}_g$; hence by Theorem A.11, $\mathcal{X}_g \in \mathcal{B}$ and $\mu(\mathcal{X}_g) = 1$ (thus $\mathcal{X}_g \neq \emptyset$). Theorem A.11 also states that for any $\sigma \in \mathcal{X}_g$ there is a unique probability $\mu(\cdot|\mathcal{B}_\infty)(\sigma)$, $\mu(\cdot|\Sigma)(\sigma)$ in the notation of Theorem A.11 (that we identify with G_σ of (2.2.2.1)), such that

$$\lim_{n\to\infty} G_{\Delta_n}\big(C|\sigma_{\Delta_n^c}\big) = \mu\big(C|\mathcal{B}_\infty\big)(\sigma).$$

By Theorem A.4 it follows that, for any $\sigma \in \mathcal{X}_g$, $G_{\Delta_n}(f|\sigma_{\Delta_n^c}) = \mu(f|\mathcal{B}_\infty)(\sigma)$ for all continuous functions.

Finally, the statement $G_\sigma \in \mathcal{G}$ is an immediate consequence of (2.2.2.1), as the latter states that G_σ is the weak limit of a sequence which is definitively in $\mathcal{G}_\Lambda$ (as soon as n is such that $\Lambda \subset \Delta_n$). Since $\mathcal{G}_\Lambda$ is weakly closed, $G_\sigma \in \mathcal{G}_\Lambda$, and by the arbitrariness of Λ it is in the intersection of all $\mathcal{G}_\Lambda$; hence it is in $\mathcal{G}$. □

The set $\mathcal{X}_g$ is good but not very good! For any $\sigma \in \mathcal{X}_g$, call

$$\Omega_\sigma = \{\sigma' \in \mathcal{X}_g : G_{\sigma'} = G_\sigma\}, \qquad \mathcal{X}_{gg} = \{\sigma \in \mathcal{X}_g : G_\sigma(\Omega_\sigma) = 1\}. \tag{2.2.2.2}$$

$\mathcal{X}_{gg}$ is the "very good" set we are looking for and which has been described in the beginning of the section, in the paragraph "Outline of main results."

Theorem 2.2.2.2 *$\mathcal{X}_{gg}$ is a non-empty Borel set and for any $\sigma \in \mathcal{X}_g$ either $\Omega_\sigma \in \mathcal{X}_{gg}$ or $\Omega_\sigma \cap \mathcal{X}_{gg} = \emptyset$. For any $\mu \in \mathcal{G}$, $\mu(\mathcal{X}_{gg}) = 1$ and for any bounded measurable function f, $G_\sigma(f)$, $\sigma \in \mathcal{X}_{gg}$, is a measurable function and*

$$\mu(f) = \int_{\mathcal{X}_{gg}} G_\sigma(f)\, \mu(d\sigma). \tag{2.2.2.3}$$

Conversely, for any probability ν on $\mathcal{X}_{gg}$, the measure μ defined by

$$\mu(f) = \int_{\mathcal{X}_{gg}} G_\sigma(f)\, \nu(d\sigma) \tag{2.2.2.4}$$

is in $\mathcal{G}$.

Proof It immediately follows from the definition (2.2.2.2) that for any $\sigma \in \mathcal{X}_g$ either $\Omega_\sigma \in \mathcal{X}_{gg}$ or $\Omega_\sigma \cap \mathcal{X}_{gg} = \emptyset$. Suppose $\mu \in \mathcal{G}$. By Theorem A.11 and the identification $G_\sigma = \mu(\cdot|\mathcal{B}_\infty)(\sigma)$, the pair $(\mathcal{X}_g, G_\sigma)$ is a version of the conditional probability of μ with respect to the σ-algebra $\mathcal{B}_\infty$, which is defined as the minimal σ-algebra which contains the sets $B \in \mathcal{B}$ which are in $\mathcal{B}_{\Delta_n^c}$ for all n (for this reason we will also write $G_\sigma = \mu(\cdot|\mathcal{B}_\infty)(\sigma)$, $\sigma \in \mathcal{X}_g$). By Theorem A.8 $(\mathcal{X}_{gg}, G_\sigma)$ is also a version of the conditional probability given $\mathcal{B}_\infty$, hence $\mathcal{X}_{gg}$ is a Borel set, $\mu(\mathcal{X}_{gg}) = 1$ and (2.2.2.3) holds. Since $\mu(\mathcal{X}_{gg}) = 1$, $\mathcal{X}_{gg}$ is non-empty. Equation (2.2.2.3) then follows from $(\mathcal{X}_{gg}, G_\sigma)$ being a conditional probability.

Let μ be as in (2.2.2.4). Since $G_\sigma \in \mathcal{G}$, $G_\sigma \in \mathcal{G}_\Lambda$, Λ bounded,

$$\begin{aligned}\mu(f) &= \int_{\mathcal{X}_{gg}} \left\{ \int_{\mathcal{X}} G_\Lambda(f|\sigma'_{\Lambda^c}) G_\sigma(d\sigma') \right\} \nu(d\sigma) \\ &= \int_{\mathcal{X}} G_\Lambda(f|\sigma'_{\Lambda^c}) \mu(d\sigma').\end{aligned} \tag{2.2.2.5}$$

Equation (2.2.2.5) shows that $\mu \in \mathcal{G}_\Lambda$ and by the arbitrariness of Λ that $\mu \in \mathcal{G}$. □

We will next improve the above analysis by establishing fine and detailed properties of the DLR measures. The analysis is more technical and heavily relies on the theory of conditional probabilities in Appendix A. In a first reading one may go directly to Sect. 2.2.6 (however, the proof of Theorem 2.2.7.2 will use some of the following results).

The sets $\mathcal{X}_g$ and $\mathcal{X}_{gg}$ depend on the choice of the sequence $\{\Delta_n\}$ used in the definition of G_σ. If we take another sequence $\{\Delta'_n\}$ we will have in general new $\mathcal{X}'_g$ and $\mathcal{X}'_{gg}$ and new measures $G'_{\sigma'}, \sigma' \in \mathcal{X}'_g$. The two families are however strictly related to each other.

Theorem 2.2.2.3 *The set* $\mathcal{X}_0 := \mathcal{X}_{gg} \cap \mathcal{X}'_{gg}$ *is non-empty,* $\Omega_\sigma \cap \mathcal{X}_0 \neq \emptyset$ *for all* $\sigma \in \mathcal{X}_{gg}$ *and* $\Omega'_\sigma \cap \mathcal{X}_0 \neq \emptyset$ *for all* $\sigma \in \mathcal{X}'_{gg}$. *Moreover,* $\mu(\mathcal{X}_0) = 1$ *for any* $\mu \in \mathcal{G}$; *for any* $\sigma \in \mathcal{X}_0$, $G_\sigma = G'_\sigma$, $\Omega_\sigma \cap \mathcal{X}_0 = \Omega'_\sigma \cap \mathcal{X}_0$ *and* $G_\sigma(\Omega_\sigma \cap \mathcal{X}_0) = 1$.

Proof By (2.2.2.3) with $\mu = G_\sigma$ and $f = \mathbf{1}_{\Omega_\sigma}$,

$$G_\sigma(\Omega_\sigma) = \int_{\mathcal{X}'_{gg}} G'_{\sigma'}(\Omega_\sigma)\, G_\sigma(d\sigma').$$

Since $G_\sigma(\Omega_\sigma) = 1$ and $G'_{\sigma'}(\Omega_\sigma) \leq 1$,

$$G_\sigma\big(\{\sigma' \in \Omega_\sigma \cap \mathcal{X}'_{gg} : G'_{\sigma'}(\Omega_\sigma) = 1\}\big) = 1.$$

Then $\Omega_\sigma \cap \mathcal{X}'_{gg} \neq \emptyset$ and if $\sigma' \in \Omega_\sigma \cap \mathcal{X}'_{gg}$, $G_\sigma = G'_{\sigma'}$. If $\Omega'_{\sigma'} \cap \Omega'_{\sigma''} = \emptyset$, then $G'_{\sigma'} \neq G'_{\sigma''}$; hence $G_\sigma \neq G'_{\sigma''}$ so that $\Omega_\sigma \cap \Omega'_{\sigma''} = \emptyset$. □

2.2.3 Pure states and extremal DLR measures

$\mathcal{G}$ is a convex set and, being weakly compact, its extremal points, whose collection is denoted by $\mathcal{G}_{\text{extr}}$, are also in $\mathcal{G}$. By the Krein–Millman theorem, see I.3.10 in the book by Naimark [175], elements in $\mathcal{G} \setminus \mathcal{G}_{\text{extr}}$ are convex combinations (in general integrals) over the extremal elements and are therefore called "mixture states," while the extremal elements are "pure states." We will see at the end of this subsection that pure states can be identified as the pure phases of the system. In the next theorem we will characterize the extremal DLR measures as the measures $G_\sigma, \sigma \in \mathcal{X}_{gg}$; then the decomposition into extremal measures is just (2.2.2.3).

Theorem 2.2.3.1 *The following two statements are equivalent*:

- $\mu \in \mathcal{G}$ *and* $\mu(\Omega_\sigma) = 1$ *for some* $\sigma \in \mathcal{X}_{gg}$.
- $\mu = G_\sigma, \sigma \in \mathcal{X}_{gg}$. (2.2.3.1)

The set $\mathcal{G}_{\text{extr}}$ *of extremal elements of* $\mathcal{G}$ *is* $\mathcal{G}_{\text{extr}} = \{G_\sigma, \sigma \in \mathcal{X}_{gg}\}$ *so that* (2.2.2.3) *is a decomposition of* μ *into extremal states.*

Proof If $\mu = G_\sigma$ for some $\sigma \in \mathcal{X}_{gg}$, then $\mu \in \mathcal{G}$ and $\mu(\Omega_\sigma) = 1$ by the definition of $\mathcal{X}_{gg}$. Suppose conversely that $\mu \in \mathcal{G}$ and $\mu(\Omega_\sigma) = 1$, then by (2.2.2.3) $\mu(f) = \int_{\Omega_\sigma} G_{\sigma'}(f)\, \mu(d\sigma') = G_\sigma(f) \int_{\Omega_\sigma} \mu(d\sigma') = G_\sigma(f)$, hence (2.2.3.1).

Let us next prove the statements about $\mathcal{G}_{\text{extr}}$. Suppose first that $\mu \in \mathcal{G}_{\text{extr}}$. Arguing by contradiction, we will show that if $\mu \neq G_\sigma(\cdot)$ for all $\sigma \in \mathcal{X}_{gg}$, then the integral decomposition in (2.2.2.3) can be reduced to a convex combination of two distinct measures in $\mathcal{G}$. We first observe that $\sigma \to G_\sigma$ on $\mathcal{X}_{gg}$ cannot be μ-a.s. constant, otherwise, by (2.2.2.3), it would be equal to μ. Then there must be f and $b \in \mathbb{R}$ such that

$$\mu\left(\left\{\sigma \in \mathcal{X}_{gg} : G_\sigma(f) \leq b\right\}\right) = \alpha, \quad \alpha \neq 0, 1. \tag{2.2.3.2}$$

Calling B the set in curly brackets, we define the probabilities

$$\begin{aligned} \mu'(\cdot) &= \alpha^{-1} \int_B G_\sigma(\cdot)\, \mu(d\sigma), \\ \mu''(\cdot) &= (1-\alpha)^{-1} \int_{B^c} G_\sigma(\cdot)\, \mu(d\sigma). \end{aligned} \tag{2.2.3.3}$$

By (2.2.2.4), both μ' and μ'' are in $\mathcal{G}$; moreover, $\mu' \neq \mu''$ because, by construction, $\mu'(f) \leq b$ and $\mu''(f) > b$. Also by construction $\mu = \alpha\mu' + (1-\alpha)\mu''$, hence the desired contradiction, which proves that $\mu = G_\sigma$ for some $\sigma \in \mathcal{X}_{gg}$.

Conversely, suppose that for $\sigma \in \mathcal{X}_{gg}$, $G_\sigma \notin \mathcal{G}_{\text{extr}}$. Then there are $\alpha \in [0, 1]$, μ' and μ'' in $\mathcal{G}$ for which $G_\sigma = \alpha\mu' + (1-\alpha)\mu''$. By applying (2.2.2.3) to $\mu = G_\sigma$, we get, for any measurable set A,

$$G_\sigma(A) = \int_{\mathcal{X}_{gg}} G_{\sigma'}(A)\, [\alpha\mu' + (1-\alpha)\mu''](d\sigma'). \tag{2.2.3.4}$$

Since for all $\sigma' \in \mathcal{X}_{gg}$, $G_{\sigma'}(\Omega_\sigma) = \mathbf{1}_{\sigma' \in \Omega_\sigma}$, (2.2.3.4) with $A = \Omega_\sigma$ yields

$$\alpha\mu'(\Omega_\sigma) + (1-\alpha)\mu''(\Omega_\sigma) = 1.$$

If $\alpha \in (0, 1)$, this implies $\mu'(\Omega_\sigma) = \mu''(\Omega_\sigma) = 1$, so that, by (2.2.3.1) which has already been proved, $\mu' = \mu'' = G_\sigma$. □

We will next argue that extremal measures have the physical interpretation of "pure phases." The idea is to start from the particular case that $\mathcal{G}$ is a singleton, in which case there is no doubt that its only element is a pure phase. We will prove for the singleton properties that in the general case are satisfied by the extremal measures of $\mathcal{G}$, thus interpreting the latter as pure phases.

Theorem 2.2.3.2 *If $\mathcal{G} = \{\mu\}$ is a singleton, then $\mathcal{X}_g = \mathcal{X}_{gg} = \mathcal{X}$, and, given any increasing sequence $\Lambda_n \to \mathbb{Z}^d$, for any continuous function f*

$$\begin{aligned} &\lim_{\Lambda_n \to \mathbb{Z}^d} G_{\Lambda_n}(f | \sigma^c_{\Lambda_n}) = \mu(f), \quad \sigma \in \mathcal{X}; \\ &\lim_{\Lambda_n \to \mathbb{Z}^d} \int |G_{\Lambda_n}(f | \sigma^c_{\Lambda_n}) - \mu(f)| \mu(d\sigma) = 0. \end{aligned} \tag{2.2.3.5}$$

Proof By compactness $G_{\Lambda_n}(\cdot|\sigma^c_{\Lambda_n})$ converges weakly by subsequences, and since $\mathcal{G}_\Delta$ is weakly compact any limit point is in $\mathcal{G}_\Delta$ and, by the arbitrariness of Δ it is in $\mathcal{G}$. Since $\mathcal{G}$ is a singleton the limit point is μ and the sequence actually converges. We have thus proved the first limit in (2.2.3.6); the second one follows from the first one by the Lebesgue dominated convergence theorem. □

If $\mathcal{G}$ is not a singleton, we have

Theorem 2.2.3.3 *If $\mu \in \mathcal{G}_{\text{extr}}$, given any increasing sequence $\Lambda_n \to \mathbb{Z}^d$ then for any continuous function f, $\mu(\{\sigma : \lim_{\Lambda_n\to\mathbb{Z}^d} G_{\Lambda_n}(f|\sigma^c_{\Lambda_n}) = \mu(f)\}) = 1$ and*

$$\lim_{\Lambda_n\to\mathbb{Z}^d} \int |G_{\Lambda_n}(f|\sigma^c_{\Lambda_n}) - \mu(f)|\mu(d\sigma) = 0. \qquad (2.2.3.6)$$

Proof Let $\mathcal{X}_{gg}$ be the set associated to the sequence Λ_n. Then by Theorem 2.2.3.1 there is $\sigma \in \mathcal{X}_{gg}$ such that $\mu = G_\sigma$; hence

$$\mu\Big(\Big\{\sigma : \lim_{\Lambda_n\to\mathbb{Z}^d} G_{\Lambda_n}(f|\sigma^c_{\Lambda_n}) = \mu(f)\Big\}\Big) \ge G_\sigma(\Omega_\sigma) = 1,$$

which proves the first statement in the theorem. Actually by taking countably many intersections, the set where there is convergence can be chosen once for all f. The limit in (2.2.3.6) then follows by the Lebesgue dominated convergence theorem. □

Extremal measures as pure phases

By comparing Theorems 2.2.3.2 and 2.2.3.3 we see that the extremal measures enjoy the same properties as the unique measure when $\mathcal{G}$ is a singleton, provided we replace "everywhere" by "almost everywhere" (i.e. with probability 1). That is, if we view the phase space from the viewpoint of an extremal measure μ then we see the same homogeneous behavior as when $\mathcal{G}$ is a singleton. For such reasons we will call the states in $\mathcal{G}_{\text{extr}}$ pure phases.

To exemplify the discussion let us refer to the $d = 2$ ferromagnetic Ising model with n.n. interactions and no magnetic field. We will prove in the next chapter that at small temperatures there exists an extremal measure, the so called plus measure, where with probability 1 all configurations have a positive magnetization. There is also another extremal measure, the minus measure, where with probability 1 all configurations have negative magnetization. There are then mixture states where in a fraction of configurations we observe a positive and in the other one a negative magnetization. Mixtures in this example are convex combinations of the two pure states.

2.2.4 Simplicial structure of DLR measures

The are two opposite types of convex sets, as far as the decomposition into their extremal points is concerned, which in the plane are visualized as circles and triangles.

In the former the decomposition into extremal elements is highly non-unique, just the opposite of what happens in the latter. $\mathcal{G}$ belongs to the latter category; we are going to show that the decomposition of an element $\mu \in \mathcal{G}$ into extremal DLR states is unique.

We want to write any $\mu \in \mathcal{G}$ as an integral over $\mathcal{G}_{\text{extr}}$. Thus the first step requires us to introduce a measurable space (Ω, Σ), i.e. a space Ω and a σ-algebra Σ, with Ω in one to one correspondence with $\mathcal{G}_{\text{extr}}$. By (2.2.3.1) we can take for Ω the space whose points ω are the sets Ω_σ, $\sigma \in \mathcal{X}_{gg}$. Call ϕ the map from $\mathcal{X}_{gg}$ onto Ω, which associates to any $\sigma \in \mathcal{X}_{gg}$ the element $\Omega_\sigma \in \Omega$. We then define Σ as the σ-algebra made of all sets $A \subset \Omega$ such that $\phi^{-1}(A)$ is measurable in $\mathcal{X}$. We also define for any $\mu \in \mathcal{G}$ a measure p_μ on (Ω, Σ) by setting $p_\mu(A) = \mu(\phi^{-1}(A))$, $A \in \Sigma$, so that (Ω, Σ, p_μ) is isomorphic to the restriction of μ to the σ-algebra $\phi^{-1}(\Sigma)$.

Theorem 2.2.4.1 *Let $\mu \in \mathcal{G}$; then there is a unique measure p on (Ω, Σ) such that for any bounded measurable function f on $\mathcal{X}$,*

$$\mu(f) = \int_\Omega G_\omega(f)\, p(d\omega),$$

where, by abuse of notation, $G_\omega \equiv G_\sigma$, $\sigma \in \phi^{-1}(\omega)$; recall that G_σ is the same for all $\sigma \in \phi^{-1}(\omega)$. The measure p is equal to p_μ.

Proof By Theorem 2.2.2.2, $\sigma \to G_\sigma(f)$ is measurable and constant on each Ω_σ; hence it is measurable on the σ-algebra $\phi^{-1}(\Sigma)$. Calling μ' the restriction of μ to $\phi^{-1}(\Sigma)$, we thus conclude from (2.2.2.3) that

$$\mu(f) = \int_{\mathcal{X}_{gg}} G_\sigma(f)\, \mu'(d\sigma) = \int_\Omega G_\omega(f)\, p_\mu(d\omega).$$

To prove uniqueness, let (Ω, Σ, q) be a probability space and

$$\mu(f) = \int_\Omega G_\omega(f)\, q(d\omega).$$

We claim that $q = p_\mu$. For $B \in \Sigma$ we have

$$\mu(\phi^{-1}(B)) = \int G_\omega(\phi^{-1}(B))\, q(d\omega) = \int_B G_\omega(\phi^{-1}(B))\, p_\mu(d\omega). \tag{2.2.4.1}$$

Since $G_{\phi^{-1}(\omega)}(\phi^{-1}(B)) = \mathbf{1}_{\omega \in B}$, (2.2.4.1) yields $q(B) = p_\mu(B)$. □

2.2.5 Sigma algebra at infinity

In the course of the proof of Theorem 2.2.2 we have introduced the σ-algebra $\mathcal{B}_\infty$ whose definition is made explicit here.

Definition The σ-algebra $\mathcal{B}_\infty = \bigcap_{\Lambda \text{ bounded}} \mathcal{B}_{\Lambda^c}$ is called the *σ-algebra at infinity*, or *tail field*.

Sets in $\mathcal{B}_\infty$ have the following property: if $A \in \mathcal{B}_\infty$ and $\sigma \in A$, then any local modification σ' of σ is also in A. Given a measure μ, a set A is μ-modulo 0 in $\mathcal{B}_\infty$ if there is a set $N \in \mathcal{B}_\infty$ such that $\mu(N) = 0$ and $A \cap N^c \in \mathcal{B}_\infty$. Functions which are μ-modulo 0 measurable at infinity are defined analogously.

The next theorem shows that measure theoretically the sets Ω_σ are the smallest ones among those measurable at infinity. Indeed, we shall see that if $\mu \in \mathcal{G}$, then any function f which is μ-modulo 0 measurable at infinity is μ almost surely constant on each set Ω_σ.

Theorem 2.2.5.1 *If $\mu \in \mathcal{G}$ and f is μ-almost surely measurable at infinity, then*

$$\mu\big(\{\sigma \in \mathcal{X}_{gg} : f(\sigma') = G_\sigma(f), \text{ for } G_\sigma\text{-almost all } \sigma' \in \Omega_\sigma\}\big) = 1. \qquad (2.2.5.1)$$

In particular for any $\mu \in \mathcal{G}_{\text{extr}}$ if A is μ-modulo 0 *in $\mathcal{B}_\infty$, then $\mu(A)$ is either* 0 *or* 1, *namely $\mathcal{B}_\infty = \{\emptyset, \mathcal{X}\}$, μ-modulo* 0.

Proof By assumption there is $N \in \mathcal{B}_\infty$ so that $\mu(N) = 0$ and $f((\sigma'_\Lambda, \sigma_{\Lambda^c})) = f((\sigma_\Lambda, \sigma_{\Lambda^c}))$, for all bounded Λ, all $\sigma \notin N$ and all σ'_Λ. By (2.2.2.3), $\mu(\{\sigma : G_\sigma(N) = 0\} = 1$, and (2.2.5.1) will be proved by showing that if $\sigma : G_\sigma(N) = 0$, then $f(\sigma') = G_\sigma(f)$ for G_σ-almost all $\sigma' \in \Omega_\sigma$.

Let $\sigma : G_\sigma(N) = 0$. Arguing by contradiction, similarly to the proof of Theorem 2.2.3 (see (2.2.3.2)–(2.2.3.4)), suppose that there are $\alpha \neq 0, 1$ and b such that $G_\sigma(B) = \alpha$, $B := \{\sigma' \in \Omega_\sigma \sqcap N^c : f(\sigma') \le b\}$. Let

$$\mu' = \alpha^{-1} \mathbf{1}_B G_\sigma, \qquad \mu'' = (1-\alpha)^{-1} \mathbf{1}_{B^c} G_\sigma,$$

so that $G_\sigma = \alpha\mu' + (1-\alpha)\mu''$ and $\mu' \neq \mu''$. The contradiction will arise once we show that μ' and μ'' are in $\mathcal{G}$, because by Theorem 2.2.3, $G_\sigma \in \mathcal{G}_{\text{extr}}$. Let us prove that $\mu' \in \mathcal{G}_\Lambda$, Λ bounded, namely that (2.1.6.5) holds with $\mu \to \mu'$. We have for any bounded measurable function h,

$$\begin{aligned}\int G_\Lambda(h|\sigma' r_{\Lambda^c})\, \mu'(d\sigma') &= \alpha^{-1} \int_{\Omega_\sigma \cap N^c} \mathbf{1}_{f(\sigma') \le b} G_\Lambda(h|\sigma'_{\Lambda^c})\, G_\sigma(d\sigma') \\ &= \alpha^{-1} \int_{\Omega_\sigma \cap N^c} G_\Lambda(h \mathbf{1}_{f \le b}|\sigma'_{\Lambda^c})\, G_\sigma(d\sigma') \\ &= \alpha^{-1} \int_{\Omega_\sigma \cap N^c} h \mathbf{1}_{f \le b}\, G_\sigma(d\sigma') = \mu'(h),\end{aligned}$$

where we have used that $\{f(\sigma') \le b\}$ does not depend on σ_Λ, and it is therefore a constant with respect to the measure $G_\Lambda(\cdot|\sigma'_{\Lambda^c}())$. Thus $\mu' \in \mathcal{G}_\Lambda$ and by the arbitrariness of Λ, $\mu' \in \mathcal{G}$; the same proof shows that $\mu'' \in \mathcal{G}$ hence the desired contradiction. □

2.2.6 *Pure phases, phase transitions*

Here we discuss another aspect of phase transitions, related to the notion of intensive variables, which will lead to a different definition of pure phases, alternative to the one used so far, where pure phases have been associated to extremal DLR measures. In the new notion pure phases still correspond to extremal DLR measures but with extremality referring to the set of translational invariant DLR measures. The new notion is related to that of intensive variables. From a physical point of view in fact the intensive variables are expected to have sharp values in a pure phase, while in mixtures they fluctuate, as their values will depend on which pure state of the mixture the system is actually in. The notion of an intensive variable is related to space translations, which have been so far ignored, and which are going to enter strongly in the game.

We start with some simple basic considerations. Observe that, while the probability of a spin configuration with energy E is just given by the Gibbs formula (and it is therefore proportional to $e^{-\beta E}$), the probability that the system has an energy E is instead not at all as simple. In fact, counting the number of states with a given energy is in general a very complex task. Let us denote their number by $e^{S_\Lambda(E)}$, Λ the bounded domain where the system is confined. $S_\Lambda(E)$ is thus an entropy, the entropy for the given values of Λ and E; see Sect. 2.3 where the notion will become central. The energy distribution is then proportional to $\exp\{-\beta E + S_\Lambda(E)\}$. Both energy and entropy are extensive quantities; the corresponding intensive quantities are $e = E/|\Lambda|$ and $s_\Lambda(e) = S_\Lambda(E)/|\Lambda|$, respectively the energy and the entropy densities. We thus get for the energy distribution $\exp\{-|\Lambda|(\beta e - s_\Lambda(e))\}$. Supposing that $s_\Lambda(e)$ has the limit $s(e)$ as $\Lambda \to \mathbb{Z}^d$, we may then conclude that the energy distribution is concentrated for large $|\Lambda|$ around the minimizers of the free energy $e - \beta^{-1}s(e)$. When there is a unique minimizer, the distribution is uni-modal and most of the mass of the distribution is in its neighborhood: the fluctuations of the energy density are thus small and disappear in the thermodynamic limit. On the other hand, when the minimizer is not unique (the distribution is then called multi-modal), the energy density has macroscopic fluctuations which survive in the limit and the system exhibits a phase transition. But phase transitions may also come from the loss of uni-modality of other intensive variables rather than the energy, as in the classical Ising model, where the relevant order parameter is the magnetization density.

As we shall see, the theory of Gibbs measures encodes the above ideas in an elegant formulation which involves the action of the group of space translations on $\mathcal{G}$. Referring to Appendix A for our notation, for any $i \in \mathbb{Z}^d$ we denote by τ_i the map on $\mathbb{Z}^d$ defined by $\tau_i(x) = x + i$ and by $\tau_i(\sigma)$, $\tau_i(f)$ and $\tau_i(\mu)$ its dual actions, respectively, on spin configurations, on functions of the spin configurations and on probabilities on the space of spin configurations. Explicitly $\tau_i(\sigma)(x) = \sigma(x-i)$, $\tau_i(f)(\sigma) = f(\tau_{-i}(\sigma))$, $\tau_i(\mu)(f) = \mu(\tau_{-i}(f))$.

Theorem 2.2.6.1 *The set of DLR measure is invariant under translations,*

$$\tau_i(\mathcal{G}) = \mathcal{G}, \quad \textit{for any } i \in \mathbb{Z}^d, \tag{2.2.6.1}$$

and its subset $\mathcal{G}^0$ of translational invariant measures is a non-empty, convex, compact set.

Proof For any bounded Δ, $\tau_i(\mathcal{G}_\Delta) = \mathcal{G}_{\tau_i(\Delta)}$, see Appendix A after (A.1.10), hence (2.2.6.1). Let Δ_n be an increasing sequence of cubes which invades $\mathbb{Z}^d$, $\mu \in \mathcal{G}$ and

$$\mu^{(n)} = \frac{1}{|\Delta_n|} \sum_{i \in \Delta_n} \tau_i(\mu). \tag{2.2.6.2}$$

For all this, $\mu^{(n)} \in \mathcal{G}$ and let ν be a weak limit point of $\{\mu^{(n)}\}$, whose existence follows from compactness. $\nu = \tau_i(\nu)$ as ν is the limit of Cesaro's averages, and $\nu \in \mathcal{G}$ because $\mathcal{G}$ is weakly closed; thus $\mathcal{G}^0 \neq \emptyset$. Convexity and compactness also easily follow. □

2.2.7 Ergodic decomposition

In the first part of this subsection we will recall the ergodic decomposition of a translational invariant measure μ into its ergodic components, and in the second part we shall use it to determine the structure of the translational invariant DLR measures. Recall that a translational invariant measure μ is ergodic if and only if $\mu(A)$ is either 0 or 1 for any measurable, translational invariant set A.

The ergodic decomposition consists in writing μ as an integral over measures which are translational invariant and ergodic. We state the results here without proofs, as they are well known from the literature; however, for the sake of completeness, a proof is given in Sect. A.9 of Appendix A.

Let Δ_n be an increasing sequence of cubes which invades $\mathbb{Z}^d$. Call

$$A^{(n)} f(\sigma) = \frac{1}{|\Delta_n|} \sum_{i \in \Delta_n} \tau_i f(\sigma),$$

and denote by $\mathcal{C}$ the set of all elementary cylinders; finally, define, in analogy with (2.2.2.2),

$$\mathcal{X}_g^0 = \Big\{\sigma : \lim_{n\to\infty} A^{(n)} \mathbf{1}_C(\sigma) \text{ exists for all } C \in \mathcal{C}\Big\}.$$

Analogously to Theorem 2.2.2.1, $\mathcal{X}_g^0$ is non-empty and for any $\sigma \in \mathcal{X}_g^0$ there is a translational invariant measure $\mathcal{A}_\sigma$ such that

$$\mathcal{A}_\sigma(f) = \lim_{n\to\infty} A^{(n)} f(\sigma) \quad \text{for all continuous functions } f. \tag{2.2.7.1}$$

The proof of the above statements uses the Birkhoff theorem. Fix arbitrarily a translational invariant measure, for instance a Bernoulli measure ν (Bernoulli measures are measures such that all spins are independent and identically distributed, hence they are translational invariant). Then by the Birkhoff theorem $\nu(\mathcal{X}_g^0) = 1$ and hence $\mathcal{X}_g^0$ is non-empty. The analogue of (2.2.2.2) is

$$\Omega_\sigma^0 = \{\sigma' \in \mathcal{X}_g^0 : \mathcal{A}_{\sigma'} = \mathcal{A}_\sigma\}, \qquad \mathcal{X}_{gg}^0 = \{\sigma \in \mathcal{X}_g^0 : \mathcal{A}_\sigma(\Omega_\sigma^0) = 1\},$$

and the analogue of Theorem 2.2.2.2 holds as well.

Theorem 2.2.7.1 *$\mathcal{X}_{gg}^0$ is non-empty and for any translational invariant measure μ, $\mu(\mathcal{X}_{gg}^0) = 1$. The measures $\mathcal{A}_\sigma$, $\sigma \in \mathcal{X}_{gg}^0$, are ergodic with respect to space translations. Moreover for any bounded measurable function f, $\mathcal{A}_\sigma(f)$ is a measurable function and for any probability ν on $\mathcal{X}_{gg}^0$, the measure μ defined by*

$$\mu(f) = \int_{\mathcal{X}_{gg}^0} \mathcal{A}_\sigma(f)\, \nu(d\sigma)$$

is translational invariant. Vice versa, if μ is translational invariant then

$$\mu(f) = \int_{\mathcal{X}_{gg}^0} \mathcal{A}_\sigma(f)\, \mu(d\sigma).$$

Finally, a translational invariant measure μ is ergodic if and only if $\mu = \mathcal{A}_\sigma$ for some $\sigma \in \mathcal{X}_{gg}^0$.

So far everything was general with no relations with the Gibbs measures. Let us now specify μ as a translational invariant DLR measure, namely $\mu \in \mathcal{G}^0$. If μ is ergodic, then for μ a.a. σ, $\mu = \mathcal{A}_\sigma$, which shows that $\mathcal{A}_\sigma \in \mathcal{G}^0$, μ almost surely. The above conclusion extends to any $\mu \in \mathcal{G}^0$.

Theorem 2.2.7.2 *If $\mu \in \mathcal{G}^0$, $\mu(\mathcal{X}_{gg}^0) = 1$ and for μ almost all $\sigma \in \mathcal{X}_{gg}^0$, $\mathcal{A}_\sigma \in \mathcal{G}^0$. Namely any translational invariant DLR measure μ is supported by configurations σ such that their ergodic averages $\mathcal{A}_\sigma$ defined in (2.2.7.1) exist and are ergodic DLR measures (so that the set of ergodic DLR measures is non-empty). As a consequence*

$$\mu(f) = \int_{\mathcal{X}_{gg}^0 \cap \mathcal{G}^0} \mathcal{A}_\sigma(f)\, \mu(d\sigma). \tag{2.2.7.2}$$

Proof The set $\mathcal{X}_g^0$ is $\mathcal{B}_\infty$ measurable because the value of $\mathcal{A}_\sigma(\cdot)$ does not change after a local modification of σ. On the other hand, by the Birkhoff theorem, $\mu(\mathcal{X}_g^0) = 1$ and by (2.2.2.3), $\mu(\{\sigma \in \mathcal{X}_{gg} : G_\sigma(\mathcal{X}_g^0) = 1\}) = 1$. Calling K the intersection of $\{\sigma \in \mathcal{X}_{gg} : G_\sigma(\mathcal{X}_g^0) = 1\}$ and $\{\sigma \in \mathcal{X}_{gg} \cap \mathcal{X}_g^0 :$ for any $C \in \mathcal{C}$, $\mathcal{A}_{\sigma'}(\mathbf{1}_C) = G_\sigma(\mathcal{A}_{\sigma''}(\mathbf{1}_C))$ for G_σ a.a. $\sigma'\}$, by (2.2.5.1) $\mu(K) = 1$. By (2.2.7.1) and the Lebesgue dominated convergence theorem, for $\sigma \in K$

$$\int_{\Omega_\sigma} \mathcal{A}_{\sigma''}(f)\, G_\sigma(d\sigma'') = \lim_{n\to\infty} \int_{\Omega_\sigma} A_{\sigma''}^{(n)}(f)\, G_\sigma(d\sigma''),$$

since $\int_{\Omega_\sigma} A^{(n)}_{\sigma''}(f)\, G_\sigma(d\sigma'') = \frac{1}{|\Delta_n|}\sum_{i\in\Delta_n} \tau_i(G_\sigma)(f) =: \nu^{(n)}(f)$, $\nu^{(n)} \in \mathcal{G}$, being a convex combination of Gibbs measures ($G_\sigma \in \mathcal{G}$ and $\tau_i(G_\sigma)$ as well by (2.2.6.1)). Thus, by definition of K, if $\sigma \in K$, for G_σ a.a. $\sigma' \in \Omega_\sigma$, $\nu^{(n)} \to \mathcal{A}_{\sigma'}$ on the cylinders and, by a density argument (see Theorem A.4)

$$\mathcal{A}_{\sigma'}(f) = \lim_{n\to\infty} \nu^{(n)}(f), \quad \text{for all continuous functions } f.$$

Then $\mathcal{A}_{\sigma'}$ is weak limit of Gibbs measures and hence $\mathcal{A}_{\sigma'} \in \mathcal{G}^0$. Since this holds for G_σ a.a. σ' and for all $\sigma \in K$, $\mathcal{A}_{\sigma'} \in \mathcal{G}^0$ for μ a.a. σ'. □

Let us next see how the theorem fits with what we have said before about phase transitions. Intensive variables are defined in physics as spatial averages of local observables, represented, for instance, by continuous functions. We then recognize in $\mathcal{A}_\sigma(f)$ the value in the state σ of the intensive variable associated to the continuous function f. Particular cases occur when $f(\sigma) = \sigma(0)$; the intensive variable then has the meaning of the magnetization density of the system. The energy density is also an intensive variable, with

$$u(\sigma) = -\frac{1}{2}\sum_{y\neq 0} J(0, y)\sigma(0)\sigma(y) - h\sigma(0). \tag{2.2.7.3}$$

The relevant quantities for phase transitions, according to what we have said before, are the probability distributions of the intensive variables $\mathcal{A}_{\sigma'}(f)$. If a measure is ergodic, by Theorem 2.2.7.1, it coincides with a measure $\mathcal{A}_\sigma$ and it is supported by a set Ω^0_σ. Then any intensive variable $\mathcal{A}_{\sigma'}(f)$ is constant in the support of the ergodic measure, i.e. when $\sigma' \in \Omega_\sigma$. By Theorems 2.2.7.2 and 2.2.7.1, any translational invariant Gibbs measure can be decomposed into Gibbs measures which are ergodic, hence in each one of these all intensive variables do not fluctuate. The absence of fluctuations of the intensive variables characterizes the pure phases; hence the ergodic Gibbs measures represent pure phases. We have thus proved that any translational invariant Gibbs measure is a mixture of pure phases with weights which are uniquely determined. The existence of intensive variables which have non-trivial fluctuations is then the indication that there is a phase transition and the variables which fluctuate can be used as order parameters to classify the transition.

We denote by $\mathcal{G}^0_{\text{extr}}$ the extremal points of $\mathcal{G}^0$, which are ergodic DLR measures and represent the pure phases of the system, described above. If the cardinality of $\mathcal{G}^0_{\text{extr}}$ is larger than 1, there are several pure phases and there is a phase transition. However, a phase transition can occur also if $|\mathcal{G}^0| = 1$, due to a break of the translational symmetry. It may in fact be that the unique element $\mu \in \mathcal{G}^0$ can be non-trivially decomposed into non-translational invariant states, if the cardinality of $\mathcal{G}_e$ is larger than 1. Such effects may be related to the appearance of crystalline structures and states of the solid phase, but also to the existence of states describing coexistence of phases, like the Dobrushin states in $d \geq 3$ ferromagnetic Ising systems at low temperatures.

2.3 Boltzmann hypothesis, entropy and pressure

In this section we completely change perspective, now focusing on thermodynamics rather than Gibbs and DLR measures. In a sense we go back to the origins as the derivation of thermodynamics was historically the first goal of statistical mechanics with the aim of establishing a quantitative link between the macroscopic thermodynamic potentials and the microscopic inter-molecular interactions.

The analysis in this section is entirely based on the postulate that "entropy is proportional to the log of the number of states," the famous Boltzmann hypothesis. We will study the Ising model starting from such an assumption and derive expressions for the thermodynamic potentials in terms of the Ising hamiltonian, in particular we will relate thermodynamic pressure to partition functions. The Gibbs assumption on the structure of the equilibrium states is not needed, nonetheless our proofs will use extensively the DLR theory but only as a technical tool.

2.3.1 An example from information theory

A simple example of what we are going to do is borrowed from information theory. Consider a channel which transmits messages with a finite alphabet Ω; *we want to compute its capacity by counting how many messages can be emitted by a source which is "sending f with frequency ϕ."* By this we mean the following: f is a real valued function on Ω, $\phi \in (\min f, \max f)$, and the "normalized" number of messages we want to count is

$$\lim_{\delta \to 0} \lim_{N \to \infty} \frac{\log K_\delta(N)}{N}, \tag{2.3.1.1}$$

where

$$K_\delta(N) = \text{card}\left\{ (\omega_1, \ldots, \omega_N) \in \Omega^N : A_N := \left| \frac{1}{N} \sum_{i=1}^{N} f(\omega_i) - \phi \right| \le \delta \right\}. \tag{2.3.1.2}$$

Instead of going into combinatorics and Stirling formulas, it is more instructive for the applications to statistical mechanics to use a probabilistic approach. We start from the identity

$$K_\delta(N) = \sum_{\omega_1, \ldots, \omega_N} \mathbf{1}_{A_N \le \delta} \frac{p(\omega_1) \cdots p(\omega_N)}{p(\omega_1) \cdots p(\omega_N)}, \tag{2.3.1.3}$$

and the whole trick is to choose properly the probability $p(\omega)$ on Ω. Calling $Z_b = \sum_{\omega \in \Omega} e^{bf(\omega)}$, we will see that the "right choice" is

$$p(\omega) = \frac{e^{bf(\omega)}}{Z_b}, \quad b \text{ such that } \sum_{\omega \in \Omega} p(\omega) f(\omega) = \phi. \tag{2.3.1.4}$$

Existence [and uniqueness] of b follows from the fact that $\sum_{\omega\in\Omega} p(\omega) f(\omega)$ is an increasing function of b which converges to $\min f$ and $\max f$ as $b \to \mp\infty$. With the choice (2.3.1.4) for $p(\cdot)$, we get from (2.3.1.3)

$$K_\delta(N) = \sum_{\omega_1,\dots,\omega_N} \mathbf{1}_{A_N\le\delta}[p(\omega_1)\cdots p(\omega_N)]e^{-\sum(bf(\omega_i)-\log Z_b)}. \tag{2.3.1.5}$$

It is now clear why (2.3.1.4) is a good choice: the sum $\sum_{i=1}^N bf(\omega_i)$ in the exponent is, by (2.3.1.2), approximately the same as the one fixed by the condition $A_N \le \delta$, while the second equality in (2.3.1.4) ensures that with probability converging to 1 the condition $A_N \le \delta$ is satisfied. Indeed, call $S := \log Z_b - b\phi$ and P_N the product probability $p(\omega_1)\cdots p(\omega_N)$, then

$$P_N(A_N \le \delta)e^{(S-b\delta)N} \le K_\delta(N) \le e^{(S+b\delta)N}. \tag{2.3.1.6}$$

By the law of large numbers, for any $\delta > 0$ $\lim_{N\to\infty} P_N(A_N \le \delta) = 1$ so that

$$\lim_{\delta\to 0}\lim_{N\to\infty} \frac{\log K_\delta(N)}{N} = S, \tag{2.3.1.7}$$

and by explicit computation

$$S = \log Z_b - b\phi = -\sum_{\omega\in\Omega} p(\omega)\log p(\omega) =: S(p). \tag{2.3.1.8}$$

In conclusion, the capacity of the channel equals the "information entropy" $S(p)$ of the associated Gibbs measure p of (2.3.1.4). In our applications $\Omega = \{-1, 1\}$, $\mathbb{N}_+$ is replaced by $\mathbb{Z}^d$, b by the inverse temperature β and f by an "energy function" u, which, however, unlike f, does not depend on a single spin. Independence will then fail and the above law of large numbers for Bernoulli measures will be replaced by ergodic theorems for DLR measures. The analysis will then identify the thermodynamic entropy with the above information entropy.

2.3.2 Boltzmann hypothesis

We will consider in the sequel the Ising hamiltonian

$$H_\Lambda(\sigma_\Lambda) = -\frac{1}{2}\sum_{x\ne y\in\Lambda} J(x,y)\sigma_\Lambda(x)\sigma_\Lambda(y) - h\sum_{x\in\Lambda}\sigma_\Lambda(x), \tag{2.3.2.1}$$

supposing that the $J(x, y)$ are translational invariant and summable; see Sect. 2.1.2. To formulate the Boltzmann hypothesis we first introduce the notion of "number of states with given energy."

Definition [Number of states with given energy density] For any bounded set $\Lambda \subset \mathbb{Z}^d$, $\delta > 0$ and $E \in \mathbb{R}$, we define

$$N_{E,\Lambda,\delta} = \text{card}\left\{\sigma_\Lambda : \left| H_\Lambda(\sigma_\Lambda) - |\Lambda|E \right| \leq \delta|\Lambda|\right\}. \tag{2.3.2.2}$$

With the above definition we have relaxed the notion of number of states with given energy by introducing the accuracy parameter δ. This is technically convenient but also natural in a lattice model where the finite volume hamiltonian has finitely many values. We will eventually let $\delta \to 0$, but only after $|\Lambda| \to \infty$. Notice that E in (2.3.2.2) has the meaning of the energy density as $H_\Lambda(\sigma_\Lambda)/|\Lambda|$ is close to E (by δ). Recalling that the Boltzmann hypothesis relates the entropy to the log of the number of states, we next introduce

$$S_{E,\Lambda,\delta} = \frac{\log N_{E,\Lambda,\delta}}{|\Lambda|}, \tag{2.3.2.3}$$

having divided by $|\Lambda|$ because we want the entropy per unit volume. $S_{E,\Lambda,\delta}$ cannot be a candidate for the Boltzmann entropy, as it is not a function of E, depending also on Λ and δ. Thus we need to let $\Lambda \to \mathbb{Z}^d$ and we want this to happen in "a regular way"; namely, we want the volume of a neighborhood of the boundary to be much smaller than the whole volume:

Definition 2.3.2.1 (van Hove sequences) A sequence Λ_n is "*van Hove*" if it is an increasing sequence of bounded regions which invades the whole $\mathbb{Z}^d$ (for any $x \in \mathbb{Z}^d$ there is n so that $x \in \Lambda_n$) and verifies the following property. Given any cube $\Delta \subset \mathbb{Z}^d$ and any partition of $\mathbb{Z}^d$ into translates of Δ, call Λ'_n the union of those cubes of the partition which are contained in Λ_n and Λ''_n of those which have non-empty intersection with Λ_n. Then

$$\lim_{n\to\infty} \frac{|\Lambda'_n|}{|\Lambda_n|} = \lim_{n\to\infty} \frac{|\Lambda''_n|}{|\Lambda_n|} = 1. \tag{2.3.2.4}$$

By default in the sequel $\Lambda \to \mathbb{Z}^d$ is meant to be taken in the van Hove sense.

We would then like to have a theorem which says that there exists $S(E)$ such that for any van Hove sequence Λ_n

$$\lim_{\delta\to 0} \lim_{n\to\infty} S_{E,\Lambda_n,\delta} = S(E). \tag{2.3.2.5}$$

With such a result we can then reasonably formulate the Boltzmann hypothesis in our Ising model by saying that "*$S(E)$ as defined by (2.3.2.5) is the thermodynamic entropy of a system whose microscopic interactions are described by (2.3.2.1).*" The definition poses consistency problems as $S(E)$ should verify the properties that entropy has in thermodynamics and a great success of the theory is that all this can indeed be rigorously established. Using the thermodynamic relations we can obtain from $S(E)$ other thermodynamic potentials, for instance the inverse temperature β

as a function of E is equal to the derivative $dS(E)/dE$. As we will see, there are also formulas for computing the pressure P_β as a function of β, once $S(E)$ as a function of E is known. We will prove that the pressure P_β obtained from $S(\cdot)$ in this way has a simple expression in terms of the Ising partition function. Let Λ_n be any van Hove sequence; then

$$P_\beta = \lim_{n\to\infty} \frac{\log Z_{\beta,\Lambda_n}}{\beta|\Lambda_n|}, \qquad Z_{\beta,\Lambda} = \sum_{\sigma_\Lambda \in \mathcal{X}_\Lambda} e^{-\beta H_\Lambda(\sigma_\Lambda)}. \tag{2.3.2.6}$$

For practical and numerical purposes the expression (2.3.2.6) is much easier to handle than (2.3.2.5); thus, while from an axiomatic viewpoint entropy is the starting point from where all the other thermodynamic potentials are derived, often in the literature (2.3.2.6) is taken as a definition of pressure and all thermodynamics can then be derived.

The equality between the pressure computed from the entropy (2.3.2.5) via thermodynamic relations and the pressure computed via (2.3.2.6) goes under the name of "equivalence of ensembles." The ensembles in the statement are the "microcanonical ensemble," which denotes the phase space reduced to the subset (2.3.2.2) and the "grand-canonical ensemble" which is the full phase space $\mathcal{X}_\Lambda$ where the partition function is computed. This reminds one of the minimization of a function in $\mathbb{R}^n$ where using Lagrange multipliers a constraint can be dropped and the problem reduced to one in the whole $\mathbb{R}^n$. The analogy is made more evident in the following outline of the proof of (2.3.2.5); the real proof is postponed to the next subsections, while (2.3.2.6) is proved in Sect. 2.3.3. We start from the trivial identity

$$\begin{aligned} N_{E,\Lambda,\delta} &= \sum_{\sigma_\Lambda \in \mathcal{X}_\Lambda} \mathbf{1}_{|H_\Lambda(\sigma_\Lambda) - |\Lambda|E| \le \delta|\Lambda|} \\ &= \sum_{\sigma_\Lambda \in \mathcal{X}_\Lambda} \mathbf{1}_{|H_\Lambda(\sigma_\Lambda) - |\Lambda|E| \le \delta|\Lambda|} \frac{e^{-\beta H_\Lambda(\sigma_\Lambda)}}{Z_{\beta,\Lambda}} \{Z_{\beta,\Lambda} e^{\beta H_\Lambda(\sigma_\Lambda)}\}. \end{aligned}$$

We proceed by writing upper and lower bounds, which will hopefully coincide in the limit, and bound the last term $e^{\beta H_\Lambda(\sigma_\Lambda)}$ in the curly brackets by $e^{\beta(E\pm\delta)|\Lambda|}$: we then find for $\frac{\log N_{E,\Lambda,\delta}}{|\Lambda|}$ the bounds

$$\begin{aligned} &\le \frac{\log Z_{\beta,\Lambda}}{|\Lambda|} + \beta(E+\delta) \\ &\ge \frac{\log Z_{\beta,\Lambda}}{|\Lambda|} + \beta(E-\delta) \\ &\quad + \frac{1}{|\Lambda|} \log G_{\beta,\Lambda}\big(\{|H_\Lambda(\sigma_\Lambda) - |\Lambda|E| \le \delta|\Lambda|\}\big), \end{aligned} \tag{2.3.2.7}$$

where $G_{\beta,\Lambda}(\sigma_\Lambda) = \frac{e^{-\beta H_\Lambda(\sigma_\Lambda)}}{Z_{\beta,\Lambda}}$ is the Gibbs measure (with zero boundary conditions). We now face two problems: the first one, the easier one, solved in the next subsection, is to prove that the limit on the r.h.s. of (2.3.2.6) exists, thus defining a function

P_β which only afterwards will be identified with the thermodynamic pressure. This yields

$$\lim_{\delta\to 0}\limsup_{\Lambda\to\mathbb{Z}^d}\frac{\log N_{E,\Lambda,\delta}}{|\Lambda|} \le \inf_{\beta>0}(\beta P_\beta + \beta E). \tag{2.3.2.8}$$

The second and more serious problem concerns the lower bound, a problem that can be avoided if we suppose that there is β^* so that

$$\lim_{\delta\to 0}\lim_{\Lambda\to\mathbb{Z}^d}\frac{1}{|\Lambda|}\log G_{\beta^*,\Lambda}\left(\left\{\left|\frac{H_\Lambda(\sigma_\Lambda)}{|\Lambda|}-E\right|\le\delta\right\}\right)=0. \tag{2.3.2.9}$$

Then

$$\lim_{\delta\to 0}\liminf_{\Lambda\to\mathbb{Z}^d}\frac{\log N_{E,\Lambda,\delta}}{|\Lambda|} \ge \beta^* P_{\beta^*}+\beta^* E, \tag{2.3.2.10}$$

which together with (2.3.2.8) yields the result that the limit in (2.3.2.5) exists and is equal to

$$S(E)=\beta^* P_{\beta^*}+\beta^* E=\inf_{\beta>0}(\beta P_\beta+\beta E), \tag{2.3.2.11}$$

which is the well known thermodynamic formula for the entropy in terms of the pressure.

The problem is that (2.3.2.9) is not true in general. As we will see it holds for E in a set $\mathcal{E}_{\rm erg}$: for any $E\in\mathcal{E}_{\rm erg}$ there is a special value of β for which (2.3.2.9) holds. We will discuss later how to proceed when $E\notin\mathcal{E}_{\rm erg}$. The terminology hints that the crucial estimate (2.3.2.9) is related to an ergodic theorem. Let $\tau_z u(\sigma)$ be the translate by $z\in\mathbb{Z}^d$ of the function $u(\sigma)$ defined in (2.2.7.3). Then

$$\mathcal{A}_{\sigma,\Lambda}(u):=\frac{1}{|\Lambda|}\sum_{x\in\Lambda}\tau_x u(\sigma) \tag{2.3.2.12}$$

is the "ergodic average" of u in Λ computed at σ. We claim that $\mathcal{A}_{\sigma,\Lambda}(u)$ is "close to" $H_\Lambda(\sigma_\Lambda)$, the latter to be thought of as a function of $\sigma\in\mathcal{X}$ (with σ_Λ the restriction of σ to Λ) and thus also denoted by $H_\Lambda(\sigma)$. Indeed

$$H_\Lambda(\sigma)\equiv H_\Lambda(\sigma_\Lambda)=|\Lambda|\mathcal{A}_{\sigma,\Lambda}(u)+\frac{1}{2}\sum_{x\in\Lambda}\sum_{y\in\Lambda^c}J(x,y)\sigma(x)\sigma(y). \tag{2.3.2.13}$$

Analogously, for any $\sigma_\Lambda\in\mathcal{X}_\Lambda$ and $\sigma_{\Lambda^c}\in\mathcal{X}_{\Lambda^c}$, denoting $\sigma=(\sigma_\Lambda,\sigma_{\Lambda^c})$

$$H_\Lambda(\sigma_\Lambda|\sigma_{\Lambda^c})=|\Lambda|\mathcal{A}_{\sigma,\Lambda}(u)-\frac{1}{2}\sum_{x\in\Lambda,y\in\Lambda^c}J(x,y)\sigma_\Lambda(x)\sigma_{\Lambda^c}(y). \tag{2.3.2.14}$$

We will see in the next subsection that if $\Lambda\to\mathbb{Z}^d$ in the van Hove sense, then

$$\lim_{\Lambda\to\mathbb{Z}^d}\frac{1}{|\Lambda|}\sum_{x\in\Lambda}\sum_{y\in\Lambda^c}|J(x,y)|=0, \tag{2.3.2.15}$$

so that the last term in (2.3.2.13)–(2.3.2.14) is a "negligible error" (once divided by $|\Lambda|$) and $H_\Lambda(\sigma)$ is "essentially" the ergodic average of u in Λ.

If μ is an ergodic measure on $\mathcal{X}$, $\mathcal{A}_{\sigma,\Lambda}(u) \to \mu(u)$ ($\mu(u)$ being the expectation of u) for μ almost all σ. Thus, if $\mu(u) = E$,

$$\lim_{\delta\to 0}\lim_{\Lambda\to\mathbb{Z}^d} \mu\left(\left\{\left|\frac{H_\Lambda(\sigma_\Lambda)}{|\Lambda|} - E\right| \le \delta\right\}\right) = 1. \tag{2.3.2.16}$$

As we shall see using (2.3.2.15) we can modify (2.3.2.7) so that instead of $G_{\beta,\Lambda}$ we can put any DLR measure at inverse temperature β with an error such that its log, divided by $|\Lambda|$ vanishes as $|\Lambda| \to \infty$. Thus (2.3.2.9) holds for all $E \in \mathcal{E}_{\text{erg}}$ where the latter is the set of all $\mu(u)$ as μ varies over all the ergodic DLR measures at $\beta > 0$ and β varies in the whole $\mathbb{R}_+$.

If all DLR measures were ergodic we will be in business, but as we have seen in the previous section this is not true in case of phase transitions, when in fact there are distinct ergodic DLR measures at the same β so that their convex combinations are DLR at β but not ergodic. In such a case the approach which starts from (2.3.2.7) must be aborted and we have to go one step back. Suppose the energy density E belongs to an interval $[E', E'']$, $E = aE' + (1-a)E''$, $a \in (0, 1)$, whose endpoints are both in $\mathcal{E}_{\text{erg}}$, namely are expectations of $u(\sigma)$ relative to two ergodic DLR measures at the same inverse temperature β, then the previous argument can be reproduced in the following way. Suppose for simplicity Λ to be a cube and suppose that it can be split into two rectangles Λ' and Λ'' such that $a = |\Lambda'|/|\Lambda|$ (approximate equality is however sufficient). By (2.3.2.15), if $|\Lambda|$ is large enough the set $\{|H_{\Lambda'}(\sigma_{\Lambda'}) - |\Lambda'|E'| \le \frac{\delta}{4}|\Lambda'|\} \cap \{|H_{\Lambda''}(\sigma_{\Lambda''}) - |\Lambda''|E''| \le \frac{\delta}{4}|\Lambda''|\}$ is contained in $\{|H_\Lambda(\sigma_\Lambda) - |\Lambda|E| \le \delta|\Lambda|\}$. Then the lower bound in (2.3.2.7) can be replaced by

$$\begin{aligned}\frac{\log N_{E,\Lambda,\delta}}{|\Lambda|} \ge \frac{\log Z_{\beta,\Lambda}}{|\Lambda|} &+ \beta(E-\delta)\\ &+ \frac{1}{|\Lambda|}\log G_{\beta,\Lambda}\left(\left\{|H_\Lambda(\sigma_{\Lambda'}) - |\Lambda'|E'| \le \frac{\delta}{4}|\Lambda'|\right\}\right.\\ &\left.\cap\left\{|H_\Lambda(\sigma_{\Lambda''}) - |\Lambda''|E''| \le \frac{\delta}{4}|\Lambda''|\right\}\right).\end{aligned} \tag{2.3.2.17}$$

Using again (2.3.2.15), we can also factorize the Gibbs measure in the sense that the log of

$$\frac{G_{\beta,\Lambda}(\{|H_{\Lambda'}(\sigma_{\Lambda'}) - |\Lambda'|E'| \le \frac{\delta}{4}|\Lambda'|\} \cap \{|H_{\Lambda''}(\sigma_{\Lambda''}) - |\Lambda''|E''| \le \frac{\delta}{4}|\Lambda''|\})}{G_{\beta,\Lambda'}(\{|H_{\Lambda'}(\sigma_{\Lambda'}) - |\Lambda'|E'| \le \frac{\delta}{4}|\Lambda'|\})G_{\beta,\Lambda''}(\{|H_{\Lambda''}(\sigma_{\Lambda''}) - |\Lambda''|E''| \le \frac{\delta}{4}|\Lambda''|\})}$$

divided by $|\Lambda$ vanishes as the cube $\Lambda \to \mathbb{Z}^d$. We can thus replace in (2.3.2.17) the $G_{\beta,\Lambda}$ expectation by the product of expectations with $G_{\beta,\Lambda'}$ and $G_{\beta,\Lambda''}$ and then reproduce the argument used when $E \in \mathcal{E}_{\text{erg}}$.

It remains for us to characterize the set $\mathcal{E}_{\text{allwd}}$ defined as the set of all E which are contained in intervals $[E', E'']$ with E' and E'' expectations of u with respect to two ergodic measures which are DLR at the same value of β. By the analysis of the previous section we know that for each $\beta > 0$ there exist ergodic DLR measures; moreover, any translational invariant DLR measure at β (their collection being denoted by $\mathcal{G}^0_\beta$) can be written as an integral over the subset of ergodic DLR measures (ergodic decomposition theorem). It then follows that

$$\mathcal{E}_{\text{allwd}} = \bigcup_{\beta>0} \{E \in \mathbb{R} : E = \mu(u), \mu \in \mathcal{G}^0_\beta\}. \tag{2.3.2.18}$$

We will next see how the sets $\{E \in \mathbb{R} : E = \mu(u), \mu \in \mathcal{G}^0_\beta\}$ are related to the pressure P_β. Together with the proof of existence of the limit (2.3.2.6) defining P_β we also have that $\pi_\beta := \beta P_\beta$ is a continuous convex function of β. Then by general theorems on convex functions, see Sect. 2.3.6, its right and left derivatives $D^\pm\pi_\beta$ exist, and $D^-\pi_\beta \le D^+\pi_\beta$. We will then prove that for any $\beta > 0$ the values $E'' = -D^-\pi_\beta$ and $E' = -D^+\pi_\beta$, $E' \le E''$ are both in $\mathcal{E}_{\text{erg}}$; thus

$$\mathcal{E} =: -\mathcal{E}_{\text{allwd}} = \bigcup_{\beta>0} [D^-\pi_\beta, D^+\pi_\beta]. \tag{2.3.2.19}$$

It will also follow from general theorems on convex functions that both $[D^-\pi_\beta, D^+\pi_\beta]$ and $\mathcal{E}_{\text{allwd}}$ are bounded intervals. The restriction to a bounded interval of energies is a consequence of the system being a lattice model in which the energy density is bounded both from above and below (in a continuum system it is generally only bounded from below). We will see in the sequel that bounded spin systems have strange properties; in particular negative temperatures can also be defined if we enlarge the set $\mathcal{E}_{\text{allwd}}$ to all E for which the limit (2.3.2.5) exists, a phenomenon which disappears if the energy density is unbounded from above.

2.3.3 *Thermodynamic limit of the pressure*

In this subsection we will prove

Theorem 2.3.3.1 *For any van Hove sequence $\Lambda \to \mathbb{Z}^d$ and any sequence of b.c. $\sigma_{\Lambda^c} \in \mathcal{X}_{\Lambda^c}$,*

$$\lim_{\Lambda \to \mathbb{Z}^d} P_{\beta,\Lambda}(\sigma_{\Lambda^c}) =: P_\beta, \tag{2.3.3.1}$$

where, recalling that $\mathcal{X}_\Lambda = \{-1,1\}^\Lambda$,

$$\begin{aligned} P_{\beta,\Lambda}(\sigma_{\Lambda^c}) &= \frac{1}{\beta|\Lambda|} \log Z_{\beta,\Lambda}(\sigma_{\Lambda^c}), \\ Z_{\beta,\Lambda}(\sigma_{\Lambda^c}) &= \sum_{\sigma_\Lambda \in \mathcal{X}_\Lambda} e^{-\beta H_\Lambda(\sigma_\Lambda|\sigma_{\Lambda^c})}. \end{aligned} \tag{2.3.3.2}$$

Before starting the proof of the theorem we state and prove the following lemma.

Lemma 2.3.3.2 *There is a constant c so that for any bounded Λ, any σ_Λ and any σ_{Λ^c}*

$$|H_\Lambda(\sigma_\Lambda|\sigma_{\Lambda^c})| \le c|\Lambda|. \tag{2.3.3.3}$$

Moreover, given any $\epsilon > 0$ for all cubes Δ large enough

$$\sum_{x\in\Delta, y\in\Delta^c} |J(x,y)| \le \epsilon|\Delta|. \tag{2.3.3.4}$$

Proof We have

$$|H_\Lambda(\sigma_\Lambda|\sigma_{\Lambda^c})| \le \sum_{x\in\Lambda}\sum_{y\neq x} |J(x,y)| + \sum_{x\in\Lambda} |h|,$$

which proves (2.3.3.3) with $c = |h| + \sum_{x\neq 0}|J(0,x)|$.
Given $R > 0$ we split the sum in the l.h.s. of (2.3.3.4) as

$$\sum_{x\in\Delta, \mathrm{dist}(x,\Delta^c)\le R}\ \sum_{y\in\Delta^c} |J(x,y)| + \sum_{x\in\Delta, \mathrm{dist}(x,\Delta^c)> R}\ \sum_{y\in\Delta^c} |J(x,y)|.$$

The first term is bounded by $cL^{d-1}R$, L being the side of Δ; the second one by

$$|\Delta| \sum_{|x|>R} |J(0,x)| < \frac{\epsilon}{2}|\Delta|,$$

if R is large enough. Given such a R we then choose L so large that $cL^{d-1}R \le (\epsilon/2)L^d$, hence (2.3.3.4). □

Proof of Theorem 2.3.3.1 In this proof we do not make explicit β in the notation. By (2.3.3.3) there is c so that

$$|P_\Lambda(\sigma_{\Lambda^c})| = \frac{1}{\beta|\Lambda|}\,|\log Z_\Lambda(\sigma_{\Lambda^c})| \le c. \tag{2.3.3.5}$$

Then by compactness there exists an increasing sequence of cubes $\Delta \to \mathbb{Z}^d$ such that

$$\lim_{\Delta\to\mathbb{Z}^d} P_\Delta =: P. \tag{2.3.3.6}$$

Now, $P_\Delta = \frac{\log Z_\Delta}{\beta|\Delta|}$, $Z_\Delta = \sum_{\sigma_\Delta} e^{-\beta H_\Delta(\sigma_\Delta)}$, $H_\Delta(\sigma_\Delta)$ as in (2.3.2.1). Let Λ be a van Hove sequence and σ_{Λ^c} a sequence of boundary conditions. We will prove that $\lim_{\Lambda\to\mathbb{Z}^d} P_\Lambda(\sigma_{\Lambda^c}) = P$, P as in (2.3.3.6). Let $\epsilon > 0$ and let Δ be a cube of the sequence in (2.3.3.6) as large as required for (2.3.3.4) to hold. Consider a partition into translates of Δ, call $\Delta(i)$ those in Λ, and Λ' their union. With $c = |h| + \sum_{x\neq 0} |J(0,x)|$ we then have, using Lemma 2.3.3.2,

$$\Big| H_\Lambda(\sigma_\Lambda|\sigma_{\Lambda^c}) - \sum_{\Delta(i)\subset\Lambda'} H_{\Delta(i)}(\sigma_{\Delta(i)}) \Big| \le \epsilon|\Lambda'| + c|\Lambda\setminus\Lambda'|. \tag{2.3.3.7}$$

From (2.3.3.7) we get

$$\Big| P_\Lambda(\sigma_{\Lambda^c}) - \frac{|\Lambda'|}{|\Lambda|} P_\Delta \Big| \le \epsilon \frac{|\Lambda'|}{|\Lambda|} + c\frac{|\Lambda\setminus\Lambda'|}{|\Lambda|}.$$

Letting $\Lambda \to \mathbb{Z}^d$ and calling $P' \le P''$ the lim inf and lim sup of $P_\Lambda(\sigma_{\Lambda^c})$

$$P_\Delta - \epsilon \le P' \le P'' \le P_\Delta + \epsilon.$$

By letting $\Delta \to \mathbb{Z}^d$ along the sequence in (2.3.3.6) we get $P - \epsilon \le P' \le P'' \le P + \epsilon$ and by the arbitrariness of ϵ, (2.3.3.1). □

2.3.4 Thermodynamic limit of the entropy

The precise statement of the results outlined in Sect. 2.3.2 is

Theorem 2.3.4.1 (Main result) *There is a non-empty interval $\mathcal{E}$ (defined in* (2.3.2.19)) *such that for any $E \in \mathcal{E}_{\text{allw}}$, $\mathcal{E}_{\text{allw}} := -\mathcal{E}$, and for any van Hove sequence $\Lambda \to \mathbb{Z}^d$*

$$\lim_{\delta\to 0} \limsup_{\Lambda\to\mathbb{Z}^d} \frac{\log N_{E,\Lambda,\delta}}{|\Lambda|} = \lim_{\delta\to 0} \liminf_{\Lambda\to\mathbb{Z}^d} \frac{\log N_{E,\Lambda,\delta}}{|\Lambda|} =: S(E), \tag{2.3.4.1}$$

with $S(E)$ a strictly increasing, concave function of E. Moreover, if P_β denotes the pressure defined in (2.3.3.1),

$$\beta P_\beta = \sup_{E\in\mathcal{E}_{\text{allw}}} \{\beta(-E) + S(E)\}, \qquad -S(E) = \sup_{\beta>0}\{\beta(-E) - \beta P_\beta\}, \tag{2.3.4.2}$$

so that the functions $s(E) = -S(-E)$, $E \in \mathcal{E}$, and $\pi_\beta = \beta P_\beta$, $\beta \in \mathbb{R}_+$, are Legendre transforms of each other. $S(E)$ and P_β are both differentiable except at countably many points, while left and right derivatives $D^\pm$ exist everywhere. We say that E and β are conjugate if $e = -E \in [D^-\pi_\beta, D^+\pi_\beta]$; in such a case $\beta \in [D^-s(e), D^+s(e)]$ and, if E and β are conjugate, then

$$P_\beta = -E + \beta^{-1} S(E). \tag{2.3.4.3}$$

Finally, $s(e)$ is linear in $[D^-\pi_\beta, D^+\pi_\beta]$ and π_β in $[D^-s(e), D^+s(e)]$.

Thermodynamic interpretation

As discussed in Sect. 2.3.2, in agreement with the Boltzmann hypothesis we interpret $S(E)$ as the thermodynamic entropy, and then P_β as given in (2.3.4.3) in terms of $S(E)$ is the thermodynamic pressure at the inverse temperature β, as thermodynamics says that entropy and pressure are related as in (2.3.4.3). In conclusion, Theorem 2.3.4.1 and the Boltzmann hypothesis completely determine the thermodynamics of the Ising model and justify the identification of the limit (2.3.3.1) as the thermodynamic pressure.

Proof of the statements of Theorem 2.3.4.1

The proof of the theorem takes most of this section. Convexity of the pressure is proved in Theorem 2.3.5.1; (2.3.4.1) is proved in Theorem 2.3.7.5; (2.3.4.2) in Theorem 2.3.8.1; the properties of $S(E)$ stated in Theorem 2.3.4.1 are proved in Theorem 2.3.8.2, while in Theorem 2.3.8.3 there is a characterization of the set $\mathcal{E}$.

Scheme of the proof of Theorem 2.3.4.1

We will first prove (2.3.4.1) only for $E \in \mathcal{E}_{\mathrm{erg}}$, $\mathcal{E}_{\mathrm{erg}} := \{E : E = \mu(u),\ \mu$ ergodic DLR at inverse temperature $\beta, \beta > 0\}$. For such special values of E we can in fact follow the heuristic argument presented in Sect. 2.3.2 to prove (2.3.4.1). The argument also proves (2.3.4.3), which is thus established for $E \in \mathcal{E}_{\mathrm{erg}}$. The extension to $E \notin \mathcal{E}_{\mathrm{erg}}$ is based on the general properties of convex functions, which are recalled together with the definition and properties of Legendre transforms in Sect. 2.3.6.

Proposition 2.3.4.2 *For any $E \in \mathcal{E}_{\mathrm{erg}}$ and for any van Hove sequence Λ_n, the limit $S(E)$ in* (2.3.4.1) *exists and $S(E)$ satisfies* (2.3.4.3).

We fix in the sequel $E \in \mathcal{E}_{\mathrm{erg}}$ so that there is an ergodic DLR measure μ at some $\beta > 0$ such that $\mu(u) = E$. We postpone the proof of Proposition 2.3.4.2 to first deal with the following three lemmas.

Lemma 2.3.4.3 *For any $\delta > 0$ and any van Hove sequence $\Lambda \to \mathbb{Z}^d$*

$$\lim_{\Lambda\to\mathbb{Z}^d} \mu\left(\frac{1}{|\Lambda|}\left|\sum_{x\in\Lambda}\{\tau_x u(\sigma) - E\}\right| \le \delta\right) = 1. \tag{2.3.4.4}$$

Proof Since μ is ergodic, for any $\epsilon > 0$ there is a cube Δ_ϵ so that for any cube Δ, $|\Delta| \ge |\Delta_\epsilon|$,

$$\mu\left(\left|\sum_{x\in\Delta}\{\tau_x u(\sigma) - E\}\right|\right) < \epsilon|\Delta|. \tag{2.3.4.5}$$

Define a partition of $\mathbb{Z}^d$ into cubes which are translates of Δ; call $\Delta(i)$ those contained in Λ, and Λ' their union. Then

$$\begin{aligned}
&\mu\left(\frac{1}{|\Lambda|}\left|\sum_{x\in\Lambda}\{\tau_x u(\sigma)-E\}\right|>\delta\right)\\
&\leq \frac{1}{\delta|\Lambda|}\mu\left(\left|\sum_{x\in\Lambda}\{\tau_x u(\sigma)-E\}\right|\right)\\
&\leq \frac{1}{\delta|\Lambda|}\left\{\sum_{\Delta(i)}\mu\left(\left|\sum_{x\in\Delta(i)}\{\tau_x u(\sigma)-E\}\right|\right)+c|\Lambda\setminus\Lambda'|\right\}, \qquad (2.3.4.6)
\end{aligned}$$

because u is bounded. By (2.3.4.5),

$$\mu\left(\frac{1}{|\Lambda|}\left|\sum_{x\in\Lambda}\{\tau_x u(\sigma)-E\}\right|>\delta\right)\leq\frac{c|\Lambda\setminus\Lambda'|}{\delta|\Lambda|}+\frac{\epsilon|\Lambda'|}{\delta|\Lambda|}.$$

In the limit $\Lambda\to\mathbb{Z}^d$, $\frac{c|\Lambda\setminus\Lambda'|}{\delta|\Lambda|}\to 0$, $\frac{\epsilon|\Lambda'|}{\delta|\Lambda|}\to\frac{\epsilon}{\delta}$ and by the arbitrariness of ϵ, (2.3.4.4) is proved. □

Lemma 2.3.4.4 *Given any $\epsilon>0$ for all cubes Δ large enough*

$$\sup_{\sigma_\Delta,\sigma_{\Delta^c}}\left|\sum_{x\in\Delta}[\tau_x u((\sigma_\Delta,\sigma_{\Delta^c}))-\tau_x u((\sigma_\Delta,0_{\Delta^c}))]\right|\leq\epsilon|\Delta|, \qquad (2.3.4.7)$$

where $u((\sigma_\Delta,0_{\Delta^c}))$ is defined by (2.2.7.3) *putting $\sigma(\cdot)=0$ on Δ^c;*

$$\sup_{\sigma_\Delta,\sigma_{\Delta^c}}\left|H_\Delta(\sigma_\Delta)-H_\Delta((\sigma_\Delta|\sigma_{\Delta^c}))\right|\leq\epsilon|\Delta|. \qquad (2.3.4.8)$$

Proof Equations (2.3.4.7)–(2.3.4.8) follow straightforwardly from (2.3.3.4). □

Lemma 2.3.4.5 *Let $\Lambda\to\mathbb{Z}^d$ be a van Hove sequence; then, given any $\epsilon>0$ for all Λ large enough, we have*

$$\sup_{\sigma_\Lambda,\sigma_{\Lambda^c}}\left|H_\Lambda(\sigma_\Lambda)-\sum_{x\in\Lambda}\tau_x u((\sigma_\Lambda,\sigma_{\Lambda^c}))\right|\leq\epsilon|\Lambda|, \qquad (2.3.4.9)$$

$$\sup_{\sigma_\Lambda,\sigma_{\Lambda^c}}\left|H_\Lambda(\sigma_\Lambda)-H_\Lambda(\sigma_\Lambda|\sigma_{\Lambda^c})\right|\leq\epsilon|\Lambda|. \qquad (2.3.4.10)$$

Proof Let $\Delta(i)$ and Λ' be defined as in the definition of the van Hove sequences; then by (2.3.2.12), the l.h.s. of (2.3.4.9) is bounded by

$$\sum_{\Delta(i)\subset\Lambda'}\left|\sum_{x\in\Delta(i)}[\tau_x u((\sigma_\Lambda,\sigma_{\Lambda^c}))-\tau_x u((\sigma_\Lambda,0_{\Lambda^c}))]\right|+c|\Lambda\setminus\Lambda'|,$$

so that (2.3.4.9) follows from (2.3.4.7), because $|\Lambda \setminus \Lambda'|/|\Lambda| \to 0$. Analogously,

$$H_\Lambda(\sigma_\Lambda|\sigma_{\Lambda^c}) = H_{\Lambda\setminus\Delta(i)}(\sigma_{\Lambda\setminus\Delta(i)}|\sigma_{\Lambda^c}) + H_{\Delta(i)}(\sigma_{\Delta(i)}|\sigma_{\Lambda\setminus\Delta(i)}, \sigma_{\Lambda^c}).$$

Hence by (2.3.4.8) for $\Delta(i)$ large enough

$$|H_\Lambda(\sigma_\Lambda|\sigma_{\Lambda^c}) - H_{\Delta(i)}(\sigma_{\Delta(i)}) - H_{\Lambda\setminus\Delta(i)}(\sigma_{\Lambda\setminus\Delta(i)}|\sigma_{\Lambda^c})| \le \frac{\epsilon}{4}|\Delta(i)|.$$

By iteration

$$\left| H_\Lambda(\sigma_\Lambda|\sigma_{\Lambda^c}) - \sum_{\Delta(i)\subset\Lambda'} H_{\Delta(i)}(\sigma_{\Delta(i)}) \right| \le c|\Lambda \setminus \Lambda'| + \frac{\epsilon}{4}|\Lambda|.$$

By the same argument

$$\left| H_\Lambda(\sigma_\Lambda) - \sum_{\Delta(i)\subset\Lambda'} H_{\Delta(i)}(\sigma_{\Delta(i)}) \right| \le c|\Lambda \setminus \Lambda'| + \frac{\epsilon}{4}|\Lambda|, \tag{2.3.4.11}$$

hence (2.3.4.10). □

Proof of Proposition 2.3.4.2 Let $\Lambda \to \mathbb{Z}^d$ be a van Hove sequence. Then by Lemmas 2.3.4.3 and 2.3.4.5 for Λ large enough,

$$\mu\big(\big|H_\Lambda(\sigma_\Lambda) - |\Lambda|E\big| \le \delta\big) \ge \frac{1}{2}, \tag{2.3.4.12}$$

and by Theorem 2.3.3.1, given any ϵ for Λ large enough and any σ_{Λ^c},

$$e^{-\beta\epsilon|\Lambda|} \le \frac{Z_\Lambda(\sigma_{\Lambda^c})}{e^{\beta P_\beta|\Lambda|}} \le e^{\beta\epsilon|\Lambda|}. \tag{2.3.4.13}$$

We are now ready for the proof of Proposition 2.3.4.2. We first write

$$\begin{aligned} N_{E,\Lambda,\delta} &= \sum_{\sigma_\Lambda} \mathbf{1}_{H_\Lambda(\sigma_\Lambda)-|\Lambda|E|\le\delta|\Lambda|}, \\ &= \int d\mu(\sigma_{\Lambda^c}) \sum_{\sigma_\Lambda} \mathbf{1}_{|H_\Lambda(\sigma_\Lambda)-|\Lambda|E|\le\delta|\Lambda|} \frac{e^{-\beta H_\Lambda(\sigma_\Lambda|\sigma_{\Lambda^c})}}{Z_\Lambda(\sigma_{\Lambda^c})} \\ &\quad \times \{Z_\Lambda(\sigma_{\Lambda^c}) e^{\beta H_\Lambda(\sigma_\Lambda|\sigma_{\Lambda^c})}\}. \end{aligned}$$

By (2.3.4.13) and (2.3.4.10) for any $\sigma_\Lambda : |H_\Lambda(\sigma_\Lambda) - |\Lambda|E| \le \delta|\Lambda|$ and all Λ large enough,

$$\{Z_\Lambda(\sigma_{\Lambda^c}) e^{\beta H_\Lambda(\sigma_\Lambda|\sigma_{\Lambda^c})}\} \le e^{\beta(P_\beta+\epsilon)|\Lambda|} e^{\beta(H_\Lambda(\sigma_\Lambda)+\epsilon)|\Lambda|} \le e^{\beta(P_\beta+E)|\Lambda|} e^{\beta(2\epsilon+\delta)|\Lambda|},$$

so that

$$N_{E,\Lambda,\delta} \le e^{\beta(P_\beta+E)|\Lambda|} e^{\beta(2\epsilon+\delta)|\Lambda|}. \tag{2.3.4.14}$$

Notice that the upper bound (2.3.4.14) is valid for any E and not only for the special value $E = \mu(u)$. For the lower bound, instead after proceeding similarly, we use (2.3.4.12) to get

$$N_{E,\Lambda,\delta} \geq e^{\beta(P_\beta+E)|\Lambda|} e^{-\beta(2\epsilon+\delta)|\Lambda|} \frac{1}{2}, \tag{2.3.4.15}$$

and Proposition 2.3.4.2 is proved. □

2.3.5 *Equivalence of ensembles*

Proposition 2.3.4.2 is a first indication of it, but it is not yet a full justification for interpreting the limit of $\log Z_\Lambda(\sigma_{\Lambda^c})/(\beta|\Lambda|)$ as the thermodynamic pressure. One reason is that the entropy $S(E)$ has only been defined for $E \in \mathcal{E}_{\rm erg}$ and therefore the identity (2.3.4.3) is only established for such values of E. Notice however that (2.3.4.3) covers all $\beta > 0$ as for any β there is an ergodic DLR measure, hence a value $E \in \mathcal{E}_{\rm erg}$ for which (2.3.4.3) holds. As we are going to explain this creates a serious consistency problem, which could make the interpretation of P_β as the pressure dubious. In fact, according to thermodynamic principles, if P_β is the thermodynamic pressure at the inverse temperature β, then the function

$$S^{\rm td}(E) := -\sup_{\beta>0}\{\beta(-E) - \beta P_\beta\} \tag{2.3.5.1}$$

is the thermodynamic entropy. We thus have two entropies, $S(E)$ defined directly from the Boltzmann hypothesis, and $S^{\rm td}(E)$ defined starting from P_β. Consistency requires that they coincide which indeed will be proved. There is also another consistency problem to check: by thermodynamic principles the pressure is related to the entropy by

$$\beta P_\beta = \sup_E\{\beta(-E) - [-S^{\rm td}(E)]\}, \tag{2.3.5.2}$$

namely $\pi_\beta = \beta P_\beta$ is the Legendre transform of $s(E) = -S^{\rm td}(-E)$; hence π_β must be a convex function of β—see the paragraph "Legendre transforms" at the beginning of Sect. 2.3.8.

We thus need to prove (1) that the limit in (2.3.4.1) is well posed also for $E \notin \mathcal{E}_{\rm erg}$; (2) that $S(E) = S^{\rm td}(E)$ for all E; (3) that βP_β is a convex function of β, a property which implies (2.3.5.2) because by (2.3.5.1) $s(E) = -S^{\rm td}(-E)$ is the Legendre transform of βP_β (see again the paragraph "Legendre transforms" in Sect. 2.3.8).

Property (3) is proved as a corollary of Theorem 2.3.5.1 below, so that the requirement (2.3.5.2) from thermodynamics is fulfilled. More serious is the consistency problem (2), i.e. that $S(E) = S^{\rm td}(E)$, which will be proved in the next subsections together with (1), leading in the end to the proof of Theorem 2.3.4.1. Problem (2) is usually referred to as "equivalence of ensembles." The equivalence is between the "grand-canonical ensemble" used in the definition of $Z_\Lambda(\sigma_{\Lambda^c})$, i.e. all $\sigma_\Lambda \in \mathcal{X}_\Lambda$, and the "micro-canonical ensemble" used in the definition of $N_{E,\Lambda,\delta}$, where σ_Λ is restricted by the energy constraint, and, as we shall see, there are other ensembles related to other variables than the energy.

We will next prove that the pressure is convex. Consider the convex family of hamiltonians $H_\Lambda(\sigma_\Lambda) + V_\Lambda(\sigma_\Lambda)$ where $H_\Lambda(\sigma_\Lambda)$ is defined in (2.3.2.13) and $V_\Lambda(\sigma_\Lambda) := \sum_{\tau_x \Delta \subset \Lambda} \tau_x v(\sigma_\Lambda)$, where v is a cylindrical function on Δ and the sum is over all translates of Δ which are in Λ. Call $Z_\Lambda(v)$ the corresponding partition function and, given any van Hove sequence, let

$$\pi(v) = \lim_{\Lambda \to \mathbb{Z}^d} \pi_\Lambda(v), \quad \pi_\Lambda(v) := \frac{\log Z_\Lambda(v)}{|\Lambda|}. \tag{2.3.5.3}$$

The existence of the above limit is proved as in Sect. 2.3.3 and it is omitted.

Theorem 2.3.5.1 *The function* $\pi(v)$ *defined in* (2.3.5.3) *is convex, i.e. for any* $a \in [0, 1]$

$$\pi\big(av^{(1)} + (1-a)v^{(2)}\big) = a\pi(v^{(1)}) + (1-a)\pi(v^{(2)}). \tag{2.3.5.4}$$

Proof Let Λ be a bounded region and

$$v = av^{(1)} + (1-a)v^{(2)}, \quad a \in [0, 1].$$

Then calling $H_\Lambda^{(1)}(\sigma_\Lambda)$ and $H_\Lambda^{(2)}(\sigma_\Lambda)$ the energies with $v^{(1)}$ and $v^{(2)}$, respectively,

$$\begin{aligned} Z_\Lambda(v) &= \sum_{\sigma_\Lambda} e^{-\beta a H_\Lambda^{(1)}(\sigma_\Lambda)} e^{-\beta(1-a)H^{(2)}(\sigma_\Lambda)} \\ &\le \Big[\sum_{\sigma_\Lambda \in \mathcal{X}_\Lambda} \big(e^{-\beta a H^{(1)}(\sigma_\Lambda)}\big)^p\Big]^{1/p} \Big[\sum_{\sigma_\Lambda \in \mathcal{X}_\Lambda} \big(e^{-\beta(1-a)H^{(2)}(\sigma_\Lambda)}\big)^q\Big]^{1/q}, \end{aligned}$$

with $1/p + 1/q = 1$. By choosing $p = a^{-1}$ and $q = (1-a)^{-1}$ we get

$$Z_\Lambda(v) \le Z_\Lambda(v^{(1)})^a \, Z_\Lambda(v^{(2)})^{1-a},$$

which yields $\pi_\Lambda(v) \le a\pi_\Lambda(v^{(1)}) + (1-a)\pi_\Lambda(v^{(2)})$. Then π_Λ is convex and $\pi(U)$ as well because of the limit of convex functions; see the paragraph "Properties of convex functions" in the next subsection. □

By Theorem 2.3.5.1, βP_β is a convex function of β and (2.3.5.2) is proved.

2.3.6 Properties of convex functions

We recall here some properties of convex functions on $\mathbb{R}$ which will often be used in the sequel, referring for their proofs to Chap. I.3 in Simon's book on Statistical Mechanics [200].

- *Definition.* $f(x)$, $x \in \mathbb{R}$, is convex if $f(\alpha x + (1-\alpha)y) \le \alpha f(x) + (1-\alpha) f(y)$ for all x and y and $\alpha \in [0, 1]$.
- *Differentiability.* If f is convex, f is differentiable at all but countably many points, the right and left derivatives, $D^+ f$ and $D^- f$, exist everywhere and

$$D^- f(x) \le D^+ f(x) \le D^- f(y) \le D^+ f(y), \quad x < y. \tag{2.3.6.1}$$

- *Limits of convex functions.* If f_n is a sequence of convex functions which converges point-wise to f, then f is convex and for any x

$$D^- f(x) \le \liminf D^- f_n(x) \le \limsup D^+ f_n(x) \le D^+ f(x). \tag{2.3.6.2}$$

- *Legendre transforms.* Let $f(x)$ be a convex function, $x \in \mathbb{R}$. Its Legendre transform $g(p)$, $p \in \mathbb{R}$, is

$$g(p) = \sup_x \{xy - f(x)\}. \tag{2.3.6.3}$$

- *Properties of the Legendre transform.* The Legendre transform $g(p)$ of a function $f(x)$ is convex and, if $f(x)$ is convex, then

$$f(x) = \sup_p \{xy - g(p)\}. \tag{2.3.6.4}$$

  In general, the Legendre transform h of the Legendre transform g of f is $h = CEf$ the convex envelope of f, namely the largest convex function $\le f$.
- *Conjugate pairs.* Let f be a convex function and g its Legendre transform. Then x and p are conjugate if

$$p \in [D^- f(x), D^+ f(x)] \text{ if and only if } g(p) = px - f(x), \tag{2.3.6.5}$$

  and

$$\text{if } p \in [D^- f(x), D^+ f(x)] \text{ then } x \in [D^- g(p), D^+ g(p)]. \tag{2.3.6.6}$$

Notice that if $D^- f(x) < D^+ f(x)$ then $g(p)$ is linear in $[D^- f(x), D^+ f(x)]$. The Legendre transform has the following geometric interpretation (see Fig. 2.1): for each p consider all the straight lines $px + c$, all with the same slope p, which are not above the graph of f, namely the set of all c so that $f(x) - [px + c] \ge 0$ for all x. Hence $c \le \inf_x \{f(x) - px\}$. Call c^* the sup of all such c, then $g(p) = -c^*$ and hence $g(p)$ is minus the intersection with the vertical axis of the highest line with slope p which is $\le f(x)$ at all x.

2.3.7 Concavity of the Boltzmann entropy

With D below denoting derivative with respect to β, define

$$\pi_\beta = \beta P_\beta, \quad \mathcal{A} = \{\beta > 0 : D^- \pi_\beta = D^+ \pi_\beta = D \pi_\beta\}, \tag{2.3.7.1}$$

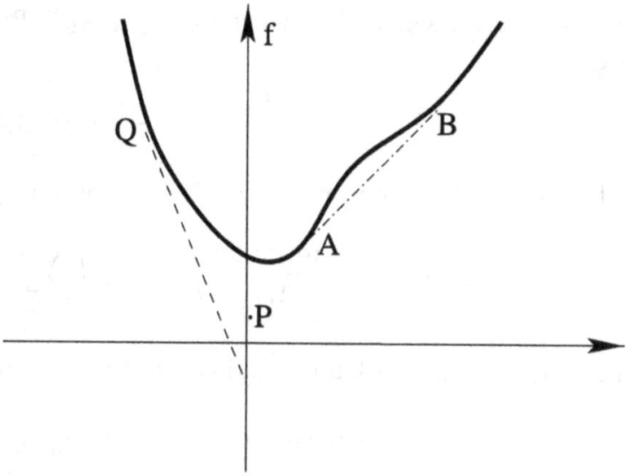

Fig. 2.1 The *dashed line* is tangent at the graph of f at Q, and the value of the Legendre transform is P. The convex envelope is obtained by joining the points A and B as shown

so that $\mathcal{A}$ covers the positive axis but for countably many points; let

$$\mathcal{G}_\beta^0 = \text{ the set of translational invariant DLR measures at } \beta. \tag{2.3.7.2}$$

Proposition 2.3.7.1 *For any $\beta > 0$ and for any $\mu \in \mathcal{G}_\beta^0$, $\mathcal{G}_\beta^0$ as in (2.3.7.2),*

$$D^- \pi_\beta \le -\mu(u) \le D^+ \pi_\beta. \tag{2.3.7.3}$$

If $\beta \in \mathcal{A}$, see (2.3.7.1), then

$$E := -D\pi_\beta \in \mathcal{E}_{\text{erg}} \quad \textit{and} \quad \textit{for any } \mu \in \mathcal{G}_\beta^0, \quad \mu(u) = E = -D\pi_\beta. \tag{2.3.7.4}$$

Proof Calling $\pi_{\beta,\Lambda}(\sigma_{\Lambda^c}) := Z_{\beta,\Lambda}(\sigma_{\Lambda^c})/|\Lambda|$, we claim that for any probability ν on $\mathcal{X}$ and any van Hove sequence $\Lambda \to \mathbb{Z}^d$,

$$\lim_{\Lambda \to \mathbb{Z}^d} \int d\nu(\sigma)\, D\pi_{\beta,\Lambda}(\sigma_{\Lambda^c}) \in [D^-\pi_\beta, D^+\pi_\beta]. \tag{2.3.7.5}$$

The proof of this claim is as follows. Since the pressure is independent of the b.c., see Sect. 2.3.3, and since $\pi_{\beta,\Lambda}(\sigma_{\Lambda^c})$ is uniformly bounded, see (2.3.3.5), by the Lebesgue dominated convergence theorem, $\lim_{\Lambda \to \mathbb{Z}^d} \int \nu(d\sigma)\, \pi_{\beta,\Lambda}(\sigma_{\Lambda^c}) = \pi_\beta$. Thus $\int \nu(d\sigma)\, \pi_{\beta,\Lambda}(\sigma_{\Lambda^c})$ is a sequence of convex, differentiable functions of β which converges to π_β and (2.3.7.5) follows from (2.3.6.2). Thus, the claim is proved.

By explicit computation,

$$D\pi_{\beta,\Lambda}(\sigma_{\Lambda^c}) = -\frac{1}{|\Lambda|} \sum_{x \in \Lambda} \frac{e^{-\beta H_\Lambda(\sigma_\Lambda|\sigma_{\Lambda^c})}}{Z_{\beta,\Lambda}(\sigma_{\Lambda^c})} H_\Lambda(\sigma_\Lambda|\sigma_{\Lambda^c}). \tag{2.3.7.6}$$

Rewrite (2.3.7.5) with (2.3.7.6) and use $\nu = \mu \in \mathcal{G}^0_\beta$. By the DLR property we then have

$$\lim_{\Lambda \to \mathbb{Z}^d} \mu\left(-\frac{H_\Lambda(\sigma_\Lambda|\sigma_{\Lambda^c})}{|\Lambda|} \right) \in [D^-\pi_\beta, D^+\pi_\beta]. \tag{2.3.7.7}$$

By Lemma 2.3.4.5 given any $\epsilon > 0$ for all Λ large enough,

$$\left| \mu\left(-\frac{H_\Lambda(\sigma_\Lambda|\sigma_{\Lambda^c})}{|\Lambda|} \right) + \frac{1}{|\Lambda|} \mu\left(\sum_{x\in\Lambda} \tau_x u \right) \right| \le 2\epsilon.$$

Since $\mu \in \mathcal{G}^0_\beta$, μ is translation invariant and $\mu(\tau_x u) = \mu(u)$, by (2.3.7.7)

$$\mu(u) \in [D^-\pi_\beta - 2\epsilon, D^+\pi_\beta + 2\epsilon],$$

which by the arbitrariness of ϵ proves (2.3.7.3).

The second statement in (2.3.7.4) is a corollary of (2.3.7.3) and we only have to prove that $\mu(u) \in \mathcal{E}_{\rm erg}$, which follows because $\mathcal{G}^0_\beta$ contains at least one ergodic measure, Theorem 2.2.7.2. □

Proposition 2.3.7.2 *Let $\beta \notin \mathcal{A}$; then there are $\mu_\pm \in \mathcal{G}^0_\beta$ so that*

$$-\mu_\pm(u) = D^\pm \pi_\beta. \tag{2.3.7.8}$$

Proof Since π_β is convex there exists an increasing sequence $\beta_n \in \mathcal{A}$ which converges to β as $n \to \infty$. By (2.3.7.4) there are ergodic measures μ_n which are DLR with respect to β_n such that $\mu_n(u) = E_n = -D\pi_{\beta_n}$. By compactness there is a subsequence n_k so that μ_{n_k} converges weakly to some probability measure μ.

We claim that μ is in $\mathcal{G}^0_\beta$. Let f be any cylindrical function, since $\mu_{n_k}(\tau_x f) = \mu_{n_k}(f)$, then also $\mu(\tau_x f) = \mu(f)$, hence μ is translational invariant. We fix arbitrarily a bounded set Λ and $\epsilon > 0$ and call $G_{\Lambda,\beta,\sigma_{\Lambda^c}}(f)$ the conditional Gibbs expectation of f at β with b.c. σ_{Λ^c}. Then, by (2.3.3.3), for all n_k large enough

$$\sup_{\sigma_{\Lambda^c}} |G_{\Lambda,\beta_{n_k},\sigma_{\Lambda^c}}(f) - G_{\Lambda,\beta,\sigma_{\Lambda^c}}(f)| \le \epsilon. \tag{2.3.7.9}$$

Moreover, since both $f(\sigma)$ and $G_{\Lambda,\beta,\sigma_{\Lambda^c}}(f)$ are continuous functions of σ, for all n_k large enough

$$\begin{aligned} \big|\mu(G_{\Lambda,\beta,\sigma_{\Lambda^c}}(f)) - \mu_{n_k}(G_{\Lambda,\beta,\sigma_{\Lambda^c}}(f))\big| &< \epsilon, \\ \big|\mu(f) - \mu_{n_k}(f)\big| &< \epsilon. \end{aligned} \tag{2.3.7.10}$$

By (2.3.7.9)–(2.3.7.11)

$$\begin{aligned} &\big|\mu(G_{\Lambda,\beta,\sigma_{\Lambda^c}}(f)) - \mu(f)\big| \\ &\quad\le \big|\mu_{n_k}(G_{\Lambda,\beta,\sigma_{\Lambda^c}}(f)) - \mu_{n_k}(f)\big| + 2\epsilon \\ &\quad\le \big|\mu_{n_k}(G_{\Lambda,\beta_{n_k},\sigma_{\Lambda^c}}(f)) - \mu_{n_k}(f)\big| + 3\epsilon. \end{aligned} \tag{2.3.7.11}$$

Since μ_{n_k} is DLR at β_{n_k}, $\mu_{n_k}(G_{\Lambda,\beta_{n_k},\sigma_{\Lambda^c}}(f)) = \mu_{n_k}(f)$; hence, by the arbitrariness of ϵ, $\mu(G_{\Lambda,\beta,\sigma_{\Lambda^c}}(f)) = \mu(f)$. As the argument applies to any cylindrical f and any bounded Λ, we may conclude the proof of the claim (namely that $\mu \in \mathcal{G}^0_\beta$). By (2.3.6.2),

$$\lim D\pi_{\beta_{n_k}} \le D^- \pi_\beta. \tag{2.3.7.12}$$

By Proposition 2.3.7.1, $\mu_{n_k}(u) = -D\pi_{\beta_{n_k}}$, so that

$$-\mu(u) \le D^- \pi_\beta. \tag{2.3.7.13}$$

By (2.3.7.3) it then follows that $-\mu(u) = D^-\pi_\beta$. By a completely analogous argument we conclude that there is a translational invariant DLR measure ν at inverse temperature β so that $-\nu(u) = D^+\pi_\beta$. □

Corollary 2.3.7.3 *If $\beta \notin \mathcal{A}$ there are ergodic DLR measures $\mu_\pm$ so that* (2.3.7.8) *holds. Moreover, for any E such that $-E \in [D^-\pi_\beta, D^+\pi_\beta]$ there is $\mu \in \mathcal{G}^0_\beta$ such that $\mu(u) = E$. The converse statement has been proved in* (2.3.7.3).

Proof By Proposition 2.3.7.2, there are translational invariant DLR measures $\nu_\pm$ for which (2.3.7.8) holds. If ν_+ is not ergodic, then by the ergodic decomposition, (2.2.7.2),

$$-D^+\pi_\beta = \nu_+(u) = \int_{\mathcal{X}^0_{gg} \cap \mathcal{G}^0} \mathcal{A}_\sigma(u)\, \nu_+(d\sigma), \tag{2.3.7.14}$$

where $\mathcal{A}_\sigma$ is an ergodic (as $\sigma \in \mathcal{X}^0_{gg}$) DLR measure (as $\sigma \in \mathcal{G}^0$). By (2.3.7.3) $-\mathcal{A}_\sigma(u) \le D^+\pi_\beta$; then by (2.3.7.14) the set of $\sigma \in \mathcal{X}^0_{gg} \cap \mathcal{G}^0$ such that $-\mathcal{A}_\sigma(u) = D^+\pi_\beta$ has ν_+ measure equal to 1 and it is therefore non-empty. The same argument applies to $D^-\pi_\beta$. Finally, if $-E \in [D^-\pi_\beta, D^+\pi_\beta]$, $-E = aD^-\pi_\beta + (1-a)D^+\pi_\beta = a\nu_+(u) + (1-a)\nu_-(u)$, $a \in [0,1]$, hence $-E = \mu(u)$, $\mu = a\nu_+ + (1-a)\nu_-$ and the corollary is proved. □

In summary, we have proved so far that

$$\{-\mu(u), \mu \in \mathcal{G}^0_\beta\} = [D^-\pi_\beta, D^+\pi_\beta], \quad \text{for any } \beta > 0, \tag{2.3.7.15}$$

which implies

$$\mathcal{E} = \Big\{-\mu(u), \mu \in \bigcup_{\beta>0} \mathcal{G}^0_\beta\Big\} = \bigcup_{\beta>0} [D^-\pi_\beta, D^+\pi_\beta]. \tag{2.3.7.16}$$

By the above corollary we also know that $\mathcal{E}_{\text{erg}}$ contains the endpoints of $[D^-\pi_\beta, D^+\pi_\beta]$ as well as those values $E = -D\pi_\beta$ for which $D\pi_\beta$ exists: recall that for $E \in \mathcal{E}_{\text{erg}}$ we can use Proposition 2.3.4.2.

Proposition 2.3.7.4 *If $E \in \mathcal{E}_{\text{erg}}$ then*

$$S^{\text{td}}(E) = S(E), \quad S^{\text{td}}(E) \text{ as in (2.3.5.1)}. \tag{2.3.7.17}$$

Proof If $E \in \mathcal{E}_{\text{erg}}$ there is β and an ergodic DLR measure μ at β such that $\mu(u) = E$. By Proposition 2.3.4.2, $S(E) = \beta(P_\beta + E)$. By Proposition 2.3.7.1, $-E \in [D^-\pi_\beta, D^+\pi_\beta]$ and therefore by (2.3.6.5), $-S^{\text{td}}(E) = \beta(-E) - \pi_\beta$, hence (2.3.7.17). □

Theorem 2.3.7.5 *For any $-E \in \mathcal{E}$ the limit* (2.3.4.1) *exists and*

$$S(E) = S^{\text{td}}(E), \quad \text{for all } -E \in \mathcal{E} \tag{2.3.7.18}$$

($S^{\text{td}}(E)$ *as in* (2.3.5.1)).

Proof We already know that $S(E) = S^{\text{td}}(E)$ for all $E \in \mathcal{E}_{\text{erg}}$, and hence for all E conjugate to $\beta \in \mathcal{A}$. By (2.3.6.5) for any E conjugate to $\beta \notin \mathcal{A}$,

$$-S^{\text{td}}(E) = \beta(-E) - \pi_\beta, \quad -E \in [-E^-, -E^+],\ -E^\pm = D^\pm\pi_\beta. \tag{2.3.7.19}$$

By Corollary 2.3.7.3, $E^\pm \in \mathcal{E}_{\text{erg}}$ and by Proposition 2.3.7.4 $S^{\text{td}}(E^\pm) = S(E^\pm)$. Theorem 2.3.7.5 then follows from the following proposition. □

Proposition 2.3.7.6 *Suppose that there are two ergodic DLR measures, μ_1 and μ_2 at the same inverse temperature β so that $\mu_1(u) = E_1 < \mu_2(u) = E_2$. Then for any E in $[E_1, E_2]$ the limit in* (2.3.4.1) *exists and*

$$S(E) - \beta E = \beta P_\beta. \tag{2.3.7.20}$$

Proof Fix $E \in (E_1, E_2)$ and let $a \in (0,1)$ be such that $E = aE_1 + (1-a)E_2$. Let $\Lambda \to \mathbb{Z}^d$ be a van Hove sequence; given any cube Δ and a partition into translates of Δ, call $\Delta(i)$, $i = 1, \ldots, N$, the cubes of the partition contained in Λ; Λ' is their union. As $\Lambda \to \mathbb{Z}^d$, $N \to \infty$ and $|1 - \frac{N|\Delta|}{|\Lambda|}| \to 0$. We then choose for any Λ a positive integer $n < N$ so that

$$\lim_{\Lambda \to \mathbb{Z}^d} \frac{n}{N} = a, \qquad \lim_{\Lambda \to \mathbb{Z}^d} \frac{N-n}{N} = 1 - a. \tag{2.3.7.21}$$

Thus for any $\zeta > 0$, as soon as $|\Lambda|$ is large enough

$$\left| \frac{n|\Delta|}{|\Lambda|} - a \right| < \zeta, \qquad \left| \frac{(N-n)|\Delta|}{|\Lambda|} - (1-a) \right| < \zeta.$$

We can now bound $N_{E,\Lambda,\delta}$ from below as follows. We choose $\delta' < \delta$ (its value will be specified later) and consider all σ_Λ such that $|H_{\Delta(i)}(\sigma_{\Delta(i)}) - |\Delta|E_1| < \delta'|\Delta|$ for

$i = 1, \ldots, n$ and $|H_{\Delta(i)}(\sigma_{\Delta(i)}) - |\Delta| E_2| < \delta' |\Delta|$ for $i = n+1, \ldots, N$. By (2.3.4.11), given any $\epsilon > 0$ if $|\Delta|$ is large enough,

$$\left| H_\Lambda(\sigma_\Lambda) - \sum_{\Delta(i) \subset \Lambda_0} H_{\Delta(i)}(\sigma_{\Delta(i)}) \right| \le c|\Lambda \setminus \Lambda'| + \frac{\epsilon}{4}|\Lambda|,$$

and therefore if ζ, ϵ and δ' are small enough and $|\Lambda|$ large enough,

$$|H_\Lambda(\sigma_\Lambda) - E|\Lambda|| \le c|\Lambda \setminus \Lambda'| + \frac{\epsilon}{4}|\Lambda| + \delta'|\Lambda| + \zeta|\Lambda| < \delta|\Lambda|. \tag{2.3.7.22}$$

We have

$$N_{E,\Lambda,\delta} \ge \left\{ \prod_{i=1}^{n} N_{E_1,\Delta(i),\delta'} \right\} \left\{ \prod_{i=n+1}^{N} N_{E_2,\Delta(i),\delta'} \right\},$$

so that, by (2.3.7.21),

$$\liminf_{\Lambda \to \mathbb{Z}^d} \frac{\log N_{E,\Lambda,\delta}}{|\Lambda|} \ge a \frac{\log N_{E_1,\Delta,\delta'}}{|\Delta|} + (1-a) \frac{\log N_{E_2,\Delta,\delta'}}{|\Delta|}.$$

We next let $|\Delta| \to \infty$ and then $\delta' \to 0$ and, since E_1 and E_2 are in $\mathcal{E}_{\text{erg}}$,

$$\liminf_{\delta \to 0} \liminf_{\Lambda \to \mathbb{Z}^d} \frac{\log N_{E,\Lambda,\delta}}{|\Lambda|} \ge aS(E_1) + (1-a)S(E_2).$$

By (2.3.4.3) the r.h.s. is equal to

$$\beta\{a[P_\beta + E_1] + (1-a)[P_\beta + E_2]\} = \beta\{P_\beta + E\}.$$

By taking $\Lambda \to \mathbb{Z}^d$ in (2.3.4.14) and then δ and ϵ to 0, we get the upper bound $\limsup_{\delta \to 0} \limsup_{\Lambda \to \mathbb{Z}^d} \frac{\log N_{E,\Lambda,\delta}}{|\Lambda|} \le \beta\{P_\beta + E\}$. □

2.3.8 *Variational principles and equivalence of ensembles*

In this subsection we discuss three variational principles. The first two, see Theorem 2.3.8.1, are just a reformulation of those stated in Theorem 2.3.4.1. The third one, see Theorem 2.3.9.1, is instead of a rather different nature.

Theorem 2.3.8.1 *For any $\beta > 0$ and with P_β and $S(E)$ as in Theorem* 2.3.4.1 *of Sect.* 2.3.4,

$$P_\beta = \sup_{-E \in \mathcal{E}} \{-(E - \beta^{-1} S(E))\}, \quad \mathcal{E} \text{ as in (2.3.7.16)}, \tag{2.3.8.1}$$

and for any $-E \in \mathcal{E}$

$$S(E) = -\sup_{\beta > 0} \{-\beta E - \beta P_\beta\}. \tag{2.3.8.2}$$

Proof As a consequence of the identification $S^{\rm td}(E) = S(E)$, $E \in \mathcal{E}_{\rm allw}$, see (2.3.7.18), we get (2.3.8.1) from (2.3.5.2) and (2.3.8.2) from (2.3.5.1). □

Remarks

The Gibbs formula and the theory of DLR states have been heavily used in the proofs but they should be regarded just as auxiliary tools. The only physical assumption in the theory has been the Boltzmann hypothesis. Equation (2.3.8.1) may also be proved directly by arguments similar to those in Sect. 2.3.3 as traditionally is done in texts on statistical mechanics. As we already had available the theory of DLR measures, it was simpler to proceed the way we did.

The variational principle (2.3.8.2) may be viewed as a method for determining the entropy in terms of the grand canonical partition function that more easily is handled because the energy constraint is dropped.

Below are some consequences of the above variational principles which have physical relevance.

Theorem 2.3.8.2 *$S(E)$ is an increasing function of E; its right and left derivatives exist everywhere and $\beta \in [D^- S(E), D^+ S(E)]$ if and only if $-E \in [D^- \pi_\beta, D^+ \pi_\beta]$. The derivative of $S(E)$ exists at all but countably many points and if $DS(E)$ exists*

$$DS(E) = \beta, \quad \beta : -E \in [D^-(\beta P_\beta), D^+(\beta P_\beta)], \tag{2.3.8.3}$$

and $S(E) = \beta E + \beta P_\beta$.

Proof Let $s(e) = -S(-e)$, $e \in \mathcal{E}$, and $\pi_\beta = \beta P_\beta$. Since $s(e)$ is the Legendre transform of π_β, see (2.3.8.2), it is convex and hence its derivative $Ds(e)$ exists at all but countably many $e \in \mathcal{E}$; we denote such a set $\mathcal{E}'$. Recalling from (2.3.7.3) that e is related to β by the relation $e \in [D^- \pi_\beta, D^+ \pi_\beta]$, then if $e \in \mathcal{E}'$, $Ds(e)$ must be equal to β by (2.3.6.6). Hence

$$s(e'') - s(e') = \int_{[e',e'']\cap\mathcal{E}'} Ds(e)de > 0, \quad e'' > e'.$$

The claimed relation between e and β is a particular case of (2.3.6.5)–(2.3.6.6). □

Remarks

Theorem 2.3.8.2 establishes the basic principle of thermodynamics that the derivative of the entropy with respect to the energy is the inverse temperature and since Theorem 2.3.8.2 proves that the entropy is an increasing function of the energy, the temperatures are positive as they should (if $S(E)$ is not differentiable, the values of β are identified with the slope of the tangent lines to the graph of S hence $\beta \in [D^- s(e), D^+ s(e)]$ and therefore $\beta > 0$ in general).

There are two main questions though, which need to be clarified: where is the limitation to $-E \in \mathcal{E}$ coming from? And what happens if we take $-E \notin \mathcal{E}$?

Theorem 2.3.8.3 *We have*

$$\sup\{E : -E \in \mathcal{E}\} = 0, \qquad \sup_{-E\in\mathcal{E}} S(E) = \ln 2. \tag{2.3.8.4}$$

Moreover, for any increasing sequence Δ of cubes which invades $\mathbb{Z}^d$, the following limit exists:

$$\lim_{\Delta\to\mathbb{Z}^d} \frac{1}{|\Delta|} \min_{\sigma_\Delta} H_\Delta(\sigma_\Delta) =: e_{\min}, \tag{2.3.8.5}$$

and the set $\mathcal{E}$ in Theorem 2.3.4.1 *is*

$$\mathcal{E} = (e_{\min}, 0). \tag{2.3.8.6}$$

Proof Recall from (2.3.7.16) that $\mathcal{E} = \bigcup_{\beta>0}[D^-\pi_\beta, D^+\pi_\beta]$; then by (2.3.6.1)

$$\sup\{E, -E \in \mathcal{E}\} = -\lim_{\beta\to 0} D^-\pi_\beta.$$

We can take the limit along a sequence β_n where $D\pi_{\beta_n}$ exists. Then there are measures $\mu_n \in \mathcal{C}^0_{\beta_n}$ so that $-D\pi_{\beta_n} = \mu_n(u)$. Using the DLR property

$$\mu_n(\sigma(0)) = \mu_n\big(\tanh\{\beta_n\kappa(\sigma)\}\big), \quad \kappa(\sigma) = h + \sum_{x\neq 0} J(0,x)\sigma(x),$$

and since $\sup_\sigma |\kappa(\sigma)| \le c$, $\lim_{n\to\infty}\mu_n(\sigma(0)) = 0$. An analogous argument proves that for any $x \neq 0$, $\lim_{n\to\infty}\mu_n(\sigma(0)\sigma(x)) = 0$ and the first half of (2.3.8.4) is proved.

By Theorem 2.3.8.2, $S(E)$ is an increasing function of E; then by the first half of (2.3.8.4), $\sup_{-E\in\mathcal{E}} S(E) = \lim_{E\nearrow 0} S(E)$. As before we take the limit along a sequence E_n such that $D\pi_{\beta_n} = -E_n$ and $\beta_n \to 0$. By (2.3.8.2), $S(E_n) = \beta_n E_n + \pi_{\beta_n}$ and the second half of (2.3.8.4) is then a consequence of the following inequality: there is $c > 0$ so that for any bounded Λ and any σ_{Λ^c}, $e^{-\beta c|\Lambda|}2^{|\Lambda|} \le Z_{\beta,\Lambda}(\sigma_{\Lambda^c}) \le e^{\beta c|\Lambda|}2^{|\Lambda|}$, which thus proves the second half of (2.3.8.4).

The proof of (2.3.8.5) now follows. Since $|H_\Delta(\sigma_\Delta)| \le c|\Delta|$, c the sup norm of u, there is a sub-sequence $\Delta_n \to \mathbb{Z}^d$ such that the limit below exists,

$$\lim_{\Delta_n\to\mathbb{Z}^d} \frac{1}{|\Delta_n|} \min_{\sigma_{\Delta_n}} H_{\Delta_n}(\sigma_{\Delta_n}) =: e. \tag{2.3.8.7}$$

By Lemma 2.3.4.4, given any $\epsilon > 0$ if n is large enough the following holds. Given any Δ, call $\Delta_n(i)$ the cubes of a partition into translates of Δ_n which are in Δ, and Δ' is their union. Then

$$\Big| H_\Delta(\sigma_\Delta) - \sum_i H_{\Delta_n(i)}(\sigma_{\Delta_n(i)})\Big| \le \epsilon|\Delta| + c|\Delta\setminus\Delta'|. \tag{2.3.8.8}$$

Thus

$$\liminf_{\Delta\to\mathbb{Z}^d}\left|\frac{1}{|\Delta|}\min_{\sigma_\Delta} H_\Delta(\sigma_\Delta) - e\right| \le \epsilon,$$

with the same inequality holding for the limsup; hence (2.3.8.5). The proof of (2.3.8.6) is omitted. □

Remarks

Notice that for any E and not only $-E \in \mathcal{E}$,

$$\lim_{\delta\to 0}\limsup_{\Lambda\to\mathbb{Z}^d}\frac{\log N_{E,\Lambda,\delta}}{|\Lambda|} \le \log 2, \tag{2.3.8.9}$$

as $2^{|\Lambda|} = \text{card}(\Lambda)$. Thus, by Theorem 2.3.8.3, if $E > 0$, i.e. $-E \notin \mathcal{E}$, then the entropy $S(E)$ (which can be proved to be well defined) is smaller than the sup of the entropy when $-E \in \mathcal{E}$. Thus, the thermodynamic law that entropy is an increasing function of E is violated, namely temperatures are negative at $E > 0$! The physical relevance of states with negative temperatures cannot be completely ruled out in systems where the energy density is bounded, as in our Ising model. Indeed if we are able to thermally isolate the system and to prepare an initial state with high energy, then it must relax to some limit, and the limit state will have by conservation of energy the same high energy as the initial one. There are indeed practical applications of such considerations. Negative temperatures are related to positive temperatures for the hamiltonian $H' = -H$, which is in general limited to spin systems as it may become pathological when variables are unbounded: the "stability condition" on the hamiltonian required for the partition function to exist is in general not satisfied after the transformation $H \to -H$.

2.3.9 A variational principle for measures

We have seen that the pressure and the thermodynamics of the Ising model can be derived under the assumption only that entropy is the log of the number of states at the given energy. Using such an assumption we have then identified the pressure in terms of the log of the partition function. On the other hand, as the partition function is the normalization factor in the Gibbs formula, it seems natural to try to push the argument further to derive the full Gibbs formula as well. The variational principle in Theorem 2.3.8.1 states that the pressure P_β, at a given inverse temperature β, is obtained by maximizing the entropy; see (2.3.8.1). But Propositions 2.3.7.1 and 2.3.7.2 complement the result by stating that the maximizing energy E is indeed the average value $\mu(u)$, with u as in (2.2.7.3), of a translational invariant measure μ DLR at inverse temperature β. Thus, the maximizer gives information on μ; it specifies the average of u. If we could repeat the argument with other functions, we would then identify other expectations of μ, and this may eventually lead to the identification of μ itself. Such a program will be carried out by a generalization of the notion of entropy which will be extended to the "entropy of probability measures," and then the assumption that equilibrium is reached when the entropy is maximal will identify the translational invariant DLR measures.

We have already found in Sect. 2.3.1 an expression for the entropy $S(\mu)$ of a measure μ, given by the formula (2.3.1.8). $S(\mu)$ in Sect. 2.3.1 counts how many "typical messages" are emitted by an independent source μ (of sequences of symbols); namely $S(\mu)$ gives a measure of the storage capacity of the system (it tells how many messages can be stored by μ). The assumption of independence used in Sect. 2.3.1 can be greatly relaxed. The Shannon–MacMillan–Breiman theorem states that if μ is ergodic (on $\{-1,1\}^{\mathbb{Z}^d}$ in our specific application), then there is a unique number $S(\mu)$ such that the following holds. Given any $\epsilon > 0$, for any cube Δ large enough we can split $\{-1,1\}^{\Delta}$ into two sets, A_1 and A_2, so that

$$\mu(A_1) < \epsilon, \qquad \left|\frac{\log \operatorname{card} A_2}{|\Delta|} - S(\mu)\right| < \epsilon,$$

and moreover for any $\sigma_\Delta \in A_2$,

$$\left|\frac{\log \mu(\sigma_\Delta)}{|\Delta|} + S(\mu)\right| < \epsilon. \tag{2.3.9.1}$$

The number $S(\mu)$, called the [information] entropy of μ, is equal to

$$-\lim_{\Delta\to\mathbb{Z}^d} \frac{1}{|\Delta|} \sum_{\sigma_\Delta} \mu(\sigma_\Delta) \log \mu(\sigma_\Delta) =: S(\mu). \tag{2.3.9.2}$$

If μ is our ergodic DLR measure of Sect. 2.3.2 at inverse temperature β, then μ concentrates on configurations with energy $(E \pm \delta)|\Delta|$, $E = \mu(u)$, the complement having vanishing measure as $\Delta \to \mathbb{Z}^d$. We have seen that the number of configurations with energy $(E \pm \delta)|\Delta|$ is $N_{E,\Delta,\delta} \approx e^{S(E)|\Delta|}$ and moreover for any configuration σ_Δ such that $H_\Delta(\sigma_\Delta) \in (E \pm \delta)|\Delta|$,

$$\frac{1}{|\Delta|} \log \mu(\{\sigma : \sigma \restriction \Delta = \sigma_\Delta\}) \approx -\beta(E \pm \delta \pm \epsilon) + C,$$

C a constant, with $\epsilon|\Delta|$ bounding the interaction with Δ^c. By taking $|\Delta| \to \infty$ this proves that $S(\mu) = S(E)$. Thus the notion of information entropy $S(\mu)$ reduces to the $S(E)$ entropy when $E \in \mathcal{E}_{\text{erg}}$. We can then accept $S(\mu)$ as the entropy of a measure without running into a conceptual conflict with our previous considerations.

Having $S(\mu)$, we can now invoke thermodynamic principles to find the equilibrium states. We start by considering the simple case of a finite region Λ. We suppose that equilibrium states are maximizers of the entropy among all states with a given energy. In the grand-canonical ensemble such a constraint is only imposed in the average by requiring that a measure μ representative of a state should verify

$$\sum_{\sigma_\Lambda} \mu(\sigma_\Lambda) H_\Lambda(\sigma_\Lambda) = E|\Lambda|. \tag{2.3.9.3}$$

The equilibrium measure is then defined as the maximizer of the entropy $S_\Lambda(\mu)$ given by (2.3.9.2) (with $\Delta = \Lambda$ and without taking the limit) under the constraint (2.3.9.3):

$$\left\{ S_\Lambda(\mu), \mu : \sum_{\sigma_\Lambda} \mu(\sigma_\Lambda) H(\sigma_\Lambda) = E|\Lambda| \right\} \longrightarrow \text{maximum.} \tag{2.3.9.4}$$

By using Lagrange multipliers we need to find the critical points of

$$S_\Lambda(\mu) - \beta \sum_{\sigma_\Lambda} \mu(\sigma_\Lambda) H(\sigma_\Lambda) - \lambda \sum_{\sigma_\Lambda} \mu(\sigma_\Lambda), \tag{2.3.9.5}$$

with β the Lagrange multiplier for the energy constraint and λ for the normalization of μ as a probability. By an elementary computation, we see that the critical point is unique, and it is given by G_Λ, the Gibbs measure at the inverse temperature β, which is a maximizer of (2.3.9.4) by the concavity of $S_\Lambda(\mu)$. Notice also that the variational problem

$$\left\{ S_\Lambda(\mu) - \frac{\beta}{|\Lambda|} \sum_{\sigma_\Lambda} \mu(\sigma_\Lambda) H(\sigma_\Lambda) \right\} \longrightarrow \text{maximum} \tag{2.3.9.6}$$

is achieved at G_Λ and that the maximum is equal to $\beta P_{\beta,\Lambda}$, $P_{\beta,\Lambda}$ being the finite volume pressure, $\beta P_{\beta,\Lambda} = \log Z_\Lambda / |\Lambda|$, namely

$$\sup_\mu \left\{ S_\Lambda(\mu) - \frac{\beta}{|\Lambda|} \mu(H_\Lambda) \right\} = \beta P_{\beta,\Lambda} = \left\{ S_\Lambda(G_\Lambda) - \frac{\beta}{|\Lambda|} G_\Lambda(H_\Lambda) \right\}. \tag{2.3.9.7}$$

Equation (2.3.9.7) extends to infinite volumes with the finite volume Gibbs measures replaced by the translational invariant DLR measures. We only state the result in our setup (its validity being more general) and refer to the literature for the proof; see for instance Theorems III.4.3, III.4.5 and III.4.9 in Simon's book on statistical mechanics [200].

Theorem 2.3.9.1 *Let $\Delta \to \mathbb{Z}^d$ be an increasing sequence of cubes and call $\mathcal{M}_0$ the set of all translational invariant probabilities on $\{-1,1\}^{\mathbb{Z}^d}$. Then we have*

(i) *Existence of entropy: for any $\mu \in \mathcal{M}_0$ the limit* (2.3.9.2) *exists.*
(ii) *Gibbs variational principle: For any $\mu \in \mathcal{M}_0$ and with u defined in* (2.2.7.3),

$$P_\beta = \sup_{\mu \in \mathcal{M}_0} \{\beta^{-1} S(\mu) - \mu(u)\}. \tag{2.3.9.8}$$

(iii) *Ruelle's theorem*: *$\mu \in \mathcal{M}_0$ is in $\mathcal{G}^0_\beta$ if and only if*

$$P_\beta = \beta^{-1} S(\mu) - \mu(u).$$

2.4 Thermodynamics and DLR measures

In this section we complete the analysis of the thermodynamics of the Ising model by proving that the translational invariant DLR measures can be identified as the functionals tangent to the graph of the pressure regarded as a function of the interaction potential. We thus have an alternative way to derive the Gibbs formula from the Boltzmann hypothesis, besides the one in Theorem 2.3.9.1, based on an extended notion of entropy of measures to which thermodynamic principles are then applied. We will conclude the section by briefly discussing (proofs are omitted) a characterization of the translational invariant DLR measures as "tangent functional to the pressure" and the relation between large deviations for Gibbs measures and thermodynamics potentials.

2.4.1 Canonical ensemble and free energy

The Boltzmann hypothesis actually states that entropy is equal to the log of the number of states *available* to the system. Supposing that the energy is a prime integral, the phase space available to the system is the energy surface relative to the energy of the system. There could, however, be other prime integrals or we can imagine that by some external action on the system the phase space available to the system is reduced. The thermodynamic potentials are then modified and new order parameters come into play.

Definitions *Grand-canonical, canonical and micro-canonical ensembles*. We will suppose that the phase space available to the system is determined by the value of an intensive quantity (also called observable). The subset of the phase space where such a value is attained is the "*canonical ensemble*" (called *micro-canonical* when the observable is the energy). Usually the notion is applied to the case where the intensive quantity is the total number of particles or, in Ising systems, the total magnetization. In the case of particle systems "micro-canonical" usually refers to the ensemble where in which both energy and total number of particles are fixed. The *grand-canonical* ensemble is instead the unrestricted phase space.

The equivalence of ensembles is a property which states the equivalence of thermodynamics computed in the canonical ensemble and the thermodynamics computed in the grand-canonical ensemble after a suitable term has been added to the hamiltonian (which is referred to as the variable conjugate to the observable defining the canonical ensemble). We have proved that the micro-canonical ensemble in which energy is fixed gives the same thermodynamic potentials as the grand-canonical ensemble with temperature the variable conjugate to the energy, (2.3.8.1)–(2.3.8.2).

Suppose that the intensive quantity defining the canonical ensemble is

$$V_\Lambda(\sigma_\Lambda) = \sum_{x\in\mathbb{Z}^d:\tau_x\Delta\subset\Lambda} \tau_x v(\sigma_\Lambda), \tag{2.4.1.1}$$

where v is a cylindrical function in a bounded set $\Delta \subset \mathbb{Z}^d$, i.e. such that $v(\sigma)$ is independent of σ_{Δ^c}. The "order parameter" is then the observable v and the values w of its ergodic averages together with the energy E parameterize the equilibrium states of the system. In such a scenario the analogue of the Boltzmann hypothesis (2.3.2.3) is that the finite volume entropy density at the values E and w of the order parameters is

$$\begin{aligned} S_{\Lambda,\delta}(E,w;v) \\ := \frac{1}{|\Lambda|}\log\left\{\sum_{\sigma_\Lambda} \mathbf{1}_{|V_\Lambda(\sigma_\Lambda)-|\Lambda|w|\le\delta|\Lambda|}\mathbf{1}_{|H_\Lambda(\sigma_\Lambda)-|\Lambda|E|\le\delta|\Lambda|}\right\}. \end{aligned} \tag{2.4.1.2}$$

We will study in the sequel a simplified scenario with only one order parameter; namely, we will suppose that the energy is not conserved, because the system exchanges energy with a "reservoir" at inverse temperature β (thermal walls), which thus fixes the temperature of the system at the value β^{-1}. We suppose however that $\sum_x \tau_x v$ is conserved and that the walls are impermeable to exchanges of this quantity, which is then the only order parameter for the system (as the temperature is fixed). In such a scenario the free energy rather than entropy is the relevant thermodynamic quantity and the analogue of (2.4.1.2) is

$$F_{\beta,\Lambda,\delta,\sigma_{\Lambda^c}}(w;v) := -\frac{1}{\beta|\Lambda|}\log\left\{\sum_{\sigma_\Lambda:|V_\Lambda(\sigma_\Lambda)-|\Lambda|w|\le\delta|\Lambda|} e^{-\beta H_\Lambda(\sigma_\Lambda|\sigma_{\Lambda^c})}\right\}. \tag{2.4.1.3}$$

The minus sign is there because we consider here the free energy and not the pressure. The available phase space, i.e. the ensemble of configurations appearing in (2.4.1.3), is called the (δ-relaxed) "canonical ensemble" relative to the variable v.

There is some arbitrariness in the definition (2.4.1.3) regarding the boundary conditions, which is fortunately unimportant because any surface correction becomes negligible in the thermodynamic limit.

As in the previous section the crucial point is an equivalence of ensemble property which says that in the infinite volume limit it is equivalent to study the above problem or another one where we drop the constraint in the phase space and add an external field which forces $V_\Lambda(\sigma_\Lambda)$ to have the desired value (in the previous section where v was u, this was achieved by choosing properly β). With the addition of the external field the new energy with "zero boundary conditions" is

$$H_{\Lambda,\lambda,v}(\sigma_\Lambda) = H_\Lambda(\sigma_\Lambda) - \lambda V_\Lambda(\sigma_\Lambda) \tag{2.4.1.4}$$

(the minus sign in front of λ is just conventional), while if the boundary condition is σ_{Λ^c},

$$H_{\Lambda;\lambda,v}(\sigma_\Lambda|\sigma_{\Lambda^c}) = H_\Lambda(\sigma_\Lambda|\sigma_{\Lambda^c}) - \lambda V_\Lambda(\sigma_\Lambda|\sigma_{\Lambda^c}), \tag{2.4.1.5}$$

where $V_\Lambda(\sigma_\Lambda|\sigma_{\Lambda^c}) := \sum_{x\in\mathbb{Z}^d:\tau_x\Delta\cap\Lambda\neq\emptyset} \tau_x v(\sigma_\Lambda, \sigma_{\Lambda^c})$. The new pressure is

$$P_{\beta,\Lambda,\sigma_{\Lambda^c}}(\lambda; v) := \frac{1}{\beta|\Lambda|} \log\left\{\sum_{\sigma_\Lambda} e^{-\beta[H_{\Lambda;\lambda,v}(\sigma_\Lambda|\sigma_{\Lambda^c})}\right\}. \tag{2.4.1.6}$$

Notice that the additional energy in the hamiltonian has a finite range, as this is equal to the diameter of the set Δ introduced above.

Theorem 2.4.1.1 *For any $\beta > 0$ and any cylindrical function v the following holds:*

- *For any van Hove sequence $\{\Lambda\}$, any sequence σ_{Λ^c} and any $\lambda \in \mathbb{R}$*

$$\lim_{\Lambda\to\mathbb{Z}^d} P_{\beta,\Lambda,\sigma_{\Lambda^c}}(\lambda; v) \text{ exists and we call it } P_\beta(\lambda; v). \tag{2.4.1.7}$$

Moreover, for any $w \in \mathcal{E}_v := \bigcup_{\lambda\in\mathbb{R}} [D^- P_\beta(\lambda; v), D^+ P_\beta(\lambda; v)]$,

$$\lim_{\delta\to 0} \liminf_{\Lambda\to\mathbb{Z}^d} F_{\beta,\Lambda,\delta,\sigma_{\Lambda^c}}(w; v) = \lim_{\delta\to 0} \limsup_{\Lambda\to\mathbb{Z}^d} F_{\beta,\Lambda,\delta,\sigma_{\Lambda^c}}(w; v) =: F_\beta(w; v). \tag{2.4.1.8}$$

- *$P_\beta(\lambda; v)$ and $F_\beta(w; v)$ are Legendre transforms of each other,*

$$P_\beta(\lambda; v) = \sup_{w\in\mathcal{E}_v} \{\lambda w - F_\beta(w; v)\}, \quad F_\beta(w; v) = \sup_{\lambda\in\mathbb{R}}\{\lambda w - P_\beta(\lambda; v)\}, \tag{2.4.1.9}$$

so that $w \in [D^- P_\beta(\lambda; v), D^+ P_\beta(\lambda; v)]$ iff $\lambda \in [D^- F_\beta(w; v), D^+ F_\beta(w; v)]$ and for such conjugate pairs,

$$P_\beta(\lambda; v) = \lambda w - F_\beta(w; v). \tag{2.4.1.10}$$

- *(λ, w) is a conjugate pair if and only if there exists a translational invariant measure μ DLR at β with hamiltonian* (2.4.1.4)–(2.4.1.5) *such that*

$$\mu(v) = w. \tag{2.4.1.11}$$

Moreover, if $w = D^\pm P_\beta(\lambda; v)$, then μ can be chosen to be ergodic and if the derivative $DP_\beta(\lambda; v)|_{\lambda=0} =: w$ exists, then all $\mu \in \mathcal{G}^0$ are such that $\mu(v) = w$.

- *Finally,*

$$P_\beta(\lambda; v) = P_\beta(1; \lambda v) =: P_\beta(\lambda v), \tag{2.4.1.12}$$

and $P_\beta(v)$ is a convex function on the space whose elements are the cylindrical functions v.

Proof The proof is completely analogous to the one for the entropy–energy conjugation and it will only be sketched. Using that the additional term in the hamiltonian due to v has a finite range the proofs are essentially unchanged. We thus start by considering all the ergodic measures μ which are DLR at (β, λ) for any $\lambda \in \mathbb{R}$. Call $w = \mu(v)$, then the analogue of Proposition 2.3.4.2 holds and proves that the limit in (2.4.1.8) exists and is equal to

$$F_\beta(w; v) = -P_\beta(\lambda; v) + \lambda w.$$

Analogously to Proposition 2.3.7.1, $\mu(v) \in [D^- P_\beta(\lambda; v), D^+ P_\beta(\lambda; v)]$ for any translational invariant measure μ DLR at β with hamiltonian (2.4.1.4). The converse is also true; its proof is given by the analogues of Propositions 2.3.7.1 and 2.3.7.2: namely for any $w \in [D^- P_\beta(\lambda; v), D^+ P_\beta(\lambda; v)]$, there is a translational invariant measure μ DLR at β with hamiltonian (2.4.1.4), such that $\mu(v) = w$. Since any such measure μ verifies

$$\mu(v) \in [D^- P_\beta(\lambda; v), D^+ P_\beta(\lambda; v)]$$

there must be ergodic ones, $\mu_\pm$, such that $\mu_\pm(v) = D^\pm P_\beta(\lambda; v)$, which is proved as in (2.3.7.8). Then, calling $w_\pm := D^\pm P_\beta(\lambda; v)$, and repeating the proof of Proposition 2.3.7.6, we conclude that the limit (2.4.1.8) exists also when $w \in (D^- P_\beta(\lambda; v), D^+ P_\beta(\lambda; v))$ and, moreover,

$$F_\beta(w; v) = a F_\beta(w_-; v) + (1-a) F_\beta(w_+; v), \quad w = a w_- + (1-a) w_+, \ a \in [0, 1].$$

Thus $F_\beta(w; v)$ is the Legendre transform of $P_\beta(\lambda; v)$ when

$$w \in \left(D^- P_\beta(\lambda; v), D^+ P_\beta(\lambda; v)\right)$$

and since for the others we have already proved that (2.4.1.10) holds (as for such w there are ergodic measures with $\mu(v) = w$), we then conclude the proof of (2.4.1.9). Equation (2.4.1.12) is obvious and the other statements in the theorem are either already proved or are a consequence of general convexity properties. □

We conclude this subsection with an interesting corollary of Theorem 2.4.1.1 which gives a characterization of the absence or presence of phase transitions, identifying their absence in the DLR context (i.e. uniqueness of translational invariant DLR measures) with the thermodynamic notion based on differentiability of the pressure.

Theorem 2.4.1.2 (Phase transitions) *There is only one invariant DLR measure at* β *with hamiltonian defined by* (2.3.2.13) *if and only if for any cylindrical function* v

$$D^+ P_\beta(\lambda; v)|_{\lambda=0} = D^- P_\beta(\lambda; v)|_{\lambda=0}. \tag{2.4.1.13}$$

Proof Suppose there is only one DLR measure μ. In the proof of Theorem 2.4.1.1 we have seen that there are ergodic measures $\mu_{\pm;\lambda,v}$ DLR at β with respect to the hamiltonian (2.4.1.4) and such that $\mu_{\pm;\lambda,v}(v) = D^{\pm}P_\beta(\lambda; v)$. Since $\mu_{\pm;0,v}$ are DLR at β with hamiltonian (2.3.2.13), $\mu_{+;0,v} = \mu_{-;0,v}$; hence (2.4.1.13). Thus $P_\beta(\lambda; v)$ is differentiable at $\lambda = 0$ for any v.

Suppose now that $P_\beta(\lambda; v)$ is differentiable at $\lambda = 0$ for any v and let μ' and μ'' be two translational invariant DLR measures with hamiltonian (2.3.2.13). We need to prove that $\mu' = \mu''$. By Theorem 2.4.1.1

$$\mu'(v) \in [D^- P_\beta(\lambda; v)|_{\lambda=0}, D^+ P_\beta(\lambda; v)|_{\lambda=0}] \Rightarrow \mu'(v) = DP_\beta(\lambda; v)|_{\lambda=0}$$

by the assumption of differentiability. The same argument applied to μ'' shows that $\mu'(v) = \mu''(v)$. Since this holds for all v, by density $\mu' = \mu''$. □

2.4.2 Tangent functionals to the pressure

There is a converse to Theorem 2.4.1.2, which shows that the DLR measures are characterized by the discontinuities in the intervals

$$[D^- P_\beta(\lambda; v)|_{\lambda=0}, D^+ P_\beta(\lambda; v)|_{\lambda=0}].$$

We need to introduce the notion of tangent functional to the graph of the pressure. We fix β and regard $P_\beta(v)$, defined in (2.4.1.12), as a function on the space $\mathcal{V}$ of all v.

Definition 2.4.2.1 A tangent functional to $P_\beta(\cdot)$ at v is a *linear functional* $\alpha(\cdot)$ on $\mathcal{V}$ such that

$$\alpha(v') \le P_\beta(v') - P_\beta(v), \quad \text{for any } v' \in \mathcal{V}. \tag{2.4.2.1}$$

Theorem 2.4.2.2 *Any translational invariant measure μ DLR at β with hamiltonian (2.3.2.13) is such that $\mu(v)$ is tangent to $P_\beta(\cdot)$ at $v = 0$. Vice versa, if α is tangent to $P_\beta(\cdot)$ at $v = 0$ then there is a unique translational invariant DLR measure μ such that $\mu(v) = \alpha(v)$ for all v.*

Proof Let μ be DLR at β with hamiltonian (2.3.2.13). Calling $P_\beta = P_\beta(0; v)$, we need to prove that $\mu(v) \le P_\beta(v) - P_\beta$ for any $v \in \mathcal{V}$. Call $\mathcal{E}'_v$ the set of all $\lambda \in \mathbb{R}$ where $DP_\beta(\lambda; v)$ exists, which is all $\mathbb{R}$ except for countably many points. Then

$$P_\beta(v) - P_\beta = \int_{\mathcal{E}'_v \cap [0,1]} DP_\beta(\lambda; v)\, d\lambda. \tag{2.4.2.2}$$

By (2.3.6.1) for $\lambda > 0$, $DP_\beta^-(\lambda; v) \ge D^+ P_\beta(\lambda; v)|_{\lambda=0}$, so that (2.4.2.2) yields

$$P_\beta(v) - P_\beta \ge DP_\beta^+(\lambda; v)|_{\lambda=0}, \tag{2.4.2.3}$$

and $DP_\beta^+(\lambda; v)|_{\lambda=0} \ge \mu(v)$: in fact, by (2.4.1.11), $(0, \mu(v))$ is a conjugate pair and $\mu(v) \in [DP_\beta^-(0; v), D^+ P_\beta(0; v)]$. For the converse statement we refer to the literature; see for instance [200]. □

Remarks

We can therefore recover the set of equilibrium states from the pressure, as the functionals tangent to the pressure are the translational invariant DLR measures. Thus ultimately the only axiom needed for the whole theory is the Boltzmann identification of the entropy in terms of the number of states.

2.4.3 Large deviations

Suppose that our Ising system is in equilibrium at an inverse temperature β for which there is a unique ergodic DLR measure μ. We then know that the ergodic average $\mathcal{A}_{\Lambda,\sigma}(u)$ defined in (2.3.2.12) is with large probability close to $\mu(u)$ so that we can read out of σ what is the temperature of the system. In principle, at least the quantity $\mathcal{A}_{\Lambda,\sigma}(u)$ can be observed experimentally, thus providing an alternative way to measure temperatures. However, $\mathcal{A}_{\Lambda,\sigma}(u)$ is close to $\mu(u)$ "only with large probability" and the question then arises of what the probability is to mistake the correct values $\mu(u)$ and β. As we will see such a probability is exponentially small with the volume and for such a reason it is called "a large deviation." As already discussed $\mathcal{A}_{\Lambda,\sigma}(u)$ is for large $|\Lambda|$ close to $H_\Lambda(\sigma_\Lambda)$. Let E be an energy density corresponding to an ergodic DLR measure with inverse temperature β'. Then

$$
\begin{aligned}
&G_{\Lambda,\beta}\big(|H_\Lambda(\sigma_\Lambda) - E|\Lambda|| \le \delta\big)\\
&\quad = \sum_{|H_\Lambda(\sigma_\Lambda)-E|\Lambda||\le\delta} \frac{e^{-\beta H_\Lambda(\sigma_\Lambda)}}{Z_{\beta,\Lambda}} \frac{e^{-(\beta'-\beta)H_\Lambda(\sigma_\Lambda)}}{Z_{\beta',\Lambda}} \{e^{(\beta'-\beta)H_\Lambda(\sigma_\Lambda)} Z_{\beta',\Lambda}\}\\
&\quad \approx \frac{Z_{\beta',\Lambda}}{Z_{\beta,\Lambda}} e^{(\beta'-\beta)E|\Lambda|} G_{\beta',\Lambda}\big(|H_\Lambda(\sigma_\Lambda) - E|\Lambda|| \le \delta\big).
\end{aligned}
$$

We thus expect that

$$
\frac{\log G_{\Lambda,\beta}\big(|H_\Lambda(\sigma_\Lambda) - E|\Lambda|| \le \delta\big)}{|\Lambda|} \approx (\beta'-\beta)E + \beta' P_{\beta'} - \beta P_\beta,
$$

in which case the rate of large deviations, namely the l.h.s., is related to thermodynamics and given by $(\beta'-\beta)E + \beta' P_{\beta'} - \beta P_\beta$. The above argument can be made rigorous and generalized as follows (proofs will be omitted).

Let μ be a DLR measure at inverse temperature β with hamiltonian as in (2.3.2.13). Let v be a cylindrical function, $\Delta \to \mathbb{Z}^d$ an increasing sequence of cubes, $w \in \mathbb{R}$ and

$$
A_{w,\delta;\Delta} := \mu\left(\frac{1}{|\Delta|}\left|\sum_{x\in\Delta}(\tau_x v - w)\right| \le \delta\right). \tag{2.4.3.1}
$$

We have

Theorem 2.4.3.1 *Let Δ be an increasing sequence of cubes; then*

$$\lim_{\delta\to 0}\lim_{\Delta\to\mathbb{Z}^d}\frac{\log A_{w,\delta;\Delta}}{\beta|\Delta|} = -[\lambda w - P_\beta(\lambda; v) + P_\beta], \tag{2.4.3.2}$$

where λ is such that $w \in [D^- P_\beta(\lambda; v), D^+ P_\beta(\lambda; v)]$. If the Hamiltonian is instead defined by $u - \lambda_0 v$ ($\lambda_0 = 0$ previously), then (2.4.3.2) becomes

$$\lim_{\delta\to 0}\lim_{\Delta\to\mathbb{Z}^d}\frac{\log A_{w,\delta;\Delta}}{\beta|\Delta|} = -[(\lambda - \lambda_0)w - P_\beta(\lambda; v) + P_\beta(\lambda_0; v)].$$

Chapter 3
The phase diagram of Ising systems

In Chap. 2 we have proved that the set of DLR measures is a simplex, namely a convex set whose elements can be uniquely decomposed into a convex combination of the extremal ones. Phase transitions have then to be related to the existence of distinct DLR measures; the extremal, translational invariant DLR measures have been interpreted as pure thermodynamical phases.

What happens to the set of DLR measures when we vary temperature, magnetic field or more generally the interaction is not known in general and it is one of the more important open questions in equilibrium statistical mechanics. We would like to have general theorems on the structure of a phase diagram, where the set of all extremal DLR measures is plotted versus temperature, magnetic field and any other relevant control parameter in the system. Like in thermodynamics we definitely expect to see regions in the diagram where only a single phase is present; such regions should then be separated from each other by regular surfaces of positive codimension, where different phases coexist. Phase transitions occur at such boundaries and since these supposedly have positive codimension, phase transitions are "exceptional."

We have seen in Sect. 2.4 that something of this picture is captured by the general theory of thermodynamic potentials, provided we enlarge the notion of phase space to phase diagrams over the space of all possible interaction potentials. Then "generically" there is only one pure phase, but such a result, unfortunately, does not give information on what happens for a single hamiltonian, when we vary β and h.

In statistical mechanics a proof of the Gibbs phase rule, also in the qualitative sense described above, is in general missing and, as said earlier, this looks like one of the most important open problems in the field. An exception is provided by some lattice models and the purpose of this chapter is to discuss the issue in the context of the Ising model, starting from what can be said in general and progressively adding assumptions which allow for a more and more detailed picture of the phase diagram.

In Sect. 3.1 we will state properties valid quite in general even though proved here only in particular cases, just for ease of presentation. We will start by proving that in $d = 1$ dimensions, phase transitions are absent, unless the interaction has a "very long range," and that the same conclusion holds in any dimensions, provided the temperature is large enough. Both results are actually valid in great generality and indeed we may say that the gaseous phase (high temperatures and low densities) is in general well understood in statistical mechanics.

The converse result, namely the existence of phase transitions, is more problematic in general. We will prove that for β large enough the ferromagnetic, n.n. Ising model with $h = 0$ and in $d \geq 2$ has a phase transition. We will show that there are two distinct DLR measures, obtained by taking the thermodynamic limit of Gibbs measures with plus and minus boundary conditions respectively. The result extends by inequalities to general ferromagnetic interactions in all $d \geq 2$ and by the Pirogov–Sinai theory also to some non-ferromagnetic interactions. The last aspect will be examined in Chap. 10 when extending the theory to models of particles in the continuum.

For ferromagnetic spin–spin interactions, see Fig. 10.1, it is proved that in the half plane (β, h) phase transitions can only occur on the line $h = 0$. Moreover, for any dimension $d \geq 2$, there is $\beta_c < \infty$ (called the inverse critical temperature), so that on the line $\{(\beta, h) : h = 0,\ \beta > \beta_c\}$ there are at least two phases, while for $\beta < \beta_c$ the phase is unique.

In the next chapter we will restrict the analysis by focusing on scale dependent interactions; first mean field and then Kac potentials, deriving in the mesoscopic (Lebowitz–Penrose) limit the van der Waals theory of the liquid–vapor phase transition. In Chap. 9 we will prove phase transitions without taking the mesoscopic limit and only supposing that the Kac parameter is small.

3.1 General results

We will first show that in $d = 1$ there is no phase transition, and then prove that the n.n. ferromagnetic Ising model at low temperatures and no magnetic field has phase transitions in $d \geq 2$ dimensions with two distinct DLR measures, introducing in the course of the proofs the notion of contours, which will have a very important role in the sequel of the book. We will then prove the absence of phase transitions at high temperatures, presenting the Dobrushin uniqueness theorem, whose proof will be given in Sect. 3.2.

3.1.1 Absence of phase transitions in one dimensions

We start with the Landau argument that in $d = 1$ the energy is not strong enough to break the spin flip symmetry. Consider the hamiltonian (2.1.2.4) with no magnetic field, $h = 0$, so that it is invariant under spin flip. The key assumption is that the interaction between two half lines is bounded; more precisely, that

$$\sup_{z\in\mathbb{Z}} \sum_{x<z} \sum_{y\geq z} |J(x, y)| =: \|J\| < \infty. \tag{3.1.1.1}$$

Then, given any finite interval $I \subset \mathbb{Z}$, we have $|H_I(\sigma_I|\sigma_{I^c}) - H_I(-\sigma_I|\sigma_{I^c})| \leq 4\|J\|$, independently of the configurations σ_I and σ_{I^c}. Thus the probability of flipping the spins in any interval, no matter how long it is, is uniformly positive; therefore, with probability 1 there will be somewhere a flip and the spin flip symmetry is not broken.

I shall only sketch the proof (the real proof of the absence of phase transitions is more abstract and is presented in full detail in Theorem 3.1.1.1 below). To make the argument simpler, suppose that $J(x, y) = J(0, y - x)$ and $J(0, x) = 0$

if $|x| \geq R$, R the range of the interaction. Fix any integer $L > R$, call the interval $[-Lk, Lk] : I_k$, k a positive integer, we want to prove that for any DLR measure μ and any string $\sigma^*_{I_1} \in \mathcal{X}_{I_1}$,

$$\mu(\{\sigma : \sigma_{I_1} = \sigma^*_{I_1}\}) = \mu(\{\sigma : \sigma_{I_1} = -\sigma^*_{I_1}\}). \tag{3.1.1.2}$$

Let T^0_k be the identity transformation on $\{-1, 1\}^{\mathbb{Z}}$ and T^1_k the transformation which flips the spins in I_k leaving the others unchanged. Given a positive integer n and $\bar{\sigma} = (\bar{\sigma}_{I_n}, \bar{\sigma}_{I^c_n})$, define the probability on $\{0, 1\}^n$

$$P_{n,\bar{\sigma}}(i_1, \ldots, i_n) = \frac{G_{I_n}(T^{i_n}_n \cdots T^{i_1}_1 \bar{\sigma}_{I_n} | \bar{\sigma}_{I^c_n})}{\sum_{j_1,\ldots,j_n} G_{I_n}(T^{j_n}_n \cdots T^{j_1}_1 \bar{\sigma}_{I_n} | \bar{\sigma}_{I^c_n})}. \tag{3.1.1.3}$$

Going back to (3.1.1.2) we see that $|\mu(\{\sigma : \sigma_{I_1} = \sigma^*_{I_1}\}) - \mu(\{\sigma : \sigma_{I_1} = -\sigma^*_{I_1}\})|$ is bounded by $2 \sup_{\sigma = (\sigma^{(k)}_{I_n}, \sigma_{I^c_n})} |P_\sigma(\{i_1 + \cdots + i_n \text{ is even}\}) - \frac{1}{2}|$, which vanishes as $n \to \infty$ because (proofs are omitted):

- there is $p_{n,\bar{\sigma};k}(i_k)$, $k = 1, \ldots, n$, so that $P_{n,\bar{\sigma}}(i_1, \ldots, i_n) = \prod_{k=1}^n p_{n,\bar{\sigma};k}(i_k)$;
- there is $\delta > 0$ so that $\delta \leq p_{n,\bar{\sigma};k}(i_k) \leq 1 - \delta$, uniformly in n, $\bar{\sigma}$ and k;
- $|P_{n,\bar{\sigma}}(\{\sum_{k=1}^n i_k \text{ is even }\}) - \frac{1}{2}| \leq \prod_{k=1}^n |2p_{n,\bar{\sigma},k}(i_k = 1) - 1|$.

The argument extends also to infinite range interactions, provided (3.1.1.1) holds. This is not the case if $J(x, y)$ decays at infinity as $|x - y|^{-2+\alpha}$, $\alpha \in [0, 1)$, and indeed for such long range interactions there are phase transitions also in $d = 1$. All of the above fails even when the range is finite in $d \geq 2$ dimensions, where the I_k are now "balls" of increasing radius. The interaction then grows proportionally to the surface of the balls and it may thus become arbitrarily expensive to flip the spins in the ball. If however the spins have continuous values the variation can be made smooth, and indeed continuous spin symmetries are not broken in $d = 2$ for a wide class of interactions.

Theorem 3.1.1.1 *Consider the general hamiltonian* (2.1.2.4) *with* $J(x, y)$ *such that we have* (3.1.1.1). *Then there is only one DLR measure at each* $\beta > 0$.

Proof By "abstract arguments" it is enough to prove that given any two configurations σ and σ', there is a constant c^* so that the following holds. For any elementary cylinder C (an elementary cylinder is a set of the form $\{\sigma \in \mathcal{X} : \sigma_\Lambda = \sigma^*_\Lambda\}$), for all n large enough

$$G_{\Delta_n}(C | \sigma_{\Delta^c_n}) \leq c^* G_{\Delta_n}(C | \sigma'_{\Delta^c_n}), \tag{3.1.1.4}$$

where $\Delta_n = [-n, n]$. At the end of the subsection we will prove the "abstract arguments" showing that (3.1.1.4) implies uniqueness of DLR measures and we proceed by proving (3.1.1.4) with $c^* = e^{8\beta c}$ independently of σ and σ'.

• Proof of (3.1.1.4). Suppose $C = \{\sigma'' : \sigma''_\Lambda = \sigma^*_\Lambda\}$ and take n so large that $\Lambda \subset \Delta_n$. Then

$$G_{\Delta_n}\left(C|\sigma_{\Delta_n^c}\right) = \frac{1}{Z_{\Delta_n}(\sigma_{\Delta_n^c})} \sum_{\sigma''_{\Delta_n}:\sigma''_\Lambda=\sigma^*_\Lambda} e^{-\beta H(\sigma''_{\Delta_n}|\sigma_{\Delta_n^c})}.$$

By (3.1.1.1),

$$\left|H(\sigma''_{\Delta_n}|\sigma_{\Delta_n^c}) - H(\sigma''_{\Delta_n}|\sigma'_{\Delta_n^c})\right| \le 4\|J\|,$$

so that

$$G_{\Delta_n}\left(C|\sigma_{\Delta_n^c}\right) \le \frac{e^{8\beta\|J\|_\infty}}{Z_{\Delta_n}(\sigma'_{\Delta_n^c})} \sum_{\sigma''_{\Delta_n}:\sigma''_\Lambda=\sigma^*_\Lambda} e^{-\beta H(\sigma''_{\Delta_n}|\sigma'_{\Delta_n^c})},$$

hence (3.1.1.4).

• The "abstract arguments." We have proved in Chap. 2 (see Theorem 2.2.3.1) that any extremal DLR measure has the form G_σ, $\sigma \in \mathcal{X}_{gg}$, and G_σ is the weak limit as $n \to \infty$ of $G_{\Delta_n}(\cdot|\sigma_{\Delta_n^c})$ (see Theorem 2.2.2.1). Moreover, G_σ is supported by all $\sigma' \in \mathcal{X}_{gg}$ such that $G_{\sigma'} = G_\sigma$ (see (2.2.2.2) and Theorem 2.2.2.2). It follows from this support property that if σ and σ' are in $\mathcal{X}_{gg}$, then either $G_\sigma = G_{\sigma'}$, or else G_σ and $G_{\sigma'}$ have mutually disjoint supports.

We thus need to prove that (3.1.1.4) implies that G_σ and $G_{\sigma'}$ do not have mutually disjoint supports. By taking the limit $n \to \infty$ in (3.1.1.4) and recalling that convergence is ensured by the condition $\sigma, \sigma' \in \mathcal{X}_{gg}$,

$$G_\sigma(C) \le c^* G_{\sigma'}(C). \tag{3.1.1.5}$$

By general arguments (see below) this implies that G_σ is absolutely continuous with respect to $G_{\sigma'}$; hence they cannot have disjoint support so that the other alternative holds and $G_\sigma = G_{\sigma'}$. By the arbitrariness of σ and σ' we then conclude that $|\mathcal{G}| = 1$, i.e. there is only one DLR measure.

• Proof of absolute continuity, namely that if (3.1.1.5) holds for all $C \in \mathcal{C}$, then it holds as well for all Borel sets and hence $m \equiv G_\sigma$ is absolutely continuous with respect to $m' = G_{\sigma'}$. Calling $\mathcal{B}$ the family of Borel sets, let $\mathcal{S} = \{A \in \mathcal{B} : m(A) \le c^* m'(A)\}$. Then $\mathcal{S} \supset \mathcal{C}$, and if we prove that $\mathcal{S}$ is a "monotone class" then by a general theorem (see Appendix A, in particular Theorem A.3), $\mathcal{S} = \mathcal{B}$. To prove that $\mathcal{S}$ is a monotone class, we need to prove two properties. (i) If $A \in \mathcal{S}$, $B \in \mathcal{S}$, $A \subset B$, then $B \setminus A \in \mathcal{S}$. (ii) If $\{A_n\}$ is an increasing sequence of elements all in $\mathcal{S}$, then $\bigcup A_n \in \mathcal{S}$. Both properties are obviously verified in our case as m and m' are probabilities; thus $\mathcal{S}$ is a monotone class and $\mathcal{S} = \mathcal{B}$. □

3.1.2 *Phase transitions at low temperatures*

In this subsection we will prove that the $d \ge 2$, n.n. ferromagnetic Ising model has a phase transition if $h = 0$ and β is large enough. The restriction to n.n. interactions is only for simplicity. The proof is based on "Peierls argument" which will be behind several other proofs in this book.

Theorem 3.1.2.1 *In the $d \geq 2$ n.n. ferromagnetic Ising model with $h = 0$, if β is large enough there are at least two distinct DLR measures.*

Proof The proof shows that the structure of the ground states persists as the temperature increases above 0. The Gibbs measures at $\beta = \infty$ are measures supported by the configurations which minimize the energy, and in our case these are the configuration with all spins equal to $+1$ and the configuration with all spins equal to -1. We shall prove that if the temperature is small enough there are two distinct DLR measures, μ^+ and μ^-, supported by configurations which look like a sea of pluses with rare and small islands of minuses and a sea of minuses with rare and small islands of pluses, respectively. This will be proved by showing that in both cases contours (namely the boundaries between $+$ and $-$) have small probability.

We consider the sequence Δ_n of cubes of side $2n+1$ and center the origin, calling $G^+_{\Delta_n}$ the Gibbs measure in Δ_n with b.c. $\mathbf{1}_{\Delta_n^c}$, which denotes the configuration identically equal to 1 on Δ_n^c. We will prove that if β is large enough, then for all n

$$G^+_{\Delta_n}\big(\sigma(0) = 1\big) \geq \alpha > \frac{1}{2}. \tag{3.1.2.1}$$

As a consequence, any limit point μ^+ of $G^+_{\Delta_n}$ will satisfy (3.1.2.1) (the existence of limit points follows by compactness, but, by using ferromagnetic inequalities, the limit itself can be shown to exist). By the spin flip symmetry of the interaction, the spin flip of a DLR measure is again DLR and since, by (3.1.2.1), μ^+ gives probability $> 1/2$ to $\sigma(0) = 1$, its spin flip μ^- will give probability $> 1/2$ to $\sigma(0) = -1$; hence $\mu^+ \neq \mu^-$. Thus, we only need to prove (3.1.2.1).

The proof uses the important notion of contours, which have here a particularly simple expression due to the n.n. assumption. To deal with more general situations in the next chapters we will modify the notion used here, which must therefore be considered as temporary. We use the shorthand $\Lambda = \Delta_n$ and by default in the sequel any configuration σ is identically equal to $+1$ outside Λ. We call C_x, $x \in \mathbb{Z}^d$, the closed unit cube in $\mathbb{R}^d$ with center x and, given σ, call $+$ a cube if the spin at its center is $+$ and $-$ otherwise. The minus region is the union of all the minus cubes and we call $\Gamma = \Gamma(\sigma)$ its boundary. Thus Γ is a surface union of unit faces, its number is denoted by $|\Gamma|$, $|\Gamma| < \infty$ because $\sigma(x) = 1$ outside Λ. We call $\mathcal{K}_\Lambda$ the family of all $\Gamma(\sigma)$ for any $\sigma = (\sigma_\Lambda, \mathbf{1}_{\Lambda^c})$, $\sigma_\Lambda \in \mathcal{X}_\Lambda$.

Definition Calling two sets *connected* if their closures have a non-empty intersection, the maximal connected components γ_i of $\Gamma \in \mathcal{K}_\Lambda$ are called "*contours*."

Each γ_i is a closed polygonal surface which may self-intersect, each face of γ_i is common to a plus and to a minus cube; therefore

$$H_\Lambda(\sigma_\Lambda|\mathbf{1}_{\Lambda^c}) = H_\Lambda(\mathbf{1}_\Lambda|\mathbf{1}_{\Lambda^c}) + 2J|\Gamma(\sigma_\Lambda, \mathbf{1}_{\Lambda^c})|.$$

Since there is a one to one correspondence between $\{\sigma : \sigma(x) = 1, x \in \Lambda^c\}$ and $\mathcal{K}_\Lambda$,

$$G^+_\Lambda(\sigma_\Lambda) = \frac{e^{-2\beta J|\Gamma(\sigma_\Lambda, \mathbf{1}_{\Lambda^c})|}}{\sum_{\Gamma' \in \mathcal{K}_\Lambda} e^{-2\beta J|\Gamma'|}}. \tag{3.1.2.2}$$

To take advantage of the representation (3.1.2.2) we need the following property of Γ.

Lemma 3.1.2.2 *If $\Gamma = \{\gamma_1, \dots, \gamma_n\}$ (arbitrarily ordered) is in $\mathcal{K}_\Lambda$, then $\Gamma' = \{\gamma_2, \dots, \gamma_n\}$ is also in $\mathcal{K}_\Lambda$.*

Proof A point $r \in \mathbb{R}^d$ is "internal" to γ_1 if any curve which connects r to infinity intersects γ_1. Denote by $\text{int}(\gamma_1)$ the set of all its internal points which are in $\mathbb{Z}^d$. Let $\sigma^{(1)}$ be the configuration obtained from σ by flipping all spins in $\text{int}(\gamma_1)$ and leaving all the others unchanged. We have $\sigma(x)\sigma(y) = \sigma^{(1)}(x)\sigma^{(1)}(y)$, $|x - y| = 1$, when both x and y are in $\text{int}(\gamma_1)$ as well as when they are both in the complement of $\text{int}(\gamma_1)$.

The remaining pairs $|x - y| = 1$ are with $x \in \text{int}(\gamma_1)$ and y in the complement of $\text{int}(\gamma_1)$. In σ the unit cubes with centers x and y are separated by a face $b(x, y) \in \gamma_1$; since $\sigma^{(1)}(x) = -\sigma(x)$ and $\sigma^{(1)}(y) = \sigma(y)$, $b(x, y)$ is no longer in $\Gamma^{(1)} := \Gamma(\sigma^{(1)})$. In conclusion, $\Gamma^{(1)}$ is strictly contained in Γ. It still has the contours $\{\gamma_2, \dots, \gamma_n\}$, because γ_i, $i \neq 1$ is either entirely in $\text{int}(\gamma_1)$ or in its complement, as $\gamma_i \cap \gamma_1 = \emptyset$. γ_1, however, may have not entirely disappeared. If there is a face in γ_1 common to two cubes in $\text{int}(\gamma_1)$, then such a face survives so that in $\sigma^{(1)}$ there may be also other contours whose union, however, is a proper subset of γ_1. By applying the previous spin flip procedure successively to such residual contours, we finally obtain a configuration with only the contours $\{\gamma_2, \dots, \gamma_n\}$. □

Let γ be a contour (it is namely a contour in some $\Gamma \in \mathcal{K}_\Lambda$), then $\{\Gamma \ni \gamma\}$ denotes the set of all $\Gamma \in \mathcal{K}_\Lambda$ such that γ is a contour in Γ; we also write $\{\sigma \Rightarrow \gamma\}$ for the set of all σ such that $\Gamma(\sigma) \in \{\Gamma \ni \gamma\}$.

Proposition 3.1.2.3 *For any contour γ*

$$G^+_\Lambda(\{\sigma \Rightarrow \gamma\}) \leq e^{-2\beta J|\gamma|}.$$

Proof We write

$$G^+_\Lambda(\{\sigma \Rightarrow \gamma\}) = \frac{\sum_{\Gamma \in \{\Gamma \ni \gamma\}} e^{-2\beta J|\Gamma|}}{\sum_{\Gamma \in \mathcal{K}_\Lambda} e^{-2\beta J|\Gamma|}} = e^{-2\beta J|\gamma|} \frac{\sum_{\Gamma \in \{\Gamma \ni \gamma\}} e^{-2\beta J|\Gamma \setminus \gamma|}}{\sum_{\Gamma \in \mathcal{K}_\Lambda} e^{-2\beta J|\Gamma|}}.$$

By the above lemma, $\sum_{\Gamma \in \{\Gamma \ni \gamma\}} e^{-2\beta J|\Gamma \setminus \gamma|} = \sum_{\Gamma' \in \mathcal{K}_\Lambda : \Gamma' \cup \gamma \in \mathcal{K}_\Lambda} e^{-2\beta J|\Gamma'|}$, which is further bounded by dropping the requirement that $\Gamma' \cup \gamma \in \mathcal{K}_\Lambda$. Therefore, the last fraction in the above equation is ≤ 1 and the proposition follows. □

Proof of (3.1.2.1) If σ is such that $\sigma(0) = -1$, then $\Gamma(\sigma)$ has a contour γ with 0 in its interior; hence

$$G^+_{\Delta_n}(\{\sigma(0) = -1\}) \leq \sum_{\gamma:0\in\text{int}(\gamma)} e^{-2\beta J|\gamma|}. \tag{3.1.2.3}$$

The proof of (3.1.2.1) follows from showing that the number of contours with given length grows exponentially with the length; then the r.h.s. of (3.1.2.3) is infinitesimal as $\beta \to \infty$ and the inequality (3.1.2.1) follows for β large enough.

In $d = 2$ the number of coordinate polygons of length n starting from a given point and without self-intersections is bounded by 3^n and in $d = 2$ one can reduce to such a case, thus proving the above statement. A bound valid also in $d > 2$ can be obtained by reducing to "trees," as we are going to see. To any face of γ as in (3.1.2.3) we associate the minus cube which has that face. The union of all such cubes, denoted by D, is a connected set which either contains 0 or has 0 in its interior. The number of cubes $|D|$ in D verifies $|D| \leq |\gamma|$, as a face of Γ cannot belong to two cubes of D, while the same cube of D may arise from more than one face, but certainly $\leq 2d$ faces, i.e. as many as there are faces of a cube. Thus, there are at most $(2^{2d})^{|D|}$ contours γ which give rise to a same set D; hence

$$G^+_{\Delta_n}(\{\sigma(0) = -1\}) \leq \sum_{D:0\in\text{int}(D)} 2^{2d|D|} e^{-2\beta J|D|}.$$

If we shift D along the x-axis by N with N larger than $|D|$, the origin which was in the interior of D is not in the interior of the shifted set; therefore, by continuity, there is a shift by $n \leq |D|$ such that the origin belongs to the shifted set. Thus

$$G^+_{\Delta_n}(\{\sigma(0) = -1\}) \leq \sum_{D:0\in D} |D| 2^{2d|D|} e^{-2\beta J|D|}.$$

We then need the following combinatorial lemma (which will be used many times in Chaps. 9 and 10:

Lemma 3.1.2.4 *Let*

$$n_d = 3^d - 1, \qquad e^{-b} 2^{n_d} < 1. \tag{3.1.2.4}$$

Then

$$\sum_{D:0\in D} e^{-b|D|} < 1, \tag{3.1.2.5}$$

where the sum is over connected sets D, unions of unit cubes centered in sites of $\mathbb{Z}^d$ (recall that two sets are connected if their closures have a non-empty intersection).

Proof The lemma is proved by reduction to a sum over trees. We thus introduce a graph which starts from a root, whose label is 0. The root is connected to n_d new elements, called individuals of the first generation, with label $01, \dots, 0n_d$. Each one of them is in its turn connected to n_d new elements, and the collection of all such new elements are the individuals of the second generation, labeled by $0i_1i_2$, $i_j \in \{1, \dots, n_d\}$; from each of them spring n_d new elements and so on, the structure repeating itself indefinitely. A tree T is a connected subset of this graph which contains the root.

We can now associate to each D in (3.1.2.5) a tree T by the following rule. We order in some arbitrary but translational invariant fashion the cubes connected to a given one. We then associate to the cube C_0 the root of the tree; we then put as elements of the first generation those which correspond to the cubes in D connected to C_0 with the pre-assigned order and then proceed iteratively excluding cubes already considered. In this way we establish a one to one correspondence between sets D in (3.1.2.5) and finite trees T. Calling $|T|$ the number of individuals in T,

$$\sum_{D:D\ni 0} e^{-b|D|} \le \sum_{T:|T|<\infty} e^{-b|T|}.$$

Call $n(T)$ the number of generations in T; then

$$\sum_{T:n(T)\le 1} e^{-b|T|} \le e^{-b}[1+e^{-b}]^{n_d} \le e^{-b}2^{n_d} < 1.$$

By induction, suppose $\sum_{T:n(T)\le N} e^{-b|T|} \le e^{-b}2^{n_d}$; then

$$\sum_{T:n(T)\le N+1} e^{-b|T|} \le e^{-b}[1+1]^{n_d} \le e^{-b}2^{n_d},$$

where the first 1 in the square brackets is when the i-th element of the first generation is absent and the second one is when it is present; in such a case it may be seen as the root of a tree with $\le N$ generations, for which the induction assumption can be used. □

Then

$$\sum_{D:0\in D} |D| 2^{2d|D|} e^{-2\beta J|D|} \le \sup_{n\ge 1}\{n 2^{2dn} e^{-(2\beta J-b)n}\},$$

hence $G^{+}_{\Lambda_n}(\{\sigma(0)=-1\}) \le 2^{2d} e^{-2\beta J+b}$ for β large. Equation (3.1.2.1) and Theorem 3.1.2.1 are proved. □

□

One can actually prove that the weak limits of $G^{\pm}_{\Lambda_n}$ as $\Lambda_n \nearrow \mathbb{Z}^d$ exist, are independent of the sequence Λ_n, define two extremal, translational invariant DLR measures $G^{\pm}$, and that we have

Theorem 3.1.2.5 *In the same context as in Theorem* 3.1.2.1, *if* μ *is a translational invariant DLR measure, then there is* $a \in [0,1]$ *such that* $\mu = aG^{+} + (1-a)G^{-}$.

We shall not prove Theorem 3.1.2.5, but in Chap. 12 we prove the analogous property for the LMP particle model.

3.1.3 Absence of phase transitions at high temperatures

At high temperatures and low densities a fluid is a gas, and phase transitions are absent. There is a mathematically rigorous theory of the phenomenon, which applies to a large variety of statistical mechanics systems. We will discuss here the theory in the context of the Ising models. At infinite temperature, $\beta = 0$, the statement becomes evident, the finite volume Gibbs measures are product measures, and they trivially converge in the thermodynamic limit to a unique DLR measure with all spins mutually independent. The system is an "ideal gas" and there is no phase transition. Approximate ideal behavior at high temperatures is proved by a perturbative argument which shows that the spins are weakly correlated. The dependence on the boundary conditions disappears in the thermodynamic limit, proving that there is a unique DLR measure.

Theorem 3.1.3.1 (High temperature uniqueness) *In any dimension and for any of the interactions considered in Sect. 2.1 there exists $\beta_0 > 0$ so that for any $\beta < \beta_0$ there is a unique DLR measure.*

Notice that the statement is meaningful also in $d = 1$ where phase transitions may occur if the interaction decays sufficiently slowly at infinity, say $|x - y|^{-\alpha}$, $1 < \alpha \leq 2$. When $\alpha > 2$ instead, there is no phase transition for any β, as proved in Sect. 3.1.1.

We shall prove Theorem 3.1.3.1 by proving that the finite volume Gibbs measures depend weakly on the boundary conditions. A function f on $\mathcal{X}$ is Lipschitz in Δ with constant c, Δ a bounded region in $\mathbb{Z}^d$, if for any σ and σ' in $\mathcal{X}$,

$$|f(\sigma) - f(\sigma')| \leq c \sum_{x \in \Delta} |\sigma(x) - \sigma'(x)|. \qquad (3.1.3.1)$$

If f is Lipschitz in Δ, then f depends only on σ_Δ, i.e. it is cylindrical in Δ; in fact, by (3.1.3.1), $f(\sigma) = f(\sigma')$ for any σ and σ' which are identical on Δ. The converse is also true: if f is cylindrical in Δ, then f is Lipschitz in Δ with constant

$$c = \frac{1}{2} \sup_{x \in \Delta} \sup_{\sigma \in \mathcal{X}} |f(\sigma) - f(\sigma^x)|,$$

where σ^x is obtained from σ by flipping the spin at x. We can write in fact $f(\sigma) - f(\sigma')$ as a telescopic sum of terms involving differences of f computed at configurations which differ by a single spin flip at a site in Δ.

We will prove that for any bounded region Δ in $\mathbb{Z}^d$ and any function f Lipschitz in Δ,

$$\lim_{\Lambda\nearrow\mathbb{Z}^d}\ \sup_{\sigma_{\Lambda^c},\sigma'_{\Lambda^c}} \left|G_\Lambda(f|\sigma_{\Lambda^c}) - G_\Lambda(f|\sigma'_{\Lambda^c})\right| = 0, \tag{3.1.3.2}$$

where $G_\Lambda(f|\sigma_{\Lambda^c})$, Λ a bounded region in $\mathbb{Z}^d$, is the expectation of f with respect to $G_\Lambda(\cdot|\sigma_{\Lambda^c})$. As the extremal DLR states are defined via the limits $G_\Lambda(f|\sigma_{\Lambda^c})$, see Theorems 2.2.2.1 and 2.2.3.1, Theorem 3.1.3.1 is a consequence of (3.1.3.2). We will actually prove a stronger result. Define for any bounded Λ, any $\sigma', \sigma'' \in \mathcal{X}$ and any x

$$\Lambda^c_{\neq}(\sigma',\sigma'') = \left\{y \in \Lambda^c : \sigma'(y) \neq \sigma''(y)\right\}. \tag{3.1.3.3}$$

Theorem 3.1.3.2 (Weak dependence on the boundaries) *Suppose that $r(x,y) := \beta|J(x,y)|$ satisfy the condition*

$$\sup_{x\in\mathbb{Z}^d} \sum_{y\neq x} r(x,y) := r < 1, \tag{3.1.3.4}$$

then for any bounded Λ and any σ', σ'' there is a non-negative function $u_{\Lambda,\sigma',\sigma''}(x)$, $x \in \Lambda$, which depends on σ', σ'' only via $\sigma'_{\Lambda^c}, \sigma''_{\Lambda^c}$ such that:

- *for any $\Delta \subset \Lambda$ and any function f Lipschitz in Δ with constant c_f,*

$$\left|G_\Lambda(f|\sigma'_{\Lambda^c}) - G_\Lambda(f|\sigma''_{\Lambda^c})\right| \le c_f \sum_{x\in\Delta} u_{\Lambda,\sigma',\sigma''}(x). \tag{3.1.3.5}$$

- *for any x*

$$\lim_{R\to\infty} \sup\left\{u_{\Lambda,\sigma',\sigma''}(x) : \mathrm{dist}(x, \Lambda^c_{\neq}(\sigma',\sigma'')) \ge R\right\} = 0. \tag{3.1.3.6}$$

- *if moreover $r(x,y)$ satisfies the uniformity condition*

$$\lim_{R\to\infty}\ \sup_{x\in\mathbb{Z}^d} \sum_{|y-x|>R} r(x,y) = 0, \tag{3.1.3.7}$$

then

$$\lim_{R\to\infty} \sup_x \sup\left\{u_{\Lambda,\sigma',\sigma''}(x) : \mathrm{dist}(x, \Lambda^c_{\neq}(\sigma',\sigma'')) \ge R\right\} = 0. \tag{3.1.3.8}$$

If in particular $r(x,y) = r(0, y-x)$ and $r(x,y) = 0$ for all $|x-y| \ge R$, then there are positive c and ω so that $u_{\Lambda,\sigma,\sigma'}(x) \le c e^{-(\omega/R)\mathrm{dist}(x,\Lambda^c)}$.

We shall see in (3.1.3.16) below that the functions $u_{\Lambda,\sigma',\sigma''}(x)$ are non-negative sub-solutions of a linear equation. In Sects. 3.2.4 and 3.2.5 we will study such equations under the assumption that (3.1.3.4) holds and prove (3.1.3.6)–(3.1.3.8); see Corollary 3.2.3.2. Notice also that we have

Corollary 3.1.3.3 *If* (3.1.3.4) *holds, then there is no phase transition, namely* $|\mathcal{G}| = 1$, $\mathcal{G}$ *the set of DLR measures, and Theorem* 3.1.3.1 *is a consequence of Theorem* 3.1.3.2.

Proof Equation (3.1.3.2) follows from (3.1.3.5) and (3.1.3.6). □

Thus Corollary 3.1.3.3 proves Theorem 3.1.3.1. Actually Theorem 3.1.3.2 proves much more than the mere decay of correlations from the boundaries and the consequent absence of phase transitions. It states in fact that the correlations decay from where the boundary conditions are varied; thus even the expectation of a cylindrical function localized close to the boundaries is weakly sensitive to variations of the boundary conditions at large distances, or, in other words, the "walls surrounding the system do not transmit correlations," a phenomenon which in principle may appear even when phase transitions are absent.

Scheme of proof Given two configurations σ^* and σ^{**} (which act as boundary conditions), let $Q_\Lambda(\sigma_\Lambda, \sigma'_\Lambda)$ be a probability on $\mathcal{X}_\Lambda \times \mathcal{X}_\Lambda$ with the property that for any σ_Λ and σ'_Λ the marginal distributions are

$$\begin{aligned}
&\sum_{\sigma''_\Lambda} Q_\Lambda(\sigma_\Lambda, \sigma''_\Lambda) = G_\Lambda(\sigma_\Lambda | \sigma^*_{\Lambda^c}), \\
&\sum_{\sigma''_\Lambda} Q_\Lambda(\sigma''_\Lambda, \sigma'_\Lambda) = G_\Lambda(\sigma'_\Lambda | \sigma^{**}_{\Lambda^c}),
\end{aligned} \tag{3.1.3.9}$$

Q_Λ is then called a joint representation or coupling of $G_\Lambda(\cdot|\sigma^*_{\Lambda^c})$ and $G_\Lambda(\cdot|\sigma^{**}_{\Lambda^c})$. The notion is non-empty because $Q_\Lambda(\sigma_\Lambda, \sigma'_\Lambda) := G_\Lambda(\sigma_\Lambda|\sigma^*_{\Lambda^c}) G_\Lambda(\sigma'_\Lambda|\sigma^{**}_{\Lambda^c})$ is a coupling. The existence of "better couplings" will be discussed below.

Let then Q_Λ be a coupling and f a Lipschitz function in Δ (as in Theorem 3.1.3.2); then

$$\begin{aligned}
&\big|G_\Lambda(f|\sigma^*_{\Lambda^c}) - G_\Lambda(f|\sigma^{**}_{\Lambda^c})\big| \\
&\quad = \Big| \sum_{\sigma_\Lambda, \sigma'_\Lambda} \{f(\sigma_\Lambda) - f(\sigma'_\Lambda)\} Q_\Lambda(\sigma_\Lambda, \sigma'_\Lambda) \Big| \\
&\quad \le c_f \sum_{\sigma_\Lambda, \sigma'_\Lambda} \Big(\sum_{x \in \Delta} |\sigma_\Lambda(x) - \sigma'_\Lambda(x)| \Big) Q_\Lambda(\sigma_\Lambda, \sigma'_\Lambda),
\end{aligned} \tag{3.1.3.10}$$

hence (3.1.3.5) with

$$u_{\Lambda,\sigma^*,\sigma^{**}}(x) := \sum_{\sigma_\Lambda, \sigma'_\Lambda} |\sigma_\Lambda(x) - \sigma'_\Lambda(x)| Q_\Lambda(\sigma_\Lambda, \sigma'_\Lambda). \tag{3.1.3.11}$$

Thus, $u_{\Lambda,\sigma^*,\sigma^{**}}(x)$ has the interpretation of the expectation of $|\sigma_\Lambda(x) - \sigma'_\Lambda(x)|$ with respect to a coupling Q_Λ, and the whole point is to find Q_Λ such that its expectation

$u_{\Lambda,\sigma^*,\sigma^{**}}(x)$ satisfies the properties stated in Theorem 3.1.3.2. The optimization under all possible couplings defines a distance between measures which is called the Vaserstein distance—a notion which will be discussed extensively in Sect. 3.2.

The beauty of the theory is that the only explicit computation which is required involves the simple conditional expectation of a single spin. Denote by

$$R\big(G_{\{x\}}(\cdot|\sigma'_{x^c}), G_{\{x\}}(\cdot|\sigma''_{x^c})\big) := \inf_{Q_{\{x\}}} \sum_{s,s' \in \{-1,1\}} Q_{\{x\}}(s,s')|s-s'|, \qquad (3.1.3.12)$$

where the inf is over all couplings $Q_{\{x\}}$ of the conditional probabilities $G_{\{x\}}(\cdot|\sigma'_{x^c})$ and $G_{\{x\}}(\cdot)|\sigma''_{x^c})$. $R(G_{\{x\}}(\cdot|\sigma'_{x^c}), G_{\{x\}}(\cdot|\sigma''_{x^c}))$ is then "the Vaserstein distance" of $G_{\{x\}}(\cdot|\sigma'_{x^c})$ and $G_{\{x\}}(\cdot)|\sigma''_{x^c})$. In the next section we shall prove

Theorem 3.1.3.4 • *Suppose that there are non-negative numbers $r(x,y)$ such that for all $x \in \mathbb{Z}^d$ and $\sigma', \sigma'' \in \mathcal{X}$*

$$R\big(G_{\{x\}}(\cdot|\sigma'), G_{\{x\}}(\cdot|\sigma'')\big) \le \sum_{y \neq x} r(x,y)|\sigma'(y) - \sigma''(y)|. \qquad (3.1.3.13)$$

Then (3.1.3.5) *holds with parameters $u_{\Lambda,\sigma',\sigma''}(x)$ which satisfy for any $x \in \Lambda$,*

$$u_{\Lambda,\sigma',\sigma''}(x) \le \sum_{y \in \Lambda : \neq x} r(x,y) u_{\Lambda,\sigma',\sigma''}(y) + \sum_{y \in \Lambda^c} r(x,y)|\sigma'(y) - \sigma''(y)|. \quad (3.1.3.14)$$

• *Parameters $r(x,y)$ which satisfy* (3.1.3.13) *always exist, a possible choice being*

$$r(x,y) := \frac{1}{2} \sup_{\sigma} R\big(G_{\{x\}}(\cdot|\sigma_{x^c}), G_{\{x\}}(\cdot|\sigma^{(y)}_{x^c})\big), \qquad (3.1.3.15)$$

where $\sigma^{(y)}(z) = \sigma(z)$ if $z \neq y$ and $= -\sigma(y)$ if $z = y$.

• *The choice* (3.1.3.15) *is optimal: if $r'(x,y)$ satisfies* (3.1.3.13), *then $r'(x,y) \ge r(x,y)$; the latter as in* (3.1.3.15).

The last statement in Theorem 3.1.3.4, namely that (3.1.3.15) gives the optimal choice for $r(x,y)$, is proved below [using Theorem 3.2.1.1 which will be stated and proved in the next section], and the other statements in Theorem 3.1.3.4 are proved in Sect. 3.2.2. To prove the last statement in Theorem 3.1.3.4, we first notice that if $\{r(x,y), y \neq x\}$ is such that (3.1.3.13) holds for all σ' and σ'' in $\mathcal{X}$, then $r(x,y)$ is larger than or equal to the r.h.s. of (3.1.3.15): just take $\sigma' = \sigma$ and $\sigma'' = \sigma^{(y)}$ in (3.1.3.13). The converse inequality, namely that (3.1.3.13) holds with $r(x,y)$ as in (3.1.3.15), follows because by Theorem 3.2.1.1 (of which it is a particular case), the Vaserstein distance satisfies the triangular inequality:

$$\begin{aligned} &R\big(G_{\{x\}}(\cdot|\sigma'), G_{\{x\}}(\cdot|\sigma'')\big) \\ &\quad \le R\big(G_{\{x\}}(\cdot|\sigma'), G_{\{x\}}(\cdot|\sigma)\big) + R\big(G_{\{x\}}(\cdot|\sigma), G_{\{x\}}(\cdot|\sigma'')\big), \end{aligned} \qquad (3.1.3.16)$$

for all $\sigma, \sigma', \sigma''$. Indeed using (3.1.3.16) we can recover any σ''_{x^c} from any σ'_{x^c} by flipping one after the other all the spins in x^c where σ'' disagrees with σ'. Then repeatedly using (3.1.3.16) we derive (3.1.3.13) with $r(x, y)$ as in (3.1.3.15).

Theorem 3.1.3.4 reduces the analysis of $|G_\Lambda(f|\sigma'_{\Lambda^c}) - G_\Lambda(f|\sigma''_{\Lambda^c})|$ to study the solutions $u_{\Lambda,\sigma',\sigma''}$ of (3.1.3.14). A particularly simple case is when the family $r(x, y)$ satisfies (3.1.3.4), namely $\sup_x \sum_{y\neq x} r(x, y) \le r < 1$, and moreover there is $R > 0$ so that $r(x, y) = 0$ for all $|x - y| \ge R$. Then by iterating (3.1.3.14) we get

$$u_{\Lambda,\sigma,\sigma'}(x) \le \sum_{n\ge N} \sum_{y\in\Lambda^c} r^n(x, y)|\sigma'(y) - \sigma''(y)|,$$

for any N such that $RN \le \mathrm{dist}(x, \Lambda^c)$. Then

$$u_{\Lambda,\sigma,\sigma'}(x) \le \sum_{n\ge N} 2r^n = \frac{2r^N}{1-r},$$

proving exponential decay from the boundaries as in the last statement in Theorem 3.1.3.2.

In Sects. 3.2.4 and 3.2.5 we will study the solutions of the inequality (3.1.3.14) and prove the following.

Theorem 3.1.3.5 (Dobrushin condition) *Suppose that*

$$\sup_x \sum_{y\neq x} r(x, y) < 1, \quad r(x, y) := \frac{1}{2} \sup_\sigma R\big(G_{\{x\}}(\cdot|\sigma_{x^c}), G_{\{x\}}(\cdot|\sigma^{(y)}_{x^c})\big), \quad (3.1.3.17)$$

then (3.1.3.6)–(3.1.3.8) *are satisfied.*

Proof of Theorem 3.1.3.2

Thus Theorem 3.1.3.2 follows from Theorem 3.1.3.5 and:

Theorem 3.1.3.6 *Equation* (3.1.3.13) *holds with* $r(x, y) \le \beta|J(x, y)|$.

Proof By an explicit computation:

$$G_x(\{\sigma(x) = u\}|\sigma'_{x^c}) = \frac{e^{ut'}}{e^{t'} + e^{-t'}}, \quad t' = \beta\Big\{h + \sum_{z\neq x} J(x, z)\sigma'(z)\Big\}. \quad (3.1.3.18)$$

$G_x(\{\sigma(x) = u\}|\sigma''_{x^c})$ is given by the same expression with $t'' = \beta\{h + \sum_{z\neq x} J(x, z)\sigma''(z)\}$. We use the shorthand notation $p'_\pm = G_x(\{\sigma(x) = \pm1\}|\sigma'_{x^c})$ and $p''_\pm = G_x(\{\sigma(x) = \pm1\}|\sigma''_{x^c})$. If for instance $-2\beta J(x, y)\sigma'(y) < 0$, then, since $e^t/(e^t + e^{-t})$ is an increasing function of t and $t'' < t'$, $p'_+ > p''_+$ and $p'_- < p''_-$. We construct a coupling by trying to put as much mass as possible on the diagonal. Thus

$Q(+,+) = p''_+$ and mass $Q(-,-) = p'_-$. By imposing that the sum over columns gives p'_- and p'_+ we get $Q(-,+) = 0$ and $Q(+,-) = p'_+ - p''_+$. It is easy to check that the sum over rows gives p''_- and p''_+, hence Q is a coupling.

The crucial feature of this choice is that the matrix $Q(u', u'')$ is triangular. It follows in fact that $u' - u'' \geq 0$ in the support of Q, hence the Q-expectation of $|u' - u''|$ is equal to that of $(u' - u'')$:

$$\sum_{u',u''=\pm} Q(u', u'')|u' - u''|$$

$$= \left| \left\{ \sum_{u=\pm 1} u G_x(\{\sigma(x) = u\}|\sigma'_{x^c}) \right\} - \left\{ \sum_{u=\pm 1} u G_x(\{\sigma(x) = u\}|\sigma''_{x^c}) \right\} \right|$$

$$= \left| \tanh(t') - \tanh(t'') \right|$$

$$\leq \sum_{y \neq x} (\beta |J(x, y)|) \, |\sigma''(y) - \sigma'(y)|.$$

□

3.2 Dobrushin uniqueness theory. I. Vaserstein distance

The theory is much more general than the applications to statistical mechanics that we discuss in this book, as a general reference see the book by Villani [208]. This is Part I because our presentation continues in Sect. 8.4 with the analysis of the spectral gap for some Perron–Frobenius operators and in Sect. 11.5 to implement the Pirogov–Sinai theory in the LMP particle model. Our starting point here is to bound the distance between two measures. The bound will be found implicitly as a solution (or better a sub-solution) of a linear system of equations, so that the final result will require as well an analysis of the linear system.

3.2.1 Couplings and Vaserstein distance

We will suppose that the space Ω below has finitely many elements, but the results extend to general complete, separable metric spaces with μ and μ' Borel probabilities on Ω. In the context of Sect. 3.1.3, $\Omega = \{-1, 1\}^\Lambda$, $\omega = \sigma_\Lambda$, $\mu(\omega) = G_\Lambda(\sigma_\Lambda|\sigma_{\Lambda^c})$ and $\mu'(\omega) = G_\Lambda(\sigma_\Lambda|\sigma'_{\Lambda^c})$.

Couplings A probability Q on $\Omega \times \Omega$, is a coupling (also called a joint representation) of μ and μ' if

$$\sum_{\omega'} Q(\omega, \omega') = \mu(\omega), \qquad \sum_{\omega} Q(\omega, \omega') = \mu'(\omega'). \tag{3.2.1.1}$$

In more general spaces sums we require $\int_{\Omega\times\Omega} f(\omega)dQ(\omega,\omega') = \int_\Omega f(\omega)d\mu(\omega)$ and $\int_{\Omega\times\Omega} f(\omega')dQ(\omega,\omega') = \int_\Omega f(\omega')d\mu'(\omega')$ for any bounded, measurable function f on Ω.

Vaserstein distance The Vaserstein distance may be viewed as a tool for a quantitative definition of a good coupling. Let d be a pseudo distance on Ω, namely a non negative, Borel measurable function on $\Omega \times \Omega$ which satisfies the triangular inequality, notice that we are not requiring that $d(a,b)=0$ implies $a=b$, indeed in the applications to Ising $d(\sigma,\sigma')$ may for instance be $|\sigma(x)-\sigma'(x)|$ or a sum over x in a finite subset.

Given a pseudo distance d (by an abuse of language) we define the Vaserstein distance of two probabilities μ and μ' relative to d as

$$R_d(\mu,\mu') := \inf_Q \sum_{\omega,\omega'} d(\omega,\omega')\, Q(\omega,\omega'), \tag{3.2.1.2}$$

where the inf is over all couplings Q of μ and μ'.

Theorem 3.2.1.1 *$R_d(\mu,\mu')$ satisfies the triangular inequality, the inf in* (3.2.1.2) *is a min; namely, there is a coupling Q such that*

$$R_d(\mu,\mu') = \sum_{\omega,\omega'} d(\omega,\omega')\, Q(\omega,\omega'). \tag{3.2.1.3}$$

Proof Since the limit of couplings is a coupling, the last statement follows by compactness, and this is the only point in the theorem where we need the fact that Ω has finitely many elements. Let Q_1 and Q_2 be the couplings which realize $R_d(\mu,\mu')$ and $R_d(\mu',\mu'')$, and let $\Omega' = \{\omega \in \Omega : \mu'(\omega)=0\}$. We will prove that

$$Q(\omega,\omega'') = \sum_{\omega'\notin\Omega'} Q_1(\omega,\omega') \frac{1}{\mu'(\omega')} Q_2(\omega',\omega'')$$

is a coupling of μ and μ''. We first observe that $Q_1(\omega,\omega')=0$ whenever $\omega' \in \Omega'$, indeed

$$0 \le Q_1(\omega,\omega') \le \sum_{\omega} Q_1(\omega,\omega') = \mu'(\omega') = 0.$$

Analogously $Q_2(\omega',\omega'')=0$ whenever $\omega' \in \Omega'$. Thus

$$\sum_{\omega''} Q(\omega,\omega'') = \sum_{\omega'\notin\Omega'} Q_1(\omega,\omega') = \sum_{\omega'} Q_1(\omega,\omega') = \mu(\omega),$$

$$\sum_{\omega} Q(\omega,\omega'') = \sum_{\omega'\notin\Omega'} Q_2(\omega',\omega'') = \sum_{\omega'} Q_2(\omega',\omega'') = \mu''(\omega''),$$

and Q is a coupling. Then $R_d(\mu, \mu'')$ is bounded by

$$\sum_{\omega,\omega''} d(\omega, \omega'')\, Q(\omega, \omega'')$$

$$\leq \sum_{\omega,\omega''} \sum_{\omega' \notin \Omega'} Q_1(\omega, \omega') \frac{1}{\mu'(\omega')} Q_2(\omega', \omega'') \{d(\omega, \omega') + d(\omega', \omega'')\}$$

$$= R_d(\mu, \mu') + R_d(\mu', \mu''). \qquad \square$$

3.2.2 *The Dobrushin method for constructing couplings*

Construction of good couplings is an art often based on probabilistic and geometric intuitions. Dobrushin has proposed an algorithm based on a local optimization strategy which constructs all by itself the coupling.

The setup

Let $\Omega \subset S^n$, S a finite state space, and $\omega = (\omega_1, \ldots, \omega_n)$ the elements of Ω. Ω may be a proper subset of S^n as it happens in Sect. 11.5 where we introduce "constraints" among the components ω_i.

We consider two probabilities μ and μ' on Ω and want to construct couplings of μ and μ'. We use $\omega = (\omega_1, \ldots, \omega_n)$ as a shorthand

$$\omega_{(i)} = \{\omega_j, j \in (\{1, \ldots, n\} \setminus \{i\})\}, \tag{3.2.2.1}$$

and denote by $\mu(\cdot|\omega_{(i)})$, $\mu'(\cdot|\omega_{(i)})$ the conditional probability on S of the component ω_i given that all the other components have the values $\omega_{(i)}$. We suppose to have already found a "good" coupling of such conditional probabilities; we call it $q_{i,\omega_{(i)},\omega'_{(i)}}(\cdot,\cdot)$ as a probability on $S \times S$. Let d be a pseudo-distance on S and if $I \subset \{1, \ldots, n\}$ $d_I(\omega, \omega') = \sum_{i \in I} d(\omega_i, \omega'_i)$. The next theorem proves that "local bounds can be made global":

Theorem 3.2.2.1 *Suppose that in the above setup there are non-negative functions* $K_i(\omega, \omega') = K_i(\omega_{(i)}, \omega'_{(i)})$, $i = 1, \ldots, n$, *such that*

$$\sum_{\omega_i,\omega'_i} d(\omega_i, \omega'_i)\, q_{i,\omega_{(i)},\omega'_{(i)}}(\omega_i, \omega'_i) \leq K_i(\omega_{(i)}, \omega'_{(i)}). \tag{3.2.2.2}$$

Then there is a coupling Q *of* μ *and* μ' *such that, for any* $i = 1, \ldots, n$,

$$\sum_{\omega,\omega'} d_i(\omega, \omega')\, Q(\omega, \omega') \equiv \sum_{\omega,\omega'} d(\omega_i, \omega'_i)\, Q(\omega, \omega') \leq \sum_{\omega,\omega'} K_i(\omega_{(i)}, \omega'_{(i)})\, Q(\omega, \omega'). \tag{3.2.2.3}$$

Proof Let P_0 be a coupling of μ and μ'. Call

$$P_0\big(\omega_{(i)},\omega'_{(i)}\big) := \sum_{\omega_i,\omega'_i} P_0\big([\omega_{(i)},\omega_i],[\omega'_{(i)},\omega'_i]\big),$$

namely $P_0(\omega_{(i)},\omega'_{(i)})$ is the marginal distribution of the components $\omega_{(i)},\omega'_{(i)}$. Define next

$$P(\omega,\omega') := q_{i,\omega_{(i)},\omega'_{(i)}}(\omega_i,\omega'_i)\, P_0\big(\omega_{(i)},\omega'_{(i)}\big). \tag{3.2.2.4}$$

We claim that $P(\omega,\omega')$ is also a coupling of μ and μ': since $q_{i,\omega_{(i)},\omega'_{(i)}}(\omega_i,\omega'_i)$ is a coupling of the conditional probabilities,

$$\begin{aligned}\sum_{\omega'} P(\omega,\omega') &= \sum_{\omega'_{(i)}}\sum_{\omega'_i} q_{i,\omega_{(i)},\omega'_{(i)}}(\omega_i,\omega'_i)\, P_0\big(\omega_{(i)},\omega'_{(i)}\big)\\ &= \sum_{\omega'_{(i)}} \mu\big(\omega_i|\omega_{(i)}\big) P_0\big(\omega_{(i)},\omega'_{(i)}\big)\\ &= \sum_{\omega'_{(i)}} \mu\big(\omega_i|\omega_{(i)}\big) \sum_{\omega''_i,\omega'_i} P_0\big([\omega_{(i)},\omega''_i],[\omega'_{(i)},\omega'_i]\big).\end{aligned}$$

We first sum over $\omega' = [\omega'_{(i)},\omega'_i]$, and, since P_0 is a coupling, we get

$$\sum_{\omega'} P(\omega,\omega') = \sum_{\omega''_i} \mu\big([\omega_{(i)},\omega''_i]\big)\mu\big(\omega_i|\omega_{(i)}\big) = \mu(\omega).$$

The same argument applies to the second component completing the proof that $P(\omega,\omega')$ is a coupling of μ and μ'. Since a convex combination of couplings is still a coupling,

$$P_1(\omega,\omega') := \frac{1}{n}\sum_{i=1}^{n} q_{i,\omega_{(i)},\omega'_{(i)}}(\omega_i,\omega'_i)\, P_0\big(\omega_{(i)},\omega'_{(i)}\big)$$

is also a couplings and by iteration

$$P_{k+1}(\omega,\omega') := \frac{1}{n}\sum_{i=1}^{n} q_{i,\omega_{(i)},\omega'_{(i)}}(\omega_i,\omega'_i)\, P_k\big(\omega_{(i)},\omega'_{(i)}\big) \tag{3.2.2.5}$$

are all couplings of μ and μ'. Then

$$Q_k(\omega,\omega') := \frac{1}{k+1}\sum_{h=0}^{k} P_h(\omega,\omega') \tag{3.2.2.6}$$

is a coupling and, by letting $k \to \infty$ along a convergent subsequence k_j (recall that the space is finite), also

$$Q(\omega, \omega') := \lim_{k_j \to \infty} Q_{k_j}(\omega, \omega') \tag{3.2.2.7}$$

is a coupling. Moreover, since $\lim_{k_j \to \infty} Q_{k_j+1}(\omega, \omega') = Q(\omega, \omega')$,

$$Q(\omega, \omega') = \lim_{k_j \to \infty} \frac{1}{n} \sum_{i=1}^{n} q_{i,\omega_{(i)},\omega'_{(i)}}(\omega_i, \omega'_i)\, Q_{k_j}\big(\omega_{(i)}, \omega'_{(i)}\big),$$

which is equal to $\frac{1}{n} \sum_{i=1}^{n} q_{i,\omega_{(i)},\omega'_{(i)}}(\omega_i, \omega'_i)\, Q(\omega_{(i)}, \omega'_{(i)})$, so that

$$\begin{aligned}
&\sum_{\omega,\omega'} d_j(\omega, \omega')\, Q(\omega, \omega') \\
&\quad = \sum_{\omega,\omega'} d_j(\omega, \omega') \frac{1}{n} \sum_{i \neq j} q_{i,\omega_{(i)},\omega'_{(i)}}(\omega_i, \omega'_i)\, Q\big(\omega_{(i)}, \omega'_{(i)}\big) \\
&\qquad + \sum_{\omega,\omega'} d_j(\omega, \omega') \frac{1}{n} q_{j,\omega_{(j)},\omega'_{(j)}}(\omega_j, \omega'_j)\, Q\big(\omega_{(j)}, \omega'_{(j)}\big).
\end{aligned} \tag{3.2.2.8}$$

We rewrite the first term on the r.h.s. of (3.2.2.8) as follows. We fix $i \neq j$ and sum over ω_i, ω'_i, getting

$$\frac{1}{n} \sum_{i \neq j} \sum_{\omega_{(i)},\omega'_{(i)}} d_j\big(\omega_{(i)}, \omega'_{(i)}\big) Q\big(\omega_{(i)}, \omega'_{(i)}\big) = \frac{1}{n} \sum_{i \neq j} \sum_{\omega,\omega'} d_j(\omega, \omega')\, Q(\omega, \omega').$$

Analogously, in the second term on the r.h.s. of (3.2.2.8) we write

$$\begin{aligned}
&\sum_{\omega_{(j)},\omega'_{(j)}} Q\big(\omega_{(j)}, \omega'_{(j)}\big) \sum_{\omega_j,\omega'_j} q_{j,\omega_{(j)},\omega'_{(j)}}(\omega_j, \omega'_j) d(\omega_j, \omega'_j) \\
&\quad \leq \sum_{\omega_{(j)},\omega'_{(j)}} Q\big(\omega_{(j)}, \omega'_{(j)}\big) K_j\big(\omega_{(j)}, \omega'_{(j)}\big) \\
&\quad = \sum_{\omega,\omega'} Q(\omega, \omega') K_j\big(\omega_{(j)}, \omega'_{(j)}\big).
\end{aligned}$$

Collecting the above estimates we get from (3.2.2.8) the result that $\sum_{\omega,\omega'} d(\omega_j, \omega'_j)\, Q(\omega, \omega')$ is equal to $\sum_{\omega,\omega'} d_j(\omega, \omega')\, Q(\omega, \omega')$, which is bounded by

$$\leq \frac{n-1}{n} \sum_{\omega,\omega'} d_j(\omega, \omega')\, Q(\omega, \omega') + \frac{1}{n} \sum_{\omega,\omega'} Q(\omega, \omega') K_j\big(\omega_{(j)}, \omega'_{(j)}\big),$$

hence (3.2.2.3). □

Corollary 3.2.2.2
Suppose that there are constants C_i and $r_{i,j}$ so that (3.2.2.2) *holds with*

$$K_i(\omega_{(i)}, \omega'_{(i)}) \le C_i + \sum_{j \neq i} r_{i,j} d(\omega_j, \omega'_j). \tag{3.2.2.9}$$

Let Q be as in Theorem 3.2.2.1 *and $v(i) := \sum_{\omega,\omega'} d_i(\omega, \omega')\, Q(\omega, \omega')$.*
Then $v(i) \le \sum_{j \neq i} r_{i,j}\, v(j) + C_i$, hence $v(\cdot) \le u(\cdot)$, where

$$u(i) = \sum_{j \neq i} r_{i,j}\, u(j) + C_i. \tag{3.2.2.10}$$

Proof of Theorem 3.1.3.4 Suppose $\{1, \dots, n\}$ to be replaced by Λ, $i \in \{1, \dots, n\}$ by $x \in \Lambda$, $\Omega = \mathcal{X}_\Lambda$, $\mu(\sigma_\Lambda) = G_\Lambda(\sigma_\Lambda|\sigma'_{\Lambda^c})$ and $\mu'(\sigma_\Lambda) = G_\Lambda(\sigma_\Lambda|\sigma''_{\Lambda^c})$. Suppose also that (3.1.3.13) holds. Then by Theorem 3.2.1.1 there is a coupling of $G_{\{x\}}(\cdot|\sigma')$ and $G_{\{x\}}(\cdot|\sigma'')$ which attains the Vaserstein distance on the l.h.s. of (3.1.3.13), and (3.2.2.9) holds with $C_x = \sum_{y \in \Lambda^c} r(x, y)|\sigma''(y) - \sigma'(y)|$ where $r_{x,y} = r(x, y)$. Theorem 3.1.3.4 then follows from Corollary 3.2.2.2. □

As we will see, the bound derived in Corollary 3.2.2.2 is useful if the parameters $r_{i,j}$ are so small that $\sum_{j \neq i} r_{i,j} \le r < 1$ for all i.

3.2.3 Total variation distance

Theorem 3.2.3.1 below generalizes the proof of Theorem 3.1.3.6 by constructing a coupling which concentrates as much mass as possible on the diagonal, which is good as the diagonal mass does not contribute to the Vaserstein distance. Let

$$A_+ = \{\omega : \mu(\omega) \ge \mu'(\omega)\}, \qquad A_- = \{\omega : \mu(\omega) < \mu'(\omega)\},$$
$$m(\omega) = \begin{cases} \mu'(\omega) & \text{if } \omega \in A_+, \\ \mu(\omega) & \text{if } \omega \in A_-. \end{cases} \tag{3.2.3.1}$$

Observe that if $A_\pm$ are empty, then $\mu = \mu'$ and the diagonal coupling $Q(\omega, \omega') = \mu(\omega)\mathbf{1}_{\omega=\omega'}$ exists.

Theorem 3.2.3.1 *If $\mu \neq \mu'$ then both $A_\pm$ are non-empty, $\mu(A_+) > \mu'(A_+)$ and $\mu'(A_-) > \mu(A_-)$. The measure Q on $\Omega \times \Omega$ defined as*

$$Q(\omega, \omega') = m(\omega)\mathbf{1}_{\omega=\omega'} + (\mu(A_+) - \mu'(A_+))\lambda(\omega)\lambda'(\omega'), \tag{3.2.3.2}$$

$$\lambda(\omega) = \frac{\mu(\omega) - \mu'(\omega)}{\mu(A_+) - \mu'(A_+)} \mathbf{1}_{\omega \in A^+},$$
$$\lambda'(\omega) = \frac{\mu'(\omega) - \mu(\omega)}{\mu'(A_-) - \mu(A_-)} \mathbf{1}_{\omega \in A_-} \tag{3.2.3.3}$$

is a coupling of μ and μ'.

Proof If $\mu \neq \mu'$ the maximal allowed mass on the diagonal is $m(\omega)$ as defined in (3.2.3.1); indeed, any additional mass would violate the condition that the total mass on rows and columns should equal the mass of μ and μ'. The theorem claims that it is possible to put mass off the diagonal in such a way that, together with the mass (3.2.3.1), we have a coupling. Since $\mu \neq \mu'$, $\mu(A_+) - \mu'(A_+) = \mu'(A_-) - \mu(A_-) > 0$ (the equality follows from $\mu(A_+) + \mu(A_-) = \mu'(A_+) + \mu'(A_-) = 1$, while the positivity follows from the assumption $\mu \neq \mu'$). Observe that λ and λ' as in (3.2.3.4) are two probabilities, and hence $\lambda(\omega)\lambda'(\omega')$ is a coupling of λ and λ'. Then Q as defined in (3.2.3.2) is a coupling of μ and μ'. In fact, the sum over ω' gives

$$\begin{aligned}\sum_{\omega'} Q(\omega,\omega') &= m(\omega) + (\mu(A_+) - \mu'(A_+))\lambda(\omega)\\ &= m(\omega) + [\mu(\omega) - \mu'(\omega)]\mathbf{1}_{\omega\in A^+} = \mu(\omega).\end{aligned}$$

The same argument applies when summing over ω, after rewriting (3.2.3.2) with the factor $(\mu(A_+) - \mu'(A_+))$ replaced by $(\mu'(A_-) - \mu(A_-))$ (we have already seen that the two are equal to each other). □

Corollary 3.2.3.2 *Let $d(\omega,\omega') = \mathbf{1}_{\omega\neq\omega'}$ and let $R_d(\mu,\mu')$ be the corresponding Vaserstein distance of μ and μ'. Then $R_d(\mu,\mu') = \mu(A_+) - \mu'(A_+)$ and*

$$2R_d(\mu,\mu') = |\mu - \mu'| := \sum_{\omega\in\Omega} |\mu(\omega) - \mu'(\omega)|, \tag{3.2.3.4}$$

$|\mu - \mu'|$ being the total variation of $\mu - \mu'$.

Proof If Q is as in (3.2.3.2), then

$$R_d(\mu,\mu') \leq \sum_{\omega\neq\omega'} Q(\omega,\omega') = \mu(A_+) - \mu'(A_+). \tag{3.2.3.5}$$

Let $P(\omega,\omega')$ be any coupling of μ and μ'; then

$$\sum_{\omega\neq\omega'} P(\omega,\omega') = 1 - \sum_{\omega} P(\omega,\omega). \tag{3.2.3.6}$$

We have $P(\omega,\omega) \leq m(\omega) := \min\{\mu(\omega), \mu'(\omega)\}$, and since $1 = \mu(\Omega) = m(A_-) + m(A_+) + \mu(A_+) - \mu'(A_+)$,

$$\sum_{\omega} m(\omega) = 1 - [\mu(A_+) - \mu'(A_+)]; \tag{3.2.3.7}$$

hence, by (3.2.3.6), $\sum_{\omega \neq \omega'} P(\omega, \omega') \geq \mu(A_+) - \mu'(A_+)$ and $R_d(\mu, \mu') \geq \mu(A_+) - \mu'(A_+)$, by the arbitrariness of P. Thus $R_d(\mu, \mu') = \mu(A_+) - \mu'(A_+)$ and (3.2.3.4) follows because $\mu(A_+) - \mu'(A_+) = |\mu - \mu'|/2$. □

3.2.4 *Equations for the first moments*

We study here the linear system of the $|\Lambda|$ equations

$$u(x) = \sum_{y \in \Lambda} r(x, y) \, u(y) + \alpha(x), \quad x \in \Lambda, \ \Lambda \subset \mathbb{Z}^d, \tag{3.2.4.1}$$

in the $|\Lambda|$ unknowns $\{u(x), x \in \Lambda\}$ and with known terms of the form

$$\alpha(x) = \alpha'(x) + \sum_{y \in \Lambda^c} r(x, y) \, \psi(y), \quad \sup |\psi(y)| =: b < \infty, \tag{3.2.4.2}$$

with $r(\cdot, \cdot)$, $\alpha'(\cdot)$ and $\psi(\cdot)$ all non-negative. We will suppose that either $\alpha'(\cdot) \equiv 0$ or that it depends on Λ in such a way as to vanish as the distance of x from Λ^c diverges.

$v(x)$, $x \in \Lambda$, is a sub-solution of (3.2.4.1) if

$$v(x) \leq \sum_{y \in \Lambda} r(x, y) \, v(y) + \alpha(x), \quad x \in \Lambda. \tag{3.2.4.3}$$

With reference to Theorem 3.1.3.2, $u_{\Lambda, \sigma^*, \sigma^{**}}(x)$ is a sub-solution of (3.2.4.1), with $\alpha'(x) = 0$ and $\psi(y) = |\sigma^*(y) - \sigma^{**}(y)|$. If the datum $\alpha(\cdot)$ has the form (3.2.4.2), then (3.2.4.1) can be interpreted as an equation with boundary conditions by taking $u(x)$ as a function on $\mathbb{Z}^d$ and writing

$$u(x) = \begin{cases} \alpha'(x) + \sum_{y \neq x} r(x, y) \, u(y) & \text{if } x \in \Lambda, \\ \psi(x) & \text{if } x \in \Lambda^c. \end{cases}$$

Theorem 3.2.4.1 (Existence and uniqueness) *Let* $r(x, y) \geq 0$ *satisfy* (3.1.3.4) *and let* Λ *be a bounded region in* $\mathbb{Z}^d$. *Then* (3.2.4.1) *has the unique solution*

$$u(x) = \sum_{y \in \Lambda} g_\Lambda(x, y) \, \alpha(y), \tag{3.2.4.4}$$

where

$$g_\Lambda(x, y) := \sum_{n \geq 0} r_\Lambda^n(x, y), \quad r_\Lambda(x, y) = r(x, y) \mathbf{1}_{x, y \in \Lambda}, \tag{3.2.4.5}$$

with $r^0_\Lambda(x,y) := \mathbf{1}_{x=y}$ *and* $r^n_\Lambda(x,y)$ *the n-convolution of* $r_\Lambda(x,y)$*. The series in* (3.2.4.5) *converges and*

$$g_\Lambda(x,y) \le \frac{1}{1-a_\Lambda}, \qquad a_\Lambda := \sup_{x\in\Lambda} \sum_{y\in\Lambda, y\ne x} r_\Lambda(x,y) \le r < 1. \tag{3.2.4.6}$$

Finally, if v *is a sub-solution of* (3.2.4.1) (*i.e. it satisfies* (3.2.4.3)), *then* $v \le u$.

Proof First we treat existence. By (3.1.3.4), $\sum_{y\in\Lambda} r_\Lambda(x.y) \le a_\Lambda \le r < 1$; then by induction on $n \ge 1$, $\sum_{y\in\Lambda} r^n_\Lambda(x,y) \le a^n_\Lambda$. Summing over n we get $g_\Lambda(x,y) \le \frac{1}{1-a_\Lambda}$, so that (3.2.4.5) and (3.2.4.4) are well defined. The function $u(x)$, $x \in \Lambda$, defined in (3.2.4.4), satisfies

$$u(x) = \alpha(x) + \sum_{y\in\Lambda}\sum_{n\ge 1} r^n_\Lambda(x,y)\alpha(y) = \alpha(x) + \sum_{y\in\Lambda} r_\Lambda(x,y)\,u(y),$$

so that $u(\cdot)$ solves (3.2.4.1).

Now consider uniqueness. Suppose that $v(x)$ is another solution of (3.2.4.1). Then $w(x) := u(x) - v(x)$ solves (3.2.4.1) with $\alpha = 0$, namely

$$w(x) = \sum_{y\in\Lambda} r_\Lambda(x,y)w(y) = \sum_{y\in\Lambda} r^n_\Lambda(x,y)w(y).$$

By (3.1.3.4), $|w(x)| \le r^n \sup_{y\in\Lambda}|w(y)|$; hence $w = 0$ by the arbitrariness of n.

Let us turn to sub-solutions. Let v be a sub-solution of (3.2.4.1). Then by induction over N, for any $x \in \Lambda$,

$$v(x) \le \sum_{y\in\Lambda}\sum_{n=0}^{N} r^n_\Lambda(y,x)\alpha(y) + \sum_{y\in\Lambda} r^n_\Lambda(y,x)v(y).$$

By letting $N \to \infty$ the second term vanishes and we have $v(x) \le u(x)$. □

3.2.5 Decay properties of the Green function

Let

$$g(x,y) := \sum_{n\ge 0} r^n(x,y).$$

By comparison with (3.2.4.5) it immediately follows that $g_\Lambda \le g$ point-wise, so that the decay properties of g immediately are reflected in the decay properties of g_Λ (the Green function in the title) and then of u.

Theorem 3.2.5.1 *Suppose* (3.1.3.4) *holds. Then for any x and for any $\epsilon > 0$, there is R (which depends on x and ϵ) so that*

$$\sum_{|y-x|\geq R} g(x,y) \leq \epsilon.$$

If (3.1.3.7) *holds, then R is independent of x.*

Proof Define recursively R_k, $k \geq 1$, with R_1 such that $\sum_{|y-x|\geq R_1} r(x,y) < \epsilon$ and

$$R_{k+1}: \qquad \sup_{z:|z-x|\leq R_1+\cdots+R_k} \sum_{|y-z|\geq R_{k+1}} r(z,y) < \epsilon.$$

If (3.1.3.7) holds, we may take R_1 independent of x and then $R_k = R_1$ for all k.

Let N be such that $r^N < \epsilon$ (r as in (3.1.3.4)). Take R in the theorem equal to $R_1 + \cdots + R_N$. Then $\sum_{|y-x|\geq R} g(x,y)$ is bounded by

$$\frac{r^N}{1-r} + \sum_{n=1}^{N} \sum_{x_1,\ldots,x_{n-1}} \sum_{|y-x|\geq R} r(x,x_1)\cdots r(x_{n-2},x_{n-1}) r(x_{n-1},y),$$

where the first term takes into account the contribution to (3.2.4.5) of $n > N$ and the second term of $n \leq N$. We split the terms in the second sum into those where $|x_1 - x| > R_1$ and those where $|x_1 - x| \leq R_1$. The former are bounded by ϵr^{n-1}, by the choice of R_1. In the latter we split the sum over x_2 into terms with $|x_2 - x_1| > R_2$ and the others. Since $|x_1 - x| \leq R_1$ the contribution of $|x_2 - x_1| > R_2$ is bounded again by $r^{n-1}\epsilon$. By iteration and after noticing that if $|x_j - x_{j-1}| \leq R_j$ for $j = 1, \ldots, n-1$, then $|y - x_{n-1}| \geq R_n$ (because $|y - x| \geq R$), and we get

$$\sum_{|y-x|\geq R} g(x,y) \leq \frac{\epsilon}{1-r} + \sum_{n=1}^{N} \epsilon r^{n-1} n \leq c\epsilon.$$

□

Corollary 3.2.5.2 *Suppose that* (3.1.3.4) *holds; then given any x and $\epsilon > 0$, there is R so that the following holds. Let u solve* (3.2.4.1), *with $\Lambda \ni x$, $\alpha'(\cdot) \equiv 0$ and* $\mathrm{dist}(x, \Lambda^c_{\neq}(\psi)) \geq R$, *where $\Lambda^c_{\neq}(\psi) = \{y \in \Lambda^c : \psi(y) \neq 0\}$.*

Then $u(x) \leq b\epsilon$; and if (3.1.3.7) *holds, then R is independent of x.*

Proof Under the assumption that $\alpha'(\cdot) \equiv 0$, by (3.2.4.4)

$$u(x) \leq \sum_{y\in\Lambda^c_{\neq}(\psi)} g(x,y)\, |\psi(y)|,$$

and the corollary follows from Theorem 3.2.5.1. □

Using stronger assumptions on the decay rate of $r(x,y)$ as $|y-x| \to \infty$, we can deduce better decay properties of $u(x)$. Suppose there is a metric $\delta(x,y)$ such that

$\delta(x,y) \geq a > 0$ for all $x \neq y$, and

$$\sup_x \sum_{y \neq x} r(x,y)\, e^{\delta(x,y)} \leq r' < 1. \tag{3.2.5.1}$$

Notice that (3.2.5.1) implies (3.1.3.4), with $r < e^{-a}$.

Theorem 3.2.5.3 (Decay rate) *Suppose that* (3.2.5.1) *holds, then for any x, any set $A \subset \mathbb{Z}^d$ and any non-negative function f on A,*

$$\sum_{y \in A} g(x,y) f(y) \leq \frac{1}{1-r'} \sup_{y \in A} \{e^{-\delta(x,y)} f(y)\}. \tag{3.2.5.2}$$

Proof Let $\pi(x,y) = r(x,y)e^{\delta(x,y)}$. Since $\delta(x,y)$ satisfies the triangular inequality,

$$r^n(x,y) \leq e^{-\delta(x,y)} \pi^n(x,y).$$

By (3.2.5.1) $\sum_y \pi(x,y) \leq r'$, $\sum_{n \geq 0} \sum_y \pi^n(x,y) \leq \frac{1}{1-r'}$; hence (3.2.5.2), recalling (3.2.4.5). □

Corollary 3.2.5.4 *Let u satisfy* (3.2.4.1); *then, if* (3.2.5.1) *holds,*

$$u(x) \leq \frac{1}{1-r'} \sup_{y \in \Lambda}\{e^{-\delta(x,y)} \alpha'(y)\} + \frac{b}{1-r'} \sup_{y \in \Lambda^c_{\neq}} e^{-\delta(x,y)}.$$

The following particular case appears frequently in the applications.

Corollary 3.2.5.5 *Suppose that* (3.1.3.4) *holds, $r(x,y) \leq ce^{-a|x-y|}$, $c > 0$, $a > 0$, and $\alpha'(x) \leq b' e^{-a\, \mathrm{dist}(x, \Lambda^c_{\neq})}$, $b' > 0$. Then there are $\kappa > 0$ and $r' \in (r,1)$, so that*

$$u(x) \leq e^{-\kappa\, \mathrm{dist}(x, \Lambda^c_{\neq})} \frac{b^*}{1-r'}, \quad b^* = \max\{b, b'\}.$$

Proof (3.2.5.1) is satisfied with $\delta(x,y) = \kappa|x-y|$, if $\kappa > 0$ is small enough. Indeed

$$\sum_{y \neq x} r(x,y) e^{\kappa|x-y|} \leq r e^{\kappa R} + \sum_{y:|y-x|>R} c e^{-(a-\kappa)|x-y|} = r' < 1,$$

after choosing R large enough to suppress the second term and then κ small enough to control the first one. Then Corollary 3.2.5.5 follows from Corollary 3.2.5.4. □

Chapter 4
Mean field, Kac potentials and the Lebowitz–Penrose limit

A sharp separation of the microscopic and macroscopic scales is the basic physical assumption which supports the derivation of thermodynamics from statistical mechanics described in the two previous chapters. We will discuss hereafter cases where a third, mesoscopic scale, intermediate between the two, enters the play.

There are in principle three main lengths in the theory, the inter-atomic distance which in the Ising model may be taken as the lattice spacing, the "effective interaction range" and, finally, the size of the spatial domain which defines the macroscopic scale. So far, we have studied systems in which the "effective interaction range" is of the same order as the lattice spacing in the sense that we have fixed once and for all a hamiltonian and imposed a summability condition, see Sect. 2.1.2; we then say that the effective interaction range is finite and hence it is of the same order as the lattice spacing. Keeping fixed the lattice spacing and the hamiltonian we then consider the limit in which the region Λ invades the whole space, so that the ratio of inter-atomic and macroscopic distances vanishes, while the ratio of inter-atomic over effective interaction range stays finite. In this limit, bulk and surface effects are sharply separated and the correct bulk thermodynamic behavior emerges. The limit has therefore been called "the thermodynamic limit."

The summability condition on the hamiltonian fails in many important cases, as when Coulomb forces or dipole–dipole interactions are taken into account. This is an important chapter in statistical mechanics, but it is not what we are aiming at. We will in fact suppose that screening effects dump the long range nature of these electromagnetic interactions which can then be replaced by short range effective potentials like those considered so far. We have rather in mind what van der Waals believed to be at the origin of the liquid–vapor phase transition; namely a small but rather long range attractive tail in the interaction, which (possibly combined with a short range repulsive force) gives rise to condensation phenomena for suitable values of the density and the temperature.

The systems we are going to consider are characterized by having the inter-atomic and the interaction lengths sharply separated. There are then two possible sub-cases: (i) the interaction and the macroscopic lengths are of the same order; these are mean field models, and (ii) both ratios, lattice spacing/interaction length and interaction length/macroscopic length, vanish; this category contains the Kac potentials which will play a central role in the book.

In Sect. 4.1 we will study the Ising model with mean field interactions. As we shall see the model can explicitly be solved, and by examining the solution we will verify the positive aspects of the theory and its shortcomings, namely the physically correct competition between energy and entropy responsible for the phase transitions, but also the failure of some basic thermodynamic inequalities, due to the long range scaling of the interaction.

In Sect. 4.2 we will refine the mean field assumption as proposed by Kac with his famous Kac potentials, which have been used by Kac, Uhlenbeck and Hemmer [147–149], and Lebowitz, Penrose [160], to derive the van der Waals theory of liquid–vapor phase transitions from a statistical mechanics context. This is the scaling regime that we will focus our attention on in the whole sequel, namely when the ratios of inter-atomic distance over effective interaction range and effective interaction range over domain size are both very small.

4.1 The mean field model

The basic physical assumption in the whole chapter is that the attractive part of the interaction has an effective range much longer than the interatomic distance, so that "statistical averages" have an important role to play in the computation of the energy.

Mean field models are those in which we also suppose that the interaction range is of the same order as the size of the region where the system is confined. We will further specialize our analysis by supposing that (i) the long range attractive force discussed so far is the only one present (such an assumption relies on the lattice nature of the model; as in continuous particle systems other forces are needed to ensure stability; see Chap. 10 where the issue will be discussed); (ii) any two spins in the system interact in the same way independently of their location. Under such assumptions the analysis of the model becomes easy, as we are going to see.

4.1.1 The finite volume, mean field free energy

The mean field hamiltonian in a bounded region Λ of $\mathbb{Z}^d$ is defined by

$$H^{\mathrm{mf}}_{\Lambda,h}(\sigma_\Lambda) = -\frac{1}{2|\Lambda|}S_\Lambda^2 - hS_\Lambda, \quad S_\Lambda = S_\Lambda(\sigma_\Lambda) = \sum_{x\in\Lambda}\sigma_\Lambda(x). \tag{4.1.1.1}$$

Equation (4.1.1.1) is (modulo the addition of a constant) the hamiltonian (2.1.2.4) with $J(x,y) = 1/|\Lambda|\mathbf{1}_{x,\in\Lambda,y\in\Lambda}$: in (2.1.2.4) the sum does not include $x = y$, which instead contributes to (4.1.1.1), but the energy due to such terms does not depend on σ_Λ, and it is an unimportant additive constant. $H^{\mathrm{mf}}_{\Lambda,h}(\sigma_\Lambda)$ is a function of the total magnetization S_Λ—this is why the mean field model is simple!

The "sharp" canonical ensemble $\mathcal{X}_{m,0,\Lambda}$ is defined as the subset of $\mathcal{X}_\Lambda$ of all σ_Λ such that $S_\Lambda(\sigma_\Lambda) = m|\Lambda|$. The range of values of S_Λ is the set

$$\{-|\Lambda|, -|\Lambda|+2, \dots, |\Lambda|\} =: |\Lambda|M_\Lambda, \tag{4.1.1.2}$$

while M_Λ is the set of values of the magnetization density $S_\Lambda/|\Lambda|$. The value $-|\Lambda|$ on the r.h.s. of (4.1.1.2) is attained at $-\mathbf{1}$, the configuration where all spins are -1; $-|\Lambda|+2$ is realized by all the configurations obtained from $-\mathbf{1}$ by flipping only one spin, and so on. Thus $\mathcal{X}_{m,0,\Lambda}=\emptyset$, unless $m\in M_\Lambda$. To have a more regular dependence on m we also introduce the ensembles $\mathcal{X}_{m,\zeta,\Lambda}$ which are canonical with accuracy $\zeta>0$:

$$\mathcal{X}_{m,\zeta,\Lambda}=\bigcup_{|m'-m|<\zeta}\mathcal{X}_{m',0,\Lambda}=\Big\{\sigma_\Lambda\in\mathcal{X}_\Lambda:\Big|\frac{S_\Lambda(\sigma_\Lambda)}{|\Lambda|}-m\Big|<\zeta\Big\}. \tag{4.1.1.3}$$

The corresponding mean field canonical partition function and mean field free energy are

$$\begin{aligned} Z^{\text{can}}_{\beta,h,m,\zeta,\Lambda}&:=\sum_{\sigma_\Lambda\in\mathcal{X}_{m,\zeta,\Lambda}}e^{-\beta H^{\text{mf}}_{\Lambda,h}(\sigma_\Lambda)},\\ F^{\text{can}}_{\beta,h,m,\zeta,\Lambda}&:=-\frac{1}{\beta|\Lambda|}\log Z^{\text{can}}_{\beta,h,m,\zeta,\Lambda}. \end{aligned} \tag{4.1.1.4}$$

We also define the "effective mean field free energy" $H^{\text{mf,eff}}_\Lambda(m)$, $m\in M_\Lambda$, as

$$\begin{aligned} \beta H^{\text{mf,eff}}_\Lambda(m)&:=-\log Z^{\text{can}}_{\beta,h,m,0,\Lambda}\\ &=|\Lambda|\Big(\Big\{-\frac{m^2}{2}-hm\Big\}-\frac{1}{\beta}I_{|\Lambda|}(m)\Big), \end{aligned} \tag{4.1.1.5}$$

where

$$I_{|\Lambda|}(m)=\frac{1}{|\Lambda|}\log|\mathcal{X}_{m,0,\Lambda}|=\frac{1}{|\Lambda|}\log\binom{|\Lambda|}{n},\quad n=|\Lambda|\frac{m+1}{2}, \tag{4.1.1.6}$$

which is the log of the number of states available to the system and thus is identified with the entropy, as discussed in Sects. 2.3 and 2.4. In mean field the entropy has the rather explicit form (4.1.1.6).

4.1.2 The infinite volume mean field free energy

By (4.1.1.5), using the Stirling formula, see (A.2.9), for any $m\in(-1,1)$,

$$\lim_{\zeta\to 0}\lim_{|\Lambda|\to\infty}F^{\text{can}}_{\beta,h,m,\zeta,\Lambda}=\phi_{\beta,h}(m), \tag{4.1.2.1}$$

$$\phi_{\beta,h}(m)=\Big\{-\frac{m^2}{2}-hm\Big\}-\frac{1}{\beta}I(m), \tag{4.1.2.2}$$

where the entropy $I(m)$ is

$$I(m) = -\frac{1-m}{2}\log\frac{1-m}{2} - \frac{1+m}{2}\log\frac{1+m}{2}. \tag{4.1.2.3}$$

Denoting by f' the derivative of f with respect to m, we have

$$\phi''_{\beta,0}(m) = \frac{1}{\beta(1-m^2)} - 1,$$

and the graph $\phi_{\beta,0}(m)$, see Fig. 4.1, has the following properties (whose proof is omitted):

Theorem 4.1.2.1 *$\phi_{\beta,0}(m)$ is a symmetric function of m, convex when $\beta \le 1$ and with a double well shape when $\beta > 1$ with minima at $\pm m_\beta$ where m_β is the unique positive solution of the mean field equation*

$$m_\beta = \tanh\{\beta m_\beta\}. \tag{4.1.2.4}$$

More specifically, $\phi''_{\beta,0}(\cdot) > 0$ for $\beta < 1$; $\phi''_{1,0}(\cdot) \ge 0$ with equality only at $m = 0$. For $\beta > 1$, the "spinodal region" where $\phi''_{\beta,0}(\cdot) < 0$ is the interval $|m| < \sqrt{1-1/\beta}$. The region where $\phi''_{\beta,0}(\cdot) > 0$ is $|m| > \sqrt{1-1/\beta}$; it splits into the "metastable region," which is the union of the two intervals $(-m_\beta, -\sqrt{1-1/\beta})$ and $(\sqrt{1-1/\beta}, m_\beta)$, and $|m| \ge m_\beta$ which is "the pure phases region."

There are both thermodynamical and dynamical justifications for the terminology "metastable" and "spinodal," but we postpone a discussion of the latter to Sect. 5.3 and restrict ourselves to the former (see Sect. 4.1.3 below). We start from "mean field thermodynamics," noticing that the mean field approach reproduces the desired competition between energy and entropy (the first and the second terms on the r.h.s. of (4.1.2.2)). It correctly indicates the existence of a critical temperature ($\beta = 1$ in the model), which separates the region where entropy wins, which is where the free energy is strictly convex, from the other region, where the energy wins and the free energy is no longer strictly convex. But mean field theory overdoes it: thermodynamics requires the free energy to be convex, losing strict convexity at the phase transition, where the graph of $F_{\beta,h}(m)$ remains convex; but it becomes flat, being

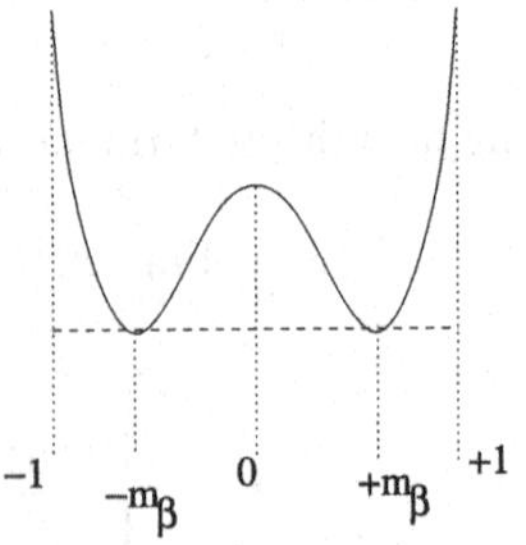

Fig. 4.1 The graph of $\phi_{\beta,0}(m)$, $\beta > 1$; the value at the extremes is $-1/2$ at the center $-\log 2/\beta$. With the *dashed line* it is the graph of the convex envelope

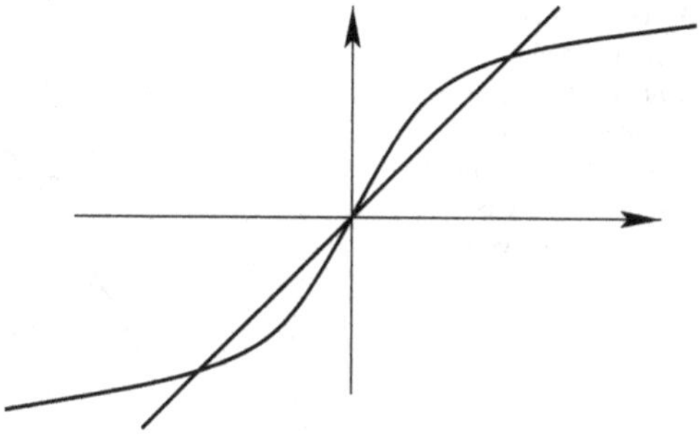

Fig. 4.2 Graph of $\tanh\beta m$, $\beta > 1$: the intersections with the diagonal are at $m = \pm m_\beta$

linear in the interval with endpoints the magnetization values of the coexisting pure phases. In mean field at $\beta > 1$ and $h = 0$, see Fig. 4.2, these are $\pm m_\beta$; thus thermodynamics predicts that $F_{\beta,h=0}(m)$ is constant in $[-m_\beta, m_\beta]$. Instead $\phi_{\beta,0}(m)$ has a double well shape, which is not thermodynamically acceptable. Indeed the free energy density of the standard hamiltonian (2.1.2.4) is convex, as proved in Sect. 2.4. The origin of the pathology lies in the mean field assumption that the hamiltonian changes with volume, the "correct thermodynamic laws" describe in fact a regime where the ratio between the effective interaction range and the macroscopic length (identified with the diameter of the region where the system is confined) vanishes. This is the thermodynamic limit where thermodynamics is valid, mean field instead describes a totally different regime. On the other hand, simplicity, i.e. the best quality of mean field, is just a byproduct of the fact that all spins interact with each other in the same way, independently of their relative positions, which is the same reason as what causes all the discrepancies with thermodynamics. Simplicity and beauty of the results, however, justify the many efforts done to find remedies, which, as we shall see, will eventually lead to Kac potentials.

4.1.3 The Maxwell equal area law

By the second principle of thermodynamics, the free energy is a convex function of m and hence the derivative of the free energy with respect to m must be a non-decreasing function of m. In the mean field approach at $\beta > 1$ this is not true, as

$$k_\beta(m) := \frac{d\phi_{\beta,0}(m)}{dm}$$

is a smooth increasing function of m in $m \le m_- = -\sqrt{1-1/\beta}$. Past m_-, it decreases till $m = m_+ = -m_-$, and, for $m > m_+$, it is again increasing. The interval $[m_-, m_+]$ is called the "spinodal region," the terminology reflecting the shape of the graph of $k_\beta(m)$; spinodal is thus the interval $[m_-, m_+]$, which is "thermodynamically unstable," because the free energy is concave instead of being convex.

To make the theory thermodynamically acceptable we must therefore modify $k_\beta(m)$ and, as argued by Maxwell, there is a "canonical way" to do it: to make $k_\beta(m)$

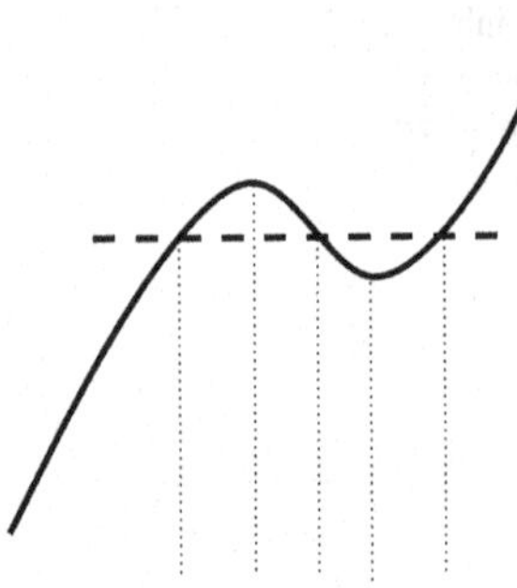

Fig. 4.3 The *vertical dashed lines* (from the left) have abscissa $m_{-1}, m_{-}, m_0, m_{+}$ and m_1. The *horizontal dashed line* cuts the graph of $k_\beta(m)$, so that the areas above and below are equal

non-decreasing we cut the graph $\{m, k_\beta(m)\}$ with a horizontal segment, see Fig. 4.3, which intersects the original graph in three points whose m-coordinates are $m_{-1} < m_0 < m_1$, $m_{-1} < m_{-}$, $m_1 > m_{+}$. Namely, we replace $k_\beta(m)$ by a function $k^*_\beta(m)$ which is the straight segment in $[m_{-1}, m_1]$ (at height $k_\beta(m_1)$) and which is equal to $k_\beta(\cdot)$ elsewhere. $k^*_\beta(m)$ is then continuous and non-decreasing, as desired. The famous "Maxwell equal area law" is a prescription for the location of the cut: the areas spanned by the original graph above and below the segment should be equal. This is the geometrical answer to the physical request that like $k_\beta(m)$ we also want the free energy $\phi_{\beta,0}(m)$ to be unchanged in the complement of the interval $[m_{-1}, m_1]$, where $k_\beta(m)$ has been modified. This is evidently true with the equal area prescription. In fact, since $k^*_\beta(m) = k_\beta(m)$ for $m \le m_{-1}$ we also also have $\phi^\star_{\beta,0}(m) = \phi_{\beta,0}(m)$ for $m \le m_{-1}$. We have $\phi_{\beta,0}(m_1) = \phi_{\beta,0}(m_{-1}) + \int_{m_{-1}}^{m_1} k_\beta(m)\,dm$ and

$$\phi^*_{\beta,0}(m_1) = \phi^*_{\beta,0}(m_{-1}) + \int_{m_{-1}}^{m_1} k^*_\beta(m)\,dm.$$

By the equal area rule, $\int_{m_{-1}}^{m_1} [k_\beta(m) - k^*_\beta(m)]\,dm = 0$, hence $\phi^*_{\beta,0}(m_1) = \phi_{\beta,0}(m_1)$ and since $k_\beta(m) = k^*_\beta(m)$ for $m \ge m_1$, also $\phi^\star_{\beta,0}(m) = \phi_{\beta,0}(m)$ for $m \ge m_1$.

The Maxwell equal-area rule has the simple geometrical interpretation

$$\phi^\star_{\beta,0}(m) = CE(\phi_{\beta,0}(m)). \tag{4.1.3.1}$$

CE is the convex envelope. Recall in fact that $CE(f)$ is the maximal convex function whose graph lies below that of f. $\phi^\star_{\beta,0}(m)$ is continuous and convex and the only region where it differs from $\phi_{\beta,0}(m)$ is the open interval (m_{-1}, m_1), where it is straight. This is the only place where we could increase $\phi^\star_{\beta,0}(m)$ keeping it below $\phi_{\beta,0}(m)$, but then we would lose convexity. In our specific case, see Fig. 4.1, $m_1 = m_\beta$ and $m_{-1} = -m_\beta$, with m_β as in (4.1.2.4), and $k_\beta(m_1) = 0$.

The region where $\phi_{\beta,0}(\cdot)$ is convex, i.e. $k_\beta(m)$ increasing, is thermodynamically acceptable; it contains, however, a part where $\phi_{\beta,0} > \phi^\star_{\beta,0}$, i.e. where the graph is above its convex envelope and yet it is convex. Such regions which are thermodynamically not "as bad as the spinodals" are called metastable. In conclusion we have a rule for changing the non-convex function $\phi_{\beta,0}(m)$ into a convex one which also allows for a thermodynamical interpretation of spinodal and metastable—but where is the rule coming from?

4.1.4 The mean field pressure

Let us see what happens if we start from the grand-canonical rather than the canonical ensemble. Recalling (4.1.1.5) we have

$$Z_{\beta,h,\Lambda} = \sum_{m\in M_\Lambda} e^{-\beta H_\Lambda^{\mathrm{mf,eff}}(m)}.$$

Set $P_{\beta,h,\Lambda} = \log Z_{\beta,h,\Lambda}/(\beta|\Lambda|)$, and since $\mathrm{card}(M_\Lambda) = |M_\Lambda| = |\Lambda| + 1$,

$$0 \le P_{\beta,h,\Lambda} - \max_{m\in M_\Lambda} \left\{hm - F^{\mathrm{can}}_{\beta,0,m,0,\Lambda}\right\} \le \frac{\log(|\Lambda|+1)}{\beta|\Lambda|}, \tag{4.1.4.1}$$

with $F_{\beta,h,m,0,\Lambda}$ as in (4.1.1.4). As the entropy $I_{|\Lambda|}(m)$, in the expression for $H_\Lambda^{\mathrm{mf,eff}}(m)$, converges uniformly to its limit value $I(m)$, see (A.2.9), we have

$$\lim_{|\Lambda|\to\infty} P_{\beta,h,\Lambda} = p_{\beta,h} = \sup_{m\in[-1,1]} [-\phi_{\beta,h}(m)] = \sup_{m\in[-1,1]} [hm - \phi_{\beta,0}(m)] \tag{4.1.4.2}$$

(the sup being a max). Thus $p_{\beta,h}$ is the Legendre transform of $\phi_{\beta,0}(m)$, and therefore the grand-canonical free energy density, which is defined as the Legendre transform of $p_{\beta,h}$, is the convex envelope of $\phi_{\beta,0}(m)$:

$$\sup_h\{hm - p_{\beta,h}\} = CE(\phi_{\beta,0}(\cdot))(m). \tag{4.1.4.3}$$

On the other hand, by (4.1.3.1) the convex envelope of $\phi_{\beta,0}(m)$ is the free energy obtained by the equal area rule, hence we have a new argument for the validity of the rule.

4.1.5 Thermodynamics of mean field

We have just seen in Sect. 4.1.4 that the grand-canonical free energy density is the convex envelope $CE(\phi_{\beta,0})(m)$ of $\phi_{\beta,0}(m)$. In agreement with the thermodynamic considerations presented in Sects. 2.3 and 2.4, the mean field grand canonical free energy indicates that

- *there are no phase transitions for* $\beta \le 1$,
- *there are phase transitions when* $\beta > 1$, *described by the "order parameter"* m,
- *when* $m \notin (-m_\beta, m_\beta)$, m_β *as in* (4.1.2.4), *there is a unique pure phase; there is no pure phase when* $m \in (-m_\beta, m_\beta)$ *and the magnetization* m *can only be realized by a superposition of the two pure phases with magnetizations* $\pm m_\beta$.

According to van der Waals, $\{|m| \le \sqrt{1-1/\beta}\}$ is the spinodal region, while the two intervals $(-m_\beta, -\sqrt{1-1/\beta})$ and $(\sqrt{1-1/\beta}, m_\beta)$ form the metastable region. All that cannot be extracted from the grand-canonical free energy, which is flat in the whole interval $[-m_\beta, m_\beta]$, but rather can be extracted from $\phi_{\beta,0}(m)$ which distinguishes spinodal and metastable. In this respect, the non-convexity of $\phi_{\beta,0}(m)$ is helpful; it gives more information and consequently has a richer structure.

4.1.6 Conclusions

With the grand-canonical ensemble we have avoided the incongruence which mean field approach exhibits in the canonical formalism. We have a reasonable phase transition theory, which even includes metastability. We have however several unsatisfactory facts.

- *We only have a thermodynamic theory without DLR states, and a thermodynamic limit mean field hamiltonian is not well defined.*
- *Equivalence of ensembles is not valid; the canonical formalism leads to a theory incompatible with thermodynamics.*
- *There are "too many" phase transitions. We do not want, in general, phase transitions in one dimension, but the theory is dimension independent. There are no geometrical considerations involved in the phase transition, which we know to be relevant in physical systems.*
- *The theory does not explain interfaces nor coexistence of phases, which are among the most important targets of our analysis.*

4.2 Kac potentials

In this section we will study systems where the lattice spacing is much smaller than the interaction range, which in turn is much smaller than the size of the domain where the system is confined, namely the three basic spatial scales of the system are well separated.

This is for instance the case of a hamiltonian with coupling constants $J(x,y) = \mathbf{1}_{|x-y|\le L}/L^d$, $L \gg 1$. Then the spin at a point, say x, interacts with those in the ball of radius L as in the mean field model, but it sees the difference with the mean field as soon as interactions with more distant spins are considered. Is the first impression of there being a mean field correct? Or does the large distance fall-off prevail?

The idea that mean field behavior might be correct is obviously the most appealing one as it may explain phase transitions as a perturbation of the mean field. In any case, however, we may expect that on distances $\ll L$ the true system may really look like a mean field, in which case its magnetization density would be either close to m_β or to $-m_\beta$ (we are considering the case where the mean field has a phase transition, with order parameter $\pm m_\beta$). We may then envision two possible scenarios; either that the same phase propagates throughout the whole domain, or else that the regions with m_β and $-m_\beta$ alternate more or less randomly in space. In the first case the energy would win, in the second one the entropy would be predominant. As we shall see in Chap. 9, the above arguments can really be pushed through, and, in $d \geq 2$ dimensions, the first alternative will be found to hold. In this way the occurrence of a phase transition will be proved in a large class of Kac interactions, which includes the above "local mean field" hamiltonian. In $d = 1$ instead there cannot be a phase transition as the interaction range is finite, and it can indeed be proved that regions with magnetization m_β and $-m_\beta$ alternate on a spatial scale of the order $e^{c\gamma^{-1}}$, $c > 0$ being a suitable constant.

The key word with Kac potentials is "scalings." We have three basic lengths: lattice distance, interaction range and size of the system. There are other lengths which are not intrinsic, but these are introduced for the sake of the analysis. They are related to coarse graining and renormalization group techniques. More lengths will come up when dynamics is considered, and correspondingly we will also have many time scales.

The difficulty with so many scales is that everybody has a favorite one and expresses everything in the corresponding units. The translation from one to the other is conceptually straightforward, but often confusing. Coming from statistical mechanics, my favorite space unit is the lattice distance, which therefore is taken equal to 1. The range of the interaction of the Kac potentials, denoted by γ^{-1}, is then very large and the size of the regions even larger, say ϵ^{-1}, ϵ much smaller than γ. In continuum mechanics, the opposite convention is mainly used: the size of the region is 1 (macroscopic units), the lattice spacing is then ϵ (thus thinking of the thermodynamic limit as a continuum limit with the lattice mesh ϵ going to 0), while the range of the interaction, $\epsilon\gamma^{-1}$, is also vanishing, but at a slower rate. The introduction of the third scale, intermediate between microscopic and macroscopic and therefore called mesoscopic, is characteristic of Kac potentials and ultimately is responsible for their nice behavior. The mesoscopic scale is the relevant one in the study of interfaces, the fine structure of the interface appears in fact in the mesoscopic scale with unit length $\epsilon\gamma^{-1}$, while in the macroscopic scale the interface looks sharp. But it is still too early for all that, so let us proceed with the analysis of the Kac potentials.

4.2.1 The Lebowitz–Penrose limit

We will study here hamiltonians of the form

$$-\frac{1}{2}\sum_{x \neq y} J_\gamma(x,y)\sigma(x)\sigma(y), \qquad J_\gamma(x,y) = \gamma^d J(\gamma x, \gamma y), \tag{4.2.1.1}$$

which satisfy the following conditions.

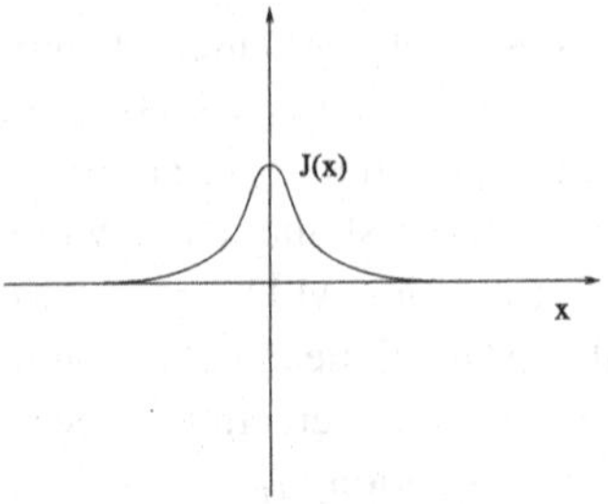

Fig. 4.4 The graph of J in $d=1$

Assumptions on the interaction

We suppose that $J(r,r') = J(r+a, r'+a)$, for all r, r' and a in $\mathbb{R}^d$; that $J(0,r)$ is a non-negative, C^2 function, supported by the unit ball and normalized as $\int_{\mathbb{R}^d} J(0,r)\,dr = 1$; see Fig. 4.4. Finally, γ in (4.2.1.1) is a positive parameter which takes values in $\{2^{-2n}, n \in \mathbb{N}\}$. The above assumptions are far from necessary and they could be strongly relaxed.

For simplicity in the following discussion we take Λ as a torus in $\mathbb{Z}^d$, calling L its side. If $\gamma^{-1} \gg L$, the interaction of any two spins $\sigma(x)$ and $\sigma(y)$ in Λ is

$$-\sum_{i \in L\mathbb{Z}^d} \gamma^d J(\gamma x, \gamma(y+i))\sigma(x)\sigma(y) \approx L^{-d}\sigma(x)\sigma(y),$$

as the l.h.s. is the Riemann sum of $\int J(0,r)\,dr$ with a mesh γL. Thus, in the regime $\gamma^{-1} \gg L$ the Kac potential reproduces the mean field model. When $\gamma^{-1} \approx L$, mesoscopic and macroscopic lengths are comparable; this is still essentially mean field but with a spatial structure. As we shall see, already in this regime there are several interesting phenomena. Finally the third regime is where γ is small and $L \gg \gamma^{-1}$. As explained in the introduction to this section, this is the regime where the lattice spacing (1 in our units), the interaction range, γ^{-1}, and the domain size, L, are all well separated.

In the third regime L is much larger than γ^{-1}, and the system may be included among those studied in Chap. 2, which is literally true if we take $L \to \infty$ keeping γ small but fixed, because in such a case the interaction has a strictly finite range. Kac, Uhlenbeck and Hemmer, and then Lebowitz and Penrose, have proposed a compromise where both ratios vanish, lattice spacing/interaction range and interaction range/domain size. This is the ideal regime where the three scales are completely separated, and the thermodynamics in this regime can be explicitly determined.

When γ is finite the interaction has a finite range, Theorem 2.3.3.1 applies, and the thermodynamic pressure $P_{\beta,h,\gamma}$ is well defined:

$$\lim_{\Lambda \to \mathbb{Z}^d} P_{\beta,h,\gamma,\Lambda}(\sigma_{\Lambda^c}) =: P_{\beta,h,\gamma}, \tag{4.2.1.2}$$

where $P_{\beta,h,\gamma,\Lambda}(\sigma_{\Lambda^c})$ is the grand-canonical pressure in Λ for the system with hamiltonian (4.2.1.1) and coupling constants (4.2.1.1); $\Lambda \to \mathbb{Z}^d$ in the van Hove sense, and σ_{Λ^c} is any sequence of boundary conditions. The dependence on β and h is here made explicit.

Theorem 4.2.1.1 (Lebowitz–Penrose) *For any $h \in \mathbb{R}$ and any $|m| < 1$,*

$$\lim_{\gamma\to 0} P_{\beta,h,\gamma} = p_{\beta,h},$$
$$\lim_{\gamma\to 0} F_{\beta,\gamma}(m) = \sup_h (hm - p_{\beta,h}) = CE\phi_{\beta,0}(m), \tag{4.2.1.3}$$

where $F_{\beta,\gamma}(m)$ is the Legendre transform of $P_{\beta,h,\gamma}$ and $p_{\beta,h}$ (the mean field pressure); $\phi_{\beta,0}(m)$ is defined in (4.1.4.2).

We shall prove Theorem 4.2.1.1 in Sect. 4.2.3. The double limit $\Lambda \to \mathbb{Z}^d$, (4.2.1.2), followed by $\gamma \to 0$, (4.2.1.3), is usually called "the Lebowitz–Penrose limit." It defines a limit procedure characterized by taking first the volume and then the range of the forces to infinity, thus fitting with the ideal regime where the three basic spatial scales are completely separated. Such a limit procedure automatically gives the correct thermodynamic behavior, as the first limit provides the convexity properties required by thermodynamics, which are then preserved by the second limit.

Theorem 4.2.1.1 proves the equivalence of ensembles. Set the magnetic field $h = 0$, call $F_{\beta,\gamma,\Lambda,\delta,\sigma_{\Lambda^c}}(m)$ the finite volume canonical free energy defined in (2.4.1.3) with the cylindrical function $v(\sigma) = \sigma(0)$ and $w = m$. We claim that

$$\lim_{\gamma\to 0} \lim_{\delta\to 0,\Lambda\to\mathbb{Z}^d} F_{\beta,\gamma,\Lambda,\delta,\sigma_{\Lambda^c}}(m) = CE\phi_{\beta,0}(m), \tag{4.2.1.4}$$

where, as before, $\Lambda \to \mathbb{Z}^d$ in the van Hove sense, and σ_{Λ^c} is any sequence of boundary conditions. In fact, by the general theory after the limits $\Lambda \to \mathbb{Z}^d$ and $\delta \to 0$, we obtain $F_{\beta,\gamma}(m)$; then (4.2.1.4) follows from the second limit in (4.2.1.3).

4.2.2 Block spins, coarse graining

The proof of Theorem 4.2.1.1 is maybe even more interesting than the result itself. Following Lebowitz–Penrose, we will in fact introduce notions as block spin, coarse graining, effective hamiltonians.... In this way we shall prove that in the limit of small γ the Gibbs probability of observing a coarse grained magnetization pattern is determined by a free energy functional (which is a non-local version of the famous Ginzburg–Landau functional) and the computation of Gibbs probabilities will be reduced to variational problems with such free energy functionals. The transition to continuum is then the key word here, and, with this in mind, we introduce some new notation which will be frequently used throughout the sequel.

The partitions $\mathcal{D}^{(\ell)}$ of $\mathbb{R}^d$

- For any $\ell \in \{2^n,\ n \in \mathbb{Z}\}$ and any $i \in \ell\mathbb{Z}^d$, we set

$$C_i^{(\ell)} = \{r \in \mathbb{R}^d : i_k \le r_k < i_k + \ell,\ k = 1, \dots, d\} \tag{4.2.2.1}$$

  (r_k and i_k being the kth coordinate of r and i) and call

$$\mathcal{D}^{(\ell)} = \{C_i^{(\ell)}, i \in \ell\mathbb{Z}^d\}, \tag{4.2.2.2}$$

  the corresponding partition of $\mathbb{R}^d$; we also denote by $C_r^{(\ell)}$ the cube of $\mathcal{D}^{(\ell)}$ which contains r. Since ℓ has the form 2^n, each cube of $\mathcal{D}^{(\ell)}$ is the union of cubes of $\mathcal{D}^{(\ell')}$ for $\ell > \ell'$; $\mathcal{D}^{(\ell)}$ is then coarser than $\mathcal{D}^{(\ell')}$ and $\mathcal{D}^{(\ell')}$ finer than $\mathcal{D}^{(\ell)}$.
- A function $f(r)$ is $\mathcal{D}^{(\ell)}$-measurable if it is constant on each cube $C_i^{(\ell)}, i \in \ell\mathbb{Z}^d$; a region Λ is $\mathcal{D}^{(\ell)}$-measurable if its characteristic function is $\mathcal{D}^{(\ell)}$-measurable, or, equivalently, it is union of cubes of $\mathcal{D}^{(\ell)}$.
- We identify a set $\Lambda \subset \mathbb{Z}^d$ with the set $\bigcup_{x\in\Lambda} C_x^{(1)}$ in $\mathbb{R}^d$, which, by abuse of notation, will also be called Λ.

The whole analysis in the sequel uses extensively coarse graining: by the scaling properties of the Kac potentials, we shall see that the relevant quantities are not the values of the spins themselves but rather their averages over suitable regions; see the third item below. Here are the main definitions.

Block spins and coarse graining

- A block spin $s(i)$, $i \in \ell\mathbb{Z}^d$, is the collection of spins $\sigma(x)$, $x \in C_i^{(\ell)}$.
- The block spin magnetization is the function on $\mathbb{R}^d$ defined by

$$\sigma^{(\ell)}(r) = \ell^{-d} \sum_{x \in C_r^{(\ell)} \cap \mathbb{Z}^d} \sigma(x), \tag{4.2.2.3}$$

  so that $\sigma^{(\ell)}(r) = ⨍_{C_r^{(\ell)}} \sigma^{(1)}(r')\,dr'$, where $⨍_\Lambda f(r)\,dr = \frac{1}{|\Lambda|}\int_\Lambda f(r)\,dr$.
- $\mathcal{M}^{(\ell)}$, $\ell \in \{2^n, n \in \mathbb{N}\}$, and $\mathcal{M}_\Lambda^{(\ell)}$, Λ a $\mathcal{D}^{(\ell)}$-measurable set, are the spaces of all $\mathcal{D}^{(\ell)}$-measurable functions on $\mathbb{R}^d$, respectively Λ, with values in

$$M^{(\ell)} = \left\{-1, -1 + \frac{2}{\ell^d}, \dots, 1 - \frac{2}{\ell^d}, 1\right\}. \tag{4.2.2.4}$$

With ℓ and Λ as in the last item, we write for any $m \in \mathcal{M}_\Lambda^{(\ell)}$

$$\{\sigma_\Lambda^{(\ell)} = m\} = \{\sigma_\Lambda \in \mathcal{X}_\Lambda \colon \sigma_\Lambda^{(\ell)}(r) = m(r), r \in \Lambda\}, \tag{4.2.2.5}$$

and recall that $\Lambda \subset \mathbb{Z}^d$ is identified with a $\mathcal{D}^{(\ell)}$-measurable set of $\mathbb{R}^d$. We then define

$$Z_{\gamma,\Lambda,\sigma_{\Lambda^c}}\big(\{\sigma_\Lambda^{(\ell)} = m\}\big) = \sum_{\sigma_\Lambda \in \{\sigma_\Lambda^{(\ell)} = m\}} e^{-\beta H_{\gamma,\Lambda}(\sigma_\Lambda|\sigma_{\Lambda^c})} \tag{4.2.2.6}$$

(the magnetic field h is not made explicit in this notation) and, for any set $\mathcal{A} \subset \mathcal{M}_\Lambda^{(\ell)}$,

$$Z_{\gamma,\Lambda}(\mathcal{A}) = \sum_{m \in \mathcal{A}} Z_{\gamma,\Lambda,\sigma_{\Lambda^c}}\big(\{\sigma_\Lambda^{(\ell)} = m\}\big). \tag{4.2.2.7}$$

The main step in the proof of the Lebowitz–Penrose theorem, Theorem 4.2.1.1, is

Theorem 4.2.2.1 *There is $c > 0$, so that for any bounded $\mathcal{D}^{(\gamma^{-1/2})}$-measurable region Λ in $\mathbb{R}^d$ and any $m_\Lambda \in \mathcal{M}^{(\gamma^{-1/2})}$*

$$\log Z_{\gamma,\Lambda,\sigma_{\Lambda^c}}\big(\{\sigma_\Lambda^{(\gamma^{-1/2})} = m_\Lambda\}\big) \le -\beta\, F_{\gamma,\Lambda}\big(m_\Lambda|\sigma_{\Lambda^c}^{(\gamma^{-1/2})}\big) + \beta c\epsilon(\gamma)|\Lambda|, \tag{4.2.2.8}$$

$$\log Z_{\gamma,\Lambda,\sigma_{\Lambda^c}}\big(\{\sigma_\Lambda^{(\gamma^{-1/2})} = m_\Lambda\}\big) \ge -\beta\, F_{\Lambda,\gamma}\big(m_\Lambda|\sigma_{\Lambda^c}^{(\gamma^{-1/2})}\big) - \beta c\epsilon(\gamma)|\Lambda|, \tag{4.2.2.9}$$

where

$$\epsilon(\gamma) = \big(\gamma^{1/2} + \gamma^{d/2}\log\gamma^{-1}\big), \tag{4.2.2.10}$$

and $F_{\gamma,\Lambda}(m_\Lambda|m_{\Lambda^c}) = F_{\gamma,\Lambda}(m_\Lambda) - \int_\Lambda\int_{\Lambda^c} J_\gamma(r,r')\,m_\Lambda(r)m_{\Lambda^c}(r')$ where

$$F_{\gamma,\Lambda}(m_\Lambda) = -\frac{1}{2}\int_\Lambda\int_\Lambda J_\gamma(r,r')\,m_\Lambda(r)m_\Lambda(r') - h\int_\Lambda m_\Lambda(r) - \frac{1}{\beta}\int_\Lambda I(m_\Lambda(r)). \tag{4.2.2.11}$$

$F_{\gamma,\Lambda}(m_\Lambda|m_{\Lambda^c})$ is "the L–P functional"; L–P after Lebowitz and Penrose, who introduced such a functional in their derivation of the van der Waals theory. We postpone the proof of Theorem 4.2.2.1 to Sect. 4.2.5 [the proof is just some combinatorics with use of the Stirling formula plus the approximation of integrals by Riemann sums] and use it in the next subsection to prove Theorem 4.2.1.1. The following corollary of Theorem 4.2.2.1 will also be very useful in the sequel.

Theorem 4.2.2.2 *There is $c > 0$, so that for any bounded $\mathcal{D}^{(\gamma^{-1/2})}$-measurable region Λ in $\mathbb{R}^d$ and any $\mathcal{A} \subset \mathcal{M}^{(\gamma^{-1/2})}$*

$$\log Z_{\gamma,\Lambda,\sigma_{\Lambda^c}}(\mathcal{A}) \le -\beta \inf_{m_\Lambda \in \mathcal{A}} F_{\gamma,\Lambda}\big(m_\Lambda|\sigma_{\Lambda^c}^{(\gamma^{-1/2})}\big) + \beta c\epsilon(\gamma)|\Lambda|, \tag{4.2.2.12}$$

$$\log Z_{\gamma,\Lambda,\sigma_{\Lambda^c}}(\mathcal{A}) \geq -\beta \inf_{m_\Lambda \in \mathcal{A}} F_{\Lambda,\gamma}\big(m_\Lambda | \sigma_{\Lambda^c}^{(\gamma^{-1/2})}\big) - \beta c \epsilon(\gamma)|\Lambda|. \quad (4.2.2.13)$$

Proof Equation (4.2.2.13) is a direct consequence of (4.2.2.9). To prove (4.2.2.12), we bound

$$\log Z_{\gamma,\Lambda,\sigma_{\Lambda^c}}(\mathcal{A}) \leq |\mathcal{A}| \sup_{m_\Lambda \in \mathcal{A}} \log Z_{\gamma,\Lambda,\sigma_{\Lambda^c}},$$

and observe that the cardinality $|\mathcal{M}_\Lambda^{(\gamma^{-1/2})}|$ of $\mathcal{M}_\Lambda^{(\gamma^{-1/2})}$ is

$$\big|\mathcal{M}_\Lambda^{(\gamma^{-1/2})}\big| = (\gamma^{-d/2} + 1)^{\gamma^{d/2}|\Lambda|}. \quad (4.2.2.14)$$

Then (4.2.2.12) follows from (4.2.2.8) (with a different constant c). □

4.2.3 Variational problems and the L–P functional

As a first consequence of Theorem 4.2.2.1 we prove here the Lebowitz–Penrose theorem, Theorem 4.2.1.1. By the general theory we are free to choose any van Hove sequence, and we can therefore restrict ourselves to an increasing sequence of $\mathcal{D}^{(\gamma^{-1})}$-measurable cubes, and Λ below is by default a $\mathcal{D}^{(\gamma^{-1})}$-measurable cube. We are going to prove upper and lower bounds on the pressure which in the limit coincide.

Upper bound on the pressure

By (4.2.2.12) with $\mathcal{A} = \mathcal{M}^{(\gamma^{-1/2})}$

$$\log Z_{\gamma,\Lambda,\sigma_{\Lambda^c}} \leq -\beta \sup_{m_\Lambda \in L^\infty(\Lambda;[-1,1])} F_{\gamma,\Lambda}\big(m_\Lambda | \sigma_{\Lambda^c}^{(\gamma^{-1/2})}\big) + \beta c \epsilon(\gamma)|\Lambda|.$$

In the Lebowitz–Penrose limit where first $\Lambda \to \mathbb{Z}^d$ and then $\gamma \to 0$, the last term vanishes, and we thus need to find lower bounds on

$$\frac{1}{|\Lambda|} \inf_{m_\Lambda \in L^\infty(\Lambda;[-1,1])} F_{\gamma,\Lambda}(m_\Lambda | m_{\Lambda^c}),$$

where $m_{\Lambda^c} \in L^\infty(\Lambda^c, [-1,1])$ (we may actually restrict ourselves to $m_{\Lambda^c} = \sigma_{\Lambda^c}^{(\gamma^{-1/2})}$). To exploit the ferromagnetic nature of the interaction, we rewrite the functional as

$$\begin{aligned} &F_{\gamma,\Lambda}(m_\Lambda | m_{\Lambda^c}) \\ &\quad = F^*_{\gamma,\Lambda}(m_\Lambda | m_{\Lambda^c}) - \frac{1}{2} \int_\Lambda \int_{\Lambda^c} J_\gamma(r,r') m_{\Lambda^c}(r')^2 \, dr' \, dr, \end{aligned} \quad (4.2.3.1)$$

where

$$F^*_{\gamma,\Lambda}(m_\Lambda|m_{\Lambda^c}) = F^*_{\gamma,\Lambda}(m_\Lambda) + \frac{1}{2}\int_\Lambda\int_{\Lambda^c} J_\gamma(r,r')[m_\Lambda(r) - m_{\Lambda^c}(r')]^2 dr'\,dr,$$

and, recalling (4.1.2.2),

$$F^*_{\gamma,\Lambda}(m_\Lambda) = \int_\Lambda \phi_{\beta,h}\big(m_\Lambda(r)\big)\,dr + \frac{1}{4}\int_\Lambda\int_\Lambda J_\gamma(r,r')[m_\Lambda(r) - m_\Lambda(r')]^2\,dr'\,dr. \qquad (4.2.3.2)$$

Since the interaction term (i.e. the one with J_γ) is non-negative, we have

$$\inf_{m_\Lambda\in L^\infty(\Lambda,[-1,1])} F^*_{\gamma,\Lambda}(m_\Lambda|m_{\Lambda^c}) \ge \inf_{m_\Lambda\in L^\infty(\Lambda,[-1,1])} \int_\Lambda \phi_{\beta,h}\big(m_\Lambda(r)\big)\,dr.$$

Since $\frac{1}{2}\int_\Lambda\int_{\Lambda^c} J_\gamma(r,r')m_{\Lambda^c}(r')^2\,dr'\,dr \le c'\gamma^{-1}L^{d-1}$, L the side of the cube Λ, c' a suitable constant, and we conclude from (4.2.3.1), recalling the definition (4.1.4.2) of $p_{\beta,h}$,

$$\inf_{m_\Lambda\in L^\infty(\Lambda,[-1,1])} F_{\gamma,\Lambda}(m_\Lambda|m_{\Lambda^c}) \ge -p_{\beta,h}|\Lambda| - c'\gamma^{-1}L^{d-1}.$$

Thus

$$\frac{\log Z_{\Lambda,\gamma}(\sigma_{\Lambda^c})}{\beta|\Lambda|} \le p_{\beta,h} + \left(c'\frac{\gamma^{-1}}{L} + c\epsilon(\gamma)\right),$$

and by taking $L\to\infty$, $P_{\beta,h,\gamma} := \lim_{\Lambda\to\mathbb{Z}^d} \frac{\log Z_{\Lambda,\gamma}(\sigma_{\Lambda^c})}{\beta|\Lambda|} \le p_{\beta,h}$ so that

$$\limsup_{\gamma\to 0} P_{\beta,h,\gamma} \le p_{\beta,h}. \qquad (4.2.3.3)$$

Lower bound on the pressure

By (4.2.2.9) and with $M^{(\gamma^{-1/2})}$ as in (4.2.2.4),

$$\log Z_{\gamma,\Lambda,\sigma_{\Lambda^c}} \ge -\beta\, F_{\gamma,\Lambda}\big(\tilde m \mathbf{1}_\Lambda|\sigma^{(\gamma^{-1/2})}_{\Lambda^c}\big) - \beta c\epsilon(\gamma)|\Lambda|$$

for any $\tilde m\in M^{(\gamma^{-1/2})}$. With L the side of the cube Λ, by (4.2.2.11)

$$F_{\gamma,\Lambda}\big(\tilde m\mathbf{1}_\Lambda|\sigma^{(\gamma^{-1/2})}_{\Lambda^c}\big) \le \phi_{\beta,h}(\tilde m)|\Lambda| + c'\gamma^{-1}L^{d-1}.$$

By taking $L\to\infty$,

$$P_{\beta,h,\gamma} \ge -\phi_{\beta,h}(\tilde m) - c\epsilon(\gamma), \quad \text{for any } \tilde m\in M^{(\gamma^{-1/2})}. \qquad (4.2.3.4)$$

Call m^* the minimizer of $\phi_{\beta,h}(\cdot)$ so that $p_{\beta,h} = -\phi_{\beta,h}(m^*)$ and denote by $[m^*]_\gamma$ the value in $M^{(\gamma^{-1/2})}$ closest to m^*. Then by (4.2.3.4)

$$P_{\beta,h,\gamma} \geq p_{\beta,h} - |\phi_{\beta,h}([m^*]_\gamma) - \phi_{\beta,h}(m^*)| - c\epsilon(\gamma), \tag{4.2.3.5}$$

and, since $[m^*]_\gamma \to m^*$ as $\gamma \to 0$,

$$\liminf_{\gamma\to 0} P_{\beta,h,\gamma} \geq p_{\beta,h}. \tag{4.2.3.6}$$

Proof of Theorem 4.2.1.1

The first limit in (4.2.1.3) follows from (4.2.3.3) and (4.2.3.6). Proof of the second limit in (4.2.1.3) now follows. Since $F_{\beta,\gamma}(m) = \sup_h(hm - P_{\beta,h})$, for any h, $\liminf_{\gamma\to 0} F_{\beta,\gamma}(m) \geq (hm - p_{\beta,h})$ and taking the sup over h,

$$\liminf_{\gamma\to 0} F_{\beta,\gamma}(m) \geq CE\phi_{\beta,0}(m).$$

For the upper bound we use (4.2.3.4) with $\tilde{m} = 1$ and -1, respectively, to prove that there is a constant c^* so that

$$P_{\beta,h,\gamma} \geq \begin{cases} h - c^* & \text{for } h > 0, \\ -h - c^* & \text{for } h < 0. \end{cases}$$

Thus, given $|m| < 1$, there is h^* so that

$$F_{\beta,\gamma}(m) = \sup_h (hm - P_{\beta,h,\gamma}) = \sup_{|h|\leq h^*} (hm - P_{\beta,h,\gamma}).$$

Then by (4.2.3.5)

$$F_{\beta,\gamma}(m) \leq \sup_h (hm - p_{\beta,h}) + \sup_{|h|\leq h^*} |\phi_{\beta,h}([m^*]_\gamma) - \phi_{\beta,h}(m^*)| + c\epsilon(\gamma),$$

the last two terms vanishing in the limit $\gamma \to 0$. Theorem 4.2.1.1 is therefore proved, pending the validity of Theorem 4.2.2.1.

4.2.4 The L–P functional as a large deviation rate function

Let us fix a square $\Lambda \subset \mathbb{R}^d$ and two continuous functions $m_\Lambda \in C(\Lambda; [-1,1])$ and $m_{\Lambda^c} \in C(\Lambda^c; [-1,1])$. We regard Λ as a "mesoscopic region," m_Λ as a "mesoscopic profile in Λ" and m_{Λ^c} as "mesoscopic boundary conditions." The microscopic counterpart of m_Λ when the micro/meso ratio is γ may be defined as the set of all spin configurations $\sigma_{\gamma^{-1}\Lambda}$ such that $\sigma_{\gamma^{-1}\Lambda}^{(\gamma^{-1/2})}(r) = [m_\Lambda(\gamma r)]_\gamma =: m_\Lambda^{(\gamma)}(r)$.

Then choosing $\bar{\sigma}$ so that $\bar{\sigma}^{(\gamma^{-1/2})}_{\gamma^{-1}\Lambda^c}(r) = [m_{\Lambda^c}(\gamma r)]_\gamma$, we may define the corresponding γ-free energy of m_Λ as

$$-\frac{\log Z_{\gamma,\gamma^{-1}\Lambda,\bar{\sigma}_{\gamma^{-1}\Lambda^c}}(\{\sigma^{(\gamma^{-1/2})}_{\gamma^{-1}\Lambda} = m^{(\gamma)}_\Lambda\})}{\beta\gamma^{-d}}.$$

By Theorem 4.2.2.1,

$$\lim_{\gamma\to 0} -\frac{\log Z_{\gamma,\gamma^{-1}\Lambda,\bar{\sigma}_{\gamma^{-1}\Lambda^c}}(\{\sigma^{(\gamma^{-1/2})}_{\gamma^{-1}\Lambda} = m^{(\gamma)}_\Lambda\})}{\beta\gamma^{-d}} = F_\Lambda(m_\Lambda|m_{\Lambda^c}),$$

where $F_\Lambda(m_\Lambda|m_{\Lambda^c})$ is given by the expression (4.2.2.11) with J_γ replaced by J. $F_\Lambda(m_\Lambda|m_{\Lambda^c})$ is then interpreted as the free energy of the mesoscopic profile m_Λ (with boundary conditions m_{Λ^c}).

Calling $G^{\text{per}}_\gamma(m_\Lambda)$ the Gibbs probability of the set $\{\sigma^{(\gamma^{-1/2})}_{\gamma^{-1}\Lambda} = m^{(\gamma)}_\Lambda\}$ when Λ is a torus, we have

$$\frac{\log G^{\text{per}}_\gamma(m_\Lambda)}{\beta\gamma^{-d}} = \frac{\log Z^{\text{per}}_{\gamma,\gamma^{-1}\Lambda}(\{\sigma^{(\gamma^{-1/2})}_{\gamma^{-1}\Lambda} = m^{(\gamma)}_\Lambda\})}{\beta\gamma^{-d}} - \frac{\log Z^{\text{per}}_{\gamma,\gamma^{-1}\Lambda}}{\beta\gamma^{-d}}.$$

The r.h.s. converges as $\gamma\to 0$ to $-F^{\text{per}}_\Lambda(m_\Lambda)+\phi_{\beta,h}(m^*)|\Lambda|$, namely to $-F^{\text{exc}}(m_\Lambda)$, where $F^{\text{exc}}(m_\Lambda)$ is equal to

$$\int_\Lambda \{\phi_{\beta,h}(m_\Lambda(r)) - \phi_{\beta,h}(m^*)\}\,dr + \frac{1}{4}\int_\Lambda\int_\Lambda J^{\text{per}}(r,r')[m_\Lambda(r) - m(r')]^2 dr'\,dr$$

(J^{per} is the periodic extension of J, and m^* the minimizer of $\phi_{\beta,h}(\cdot)$).

$F^{\text{exc}}(m_\Lambda)$ is the excess free energy of m_Λ as it is normalized so that its minimum is 0. Since $G^{\text{per}}_\gamma(m_\Lambda) \approx e^{-\beta\gamma^{-d}F^{\text{exc}}(m_\Lambda)}$, the excess free energy of m_Λ is the rate function for large deviations of the Gibbs measure (as the scaling parameter $\gamma\to 0$).

4.2.5 *Proof of Theorem 4.2.2.1*

We start by coarse graining the energy. Define

$$U_{\gamma,\Lambda}(m_\Lambda|m_{\Lambda^c}) = F_{\gamma,\Lambda}(m_\Lambda|m_{\Lambda^c}) + \frac{1}{\beta}\int_\Lambda I(m_\Lambda(r)), \tag{4.2.5.1}$$

with $F_{\gamma,\Lambda}(m_\Lambda|m_{\Lambda^c})$ as in (4.2.3.1), namely by subtracting from the free energy the contribution of the entropy. Let

$$J_\gamma^{(\ell)}(r_1,r_2) = \fint C_{r_1}^{(\ell)} \times C_{r_2}^{(\ell)} J_\gamma(r_1',r_2')\,dr_1'\,dr_2',$$
$$J_\gamma(r,r') = \gamma^d J(\gamma r, \gamma r'). \tag{4.2.5.2}$$

Notice that $J_\gamma^{(\ell)}(r,r')$ is $\mathcal{D}^{(\ell)}$-measurable in both variables, r and r'.

Lemma 4.2.5.1 *There is c so that for all γ, all bounded $\mathcal{D}^{(\gamma^{-1/2})}$-measurable regions Λ, and all $\sigma_\Lambda, \sigma_{\Lambda^c}, m_\Lambda, m_{\Lambda^c}$*

$$\big|H_{\gamma,\Lambda}(\sigma_\Lambda|\sigma_{\Lambda^c}) - U_{\gamma,\Lambda}\big(\sigma_\Lambda^{(\gamma^{-1/2})}|\sigma_{\Lambda^c}^{(\gamma^{-1/2})}\big)\big|$$
$$\le c\gamma^{1/2}|\Lambda|, \tag{4.2.5.3}$$
$$\big|U_{\gamma,\Lambda}(m_\Lambda|m_{\Lambda^c}) - U_{\Lambda,\gamma}\big(m_\Lambda^{(\gamma^{-1/2})}|m_{\Lambda^c}^{(\gamma^{-1/2})}\big)\big|$$
$$\le c\gamma^{1/2}|\Lambda|. \tag{4.2.5.4}$$

Proof By (4.2.5.2) and (4.2.1.1), for any x, y in $\mathbb{Z}^d$ and any γ there is c' so that

$$\big|J_\gamma(x,y) - J_\gamma^{(\gamma^{-1/2})}(x,y)\big| \le 2c'\gamma^{1/2+d}\mathbf{1}_{|x-y|\le 2\gamma^{-1}}. \tag{4.2.5.5}$$

The factor 2 in the characteristic function is not optimal; c' above can be estimated as $\sqrt{d}\|DJ\|_\infty$, $\|DJ\|_\infty$ being the sup norm of the first derivatives of $J(0,r)$. Let C and C' be cubes of $\mathcal{D}^{(\gamma^{-1/2})}$ and r and r' points in C and C'. Then

$$\Bigg|\sum_{x\in C\cap\mathbb{Z}^d}\sum_{y\in C'\cap\mathbb{Z}^d}\mathbf{1}_{x\ne y}J_\gamma(x,y)\sigma(x)\sigma(y)$$
$$-\sum_{x\in C\cap\mathbb{Z}^d}\sum_{y\in C'\cap\mathbb{Z}^d}\mathbf{1}_{x\ne y}J_\gamma^{(\gamma^{-1/2})}(x,y)\sigma(x)\sigma(y)\Bigg|$$
$$\le c'|C|^2\gamma^{1/2+d}\mathbf{1}_{|r-r'|\le 3\gamma^{-1}}.$$

Let $C\ne C'$, $r\in C$, $r'\in C'$; then $\sum_{x\in C\cap\mathbb{Z}^d}\sum_{y\in C'\cap\mathbb{Z}^d}J_\gamma^{(\gamma^{-1/2})}(x,y)\sigma(x)\sigma(y)$ equals

$$|C|^2 J_\gamma^{(\gamma^{-1/2})}(r,r')\sigma^{(\gamma^{-1/2})}(r)\sigma^{(\gamma^{-1/2})}(r')$$
$$=\int_C dr\int_{C'} dr'\,J_\gamma(r,r')\sigma^{(\gamma^{-1/2})}(r)\sigma^{(\gamma^{-1/2})}(r'),$$

while if $C = C'$ there is a correction due to the pairs x, y with $x = y$:

$$\Big| \sum_{x\in C\cap\mathbb{Z}^d} \sum_{y\in C'\cap\mathbb{Z}^d} \mathbf{1}_{x\neq y} J_\gamma^{(\gamma^{-1/2})}(x,y)\sigma(x)\sigma(y) - \int_C dr \int_{C'} dr' J_\gamma(r,r')\sigma^{(\gamma^{-1/2})}(r)\sigma^{(\gamma^{-1/2})}(r') \Big| \le \gamma^d \|J\|_\infty |C|.$$

After collecting all these bounds, we get

$$\big|H_{\gamma,\Lambda}(\sigma_\Lambda|\sigma_{\Lambda^c}) - U_{\gamma,\Lambda}\big(\sigma_\Lambda^{(\gamma^{-1/2})}|\sigma_{\Lambda^c}^{(\gamma^{-1/2})}\big)\big| \le c|\Lambda|(\gamma^{1/2} + \gamma^d\|J\|_\infty),$$

which proves (4.2.5.3). The proof of (4.2.5.4) is similar, and we omit it. □

Recalling (4.1.1.5), the l.h.s. of (4.2.2.8) is then bounded by

$$\le \beta c|\Lambda|(\gamma^{1/2} + \gamma^d\|J\|_\infty) - \beta\, U_{\gamma,\Lambda}\big(m_\Lambda|\sigma_{\Lambda^c}^{(\gamma^{-1/2})}\big) + \gamma^{-d/2} \sum_{i\in\Lambda\cap\gamma^{-1/2}\mathbb{Z}^d} I_{\gamma^{-d/2}}(m_\Lambda(i)).$$

By (A.2.9), $|I_{\gamma^{-d/2}}(m_\Lambda(r)) - I(m_\Lambda(r))| \le c'\gamma^{d/2}\log\gamma^{-d/2}$, so that

$$\log Z_{\gamma,\Lambda}(\{\sigma_\Lambda^{(\gamma^{-1/2})} = m_\Lambda\}; \sigma_{\Lambda^c}) \le -\beta F_{\gamma,\Lambda}\big(m_\Lambda|\sigma_{\Lambda^c}^{(\gamma^{-1/2})}\big) + \beta c\epsilon(\gamma)|\Lambda|, \quad (4.2.5.6)$$

and (4.2.2.12) is proved. Equation (4.2.2.12) is proved similarly, so that Theorem 4.2.2.1 is proved and with it Theorem 4.2.1.1 is also proved; see the end of Sect. 4.2.3.

4.2.6 Summary and final comments

Even though Theorem 4.2.1.1 may look as the nth derivation in this section of the mean field thermodynamics, it should instead be regarded, at least conceptually, as a giant step! Theorem 4.2.1.1 in fact states that there exist systems with purely finite range interactions whose "correct thermodynamics" (as derived in Chap. 2 by only using thermodynamic limit procedures) is as close as desired to mean field thermodynamics. In the mean field models such a statement is missing, because the finite volume approximants are systems whose interaction range is equal to the size of the region, an assumption which is unphysical at least for Lennard-Jones intermolecular forces and macroscopic domains. Kac potentials show that the mean field assumption is not necessary. When we talk of mean field properties, we should always think of an idealization of properties in a system with a Kac potential with γ maybe very small but positive, and therefore of a system with finite range interactions.

So far the good news, but unfortunately there is bad news too. If we regard mean field in the way stressed above, then we are lost, or better, we may have lost the phase transitions! The straight segment in the diagram of the mean field free energy may not be found in any free energy of the Kac approximants, as strictly convex functions may approximate arbitrarily well a straight segment. The whole business of the mean field approach was to produce phase transitions, and, in order to avoid the related unphysical features, we have gone to Kac potentials and with that maybe missed the final goal. I guess this was one of the main reasons why, after great excitement, Kac potentials have essentially been abandoned.

Observe that the problem is real; it is not just a technicality. In $d = 1$, by Theorem 3.1.1.1, any Kac approximant has no phase transitions, since the interaction has finite range. It does happen, as it should in $d = 1$, that the straight segment in the graph of the free energy is there only in the limit $\gamma \to 0$ and never at $\gamma > 0$. (We had already remarked that mean field has too many phase transitions, so it is good that some of them can be ruled out, but not all of them of course!) So we really have to look at the system with fixed positive γ, without taking the limit $\gamma \to 0$. It thus seems that we are back to square one, with the problem of proving phase transitions for finite range interactions. The big advantage however is that we have a limit theory with phase transitions and we may look for a perturbative approach in the same spirit as the Peierls argument of Sect. 3.1.2. Instead of perturbing ground states we would then perturb mean field minimizers, small temperatures being replaced by small values of γ. Such ideas have been extremely successful, and as we shall see in Part III of the book they apply to Ising systems with Kac potentials and also to some continuum particle models, leading to a proof of the existence of liquid–vapor phase transitions in the continuum, a long standing, open problem in statistical mechanics.

The whole approach requires very detailed knowledge of the limit theory, which in turns means a detailed study of the non-local functional which appears in the proof of Theorem 4.2.1.1. This is required now much more than the mere estimate of its minimum, as in the previous subsection; see (4.2.3.2). The analysis brings in all the geometric structures hidden when deriving the thermodynamic potentials, and this will be the subject of Part II of this book. We shall study interfaces and surface tension associated to the above free energy functional, and the results are of the same type as in the Ginzburg–Landau functional approach, but the non-local structure of our functionals will give rise to several new interesting facts, as we shall see.

Chapter 5
Stochastic Dynamics

This chapter is devoted to dynamics. In the finite volume Ising model where the phase space is finite, non-trivial dynamics is necessarily stochastic and is usually described by Markov processes like Glauber, spin flip Markov dynamics. Historically Glauber dynamics was devised as a tool to compute partition functions, with the idea that by running dynamics we select the most relevant configurations in the full phase space. In spirit this is similar to gradient flow dynamics when it is viewed as a tool to search the minima of a function.

In Sect. 5.1 we shall define the spin flip dynamics as a Markov semigroup for general Ising interactions. We shall then specialize to the mean field model and Kac potentials and determine their scaling limit behavior, proving for the latter convergence to a deterministic, non-local evolution equation which will be extensively studied in Part II of this book. In Sect. 5.2 we shall define the evolution (previously defined in terms of Markov semigroups) as a continuous time Markov process and finally, in Sect. 5.3, we shall study in the simpler context of the mean field model phenomena in which stochastic effects persist after taking the thermodynamic limit.

5.1 The Glauber semigroup

In this section we introduce and study spin flip semigroups, eventually concentrating on the Glauber semigroup for the mean field approach and Kac interactions for which we shall derive deterministic evolutions in the macroscopic limit.

5.1.1 Spin flip dynamics

In this subsection we shall introduce a spin flip Markov semigroup for Ising models in bounded domains.

Semigroups and generators

Let L be a linear operator on $L^\infty(\mathcal{X}_\Lambda)$ and $\{e^{Lt}, t \ge 0\}$ the semigroup generated by L. $L^\infty(\mathcal{X}_\Lambda)$ may be identified to $\mathbb{R}^N, N = 2^{|\Lambda|}$, and linear operators A on $L^\infty(\mathcal{X}_\Lambda)$ by $N \times N$ matrices by setting $Af(\sigma_\Lambda) = \sum_{\sigma'_\Lambda} A(\sigma_\Lambda, \sigma'_\Lambda) f(\sigma'_\Lambda)$, where $A(\sigma_\Lambda, \sigma'_\Lambda) = (A\mathbf{1}_{\sigma'_\Lambda})(\sigma_\Lambda)$ and $\mathbf{1}_{\sigma'_\Lambda}(\sigma_\Lambda) = 1$ if $\sigma_\Lambda = \sigma'_\Lambda$ and $= 0$ otherwise. e^{Lt} is then the exponential of the matrix Lt (defined as a power series). L is a spin flip generator if for any $f \in L^\infty(\mathcal{X}_\Lambda)$

$$Lf(\sigma_\Lambda) = \sum_{x\in\Lambda} c(x,\sigma_\Lambda)\big(f(\sigma_\Lambda^x) - f(\sigma_\Lambda)\big), \quad c(x,\sigma_\Lambda) > 0, \tag{5.1.1.1}$$

where $\sigma_\Lambda^x(y) = \sigma_\Lambda(y)$ for $y \neq x$ and $\sigma_\Lambda^x(x) = -\sigma_\Lambda(x)$; $c(x,\sigma_\Lambda)$ is called the spin flip intensity at x when the state is σ_Λ. As we shall see $c(x,\sigma_\Lambda)dt$ is "the probability that the spin at x flips in the time interval $[t, t+dt]$, knowing that at time t the configuration is σ_Λ."

While e^{Lt} as an operator on $L^2(\mathcal{X}_\Lambda)$ is positive being a square, $e^{Lt} = e^{Lt/2}e^{Lt/2}$, its matrix elements $e^{Lt}(\sigma_\Lambda, \sigma'_\Lambda)$ are not in general positive. When they are all non-negative and normalized to a probability (i.e. for each σ_Λ and t, the sum of $e^{Lt}(\sigma_\Lambda, \sigma'_\Lambda)$ over σ'_Λ is equal to 1) then the semigroup is called Markov. This is the case for the generator in (5.1.1.1):

Proposition 5.1.1.1 *If L has the form* (5.1.1.1), *then, for any $t \geq 0$ and $\sigma_\Lambda^* \in \mathcal{X}_\Lambda$,*

$$P_t(\sigma_\Lambda|\sigma_\Lambda^*) := e^{Lt}(\sigma_\Lambda^*, \sigma_\Lambda) \geq 0, \quad \sum_{\sigma_\Lambda} P_t(\sigma_\Lambda|\sigma_\Lambda^*) = 1. \tag{5.1.1.1}$$

Thus, $\sigma_\Lambda \to P_t(\sigma_\Lambda|\sigma_\Lambda^*)$ is for any $t \geq 0$ and any σ_Λ^* a probability on $\mathcal{X}_\Lambda$.

Proof Since $e^{Lt}(\sigma_\Lambda^*, \sigma_\Lambda) = (e^{Lt}\mathbf{1}_{\sigma_\Lambda})(\sigma_\Lambda^*)$ and $\frac{de^{Lt}}{dt} = e^{Lt}L = Le^{Lt}$,

$$\frac{d}{dt}P_t(\sigma_\Lambda|\sigma_\Lambda^*) = [e^{Lt}L\mathbf{1}_{\sigma_\Lambda}](\sigma_\Lambda^*). \tag{5.1.1.2}$$

By summing over σ_Λ we get

$$\frac{d}{dt}\sum_{\sigma_\Lambda} P_t(\sigma_\Lambda|\sigma_\Lambda^*) = \Big[e^{Lt}L\sum_{\sigma_\Lambda}\mathbf{1}_{\sigma_\Lambda}\Big](\sigma_\Lambda^*) = 0,$$

because $\sum_{\sigma_\Lambda}\mathbf{1}_{\sigma_\Lambda} \equiv 1$ and $L1 = 0$. As a consequence

$$\sum_{\sigma_\Lambda} P_t(\sigma_\Lambda|\sigma_\Lambda^*) = \sum_{\sigma_\Lambda} P_0(\sigma_\Lambda|\sigma_\Lambda^*) = 1,$$

because $P_0(\sigma_\Lambda|\sigma_\Lambda^*) = (e^0\mathbf{1}_{\sigma_\Lambda})(\sigma_\Lambda^*) = \mathbf{1}_{\sigma_\Lambda}(\sigma_\Lambda^*)$. Call

$$c(\sigma_\Lambda) = \sum_{x\in\Lambda} c(x,\sigma_\Lambda), \qquad \pi_{\sigma_\Lambda}(x) = \frac{c(x,\sigma_\Lambda)}{c(\sigma_\Lambda)} \tag{5.1.1.3}$$

($\pi_{\sigma_\Lambda}(x)$ is for future reference). Fix σ_Λ^*, call $g = L\mathbf{1}_{\sigma_\Lambda}$, then by (5.1.1.2) $\frac{d}{dt}P_t(\sigma_\Lambda|\sigma_\Lambda^*) = \sum_{\eta_\Lambda} P_t(\eta_\Lambda|\sigma_\Lambda^*)g(\eta_\Lambda)$, i.e. the "forward Kolmogorov equation"

$$\frac{d}{dt}P_t(\sigma_\Lambda|\sigma_\Lambda^*) = -c(\sigma_\Lambda)P_t(\sigma_\Lambda|\sigma_\Lambda^*) + \sum_{x\in\Lambda} c(x,\sigma_\Lambda^x)P_t(\sigma_\Lambda^x|\sigma_\Lambda^*) \tag{5.1.1.4}$$

("forward" because we have fixed the starting configuration σ_Λ^*, if we instead fix the last configuration σ_Λ and regard $P_t(\sigma_\Lambda|\sigma_\Lambda^*)$ as a function of the "old one," σ_Λ^*, $g(\sigma_\Lambda^*) = P_t(\sigma_\Lambda|\sigma_\Lambda^*)$, then we get the "backward Kolmogorov equation" $\frac{dg}{dt} = Lg$). Rewriting (5.1.1.4) in integral form, we get

$$P_t(\sigma_\Lambda|\sigma_\Lambda^*) = e^{-c(\sigma_\Lambda)t} P_0(\sigma_\Lambda|\sigma_\Lambda^*) + \int_0^t e^{-c(\sigma_\Lambda)(t-s)} \sum_{x\in\Lambda} c(x, \sigma_\Lambda^x) P_s(\sigma_\Lambda^x|\sigma_\Lambda^*).$$

As an equation in the unknowns $\{P_s(\sigma_\Lambda|\sigma_\Lambda^*), \sigma_\Lambda \in \mathcal{X}_\Lambda, s \in (0,t]\}$ it can be solved by iteration giving rise to a convergent series. The inequality in (5.1.1.1) follows because all the terms of the series are non-negative. □

The spin flip dynamics is a particular case of a more general class of jump processes where (5.1.1.1) is replaced by

$$Lf(\sigma_\Lambda) = \sum_{\sigma'_\Lambda} c(\sigma_\Lambda, \sigma'_\Lambda)\big(f(\sigma'_\Lambda) - f(\sigma_\Lambda)\big),$$

with $c(\sigma_\Lambda, \sigma'_\Lambda) \geq 0$ being the intensity to jump from σ_Λ to σ'_Λ. Of physical importance are the so called exchange dynamics, where

$$Lf(\sigma_\Lambda) = \sum_{x,y\in\Lambda : x\neq y} c(\sigma_\Lambda, \sigma_\Lambda^{x,y})\big(f(\sigma_\Lambda^{x,y}) - f(\sigma_\Lambda)\big), \tag{5.1.1.5}$$

and $\sigma_\Lambda^{x,y}(z) = \sigma_\Lambda(z)$ if $z \neq x, y$; $\sigma_\Lambda^{x,y}(x) = \sigma_\Lambda(y)$ and $\sigma_\Lambda^{x,y}(y) = \sigma_\Lambda(x)$, i.e. the spins at $x \neq y$ exchange at rate $c(\sigma_\Lambda, x, y)$. Usually one restricts to nearest neighbor exchanges ($c(\sigma_\Lambda, x, y) = 0$ if $|x - y| > 1$). As we shall see, for a special choice of the intensities (related to the hamiltonian of the system) the exchange dynamics is called Kawasaki dynamics.

5.1.2 Invariant measures

A measure ν on $\mathcal{X}_\Lambda$ evolves under the semigroup e^{Lt} by duality, namely for any $f \in \mathcal{X}_\Lambda$,

$$\nu_t(f) = \nu(e^{Lt} f), \tag{5.1.2.1}$$

where $\nu(f)$, the ν-expectation of f, is $\sum_{\sigma_\Lambda\in\mathcal{X}_\Lambda} \nu(\sigma_\Lambda) f(\sigma_\Lambda)$. μ is invariant under the semigroup e^{Lt} if for any $t > 0$

$$\mu(f) = \mu(e^{Lt} f), \tag{5.1.2.2}$$

or, equivalently, if for any σ_Λ and recalling that $P_t(\sigma_\Lambda|\sigma_\Lambda^*)$ is as in (5.1.1.1),

$$\sum_{\sigma_\Lambda^*} \mu(\sigma_\Lambda^*) P_t(\sigma_\Lambda|\sigma_\Lambda^*) = \mu(\sigma_\Lambda). \tag{5.1.2.3}$$

Theorem 5.1.2.1 *μ is invariant if and only if*

$$\mu(Lf) = 0, \quad \text{for all } f \in L^\infty(\mathcal{X}_\Lambda), \tag{5.1.2.4}$$

or equivalently

$$\sum_{x \in \Lambda} \mu(\sigma_\Lambda^x)\, c(x, \sigma_\Lambda^x) = c(\sigma_\Lambda)\mu(\sigma_\Lambda), \quad \text{for all } \sigma_\Lambda. \tag{5.1.2.5}$$

Proof Suppose μ invariant. By taking the time derivative of (5.1.2.3) and then setting $t = 0$ we get (5.1.2.4). Vice versa, suppose (5.1.2.4) holds. Then

$$\frac{d}{dt} \sum_{\sigma_\Lambda^*} \mu(\sigma_\Lambda^*)(e^{Lt} f)(\sigma_\Lambda^*) = \sum_{\sigma_\Lambda^*} \mu(\sigma_\Lambda^*)(Le^{Lt} f)(\sigma_\Lambda^*).$$

Call $g := e^{Lt} f$ then by (5.1.2.4)

$$\frac{d}{dt} \sum_{\sigma_\Lambda^*} \mu(\sigma_\Lambda^*)(e^{Lt} f)(\sigma_\Lambda^*) = \sum_{\sigma_\Lambda^*} \mu(\sigma_\Lambda^*)(Lg)(\sigma_\Lambda^*) = 0,$$

so that $\sum_{\sigma_\Lambda^*} \mu(\sigma_\Lambda^*) \sum_{\sigma_\Lambda} P_t(\sigma_\Lambda|\sigma_\Lambda^*) f(\sigma_\Lambda) = \sum_{\sigma_\Lambda^*} \mu(\sigma_\Lambda^*) \sum_{\sigma_\Lambda} P_0(\sigma_\Lambda|\sigma_\Lambda^*) f(\sigma_\Lambda)$, hence (5.1.2.3). Writing (5.1.2.4) with $f = \mathbf{1}_{\sigma_\Lambda}$:

$$\mu(\eta_\Lambda) \sum_{x \in \Lambda} c(x, \eta_\Lambda)[\mathbf{1}_{\sigma_\Lambda}(\eta_\Lambda^x) - \mathbf{1}_{\sigma_\Lambda}(\eta_\Lambda)] = 0, \tag{5.1.2.6}$$

which is (5.1.2.5). Vice versa if (5.1.2.5) holds then (5.1.2.6) holds as well and (5.1.2.4) then follows by multiplying (5.1.2.6) by $f(\sigma_\Lambda)$ and then summing over σ_Λ. □

Remarks By the assumption that all spin flip intensities $c(x, \sigma_\Lambda) > 0$, it follows that for any $t > 0$

$$\inf_{\sigma_\Lambda^*, \sigma_\Lambda} P_t(\sigma_\Lambda|\sigma_\Lambda^*) > 0. \tag{5.1.2.7}$$

This is known as the Döblin condition (see for instance Foguel [122]): it ensures uniqueness of the invariant measure as well as exponential convergence as $t \to \infty$. We shall prove in Sect. 8.4 extensions to Markov chains in unbounded spaces, and, as we shall see, there is a strict relation with the Dobrushin uniqueness theorem and the Perron–Frobenius theorem. See Revuz [188] for a general reference on Markov chains.

5.1.3 Glauber spin flip rates

The Glauber dynamics is defined by a special choice of the intensities of the generator of the spin flip semigroup. Let Λ be a bounded region, σ_{Λ^c} a fixed boundary condition, $H_\Lambda(\sigma_\Lambda|\sigma_{\Lambda^c})$ the hamiltonian and β the inverse temperature. The Glauber dynamics in this setup is the spin flip process with generator (5.1.1.1) if $c(x,\sigma_\Lambda)$ has the form

$$c(x,\sigma_\Lambda) = c_0(x,\sigma_{\Lambda\setminus x})\, e^{-(\beta/2)[H_\Lambda(\sigma_\Lambda^x|\sigma_{\Lambda^c}) - H_\Lambda(\sigma_\Lambda|\sigma_{\Lambda^c})]}. \tag{5.1.3.1}$$

$c_0(x,\sigma_{\Lambda\setminus x})$ may be any strictly positive function of $\sigma_{\Lambda\setminus x}$, it is important that it does not depend on the spin at x; $c_0(x,\sigma_{\Lambda\setminus x})$ is a "mobility coefficient."

Theorem 5.1.3.1 *The Gibbs measure* $\mu(\sigma_\Lambda) \equiv G_\Lambda(\sigma_\Lambda|\sigma_{\Lambda^c})$ *at inverse temperature* β *with hamiltonian* $H_\Lambda(\sigma_\Lambda|\sigma_{\Lambda^c})$ *is invariant under the Glauber dynamics (with the same* β *and* $H_\Lambda(\sigma_\Lambda|\sigma_{\Lambda^c})$*), i.e. it satisfies* (5.1.2.4)–(5.1.2.5). *Moreover,*

$$\mu(\sigma_\Lambda)\,c(x,\sigma_\Lambda) = \mu(\sigma_\Lambda^x)\,c(x,\sigma_\Lambda^x), \quad \text{for all } \sigma_\Lambda \text{ and all } x\in\Lambda, \tag{5.1.3.2}$$

$$\mu(gLf) = \mu(fLg), \quad \text{for all } f \text{ and } g \text{ in } L^\infty(\mathcal{X}_\Lambda). \tag{5.1.3.3}$$

Proof Invariance follows from (5.1.3.3) (taking $g=1$), which in turns follows from (5.1.3.2), which is a direct consequence of (5.1.3.1). □

Remarks The condition (5.1.3.2) (which is stronger than mere invariance) is known in the physical literature as "detailed balance." It states in fact that the measure is invariant even for each elementary transitions between σ_Λ and σ_Λ^x and not by compensations after summing over all possibilities as in (5.1.2.5). Such a strong property is then responsible for (5.1.3.3), which states that L as an operator on $L^2(\mathcal{X}_\Lambda,\mu)$ is self-adjoint. If L is self-adjoint the Markov semigroup e^{Lt} with invariant measure μ is "reversible," and the reason for the name is that the trajectories of the corresponding Markov process (see Sect. 5.2) have the same law both if they are read forward or backwards.

The Kawasaki dynamics is defined by (5.1.1.5) by setting

$$c(\sigma_\Lambda,x,y) = c_0(x,y,\sigma_{\Lambda\setminus\{x,y\}})\, e^{-(\beta/2)[H_\Lambda(\sigma_\Lambda^{x,y}|\sigma_{\Lambda^c}) - H_\Lambda(\sigma_\Lambda|\sigma_{\Lambda^c})]}, \tag{5.1.3.4}$$

for $|x-y|=1$ and $=0$ otherwise, so that also for the Kawasaki dynamics the Gibbs measure is invariant and reversible. Of particular importance in the literature of stochastic interacting particle systems is the Kawasaki dynamics at $\beta=0$ (infinite temperature, ideal gas) the generator then coming with the choice $c_0 = 1/(2d)$,

$$Lf(\sigma_\Lambda) = \frac{1}{2d}\sum_{x,y\in\Lambda:|x-y|=1} \big(f(\sigma_\Lambda^{x,y}) - f(\sigma_\Lambda)\big), \tag{5.1.3.5}$$

which is called "the stirring process," as it is an independent exchange of the content of neighboring sites. By introducing the particle variables $\eta_\Lambda(x) = (1+\sigma_\Lambda(x))/2$ (and interpreting $\eta_\Lambda(x)=1$ as a particle being at x, while if $\eta_\Lambda(x)=0$, then x is empty), (5.1.3.5) becomes

$$Lf(\eta_\Lambda) = \sum_{x\in\Lambda}\sum_{y\in\Lambda} p(x,y)\eta_\Lambda(x)(1-\eta_\Lambda(y))\big(f(\eta_\Lambda - 1_x + 1_y) - f(\eta_\Lambda)\big). \tag{5.1.3.6}$$

$\eta_\Lambda - 1_x + 1_y$ means subtracting from η_Λ the particle at x and adding one at y; $p(x,y)$ in general is a symmetric transition probability, to identify with (5.1.3.5) we should take $p(x,y) = 1/(2d)$ if $|y-x| = 1$ and $= 0$ otherwise. In words (5.1.3.6) means that a particle at x decides to jump at rate 1, then chooses y with probability $p(x,y)$ and moves there if y was empty. The process is then called the symmetric simple exclusion. Notice that even at infinite temperature the process feels an interaction, the exclusion interaction (which is, however, not too difficult to handle).

Choice of the mobility coefficient

It is convenient to write $c(x,\sigma_\Lambda)$ in (5.1.3.1) as

$$c(x,\sigma_\Lambda) = c_{\sigma_\Lambda(x)}(x,\sigma_{\Lambda\setminus x}), \tag{5.1.3.7}$$

where

$$c_\pm(x,\sigma_{\Lambda\setminus x}) = c_0(h_x)\,e^{\mp\beta h_x}, \quad h_x(\sigma_{\Lambda\setminus x}) = h + \sum_{y\neq x} J(x,y)\sigma_\Lambda(y). \tag{5.1.3.8}$$

h_x is called the "molecular magnetic field" at $x\in\Lambda$; it is the sum of the external magnetic field h and the effective field at x produced by the other spins. For the sake of definiteness we take

$$c_0(h_x) = \frac{1}{e^{-\beta h_x} + e^{\beta h_x}}. \tag{5.1.3.9}$$

Another choice which often appears in the literature is $c_0 = 1/2$. We shall proceed by examining two particular cases, the mean field Glauber dynamics and the Glauber dynamics with Kac potentials.

5.1.4 Mean field dynamics

The mean field Glauber dynamics is defined by the spin flip generator L with rates (5.1.3.8) with $J(x,y) = \mathbf{1}_{x\neq y}/|\Lambda|$ and $c_0(h_x)$ as in (5.1.3.9). We are however specifically interested only in the total magnetization density

$$m(\sigma_\Lambda) = \frac{1}{|\Lambda|}\sum_{x\in\Lambda}\sigma_\Lambda(x), \tag{5.1.4.1}$$

which means that we want only to look at the action of the semigroup on functions $f \in L^\infty(\mathcal{X}_\Lambda)$ which are constant on the levels $\{\sigma_\Lambda \in \mathcal{X}_\Lambda : m(\sigma_\Lambda) = m\}$, $m \in M_\Lambda$, where

$$M_\Lambda := \left\{ -1, -1 + \frac{2}{|\Lambda|}, \dots, 1 - \frac{2}{|\Lambda|}, 1 \right\}. \tag{5.1.4.2}$$

Such functions have then the form $f(\sigma_\Lambda) = g(m(\sigma_\Lambda))$ with g a function on M_Λ, $g \in L^\infty(M_\Lambda)$.

Theorem 5.1.4.1 *There is a Markov generator $\mathcal{L}$ on $L^\infty(M_\Lambda)$ so that for any $g \in L^\infty(M_\Lambda)$ and any $\sigma_\Lambda \in \mathcal{X}_\Lambda$*

$$e^{Lt} f(\sigma_\Lambda) = e^{\mathcal{L}t} g\big(m(\sigma_\Lambda)\big), \quad \textit{where } f(\sigma_\Lambda) := g\big(m(\sigma_\Lambda)\big), \tag{5.1.4.3}$$

and L is the mean field generator.

Proof We shall define a generator $\mathcal{L}$ and then check that (5.1.4.3) is satisfied. The generator $\mathcal{L}$ acts on $g \in L^\infty(M_\Lambda)$ by

$$\begin{aligned} \mathcal{L}g(m) = |\Lambda| \Bigg(& c_{+,\Lambda}(m) \left\{ g\left(m - \frac{2}{|\Lambda|}\right) - g(m) \right\} \\ & + c_{-,\Lambda}(m) \left\{ g\left(m + \frac{2}{|\Lambda|}\right) - g(m) \right\} \Bigg), \end{aligned} \tag{5.1.4.4}$$

where

$$c_{\pm,\Lambda}(m) = \frac{1 \pm m}{2} \, c_{0,\Lambda}(m,h) \, e^{\mp\beta(h+m\mp 1/|\Lambda|)}. \tag{5.1.4.5}$$

$c_{0,\Lambda}(m,h)$ is the mobility coefficient, which, corresponding to the choice (5.1.3.9), is given by

$$c_{0,\Lambda}(m,h) = \frac{1}{e^{-\beta(h+m\mp 1/|\Lambda|)} + e^{\beta(h+(m\mp 1/|\Lambda|)}}. \tag{5.1.4.6}$$

When $f(\sigma_\Lambda) = g(m(\sigma_\Lambda))$ then $Lf(\sigma_\Lambda) = \mathcal{L}g(m(\sigma_\Lambda))$. By induction on n, $L^n f(\sigma_\Lambda) = \mathcal{L}^n g(m(\sigma_\Lambda))$, and expanding $e^{Lt} f$ in a power series, we finally get (5.1.4.3). □

We shall sometimes replace $\mathcal{L}$ by the generator with intensities

$$\begin{aligned} c_\pm(m) &= \frac{1 \pm m}{2} \, c_0(m,h) \, e^{\mp\beta(h+m)}, \\ c_0(m,h) &= \big(e^{-\beta(h+m)} + e^{\beta(h+m}\big)^{-1}. \end{aligned} \tag{5.1.4.7}$$

5.1.5 Scaling limit in mean field dynamics

As we shall see the limit behavior of the mean field Glauber evolution, with the choice (5.1.3.9) for the mobility coefficient, in the limit $|\Lambda| \to \infty$ is described by a deterministic evolution equation:

$$\frac{du}{dt} = -u + \tanh\{\beta(u+h)\} =: V(u) = V_-(u) - V_+(u), \tag{5.1.5.1}$$

where $V_\mp(u) = (1 \mp u)\,\frac{e^{\mp\beta(h+u)}}{e^{\beta(h+u)}+e^{-\beta(h+u)}}$.

Let $m' \in M_\Lambda$ and

$$P_{t,\Lambda,m'}(m) := e^{\mathcal{L}t}(m', m). \tag{5.1.5.2}$$

$P_{t,\Lambda,m}(\cdot)$ is a probability on M_Λ, interpreted as the mean field probability that the magnetization at time t is m if at time 0 it is m'. Next theorem proves that as $|\Lambda| \to \infty$ the probability concentrates around $u(t)$.

Theorem 5.1.5.1 *Fix $u \in [-1, 1]$ and let $m' \in M_\Lambda$ be such that $m' \to u$ as $|\Lambda| \to \infty$. Call $u(t)$ the solution starting from u of* (5.1.5.1), *then for any $\zeta > 0$ and any $t > 0$,*

$$\lim_{|\Lambda|\to\infty} P_{t,\Lambda,m'}\big(|m - u(t)| \ge \zeta\big) = 0. \tag{5.1.5.3}$$

Proof Define

$$\frac{du_\Lambda}{dt} = V_{-,\Lambda}(u_\Lambda) - V_{+,\Lambda}(u_\Lambda) =: V_\Lambda(u_\Lambda), \quad u_\Lambda(0) = m', \tag{5.1.5.4}$$

where

$$V_{\pm,\Lambda}(u) = (1 \mp u)c_{0,\Lambda}(u,h)e^{\mp\beta(h+u\mp 1/|\Lambda|)}. \tag{5.1.5.5}$$

There is a constant c so that for all u and $|\Lambda|$,

$$|V_{\pm,\Lambda}(u) - V_\pm(u)| \le \frac{c}{|\Lambda|}, \tag{5.1.5.6}$$

and it is readily seen that for any $T > 0$,

$$\lim_{|\Lambda|\to\infty} \sup_{t\le T} |u_\Lambda(t) - u(t)| = 0. \tag{5.1.5.7}$$

Denote by $\langle\cdot\rangle$ expectation with respect to $P_{t,\Lambda,m'}$, then writing u_Λ for $u_\Lambda(t)$ and $\dot u_\Lambda$ for $du_\Lambda/dt = V_\Lambda(u_\Lambda)$,

$$\frac{d}{dt}\big\langle(m-u_\Lambda)^2\big\rangle = \big\langle \mathcal{L}m^2 - 2u_\Lambda\mathcal{L}m - 2m\dot u_\Lambda + 2u_\Lambda\dot u_\Lambda\big\rangle. \tag{5.1.5.8}$$

By an explicit computation

$$\mathcal{L}m = V_\Lambda(m), \qquad \mathcal{L}m^2 - 2m\mathcal{L}m =: R_\Lambda, \qquad \|R_\Lambda\|_\infty \le \frac{c}{|\Lambda|}. \tag{5.1.5.9}$$

Thus

$$\frac{d}{dt}\big\langle (m-u_\Lambda)^2\big\rangle = 2\big\langle (m-u_\Lambda)(V_\Lambda(m) - V_\Lambda(u_\Lambda)) + R_\Lambda\big\rangle, \tag{5.1.5.10}$$

and since V_Λ is Lipschitz,

$$\frac{d}{dt}\big\langle (m-u_\Lambda)^2\big\rangle \le c'\big\langle (m-u_\Lambda)^2\big\rangle + \frac{c}{|\Lambda|}, \tag{5.1.5.11}$$

hence (5.1.5.3). □

Remarks For the sake of definiteness let us refer to (5.1.5.1). We can distinguish three factors in $V_\pm$, the first one is $(1 \pm m)/2$, it is the density of plus, respectively minus spins when the magnetization is m; $\frac{e^{\mp\beta(h+m)}}{e^{\beta(h+m)}+e^{-\beta(h+m)}}$ is the flip intensity of a plus, respectively minus spin; the third factor is 2 and it is the change due to a spin flip. Since V_- refers to minus spins it produces a positive change in the magnetization, hence it appears with a positive sign while V_+ with a negative one.

The limit mean field free energy $\phi_{\beta,h}(m)$, see (4.1.2.2), is a non-increasing function of time, namely a Lyapunov function for (5.1.5.1):

$$\frac{d\phi_{\beta,h}(m(t))}{dt} \le 0, \quad \text{if } m(t) \text{ satisfies (5.1.5.1).} \tag{5.1.5.12}$$

Glauber dynamics plus the thermodynamic limit prove the thermodynamic law that the free energy does not increase. Actually the inequality in (5.1.5.12) is strict unless

$$m = \tanh\{\beta(m+h)\}. \tag{5.1.5.13}$$

Thus the thermodynamical pure phases, i.e. the values of m which minimize $\phi_{\beta,h}(m)$, are as they should time invariant. Moreover if m is a critical point of $\phi_{\beta,h}(m)$ which is in the spinodal region, it is unstable for (5.1.5.1), i.e. small perturbations produce departing orbits. If instead the stationary point is metastable, then it is stable but only for "small perturbations." We shall come back on the issue later with more details and just observe that dynamics has indeed confirmed the thermodynamical considerations of the previous section with a deeper justification of the spinodal and metastable regimes.

5.1.6 Limit evolution of dynamics with Kac potentials

We denote by L_γ the generator L of the semigroup of Sect. 5.1.3 when $J(x,y) = J_\gamma(x,y)$, the latter as in (4.2.1.1), and when the mobility coefficient is given by

(5.1.3.9). We shall prove that in the mesoscopic limit $\gamma \to 0$ the magnetization density profile m_Ω in a regular "mesoscopic region" Ω, with fixed boundary conditions m_{Ω^c} outside Ω evolves according to the deterministic non-local evolution equation $(J * m(r) = \int J(r,r')m(r')dr')$

$$\frac{dm_\Omega}{dt} = -m_\Omega + \tanh\{\beta(J * (m_\Omega + m_{\Omega^c}) + h)\}. \tag{5.1.6.1}$$

The setup

- $\Omega \subset \mathbb{R}^d$ is a bounded, "regular" region (Ω should be regarded as "mesoscopic"). Initial datum m^*_Ω and "boundary condition" $\bar{m}$ are given continuous functions, respectively on Ω and on $\mathbb{R}^d$, with values in $[-1, 1]$,
- $\Omega_\gamma \subset \mathbb{Z}^d$, the "microscopic image" of Ω, is defined as

$$\Omega_\gamma = \bigcup_{C_i^{(\gamma^{-1/2})} \subset \gamma^{-1}\Omega} C_i^{(\gamma^{-1/2})} \cap \mathbb{Z}^d. \tag{5.1.6.2}$$

We have $C_i^{(\gamma^{-1/2})} \in \mathcal{D}^{(\gamma^{-1/2})}$, see Sect. 4.2.2. In Ω_γ we assign an initial configuration $\sigma^*_{\Omega_\gamma}$ and a boundary condition $\bar{\sigma}_{\Omega^c_\gamma}$ which "approximate" m^*_Ω and $\bar{m}$ as in (5.1.6.4) below.

- We denote by $|f|_{\Omega_\gamma} = \sup_{x \in \Omega_\gamma} |f^{(\gamma^{-1/2})}(x)|$, f a function on Ω_γ, where

$$f^{(\gamma^{-1/2})}(x) = \frac{1}{|C_x^{(\gamma^{-1/2})}|} \sum_{y \in C_x^{(\gamma^{-1/2})}} f(y). \tag{5.1.6.3}$$

$|f|_{\Omega^c_\gamma}$ is defined similarly. We then suppose that

$$\lim_{\gamma \to 0} |\sigma^*_{\Omega_\gamma}(\cdot) - m^*_\Omega(\gamma\,\cdot)|_{\Omega_\gamma} = \lim_{\gamma \to 0} |\bar{\sigma}_{\Omega^c_\gamma}(\cdot) - \bar{m}(\gamma\,\cdot)|_{\Omega^c_\gamma} = 0. \tag{5.1.6.4}$$

- $u_\gamma(x,t)$, $x \in \mathbb{Z}^d$, $t \geq 0$, denotes the function which for $x \in \Omega_\gamma$ satisfies

$$\frac{du_\gamma(x,t)}{dt} = -u_\gamma(x,t) + \tanh\left\{\beta\left(\sum_{y \neq x} J_\gamma(x,y)u_\gamma(y,t) + h\right)\right\}, \tag{5.1.6.5}$$

while for $x \in \Omega^c_\gamma$ and $t \geq 0$, $u_\gamma(x,t) = \sigma^{*\,(\gamma^{-1/2})}_{\Omega_\gamma}(x)$. Finally $u_\gamma(x,0) = \sigma^{*\,(\gamma^{-1/2})}_{\Omega_\gamma}(x)$, for $x \in \Omega_\gamma$.

We omit the proof that if Ω is sufficiently regular then for any $t \geq 0$,

$$\lim_{\gamma \to 0} |u_\gamma(\cdot,t) - m_\Omega(\gamma\,\cdot,t)|_{\Omega_\gamma} = 0, \tag{5.1.6.6}$$

namely that the discrete evolution equation (5.1.6.5) approximates the continuum equation (5.1.6.1) in the $|\cdot|_{\Omega_\gamma}$ sense.

Convergence to a limit equation will be proved in terms of the semi-norm

$$\|f\|_\gamma^2 := \frac{1}{|\Omega_\gamma|}\sum_{x\in\Omega_\gamma} f^{(\gamma^{-1/2})}(x)^2. \tag{5.1.6.7}$$

Calling $P_{\gamma,t}(f)=\sum_{\sigma_{\Omega_\gamma}} e^{L_\gamma t}(\sigma^*_{\Omega_\gamma},\sigma_{\Omega_\gamma})f(\sigma_{\Omega_\gamma})$, L_γ the generator of the Markov semigroup when the boundary conditions outside Ω_γ are given by $\bar\sigma_{\Omega^c_\gamma}$.

Theorem 5.1.6.1 *Let $m_\Omega(r,t)$, $r\in\Omega$, $t\ge 0$, solve* (5.1.6.1) *with boundary conditions $\bar m_{\Omega^c}$ and initial datum m^*_Ω. Then for any $t>0$,*

$$\lim_{\gamma\to 0} P_{\gamma,t}\big(\|(\sigma_{\Omega_\gamma}(\cdot)-m_\Omega(\gamma\,\cdot,t))\|_\gamma^2\big)=0. \tag{5.1.6.8}$$

Proof By (5.1.6.6) it suffices to prove that for any $t>0$,

$$\lim_{\gamma\to 0} P_{\gamma,t}\big(\|(\sigma_{\Omega_\gamma}(\cdot)-u_\gamma(\cdot,t))\|_\gamma^2\big)=0. \tag{5.1.6.9}$$

We proceed as in the proof of Theorem 5.1.5.1. Write $\langle\cdot\rangle$ for the expectation with respect to $P_{\gamma,t}$, $u_\gamma(\cdot)$ for $u_\gamma(\cdot,t)$ and $du_\gamma/dt=V_\gamma(u_\gamma)$ as given by the r.h.s. of (5.1.6.5), then

$$\begin{aligned}&\frac{d}{dt}\Big\langle\big(\sigma_{\Omega_\gamma}^{(\gamma^{-1/2})}-u_\gamma^{(\gamma^{-1/2})}\big)^2\Big\rangle\\ &\quad=\big\langle L_\gamma(\sigma_{\Omega_\gamma}^{(\gamma^{-1/2})})^2-2u_\gamma^{(\gamma^{-1/2})}L_\gamma\sigma_{\Omega_\gamma}^{(\gamma^{-1/2})}\\ &\qquad-2\sigma_{\Omega_\gamma}^{(\gamma^{-1/2})}V_\gamma(u_\gamma)^{(\gamma^{-1/2})}+2u_\gamma^{(\gamma^{-1/2})}V_\gamma(u_\gamma)^{(\gamma^{-1/2})}\big\rangle.\end{aligned} \tag{5.1.6.10}$$

Let $V_\gamma(\sigma)$ be the r.h.s. of (5.1.6.5) with u_γ replaced by $\sigma=(\sigma_{\Omega_\gamma},\bar\sigma_{\Omega^c_\gamma})$, then

$$L_\gamma\sigma_{\Omega_\gamma}^{(\gamma^{-1/2})}=[V_\gamma(\sigma)]^{(\gamma^{-1/2})},\qquad L_\gamma(\sigma_{\Omega_\gamma}^{(\gamma^{-1/2})})^2=2\sigma_{\Omega_\gamma}^{(\gamma^{-1/2})}L_\gamma\sigma_{\Omega_\gamma}^{(\gamma^{-1/2})}+R_\gamma,$$

with $\|R_\gamma\|_\infty\le c\gamma^{d/2}$. Hence $\frac{d}{dt}\langle(\sigma_{\Omega_\gamma}^{(\gamma^{-1/2})}-u_\gamma^{(\gamma^{-1/2})})^2\rangle$ is equal to

$$\big\langle 2\{\sigma_{\Omega_\gamma}^{(\gamma^{-1/2})}-u_\gamma^{(\gamma^{-1/2})}\}\{[V_\gamma(\sigma)]^{(\gamma^{-1/2})}-[V_\gamma(u_\gamma)]^{(\gamma^{-1/2})}\}+R_\gamma\big\rangle. \tag{5.1.6.11}$$

By (4.2.5.5), $|J_\gamma(r,r')-J^{(\gamma^{-1/2})}(r,r')|\le c\gamma^{d+1/2}\mathbf{1}_{|r-r'|\le 2\gamma^{-1}}$ so that

$$\begin{aligned}\Bigg|V_\gamma(u_\gamma)(x)-V_\gamma(\sigma)(x)-\Bigg\{\sigma(x)-u_\gamma(x)+\tanh\Bigg(\beta\Bigg[\sum_y J^{(\gamma^{-1/2})}(x,y)u_\gamma(y)\\ +h\Bigg]\Bigg)-\tanh\Bigg(\beta\Bigg[\sum_y J^{(\gamma^{-1/2})}(x,y)\sigma(y)+h\Bigg]\Bigg)\Bigg\}\Bigg|\le 2\beta c\gamma^{1/2}.\end{aligned}$$

Since $\sum_y J^{(\gamma^{-1/2})}(x,y)u_\gamma(y) = \sum_y J^{(\gamma^{-1/2})}(x,y)u_\gamma^{(\gamma^{-1/2})}(y)$ and since an analogous formula holds with $\sigma(y)$, then

$$\begin{aligned}
&\big|V_\gamma(u_\gamma)^{(\gamma^{-1/2})}(x) - V_\gamma(\sigma)^{(\gamma^{-1/2})}(x)\big| \\
&\leq \big|u_\gamma^{(\gamma^{-1/2})}(x) - \sigma^{(\gamma^{-1/2})}(x)\big| \\
&\quad + c\Big(\gamma^{1/2} + \gamma^{d/2} \sum_{z\in C_x^{(\gamma^{-1/2})}} \sum_{y\in\Omega_\gamma} J^{(\gamma^{-1/2})}(z,y)\big|u_\gamma^{(\gamma^{-1/2})}(y) - \sigma^{(\gamma^{-1/2})}(y)\big|\Big).
\end{aligned}$$

In conclusion, we get from (5.1.6.11),

$$\frac{d}{dt}\big\langle \|\sigma_{\Omega_\gamma} - u_\gamma\|_\gamma^2\big\rangle \leq c\big\langle \|u_\gamma - \sigma_{\Omega_\gamma}\|_\gamma^2\big\rangle + c'\gamma^{1/2},$$

which yields (5.1.6.9). □

5.2 Markov processes

We have mainly used so far analytic techniques to investigate spin flip Markov semigroups; in the sequel, methods, ideas and intuitions from probability theory will instead have greater relevance. By Proposition 5.1.1.1 we can interpret $e^{Lt}(\sigma^*_\Lambda, \sigma_\Lambda)$ as the probability that starting from σ^*_Λ at time 0 the state at time $t \geq 0$ is σ_Λ. This is a "single time transition probability" and our aim in this section is to extend the notion to the "probability of a time trajectory." This will lead to the introduction of Markov chains when time is discrete and Markov processes when time is continuous. Poisson processes and exponential distributions are going to play an important role. All such notions will be applied extensively in the next section to investigate the long time behavior of the mean field model.

5.2.1 The spin flip Markov chains

Markov chains are Markov processes in discrete times. Thus, times are $t \in \delta\mathbb{N}$, $\delta > 0$ the time mesh and trajectories elements in $\mathcal{X}_\Lambda^{\delta\mathbb{N}}$, i.e. sequences $\{\sigma_\Lambda(t), t\in\delta\mathbb{N}\}$. We fix an initial value σ^*_Λ at time 0 and for any $n \in \mathbb{N}_+$ define a probability $\mathcal{P}_{\sigma^*_\Lambda}$ on the space of $n+1$-step trajectories $\{\sigma_\Lambda(t), t = 0, \delta, \dots, \delta n\}$ as

$$\mathcal{P}_{\sigma^*_\Lambda}\big(\sigma_\Lambda(\delta k) = \eta_\Lambda^{(k)}, k = 0,\dots,n\big) = \prod_{k=1}^n e^{L\delta}(\eta_\Lambda^{(k-1)}, \eta_\Lambda^{(k)}) \mathbf{1}_{\eta_\Lambda^{(0)} = \sigma^*_\Lambda}. \qquad (5.2.1.1)$$

By Proposition 5.1.1.1 this is indeed a probability. By varying n we obtain a compatible family of probabilities, in the sense that

$$\sum_{\eta_\Lambda^{(n)}} \mathcal{P}_{\sigma_\Lambda^*}\big(\sigma_\Lambda(\delta k)=\eta_\Lambda^{(k)},\ k=0,\dots,n\big)=\mathcal{P}_{\sigma_\Lambda^*}\big(\sigma_\Lambda(\delta k)=\eta_\Lambda^{(k)},\ k=0,\dots,n-1\big).$$

Using the extension theorems of Sect. A.5 it can then be proved (details are omitted) that (5.2.1.1) uniquely extends to a probability on $\mathcal{X}^{\delta\mathbb{N}}$. The family of variables $\{\sigma_\Lambda(t), t\in\delta\mathbb{N}\}$ with law $\mathcal{P}_{\sigma_\Lambda^*}$ is called "a Markov chain" with transition probability $e^{L\delta}(\cdot,\cdot)$ and initial state σ_Λ^*. Call $\mathcal{F}_t$ the σ-algebra generated by $\{\sigma_\Lambda(s), s\le t\}$, then by (5.2.1.1) and using the semigroup property, for any $\eta_\Lambda\in\mathcal{X}_\Lambda$ and $s,t\in\delta\mathbb{N}$, $s<t$,

$$\mathcal{P}_{\sigma_\Lambda^*}\big(\sigma_\Lambda(t)=\eta_\Lambda\,|\mathcal{F}_s\big)=e^{L(t-s)}(\sigma_\Lambda(s),\eta_\Lambda) \tag{5.2.1.2}$$

$\sigma_\Lambda(s)$ being specified by $\mathcal{F}_s$. With the Markov chain we can not only talk of the probability of going from $\sigma_\Lambda(s)$ at time s to $\sigma_\Lambda(t)$ at time t, which is given by (5.2.1.2), but also of the way in which this is done, by specifying the probability of the whole path. As we shall see this may be of great help in computing transition probabilities. We will next discuss how all that can be generalized to continuous times.

5.2.2 *The spin flip Markov process*

The extension of the chain to continuous times is called a Markov process, the main point is the request that (5.2.1.2) should hold for all s and all $t\ge s$ in $\mathbb{R}_+$. To make it precise we shall say that the Markov process with generator L (and some fixed or random initial state) is a process $\sigma_\Lambda(t)$ on a probability space $(\Omega,\mathcal{P})$ where there is an increasing sequence of σ-algebras $\mathcal{F}_t$, $t\ge 0$, each one contained in the σ-algebra of Ω, with $\sigma_\Lambda(t)$ measurable with respect to $\mathcal{F}_t$, $\sigma_\Lambda(t)$ is then said to be "adapted to" $\mathcal{F}_t$. We then demand that for all $s\ge 0$ and all $t\ge s$,

$$\mathcal{P}\big(\sigma_\Lambda(t)=\sigma_\Lambda\;\big|\;\mathcal{F}_s\big)=e^{L(t-s)}(\sigma_\Lambda(s),\sigma_\Lambda) \tag{5.2.2.1}$$

holds. The definition of Markov process therefore is not really a definition but rather an equation on the law $\mathcal{P}$ which is requested to satisfy the family of equations (5.2.2.1) where the conditional probability with respect to $\mathcal{F}_s$ are assigned by the r.h.s. of (5.2.2.1), same scenario as in Chap. 2 for the DLR equations which in fact may be viewed as an extension of the notion of Markov processes to d-dimensional times. The problem here will be solved by an explicit construction of the process. We thus start by defining what we want a continuous time, spin flip, Markov process to be, we shall be more demanding than necessary with several extra properties required to hold; we shall then show that the definition is non-empty. As we shall eventually apply the notion to several other systems, we give the definition in a slightly more abstract way to include all of them. In the present application the state space Ξ below will be $\mathcal{X}_\Lambda$, M_Λ in the mean field case. With this in mind we introduce a finite space Ξ and a family $\{c(\xi,\xi') \ge 0, \xi \ne \xi' \in \Xi\}$ of "intensities." Given such a family we then define a Markov generator L on $L^\infty(\Xi)$ as

$$Lf(\xi) = \sum_{\xi'} c(\xi,\xi')\big(f(\xi') - f(\xi)\big). \tag{5.2.2.2}$$

When $\Xi = \mathcal{X}_\Lambda$, $c(\xi,\xi') = c(x,\sigma_\Lambda)$ with the identification $\xi = \sigma_\Lambda$ and $\xi' = \sigma^x_\Lambda$. When $\Xi = M_\Lambda$, $c(\xi,\xi') = c_{\pm,\Lambda}(m)$. $\xi = m$, $\xi' = m \pm 2/|\Lambda|$. To define the Poisson process we shall also consider the case when Ξ is a doubleton with $c(\xi,\xi') = c(\xi',\xi) = c > 0$.

Definition 5.2.2.1 The collection $(\Omega, \mathcal{P}, (\mathcal{F}_t)_{t\ge 0}, \underline{t}, \xi(\cdot))$ is a *realization of the semigroup* with intensities $c(\xi,\xi')$ if: $(\Omega,\mathcal{P})$ is a probability space; $\mathcal{F}_t$ an increasing sequence of σ-algebras on Ω (contained in the σ-algebra of Ω); $\underline{t} = (t_1, t_2, \dots)$ an increasing, divergent sequence of times; $\xi(t)$ a Ξ-valued function which verifies the following properties:

- For any $t \ge 0$ call $n(t) = \max\{i \in \mathbb{N} : t_i \le t\}$, then for any $t > 0$, $n(t)$ is $\mathcal{F}_t$-measurable, i.e. $n(t)$ is adapted to $\mathcal{F}_t$. $\xi(t)$ is also adapted to $\mathcal{F}_t$ and for any $i \ge 0$, $\xi(t) = \xi(t_i)$ for $t \in [t_i, t_{i+1})$, $t_0 = 0$.
- For any $s \ge 0$, $t > s$ and $\xi \in \Xi$,

$$\mathcal{P}\big(n(t) = n(s) \,\big|\, \mathcal{F}_s\big) = e^{-c(\xi(s))(t-s)}, \quad c(\xi) = \sum_{\xi'} c(\xi,\xi'), \tag{5.2.2.3}$$

and for any $s \ge 0$, $t > s$ and $\xi \in \Xi$

$$\begin{aligned}&\mathcal{P}\big(n(t) = n(s)+1, \xi_{n(s)+1} = \xi \,\big|\, \mathcal{F}_s\big)\\ &\quad = \int_s^t e^{-c(\xi(s))(s'-s)-c(\xi)(t-s')} c(\xi(s),\xi)\,ds'.\end{aligned} \tag{5.2.2.4}$$

It is not a priori clear (but true a posteriori!) that Definition 5.2.2.1 is non-empty; in any case it defines what we are looking for:

Theorem 5.2.2.2 *Suppose* $(\Omega, \mathcal{P}, (\mathcal{F}_t)_{t\ge 0}, \underline{t}, \xi(\cdot))$ *is a realization of the semigroup with intensities* $c(\xi,\xi')$*, in the sense of Definition* 5.2.2.1. *Then for any* $s \ge 0$, $t > s$ *and* $\xi \in \Xi$,

$$\mathcal{P}\big(\xi(t) = \xi \,\big|\, \mathcal{F}_s\big) = e^{L(t-s)}(\xi(s),\xi). \tag{5.2.2.5}$$

Proof We decompose

$$\{\xi(t) = \xi\} = \{\xi(t) = \xi, n(t) = n(s)\} \cup \{\xi(t) = \xi, n(t) > n(s)\}.$$

Since the two sets are disjoint, by (5.2.2.3), and calling $\eta = \xi(s)$,

$$\text{l.h.s. of (5.2.2.5)} = e^{-c(\eta)(t-s)}\mathbf{1}_{\eta=\xi} + \mathcal{P}\big(\{\xi(t) = \xi, n(t) > n(s)\} \,\big|\, \mathcal{F}_s\big). \tag{5.2.2.6}$$

We split the interval $(s,t]$ into equal intervals $I_k = (t'_k, t''_k]$ and decompose $\{\xi(t) = \xi, n(t) > n(s)\}$ by specifying the interval where the last jump occurs:

$$\{\xi(t) = \xi, n(t) = n(t''_k) = n(t'_k) + 1\} \cup \{\xi(t) = \xi, n(t) = n(t''_k) > n(t'_k) + 1\}.$$

The first set $\{\xi(t) = \xi, n(t) = n(t''_k) = n(t'_k) + 1\}$ is equal to

$$\bigcup_{\xi'} \{\xi(t) = \xi, \xi(t'_k) = \xi', n(t) = n(t''_k) = n(t'_k) + 1\}.$$

We first condition to $\mathcal{F}_{t''_k}$ and use (5.2.2.3); then to $\mathcal{F}_{t'_k}$ and use (5.2.2.4). We get that $\mathcal{P}(\{\xi(t) = \xi, n(t) > n(s)\} \mid \mathcal{F}_s)$ is the sum of

$$\sum_k \sum_{\xi'} \int_{I_k} e^{-c(\xi)(t-s')-c(\xi')(s'-t'_k)} c(\xi', \xi) \mathcal{P}\big(\xi(t'_k) = \xi' \mid \mathcal{F}_s\big), \quad (5.2.2.7)$$

and of

$$\sum_k \mathcal{P}\big(\{\xi(t) = \xi, n(t) = n(t''_k) > n(t'_k) + 1\} \mid \mathcal{F}_s\big). \quad (5.2.2.8)$$

Calling $\epsilon = t''_k - t'_k$, we claim that there is c so that for all $\epsilon > 0$,

$$\mathcal{P}\big(\{n(t''_k) > n(t'_k) + 1\} \mid \mathcal{F}_s\big) \le c\epsilon^2.$$

Indeed the l.h.s. is equal to $1 - \mathcal{P}(\{n(t''_k) = n(t'_k)\} \mid \mathcal{F}_s) - \mathcal{P}(\{n(t''_k) = n(t'_k) + 1\} \mid \mathcal{F}_s)$ and the claim follows using (5.2.2.3) and (5.2.2.4). Then the expression (5.2.2.8) is bounded by $\le \frac{t-s}{\epsilon} c\epsilon^2$. Postponing the proof that $\mathcal{P}(\xi(t) = \xi \mid \mathcal{F}_s)$ depends continuously on t, we let $\epsilon \to 0$ and obtain

$$\begin{aligned}\mathcal{P}\big(\xi(t) = \xi \mid \mathcal{F}_s\big)\\ = e^{-c(\xi(s))(t-s)} \mathbf{1}_{\eta=\xi} + \sum_{\xi'} \int_s^t e^{-c(\xi)(t-s')} c(\xi', \xi) \mathcal{P}\big(\xi(s') = \xi' \mid \mathcal{F}_s\big), \quad (5.2.2.9)\end{aligned}$$

which is equal to $e^{L(ts)}(\eta, \xi)$ by the analogue of (5.1.1.4).

Since $|\mathcal{P}(\xi(t') = \xi \mid \mathcal{F}_s) - \mathcal{P}(\xi(t) = \xi \mid \mathcal{F}_s)|$, $t' > t$, is bounded by $\mathcal{P}(n(t') > n(t) \mid \mathcal{F}_s)$ which in turns is bounded by $1 - e^{c(t'-t)}$, $c = \sup_\xi c(\xi)$, then $\mathcal{P}(\xi(t') = \xi \mid \mathcal{F}_s)$ is a continuous function of t. □

5.2.3 *Poisson processes and exponential times*

"The Poisson process with intensity c" can be realized as the Markov process with $\Xi = \{\xi, \xi'\}$ a doubleton, $c(\xi, \xi') = c(\xi', \xi) = c$ and an initial condition fixed once for all.

The values $\xi(t)$ of the process are then uniquely determined by the sequence $\underline{t}$ of jumps and $(\Omega, \mathcal{P}, (\mathcal{F}_t)_{t\geq 0}, \underline{t}, \xi(\cdot)\})$ reduces to $(\Omega, \mathcal{P}, (\mathcal{F}_t)_{t\geq 0}, \underline{t})$. $(\Omega, \mathcal{P}, (\mathcal{F}_t)_{t\geq 0}, \underline{t})$ is a realization of the Poisson process with intensity c. The conditions (5.2.2.3) and (5.2.2.4) on the law $\mathcal{P}$ have the simpler form: for any $s \geq 0$ and $t > s$,

$$\mathcal{P}\big(n(t) = n(s)\,\big|\mathcal{F}_s\big) = e^{-c(t-s)}, \tag{5.2.3.1}$$

$$\mathcal{P}\big(n(t) = n(s) + 1\,\big|\,\mathcal{F}_s\big) = e^{-c(t-s)} c(t-s). \tag{5.2.3.2}$$

The key ingredient in the construction of a Poisson process is the following.

Definition 5.2.3.2 (Exponential random times) A random variable $S \geq 0$ has *exponential distribution* P_c with intensity $c > 0$ if for any $t \geq 0$, $P_c(S > t) = e^{-ct}$.

It can easily be seen that for any $f \in L^\infty$,

$$\int f\,dP_c = \int_0^\infty f(t) e^{-ct} c\,dt.$$

Notice also that exponential laws are unchanged by conditioning, in the sense that

$$P_c\big(S > s + t\,\big|\,S > s\big) = e^{-ct}, \tag{5.2.3.3}$$

which is the well known paradox that if, waiting at a bus stop, you are told that it has been a long time since the last bus passed by, you should not raise your hopes.

Realization of the Poisson processes

We shall next construct an explicit realization of the Poisson process, which thus proves that the above definition is non-empty. Consider the probability space $(\mathbb{R}_+^{\mathbb{N}}, \mathcal{P})$ equal to the product $(\mathbb{R}_+, P_c)^{\mathbb{N}}$ and define on such a space the variables

$$t_n = t_n(\underline{s}) = s_1 + \cdots + s_n, \quad \underline{s} = (s_1, s_2, \ldots) \in \mathbb{R}_+^{\mathbb{N}}.$$

Define $n(t) = \min\{k : t_k \leq t\}$, and call $\mathcal{F}_t$ the σ-algebra generated by the variables $n(s)$, $s \leq t$. Then by an explicit computation for any $s \geq 0$, $t > s$ and $k \in \mathbb{N}$,

$$\mathcal{P}\big(n(t) = n(s) + k\,\big|\mathcal{F}_s\big) = e^{-c(t-s)} \frac{(c(t-s))^k}{k!}. \tag{5.2.3.4}$$

By setting $k = 0$ and $k = 1$ in (5.2.3.4) we see that (5.2.3.1) and (5.2.3.2) are verified. Thus the above is a realization of the Poisson process with intensity c, whose existence is therefore proved.

Markov chains with random times and universal couplings

We shall realize the Markov process with jump intensities $c(\xi,\xi')$ in terms of the Markov chain with transition probability $P(\xi,\xi')=\pi_\xi(\xi')=\frac{c(\xi,\xi')}{c(\xi)}$, $c(\xi)=\sum_{\xi'}c(\xi,\xi')$, by replacing the deterministic times between transitions by suitable random times. They will be exponential times of intensity $c(\xi)$ whenever the transition goes from ξ to any other $\xi'\in\Xi$. We shall actually define the process by imbedding it into a larger process where we can realize simultaneously all Markov processes with any initial condition, thus implicitly defining a coupling between processes with different initial conditions, "universal couplings."

The Markov chain with transition probability $\pi_\xi(\xi')$ to go from $\xi\to\xi'$ can be realized in the following way. Consider for each ξ the (Bernoulli) product probability space $(\Xi,\pi_\xi(\cdot))^{\mathbb{N}}$ whose elements are denoted by $\underline{\xi}'(\xi)=\xi'_n(\xi), n\in\mathbb{N}$. We then call $(\Omega',\mathcal{P}')$ the product over all $\xi\in\Xi$ of the above Bernoulli measures. Given an initial condition $\xi_0\in\Xi$ we then construct the trajectory $\xi_n, n\in\mathbb{N}$ on Ω' by setting recursively $\xi_{n+1}=\xi'_{n+1}(\xi_n)$ observing that under $\mathcal{P}'$ the process $\xi_n\in\mathbb{N}$ has the law of the Markov chain with transition probability $\pi_\xi(\xi')$ which starts at ξ_0. Notice also that we can run with the same element of Ω' any other initial datum thus obtaining a coupling for the chains with different initial data (universal couplings). The coupled processes move independently of each other till they meet, after then they stick together, thus the meeting time gives an estimate of the correlation length in the chain.

In continuous time the construction is very similar. We introduce for each $\xi\in\Xi$ a Poisson process with intensity $c(\xi)$, its law is denoted by $P^{(\xi)}$ and the realizations of the process by $\underline{t}(\xi)=\{t_n(\xi), n\ge 1\}$. We call $(\Omega,\mathcal{P})$ the probability space product of the space $(\Omega',\mathcal{P}')$ and the product of the above Poisson processes over all $\xi\in\Xi$. Let $n(t;\xi)$ the variable $n(t)$ defined on the Poisson process labeled by ξ. With probability 1 all $n(t;\xi)$ are finite and no two elements $t_{n'}(\xi')$ and $t_n(\xi)$ are equal to each other. The following definition is given in such a subset. Let $\{\underline{t}(\xi),\underline{\xi}'(\xi),\xi\in\Xi\}$ be one of its elements. For each ξ draw a vertical semi-axis interpreted as the semi-axis of positive times. At each $t_n(\xi)$ draw an arrow from ξ to ξ' if $\xi'_n(\xi)=\xi'$. Given an initial configuration $\eta\in\Xi$, which we regard as fixed once for all, define $\xi(s), s\ge 0$, by following the arrows starting from η. Namely $\xi(s)=\eta$ for all $s<t_1$, $t_1:=t_1(\eta)$. We then set $\xi(t_1)=\eta'$, $\eta'=\xi'_1(\eta)$. $\xi(s)=\xi(t_1)$ for $s\in[t_1,t_2)$ where t_2 is the first time after t_1 when an arrow starts from $\xi(t_1)$. At such a time $\xi(s)$ jumps to the new configuration indicated by the arrow and so on.

Finally, define $\mathcal{F}_t$ as the σ-algebra generated by the variables $\{n(s;\xi),\xi\in\Xi$, $s\le t\}$ and by all $\{\xi'_k(\xi), k\le n(t;\xi)\}$, thus $\mathcal{F}_t$ specifies the whole trajectory $\xi(s), s\in[0,t]$.

Theorem 5.2.3.3 *The collection $(\Omega,\mathcal{P},(\mathcal{F}_t)_{t\ge 0},\underline{t},\xi(\cdot))$ defined above is a realization of the semigroup with intensities $c(\xi,\xi')$.*

Proof As all the other properties are clearly satisfied we just need to check (5.2.2.3)–(5.2.2.4). Once we condition on $\mathcal{F}_s$ we specify the trajectory $\xi(s')$ for all

$s' \in [0, s]$, call $\xi = \xi(s)$. (5.2.2.3) requires to compute the probability that $\xi(\cdot)$ does not change in $(s, t]$, in Ω this means that $n(t, \xi) = n(s, \xi)$. Since $\underline{t}(\xi)$ is a Poisson process,

$$\mathcal{P}\big(n(t,\xi) = n(s,\xi) \mid \mathcal{F}_s\big) = e^{-c(\xi)(t-s)},$$

and (5.2.2.3) is proved.

The expression of the event in (5.2.2.4) in the space Ω is more complex. Calling $n = n(s, \xi)$ this is

$$A := \{t_{n+1}(\xi) \in (s,t]\} \cap \{\xi'_{n+1}(\xi) = \xi'\} \cap \{n(t_{n+1}(\xi), \xi') = n(t, \xi')\}.$$

We are going to prove that

$$\mathcal{P}(A \mid \mathcal{F}_s) = \int_s^t e^{-c(\xi)(t'-s)} c(\xi) \pi_\xi(\xi') e^{-c(\xi')(t-t')} dt', \tag{5.2.3.5}$$

from which (5.2.2.4) follows. (5.2.3.5) is at least heuristically clear: the factor $e^{-c(\xi)(t'-s)}c(\xi)$ is the distribution density of the exponential time $t_{n+1}(\xi) - s$, $\pi_\xi(\xi')$ is the probability that $\xi'_{n+1}(\xi) = \xi'$ and $e^{-c(\xi')(t-t')}$ is the probability that conditioned on the value $t_{n+1}(\xi) = t'$, $\underline{t}(\xi') \cap [t', t] = \emptyset$. To make this precise, we split the interval $[s, t]$ into N equal intervals of length ϵ: $\epsilon N = t - s$. We shall eventually let $\epsilon \to 0$. A lower bound for $\mathcal{P}(A \mid \mathcal{F}_s)$ is obtained by requiring that when $t_{n+1}(\xi) \in [\epsilon k, \epsilon(k+1))$ then $t_{n+2}(\xi) > \epsilon(k+1)$ and $n(\epsilon k, \xi') = n(t, \xi')$. This gives the lower bound

$$\mathcal{P}(A \mid \mathcal{F}_s) \ge \sum_k e^{-\epsilon k c(\xi)} e^{-\epsilon c(\xi)} \epsilon c(\xi) \pi_\xi(\xi') e^{-c(\xi')(t-\epsilon k)},$$

which converges to (5.2.3.5) as $\epsilon \to 0$. For the upper bound we use the relaxed constraint $n(\epsilon(k+1), \xi') = n(t, \xi')$ then getting

$$\mathcal{P}(A \mid \mathcal{F}_s) \le \sum_k e^{-\epsilon k c(\xi)} [1 - e^{-\epsilon c(\xi)}] \pi_\xi(\xi') e^{-c(\xi')(t-\epsilon(k+1))},$$

which also converges to (5.2.2.4) as $\epsilon \to 0$. □

The arrows process and the graphical construction

There is an alternative (actually isomorphic) realization of the Markov process with intensities $c(\xi, \xi')$ which has a nice graphical representation and for this reason it is also referred to in the literature as the "graphical construction."

For any ordered pair (ξ, ξ') we consider a Poisson process $(\Omega^{(\xi,\xi')}, \mathcal{P}^{(\xi,\xi')})$ with intensity $c(\xi, \xi')$, whose elements are denoted by $\underline{t}(\xi, \xi') = \{t_n(\xi, \xi'), n \in \mathbb{N}\}$. $(\Omega, \mathcal{P})$ is the product of $(\Omega^{(\xi,\xi')}, \mathcal{P}^{(\xi,\xi')})$ over all (ξ, ξ'). The graphical representation is based on drawing an arrow joining ξ and ξ' at the times $t_n(\xi, \xi')$ whenever "the random clock (ξ, ξ') rings." Given an element in Ω and an initial datum ξ_0 we then construct $\xi(t)$ by following the arrows: namely if $\xi(s) = \xi$ then $\xi(t) = \xi$ for all $t \in [s, \tau)$, $\tau = \min\{t_n(\xi, \xi') \geq s, n \in \mathbb{N}, \xi' \in \Xi\}$ and $\xi(\tau) = \xi'$ if $t_n(\xi, \xi')$ is where the above minimum is reached (the definition is well posed with probability 1). Calling $\mathcal{F}_t$ the σ-algebra generated by $n(s, \xi, \xi')$, $s \leq t$, we then have (the proof is completely analogous to that of Theorem 5.2.3.3 and omitted).

Theorem 5.2.3.4 *The process $\xi(t)$ defined on the space $(\Omega, \mathcal{P})$ as explained above obeys the law of the Markov process with intensities $c(\xi, \xi')$.*

5.2.4 Realizations of the mean field process

The continuous time Markov process associated to the mean field semigroup $e^{\mathcal{L}t}$ is defined as in Definition 5.2.2.1 with $\Xi = M_\Lambda$. Theorem 5.2.2.2 then reads:

Theorem 5.2.4.1 *If $(\Omega, \mathcal{P}, (\mathcal{F}_t)_{t \geq 0}, (\hat{m}(t))_{t \geq 0})$ realizes the semigroup $e^{\mathcal{L}t}$, then for any $s \geq 0$, $t > 0$ and $m \in M_\Lambda$,*

$$\mathcal{P}(\hat{m}(t) = m | \mathcal{F}_s\}) = e^{\mathcal{L}(t-s)}(\hat{m}(s), m).$$

The process can be visualized as in the graphical representations of Sect. 5.2.3 by simply replacing Ξ by M_Λ and $c(\xi, \xi')$ by $c_{\pm,\Lambda}(m)$, $\xi' = m \pm 2/|\Lambda|$.

5.3 Persistence of stochasticity in the macroscopic limit

In this section we discuss phenomena which contradict the limit evolution equation (5.1.5.1) (derived in Theorem 5.1.5.1). Their origin is in the difference between the actual magnetization $m(t)$ (when $|\Lambda|$ is still finite) and its deterministic limit. In the proof of Theorem 5.1.5.1 such a difference is considered as "an error" which in fact vanishes as $|\Lambda| \to \infty$. The phenomena we are going to study however involve time intervals which diverge as $|\Lambda| \to \infty$ and for which the proof of Theorem 5.1.5.1 does not apply, the bounds on the errors in fact grow exponentially in time. The key point will then be to extract from what was considered as error a term which will be the main contribution to the phenomena we want to study and which, by its very origin, is not contained in the limit deterministic equation (5.1.5.1).

5.3.1 Spinodal decomposition

In this subsection we study the "spinodal decomposition" in the context of mean field Glauber dynamics; we thus suppose $\beta > 1$, $h = 0$ and the initial state $m_0 = 0$.

We are modeling a quenching experiment where the temperature is suddenly dropped from above to below the critical value β_c ($= 1$ in our mean field model). Thus, at times $t < 0$ the system is in thermal equilibrium at $\beta < 1$ and its magnetization is therefore $m_\beta = 0$. At $t = 0$ β increases so fast to a value larger than 1 that the state of the system and thus its magnetization is essentially unchanged. The quenched state $m_0 = 0$ is therefore no longer in thermodynamic equilibrium as now $\beta > 1$, yet it is stationary for the limit dynamics (5.1.5.1) (recall $h = 0$). "Spinodal decomposition" refers to the phenomena which describe the way the system leaves the quenched state $m_0 = 0$ and which cannot be explained by the evolution (5.1.5.1) alone which instead predicts $m(t) = 0$ forever.

The argument often used to explain the paradox is that a system cannot be perfectly isolated, so that the initial state has unavoidably small non-zero fluctuations around $m = 0$. Since $m = 0$ is unstable under (5.1.5.1), if we perturb even slightly the initial state the successive orbit is attracted by the equilibrium values $\pm m_\beta$ (according to the sign of the perturbation). Thus the final state is a probability supported by $\pm m_\beta$ with weights given by the probabilities that the initial state is positive, respectively negative.

The argument is not completely satisfactory because it needs a deus ex machina, i.e. the initial perturbation, and the final state depends on the structure of the perturbation (the probability that the initial magnetization is positive/negative). Same considerations apply if we interpret the fact that the system is not isolated by adding to the limit equation a stochastic noise.

Glauber dynamics gives instead a self-contained explanation of the spinodal decomposition, despite the apparent contradiction that the limit dynamics we have derived from the Glauber dynamics in the thermodynamic limit does not account for the effect. There is indeed a clear inversion of limits as we are using (5.1.5.1) to read the asymptotic behavior of the system (in particular convergence to thermodynamic equilibrium) while the equation has been derived only for compact time intervals in the limit $|\Lambda| \to \infty$.

We shall prove that if we instead consider the finite volume Glauber dynamics for long enough times, then it is of the same stochastic nature as the dynamics which displaces the system from its initial unstable equilibrium, and after that the randomness is overcome by the deterministic drift present in (5.1.5.1) at $m \neq 0$. As we shall see, the relevant time scale is $\log|\Lambda|$ and in fact after a time of order $\log|\Lambda|$ the state will be in equilibrium, i.e. close to $\pm m_\beta$. If instead we fix t and let $|\Lambda| \to \infty$ we always see $m = 0$, as predicted by (5.1.5.1).

Theorem 5.3.1.1 *Let $h = 0$, $\beta > 1$ and $\mathcal{P}_\Lambda$ the law of the Glauber process $m(\cdot)$ which starts from an initial configuration with 0 magnetization (we must then suppose that $|\Lambda|$ is even). Call $\alpha := \beta - 1$ and $\tau_c = 1/(2\alpha)$. Then:*

- *Escape time. For any $\delta > 0$*

$$\lim_{|\Lambda|\to\infty} \mathcal{P}_\Lambda\big(|m(\tau\log|\Lambda|)| \geq \delta\big) = 0, \quad \tau < \tau_c, \tag{5.3.1.1}$$

$$\lim_{|\Lambda|\to\infty} \mathcal{P}_\Lambda\big(|m(\tau\log|\Lambda|) \mp m_\beta| \leq \delta\big) = \frac{1}{2}, \quad \tau > \tau_c. \tag{5.3.1.2}$$

- *Self-similar growth. For any $\tau \in (0, \tau_c)$ the distribution of the variable $X(\tau) := (e^{\alpha\tau \log|\Lambda|}|\Lambda|^{-1/2})^{-1} m(\tau \log|\Lambda|)$ converges to [the same] gaussian distribution with mean zero and covariance $C > 0$.*

The proof of Theorem 5.3.1.1 is given in a complementary Sect. 5.4, we shall sketch below the main ideas after some brief comments on the statements in Theorem 5.3.1.1 about "the strange interplay" between stochastic and deterministic effects. Despite the whole phenomenon being random, yet (in the relevant $\log|\Lambda|$ time units) the escape time is deterministic and it occurs "precisely" at time $\tau_c = 1/(2\alpha)$. Moreover, after an initial time layer (infinitesimal in $\log|\Lambda|$-time units) the distribution of the "normalized magnetization" is self-similar for $\tau < \tau_c$, namely: the magnetization in $e^{\alpha\tau \log|\Lambda|}|\Lambda|^{-1/2}$ units has a distribution (in the limit $|\Lambda| \to \infty$) independent of τ.

Sketch of proof The proof uses for the first time the Markov process structure of the dynamics. Even though the statements in (5.3.1.1)–(5.3.1.2) refer to single time events and can then be expressed in terms of the semi-group $e^{\mathcal{L}t}$, it is more convenient to look at them in terms of the whole trajectories of the process. The function $u_\Lambda(t)$ defined as the solution of $u_\Lambda(t) = \int_0^t V_\Lambda(u_\Lambda(s))$ (which is (5.1.5.4) in integral form, since $m_0 = 0$), is close to the actual magnetization $m(t)$ (the error vanishing as $|\Lambda| \to \infty$ and t fixed), as shown in the proof of Theorem 5.1.5.1. Thus, $m(t)$ is "almost" a solution of the above integral equation and the random variable

$$b(t) := m(t) - \int_0^t V_\Lambda(m(s)) \tag{5.3.1.3}$$

is a measure of the deviations from the deterministic evolution and of the randomness of the system. We now change perspective and regard $b(t)$ as a known quantity; (5.3.1.3) written as $m(t) = \int_0^t V_\Lambda(m(s)) + b(t)$ is now regarded as an integral equation in the unknown $m(t)$ with $b(t)$ an additional stochastic forcing. It is thus crucial to control $b(t)$ and here we shall exploit that $b(t)$ is a martingale. We thus interrupt for a moment the sketch of the proof to recall some basic facts about martingales.

Martingales in Markov processes

A process $\{M(t), t \geq 0\}$ is a martingale relative to $\{\mathcal{F}_t, t \geq 0\}$ if $\mathcal{F}_t$ is an increasing family of σ-algebras; for any $t \geq 0$ M_t is $\mathcal{F}_t$-measurable and for any $s > 0$ the conditional expectation

$$\mathcal{E}\big(M(t+s)|\mathcal{F}_t\big) = M(t). \tag{5.3.1.4}$$

In a Markov process $\{X(t), t \geq 0\}$, there are natural martingales. Let f be a function on the (finite) state space of the process, call $f(t) := f(X(t))$, L the generator of the Markov process and $\mathcal{F}_t$ the σ-algebra generated by $\{X(s), s \le t\}$. Then

$$M(t) := f(t) - f(0) - \int_0^t Lf(s) \tag{5.3.1.5}$$

is a martingale; see Theorem 5.3.1.2 below. The integrand in the last term is called "the compensator" of the process $f(t)$. Moreover, see again Theorem 5.3.1.2 below,

$$M(t)^2 - \int_0^t \{Lf^2 - 2fLf\} =: N(t) \tag{5.3.1.6}$$

is also a martingale and the integrand is called "the second compensator" of f or the quadratic variation of the martingale.

Theorem 5.3.1.2 *The above processes $M(t)$ and $N(t)$ are martingales.*

Proof Let $t_0 \ge 0$ and shorthand $\mathcal{E}_{t_0}(\cdot) = \mathcal{E}(\cdot|\mathcal{F}_{t_0})$. We need to prove that $\mathcal{E}_{t_0}(M(t)) = M_{t_0}$, namely that $\mathcal{E}_{t_0}(f(t) - \int_{t_0}^t Lf(s)) = f(t_0)$. By the Fubini theorem and using (5.2.2.1), this is equal to

$$\mathcal{E}_{t_0}\big(f(t)\big) - \int_{t_0}^t \mathcal{E}_{t_0}\big(Lf(s)\big) = \{e^{L(t-t_0)}f\}(X(t_0)) - \int_{t_0}^t \{e^{L(s-t_0)}Lf\}(X(t_0)).$$

Since $e^{L(s-t_0)}Lf = \frac{d}{ds}e^{L(s-t_0)}f$, the last expression is equal to $f(X(t_0))$, thus completing the proof that $M(t)$ is a martingale. In order to prove that $N(t)$ is a martingale, we need to show that

$$\mathcal{E}_{t_0}\left(M(t)^2 - \int_{t_0}^t \{Lf^2 - 2fLf\}\right) = M(t_0)^2. \tag{5.3.1.7}$$

Since $M(t) - M(t_0) = f(t) - f(t_0) - \int_{t_0}^t Lf(s)$, $[M(t) - M(t_0)]^2$ is equal to

$$[f(t) - f(t_0)]^2 + \int_{t_0}^t \int_{t_0}^t \{Lf(s)\}\{Lf(s')\} - 2\int_{t_0}^t [f(t) - f(t_0)]Lf(s).$$

Thus $\mathcal{E}_{t_0}(M(t)^2) - M(t_0)^2 = \mathcal{E}_{t_0}([M(t) - M(t_0)]^2)$ which is equal to

$$\mathcal{E}_{t_0}\left([f(t) - f(t_0)]^2 + \int_{t_0}^t \int_{t_0}^t \{Lf(s)\}\{Lf(s')\} - 2\int_{t_0}^t [f(t) - f(t_0)]Lf(s)\right).$$

By (5.3.1.5) $\mathcal{E}_{t_0}([f(t) - f(t_0)]^2) = \mathcal{E}_{t_0}(\int_{t_0}^t L[f(s) - f(t_0)]^2)$, so that

$$\begin{aligned}\mathcal{E}_{t_0}(M(t)^2) - M(t_0)^2 = \mathcal{E}_{t_0}\Bigg(&\int_{t_0}^t \int_{t_0}^t \{Lf(s)\}\{Lf(s')\} - 2\int_{t_0}^t [f(t) - f(t_0)]Lf(s) \\ &- 2\int_{t_0}^t f(t_0)Lf(s) + \int_{t_0}^t Lf(s)^2\Bigg).\end{aligned} \tag{5.3.1.8}$$

Since $\mathcal{E}_{t_0}(f(t)Lf(s)) = \mathcal{E}_{t_0}(\int_s^t \{Lf(s')\}\{Lf(s)\}) + \mathcal{E}_{t_0}(f(s)Lf(s))$ we then get (5.3.1.7). □

Going back to the "sketch of the proof," we observe that indeed $b(t)$ is a martingale because, by (5.1.5.9), (5.3.1.3) is (5.3.1.5) with $f(t) = m(t)$ and $b(t) = M(t)$. (5.3.1.6) then yields

$$b(t)^2 = \int_0^t \{\mathcal{L}m^2 - 2m\mathcal{L}m\} + N(t). \tag{5.3.1.9}$$

$N(t)$ a martingale. Being a martingale,

$$\mathcal{E}_\Lambda(N(t)) = \mathcal{E}_\Lambda(N(0)) = \mathcal{E}_\Lambda(b(0)^2) = 0. \tag{5.3.1.10}$$

By Doob's inequality, for any $T > 0$, $\mathcal{E}_\Lambda(\{\sup_{t\le T} b(t)^2\}) \le 4\mathcal{E}_\Lambda(b(T)^2)$. By (5.3.1.10), and since, by (5.1.5.9), $\|\mathcal{L}m^2 - 2m\mathcal{L}m\|_\infty \le \frac{c}{|\Lambda|}$,

$$\mathcal{E}_\Lambda\Big(\sup_{t\le T} b(t)^2\Big) \le 4\mathcal{E}_\Lambda(b(T)^2) \le 4\frac{cT}{|\Lambda|}. \tag{5.3.1.11}$$

Then, for any $\epsilon > 0$, $\mathcal{P}_\Lambda(\sup_{t\le T} b(t)^2 \ge \epsilon) \le \frac{4cT\epsilon^{-2}}{|\Lambda|}$. We fix $\tau < \tau_c = \frac{1}{2\alpha}$, call $T = \tau\log|\Lambda|$ and choose $\epsilon = |\Lambda|^{-\theta}$, $\theta \in (\alpha\tau, 1/2)$. Then

$$\mathcal{P}_\Lambda\Big(\sup_{t\le\tau\log|\Lambda|} |b(t)| \le |\Lambda|^{-\theta}\Big) \ge 1 - 4(c\tau\log|\Lambda|)|\Lambda|^{2\theta-1}. \tag{5.3.1.12}$$

Since $\theta < 1/2$ the r.h.s. goes to 1 as $|\Lambda| \to \infty$. We can then suppose $|b(t)| \le |\Lambda|^{-\theta}$ in (5.3.1.3) regarded as an integral equation for $m(t)$ with $b(t)$ a "known" term. If we linearize around $m = 0$ we get $m^0(t) = \int_0^t \alpha m^0(s) + b(t)$, so that

$$m^0(t) = \int_0^t e^{\alpha(t-s)}b(s), \qquad |m^0(t)| \le \alpha^{-1}e^{\alpha t}|\Lambda|^{-\theta},\ t \le \tau\log|\Lambda|,$$

which vanishes for $\alpha\tau < \theta$. By a perturbative argument whose details are omitted, the solution $m(t)$ of (5.3.1.3) has the same property. For any $\alpha' > \alpha$ and $\alpha'\tau < \theta$ and, correspondingly, for all $|\Lambda|$ large enough, $|m(t)| \le e^{\alpha' t}|\Lambda|^{-\theta}$ for all $t \le \tau\log|\Lambda|$, so that for $\theta \in (\alpha'\tau, 1/2)$

$$\mathcal{P}_\Lambda\Big(\sup_{t\le\tau\log|\Lambda|} e^{-\alpha' t}|m(t)| \le |\Lambda|^{-\theta}\Big) \ge 1 - 4(c\log|\Lambda|)|\Lambda|^{2\theta-1}, \tag{5.3.1.13}$$

and (5.3.1.1) is proved in its stronger version (5.3.1.13). The remaining statements in Theorem 5.3.1.1 are proved in Sect. 5.4.

5.3.2 Large deviations in the mean field dynamics

The analysis of the spinodal decomposition can be extended in various directions to include as a first case fluctuations of the initial state which in the previous analysis have been neglected. "Normal" fluctuations have order $1/\sqrt{|\Lambda|}$ (namely they scale as the square root of the number of spins), it is thus natural to consider the case when $m(0)$ fluctuates by $1/\sqrt{|\Lambda|}$. Even such small fluctuations may produce a non-trivial effect. If they are non-symmetric around $m=0$ the limit distribution at times $\tau \log|\Lambda|$, $\tau > 1/(2\alpha)$, may change and the weights of the limit states $\pm m_\beta$ altered.

Another interesting case is when a magnetic field is present. Let $h \in (0, h_c)$, where $h_c > 0$ is the critical field: namely for $h \in (0, h_c)$ the limit free energy $\phi_{\beta,h}(m)$ has three critical points, $m^+_{\beta,h}$, $m^0_{\beta,h}$ and $m^-_{\beta,h}$. Only $m^+_{\beta,h}$, however, is a minimum, $m^-_{\beta,h}$ is metastable being only a local minimum, while $m^0_{\beta,h} < 0$ is spinodal, i.e. it is stationary but unstable. If the initial state of the dynamics (which now contains h as well) is exactly $m^0_{\beta,h}$ the analogue of Theorem 5.3.1.1 holds and calling $\alpha = \alpha(h)$ the new coefficient of the linear instability, the distribution at time $\tau \log|\Lambda|$, $\tau > 1/(2\alpha)$, is peaked around $m^+_{\beta,h}$ and $m^-_{\beta,h}$ with equal weights. However this is not the true equilibrium distribution which is instead supported only by $m^+_{\beta,h}$ (in the limit $|\Lambda| \to \infty$). The escape from the unstable equilibrium $m^0_{\beta,h}$ is indeed shortsighted, it only sees the local curvature at $m^0_{\beta,h}$ where everything looks symmetric. After leaving the top, the system feels the difference between right and left, but it is too late as it is already falling in the chosen well.

True equilibrium in the two previous examples is reached after longer times, when with large probability there are fluctuations so strong to make the state jump from one well to the other. This is the tunneling phenomenon, here we shall discuss physical origin and effects of the strong fluctuations, quoting from the master reference in the field, the Statistical Physics book in the Landau and Lifschitz series. They explain the approach to equilibrium in a sub-system Λ of a macroscopic system as due to the fact that the other parts of the system being extremely large, effectively act as a reservoir on Λ which is eventually able to produce any forcing pattern. As a consequence any point of the phase space of the subsystem becomes reachable and memory of the initial state is lost.

We can think of the large variety and unpredictability of the interactions with the reservoir as modeled by the stochastic nature of the Glauber dynamics. If the sub-system in the Landau–Lifschits argument is large, $|\Lambda|$ large, the self-interaction in the system is in general dominant and the external action of the reservoir is typically comparatively small. This is well in agreement with our scheme where we have convergence to a limit deterministic evolution as $|\Lambda| \to \infty$. As a consequence, if the system has a negative magnetization it is then initially attracted by the stable value $-m_\beta$ around which it stays with small random fluctuations. However even if improbable the reservoir may and eventually will produce all sort of forces, for instance a sufficiently strong, positive magnetic field for a sufficiently long time interval to make $-m_\beta$ no longer an equilibrium value: the system is then driven to the other equilibrium $+m_\beta$. But the same effect may be obtained in many other ways, for instance by a magnetic field maybe initially stronger and smaller afterwards, or .... There are in fact many different patterns which lead in the end to a same outcome, each one, being very improbable, has a very small probability (exponentially small in $|\Lambda|$ as we shall see). It is then natural to expect that the one which appears first is the one with the largest probability and that the time to expect for such an event is proportional to the inverse of its probability.

The above considerations raise two main questions: estimate the probability of any macroscopic fluctuation which produces a desired effect (as for instance tunneling to the other well), find the one with largest probability. Large deviations and variational problems thus enter massively into play and with them probability and optimal control theory. The structures we are investigating are undoubtedly very rich and already non-trivial even in the over-simplified mean field model. We thus leave for the moment the analysis of the tunneling phenomenon and study large deviations.

5.3.3 Rate function of large deviations

Fix arbitrarily $\tau > 0$ and $u_0 \in [-1, 1]$ and consider for each Λ the Glauber dynamics in the time interval $[0, \tau]$ starting from a magnetization density $m \in M_\Lambda$ defined (for instance) as the closest value to u_0 such that $|m| \le |u_0|$. We want to imbed all possible trajectories $\{m(t), t \in [0, \tau]\}$ in the product space

$$X := \prod_{\delta,t}[-2, 2], \quad \delta = 2^{-n},\ n \in \mathbb{N},\ t = \delta k,\ k \in \mathbb{N} \cap [0, \delta^{-1}\tau]. \tag{5.3.3.1}$$

Denoting the elements of X by $\{D(t, \delta)\}$ and given a trajectory $m(\cdot)$ we set for each (δ, t) as in (5.3.3.1) $D(t, \delta) = m(t + \delta) - m(t)$. In this way $m(\cdot)$ identifies an element of X. We equip X with the product topology (and hence the space of magnetization density trajectories). Recalling that a function $u(t), t \in [0, \tau]$, is absolutely continuous if there is a $L^1([0, \tau])$ function $\rho(t)$ such that for all $t \in [0, \tau]$,

$$u(t) = u(0) + 2\int_0^t \rho(s)\,ds \tag{5.3.3.2}$$

(it is convenient to factorize a 2 because each jump produces a change of magnetization $\pm 2/|\Lambda|$). We state without proof the following:

Theorem 5.3.3.1 (Large deviations) *Let B and $\mathcal{O}$ be a closed and respectively open set in X, then*

$$\limsup_{|\Lambda|\to\infty} \frac{1}{|\Lambda|} \log P_{\Lambda,m}(m \in B) \le -\inf_{u\in B} I_\tau^0(u), \tag{5.3.3.3}$$

$$\liminf_{|\Lambda|\to\infty} \frac{1}{|\Lambda|} \log P_{\Lambda,m}(m \in \mathcal{O}) \ge -\inf_{u\in \mathcal{O}} I_\tau^0(u), \tag{5.3.3.4}$$

where $I_\tau(u) = +\infty$ *if* u *is not absolutely continuous, otherwise let* $\rho(\cdot)$ *as in* (5.3.3.2), *then*

$$I_\tau^0(u) := \int_0^\tau w\big(u(t), 2\rho(t)\big), \quad u(0) = u_0, \tag{5.3.3.5}$$

where calling for each $t \in [0, \tau]$, $\rho_\pm(t)$ *the solutions of*

$$\rho_-(t)\rho_+(t) = c_-(u(t))c_+(u(t)), \qquad \rho_-(t) - \rho_+(t) = \rho(t). \tag{5.3.3.6}$$

We have $w = W(\rho_+(t)|c_+(u(t))) + W(\rho_-(t)|c_-(u(t)))$ *and*

$$W(x|a) = x\log\frac{x}{a} - x + a = aW\left(\frac{x}{a}\right), \qquad W(x) \equiv W(x|1). \tag{5.3.3.7}$$

The proof of Theorem 5.3.3.1 is technical and omitted and just refer to the literature, see [85]. We shall just explain the heuristics behind $I_\tau^0(u)$. One of the possible realizations of the mean field Glauber dynamics is given in Theorems 5.2.3.4 and 5.2.4.1 by constructing for each $m \in M_\Lambda$ two independent Poisson processes, $\underline{t}_\pm(m)$ called "random times," and then taking the product over all m, we thus have a large product space where all these processes live and are mutually independent. The process $m(t), t \geq 0$, is constructed in this product space as follows. If at time $s \geq 0$ the process is in m then it remains in m till the minimum between the two times $t_\pm$ defined as $t_\pm = \min\{\underline{t}_\pm(m) \geq s\}$. If $t_- < t_+$ then at time t_- the magnetization jumps to $m - (2/|\Lambda|)$. If instead $t_- > t_+$ then the magnetization increases by $(2/|\Lambda|)$ at time t_+, the case $t_- = t_+$ has probability 0.

While the random times processes $\{\underline{t}_\pm(m)\}$ are simple objects to handle the way $m(t)$ is constructed is more complex as we need to know how random times are inter-related. The analysis would be drastically simpler if the intensities of the random times were only two, one for the plus jumps, say c_+, and one for the minus jumps, say c_-, thus neglecting their dependence on m. In such a case the change of magnetization in any time interval $[0, t]$ is just equal to $2(N_-(t) - N_+(t))$ where $N_\pm(t)$ is the number of plus/minus random times in $[0, t]$. Before worrying on how to reduce to such a case let us just see what are the large deviations for $N_\pm(t)$.

Proposition 5.3.3.2 *Let* $P_{|\Lambda|a}$ *be the law of a Poisson process* $\{\underline{t} = (0 \leq t_1 \leq \cdots \leq t_n \leq \cdots)\}$ *with intensity* $|\Lambda|a > 0$, *let* $x > 0$ *and* $\tau > 0$. *Then, calling* $N(t) = |\underline{t} \cap [0, t]|$ *the random number of points of the process in* $[0, t]$,

$$\lim_{\zeta\to 0}\lim_{|\Lambda|\to\infty}\frac{1}{|\Lambda|}\log P_{|\Lambda|a}\left(\left|\frac{N(\tau)}{|\Lambda|} - x\tau\right| < \zeta\right) = -\tau W(x|a), \tag{5.3.3.8}$$

with $W(x|a)$ *as in* (5.3.3.7).

Proof Recall first that

$$P_{|\Lambda|a}(N_\tau = n) = e^{-|\Lambda|a\tau}\frac{(|\Lambda|a\tau)^n}{n!},$$

and then use the Stirling formula, (A.2.5). □

Remarks Proposition 5.3.3.2 proves that a Poisson process with intensity a may look like a Poisson process with a different intensity x for a time t, the "cost" depends on how much x differs from a and it is proportional to t.

Let us now go back to our constant-intensity approximation of the mean field process which involves two Poisson processes one for the plus and the other for the minus jumps. Calling $b = c_+|\Lambda|$ and $a = c_-|\Lambda|$ the minus and the plus jump intensities (+ refers to a plus spin flipping which thus decreases the magnetization), the rate of magnetization change is then equal to $2(a - b)$. The next proposition estimates the optimal way to produce a different rate, say $\dot{m}$.

Proposition 5.3.3.3 *Let $\nu = P_{|\Lambda|a} \times P_{|\Lambda|b}$ be the law of two independent Poisson processes with intensity $|\Lambda|a > 0$ and respectively $|\Lambda|b > 0$. Call $N_-(t)$ and $N_+(t)$ the random number of points of the two processes in $[0, t]$. For any $\tau > 0$ and any $\dot{m}$*

$$\lim_{\zeta\to 0}\lim_{|\Lambda|\to\infty}\frac{1}{|\Lambda|}\log\nu\left(\left|\frac{2}{|\Lambda|}(N_-(\tau) - N_+(\tau)) - \dot{m}\tau\right| < \zeta\right),$$
$$= -\tau\big(W(x_-|a) + W(x_+|b)\big) \tag{5.3.3.9}$$

where

$$x_\pm: \quad x_-x_+ = ab, \qquad 2(x_- - x_+) = \dot{m}. \tag{5.3.3.10}$$

Proof To leading orders $-\tau(W(x_-|a) + W(x_+|b))$ is the log of the probability of having $N_-(\tau)/|\Lambda|\tau$ and $N_+(\tau)/|\Lambda|$ respectively close to $x_-\tau$ and $x_+\tau$ (as $|\Lambda| \to \infty$), as it follows from (5.3.3.8). By optimizing on $x_\pm$ with the constraint $2(x_- - x_+) = \dot{m}$ we find the pair defined in (5.3.3.10). □

Remarks Under the constraint $2(x_- - x_+) = \dot{m}$ the minimizer of $W(x_-|a) + W(x_+|b)$ has the form $x_- =: e^{\theta^*}a$ and $x_+ =: e^{-\theta^*}b$. As we are going to argue, $\theta = \beta^{-1}\theta^*$ has the meaning of a "large deviations" magnetic field. Indeed, to apply the above to the mean field Glauber dynamics we take

$$\begin{aligned} a &= |\Lambda|c_-(m; h), \\ b &= |\Lambda|c_+(m; h); \quad c_\pm(m; h) \text{ as in (5.1.4.7).} \end{aligned} \tag{5.3.3.11}$$

The true intensities are instead $c_{\pm,\Lambda}$, but "the error is small" as $c_{\pm,\Lambda}/|\Lambda| \to c_\pm$ as $|\Lambda| \to \infty$ and with the choice (5.3.3.11) we have simpler expressions. Calling $\theta^* = \beta\theta$ and recalling that c_0 denotes the mobility coefficient,

$$\dot{m} = 2c_0(m; h)\big(\sinh\{\beta[m + h + \theta]\} - m\cosh\{\beta[m + h + \theta]\}\big), \tag{5.3.3.12}$$

which can also be written as

$$\dot{m} = 2\frac{c_0(m,h)}{c_0(m,h+\theta)}\big(c_-(m;h+\theta) - c_+(m;h+\theta)\big)$$
$$= \frac{c_0(m,h)}{c_0(m,h+\theta)} V(m,h+\theta). \tag{5.3.3.13}$$

Thus the large deviation velocity $\dot{m}$ is produced by adding a magnetic field θ but only in the Gibbs factor $e^{\pm\beta(m+h)}$ and not in the mobility c_0 which thus remains unchanged. If we had chosen the second alternative in (5.1.4.7), namely $c_0 = 1/2$, then the ratio of the mobility coefficients in (5.3.3.13) is equal to 1 and

$$\dot{m} = V(m,h+\theta) = \sinh\{\beta[m+\theta]\} - m\cosh\{\beta[m+\theta]\},$$

so that the new rates for the large deviation profile are exactly equal to the old ones, except for the additional magnetic field θ.

The rate function for large deviations in Proposition 5.3.3.3 is the same as the integrand of the rate function $I^0_\tau(u)$ in Theorem 5.3.3.1 and clearly hints at a proof of the latter by splitting the whole time interval $[0,\tau]$ in short subintervals $[k\delta,(k+1)\delta)$, $k \in \mathbb{N}$, and approximating in each interval the plus/minus jump intensities with constant ones. Indeed this strategy works and in the limit where we first let $|\Lambda| \to \infty$ and then $\delta \to 0$ we obtain Theorem 5.3.3.1. The proof is complicated by the fact that when the magnetization m goes to ± 1 then the minus, respectively plus intensities vanish; another technical point is to introduce cutoff for very large number of jumps which would make the magnetization vary too much and interfere with our discretization scheme.

5.3.4 Cost functional and optimal cost problems

The large deviations rate function $I^0_\tau(u)$ allows to introduce in the macroscopic theory dynamical behaviors which are in contrast with the laws of dynamics. There are indeed several phenomena with metastability and tunneling among the most famous ones, which cannot be explained by the basic principle of the macroscopic dynamics namely that time evolution is described by the gradient flow of the free energy functional. The above phenomena however can be included in the theory if we relax the gradient flow assumption by saying that all trajectories are possible but they have a "non-negative cost," those which are produced by the gradient flow equations have a zero cost. We then introduce as a primitive notion the functional

$$I_{T,u_0}(u) := \beta^{-1}\int_0^T w\left(u(t), \frac{du(t)}{dt}\right), \quad u(0) = u_0, \tag{5.3.4.1}$$

on the space of trajectories u with the meaning that $I_{T,u_0}(u)$ is the cost of u. Of course we must choose w in such a way that the cost of gradient flow trajectories is 0. If we have an underlying microscopic model then the choice of $I_{T,u_0}(u)$ is dictated by the large deviations rate function, but, as said, in a macroscopic theory $I_{T,u_0}(u)$ must be considered as a primitive notion.

Depending on the specific application, we may also want to add to $I_{T,u_0}(u)$ the cost, not included in $I_{T,u_0}(u)$, of being in u_0 at time 0. If $\phi_{\beta,0}(\cdot)$ denotes the equilibrium free energy (using the mean field notation), we then define the total cost of a trajectory as

$$F_T(u) := \phi_{\beta,0}(u(0)) + I_{T,u(0)}(u). \tag{5.3.4.2}$$

Optimal cost problems refer to the minimization of $I_{T,u_0}(u)$ (or $F_T(u)$) under given constraints, for instance in the tunneling problem we minimize $I_{T,u_0}(u)$, $u_0 = -m_\beta$ under the request that $u(T) = m_\beta$. We shall study this problem in Theorem 5.3.4.4 below leaving T (the time to tunnel) free. We shall use crucially in our analysis the following property:

Definition 5.3.4.1 (Reversibility) The cost functional $I_{T,u(0)}$ and the "lagrangian" $w(u,\dot u)$ in (5.3.4.1) are *"reversible" relative* to $\phi_{\beta,0}$ if

$$w(u,\dot u) - w(u,-\dot u) = \beta \frac{d\phi_{\beta,0}(u)}{du}\,\dot u. \tag{5.3.4.3}$$

An important consequence of reversibility in variational problem is that if $I_{T,u(0)}$ is reversible, then, calling $v(t) = u(T-t)$, $0 \le t \le T$, it immediately follows from (5.3.4.3) that

$$\begin{aligned} I_{T,u(0)}(u) &= I_{T,u(T)}(v) + \phi_{\beta,0}(u(T)) - \phi_{\beta,0}(u(0)),\\ F_T(u) &= F_T(v), \end{aligned} \tag{5.3.4.4}$$

which implies that the cost of a path u which goes against the gradient flow, i.e. with v above a gradient flow trajectory, is simply the difference of the free energies at $u(T)$ and $u(0)$.

In classical mechanics $w(u,\dot u) = w(u,-\dot u)$ so that the mechanical action is reversible and indeed if a trajectory $r(t)$ solves the Newton equations then it minimizes the action with constraints $r(0)$ and $r(T)$ (at least if T is small) and its time reversed is also a solution of the Newton equations minimizing the action with the inverted constraints.

Theorem 5.3.4.2 *The large deviations rate function I^0_τ of* (5.3.3.5) *is reversible.*

Proof The pair $\rho_\pm(u,\rho)$ associated to (u,ρ) by (5.3.3.6) is such that $\rho_\pm(u,\rho) = \rho_\mp(u,-\rho)$, hence

$$\begin{aligned} w(u,\dot u) - w(u,-\dot u) &= -(\rho_- - \rho_+)\log\frac{c_-(u)}{c_+(u)} = \rho\,\log\frac{c_+(u)}{c_-(u)}\\ &= \frac{\dot u}{2}\left(\log\frac{1+u}{1-u} - 2\beta u\right), \end{aligned}$$

which proves (5.3.4.3). □

From (5.3.4.4) it then follows that

Corollary 5.3.4.3 *Let* $[t_0, t_1] \subset [0, T]$, *call* $v(s) = u(t_0 + s)$, $0 \le s \le t_1 - t_0$, *then*

$$I_{T,u(0)}(u) \ge I_{t_1-t_0,u(t_0)}(v) \ge \big(\phi_{\beta,0}(u(t_1)) - \phi_{\beta,0}(u(t_0))\big), \qquad (5.3.4.5)$$

with equality if and only if $S_{t_1-t_0}(u(t_1)) = u(t_0)$ (S_t *the mean field evolution*).

Corollary 5.3.4.3 is one of the main ingredients in the proof of tunneling:

Theorem 5.3.4.4 (Tunneling) *The cost of tunneling, defined by the l.h.s. below, is*

$$\inf_{T>0} \inf\{I_{T,-m_\beta}(u) \mid u(T) = m_\beta\} = \phi_{\beta,0}(0) - \phi_{\beta,0}(m_\beta) =: \phi. \qquad (5.3.4.6)$$

Proof Lower bound. Since in any tunneling orbit $u(t), 0 \le t \le T$, there are t_0 and t such that $u(t_0) = -m_\beta + \epsilon$ and $u(t_0 + t) = -\epsilon$, $\epsilon > 0$, then by Corollary 5.3.4.3,

$$I_{T,-m_\beta}(u) \ge \phi_{\beta,0}(-\epsilon) - \phi_{\beta,0}(-m_\beta + \epsilon).$$

By the arbitrariness of ϵ this proves that the r.h.s. of (5.3.4.6) is $\ge \phi$.

Upper bound. We construct a minimizing sequence. We shall define $u(t)$ at the times when it is negative as the time reversal of the flow S_t while we follow the forward evolution when positive; all that except around the critical points 0 and $\pm m_\beta$. More precisely let $\epsilon > 0$ and $u(t) = -m_\beta + t/2$, $0 \le t \le 2\epsilon$. With τ such that $S_\tau(-\epsilon) = -m_\beta + \epsilon$ we then define $u(2\epsilon + t) = S_{\tau-t}(-\epsilon)$, $t \le \tau$. We set $u(2\epsilon + \tau + t) = -\epsilon + t/2$ till $t = 4\epsilon$. After that, $u(\tau + 6\epsilon + t) = S_t(\epsilon)$, $t \le \tau$. Finally, $u(2\tau + 6\epsilon + t) = m_\beta - \epsilon + t/2$, $t \le 2\epsilon$. For such u

$$I_{T,-m_\beta}(u) \le \phi_{\beta,0}(-\epsilon) - \phi_{\beta,0}(-m_\beta + \epsilon) + c\epsilon, \quad T = 2\tau + 8\epsilon.$$

c a suitable constant. Hence the upper bound by letting $\epsilon \to 0$. □

It is also possible to characterize the minimizing sequences, we state without proof the following theorem:

Theorem 5.3.4.5 (Optimizing sequences) *In the same context of Theorem 5.3.4.4, if* u_n *is a minimizing sequence and* $[0, T_n]$ *its time span, then* $T_n \to \infty$ *and for any* $\epsilon > 0$ *and all* n *large enough the following holds. Call* t_0, t_1 *the first times when* $u_n(\cdot)$ *is respectively equal to* $-m_\beta + \epsilon$ *and* ϵ. *Then with* τ *such that* $S_\tau(\epsilon) = m_\beta - \epsilon$, $T_n > t_1 + \tau$ *and*

$$\sup_{0 \le t \le \tau} |u_n(t_0 + t) - S_{\tau-t}(-\epsilon)| < \epsilon, \qquad \sup_{0 \le t \le \tau} |u_n(t_1 + t) - S_t(\epsilon)| < \epsilon.$$

As argued so far the macroscopic theory can include phenomena where trajectories deviate from the gradient flow dynamics by assigning to any such trajectory a cost. We have not yet discussed though the physical interpretation of such a cost which is in fact related to the time the system has typically to wait before the given event appears. This interpretation is validated by the analysis of tunneling in the mean field Glauber dynamics. For notational simplicity we take $h = 0$ so that tunneling refers to transitions from $-m_\beta$ to $+m_\beta$ and vice versa. The results are stated in the following theorem, we refer for its proof to the literature, see [177].

Theorem 5.3.4.6 *Let $h = 0$, $\beta > 1$ and denote by $\mathcal{P}_{\Lambda,m_0}$ the law of the Glauber process which starts from an initial configuration with magnetization $m_0 \le -m_\beta$. Let T^* be the [random] first time when the magnetization is $\ge m_\beta/2$. Call $\phi := \phi_{\beta,0}(0) - \phi_{\beta,0}(m_\beta) > 0$, the free energy difference between unstable and stable equilibria, then for any $\delta > 0$ and any $m_0 \le -m_\beta$*

$$\lim_{|\Lambda|\to\infty} \mathcal{P}_{\Lambda,m_0}\big(e^{[\phi-\delta]|\Lambda|} \le T^* \le e^{[\phi+\delta]|\Lambda|}\big) = 1. \tag{5.3.4.7}$$

5.4 Complements to Sect. 5.3

In this section we shall complete the proof of Theorem 5.3.1.1 ((5.3.1.1) has already been proved in Sect. 5.3.1).

5.4.1 Self similar growth

We renormalize $m(t)$ by dividing by its linear growth:

$$X(t) = m(t)[e^{\alpha t}|\Lambda|^{-1/2}]^{-1}, \qquad t \le \tau \log|\Lambda| =: t^*,\ \tau \in \left(0, \frac{1}{2\alpha}\right). \tag{5.4.1.1}$$

Analogously to (5.3.1.5) and (5.3.1.6) there are martingale relations for $X(t)$ as well (proofs are omitted)

$$X(t) - \int_0^t e^{-\alpha s}|\Lambda|^{1/2}\{\mathcal{L}m(s) - \alpha m(s)\} =: M(t) \quad \text{is a martingale,} \tag{5.4.1.2}$$

$$M(t)^2 - \int_0^t e^{-2\alpha s}|\Lambda|\{\mathcal{L}m^2 - 2m\mathcal{L}m\}(s) =: N(t) \quad \text{is a martingale.} \tag{5.4.1.3}$$

Lemma 5.4.1.1 *There is c and for any $\tau \in (0, \frac{1}{2\alpha})$ there are $a > 0$ and $\theta \in (0, 1/2)$ so that for all $|\Lambda|$ large enough,*

$$\mathcal{P}_\Lambda\Big(\sup_{t\le\tau\log|\Lambda|} |X(t) - M(t)| \le |\Lambda|^{-a}\Big) \ge 1 - 4(c\log|\Lambda|)|\Lambda|^{2\theta-1}. \tag{5.4.1.4}$$

Proof By (5.1.5.6) for any $t \le t^* := \tau \log|\Lambda|$

$$|X(t) - M(t)| \le t\Big(\frac{c}{|\Lambda|^{1/2}} + c'|\Lambda|^{1/2} \sup_{s\le t^*} e^{-\alpha s} m(s)^2\Big).$$

To bound the term $e^{-\alpha s}m(s)^2$ we are going to use (5.3.1.13) with α' and θ chosen in the following way. Since $\alpha\tau < 1/2$ there is θ so that $\alpha\tau + \frac{1}{2} < 2\theta < 1$. We then choose $\alpha' \in (\alpha, \frac{1}{2\tau})$ in such a way that $(2\alpha' - \alpha)\tau + \frac{1}{2} < 2\theta$. In this way the condition $\alpha'\tau < \theta$ in (5.3.1.13) is verified because

$$2\theta > (2\alpha' - \alpha)\tau + \frac{1}{2} > \alpha'\tau + \frac{1}{2} > 2\alpha'\tau. \tag{5.4.1.5}$$

Then by (5.3.1.13) with probability $\ge 1 - 4(c\log|\Lambda|)|\Lambda|^{2\theta-1}$, for any $\sigma \le \tau$,

$$\begin{aligned}&|\Lambda|^{1/2} e^{-\alpha\sigma\log|\Lambda|} m(\sigma\log|\Lambda|)^2\\ &\quad\le e^{[1/2 - \alpha\sigma + 2(\alpha'\sigma - \theta)]\log|\Lambda|} \le e^{[(2\alpha'-\alpha)\tau - 2\theta + 1/2]\log|\Lambda|},\end{aligned}$$

and (5.4.1.4) holds with $a := 2\theta - (2\alpha' - \alpha)\tau - 1/2 > 0$, see (5.4.1.5). □

Lemma 5.4.1.2 *In the same context of Lemma* 5.4.1.1 *and with* $\alpha'\tau < \theta$ *as in its proof,*

$$\begin{aligned}&\mathcal{P}_\Lambda\Big(\sup_{t\le\tau\log|\Lambda|} \big||\Lambda|\{\mathcal{L}m^2 - 2m\mathcal{L}m\} - 2\big| \le c\big(|\Lambda|^{-1} + |\Lambda|^{2(\alpha'\tau-\theta)}\big)\Big)\\ &\quad\ge 1 - 4(c\log|\Lambda|)|\Lambda|^{2\theta-1}.\end{aligned} \tag{5.4.1.6}$$

Proof Since $\mathcal{L}m^2 - 2m\mathcal{L}m = \frac{4}{|\Lambda|}(c_{+,\Lambda} + c_{-,\Lambda})$, by (5.1.4.4)

$$\begin{aligned}|\Lambda|\{\mathcal{L}m^2 - 2m\mathcal{L}m\} = 4\Big\{&\frac{1-m}{2}\,\frac{e^{\beta[m+1/|\Lambda|]}}{e^{-\beta[m+1/|\Lambda|]} + e^{\beta[m+1/|\Lambda|]}}\\ &+ \frac{1+m}{2}\,\frac{e^{-\beta[m-1/|\Lambda|]}}{e^{-\beta[m-1/|\Lambda|]} + e^{\beta[m-1/|\Lambda|]}}\Big\} \le c,\end{aligned} \tag{5.4.1.7}$$

where c is a constant, independent of $|\Lambda|$ and m.

By (5.4.1.7), $\big||\Lambda|\{\mathcal{L}m^2 - 2m\mathcal{L}m\} - 2\big| \le c(\frac{1}{|\Lambda|} + m^2)$. By (5.3.1.13) with probability $\ge 1 - 4(c\log|\Lambda|)|\Lambda|^{2\theta-1}$, $\sup_{t\le t^*}|m(t)| \le |\Lambda|^{\alpha'\tau-\theta}$ hence (5.4.1.6). □

Proposition 5.4.1.3 *For any fixed t the law of $M(t)$ converges in distribution to a gaussian law of mean zero and covariance C_t, namely with density*

$$\rho(x,t) = \frac{e^{-x^2/(2C_t)}}{(2\pi C_t)^{1/2}}, \quad C_t = \int_0^t 2e^{-2\alpha s}. \tag{5.4.1.8}$$

Proof The proof follows from general theorems on martingales which actually prove convergence of the process $\{M(s), s \le t\}$. As said, we need to verify a few conditions in our model and then general theorems apply. The conditions to verify are:

- The integrand $e^{-2\alpha s}|\Lambda|\{\mathcal{L}m^2 - 2m\mathcal{L}m\}(s)$ in the second martingale relation (5.4.1.3) is uniformly bounded. This follows from (5.4.1.7).
- Call $[M(s)] = \lim_{\epsilon\to 0}\{M(s+\epsilon) - M(s-\epsilon)]$, then

$$\lim_{|\Lambda|\to\infty} \Big\| \sup_{s\le t}[M(s)] \Big\|_{\infty} = 0,$$

  which follows because $[M(s)] = [X(s)] = e^{-\alpha s}|\Lambda|^{1/2}[m(s)]$ and $[m(s)] \le 2/|\Lambda|$.
- $\lim_{|\Lambda|\to\infty} \mathcal{E}_\Lambda(e^{-2\alpha s}|\ |\Lambda|\{\mathcal{L}m^2 - 2m\mathcal{L}m\}(s) - 2|) = 0$, which follows from (5.4.1.6).

We just sketch the steps which using the above properties lead to the proof of the proposition, referring to the literature for details. In Sects. 2.6–2.7 of [89] the reader may find a more detailed discussion on which conditions imply given properties and reference to the literature for their proofs.

We regard $\{M(s), s \in [0,t]\}$, $t > 0$, as a process on $D([0,t],\mathbb{R})$, the Skorohod space of trajectories (which may have discontinuities) and denote by P_Λ its law (inherited by $\mathcal{P}_\Lambda$). By the first condition above $\{P_\Lambda\}$ is a tight family and converges weakly by subsequence to a limit law P on $D([0,t],\mathbb{R})$. P has support on $C([0,t],\mathbb{R})$, where the trajectories are continuous. This follows from the second condition above.

Under P, $M(s)$ is a martingale as well as $N(s) = M(s)^2 - 2\int_0^s e^{-\alpha s'}$, this follows from the third condition above.

P is uniquely determined by having support on $C([0,t],\mathbb{R})$ and by $M(s)$ and $N(s)$ being P-martingales. Define

$$\tilde{M}(s) = \sqrt{2}\int_0^s e^{-\alpha s'} db(s'), \qquad b(\cdot) \text{ the standard Brownian motion.} \tag{5.4.1.9}$$

$\tilde{M}(\cdot)$ is a martingale as well as $\tilde{M}(s)^2 - 2\int_0^s e^{-\alpha s'}$, P is then identified as the law of $\{\tilde{M}(s), s \le t\}$ as defined in (5.4.1.9). Then the P-distribution of $\tilde{M}(t)$ is as in (5.4.1.8). □

Theorem 5.4.1.4 *For any $\tau \in (0, \frac{1}{2\alpha})$, see* (5.4.1.1), *the law of $X(\tau \log|\Lambda|)$ converges weakly to the gaussian law of density*

$$\rho(x) := \frac{e^{-x^2/(2C)}}{(2\pi C)^{1/2}}, \qquad C = \int_0^\infty 2e^{-2\alpha s}. \tag{5.4.1.10}$$

Proof Let $\rho(x,t)$ be as in (5.4.1.8) and $t^* = \tau\log|\Lambda|$. There is c so that

$$\int (1+|x|)\,|\rho(x) - \rho(x,t)| \le ce^{-2\alpha t}. \tag{5.4.1.11}$$

On the other hand, by the martingale relation (5.4.1.3) and (5.4.1.7), for any $t \le t^*$

$$\mathcal{E}_\Lambda\big((M(t^*) - M(t))^2\big) = \mathcal{E}_\Lambda\big(M(t^*)^2 - M(t)^2\big) \le c \int_t^{t^*} e^{-2\alpha s}.$$

By compactness the joint distribution of $M(t^*)$ and $M(t)$ converges weakly by subsequences to a limit law $dQ(x,x')$ and

$$\int |x - x'|^2 dQ \le c \int_t^\infty e^{-2\alpha s}.$$

The marginal of Q over x' is the gaussian law P_t with density $\rho(x,t)$, the marginal over x is some law that we denote by P^*. The Vaserstein distance between P_t and P^* is then

$$R(P_t, P^*) := \inf_{Q} \int |x - x'| dQ \le \left(c \int_t^\infty e^{-2\alpha s} \right)^{1/2},$$

the inf being over all joint representations of P_t and P^*. Calling P_∞ the gaussian law with density (5.4.1.10),

$$R(P_t, P_\infty) \le \int |x|\, |\rho(x) - \rho(x,t)|.$$

The inequality follows by using the following joint representation of the two measures: $f(x)\delta(x-x') + (\rho(x,t) - f(x))\mathbf{1}_{x\in A}(\rho(x') - f(x'))\mathbf{1}_{x\in B}$ where

$$f(x) = \min\{\rho(x,t), \rho(x)\},$$
$$A = \{x : \rho(x,t) > f(x)\}, \qquad B = \{x : \rho(x) > f(x)\}.$$

By the triangular inequality

$$R(P_\infty, P^*) \le \left(c \int_t^\infty e^{-2\alpha s} \right)^{1/2} + c e^{-2\alpha t},$$

and by the arbitrariness of t that P^* is the gaussian (5.4.1.10). The theorem then follows from Lemma 5.4.1.1. □

5.4.2 *Proof of* (5.3.1.2)

Calling $t^* = \log|\Lambda|/(8\alpha)$, by Theorem 5.4.1.4 for any $\epsilon > 0$ there is $\delta_\epsilon > 0$ so that for all $|\Lambda|$ large enough

$$\mathcal{P}_\Lambda\left(X(t^*) \ge \frac{\delta_\epsilon}{4} \right) > \frac{1}{2} - 4\epsilon, \qquad \mathcal{P}_\Lambda\left(X(t^*) \le -\frac{\delta_\epsilon}{4} \right) > \frac{1}{2} - 4\epsilon. \qquad (5.4.2.1)$$

Then recalling the relation (5.4.1.1) between X and m, for any $\epsilon > 0$ there is $\delta > 0$ so that with probability $> 1/2 - \epsilon$, $m(t^*) \geq |\Lambda|^{-3/8}\delta$. Same for the negative values: with probability $> 1/2 - \epsilon$, $m(t^*) \leq -|\Lambda|^{-3/8}\delta$. By the Markov property, see (5.2.2.1), the evolution of $m(t)$ for $t \geq t^*$ has the same law as starting at time 0 with $m(t^*)$. By the symmetry under change of sign $m \to -m$ we may restrict to $\mathcal{P}_{\Lambda,m_0}$, with $m_0 \geq |\Lambda|^{-3/8}\delta$. Given any $\tau > 0$ call $T = \tau \log|\Lambda|$, then by (5.3.1.12) with $\theta \in (3/8, 1/2)$,

$$m(t) = m_0 + \int_0^t V_\Lambda(m(s)) + b(t).$$

$$\mathcal{P}_{\Lambda,m_0}\Big(\sup_{t\leq T}|b(t)| \geq |\Lambda|^{-\theta}\Big) \leq 4cT|\Lambda|^{2\theta-1}.$$

Thus with probability $\geq 1 - 4cT|\Lambda|^{2\theta-1}$ for all $|\Lambda|$ large enough,

$$m(t) = m_0 + \int_0^t V(m(s)) + b'(t), \quad \sup_{t\leq T}|b'(t)| \leq 2|\Lambda|^{-\theta}, \tag{5.4.2.2}$$

where we have used (5.1.5.6) to replace V_Λ by V, $V(m) = -m + \tanh\{\beta m\}$. Recalling that $m_0 \geq |\Lambda|^{-3/8}\delta$, $\delta > 0$ independent of Λ, and that $\theta > 3/8$ it follows from a perturbative analysis of (5.4.2.2) considering $b'(t)$ as the small perturbation, that $m(\tau \log|\Lambda|) \to m_\beta$ as $|\Lambda| \to \infty$ if $e^{\alpha\tau \log|\Lambda|}|\Lambda|^{-3/8} = |\Lambda|^{\alpha\tau - 3/8}$ diverges, i.e. $\alpha\tau > 3/8$. Recalling that the initial value m_0 corresponds in the original process to the value of the magnetization at time $t^* = \frac{1}{8\alpha}\log|\Lambda|$, then for $|\Lambda$ large enough, $m(t)$ in the original process is with probability approaching $1/2$ either close to $+m_\beta$ or to $-m_\beta$ if $t = \tau \log|\Lambda|$ and $\tau > 1/(2\alpha)$.

5.5 Notes and reference to Part I

A survey on equilibrium statistical mechanics was beyond my purposes, I have just presented some of the most basic properties selected in view of the successive Parts II and III. I will only give here few specific references of what used in the text. In the theory of DLR measures of Sects. 2.1–2.2 I have followed Georgii [135] and Bonetto, Gentile, Gallavotti [43]. The theory of conditional probabilities presented in Appendix A follows closely Rohlin [194, 195]; I have also used in this and in other parts of the book the basic texts in probability by Billingsley [31, 32], Dellacherie and Meyer [104], Doob, [113], Feller [118], Parthasarathy [179], Revuz and Yor [189], and the book of Dunford–Schwartz [114]. In Sects. 2.3–2.4 I have followed Simon [200], and Khinchin [154]. The existence of the thermodynamic limit for the entropy can be derived directly without going through the analysis of DLR measures as I did, see for instance Ruelle [197], and Simon [200].

Besides the original papers of Peierls [180], and Dobrushin [106], in Sect. 3.1 I have followed Georgii [135], Gallavotti [128]; in Sect. 3.2 the original papers by Dobrushin [107, 109] and the survey by Villani [208]. In Sect. 4.2 I have adapted the original work by Lebowitz–Penrose [160] to the much simpler case of the Ising model without "reference interactions," namely without an unscaled spin–spin interaction which is instead included in [160], see also [131–133]. In Chap. 5 I have used [89], in Sect. 5.3 I have adapted to the simpler case of the mean field model the analysis of spinodal decomposition in De Masi, Orlandi, Presutti, Triolo [98] and of large deviations in Comets [85].

A discussion on the foundations of statistical mechanics can be found in Penrose [181]. The relation with large deviations is underlined in Pfister [184]. I did not consider at all the quantum case, for which I refer to G.L. Sewell [199], and the random interactions, spin glass, random magnetic fields, ..., see Bovier [44], for a recent survey on the subject.

One of the issues raised in Chap. 5 is the macroscopic limits for stochastic dynamics and the persistence of stochasticity after taking the limit. Derivation of macroscopic equations (the hydrodynamic limit) from Glauber and/or Kawasaki dynamics is much studied in the literature, as general references see the books by H. Spohn [202], and C. Kipnis, C. Landim [155], and the lecture notes by J. Fritz [123]. Persistence of stochasticity in macroscopic systems is a very intriguing subject both for its theoretical and applicative implications. Here are some references: a survey on the subject is in Giacomin–Lebowitz–Presutti [140]. Interface fluctuations in a mechanical model [187], in simple exclusion processes [90, 119, 120]. Interface fluctuations in a sharp interface limit of stochastic PDEs: [28–30, 51, 52, 126]. Spinodal decomposition has been studied in [64, 95] in the context of stochastic systems; spinodal decomposition in the Cahn–Hilliard equation: [37, 38, 63, 143, 166–168]. Derivation of a limit process described by nonlinear stochastic equations: [24–26, 124]. A nice survey on experimental evidence of macroscopic randomness is in Ahlers [1]. Persistence of randomness has particular relevance in biological systems where often the initial stage of the evolution involves few individuals. The fluctuations are therefore large leading to possibly drastically different macroscopic outcomes, see for instance [10] and [205].

Metastability is maybe the most famous example of randomness acting at a macroscopic level. There are many rigorous derivations of metastability from microscopic systems and I will just refer to the recent survey by Olivieri–Vares [177], where in particular tunneling in the mean field (Curie–Weiss) model can also be found. Metastability with Kac potentials is studied in [11, 85], metastability and tunneling in a variational setting are studied in [22, 101, 156, 157].

Part II
Mesoscopic Theory, non-local functionals

Chapter 6
Non-local, free energy functionals

In this chapter we "move up to the mesoscopic level," which is Part II of the book. Recalling the discussion of Sect. 1.2, Postulate 0 of the theory asserts that states are order parameter valued functions on the spatial domain Λ occupied by the system. We shall suppose again that the order parameter is the magnetization density (in some cases, however, we shall take a particle density). Thus, a state is a function $m(r), r \in \Lambda$. $m(r)$ has the meaning of an average of spins (or of molecule positions) on the underlying microscopic level; the size of the box where averages are taken defines the spatial scale of the mesoscopic level (in terms of the microscopic one). While such an interpretation is essential to establish the relation with microscopics, it is, however, inconsequential for what follows. The mesoscopic theory that we present is in fact entirely based upon a free energy functional, which is a primitive notion in the theory, Postulate 1, say. The functional specifies the free energy of all possible mesoscopic states and based on it all the thermodynamic properties of the system can be derived.

We shall eventually restrict our analysis to the L–P free energy functionals of Sect. 4.2.3, because of their relations with the Ising model with Kac potentials. However, we want to stress again that their microscopic origin is for the moment inconsequential, and the free energy functionals are here regarded as primitive notion. They will be the starting point of a self-consistent theory, whose roots go back to the original work of van der Waals.

The chapter is organized as follows. We start in Sect. 6.1 by recalling the basic postulates of the mesoscopic theory, which are then applied to the study of thermodynamics: we prove the existence of pressure and free energy density in the thermodynamic limit and "equivalence of ensembles." In Sect. 6.2, we introduce and study dynamics, which is defined by non-local evolution equations. We mainly have in mind applications to variational problems, as the dynamics is defined in such a way as to make the free energy decrease (or more precisely, not increase). The first applications are seen right after this, in Sect. 6.3, where dynamics is used to study the structure of states in "local equilibrium." Finally, in Sect. 6.4 we study the "large deviations from equilibrium"; namely, we estimate the free energy cost of profiles which deviate from local equilibrium. The analysis is here centered around the notion of contours, and on the proof of lower bounds on their free energy, called Peierls estimates, in analogy with those in the Ising model. Such bounds will be fundamental in many successive applications. The reader may notice the analogies with the analysis in Part I of the Ising model.

6.1 Thermodynamics

6.1.1 The basic axioms

The mesoscopic theory is structurally similar to the microscopic theory of DLR states. The two refer to different spatial scales, and their elementary constituents are accordingly distinct as are the specific techniques used in their analysis, but the general procedures are structurally analogous.

The starting notion is the specification of all the possible states of the system, whose ensemble, the phase space, reflects the nature of the system. In this and in the next chapters we restrict ourselves to magnetic systems. Later on we shall consider systems of particles in the continuum. In the mesoscopic theory the states of magnetic systems are specified by measurable functions $m(r)$, defined on some spatial domain, a region Λ or the whole $\mathbb{R}^d$, according to what the case is. Thus, $m(r) \in \mathbb{R}$ has the meaning of a magnetization density (along some axis of the body) at the point r. The reader may refer to Sect. 4.2 to have in mind a specific example of how such a continuum description may arise from lattice spin systems, but, as already remarked, the underlying microscopic structure is not going to play a role: the elementary quantities are the phase space, $L^\infty(\mathbb{R}^d)$ or $L^\infty(\Lambda)$, and its elements.

As mentioned, in the last part of the book we shall study particle systems in the continuum; in such a case the mesoscopic states are non-negative, measurable functions $\rho(r)$, expressing the density of molecules at r. These are the simplest cases; the reader may easily imagine various extensions: several kinds of particles, spins with values in $S^n, \ldots,$ which can all be accommodated by generalizing the space where our densities take values. We may even conceive of more complex cases, in which, for instance, at each point the state of the system is specified by configurations of particles on (approximate) crystal structures if we want to model solid bodies, ... As explained in the introduction to the chapter, the mesoscopic theory is a level up from microscopics, describing collective properties of the system and neglecting all statistical fluctuations. Thus, $m(r)$ is a local average of spins (present in some microscopically large region around the point r, which, in mesoscopic units, must be regarded as infinitesimal and in the end not distinguishable from a point). The basic quantity in the theory is the free energy functional $F(m)$, which specifies the free energy of the state $m(\cdot)$. Together with $F(m)$ we also define its restrictions to sub-domains Λ of $\mathbb{R}^d$, in which cases the free energy functional may also depend on the boundary conditions. As a rule we shall use the symbol $\mathcal{F}(m)$ to denote the excess free energy functional, which differs from the previous one by an additive constant, chosen in such a way that the minimum of the latter is 0; this is why it is called an "excess free energy." Together with the free energy functional comes the physical assumption that the equilibrium states are the minimizers of the free energy, in the whole space [where the functional is defined] or in a sub-domain if external constraints are imposed upon the system, as for instance when the total magnetization is prescribed (the Wulff problem).

The mesoscopic theory, like statistical mechanics, has two main objectives, namely to characterize the equilibrium states and the thermodynamics potentials.

The former are defined as the minimizers of the free energy functional in the thermodynamic limit when the system occupies the whole $\mathbb{R}^d$; the thermodynamic potentials are related to the inf of the free energy functional in the thermodynamic limit.

Phase transitions correspond to degeneracy of minimizers, namely when there is more than one equilibrium state. In the models that we shall consider, phase transitions occur when the graph of the thermodynamic pressure versus magnetic field has a cusp, while the graph of the free energy versus magnetization has a straight segment. The detailed analysis of phase transitions and the coexistence of phases will be carried out for the particular class of L–P functionals, whose definition will be recalled in Sect. 6.1.3 below. Statements on the basic properties of the thermodynamic potentials will instead be proved for a more general class of functionals; see Sect. 6.1.4 below. We start with a famous example.

6.1.2 Example 1. The Ginzburg–Landau functional

The Ginzburg–Landau excess free energy functional,

$$\mathcal{F}^{\mathrm{gl}}(m) = \int_{\mathbb{R}^d} w\big(m(r)\big) + C|\nabla m(r)|^2\, dr, \tag{6.1.2.1}$$

is the prototype to have in mind. The simplest example for phase transitions is when $w(s) = (s^2-1)^2$. More generally $w(s)$, $s \in \mathbb{R}$, is a smooth function whose minimal value is 0, according to the interpretation of the functional as a free energy excess (as energy is defined modulo an additive constant, the assumption really is about the existence of a minimum). C in (6.1.2.1) is a positive constant. $m(r)$ is required to be differentiable such that the integral in (6.1.2.1) is well defined; since the interest is in minimization problems, often the domain of definition is extended, allowing for the functional to have also the value $+\infty$ and thus dropping the smoothness assumption.

As before, we regard the order-parameter valued function $m(r)$ as a magnetization density. $m(\cdot)$ is the generic non-equilibrium profile and $\mathcal{F}^{\mathrm{gl}}(m)$ quantifies the distance of m from equilibrium: the smaller $\mathcal{F}^{\mathrm{gl}}(m)$, the closer to equilibrium.

With their functional, Ginzburg and Landau had in mind two mechanisms to penalize departures from equilibrium: the first one is ruled by a "mean field" free energy density, described in (6.1.2.1) by $w(m)$. Any value of $m(r)$ which is not a minimizer of $w(\cdot)$ (i.e. a non-equilibrium value of the order parameter) contributes to the total free energy $\mathcal{F}^{\mathrm{gl}}(m)$ proportionally to the volume of where it is attained. This term, therefore, favors profiles supported by the minimizers of $w(\cdot)$. There must, however, be also a penalty for changing a minimizer, which, in the Ginzburg–Landau theory, is the last term in (6.1.2.1) which penalizes variations of m. The coefficient C will be related to the second moment of the interaction potential of the underlying microscopic model.

The functional (6.1.2.1) with w a double well describes a system with a phase transition. Indeed, the global minimizers of $\mathcal{F}^{gl}(\cdot)$ are constant functions $m(r)$ identically equal to minimizers of $w(\cdot)$, and by the double well assumption on $w(s)$ there are two equilibrium states. Their excess free energy is 0, and they are therefore interpreted as pure phases. All the other profiles have a non-zero free energy and are therefore non-equilibrium profiles.

6.1.3 Example 2. The L–P functional

This is the functional to which we shall eventually restrict ourselves in this chapter. It arises from scaling Ising models with Kac potentials as discussed in Chap. 4. L–P stands for Lebowitz and Penrose. Its specific expression when written as an excess free energy in the whole $\mathbb{R}^d$ is

$$\mathcal{F}_{\beta,h}(m) = \int_{\mathbb{R}^d} f_{\beta,h}\big(m(r)\big)\,dr + \frac{1}{4}\int_{\mathbb{R}^d\times\mathbb{R}^d} J(r,r')[m(r)-m(r')]^2\,dr\,dr', \qquad (6.1.3.1)$$

$f_{\beta,h}$ and J being non-negative functions. $\mathcal{F}(\cdot)$ is regarded as a functional on $L^\infty(\mathbb{R}^d,[-1,1])$ with values in $[0,\infty]$. We postpone for a while the precise definition of the terms appearing in (6.1.3.1), as for the moment it just suffices to say that $f_{\beta,h}(m)$ (when $\beta>1$ and $h=0$) has a double well shape with minimum equal to 0 and that J is a smooth probability kernel. To simplify the notation, we shall often drop β and h from the notation, resuming the full notation $\mathcal{F}_{\beta,h}(m)$ when convenient.

Relation with the Ginzburg–Landau functional

The two functionals (6.1.2.1) and (6.1.3.1) are similar. The second terms in (6.1.2.1) and (6.1.3.1) both penalize the variations of $m(\cdot)$, while the first ones push toward the minimizers. The Ginzburg–Landau analogue of (6.1.3.1) obviously has $w(m) = f_{\beta,h}(m)$. In the L–P functional of Chap. 4, $f_{\beta,h}(m)$ is the mean field free energy of the Ising model, not necessarily convex; hence the interpretation of the potential $w(m)$ in the Ginzburg–Landau theory as a mean field free energy density. Moreover, if we compute (6.1.3.1) on slowly varying functions, say $m(r)=m^\star(\epsilon r)$, $m^\star$ being a smooth function independent of ϵ, then, to leading orders in ϵ as $\epsilon\to 0$, the second term in (6.1.3.1) becomes

$$\frac{\epsilon^2}{2}\int_{\mathbb{R}^d}\langle\nabla m^\star, D\nabla m^\star\rangle\,dr,\quad D=(D_{i,j}),\ D_{i,j}=\frac{1}{2}\int_{\mathbb{R}^d} J(0,r)r_i r_j dr, \qquad (6.1.3.2)$$

assuming $J(r,r') = J(0,r'-r)$. The second term in (6.1.2.1) with $m(r) = m^\star(\epsilon r)$ is equal to $\frac{C\epsilon^2}{2}\int_{\mathbb{R}^d} |\nabla m^\star|^2(\epsilon r)dr$. The two are equal if D is a multiple of the identity (as when J is isotropic) and each term on the diagonal is equal to C, thus justifying the interpretation of the coefficient C as the second moment of the interaction. Thus, the Ginzburg–Landau functional may be thought of as a first order expansion of $\mathcal{F}_{\beta,h}(m)$, taking $w(m) = f_{\beta,h}(m)$ and $C = D$. As we shall see, the bulk, thermodynamic properties of the two functionals are the same, as they both rely entirely on the one body term $f_{\beta,h}(m)$; the interface properties, however, are only qualitatively, but not quantitatively the same. The reason is that the relevant profiles m which describe the coexistence of two phases vary at the interface between the two phases on the same scale as J, so that the approximation (6.1.3.2) becomes meaningless and the choice $C = D$ becomes inconsequential.

Choice of $f_{\beta,h}(m)$ and of $J(r,r')$

In this and in the following chapters we shall concentrate on the case that $f_{\beta,h}(m)$ in (6.1.3.1) is equal to

$$f_{\beta,h}(s) = \phi_{\beta,h}(s) + p_{\beta,h}, \quad p_{\beta,h} = -\min_{s\in[-1,1]} \phi_{\beta,h}(s), \tag{6.1.3.3}$$

where $\phi_{\beta,h}$ is the function defined in (4.1.2.2). The interaction $J(r,r')$ in (6.1.3.2) verifies the same conditions as stated in the paragraph *Assumptions on the interaction* of Sect. 4.2.1, so that $J(r,r') = J(0,r'-r)$, and it is a smooth, symmetric, probability kernel supported by $|r-r'| \le 1$. As $\mathcal{F}_{\beta,h}$ depends symmetrically on $J(r,r')$, there is no loss of generality on making the assumption $J(r,r') = J(r',r)$ and, equivalently, $J(0,r) = J(0,-r)$.

Pure phases

The equilibrium states, also called pure phases, are defined as the minimizers of $\mathcal{F}_{\beta,h}(\cdot)$. They are therefore constant functions identically equal to the minimizers of $\phi_{\beta,h}(s)$. When $h \neq 0$, $\phi_{\beta,h}(s)$ has a unique minimizer called $m_{\beta,h}$. The minimizer is also unique, and equal to 0, when $h = 0$ and $\beta \le 1$. At the coexistence curve, $\{(\beta,h) : h = 0, \beta > 1\}$, there are instead two minimizers $\pm m_\beta$, where m_β is the positive solution of the mean field equation, see Fig. 4.2 in Sect. 4.1.2, and

$$m_\beta = \tanh\{\beta m_\beta\}. \tag{6.1.3.4}$$

Any magnetization m in the interval $(-m_\beta, m_\beta)$ is "forbidden," in the sense that for no magnetic field h there is an equilibrium state at inverse temperature β with magnetization m (in the microscopic theory instead, all m can be attained by considering the DLR measures obtained as convex combinations of those corresponding to the pure phases). If instead $s \notin [-m_\beta, m_\beta]$, there exists a [unique] value h_s of the magnetic field h such that $m(r) = m_{\beta,h_s}$ is the unique minimizer of $\mathcal{F}_{\beta,h_s}(\cdot)$, and it is therefore an equilibrium state.

Excess free energy in bounded domains

Before turning to other non-local functionals, we define the excess free energy in bounded domains of $\mathbb{R}^d$. Let $m \in L^\infty(\mathbb{R}^d, [-1, 1])$, Λ a Borel set in $\mathbb{R}^d$ and m_Λ, m_{Λ^c} the restrictions of m to Λ and to Λ^c. We then write

$$\mathcal{F}(m) = \mathcal{F}_\Lambda(m_\Lambda | m_{\Lambda^c}) + \mathcal{F}_{\Lambda^c}(m_{\Lambda^c}), \tag{6.1.3.5}$$

where $\mathcal{F}_\Lambda(m_\Lambda | m_{\Lambda^c})$ is the excess free energy in Λ with boundary conditions m_{Λ^c}, and $\mathcal{F}_\Lambda(m_\Lambda)$ the excess free energy in Λ, without interactions with Λ^c:

$$\begin{aligned}\mathcal{F}_\Lambda(m_\Lambda) &= \int_\Lambda f_{\beta,h}\big(m_\Lambda(r)\big)\, dr \\ &\quad + \frac{1}{4}\int_\Lambda\int_\Lambda J(r,r')[m_\Lambda(r) - m_\Lambda(r')]^2\, dr'\, dr, \\ \mathcal{F}_\Lambda(m_\Lambda | m_{\Lambda^c}) &= \mathcal{F}_\Lambda(m_\Lambda) \\ &\quad + \frac{1}{2}\int_\Lambda\int_{\Lambda^c} J(r,r')[m_\Lambda(r) - m_{\Lambda^c}(r')]^2\, dr'\, dr.\end{aligned} \tag{6.1.3.6}$$

Observe that by (6.1.3.5) and (6.1.3.6), $\mathcal{F}(m) \ge \mathcal{F}_\Lambda(m_\Lambda) + \mathcal{F}_{\Lambda^c}(m_{\Lambda^c})$.

Free energy in bounded domains

If we expand the square in the double integral on the r.h.s. of (6.1.3.1), we have a simplification between the term with $m(r)^2$ and the $m(r)^2$ which is in $f_{\beta,h}(m(r))$; see (6.1.3.3). The argument is, however, formal, as the integrals are over unbounded domains. But it can be made rigorous when the functional is defined in bounded domains with boundary conditions. We then have

$$\mathcal{F}_\Lambda(m_\Lambda | m_{\Lambda^c}) = F_\Lambda(m_\Lambda | m_{\Lambda^c}) - R_\Lambda(m_{\Lambda^c}), \tag{6.1.3.7}$$

where

$$F_\Lambda(m_\Lambda | m_{\Lambda^c}) = F_\Lambda(m_\Lambda) - \int_\Lambda\int_{\Lambda^c} J(r,r') m_\Lambda(r) m_{\Lambda^c}(r')\, dr' dr, \tag{6.1.3.8}$$

$$\begin{aligned}F_\Lambda(m_\Lambda) &= \int_\Lambda -h m_\Lambda(r) - \frac{I(m_\Lambda(r))}{\beta}\, dr \\ &\quad - \frac{1}{2}\int_\Lambda\int_\Lambda J(r,r') m_\Lambda(r) m_\Lambda(r')\, dr' dr,\end{aligned} \tag{6.1.3.9}$$

$$R_\Lambda(m_{\Lambda^c}) = -p_{\beta,h}|\Lambda| - \frac{1}{2}\int_\Lambda\int_{\Lambda^c} J(r,r') m_{\Lambda^c}(r')^2\, dr'\, dr. \tag{6.1.3.10}$$

$I(m)$ is defined in (4.1.2.3) and $p_{\beta,h}$ in (6.1.3.3). In the sequel we shall often study variational problems with m_{Λ^c} a fixed boundary condition. Then $R_\Lambda(m_{\Lambda^c})$ is a constant and, by (6.1.3.7), the variational problems for $\mathcal{F}_\Lambda(m_\Lambda | m_{\Lambda^c})$ and $F_\Lambda(m_\Lambda | m_{\Lambda^c})$

are equivalent. $F_\Lambda(m_\Lambda|m_{\Lambda^c})$ is the L–P functional derived in Theorem 4.2.2.1, in the analysis of Ising spins with Kac potentials, which thus provides an extra justification for $F_\Lambda(m_\Lambda|m_{\Lambda^c})$ and $\mathcal{F}_\Lambda(m_\Lambda|m_{\Lambda^c})$ being the free energy functionals with boundary conditions m_{Λ^c}.

6.1.4 Generalizations of the L–P functional

We consider here a class of non-local free energy functionals which generalizes in various directions the L–P functional. They can be written as

$$F_{\beta,h,\Lambda}(m_\Lambda|m_{\Lambda^c}) = \int_\Lambda \psi_\beta(m_\Lambda) - m_\Lambda\left(h + \frac{J * m_\Lambda}{2} + J * m_{\Lambda^c}\right), \qquad (6.1.4.1)$$

where Λ is a bounded, measurable region in $\mathbb{R}^d$, and m_Λ and m_{Λ^c} are measurable functions with support on Λ and Λ^c, respectively, and values in $[-1,1]$. $\psi_\beta(s)$ is a continuous function on $[-1,1]$, differentiable in $(-1,1)$ with derivative $\psi'_\beta(s)$ such that $\lim_{s\pm 1}\psi'_\beta(s) = \pm\infty$. We also suppose that $\psi_\beta(s)$ is convex in a neighborhood of $s=-1$ and $s=1$. As usual $J(r,r') = J(r',r) = J(0,r'-r)$, $J(0,r)=0$ for $|r|\geq 1$ and J is smooth, but we do not require that $J(r,r')\geq 0$. Finally, $J * f(r) = \int J(r,r')f(r')$. This class has a rich and interesting structure, as becomes evident by the few examples that we discuss below.

- ψ_β convex; J a probability kernel.

Such a class includes the L–P functional if $\psi_\beta(s) = -I(s)/\beta$, which, for what was proved in Chap. 4, arises from Ising systems with a ferromagnetic Kac interaction. If we add to such a system a fixed interaction (i.e. independent of γ) then—see [160]—the free energy functional derived in the limit $\gamma\to 0$ has the expression (6.1.4.1) with $\psi_\beta(s)$ the free energy density at magnetization s and inverse temperature β of the system with the Kac potential removed, which in the Lebowitz and Penrose terminology is called the "reference system." As proved in Chap. 2, the reference free energy $\psi_\beta(s)$ is convex but not necessarily strictly convex, as $\psi_\beta(s)$ may itself have a phase transition, its graph having a straight segment which will persist and maybe widen after the addition of the ferromagnetic Kac potential. See [40] for an extension of the analysis when γ is positive and "small."

- ψ_β strictly convex; J has also negative values.

Such functionals include those derived from Ising systems with Kac potentials $J_\gamma(r,r') = \gamma^d J(\gamma r,\gamma r')$ when $\psi_\beta(s) = -I(s)/\beta$. If there is also a reference interaction then ψ is the reference free energy [131–133]. If the Fourier transform of J has a maximum at a non-zero wave number, then the constants lose stability for β large enough and the minimizers are no longer constant functions. Pure phases for functionals in this class may have non-trivial spatial patterns; for instance, they may be periodic or quasi-periodic, depending on the form of $J(r,r')$; see [86] for the discussion of the $d=1$ case with the L–P functional with J also negative, but even in such a case the theory is far from complete. Another nice example in $d=1$ has the form

$$F_\Lambda(\rho_\Lambda)=\frac{1}{2}\int_\Lambda \rho(x)\int_{x-1}^{x+1}\rho(y)-\int_\Lambda \frac{-\rho}{\beta}(\log\rho-1)+\lambda\rho, \qquad (6.1.4.2)$$

where Λ is a circle of length >1 and ρ is a non-negative, periodic function. ρ has the meaning of a particle density and the model describes a two body, constant, repulsive interaction when particles are at (mesoscopic) distance ≤ 1, and particles at larger distance do not interact. The ground states ($\beta=\infty$) have been fully characterized (Buttà, Caglioti, Presutti, unpublished). They are density pockets at mutual distance 1, but the analysis at $\beta<\infty$ is still incomplete. The spatial patterns described above are on a mesoscopic scale length, and therefore they should not be confused with crystalline structures which appear instead on an inter-atomic microscopic scale.

- $\psi_\beta(s)$ has a double well shape with minima equal to 0 at $\pm m^*$; J is a probability kernel.

Such a case may be interpreted as a system with two Kac potentials with different scaling parameters. The one with shorter range is then approximated by a local function $\psi_\beta(s)$ which for a suitable choice of β has a double well shape. The non-convexity of $\psi_\beta(s)$ has interesting implications on the structure and stability of interfaces [16, 17].

The last model we discuss here is not exactly in the above class. Let Λ be a torus in $\mathbb{R}^d$ of side $L=\epsilon^{-1}$,

$$F_{\beta,\Lambda}(m_\Lambda)=\int_\Lambda\Big(-\frac{I(m_\Lambda)}{\beta}-m_\Lambda\frac{J*m_\Lambda}{2}+m_\Lambda K_\epsilon*m_\Lambda\Big),$$

where J and K are smooth, symmetric probability kernels with range 1, and $K_\epsilon(r,r')=\epsilon^d K(\epsilon r,\epsilon r')$, K a positive definite kernel. Without the last term this is the L–P functional and since we suppose $\beta>1$, its minimizers are constant functions equal to m_β or to $-m_\beta$. The last term instead favors $m=0$, however, due to the convolution structure $m=0$ should be understood only in a weak sense, and we may exploit this to "satisfy" the first unscaled part of the functional by doing fast oscillations (on the scale of Λ) between the two values $\pm m_\beta$; hence the emergence of micro-structures. Interesting problems then arise like determining the period (if it in fact is periodic) and the shape of the optimizing sequences. Such questions have been studied in similar problems with free energy functionals arising in the study of elastic and magnetic bodies [82, 84, 174] and in the analysis of ground states in $d=1$ Ising models with ferromagnetic n.n. interaction plus long range negative potentials [141]. See also the $d=2$ analysis in [165].

6.1.5 Example 3. The LMP model

The LMP model (LMP stands for Lebowitz, Mazel, Presutti) is a system of particles in $\mathbb{R}^d$ which will be studied in Part III. Here we describe its mesoscopic version as a further example of non-local free energy functionals. The state space is now $L_0^\infty(\mathbb{R}^d;\mathbb{R}_+)$, namely the set of all bounded, measurable, non-negative functions ρ with compact support. $\rho(r)$ is interpreted as the particle density at $r \in \mathbb{R}^d$ and the compact support condition comes from considering systems confined to bounded regions. The free energy of a state ρ in LMP is

$$F(\rho) = \int_{\mathbb{R}^d} \Big\{ e_\lambda(J*\rho) - \frac{1}{\beta} S(\rho) \Big\}, \qquad S(\rho) = -\rho(\log\rho - 1), \tag{6.1.5.1}$$

where J is a smooth, symmetric, translational invariant probability kernel and $J*\rho$ the convolution of J and ρ. $e_\lambda(s)$ is the energy density when the particle density is s. The specific choice in Chap. 10 is $e_\lambda(s) = s^4/4! - s^2/2 - \lambda s$; λ is interpreted as the chemical potential. The mesoscopic nature of the model is reflected by the fact that the particle density is "diffuse"; namely, the contribution to the energy of a point r is not $e_\lambda(\rho(r))$ but rather $e_\lambda(J*\rho(r))$. Finally, $S(\rho)$ is an entropy term, and its expression is similar to that of $I(\cdot)$ in (4.1.2.3), from which it differs because the values of the order parameter are here in $[0,\infty)$, while in the magnetic systems they are in $[-1,1]$.

A difference from the previous models in (6.1.5.1) is that there is no explicit penalty for variation of the order parameter, but the joint action of entropy and energy with the concavity property of the latter will indeed imply that minimizers (on the torus) are constant functions. We refer to Chap. 10 for more details on the LMP model, which will not be considered anymore in this chapter.

6.1.6 Thermodynamic potentials

We restrict ourselves in the sequel to the functionals $F_{\beta,h,\Lambda}(m_\Lambda|m_{\Lambda^c})$ of (6.1.4.1) with ψ_β smooth and J not necessarily ferromagnetic; see Sect. 6.1.4.

Theorem 6.1.6.1 *There exist functions $a_\beta(s)$, $s\in[-1,1]$, and $p_\beta(h)$, $h\in\mathbb{R}$, one the Legendre transform of the other:*

$$a_\beta(s) = \sup_h \big(hs - p_\beta(h)\big), \qquad p_\beta(h) = \sup_s \big(hs - a_\beta(s)\big) \tag{6.1.6.1}$$

such that for any van Hove sequence $\Lambda\to\mathbb{R}^d$ (see Definition 2.3.2.1) and any $m_{\Lambda^c}\in L^\infty(\Lambda^c,[-1,1])$,

$$\lim_{\Lambda\to\mathbb{R}^d} \frac{1}{|\Lambda|} \inf_{m_\Lambda\in\mathcal{X}_{\Lambda,s}} F_{\beta,0,\Lambda}(m_\Lambda|m_{\Lambda^c}) = a_\beta(s), \quad \textit{for any } s\in[-1,1], \tag{6.1.6.2}$$

where $\mathcal{X}_{\Lambda,s} = \{m_\Lambda \in L^\infty(\Lambda,[-1,1]) : \fint_\Lambda m_\Lambda = s\}$; *while for any* $h \in \mathbb{R}$

$$\lim_{\Lambda \to \mathbb{R}^d} \frac{1}{|\Lambda|} \inf_{m_\Lambda \in L^\infty(\Lambda,[-1,1])} F_{\beta,h,\Lambda}(m_\Lambda|m_{\Lambda^c}) = -p_\beta(h). \tag{6.1.6.3}$$

Proof The proof will follow from the three claims below. Let C_n be a cube of side 2^n, $n \in \mathbb{N}_+$, $0_\Delta(r)$ the function identically equal to 0 in Δ,

$$a_{\beta,C_n}(s) = \inf_{m_{C_n} \in \mathcal{X}_{C_n,s}} \frac{F_{\beta,0,C_n}(m_{C_n}|0_{C_n^c})}{|C_n|}. \tag{6.1.6.4}$$

Claim 1 *For any* $\epsilon > 0$ *there is* $\delta > 0$ *so that*

$$\sup_n \sup_{|s-s'|\le\delta} |a_{\beta,C_n}(s) - a_{\beta,C_n}(s')| \le \epsilon. \tag{6.1.6.5}$$

Proof Without loss of generality we may suppose $s' \in [s-\delta, s)$. If $s \ge 0$, we write $s' = a(s+1) - 1$ with a in $[\frac{1+s-\delta}{1+s}, 1)$. If instead $s < 0$, we write $s = 1 - b(1-s')$ with b in $(1, \frac{1-s'+\delta}{1-s'}]$. Correspondingly, let $T_a : \mathcal{X}_{C_n,s} \to \mathcal{X}_{C_n,s'}$ and $S_b : \mathcal{X}_{C_n,s'} \to \mathcal{X}_{C_n,s}$ where for $m \in \mathcal{X}_{C_n,s}$ and $m' \in \mathcal{X}_{C_n,s'}$,

$$T_a(m)(r) = a(m(r)+1) - 1, \qquad S_b(m')(r) = 1 - b(1 - m'(r)). \tag{6.1.6.6}$$

Consider the case $s \ge 0$; then $a \in [1-\delta, 1]$, and for any $\epsilon > 0$ there is $\delta > 0$ so that $|F_{\beta,0,C_n}(m_{C_n}|0_{C_n^c}) - F_{\beta,0,C_n}(T_a(m_{C_n})|0_{C_n^c})| \le \epsilon|C_n|$, and hence $|a_{\beta,C_n}(s) - a_{\beta,C_n}(s')| \le \epsilon$. An analogous argument with S_b instead of T_a applies when $s < 0$. Claim 1 is proved. □

Claim 2 $a_{\beta,C_n}(s)$ *converges uniformly as* $n \to \infty$ *and the limit* $a_\beta^*(s)$ *is a convex function.*

Proof Since $\mathcal{X}_{C_n,s}$ contains the set of all m_{C_n} whose restriction to the cubes $C_{n-1}(i)$ contained in C_n are all in $\mathcal{X}_{C_{n-1}(i),s}$, there is a constant c so that $a_{\beta,C_n}(s) \le a_{\beta,C_{n-1}}(s) + c2^{-n}$, because $c2^{n(d-1)}$ bounds the interaction among the different sub-cubes. Since $c2^{-n} = c2^{-(n-1)} - c2^{-n}$, $b_n \le b_{n-1}$ where $b_n := a_{\beta,C_n}(s) - c2^{-n}$. b_n is bounded and monotone; it has therefore a limit and $a_{\beta,C_n}(s)$ converges to some $a_\beta^*(s)$. Uniformity follows from Claim 1. By the continuity of a_β for convexity it is sufficient to prove that

$$a_\beta\big(\alpha s + (1-\alpha)s'\big) \le \alpha a_\beta(s) + (1-\alpha)a_\beta(s'), \tag{6.1.6.7}$$

with $\alpha = 2^{-pd}q$, p and q positive integers, $q \in [0, 2^{pd}]$. For $n > p$ we divide C_n into 2^{pd} cubes $C(i)$ of side 2^{n-p}. We define functions m in $\mathcal{X}_{s^*,C_n}$ in the following way. We choose q cubes among the 2^{pd} cubes $C(i)$ and impose that the restriction of m to each one of the $C(i)$ is in $\mathcal{X}_{s,C(i)}$. In the remaining cubes $C(j)$, we instead impose that $m_{C(j)} \in \mathcal{X}_{C(j),s'}$. Then $m \in \mathcal{X}_{C_n,s^*}$ and by optimizing the choices in $\mathcal{X}_{s,C(i)}$ and $\mathcal{X}_{C(j),s'}$:

$$a_{\beta,C_n}(s^*) \le \alpha a_{\beta,C_{n-p}}(s) + (1-\alpha) a_{\beta,C_{n-p}}(s') + c2^{-(n-p)}. \tag{6.1.6.8}$$

By letting $n \to \infty$ we then obtain (6.1.6.7). Claim 2 is proved. □

Having in mind (6.1.6.3) we define $p^*_\beta(h) := \lim_{n\to\infty} -\frac{1}{|C_n|} \times \inf_{m_{C_n}\in L^\infty(C_n,[-1,1])} F_{\beta,h,C_n}(m_{C_n}|0_{C_n^c})$. By Claim 1 and Claim 2 the limit exists and

$$p^*_\beta(h) := \lim_{n\to\infty} \sup_{s\in[-1,1]} \{hs - a_{\beta,C_n}(s)\} = \sup_{s\in[-1,1]} \big(hs - a^*_\beta(s)\big); \tag{6.1.6.9}$$

Since by Claim 2 $a^*_\beta(s)$ is convex, we conclude from (6.1.6.9) that $a^*_\beta(s)$ is the Legendre transform of $p^*_\beta(h)$ and hence

$$a^*_\beta(s) = \sup_h \big(hs - p^*_\beta(h)\big). \tag{6.1.6.10}$$

Claim 3 *Equations* (6.1.6.2)–(6.1.6.3) *hold with* $a_\beta(s) = a^*_\beta(s)$ *and* $p_\beta(h) = p^*_\beta(h)$.

Proof Write $p_{\beta,n}(h) = -\frac{1}{|C_n|} \inf_{m_{C_n}\in L^\infty(C_n,[-1,1])} F_{\beta,h,C_n}(m_{C_n}|0_{C_n^c})$, so that $\lim_{n\to\infty} p_{\beta,n}(h) = p^*_\beta(h)$. Consider a partition of $\mathbb{R}^d$ into translates of C_n and call $\Lambda' \subset \Lambda$ the union of the cubes in the partition contained in Λ. Then

$$\left| \inf_{m_\Lambda\in L^\infty(\Lambda,[-1,1])} F_{\beta,h,\Lambda}(m_\Lambda|m_{\Lambda^c}) - p_{\beta,n}(h)\frac{|\Lambda'|}{|\Lambda|}\right| \le c\frac{|\Lambda|-|\Lambda'|}{|\Lambda|} + c'2^{-n}.$$

By letting $\Lambda \to \mathbb{R}^d$ and then $n \to \infty$ we have (6.1.6.3) with $p_\beta(h) = p^*_\beta(h)$. We first prove (6.1.6.2) with $\ge$. We write

$$\frac{1}{|\Lambda|} \inf_{m_\Lambda\in\mathcal{X}_{\Lambda,s}} F_{\beta,0,\Lambda}(m_\Lambda|m_{\Lambda^c}) \ge hs + \frac{1}{|\Lambda|} \inf_{m_\Lambda\in L^\infty(\Lambda,[-1,1])} F_{\beta,h,\Lambda}(m_\Lambda|m_{\Lambda^c});$$

hence, for what was proved so far,

$$\text{l.h.s. of (6.1.6.2)} \ge \sup_h \big(hs - p^*_\beta(h)\big) = a^*_\beta(s).$$

For the upper bound we use again the partition into translates of C_n and choose m_Λ so that in each cube in Λ' it is the minimizer of the free energy in $\mathcal{X}_{C_n,s}$ (or an optimizing sequence). In $\Lambda \setminus \Lambda'$ we put $m_\Lambda = s$ so that $m_\Lambda \in \mathcal{X}_{\Lambda,s}$ and

$$\frac{1}{|\Lambda|} \inf_{m_\Lambda\in\mathcal{X}_{\Lambda,s}} F_{\beta,0,\Lambda}(m_\Lambda|m_{\Lambda^c}) \le a_{\beta,n}(s)\frac{|\Lambda'|}{|\Lambda|} + c\frac{|\Lambda|-|\Lambda'|}{|\Lambda|} + c'2^{-n}.$$

Taking first $\Lambda \to \mathbb{R}^d$ and then $n \to \infty$ we see that the l.h.s. of (6.1.6.2) is $\le a_\beta^*(s)$. Claim 3 is proved. □

Claim 1, Claim 2 and Claim 3 conclude the proof of Theorem 6.1.6.1. □

Physical interpretation

Recall that in the macroscopic theory the equilibrium states are defined as the minimizers of the free energy functional in the class of states available to the system. In (6.1.6.2) the only available states are those which have magnetization density equal to s; thus the limit is interpreted as the equilibrium free energy density of the system (with the dependence on Λ and m_{Λ^c} which disappears in the limit). $a_\beta(\cdot)$ is called the "canonical free energy density," because it is computed using the "canonical ensembles" $\mathcal{X}_{\Lambda,s}$. By the laws of thermodynamics the Legendre transform of $a_\beta(\cdot)$ is the pressure; thus, $p_\beta(h)$ is the thermodynamic pressure. It is also equal to minus the free energy of the equilibrium state when there is an external magnetic field h and the free energy functional is $F_{\beta,h,\Lambda}$. In such a case the minimization is over the grand canonical ensemble $\mathcal{X}_\Lambda$ and Theorem 6.1.6.1 may interpreted as "equivalence of ensembles."

Pressure and free energy have an explicit expression in the ferromagnetic case:

Theorem 6.1.6.2 *Let $J(r,r')$ be a probability kernel, $\phi_\beta(s) = \psi_\beta(s) - s^2/2$ and*

$$\pi_\beta(h) = \sup_{|s|\le 1}\big(hs - \phi_\beta(s)\big), \qquad \hat{\phi}_\beta(s) = \sup_h\big(hs - \pi_\beta(h)\big) \tag{6.1.6.11}$$

($\hat{\phi}_\beta = CE\phi_\beta$ being the convex envelope of ϕ_β). Then $p_\beta(h) = \pi_\beta(h)$ and $a_\beta(s) = \hat{\phi}_\beta(s)$.

Proof By Theorem 6.1.6.1 we may restrict ourselves to a sequence of cubes with boundary conditions 0_{Λ^c}, thus considering $F_{\beta,h,\Lambda}(m_\Lambda|0_{\Lambda^c})$ with L the side of the cube Λ. We have $|F_{\beta,h,\Lambda}(m_\Lambda|0_{\Lambda^c}) - G(m_\Lambda)| \le cL^{d-1}$, where

$$G(m_\Lambda) = \int_\Lambda \big(-hm_\Lambda + \phi_\beta(m_\Lambda)\big) + \frac{1}{4}\int_\Lambda\int_\Lambda J(r,r')\big(m_\Lambda(r) - m_\Lambda(r')\big)^2.$$

$\inf_{m_\Lambda} G(m_\Lambda) = |\Lambda| \inf_s(-hs + \phi_\beta(s))$, and hence $p_\beta(h) = \pi_\beta(h)$. Then by (6.1.6.1), $a_\beta(s) = \sup_h(hs - \pi_\beta(h)) = \hat{\phi}_\beta(s)$. □

6.2 Properties of the L–P free energy functional

In the sequel we restrict ourselves to the L–P free energy functional and prove some basic properties which will then be used to study several variational problems. One of the main tools will be dynamics. Time evolution in the mesoscopic theory is usually defined as the gradient flow of the free energy functional, so that by its very definition, the free energy decreases in time. The evolution can then be used to prove lower bounds on the free energy of states, as the free energy at time $t > 0$ is not larger than that at time 0. Gradient flow dynamics, however, has its own interest, and the next chapters will contain interesting properties of time evolution.

6.2.1 Positivity and lower semi-continuity

The typical properties of a functional when variational problems are concerned are positivity, lower semi-continuity and coercivity. The L–P excess free energy functional $\mathcal{F}(m)$ of (6.1.3.1) is clearly positive; lower semi-continuity is proved in Theorem 6.2.1.1 below, but coercivity is missing. Coercivity is a property which ensures that profiles with bounded free energy are "regular"; namely, the sub-levels of the functional have some regularity properties. Since the L–P functional restricted to a bounded domain, see Sect. 6.1.3, is bounded, no regularity can be gained by requiring that the free energy of a function is finite. On the contrary the request that a function has its Ginzburg–Landau free energy finite implies that the L^2 norm of its derivative is finite. Thus, an ingredient which makes $\mathcal{F}(m)$ in (6.1.3.1) simple, namely that it has no divergencies, has its counterparts. As we shall see in this chapter, deviations from "good patterns," however, have a cost, and this will be enough to regain the coercivity needed, for instance, in the analysis of the sharp interface limit in Chap. 7.

Theorem 6.2.1.1 *If $m_n \in L^\infty(\mathbb{R}^d, [-1,1])$ converges to m uniformly on the compacts, i.e. $\|m_n - m\|_{L^\infty(\Delta)} \to 0$ for any compact $\Delta \subset \mathbb{R}^d$, then*

$$\liminf_{n\to\infty} \mathcal{F}(m_n) \ge \mathcal{F}(m). \tag{6.2.1.1}$$

Let $\Lambda \subset \mathbb{R}^d$ be a bounded Borel set, $m_n \in L^\infty(\Lambda, [-1,1])$, $m_{\Lambda^c} \in L^\infty(\Lambda^c, [-1,1])$. Then if m_n converges weakly in $L^2(\Lambda)$ to m,

$$\liminf_{n\to\infty} \mathcal{F}_\Lambda(m_n|m_{\Lambda^c}) \ge \mathcal{F}_\Lambda(m|m_{\Lambda^c}), \tag{6.2.1.2}$$

while, if $m_n \to m$ almost everywhere, $\mathcal{F}_\Lambda(m_n|m_{\Lambda^c}) \to \mathcal{F}_\Lambda(m|m_{\Lambda^c})$.

Proof If $\lim_{n\to\infty} \|m_n - m\|_{L^\infty(\Delta)} = 0$ for any compact $\Delta \subset \mathbb{R}^d$, the function

$$g_n(r) := f_{\beta,h}\big(m_n(r)\big) + \frac{1}{4}\int J(r,r')[m_n(r) - m_n(r')]^2\, dr' \tag{6.2.1.3}$$

converges almost everywhere to $g(r)$, defined by (6.2.1.3) with m_n replaced by m. Supposing, without loss of generality, that $\mathcal{F}(m_n) < \infty$, by Fatou's lemma, the limit of the integrals of g_n is larger than the integral of the limit, hence we have (6.2.1.1).

By using the expression (6.1.3.8) we readily see that $F_\Lambda(m_\Lambda|m_{\Lambda^c})$, Λ bounded, is lower semi-continuous also in the weak $L^2(\Lambda)$-topology, because the entropy $I(\cdot)$ is concave and the double integral in (6.1.3.9) is continuous in $L^2(\Lambda)$. Equation (6.2.1.2) then follows while the last statement is proved using the Lebesgue dominated convergence theorem. □

6.2.2 Gradient flows, the non-local L–P evolution

The gradient flow dynamics for a free energy functional F is defined by a velocity field anti-parallel to the gradient,

$$\frac{dm}{dt} = -DF(m). \tag{6.2.2.4}$$

The evolution (6.2.2.4) has built in the correct thermodynamic property that the free energy does not increase:

$$\frac{dF(m)}{dt} = -\int |DF(m)|^2\, dr \le 0.$$

Of course, (6.2.2.4) is not the only one with such a property, as we can trivially multiply the r.h.s. of (6.2.2.4) by a positive function keeping the right monotonicity property of the free energy. The multiplying factor is interpreted as a "mobility coefficient," which in (6.2.2.4) has been set equal to 1.

By applying (6.2.2.4) to the L–P functional (6.1.3.1), we get the non-local evolution equation

$$\frac{dm}{dt} = J\star m + h - \frac{1}{2\beta}\log\frac{1+m}{1-m} = J\star m + h - \frac{1}{\beta}\operatorname{arctanh}(m), \tag{6.2.2.5}$$

while, applied to (6.1.2.1), we find the well known Allen–Cahn equation

$$\frac{dm}{dt} = C\Delta m - w'(m). \tag{6.2.2.6}$$

Like the functionals from which they are derived, the two equations have a similar structure: $J\star m - m$ plays the role of Δm in (6.2.2.6), while $\frac{1}{2\beta}\log\frac{1+m}{1-m} - m + h$ is the analogue of $w'(m)$.

By Theorem 5.1.6.1 the Glauber dynamics in Ising systems with Kac potentials converges as $\gamma \to 0$ (γ being the Kac scaling parameter) to

$$\frac{dm(r,t)}{dt} = -m(r,t) + \tanh\{\beta[J\star m(r,t) + h]\} \tag{6.2.2.7}$$

(the result in Theorem 5.1.6.1 refers to finite volumes, but the proof extends to $\mathbb{R}^d$). The two evolutions (6.2.2.5) and (6.2.2.7) are similar. Because of its relation with the spins, in the sequel we shall only refer to the latter.

6.2.3 *Partial dynamics*

Besides (6.2.2.7) which refers to the evolution of infinite systems (i.e. in the whole $\mathbb{R}^d$) it is also convenient to study dynamics in bounded regions. Given a bounded set Λ in $\mathbb{R}^d$ (by default all sets in the sequel are Borel), we define a "partial dynamics," partial because it evolves only m_Λ, the restriction of a function $m \in L^\infty(\mathbb{R}^d, [-1,1])$ to Λ, leaving unchanged m_{Λ^c}, the restriction of m to Λ^c:

$$\frac{dm_\Lambda}{dt} = -m_\Lambda + \tanh\big\{\beta\big(J \star (m_\Lambda + m_{\Lambda^c}) + h\big)\big\}, \quad \text{in } \Lambda, \tag{6.2.3.1}$$

with m_{Λ^c} independent of time. Complemented by an initial condition, (6.2.3.1) defines a Cauchy problem for the unknown $m_\Lambda(r,t)$, $r \in \Lambda$, $t \ge 0$.

Integral equations

Integral versions of (6.2.2.7) and (6.2.3.1) are

$$m(r,t) = e^{-t} m(r) + \int_0^t e^{-(t-s)} \tanh\{\beta[J \star m(r,s) + h]\}\, ds, \tag{6.2.3.2}$$

$$m_\Lambda = e^{-t} m_\Lambda + \int_0^t e^{-(t-s)} \tanh\big\{\beta\big(J \star (m_\Lambda + m_{\Lambda^c}) + h\big)\big\}\, ds. \tag{6.2.3.3}$$

Theorem 6.2.3.1 (Existence and uniqueness) *Let $m \in L^\infty(\mathbb{R}^d, [-1,1])$; then there is a unique function $m(r,t) \in L^\infty(\mathbb{R}^d \times \mathbb{R}_+, [-1,1])$ which satisfies* (6.2.3.2) *with $m(\cdot,0) = m(\cdot)$, $m(r,t)$ is continuously differentiable in t for any r and solves* (6.2.2.7). *Analogous statements hold for the partial dynamics.*

Proof Let $\tau > 0$ and Ψ the map on $L^\infty(\mathbb{R}^d \times [0,\tau];\ [-1,1]))$ defined by

$$[\Psi(u)](r,t) = e^{-t} m(r) + \int_0^t e^{-(t-s)} \tanh\{\beta[J \star u(r,s) + h]\}\, ds. \tag{6.2.3.4}$$

The image of Ψ is in $L^\infty(\mathbb{R}^d \times [0,\tau];\ [-1,1]))$ because $\tanh z < 1$ for all z. $[\Psi(u)](r,t)$ is obviously differentiable in t, and for $\tau > 0$ small enough, Ψ is a contraction (because the r.h.s. of (6.2.3.4) is uniformly Lipschitz). Hence it has a unique fixed point. Then the sequence $u_n = \Psi(u_{n-1})$, $u_0(r,t) = m(r)$, converges to the fixed point of Ψ and the integral equation (6.2.3.2) has a unique solution in the time interval $[0,\tau]$. By the arbitrariness of the initial value the argument can be iterated to all times proving global existence and uniqueness. By differentiating (6.2.3.2) we obtain (6.2.2.7), and the same arguments apply to the partial dynamics and the theorem is proved. □

Definition 6.2.3.2 (Evolution semigroups) Let $T_t(\cdot)$ and $T_t^\Lambda(\cdot)$, $t \ge 0$, be the semigroups which solve (6.2.2.7) and (6.2.3.1), respectively, with initial value m. The restriction of $T_t^\Lambda(m)$ to Λ will be denoted $[T_t^\Lambda(m)]_\Lambda$.

Theorem 6.2.3.3 (Regularity) *For any $t > 0$, the functions $T_t(m) - e^{-t}m$ on $\mathbb{R}^d$ and $T_t^\Lambda(m) - e^{-t}m$ on Λ are differentiable and their gradient is uniformly bounded. Moreover, for any positive integer k, T_t and T_t^Λ map $C^k(\mathbb{R}^d)$ and $C^k(\Lambda)$, respectively, into themselves.*

Proof Let $m(r,t) = T_t(m)(r)$; we are going to prove that

$$\sup_{t\geq 0} \left\| \nabla \left(m(r,t) - e^{-t}m(r,0) \right) \right\|_\infty \leq \beta \|\nabla J\|_1. \tag{6.2.3.5}$$

Let $u(r,t) := m(r,t) - e^{-t}m(r,0)$. Then

$$\begin{aligned} |\nabla u(r,t)| &\leq \int_0^t e^{-(t-s)} \left| \nabla \tanh\{\beta(J \star m)(r,s)\} \right| ds \\ &\leq \int_0^t e^{-(t-s)} \beta(|\nabla J| \star |m|)(r,s)\, ds. \end{aligned}$$

Since $|m(r,t)| \leq 1$, we have $|\nabla u(r,t)| \leq \beta \|\nabla J\|_1$. The same argument applies to $T_t^\Lambda(m)$.

The statements about the invariance of the spaces C^k are proved by differentiating (6.2.3.2)–(6.2.3.3). □

Theorem 6.2.3.4 (Continuity on the initial datum) *Let u and v both be in $L^\infty(\mathbb{R}^d, [-1,1])$; then for any $t > 0$, $\|T_t(u) - T_t(v)\|_\infty \leq e^{(\beta-1)t}\|u - v\|_\infty$ and for any bounded Borel set Λ in $\mathbb{R}^d$,*

$$\|[T_t^\Lambda(u)]_\Lambda - [T_t^\Lambda(v)]_\Lambda\|_\infty \leq e^{(\beta-1)t}\|u - v\|_\infty. \tag{6.2.3.6}$$

Proof By (6.2.3.2) and since $\int J(r,r')dr' = 1$

$$|T_t(u) - T_t(v)| \leq e^{-t}|u - v| + \int_0^t e^{-(t-s)} \|T_s(u) - T_s(v)\|_\infty ds.$$

Hence $\|T_t(u) - T_t(v)\|_\infty \leq e^{(\beta-1)t}\|u - v\|_\infty$. (6.2.3.6) is obtained similarly. □

6.2.4 *Local mean field equations*

The equilibrium phases are invariant under the evolutions T_t and T_t^Λ, but in many cases there are other stationary profiles as well. The equations they satisfy are called "local mean field equations" and have the form

$$m(r) = \tanh\big\{\beta\big(J \star m(r) + h\big)\big\}, \quad \text{for all } r \in \mathbb{R}^d, \tag{6.2.4.1}$$

while for the dynamics (6.2.3.1) the stationary profiles are solutions of the local mean field equation

$$m_\Lambda = \tanh\{\beta(J \star (m_\Lambda + m_{\Lambda^c}) + h)\}, \quad \text{on } \Lambda. \tag{6.2.4.2}$$

Notice that the solutions of (6.2.4.1), respectively (6.2.4.2), are critical points (i.e. solutions of the Euler–Lagrange equations) of the functional $\mathcal{F}(m)$ and $F(m_\Lambda|m_{\Lambda^c})$, respectively.

6.2.5 Free energy dissipation for partial dynamics

The main property of dynamics stemming directly from its definition is that it decreases the free energy, namely the free energy functional computed along an orbit is a non-increasing function of time. The property holds both for infinite and finite volume dynamics. As we shall use the statement for the infinite volume dynamics only later, we split the proofs and examine first the simpler case of the finite volume dynamics. The "free energy dissipation rate" for the partial dynamics in Λ will turn out to be the functional $\mathcal{I}^{(\Lambda)}(m)$, $m \in L^\infty(\mathbb{R}^d, [-1,1])$, defined by

$$\mathcal{I}^{(\Lambda)}(m) = \int_\Lambda \left(-J \star m - h + \frac{1}{2\beta} \log \frac{1+m}{1-m}\right) \times \left(-m + \tanh\{\beta[J \star m + h]\}\right) dr, \tag{6.2.5.1}$$

and $\mathcal{I}^{(\Lambda)}(m) = -\infty$ if the integrand is not in L^1. $\mathcal{I}^{(\Lambda)}(m)$ is indeed a dissipation rate, namely $\mathcal{I}^{(\Lambda)}(m) \le 0$; in fact, if the first bracket, on the r.h.s. of (6.2.5.1) is non-negative, i.e. $\beta(J \star m + h) \le \frac{1}{2} \log \frac{1+m}{1-m} = \operatorname{arctanh} m$, then the second one is non-positive, because, by taking the hyperbolic tangent of the two sides of the above inequality, we get $m \ge \tanh\{\beta[J \star m + h]\}$. Moreover, equality holds if and only if m satisfies (6.2.4.2).

We shall show in the next theorem that $\mathcal{I}^{(\Lambda)}(m)$ is the rate at which the free energy $F_\Lambda([T_t^\Lambda(m)]_\Lambda|m_{\Lambda^c})$ (F_Λ the L–P free energy functional defined in (6.1.3.8)) is dissipated along the orbit $T_t^\Lambda(m)$.

Theorem 6.2.5.1 *Let Λ be a bounded Borel set in $\mathbb{R}^d$, $m \in L^\infty(\mathbb{R}^d, [-1,1])$; then*

$$F_\Lambda\big([T_t^\Lambda(m)]_\Lambda|m_{\Lambda^c}\big) - F_\Lambda(m|m_{\Lambda^c}) = \int_0^t \mathcal{I}^{(\Lambda)}(T_s^\Lambda(m))\, ds. \tag{6.2.5.2}$$

Proof As an application of the maximum principle, we shall prove in Lemma 6.2.10 below that for any $m \in L^\infty(\mathbb{R}^d, [-1,1])$ and any $s > 0$, $T_t^\Lambda(m)$, $t \ge s$, is in Λ uniformly bounded away from ± 1. More explicitly, it is proved that for any $t \ge 0$ and any $r \in \Lambda$

$$\big|T_t^\Lambda(m)(r)\big| \le \big(\|m_\Lambda\|_\infty - \tanh\{\beta(|h|+1)\}\big)e^{-t} + \tanh\{\beta(|h|+1)\}, \tag{6.2.5.3}$$

which is $= 1-(1-x)(1-e^{-t}) < 1$, $x = \tanh\{\beta(|h|+1)\}$. Thus, for $t \ge s > 0$,

$$\frac{d}{dt} F_\Lambda\big([T_t^\Lambda(m)]_\Lambda | m_{\Lambda^c}\big) = \mathcal{I}^{(\Lambda)}(T_t^\Lambda(m)),$$

so that

$$F_\Lambda\big([T_t^\Lambda(m)]_\Lambda | m_{\Lambda^c}\big) - F_\Lambda\big([T_s^\Lambda(m)]_\Lambda | m_{\Lambda^c}\big) = \int_s^t \mathcal{I}^{(\Lambda)}(T_{t'}^\Lambda(m))\, dt'. \quad (6.2.5.4)$$

We have already proved that $\mathcal{I}^{(\Lambda)}(m) \le 0$; hence the r.h.s. of (6.2.5.4) is a monotone non-decreasing function of s. It thus converges as $s \to 0$ and the limit is the r.h.s. of (6.2.5.2). On the other hand, $|T_s^\Lambda(m) - m| \to 0$ uniformly as $s \to 0$ and by the Lebesgue dominated convergence theorem, $F_\Lambda([T_s^\Lambda(m)]_\Lambda | m_{\Lambda^c}) \to F_\Lambda(m_\Lambda | m_{\Lambda^c})$ as $s \to 0$. Thus (6.2.5.2) follows from (6.2.5.4). □

6.2.6 Limit points of orbits for partial dynamics

We shall prove that $T_t^\Lambda(m)$ converges by subsequences as $t \to \infty$ and that, under suitable conditions, the limit points satisfy the local mean field equation. The analogous statement for the infinite volume dynamics $T_t(m)$ is proved in Sect. 6.2.9 below.

Theorem 6.2.6.1 (Compactness of orbits) *Let $m \in L^\infty(\mathbb{R}^d, [-1,1])$ and Λ a bounded Borel set; then for any sequence $t_n \to \infty$, there are a subsequence $t_{n_k} \to \infty$ and $u \in C(\Lambda, [-1,1])$ such that*

$$\lim_{k\to\infty} T_{t_{n_k}}^\Lambda(m) = u. \quad (6.2.6.1)$$

Proof By (6.2.3.3) $T_t^\Lambda(m) - e^{-t}m$ and its gradient are uniformly bounded. The theorem then follows from the Ascoli–Arzelà theorem. □

Theorem 6.2.6.2 *In the same context as Theorem* 6.2.6.1 *any limit point u of $T_t^\Lambda(m)$ satisfies* (6.2.4.2), $|\nabla u|$ *is bounded in Λ and*

$$F_\Lambda(u_\Lambda | m_{\Lambda^c}) \le F_\Lambda(m_\Lambda | m_{\Lambda^c}). \quad (6.2.6.2)$$

Proof Let u be a limit point of $T_t^\Lambda(m)$; namely, suppose there is an increasing sequence $t_n \to \infty$ such that

$$\lim_{n\to\infty} \|T_{t_n}^\Lambda(m) - u\|_\infty = 0. \quad (6.2.6.3)$$

Suppose by contradiction that u does not satisfy the mean field equation (6.2.4.2); then by (6.2.5.2) the free energy difference is

$$A(u) := F\big([T_1^\Lambda(u)]_\Lambda | m_{\Lambda^c}\big) - F(u_\Lambda | m_{\Lambda^c}) = \int_0^1 \mathcal{I}^{(\Lambda)}(T_t^\Lambda(u))\, dt < 0. \quad (6.2.6.4)$$

We claim that, as a consequence of (6.2.6.4),

$$\lim_{t\to\infty}\int_0^t \mathcal{I}^{(\Lambda)}(T_s^{\Lambda}(m))\,ds = -\infty, \tag{6.2.6.5}$$

which leads to a contradiction, because

$$\int_0^t \mathcal{I}^{(\Lambda)}(T_s^{\Lambda}(m))\,ds = F_\Lambda([T_t^{\Lambda}(m)]_\Lambda|m_{\Lambda^c}) - F_\Lambda(m_\Lambda|m_{\Lambda^c}),$$

and $F_\Lambda(\cdot|m_{\Lambda^c})$ is bounded.

Proof of (6.2.6.5). By the continuity of the semigroup $T_t^{\Lambda}(\cdot)$ on the initial value, Theorem 6.2.3.4,

$$\lim_{n\to\infty}\left|T_t^{\Lambda}(u) - T_t^{\Lambda}\big(T_{t_n}^{\Lambda}(m)\big)\right| = 0, \quad t\in[0,1].$$

Then, by the Lebesgue dominated convergence theorem, recalling from the proof of Theorem 6.2.5.1 that the functions under consideration are bounded away from 1 in Λ,

$$\lim_{n\to\infty} A\big(T_{t_n}^{\Lambda}(m)\big) = A(u),$$

and since $A(u) < 0$, there is n^* so that, for all $n \ge n^*$,

$$A\big(T_{t_n}^{\Lambda}(m)\big) \le \frac{A(u)}{2}.$$

Assuming, without loss of generality, that $t_n - t_{n-1} > 1$ (otherwise we would take a subsequence of $\{t_n\}$ with such a property), we have

$$\int_0^t \mathcal{I}^{(\Lambda)}(T_s^{\Lambda}(m))\,ds \le \frac{A(u)}{2}N(t), \qquad N(t) = \mathrm{Card}\{n \ge n^* : t_n \le t-1\},$$

and (6.2.6.5) follows because $N(t)\to\infty$ as $t\to\infty$. Equation (6.2.6.2) follows from Theorem 6.2.1.1. □

6.2.7 Barrier Lemma

The next theorem proves that in a weak sense the velocity of propagation is finite:

Theorem 6.2.7.1 (Barrier Lemma) *There is a constant $C > 0$ so that for any $V \ge e^2\beta$, for any $t > 0$, for any $m_i \in L^\infty(\mathbb{R}^d, [-1,1])$, $i = 1, 2$, and writing $m_i(\cdot, t) = T_t(m_i)$,*

$$\begin{aligned}\big|m_1(0,t) - m_2(0,t)\big| &\le e^{(\beta-1)t}\sup_{|r|\le Vt}\big|m_1(r,0) - m_2(r,0)\big| \\ &\quad + C\exp\left\{-tV\log\frac{V}{e\beta}\right\}.\end{aligned} \tag{6.2.7.1}$$

Moreover, for any bounded Borel set Λ, any $m \in L^\infty(\mathbb{R}^d, [-1,1])$ and any $r \in \Lambda$

$$|T_t(m)(r) - T^\Lambda(m)(r)| \le C \exp\left\{-d_\Lambda(r)\log\left(\frac{d_\Lambda(r)}{e\beta t}\right)\right\},$$
$$d_\Lambda(r) = \mathrm{dist}(r, \Lambda^c). \tag{6.2.7.2}$$

Proof From (6.2.3.2), we get

$$\left|m_1(r,t) - m_2(r,t)\right| \le e^{-t}\left|m_1(r,0) - m_2(r,0)\right| + \int_0^t e^{-(t-s)}\beta \sup_{|r'-r|\le 1} \left|m_1(r',s) - m_2(r',s)\right| ds,$$

as $J(r,r')$ is supported by $|r - r'| \le 1$. Iterating the inequality and denoting by N the smallest integer larger than or equal to Vt, we get

$$\left|m_1(0,t) - m_2(0,t)\right| \le \sum_{n<N} e^{-t}\frac{(\beta t)^n}{n!} \sup_{|r|\le Vt} \left|m_1(r,0) - m_2(r,0)\right| + 2\frac{(\beta t)^N}{N!},$$

as $|m_i(r,t)| \le 1$ for all r and t. By the Stirling formula, (A.2.1), the last term is bounded by

$$\le C\exp\{N\log\beta t - N(\log N - 1)\} \le C\exp\left\{-N\log\frac{N}{e\beta t}\right\}.$$

As $N \ge Vt$ we get the last term on the r.h.s. of (6.2.7.1).

For (6.2.7.2) we proceed in the same way, iterating N times, with N the smallest integer larger or equal to $d_\Lambda(r)$. The theorem is proved. □

6.2.8 Free energy dissipation in infinite volume dynamics

Here we extend the analysis of Sect. 6.2.5 to the dynamics defined in the whole space.

Theorem 6.2.8.1 *Let $m \in L^\infty(\mathbb{R}^d, [-1,1])$ and $\mathcal{F}(m) < \infty$; then*

$$\mathcal{F}(T_t(m)) - \mathcal{F}(m) \le \int_0^t \mathcal{I}(T_s(m))\,ds, \tag{6.2.8.1}$$

with $\mathcal{I}(m)$ given by (6.2.5.1) with $\mathbb{R}^d$ instead of Λ. Equality in (6.2.8.1) holds only if $T_t(m) = m$ for all t.

Proof Let Δ_n be the cube of side $2n+1$ and center the origin, m_{Δ_n} and $m_{\Delta_n^c}$ the restrictions of m to Δ_n and to Δ_n^c. It follows from (6.2.5.2) that

$$\mathcal{F}(T_t^{\Delta_n}(m)) - \mathcal{F}(m) = \int_0^t \mathcal{I}^{(\Delta_n)}(T_s^{\Delta_n}(m))\,ds.$$

By the Barrier Lemma, Theorem 6.2.7.1, $T_t^{\Delta_n}(m) \to T_t(m)$, uniformly on the compacts as $n \to \infty$. Then, by lower semi-continuity, Theorem 6.2.1.1,

$$\liminf_{n\to\infty}[\mathcal{F}(T_t^{\Delta_n}(m)) - \mathcal{F}(m)] \geq \mathcal{F}(T_t(m)) - \mathcal{F}(m).$$

By Fatou's lemma,

$$\limsup_{n\to\infty} \int_0^t \mathcal{I}^{(\Delta_n)}(T_s^{\Delta_n}(m))\,ds \leq \int_0^t \mathcal{I}(T_s(m))\,ds;$$

hence we have (6.2.8.1). □

6.2.9 Limit points of orbits in infinite volume dynamics

We shall next prove that $T_t(m)$ converges by subsequences as $t \to \infty$; the same statement has been proved in Sect. 6.2.6 for finite volume dynamics.

Theorem 6.2.9.1 *Let $m \in L^\infty(\mathbb{R}^d, [-1,1])$ be such that $\mathcal{F}(m) < \infty$. Then for any sequence $t_n \to \infty$ there is a subsequence t_{n_k} such that $T_{t_{n_k}}(m)$ converges uniformly on the compacts, and any limit point u satisfies* (6.2.4.1) *and $\mathcal{F}(u) \leq \mathcal{F}(m)$.*

Proof By Ascoli–Arzelà $T_t(m)$ converges by subsequences uniformly on the compacts and by a diagonalization procedure over the whole $\mathbb{R}^d$. Let u be a limit point of $T_t(m)$; namely, suppose there is an increasing sequence $t_n \to \infty$ such that, for any $R > 0$:

$$\lim_{n\to\infty} \sup_{|r|\leq R} \left|T_{t_n}(m)(r) - u(r)\right| = 0. \tag{6.2.9.1}$$

Suppose by contradiction that u does not satisfy the mean field equation, i.e. $u \neq \tanh\{\beta[J \star u + h]\}$. Then there is a cube Δ so that

$$A(u) := \int_0^1 \mathcal{I}^{(\Delta)}(T_t(u))\,dt < 0. \tag{6.2.9.2}$$

We are going to show that if all that happens, then

$$\lim_{t\to\infty} \int_0^t \mathcal{I}^{(\Delta)}(T_s(m))\,ds = -\infty, \tag{6.2.9.3}$$

which contradicts the inequality (6.2.8.1), because

$$\int_0^t \mathcal{I}^{(\Delta)}(T_s(m))\,ds \geq \int_0^t \mathcal{I}(T_s(m))\,ds \geq \mathcal{F}(T_t(m)) - \mathcal{F}(m) \geq -\mathcal{F}(m) > -\infty.$$

Proof of (6.2.9.3). By the Barrier Lemma and (6.2.9.1), calling $\Delta' = \{r : \mathrm{dist}(r, \Delta) \leq 1\}$

$$\lim_{n\to\infty} \sup_{r\in\Delta'} \sup_{t\leq 1} \left|T_t(u)(r) - T_t\big(T_{t_n}(m)\big)(r)\right| = 0.$$

Then, by Fatou's lemma,

$$\lim_{n\to\infty} A\big(T_{t_n}(m)\big) \leq A(u),$$

and the remaining arguments are the same as in the proof of Theorem 6.2.6.1 and are omitted. The theorem is proved. □

6.2.10 *Comparison Theorem*

We shall consider explicitly only (6.2.2.7), which describes evolution on the whole $\mathbb{R}^d$; the case of the dynamics in a bounded region Λ with boundary conditions m_{Λ^c} in (6.2.3.1) can be recovered by reading in the sequel $r \in \Lambda$ and replacing h by $h(r)$, $r \in \Lambda$, where

$$h(r) = h + \int_{\Lambda^c} J(r, r') m_{\Lambda^c}(r')\,dr'. \tag{6.2.10.1}$$

Definition 6.2.10.1 $v(r,t)$ is a *sub-solution* of the Cauchy problem (6.2.2.7) with initial value $v(\cdot, 0)$ if for any $t > 0$

$$\sup_{s\leq t} \|v(\cdot, s)\|_\infty < \infty, \qquad \sup_{s\leq t} \left\|\frac{dv(\cdot, s)}{ds}\right\|_\infty < \infty, \tag{6.2.10.2}$$

and for all r and $t > 0$

$$\frac{dv(r,t)}{dt} \leq -v(r,t) + \tanh\{\beta(J \star v)(r,t) + \beta h\}. \tag{6.2.10.3}$$

Analogously, $w(r,t)$ is a *super-solution* of (6.2.2.7) if $w(\cdot)$ satisfies (6.2.10.2) and, with the reverse inequality (6.2.10.3).

Theorem 6.2.10.2 *Let $v(r,t)$, $w(r,t)$ and $m(r,t)$ be respectively a sub-solution, a super-solution and the solution of* (6.2.2.7) *with initial values $v(r,0) \leq m(r,0) \leq w(r,0)$. Then for all r and all $t \geq 0$:*

$$v(r,t) \leq m(r,t) \leq w(r,t). \tag{6.2.10.4}$$

Proof We shall only prove the first inequality in (6.2.10.4); the second one is similar and we omit it. Define $b(r,t)$ as

$$b(r,t) = \frac{dv(r,t)}{dt} - \big(-v(r,t) + \tanh\{\beta(J \star v)(r,t) + \beta h\}\big) \le 0. \qquad (6.2.10.5)$$

By assumption for any $t > 0$, $b \in L^\infty(\mathbb{R}^d \times [0,t])$. We shall regard b as a "known function" of r and t and, for any $\epsilon > 0$, we define $v_\epsilon(r,t)$ as the solution starting from $v(\cdot,0)$ of

$$\frac{dv_\epsilon(r,t)}{dt} = -v_\epsilon(r,t) + \tanh\{\beta(J \star v_\epsilon)(r,t) + \beta h\} + b(r,t) - \epsilon. \qquad (6.2.10.6)$$

Indeed, since $\|b(r,t)\|_\infty < \infty$, an existence and uniqueness theorem holds for (6.2.10.6), and v_ϵ is therefore well defined. Moreover, v satisfies (6.2.10.6) with $\epsilon = 0$; hence we have $v_0 = v$. Since the solution to (6.2.10.6) depends continuously on ϵ, for any (r,t)

$$\lim_{\epsilon \to 0} v_\epsilon(r,t) = v(r,t). \qquad (6.2.10.7)$$

Let $v_\epsilon^{(n)}$ be the solution of the equation obtained from (6.2.10.6) by restricting r to the ball $\{|r| \le n\}$ and by imposing the boundary conditions $v_\epsilon^{(n)}(r,t) = 0$ identically in $\{|r| > n, t \ge 0\}$ and $v_\epsilon^{(n)}(r,0) = v(r,0)$ for $|r| \le n$. Then by the Barrier Lemma which applies to (6.2.10.6) as well (as $b(r,t)$ is bounded)

$$v_\epsilon(r,t) = \lim_{n \to \infty} v_\epsilon^{(n)}(r,t), \quad \text{for any } (r,t). \qquad (6.2.10.8)$$

Analogously, $m(r,t) = \lim_{n\to\infty} m^{(n)}(r,t)$, where $m^{(n)}(r,t)$ is the solution of the equation obtained from (6.2.2.7) by restricting r to $\{|r| \le n, t \ge 0\}$ and by imposing the boundary conditions $m^{(n)}(r,t) = 0$ in $|r| > n$ and $m^{(n)}(r,0) = m(r,0)$ in $|r| \le n$.

We shall show in Claim 3 below that the function $\psi_\epsilon^{(n)}(r,t) = v_\epsilon^{(n)}(r,t) - m^{(n)}(r,t) \le 0$ for all r and all $t \ge 0$. After that it will be easy to conclude the proof of the theorem. We call $X_\epsilon^{(n)}(t) = \sup_{|r|\le n} \psi_\epsilon^{(n)}(r,t)$ and observe that there is a constant C so that $|d\psi_\epsilon^{(n)}(r,t)/dt| \le C$ for all $|r| \le n$; recall $\psi_\epsilon^{(n)}(r,t) = 0$ for $|r| > n$ and all $t \ge 0$.

Claim 1 *$X_\epsilon^{(n)}(t)$ is a Lipschitz continuous function of $t \ge 0$ with constant C, i.e. $|X_\epsilon^{(n)}(t) - X_\epsilon^{(n)}(s)| \le C|t-s|$, and (by assumption) $X_\epsilon^{(n)}(0) \le 0$.*

Proof Assume without loss of generality that $t > s$ and $X_\epsilon^{(n)}(t) > X_\epsilon^{(n)}(s)$. Let r_k be such that $\lim_{k\to\infty} \psi_\epsilon^{(n)}(r_k,t) = X_\epsilon^{(n)}(t)$. On the other hand $\psi_\epsilon^{(n)}(r_k,s) \ge \psi_\epsilon^{(n)}(r_k,t) - C(t-s)$, so that $X_\epsilon^{(n)}(s) \ge X_\epsilon^{(n)}(t) - C(t-s)$. Claim 1 is proved. □

Claim 2 *There are δ and ζ both positive so that if at some time $t \ge 0$, $X_\epsilon^{(n)}(t) \le \delta$ and $\psi_\epsilon^{(n)}(r,t) = 0$ for some $|r| \le n$, then $\frac{d\psi_\epsilon^{(n)}(r,t)}{dt} < -\zeta$.*

Proof Since $\psi_\epsilon^{(n)}(r,t)=0$, $\frac{d\psi_\epsilon^{(n)}(r,t)}{dt}$ is equal to

$$b(r,t)-\epsilon+\big(\tanh\{\beta(J\star v_\epsilon^{(n)})(r,t)+\beta h\}$$
$$-\tanh\{\beta(J\star m_\epsilon^{(n)})(r,t)+\beta h\}\big), \tag{6.2.10.9}$$

$\tanh\{\beta(J\star v_\epsilon^{(n)})(r,t)+\beta h\}-\tanh\{\beta(J\star m_\epsilon^{(n)})(r,t)+\beta h\}\le\beta\delta$ and $b\le 0$; then, for all $|r|\le n$ such that $\psi_\epsilon^{(n)}(r,t)=0$, $\frac{d\psi_\epsilon^{(n)}(r,t)}{dt}\le-\epsilon+\beta\delta$. Hence Claim 2 with $\delta=\frac{\epsilon}{2\beta}$ and $\zeta=\frac{\epsilon}{2}$. □

Claim 3 $X_\epsilon^{(n)}(t)\le 0$ *for all* $t\ge 0$.

Proof Consider the time interval $[0,\tau]$ with $\tau=C^{-1}\delta$, C as in Claim 1 and δ as in Claim 2. By Claim 1, $X_\epsilon^{(n)}(t')\le\delta$ for all $t'\in[0,\tau]$ and by Claim 2 if $\psi_\epsilon^{(n)}(r,t')=0$, $|r|\le n$, then $\frac{d\psi_\epsilon^{(n)}(r,t')}{dt}\le-\zeta$. Then for all $|r|\le n$ and $t'\in(0,\tau]$, $\psi_\epsilon^{(n)}(r,t')<0$, and therefore $X_\epsilon^{(n)}(t)\le 0$ for all $t\in[0,\tau]$. Claim 3 follows by iteration. □

Conclusions By the Barrier Lemma, the limit as $n\to\infty$ exists; then for any r and t, $v_\epsilon(r,t)-m(r,t)=\lim_{n\to\infty}(v_\epsilon^{(n)}(r,t)-m^{(n)}(r,t))\le 0$, which by (6.2.10.7), yields $v(r,t)\le m(r,t)$. The proof of the upper bound for the super-solution is similar and we omit it. □

The notion of sub- and super-solutions can be extended to the integral version of (6.2.2.7) and as we shall see the same conclusions apply. But while the validity of (6.2.10.4) extends to equations with more general local terms (in fact in (6.2.10.9) the local term has disappeared), for the integral version our proof uses in an essential way that the local term is linear.

We say that v is a sub-solution of (6.2.3.2) if

$$v(r,t)\le e^{-t}v(r,0)+\int_0^t e^{-(t-s)}\tanh\{\beta[J\star v(r,s)+h]\}\,ds,$$
$$(r,t)\in\mathbb{R}^d\times\mathbb{R}_+. \tag{6.2.10.10}$$

Super-solutions $w(r,t)$ are defined with the reversed inequality.

Theorem 6.2.10.3 *Let* $v(r,t)$, $w(r,t)$ *and* $m(r,t)$ *be respectively a sub-solution, a super-solution and the solution of* (6.2.3.2), *with initial values* $v(r,0)\le m(r,0)\le w(r,0)$. *Then* (6.2.10.4) *holds for all* r *and all* $t\ge 0$.

Proof Given $T>0$, let $X=C([0,T];L^\infty(\mathbb{R}^d,[-1,1]))$ and denote by $u(r,t)$ its elements. Let Ψ be the map from X into itself defined by

$$\Psi(u)(r,t)=e^{-t}u(r,0)+\int_0^t e^{-(t-s)}\tanh\{\beta(J\star u)(r,s)+\beta h\}\,ds. \tag{6.2.10.11}$$

With this notation, (6.2.3.2) is the fixed point equation for Ψ and

$$v \le \Psi(v), \qquad m = \Psi(m). \tag{6.2.10.12}$$

As in the proof of Theorem 6.2.10.2 we shall use in an essential way the monotonicity of Ψ: $\Psi(u) \ge \Psi(u')$ if $u \ge u'$, point-wise in $\mathbb{R}^d \times [0, T]$. This property, which obviously follows from (6.2.10.11), rests on two facts: the positivity of J and the property of increasing of the hyperbolic tangent. Both conditions may fail in more general systems, where it may happen that the interactions have different signs and the hyperbolic tangent is replaced by functions with more complex diagrams.

Going back to our case, we also observe that $(\Psi(u))(r, 0) = u(r, 0)$ and that for $\beta T < 1$, Ψ is a contraction on subsets of functions of X with the same values at $t = 0$. Thus if $m = \Psi(m)$,

$$m = \lim_{n\to\infty} \Psi^n(m^0), \quad m^0(r, t) = m(r, 0) \text{ in } \mathbb{R}^d \times [0, T]. \tag{6.2.10.13}$$

We are going to prove that

$$v \le \lim_{n\to\infty} \Psi^n(v^0), \quad v^0(r, t) = v(r, 0) \text{ in } \mathbb{R}^d \times [0, T], \tag{6.2.10.14}$$

and since $v(\cdot, 0) \le m(\cdot, 0)$, $\Psi^n(v^0) \le \Psi^n(m^0)$; hence $v \le m$. To prove (6.2.10.14) we observe that, since $v \le \Psi(v)$, $v \le \Psi^n(v)$ and

$$v \le z, \quad \text{where } z = \lim_{n\to\infty} \Psi^n(v). \tag{6.2.10.15}$$

Moreover, $z = \Psi(z)$ because, by the continuity of Ψ,

$$\Psi(z) = \Psi\Big(\lim_{n\to\infty} \Psi^n(v)\Big) = \lim_{n\to\infty} \Psi^{n+1}(v) = z. \tag{6.2.10.16}$$

Since $z = \Psi(z)$, by (6.2.10.13),

$$z = \lim_{n\to\infty} \Psi^n(z^0), \quad z^0(x, t) = z(x, 0) \text{ in } \mathbb{R}^d \times [0, T]. \tag{6.2.10.17}$$

Recalling that $z(\cdot, 0) = v(\cdot, 0)$, we have $z \le \lim_{n\to\infty} \Psi^n(v^0)$. Hence (6.2.10.14) follows from (6.2.10.15). We have thus proved that $v \le m$.

The same argument applies to super-solutions; thus (6.2.10.4) is proved for $0 \le t \le T$. But the same argument applies to $[T, 2T]$, because the proof did not depend on the initial value; hence, by iteration, we get (6.2.10.4) at all times. □

We conclude this subsection by proving a statement used in the proof of Theorem 6.2.5.1.

Lemma 6.2.10.4 *Let $m \in L^\infty(\mathbb{R}^d, [-1, 1]$, Λ a Borel set in $\mathbb{R}^d$ (possibly $\Lambda = \mathbb{R}^d$), $m(\cdot, t) = T_t^\Lambda(m)$ and $v(0) := \sup_{r\in\Lambda} |m(r)|$. Then, for all $r \in \Lambda$ and $t \ge 0$,*

$$|m(r, t)| \le [v(0) - \tanh\{\beta(|h| + 1)\}]e^{-t} + \tanh\{\beta(|h| + 1)\}. \tag{6.2.10.18}$$

Proof Let $v_\pm(t)$, $t \ge 0$, solve $\frac{dv_\pm(t)}{dt} = -v_\pm(t) + \tanh\{\beta(h \pm 1)\}$ with initial condition $\pm v(0)$. Then

$$v_\pm(t) = [\pm v(0) - \tanh\{\beta(h \pm 1)\}]e^{-t} + \tanh\{\beta(h \pm 1)\}.$$

Write $w_\pm(r,t) = v_\pm(t)$, $r \in \Lambda$, $t \ge 0$, and $w_\pm(r,t) = m_{\Lambda^c}$, $r \in \Lambda^c$, $t \ge 0$. Then w_+ is a super-solution of (6.2.3.1) because for $r \in \Lambda$ and $t \ge 0$,

$$\frac{dw_+}{dt} \ge -w_+ + \tanh\{\beta(J \star w_+ + h)\}$$

(since $w_+ \le 1$).

Analogously, w_- is a sub-solution of (6.2.3.1). Then, by the Comparison Theorem, for all r and t,

$$\begin{aligned}&[-v(0) - \tanh\{\beta(h-1)\}]e^{-t} + \tanh\{\beta(h-1)\}\\ &\le m(r,t) \le [v(0) - \tanh\{\beta(h+1)\}]e^{-t} + \tanh\{\beta(h+1)\},\end{aligned}$$

which is (6.2.10.18). □

6.3 Local equilibrium

In this and in the next section we keep $h = 0$ and $\beta > 1$ fixed. Recall from Sect. 6.1.3 that for such values of the parameters there is a phase transition, namely the L–P excess free energy functional $\mathcal{F}(m)$ has two minimizers. The two minimizers are two constant functions, respectively equal to $\pm m_\beta$, $m_\beta > 0$, and they are identified as the two pure equilibrium phases of the system. In this section we introduce a notion of local equilibrium, and we study the structure of profiles which are in local equilibrium; in the next section instead we shall study profiles in regions where they deviate from local equilibrium.

6.3.1 *A definition of local equilibrium*

The notion of "local equilibrium" will be such that profiles in local equilibrium are "locally but not point-wise" close to m_β, or to $-m_\beta$. An important consequence of the definition is that the free energy of a profile grows proportionally to the volume of the region where it deviates from local equilibrium, as we shall see in the next section. Here we study the structure of profiles which are in local equilibrium, proving that those with minimal free energy and for given "boundary conditions" are smooth and point-wise close to m_β, or to $-m_\beta$; see the remarks at the end of this subsection.

Partitions of the space

$\mathcal{D}^{(\ell)}$, $\ell \in \{2^n, n \in \mathbb{Z}\}$, is the partition of $\mathbb{R}^d$ into cubes $C^{(\ell)}$ of side ℓ defined in Sect. 4.2.2. $C_r^{(\ell)}$ denotes the cube of $\mathcal{D}^{(\ell)}$ which contains r and $C_r^{(\ell)} \subset C_r^{(\ell')}$ if $\ell < \ell'$.

Coarse graining and phase indicators

By "coarse graining with mesh ℓ" we mean the projection of L^∞ into itself which associates to a function m its averages

$$m^{(\ell)}(r) = \fint_{C_r^{(\ell)}} m(r')dr'. \tag{6.3.1.1}$$

For any $\zeta > 0$ and $\ell \in \{2^{-n}, n \in \mathbb{N}\}$ (thus ℓ is smaller than the range of the interaction) we define the phase indicator

$$\eta^{(\zeta,\ell)}(m;r) = \begin{cases} \pm 1 & \text{if } |m^{(\ell)}(r) \mp m_\beta| \le \zeta, \\ 0 & \text{otherwise.} \end{cases} \tag{6.3.1.2}$$

ζ is "the accuracy parameter." Later on we shall need two more phase indicators, but for the present considerations $\eta^{(\zeta,\ell)}$ is enough.

Local equilibrium

Given ζ and ℓ, $m \in L^\infty(\mathbb{R}^d, [-1,1])$ is in plus local equilibrium at r if $\eta^{(\zeta,\ell)}(m; r') = 1$ for all r' such that $\text{dist}(r, C_{r'}^{(\ell)}) \le 1$, and it is in plus local equilibrium in Λ if it is in plus local equilibrium at all $r \in \Lambda$. Minus local equilibrium is defined analogously. In this section we shall denote by $M^{\pm}_{\Lambda;\zeta,\ell}$ the set of all m in $\pm$ local equilibrium in Λ.

Remarks

Having defined a pure phase as a function constantly equal to m_β [or to $-m_\beta$], it may look more natural to call Λ in $\pm$ equilibrium if $m(r) = \pm m_\beta$, $r \in \Lambda$. This is, however, far too restrictive, and so is the relaxed notion $|m(r) \mp m_\beta| \le \zeta$ for all $r \in \Lambda$, because both conditions can be violated with an arbitrarily small free energy cost: the free energy in fact is expressed by integrals and therefore it is not sensitive to sup norms. A reasonable compromise leads to the above definition, which still requires closeness in sup norm, but only for the coarse grained image of m rather than for m itself. The introduction of coarse graining is in any case natural, once we remember the microscopic origin of the magnetization density profiles $m(r)$ as averages of spins over suitable boxes.

The definition of local equilibrium ensures the existence of a "safety zone" around Λ, because $m^{(\ell)}(r)$ (the coarse grained image of m) is required to be close to equilibrium not only in Λ, but also at all points within the interaction range from Λ. As we shall see, the definition is structured in such a way that if ℓ and ζ are small enough, then the minimizer of the free energy in $M^{\pm}_{\Lambda;\zeta,\ell}$ with fixed restriction to Λ^c is point-wise close in Λ to m_β [or to $-m_\beta$], with an exponential decay from the boundaries of Λ. By symmetry, it is sufficient to prove the statements for the + ensemble, to which in the sequel we restrict ourselves.

6.3.2 Time invariance of local equilibrium ensembles

We call m_{Λ^c} a plus boundary condition [for the bounded $\mathcal{D}^{(\ell)}$-measurable region Λ] if $\eta^{(\zeta,\ell)}(m_{\Lambda^c};r) = 1$ for all $r \in \Lambda^c$ with $\mathrm{dist}(C_r^{(\ell)}, \Lambda) \le 1$. As mentioned at the end of the previous subsection, we want to study the following variational problem: given any Λ as above and any "plus boundary condition" m_{Λ^c}, find the minimum and minimizer of $\mathcal{F}_\Lambda(m_\Lambda|m_{\Lambda^c})$ over all m_Λ such that $\eta^{(\zeta,\ell)}(m_\Lambda;r) = 1$ for all $r \in \Lambda$, i.e. $(m_\Lambda, m_{\Lambda^c}) \in M^+_{\Lambda;\zeta,\ell}$.

We shall use the partial dynamics T_t^Λ defined in Sect. 6.2.3. Recall in fact that T_t^Λ decreases the free energy functional and leaves invariant the function outside Λ. The possibility of using T_t^Λ in the above variational problem rests on the non-trivial fact that T_t^Λ also leaves $M^+_{\Lambda;\zeta,\ell}$ invariant, and such a crucial property will be established in this subsection and used in the next one to solve the above variational problem. We start with a preliminary lemma:

Lemma 6.3.2.1 *There are ζ_0' and κ_0 positive, $\kappa_0 c < 1$, $c := 4^d\sqrt{d}\|\nabla J\|_\infty$, so that if $\zeta < \zeta_0'$ and $\ell < \kappa_0\zeta$, then for any $r \in \mathbb{R}^d$ and any $m \in M^+_{r;\zeta,\ell}$,*

$$\left|J \star m(r) - m_\beta\right| \le (1 + c\kappa_0)\zeta < 2\zeta, \qquad (6.3.2.1)$$

$$\left|\tanh\{\beta J \star m(r)\} - m_\beta\right| \le \zeta - \epsilon_0, \quad \epsilon_0 = \kappa_0\zeta. \qquad (6.3.2.2)$$

Proof Outline: by the regularity of J we can approximate the convolution $J * m$ by a Riemann sum on the cubes of $\mathcal{D}^{(\ell)}$. By exploiting the definition of $M^+_{\Lambda;\zeta,\ell}$ we then get (6.3.2.1), and since $s \to \tanh\{\beta s\}$ is a contraction in a neighborhood of m_β, we also get (6.3.2.2). Details are given below.

Write $J^{(\ell)}(r,r') = \fint_{C_{r'}^{(\ell)}} J(r,r'')dr''$; then for all ℓ small enough,

$$\left|J(r,r') - J^{(\ell)}(r,r')\right| \le c'\ell \mathbf{1}_{|r-r'|\le 2}, \quad c' := \sqrt{d}\|\nabla J\|_\infty < \infty, \qquad (6.3.2.3)$$

d the space dimensions, so that $|J \star m - J^{(\ell)} \star m| \le c\ell$, $c = c'4^d$, because $|m| \le 1$. Since $J^{(\ell)} \star m = J^{(\ell)} \star m^{(\ell)} = J \star m^{(\ell)}$, $|J \star m - J \star m^{(\ell)}| \le c\ell$. On the other hand, by assumption, $|m^{(\ell)}(r') - m_\beta| \le \zeta$, for all $|r'-r| \le 1$; hence $|J \star m^{(\ell)}(r) - m_\beta| \le \zeta$, which proves the first inequality (6.3.2.1). The second one follows from the choice $c\kappa_0 < 1$.

Since $\frac{d}{ds}\tanh\{\beta s\}|_{s=m_\beta} < 1$ (see Fig. 4.2), there are $a \in (0,1)$ and $\zeta^* > 0$ so that $\frac{d}{ds}\tanh\{\beta(m_\beta + s)\} < a$ for $|s| \le \zeta^*$. Hence

$$\big|\tanh\{\beta(m_\beta + s)\} - m_\beta\big| \le a|s|, \quad \text{for } |s| \le \zeta^*. \tag{6.3.2.4}$$

We take ζ_0' so small that $2\zeta_0' < \zeta^*$; then for all $|s| \le \zeta_0'$

$$\big|\tanh\{\beta[m_\beta + (1 + c\kappa_0)s]\} - m_\beta\big| \le a(1 + c\kappa_0)|s| \le (1 - \kappa_0)|s| \tag{6.3.2.5}$$

if κ_0 is so small that $\kappa_0 \le \frac{1-a}{1+ac}$. By (6.3.2.1),

$$\big|\tanh\{\beta J \star m(r)\} - m_\beta\big| \le \big|\tanh\{\beta(m_\beta + (1 + c\kappa_0)\zeta\} - m_\beta\big|,$$

and (6.3.2.2) then follows from (6.3.2.5). □

For later reference we observe that we have also proved the following:

Statement

Let ζ and ℓ be as in Lemma 6.3.2.1, and suppose $\eta^{(\zeta,\ell)}(m;r) = 1$ for $r \in \Delta$, Δ a $\mathcal{D}^{(\ell)}$-measurable region. Then

$$\|J \star (m\mathbf{1}_\Delta) - J \star (m_\beta \mathbf{1}_\Delta)\|_\infty \le (1 + c\kappa_0)\zeta < 2\zeta. \tag{6.3.2.6}$$

Theorem 6.3.2.2 *Let ζ and ℓ be as in Lemma* 6.3.2.1; *then for any bounded $\mathcal{D}^{(\ell)}$-measurable region Λ*

$$T_t^\Lambda(M^+_{\Lambda;\zeta,\ell}) \subset M^+_{\Lambda;\zeta,\ell}, \quad \textit{for any } t \ge 0. \tag{6.3.2.7}$$

Proof For $\tau > 0$ and $m_0 \in M^+_{\Lambda;\zeta,\ell}$, we set

$$X_{\tau,m_0} = \big\{m \in L^\infty\big(\mathbb{R}^d \times [0,\tau]; [-1,1]\big) : m(r,0) = m_0(r), r \in \mathbb{R}^d;$$
$$m(r,t) = m_0(r),\ (r,t) \in \Lambda^c \times [0,\tau]\big\}.$$

Let $\Psi : X_{\tau,m_0} \to X_{\tau,m_0}$ be defined for $t \in [0,\tau]$ and $r \in \Lambda$ by

$$[\Psi(m)](r,t) = e^{-t} m_0(r) + \int_0^t e^{-(t-s)} \tanh\{\beta[J * m(\cdot,s)](r)\}\, ds.$$

Ψ is a contraction on X_{τ,m_0} (equipped with the sup norm) if $\tau > 0$ is small enough. For any $u \in X_{\tau,m_0}$, $\lim_{n\to\infty} \Psi^n(u) = (T_t^\Lambda(m_0))_{0\le t\le\tau}$; see the proof of Theorem 6.2.3.1. By (6.3.2.2), the set

$$Y = \big\{u \in X_{\tau,m_0} : u(\cdot,t) \in M^+_{\Lambda;\zeta,\ell} \text{ for all } t \in [0,\tau]\big\} \tag{6.3.2.8}$$

is invariant under the map Ψ. Since Y is closed, $\lim_{n\to\infty} \Psi^n(u) \in Y$ if $u \in Y$; hence $T_t^\Lambda(m_0) \in M^+_{\Lambda;\zeta,\ell}$, $0 \le t \le \tau$. By induction the statement remains valid for all t. □

6.3.3 Minimization of local equilibrium profiles

We shall use here Theorem 6.3.2.2 to solve the variational problem stated at the beginning of Sect. 6.3.2, to which we refer for notation.

Theorem 6.3.3.1 *There are $\zeta_0 < \zeta_0'$ (ζ_0' and κ_0 as in Lemma 6.3.2.1), ω and c_ω all positive, so that for any $\zeta < \zeta_0$, $\ell < \kappa_0\zeta$, any bounded $\mathcal{D}^{(\ell)}$-measurable region Λ and any plus boundary condition m_{Λ^c} the following holds.*

- *There is a unique ψ_Λ such that $\eta^{(\zeta,\ell)}(\psi_\Lambda;\cdot) = 1$ on Λ and*

$$\inf_{m_\Lambda:\eta^{(\zeta,\ell)}(m_\Lambda;r)=1, r\in\Lambda} \mathcal{F}_\Lambda(m_\Lambda|m_{\Lambda^c}) = \mathcal{F}_\Lambda(\psi_\Lambda|m_{\Lambda^c}). \tag{6.3.3.1}$$

- *ψ_Λ is the unique solution of the mean field equation (6.2.4.2) in $\{m_\Lambda : \eta^{(\zeta,\ell)}(m_\Lambda;r) = 1, r \in \Lambda\}$, $\psi_\Lambda \in C^\infty(\Lambda, [m_\beta - \zeta + \kappa_0\zeta, m_\beta + \zeta - \kappa_0\zeta])$, $\sup_{r\in\Lambda} |\nabla\psi_\Lambda(r)| \le \beta\|\nabla J\|_\infty$ and writing $\Lambda^c_{\neq} = \{r \in \Lambda^c : \operatorname{dist}(r,\Lambda) \le 1;\ m_{\Lambda^c}(r) \ne m_\beta)\}$,*

$$|\psi_\Lambda(r) - m_\beta| \le c_\omega e^{-\omega \operatorname{dist}(r,\Lambda^c_{\neq})}. \tag{6.3.3.2}$$

- *Let m'_{Λ^c} be a plus boundary condition and ψ'_Λ the corresponding minimizer. Write (by abuse of notation) $\Lambda^c_{\neq} = \{r \in \Lambda^c : \operatorname{dist}(r,\Lambda) \le 1;\ m'_{\Lambda^c}(r) \ne m_{\Lambda^c}(r)\}$; then*

$$|\psi'_\Lambda(r) - \psi_\Lambda(r)| \le c_\omega e^{-\omega \operatorname{dist}(r,\Lambda^c_{\neq})}. \tag{6.3.3.3}$$

Proof Let m_Λ be as on the l.h.s. of (6.3.3.1); then by Theorem 6.3.2.2 $T_t^\Lambda(m_\Lambda, m_{\Lambda^c}) =: (m_\Lambda(t), m_{\Lambda^c}) \in M^+_{\Lambda;\zeta,\ell}$. By Theorem 6.2.6.1 $m_\Lambda(t)$ converges in sup norm by subsequences as $t \to \infty$ to some ψ_Λ, which solves the mean field equation (6.2.4.2). Moreover, by continuity, for all $r \in \Lambda$, $\eta^{(\zeta,\ell)}(\psi_\Lambda; r) = 1$ and by Theorem 6.2.6.2,

$$\mathcal{F}_\Lambda(m_\Lambda|m_{\Lambda^c}) \ge \mathcal{F}_\Lambda(\psi_\Lambda|m_{\Lambda^c}). \tag{6.3.3.4}$$

Since ψ_Λ solves (6.2.4.2) it is in C^∞ and by differentiating the mean field equation, recalling that $|\psi_\Lambda| \le 1$, we also get the bound $|\nabla\psi_\Lambda(r)| \le \beta\|\nabla J\|_\infty$, for all r in the interior of Λ. Moreover, by (6.3.2.2), ψ_Λ has values in $[m_\beta - \zeta + \kappa_0\zeta, m_\beta + \zeta - \kappa_0\zeta]$.

We shall next show that if ζ is small enough, then there is a unique function ϕ_Λ which solves (6.2.4.2). It is such that $\eta^{(\zeta,\ell)}(\phi_\Lambda; r) = 1$, $r \in \Lambda$; thus $\psi_\Lambda = \phi_\Lambda$ is the strict minimizer of the l.h.s. of (6.3.3.1). Suppose by contradiction that there are two distinct ϕ_Λ and ϕ'_Λ as above. Call ϕ and ϕ' their extension to $\mathbb{R}^d$ obtained by setting them equal to m_{Λ^c} outside Λ. Then, by (6.3.2.1), $J * \phi(r)$ and $J * \phi'(r)$, $r \in \Lambda$, are in $[m_\beta - 2\zeta, m_\beta + 2\zeta]$, so that, for ζ small enough, $\tanh\{\beta J * \phi(r)\} - \tanh\{\beta J * \phi'(r)\}|$ is bounded by

$$\frac{\beta}{\cosh^2\{\beta(m_\beta - 2\zeta)\}}\left(\int_\Lambda J(r,r')|\phi'(r') - \phi(r')|\,dr'\right).$$

Since $\beta \cosh^{-2}(\beta m_\beta) < 1$, we can choose ζ_0 so small that

$$e^{-2\omega} := \frac{\beta}{\cosh^2\{\beta(m_\beta - 2\zeta_0)\}} < 1. \tag{6.3.3.5}$$

Then, for $\zeta \le \zeta_0$ and for all $r \in \Lambda$,

$$|\tanh\{\beta J * \phi'(r)\} - \tanh\{\beta J * \phi(r)\}| \le e^{-2\omega} \sup_{r' \in \Lambda} |\phi'(r') - \phi(r')|,$$

which implies $\phi'_\Lambda = \phi_\Lambda$.

Finally, let ψ'_Λ and ψ_Λ be as in the last statement of the theorem, $\psi' = (\psi'_\Lambda, m'_{\Lambda^c})$ and $\psi = (\psi_\Lambda, m_{\Lambda^c})$; then for $r \in \Lambda$

$$|\psi'(r) - \psi(r)| \le e^{-2\omega}\left(\int_\Lambda J(r,r')|\psi'(r') - \psi(r')|\,dr' + \int_{\Lambda^c_{\neq}} J(r,r')|m'_{\Lambda^c}(r') - m_{\Lambda^c}(r')|\,dr'\right). \tag{6.3.3.6}$$

By iterating (6.3.3.6) and calling n_0 the smallest integer larger or equal to $\mathrm{dist}(r, \Lambda^c_{\neq})$, we get $|\psi'_\Lambda(r) - \psi_\Lambda(r)| \le \sum_{n \ge n_0} e^{-2\omega n} 2 \le (2\sum_{n\ge 0} e^{-\omega n})\, e^{-\omega n_0}$, which yields (6.3.3.3) with $c_\omega := \frac{2}{1-e^{-\omega}}$. By choosing $m'(r) = m_\beta$, we have $\psi_\Lambda = m_\beta$, and the last inequality in (6.3.3.2) follows from (6.3.3.3). □

6.4 Large deviations

In this section we shall prove that the deviations from local equilibrium have a free energy cost which grows proportionally to the volume of the region where they occur. Such estimates are called "Peierls bounds" in analogy with those proved in Sect. 3.1.2 for Ising systems at low temperatures. The setup and techniques are different, as we deal here with functionals rather than with probabilities, but in spirit they are similar. The analogy will become closer in Chap. 9, when we use the estimates of this section to prove the existence of phase transitions in Ising systems with Kac potentials. Another application of the results of this section can be found in Chap. 7, in the analysis of surface tension and interfaces.

6.4.1 The multi-canonical constraint

We first make a reduction to regular profiles and then prove for these the Peierls estimates. Regularity can be gained by running dynamics for an infinite time as we have already seen in Theorem 6.3.3.1, but this is useless here, because the free energy decreases in time. It may thus happen that once we have reached the desired regularity we have also consumed the whole initial free energy excess that we wanted to estimate. Thus dynamics does not seem useful when there are constraints which are not time-invariant, like the deviations from local equilibrium that we want to study.

We shall first prove that a minimizer ϕ_Λ of the free energy functional under the constraint "$m_\Lambda^{(\ell)}(\cdot)$ is given" has "good" regularity properties; since $\eta^{(\zeta,\ell)}(m_\Lambda; r)$ is constant on the set "$m_\Lambda^{(\ell)}(\cdot)$ is given," we can replace m_Λ by ϕ_Λ without changing $\eta^{(\zeta,\ell)}$ and thus make a reduction to estimate the free energy cost of smooth profiles which deviate from equilibrium. The first step of this program is discussed in the present subsection.

Theorem 6.4.1.1 *Let $\ell \in \{2^{-n}, n \in \mathbb{N}\}$ be such that $\beta \|J\|_\infty \ell^d < \frac{1}{4}$. Denote by Λ either the cube $C \in \mathcal{D}^{(\ell)}$ or $C \cup C'$, $C' \in \mathcal{D}^{(\ell)}$ connected to C (i.e. the closures of C and C' have non-empty intersection). Let finally $s \in [-1,1]$ and $m_{\Lambda^c} \in L^\infty(\Lambda^c, [-1,1])$. Then there is a unique u_Λ such that $\fint_\Lambda u_\Lambda = s$ and*

$$\mathcal{F}_\Lambda(m_\Lambda|m_{\Lambda^c}) \geq \mathcal{F}_\Lambda(u_\Lambda|m_{\Lambda^c}), \quad \text{for any } m_\Lambda \text{ such that } \fint_\Lambda m_\Lambda = s \tag{6.4.1.1}$$

Moreover, for any $r \in \Lambda$, $|u_\Lambda(r) - s| \leq \beta \|\nabla J\|_\infty \mathrm{diam}(\Lambda)$, $\mathrm{diam}(\Lambda) \leq 2d\,\ell$.

Proof If $s = \pm 1$ and $\fint_\Lambda m_\Lambda = \pm 1$, since $m_\Lambda \in L^\infty(\Lambda, [-1,1])$, then $m_\Lambda \equiv \pm 1$ almost everywhere and the statements in the theorem trivially follow. Let then $|s| < 1$. We are going to prove the theorem using Lagrange multipliers. To this end we define for any $h \in \mathbb{R}$ and $\psi_\Lambda \in L^\infty(\Lambda, [-1,1])$

$$\mathcal{F}_{\Lambda,h}(\psi_\Lambda|m_{\Lambda^c}) = \mathcal{F}_\Lambda(\psi_\Lambda|m_{\Lambda^c}) - h \int_\Lambda \psi_\Lambda(r)\,dr. \tag{6.4.1.2}$$

Let $A_h(\psi_\Lambda) = \tanh\{\beta(J \star (\psi_\Lambda + m_{\Lambda^c}) + h)\}$; then by Theorem 6.2.6.2 the inf of $\mathcal{F}_{\Lambda,h}(\psi_\Lambda|m_{\Lambda^c})$ is a minimum reached on functions ψ_Λ such that $A_h(\psi_\Lambda) = \psi_\Lambda$. Thus

$$X^0_{h,m_{\Lambda^c}} = \{\phi_\Lambda \in L^\infty(\Lambda, [-1,1]) : \phi_\Lambda = A_h(\phi_\Lambda)\} \tag{6.4.1.3}$$

is non-empty, and

$$\mathcal{F}_{\Lambda,h}(\psi_\Lambda|m_{\Lambda^c}) > \inf_{\phi_\Lambda \in X^0_{h,m_{\Lambda^c}}} \mathcal{F}_{\Lambda,h}(\phi_\Lambda|m_{\Lambda^c}),$$

$$\text{for any } \psi_\Lambda \notin X^0_{h,m_{\Lambda^c}}; \tag{6.4.1.4}$$

$X^0_{h,m_{\Lambda^c}}$ is actually a singleton, because (recall that $|\Lambda| \le 2\ell^d$)

$$\|A_h(\phi_\Lambda) - A_h(\psi_\Lambda)\|_\infty \le \beta \|J\|_\infty 2\ell^d \|\phi_\Lambda - \psi_\Lambda\|_\infty,$$

so that, by the assumption $\beta\|J\|_\infty \ell^d < \frac{1}{4}$, A_h is a contraction in $L^\infty(\Lambda, [-1,1])$ and its unique fixed point $\phi^{(h)}_\Lambda$ is (for instance)

$$\phi^{(h)}_\Lambda = \lim_{n\to\infty} A_h(u_n), \quad u_n = A_h(u_{n-1}),\ u_0 = s\mathbf{1}_\Lambda.$$

The convergence is in sup norm and it is also uniform in h. We are going to prove by induction on n that u_n is differentiable in h, that its derivative Du_n is

$$Du_n = p_n\{J * Du_{n-1} + 1\}, \tag{6.4.1.5}$$

$p_n = \beta \cosh^{-2}\{\beta(J * (u_{n-1} + m_{\Lambda^c}) + h)\}$, and finally that $\|Du_n\|_\infty \le 2\beta$.

Indeed $Du_0 = 0$ and if u_{n-1} is differentiable then Du_n exists and it is given by (6.4.1.5). Assume $\|Du_{n-1}\|_\infty \le 2\beta$; then

$$|Du_n| \le \beta\big(\|J\|_\infty 2\ell^d (2\beta) + 1\big) \le 2\beta, \quad \text{because } \beta\|J\|_\infty \ell^d < \frac{1}{4},$$

which completes the induction proof. Then $\phi^{(h)}_\Lambda$ is differentiable in h and $D\phi^{(h)}_\Lambda$ is the limit of Du_n. Thus writing $p = \beta\cosh^{-2}\{\beta(J * (\phi^{(h)}_\Lambda + m_{\Lambda^c}) + h)\}$,

$$D\phi^{(h)}_\Lambda = p\{J * D\phi^{(h)}_\Lambda + 1\} \ge p\{-2\beta\|J\|_\infty \ell^d + 1\} \ge \frac{p}{2}.$$

Then $a(h) := \fint_\Lambda \phi^{(h)}_\Lambda(r)\,dr$ is a strictly increasing, continuous function of h; since $a(h) \to \pm 1$ as $h \to \pm\infty$, there is h^* such that, $a(h^*) = s$. Let $\fint_\Lambda \psi_\Lambda = s$; then, unless $\psi_\Lambda = \phi^{(h^*)}_\Lambda$ almost everywhere,

$$\begin{aligned}\mathcal{F}_\Lambda(\psi_\Lambda | m_{\Lambda^c}) &= \mathcal{F}_{\Lambda,h^*}(\psi_\Lambda | m_{\Lambda^c}) + h^*|\Lambda| s > \mathcal{F}_{\Lambda,h^*}(\phi^{(h^*)}_\Lambda | m_{\Lambda^c}) + h^*|\Lambda| s \\ &= \mathcal{F}_\Lambda(\phi^{(h^*)}_\Lambda | m_{\Lambda^c}).\end{aligned}$$

Thus $\phi^{(h^*)}_\Lambda$ is the unique minimizer of the variational problem (6.4.1.1) and, by differentiating $\phi^{(h^*)}_\Lambda = A_h(\phi^{(h^*)}_\Lambda)$, $|\nabla\phi^{(h^*)}_\Lambda| \le \beta\|\nabla J\|_\infty$, so that $|\phi^{(h^*)}_\Lambda - s| \le \beta\|\nabla J\|_\infty \mathrm{diam}(\Lambda)$. □

6.4.2 Extensive bounds

In this subsection we shall bound from below the "free energy of profiles which deviate from equilibrium." The following definition identifies the region (denoted by $\mathcal{C}$) where the deviations occur.

Definition 6.4.2.1 Given $\zeta > 0$, $\ell \in \{2^{-n}, n \in \mathbb{N}\}$, a bounded $\mathcal{D}^{(\ell)}$-measurable region Λ and $m_\Lambda \in L^\infty(\Lambda, [-1,1])$, we call $\mathcal{C}_0$ a *collection of cubes* $C \in \mathcal{D}^{(\ell)}$ where $\eta^{(\zeta,\ell)}(m_\Lambda; \cdot) = 0$ and $\mathcal{C}_{\neq}$ a *collection of pairs* (C_i', C_i'') *of cubes* in $\mathcal{D}^{(\ell)}$ where $\eta^{(\zeta,\ell)}(m_\Lambda; \cdot) = \pm 1$, respectively, and such that no cube appears more than once. $\mathcal{C}$ stands for the union of all cubes in $\mathcal{C}_0$ and $\mathcal{C}_{\neq}$.

$\mathcal{C}_0$ is "*maximal*" if it contains all the cubes $C^{(\ell)}$ in Λ where $\eta^{(\zeta,\ell)}(m; r) = 0$; $\mathcal{C}_{\neq}$ is maximal if any pair (C', C'') where $\eta^{(\zeta,\ell_-)}(m; r) = \pm 1$ is such that at least one among C' and C'' is contained in $\mathcal{C}$.

As in the Ginzburg–Landau functional the energy excess is due to deviations from the equilibrium values, here represented by $\mathcal{C}_0$, and from large gradients, here represented by $\mathcal{C}_{\neq}$. The next lemma, which is a corollary of Theorem 6.4.1.1, shows that we can make a reduction to smooth functions.

Lemma 6.4.2.2 *There are* ζ_1 *and* κ_1 *positive so that for any* $\zeta \in (0, \zeta_1)$, $\ell < \kappa_1 \zeta$ *the following holds. Let* Λ, m_Λ, $\mathcal{C}_0$ *and* $\mathcal{C}_{\neq}$ *as in Definition* 6.4.2.1 *and* $m_{\Lambda^c} \in L^\infty(\Lambda^c, [-1,1])$. *Then there is* $\phi_\Lambda \in L^\infty(\Lambda, [-1,1])$, $\phi_\Lambda(r) = m_\Lambda(r)$ *in* $\Lambda \setminus \mathcal{C}$, *such that* $\mathcal{F}_\Lambda(m_\Lambda|m_{\Lambda^c}) \ge \mathcal{F}_\Lambda(\phi_\Lambda|m_{\Lambda^c})$ *and, denoting by* Δ *either* $C \in \mathcal{C}_0$ *or* $C' \cup C''$ *with* $(C', C'') \in \mathcal{C}_{\neq}$,

$$\fint_\Delta m_\Lambda = \fint_\Delta \phi_\Lambda, \qquad \sup_{r \in \Delta} \left| \phi_\Lambda(r) - \fint_\Delta \phi_\Lambda \right| \le \frac{\zeta}{4}, \tag{6.4.2.1}$$

$$\eta^{(\zeta,\ell)}(\phi_\Lambda; r) = 0, \qquad \big|\, |\phi_\Lambda(r)| - m_\beta \big| \ge \frac{\zeta}{2}, \quad \textit{for all } r \in \mathcal{C}. \tag{6.4.2.2}$$

Proof We apply successively Theorem 6.4.1.1 to the sets $C_i' \cup C_i''$, $(C_i', C_i'') \in \mathcal{C}_{\neq}$ and then to the sets $C_i \in \mathcal{C}_0$, obtaining in the end a function ϕ_Λ such that $\mathcal{F}_\Lambda(m_\Lambda|m_{\Lambda^c}) \ge \mathcal{F}_\Lambda(\phi_\Lambda|m_{\Lambda^c})$; $\phi_\Lambda(r) = m_\Lambda(r)$ in $\Lambda \setminus \mathcal{C}$. The equality in (6.4.2.1) follows by construction, the inequality from Theorem 6.4.1.1 because $\sqrt{d}\, 2\ell$ bounds the diameter of Δ. Then the l.h.s. is bounded by $\beta \|\nabla J\|_\infty \sqrt{d}\, 2\ell$. By choosing $\kappa_1 < (\beta \|\nabla J\|_\infty \sqrt{d}\, 8)^{-1}$ we then get the inequality in (6.4.2.1).

By the equality in (6.4.2.1) it follows that $\eta^{(\zeta,\ell)}(\phi_\Lambda; r) = \eta^{(\zeta,\ell)}(m_\Lambda; r)$, so that $\eta^{(\zeta,\ell)}(\phi_\Lambda; r) = 0$ on any $C \in \mathcal{C}_0$, and, by the inequality (6.4.2.1), $\big|\, |\phi_\Lambda(r)| - m_\beta \big| \ge \zeta/2$ for all $r \in C$, $C \in \mathcal{C}_0$. Let $(C', C'') \in \mathcal{C}_{\neq}$,

$$\fint_{C' \cup C''} \phi_\Lambda = \frac{1}{2}\left(\fint_{C'} m_\Lambda + \fint_{C''} m_\Lambda \right),$$

and since $\eta^{(\zeta,\ell)}(m_\Lambda; r) = \pm 1$ on C' and C'', $|\fint_{C' \cup C''} \phi_\Lambda| \le \zeta$. By the inequality in (6.4.2.1)

$$\left| \fint_{C'} \phi_\Lambda - \fint_{C' \cup C''} \phi_\Lambda \right| \le \frac{\zeta}{4}, \qquad \left| \phi_\Lambda(r) - \fint_{C'} \phi_\Lambda \right| \le \frac{\zeta}{2}. \tag{6.4.2.3}$$

By the first inequality in (6.4.2.3), $|f_{C'}\phi_\Lambda| \le \zeta + \frac{\zeta}{4} < m_\beta - \zeta$ for ζ_0 small enough; hence $\eta^{(\zeta,\ell)}(\phi_\Lambda; r) = 0, r \in C'$. By (6.4.2.3) $|\phi_\Lambda(r)| \le \zeta(1 + \frac{1}{4} + \frac{1}{2})$; hence we have (6.4.2.2) for ζ_0 small enough. □

Theorem 6.4.2.3 *There is $c_1' > 0$ so that for any $\zeta \in (0, \zeta_1)$ and $\ell < \kappa_1 \zeta$, ζ_1 and κ_1 as in Lemma* 6.4.2.2, *the following holds. Let Λ, m_{Λ^c}, m_Λ, C_0 and $C_{\neq}$ as in Lemma* 6.4.2.2, *then calling $2n$ and p the number of cubes in $C_{\neq}$ and C_0*

$$\mathcal{F}_\Lambda(m_\Lambda | m_{\Lambda^c}) \ge c_1' \zeta^2 \, \ell^d (2n + p). \tag{6.4.2.4}$$

Proof We write $\mathcal{F}_\Lambda(m_\Lambda | m_{\Lambda^c}) \ge \mathcal{F}_\Lambda(\phi_\Lambda | m_{\Lambda^c})$ with ϕ_Λ as in Lemma 6.4.2.2. By (6.4.2.2) $\int_C f_{\beta,0}(\phi_\Lambda)\, dr \ge \frac{c\zeta^2}{4} \ell^d \,(2n + p)$ because $f_{\beta,0}(s) \ge c[|s| - m_\beta]^2$. All the other terms in $\mathcal{F}_\Lambda(\phi_\Lambda | m_{\Lambda^c})$ are non-negative; hence (6.4.2.4). □

6.4.3 Topological properties of $\mathcal{D}^{(\ell)}$-sets

We shall extensively use in the sequel some topological notions that we collect here. We fix $\ell \in \{2^n, n \in \mathbb{Z}\}$ and by default in this subsection all sets will be $\mathcal{D}^{(\ell)}$-measurable.

Connected sets

Two sets are connected if their closures have non-empty intersection; thus two cubes with a common vertex are connected (for instance the two + sets in Fig. 6.2 of Sect. 6.5) and connection here is what it is usually called $*$-connection when working in $\mathbb{Z}^d$. A maximal connected component B of a set A is a connected subset of A which is "maximal" in the sense that if $C \subset A$ is connected and $C \supset B$, then $C = B$. Any set is the disjoint union of its maximal connected components. A connected set A is simply connected if its complement is connected.

Outer and inner boundaries of a set

The outer boundary $\delta^\ell_{\text{out}}[\Lambda]$ of a $\mathcal{D}^{(\ell)}$-measurable region Λ is the union of all the cubes $C \in \mathcal{D}^{(\ell)}$ not in Λ but connected to Λ. The inner boundary $\delta^\ell_{\text{in}}[\Lambda]$ of Λ is the outer boundary of Λ^c. With reference to Fig. 6.6 in Sect. 6.5, if Λ is the region in the interior of the thick line, $\delta^\ell_{\text{out}}[\Lambda]$ is the region between the thick and the dashed external lines; $\delta^\ell_{\text{in}}[\Lambda]$ between the thick and the internal dashed lines.

Theorem 6.4.3.1 *Let Λ be a bounded, connected $\mathcal{D}^{(\ell)}$-measurable region. Then we have the following.*

- *The maximal connected components of Λ^c are all bounded except one which is unbounded.*
- *Any bounded, maximal connected component of Λ^c is simply connected.*
- *If Λ is bounded and simply connected, then $\delta^{\ell}_{\text{out}}[\Lambda]$ and $\delta^{\ell}_{\text{in}}[\Lambda]$ are both connected.*

6.4.4 Contours

The definition of contours requires the introduction of some new notions.

The three scales

The three basic scales are ℓ_-, ℓ_+ and the interaction range, and they are all supposed to be in $\{2^n, n \in \mathbb{Z}\}$. ℓ_- will always be much smaller and ℓ_+ much larger than the interaction range. In this chapter the interaction range is equal to 1; thus $\ell_- \ll 1 \ll \ell_+$. In Chaps. 9 and 10 the range of the interaction is γ^{-1}, respectively $2\gamma^{-1}$; $\ell_\pm$ will then depend on γ and will be called $\ell_{\pm,\gamma}$. They will verify the inequalities $\ell_{-,\gamma} \ll \gamma^{-1} \ll \ell_{+,\gamma}$. Moreover, $\ell_{-,\gamma} \gg 1$, 1 being the lattice spacing in Chap. 9 and [of the order of] the inter-particle distance in Chap. 10. This condition ensures that statistical fluctuations are small on the scale $\ell_{-,\gamma}$. In the present context we have already invoked a mesoscopic limit, damping off all microscopic fluctuations, so that the lower bound on ℓ_- is just $\ell_- > 0$.

The three phase indicators

We introduce an accuracy parameter $\zeta > 0$ whose value is related in the applications to the choice of ℓ_-, and we define the phase indicators in terms of the triple (ζ, ℓ_-, ℓ_+). In Chaps. 9 and 10 ζ will depend on γ.

By default, in the sequel m denotes a function in $L^\infty(\mathbb{R}^d, [-1,1])$. Given m we shall define three functions on $\mathbb{R}^d$ denoted by $\eta^{(\zeta,\ell_-)}(m;r)$, $\theta^{(\zeta,\ell_-,\ell_+)}(m;r)$ and $\Theta^{(\zeta,\ell_-,\ell_+)}(m;r)$. The values of the three functions will describe with increasing degree of accuracy the local phase of the system. As we shall see $\theta^{(\zeta,\ell_-,\ell_+)}(m;r)$ and $\Theta^{(\zeta,\ell_-,\ell_+)}(m;r)$ depend on m only via $\eta^{(\zeta,\ell_-)}(m;\cdot)$, so that once the function $\eta^{(\zeta,\ell_-)}(m;\cdot)$ is specified the other two phase indicators are completely determined. We may thus use only $\eta^{(\zeta,\ell_-)}(m;\cdot)$, but the other indicators hopefully make the definitions more clear. $\eta^{(\zeta,\ell_-)}(m;r)$, see (6.3.1.2), depends on m via $\fint_{C_r^{(\ell_-)}} m(r')dr'$. If the latter is close to $\pm m_\beta$ by ζ, then $\eta^{(\zeta,\ell_-)}(m;r) = \pm 1$, and we call the cube $C_r^{(\ell_-)}$ in $\pm$ equilibrium. Otherwise $\eta^{(\zeta,\ell_-)}(m;r) = 0$ and in $C_r^{(\ell_-)}$ the system is off equilibrium. Thus $\eta^{(\zeta,\ell_-)}(m;r)$ indicates the phase, or its absence, on the small scale ℓ_-. By replacing the average of m by the empirical average of a spin configuration in $C_r^{(\ell_{-,\gamma})} \cap \mathbb{Z}^d$ we shall extend in Chap. 9 the definition of this phase indicator to

Ising spins. In Chap. 10 the definition is extended to a configuration q of particles, and $\eta^{(\zeta,\ell_{-,\gamma})}(q;r)$ will indicate if the density $|q \cap C_r^{(\ell_{-,\gamma})}|/\ell_{-,\gamma}^d$ of particles of q which are in $C_r^{(\ell_{-,\gamma})}$ is close to the [mean field] equilibrium values $\rho_{\beta,+}$ or $\rho_{\beta,-}$.

Definition 1 $\theta^{(\zeta,\ell_-,\ell_+)}(m;r)$ and $\Theta^{(\zeta,\ell_-,\ell_+)}(m;r)$ are defined in terms of $\eta^{(\zeta,\ell_-)}(m;\cdot)$ as:

- $\theta^{(\zeta,\ell_-,\ell_+)}(m;r) = \pm 1$ if $\eta^{(\zeta,\ell_-)}(m;r') = \pm 1$ constantly for all r' in $C_r^{(\ell_+)}$. $\theta^{(\zeta,\ell_-,\ell_+)}(m;r) = 0$ otherwise.
- $\Theta^{(\zeta,\ell_-,\ell_+)}(m;r) = \pm 1$ if $\eta^{(\zeta,\ell_-)}(m;r') = \pm 1$ constantly for all r' in $C_r^{(\ell_+)} \cup \delta_{\text{out}}^{\ell_+}[C_r^{(\ell_+)}]$. $\Theta^{(\zeta,\ell_-,\ell_+)}(m;r) = 0$ otherwise.

$\theta^{(\zeta,\ell_-,\ell_+)}(m;r) = 1$ means that the phase indicated by $\eta^{(\zeta,\ell_-)}(m;\cdot)$ on the small scale ℓ_- is constantly the plus phase in the whole [large] cube $C_r^{(\ell_+)}$. Points r close to the boundary of $C_r^{(\ell_+)}$, however, may not be in local equilibrium, according to the definition of Sect. 6.3.1, but if $\Theta^{(\zeta,\ell_-,\ell_+)}(m;r) = 1$ then $\theta^{(\zeta,\ell_-,\ell_+)}(m;\cdot) = 1$ in all cubes of $\mathcal{D}^{(\ell_+)}$ connected to $C_r^{(\ell_+)}$, and since $\ell_+ > 1$ all points of $C_r^{(\ell_+)}$ are in local equilibrium. Thus, while $\eta^{(\zeta,\ell_-)}(m;r)$ and $\theta^{(\zeta,\ell_-,\ell_+)}(m;r)$ are "local" in the sense that their values depend only on the restriction of m to $C_r^{(\ell_-)}$ and $C_r^{(\ell_+)}$, respectively, this is no longer true for $\Theta^{(\zeta,\ell_-,\ell_+)}(m;r)$, which is "non-local" as it depends on the restriction of m to $C_r^{(\ell_+)} \cup \delta_{\text{out}}^{\ell_+}[C_r^{(\ell_+)}]$. See Sect. 6.5 for examples and pictures.

Definition 2 A point r is correct (for the profile m) if $\Theta^{(\zeta,\ell_-,\ell_+)}(m;r) \neq 0$ and called *plus/minus (correct)* if $\Theta^{(\zeta,\ell_-,\ell_+)}(m;r) = \pm 1$. The plus/minus phases of m are the sets of its plus/minus correct points; they are mutually disconnected and the regions in between are "the spatial supports of the contours."

Definition 3 A *contour of* m is a pair $\Gamma = (\text{sp}(\Gamma), \eta_\Gamma)$, where $\text{sp}(\Gamma)$, "the spatial support of Γ," is a maximal connected component of $\{r \in \mathbb{R}^d : \Theta^{(\zeta,\ell_-,\ell_+)}(m;r) = 0\}$ and η_Γ is the restriction to $\text{sp}(\Gamma)$ of $\eta^{(\zeta,\ell_-)}(m;\cdot)$.

$\Gamma = (\text{sp}(\Gamma), \eta_\Gamma)$ is "an abstract contour" if it is a contour of some m.

Geometry of contours

In the sequel we restrict to bounded contours, meaning that their spatial support is bounded. Definitions and properties stated in this paragraph are exemplified and visualized in Sect. 6.5.

The exterior, $\text{ext}(\Gamma)$, of Γ is the unbounded, maximal connected component of $\text{sp}(\Gamma)^c$. The interior is the set $\text{int}(\Gamma) = \text{sp}(\Gamma)^c \setminus \text{ext}(\Gamma)$; we denote by $\text{int}_i(\Gamma)$ the maximal connected components of $\text{int}(\Gamma)$. Let

$$c(\Gamma) = \text{sp}(\Gamma) \cup \text{int}(\Gamma); \tag{6.4.4.1}$$

then, by Theorem 6.4.3.1, $\text{int}_i(\Gamma)$ and $c(\Gamma)$ are both simply connected.

The outer boundaries of Γ are the sets

$$A(\Gamma) := \delta_{\text{out}}^{\ell_+,\gamma}[\text{sp}(\Gamma)] \cap \text{int}(\Gamma), \qquad A_{\text{ext}}(\Gamma) := \delta_{\text{out}}^{\ell_+,\gamma}[c(\Gamma)]. \tag{6.4.4.2}$$

We shall also call $A_i(\Gamma) = A(\Gamma) \cap \text{int}_i(\Gamma)$. The inner boundaries of Γ are the sets

$$B_i(\Gamma) = \delta_{\text{out}}^{\ell_+}[A_i(\Gamma)] \cap \text{sp}(\Gamma), \qquad B_{\text{ext}}(\Gamma) = \delta_{\text{in}}^{\ell_+}[c(\Gamma)]. \tag{6.4.4.3}$$

An easy but important consequence of the above definitions is the following theorem (the reader may check its validity in the case of Fig. 6.1).

Theorem 6.4.4.1 *If m has a bounded contour Γ, $\Theta^{(\zeta,\ell_-,\ell_+)}(m;r)$ is a non-zero constant on any $A_i(\Gamma)$ and $A_{\text{ext}}(\Gamma)$.*

Proof $\Theta^{(\zeta,\ell_-,\ell_+)}(m;r) \neq 0$ for any $r \in A$, A either $A_{\text{ext}}(\Gamma)$ or $A_i(\Gamma)$, because A is outside $\text{sp}(\Gamma)$ and, being connected to $\text{sp}(\Gamma)$, cannot intersect the spatial support of any other contour. We shall next prove that $\Theta^{(\zeta,\ell_-,\ell_+)}(m;r)$ is constant and non-zero on A. Since A is connected (by Theorem 6.4.3.1), it will suffice to prove that if two cubes C and C' of $\mathcal{D}^{(\ell_+)}$, are connected to each other and both in A, then $\Theta^{(\zeta,\ell_-,\ell_+)}(m;r)$ has a constant value on the union $C \cup C'$. Indeed, suppose for instance that $\Theta^{(\zeta,\ell_-,\ell_+)}(m;r) = 1$, $r \in C$; then $\theta^{(\zeta,\ell_-,\ell_+)}(m;r') = 1$ for all $r' \in \delta_{\text{out}}^{\ell_+}[C]$ and hence for all $r' \in C'$; thus $\Theta^{(\zeta,\ell_-,\ell_+)}(m;r') = 1$ on C' (as we already know that $\Theta^{(\zeta,\ell_-,\ell_+)}(m;r') \neq 0$ on C'). □

Definition 4 Γ is a *plus, minus, contour* if $\eta_\Gamma(\cdot) = \pm 1$ on $B_{\text{ext}}(\Gamma)$.

We add a superscript $\pm$ to $A_i(\Gamma)$ writing $A_i^\pm(\Gamma)$, $i = 1, \dots, n_\pm$, with $\pm$ chosen so that $\eta_\Gamma = 1$ on $B_i^+(\Gamma)$ and $\eta_\Gamma = -1$ on $B_i^-(\Gamma)$. We also write

$$A^+(\Gamma) = \bigcup_{i=1}^{n_+} A_i^+(\Gamma), \qquad A^-(\Gamma) = \bigcup_{i=1}^{n_-} A_i^-(\Gamma); \tag{6.4.4.4}$$

$\text{int}_i^\pm(\Gamma)$ if $\text{int}_i(\Gamma)$ contains $A_i^\pm(\Gamma)$,

$$\text{int}^\pm(\Gamma) = \bigcup_{i=1}^{n_\pm} \text{int}_i^\pm(\Gamma), \qquad \text{int}(\Gamma) = \text{int}^+(\Gamma) \cup \text{int}^-(\Gamma). \tag{6.4.4.5}$$

Remarks

We shall choose ℓ_- "very small" and ℓ_+ "very large," so that "correct points" (where $\Theta^{(\zeta,\ell_-,\ell_+)}(m;\cdot) \neq 0$) are always in the interior of a "large" set where $\eta^{(\zeta,\ell_-)}(m;\cdot)$ is constant and non-zero. Moreover, the correct region and the "red zone" $\mathcal{C}$ (see Definition 6.4.2.1) are separated by a "safety zone" made of the sets $B_i(\Gamma)$ and

$B_{\text{ext}}(\Gamma)$, where, on each of them, $\eta^{(\zeta,\ell_-)}(m;r)$ is non-zero and constant. When he explained to me the reason why he devised this definition, Milos Zahradnik told me he had in mind a boat in the sea and the idea was to put some oil at the contact between the wood of the boat and the salty water of the sea.

6.4.5 Free energy of contours

The free energy of a contour will be bounded from below by using Theorem 6.4.2.3 after choosing the collections $\mathcal{C}_0$ and $\mathcal{C}_{\neq}$ of Definition 6.4.2.1 as maximal. In the next theorem, ζ_1, κ_1, and c'_1 are the parameters defined in Theorem 6.4.2.3 and

$$N_\Gamma = \frac{|\text{sp}(\Gamma)|}{\ell_+^d} = \text{number of } \mathcal{D}^{(\ell_+)}\text{-cubes in sp}(\Gamma)\,. \tag{6.4.5.1}$$

Theorem 6.4.5.1 *Let $\zeta < \zeta_1$, $\ell_- < \kappa_1\zeta$, $m \in L^\infty(\mathbb{R}^d,[-1,1])$, Λ a bounded $\mathcal{D}^{(\ell_+)}$-measurable region. Assume m has n distinct, (ζ,ℓ_-,ℓ_+) contours Γ_i, $i=1,\dots,n$, all in Λ (i.e. $\text{sp}(\Gamma_i)\subset\Lambda$). Then*

$$\mathcal{F}_\Lambda(m_\Lambda|m_{\Lambda^c}) \ge c''_1\,(\zeta^2\ell_-^d)\sum_{i=1}^n N_{\Gamma_i}, \quad c''_1 = 3^{-d}c'_1. \tag{6.4.5.2}$$

Proof For each Γ_i let $\mathcal{C}_0^{(i)}$ and $\mathcal{C}_{\neq}^{(i)}$ be maximal collections in $\text{sp}(\Gamma_i)$ in the sense of Definition 6.4.2.1. Denote by $2n_i$ and p_i the number of cubes in $\mathcal{C}_{\neq}^{(i)}$ and $\mathcal{C}_0^{(i)}$. Then by Theorem 6.4.2.3 $\mathcal{F}_\Lambda(m_\Lambda|m_{\Lambda^c}) \ge c'_1\,(\zeta^2\ell_-^d)\sum_{i=1}^n(2n_i+p_i)$, so that we just need to show that $(2n_i+p_i)\ge 3^{-d}N_{\Gamma_i}$. Let us fix Γ_i and drop i from the notation. Writing $\mathcal{C}^+$ for the collection of cubes of $\mathcal{D}^{(\ell_+)}$ which intersect $\mathcal{C}$ and N for the number of elements in $\mathcal{C}^+$, we have $(2n+p)\ge N$. We shall prove that

$$\bigcup_{C\in\mathcal{C}^+}\left(C\cup\delta_{\text{out}}^{\ell_+}[C]\right)\supseteq \text{sp}(\Gamma); \tag{6.4.5.3}$$

By (6.4.5.3), $\ell_+^d N_\Gamma$ (which is the volume of $\text{sp}(\Gamma)$) is $\le N3^d\ell_+^d$ (because the number of cubes connected to a given one is 3^d-1); hence $(2n+p)\ge 3^{-d}N_\Gamma$. Thus the theorem is reduced to (6.4.5.3) which is proved next. Let $C_x^{(\ell_+)}\subset\text{sp}(\Gamma)$, $x\in\ell_+\mathbb{Z}^d$. We have the following cases.

Case 1: $\theta^{(\zeta,\ell_-,\ell_+)}(m;x)=0$ splits into Case 1a: (see Fig. 6.8) there is $y\in\ell_-\mathbb{Z}^d\cap C_x^{(\ell_+)}$ such that $\eta^{(\zeta,\ell_-)}(m;y)=0$; if no such y exists, Case 1b, (see Fig. 6.9) there are $z,z'\in\ell_-\mathbb{Z}^d\cap C_x^{(\ell_+)}$ such that $C_z^{(\ell_-)}$ and $C_{z'}^{(\ell_-)}$ are connected and $\eta^{(\zeta,\ell_-)}(m;z)=-\eta^{(\zeta,\ell_-)}(m;z')\ne 0$. In Case 1a $C_y^{(\ell_-)}\in\mathcal{C}_0$; in Case 1b at least one of the two cubes $C_z^{(\ell_-)}$ and $C_{z'}^{(\ell_-)}$ is in $\mathcal{C}_{\neq}$ by the maximality of $\mathcal{C}_{\neq}$. Thus in Case 1, $C_x^{(\ell_+)}$ is contained in the set on the l.h.s. of (6.4.5.3).

Case 2 (visualize the various subcases in Figs. 6.8 and 6.9): $\theta^{(\zeta,\ell_-,\ell_+)}(m;x)=1$. Case 2a: there is $y\in\ell_+\mathbb{Z}^d\cap\delta_{\rm out}^{\ell_+}[C_x^{(\ell_+)}]$ such that $\theta^{(\zeta,\ell_-,\ell_+)}(m;y)=0$; then by the definition of contours $C_y^{(\ell_+)}\subset \mathrm{sp}(\Gamma)$ and by the analysis of Case 1, $C_y^{(\ell_+)}\in\mathcal{C}^+$ so that $C_x^{(\ell_+)}$ is contained in the set on the l.h.s. of (6.4.5.3). Case 2b: if Case 2a does not hold; by the definition of contours there is $y\in\ell_+\mathbb{Z}^d\cap\delta_{\rm out}^{\ell_+}[C_x^{(\ell_+)}]$ such that $\theta^{(\zeta,\ell_-,\ell_+)}(m;y)=-\theta^{(\zeta,\ell_-,\ell_+)}(m;x)$ and they are both in $\mathrm{sp}(\Gamma)$. Then there are $z\in\ell_-\mathbb{Z}^d\cap C_x^{(\ell_+)}$ and $z'\in\ell_-\mathbb{Z}^d\cap C_y^{(\ell_+)}$ such that $\eta^{(\zeta,\ell_-)}(m;z)=-\eta^{(\zeta,\ell_-)}(m;z')\neq 0$. $C_z^{(\ell_-,\gamma)}$ and $C_{z'}^{(\ell_-)}$ are connected and by maximality (the same argument as in Case 1b) at least one among $C_z^{(\ell_-,\gamma)}$ and $C_{z'}^{(\ell_-)}$ is in $\mathcal{C}_{\neq}$. Hence at least one among $C_x^{(\ell_+)}$ and $C_y^{(\ell_+)}$ is in $\mathcal{C}^+$, so that $C_x^{(\ell_+)}$ is contained in the set on the l.h.s. of (11.8.1.5).

Case 3: $\theta^{(\zeta,\ell_-,\ell_+)}(m;x)=-1$ is just as Case 2 and the proof is complete. □

6.4.6 Cut and paste techniques

The lower bound in Theorem 6.4.5 is not very satisfactory, because the true excess free energy when there are boundary conditions is

$$\mathcal{F}_\Lambda(m_\Lambda|m_{\Lambda^c})-\inf_{\psi_\Lambda\in L^\infty(\Lambda,[-1,1])}\mathcal{F}_\Lambda(\psi_\Lambda|m_{\Lambda^c}) \tag{6.4.6.1}$$

(the second term may not vanish due to the interaction with Λ^c). We would like a lower bound of the form: there exists $\psi_\Lambda\in L^\infty(\Lambda,[-1,1])$ so that

$$\mathcal{F}_\Lambda(m_\Lambda|m_{\Lambda^c})-\mathcal{F}_\Lambda(\psi_\Lambda|m_{\Lambda^c})\geq c(\zeta^2\ell_-^d)\sum_i N_{\Gamma_i} \tag{6.4.6.2}$$

($c>0$ and $\{\Gamma_i\}$ the contours of m in Λ). The operation of replacing in a domain of the space the original profile by a new one, leaving it unchanged elsewhere, is very common in statistical mechanics where it goes under the name of "surgery" or "cut and paste."

Equation (6.4.6.2) is too general to be true. We shall prove here a weaker version where $\Lambda=c(\Gamma)$, $c(\Gamma)$ as in (6.4.4.1), and the r.h.s. of (6.4.6.2) replaced by $c(\zeta^2\ell_-^d)N_\Gamma$. We shall thus change the original profile inside $c(\Gamma)$ and leave it unchanged elsewhere. The change will be such that the new profile, supposing Γ to be a plus contour, is equal to m_β on $c(\Gamma)\setminus B_{\rm ext}(\Gamma)$, $B_{\rm ext}(\Gamma)$ the layer in $\mathrm{sp}(\Gamma)$ close to $c(\Gamma)^c$; see (6.4.4.3). The delicate point in the surgery is to control the free energy cost of the interpolation between the region $c(\Gamma)\setminus B_{\rm ext}(\Gamma)$ where the new function has been set equal to m_β and the complement of $c(\Gamma)$ where the function has not been changed, a problem which is solved by exploiting that • in the interpolating region $\eta^{(\zeta,\ell_-)}$ is constant and non-zero, and • the region can be made large by appropriately choosing the parameter ℓ_+. This is the first time that we exploit the freedom in the choice of ℓ_+; so far all proofs have been uniform in ℓ_+. By the symmetry under change of sign, we are restricting ourselves without loss of generality to + contours, which we shall keep doing unless otherwise stated.

Any triple (ζ, ℓ_-, ℓ_+) with $\zeta > 0$ small enough, $\ell_- > 0$ correspondingly small and then ℓ_+ correspondingly large, meets the requests below. However, the Peierls bounds do get worse when ζ and ℓ_- decrease and ℓ_+ increases, and for this reason it is convenient to give a more quantitative definition:

Definition [Choice of parameters]

- The pair (ζ, ℓ_-) is "*good*" if $\zeta < \zeta^*/2$ and $\ell_- < \kappa^*\zeta$, where

$$\zeta^* = \min\{\zeta_0, \zeta_1\}, \qquad \kappa^* = \min\{\kappa_0, \kappa_1\}, \tag{6.4.6.3}$$

  with ζ_0 as in Theorem 6.3.3.1; ζ_1 as in Lemma 6.4.2.2, κ_0 as in Lemma 6.3.2.1 and κ_1 as in Theorem 6.4.2.3.
- The triple (ζ, ℓ_-, ℓ_+) is "*good*" if the pair (ζ, ℓ_-) is good, if $\ell_+ > 100$,

$$c_1''\ell_-^d\zeta^2 \ge 4c_\omega \ell_+^d e^{-\ell_+\omega/8}, \tag{6.4.6.4}$$

  with c_1'' as Theorem 6.4.5.1, ω and c_ω as in Theorem 6.3.3.1. We also require ℓ_+ to be so large that $(c_\omega e^{-\omega[(\ell_+/8)-2]})^2 \le c_\omega e^{-\omega\ell_+/8}$ and such that the function $x^d e^{-x\omega/8}$ is a decreasing function of x for $x \ge \ell_+/2$.

Theorem 6.4.6.1 *Let (ζ, ℓ_-, ℓ_+) be a good triple. Let $m \in L^\infty(\mathbb{R}^d, [-1, 1])$ have a bounded plus (ζ, ℓ_-, ℓ_+)-contour Γ. Then there exists a function $\psi_{c(\Gamma)} \in L^\infty(c(\Gamma); [m_\beta - \epsilon, m_\beta + \epsilon])$, $\epsilon = (1 - \kappa_0)\zeta$, κ_0 as in Lemma* 6.3.2.1, *such that $\psi_{c(\Gamma)}(r) = m_\beta$ for $r \in c(\Gamma) \setminus B_{\text{ext}}(\Gamma)$ and, writing $c_1 = \frac{c_1''}{2}$, c_1'' as in Theorem* 6.4.5.1,

$$\mathcal{F}_{c(\Gamma)}(m_{c(\Gamma)}|m_{c(\Gamma)^c}) \ge \mathcal{F}_{c(\Gamma)}(\psi_{c(\Gamma)}|m_{c(\Gamma)^c}) + c_1\,(\zeta^2\ell_-^d)\,N_\Gamma. \tag{6.4.6.5}$$

Proof We assume without loss of generality that $\mathcal{F}_{c(\Gamma)^c}(m_{c(\Gamma)^c}) < \infty$ (as $\mathcal{F}_{c(\Gamma)}(m_{c(\Gamma)}|m_{c(\Gamma)^c})$ is independent of the values of $m(r)$, $\text{dist}(r, \Lambda) > 1$). Let $\psi \in L^\infty(\mathbb{R}^d)$ be the function equal to $m_{c(\Gamma)^c}$ and to $\psi_{c(\Gamma)}$ on $c(\Gamma)^c$ and $c(\Gamma)$, respectively; then (6.4.6.5) is the same as

$$\mathcal{F}(m) \ge \mathcal{F}(\psi) + c_1\,(\zeta^2\ell_-^d)\,N_\Gamma. \tag{6.4.6.6}$$

- Let $B = \Delta_1 \cup \Delta_2 \cup \Delta_3$ (see Sect. 6.5 and Fig. 6.10), where writing $\ell = \ell_+/8$, $\Delta_1 = \delta_{\text{out}}^\ell[c(\Gamma)^c]$, $\Delta_2 = \delta_{\text{out}}^\ell[c(\Gamma)^c \cup \Delta_1]$ and $\Delta_3 = \delta_{\text{out}}^\ell[c(\Gamma)^c \cup \Delta_1 \cup \Delta_2]$. We rename $\Sigma_0 = \Delta_2$. $\mathcal{F}(m) = \mathcal{F}_B(m_B|m_{B^c}) + \mathcal{F}_{B^c}(m_{B^c})$; by definition of contours, $m \in M^+_{B,\zeta,\ell_-}$ (i.e. it is in local equilibrium in B; see Sect. 6.3.1). Then by Theorem 6.3.3.1 $\mathcal{F}_B(m_B|m_{B^c}) \ge \mathcal{F}_B(\phi_B^*|m_{B^c})$, where $\phi_B^*(\cdot) \in [m_\beta - \epsilon, m_\beta + \epsilon]$ and $|\phi^*(r) - m_\beta| \le c_\omega e^{-\omega\,\text{dist}(r, B^c)}$. Let $\phi^* = \phi_B^*$ on B and $= m$ elsewhere; then $\mathcal{F}(m) \ge \mathcal{F}(\phi^*)$.

• $\mathcal{F}(\phi^*) \ge \mathcal{F}_{\Sigma_0^c}(\phi^*_{\Sigma_0^c})$, $\mathcal{F}_{\Sigma_0}(m_\beta 1_{\Sigma_0}|\phi^*_{\Sigma_0^c}) \le \frac{1}{2}\int_{\Sigma_0^c}\int_{\Sigma_0} J(r,r')[\phi^*(r) - m_\beta]^2$. Since $\text{dist}(r, \Sigma_0) \le 1$ (otherwise $J(r,r') = 0$), $\text{dist}(r, B^c) \ge (\ell_+/8) - 2$, calling $\delta\Sigma_0 = \{r' \in \Sigma_0 : \text{dist}(r', \Sigma_0^c) \le 1\}$, by the choice of ℓ_+,

$$\mathcal{F}_{\Sigma_0}(m_\beta 1_{\Sigma_0}|\phi^*_{\Sigma_0^c}) \le (c_\omega e^{-\omega[(\ell_+/8)-2]})^2|\delta\Sigma_0| \le c_\omega e^{-\omega(\ell_+/8)}|\delta\Sigma_0|. \quad (6.4.6.7)$$

Hence $\mathcal{F}(\phi^*) \ge \mathcal{F}_{\Sigma_0^c}(\phi^*_{\Sigma_0^c}) + \mathcal{F}_{\Sigma_0}(m_\beta 1_{\Sigma_0}|\phi^*_{\Sigma_0^c}) - c_\omega e^{(\ell_+/8)}|\delta\Sigma_0|$. Thus calling ϕ the function equal to m_β on Σ_0 and to ϕ^* elsewhere,

$$\mathcal{F}(m) \ge \mathcal{F}(\phi) - c_\omega|\text{sp}(\Gamma)|e^{-\omega\ell_+/8}. \quad (6.4.6.8)$$

• Write $\Lambda_0 = c(\Gamma)\setminus\{\Delta_1 \cup \Delta_2\}$ ($\Delta_2 \equiv \Sigma_0$) and write $\mathcal{F}(\phi) = \mathcal{F}_{\Lambda_0}(\phi_{\Lambda_0}|\phi_{\Lambda_0^c}) + \mathcal{F}_{\Lambda_0^c}(\phi_{\Lambda_0^c})$. By Theorem 6.4.5.1, $\mathcal{F}(\phi) \ge \mathcal{F}_{\Lambda_0^c}(\phi_{\Lambda_0^c}) + c_1''\zeta^2\ell_-^d N_\Gamma$ so that by (6.4.6.8),

$$\mathcal{F}(m) \ge \mathcal{F}_{\Lambda_0^c}(\phi_{\Lambda_0^c}) + c_1''\zeta^2\ell_-^d N_\Gamma - c_\omega|\text{sp}(\Gamma)|e^{-\omega\ell_+/8}, \quad (6.4.6.9)$$

and by (6.4.6.4), $\mathcal{F}(m) \ge \mathcal{F}_{\Lambda_0^c}(\phi_{\Lambda_0^c}) + c_1\zeta^2\ell_-^d N_\Gamma$; recall $c_1 := c_1''/2$. Set $\psi^* = \phi$ on Λ_0^c and $= m_\beta$ on Λ_0. Then, trivially, $\mathcal{F}_{\Lambda_0^c}(\phi_{\Lambda_0^c}) \equiv \mathcal{F}_{\Lambda_0^c}(\psi^*_{\Lambda_0^c})$, while $\mathcal{F}_{\Lambda_0}(\psi^*_{\Lambda_0}|\psi^*_{\Lambda_0^c}) = 0$ because $f_\beta(\psi^*_{\Lambda_0}) = 0$ as $\psi^*_{\Lambda_0} \equiv m_\beta$ and the interaction with $\psi^*_{\Lambda_0^c}$ is also equal to 0, because $\psi^*_{\Lambda_0^c} = m_\beta$ on Σ_0. □

The following corollary of Theorem 6.4.6.1 will be used in Chap. 7.

Corollary 6.4.6.2 *Let (ζ, ℓ_-, ℓ_+) be a good triple; Λ and $\Delta \subset \Lambda$ two bounded, $\mathcal{D}^{(\ell_+)}$-measurable regions; $m \in L^\infty(\mathbb{R}^d, [-1,1])$ with $\eta^{(\zeta,\ell_-)}(m; r) = 1$, $r \in \delta^{\ell_+}_{\text{out}}[\Lambda] \cup \delta^{\ell_+}_{\text{in}}[\Lambda]$. Then there is ϕ in $L^\infty(\mathbb{R}^d)$, $\phi = m$ on Λ^c, $\phi = m_\beta$ on Δ, $\eta^{(\zeta,\ell_-)}(\phi; \cdot) = 1$ on $\Lambda \setminus \Delta$ and*

$$\mathcal{F}_\Lambda(m_\Lambda|m_{\Lambda^c}) \ge \mathcal{F}_\Lambda(\phi|m_{\Lambda^c}) - c_\omega|\delta\Delta|e^{-\omega\,\text{dist}(\Delta,\Lambda^c_{\neq})}, \quad (6.4.6.10)$$

where $\delta\Delta = \{r \in \Delta : \text{dist}(r, \Delta^c) \le 1\}$ and $\Lambda^c_{\neq} = \{r \in \Lambda^c, m(r) \neq m_\beta, \text{dist}(r, \Lambda^c) \le 1\}$.

Proof Since the values of m on $\Lambda^c \setminus \delta^{\ell_+}_{\text{out}}[\Lambda]$ are irrelevant, without loss of generality we shall suppose that on such a region $m = m_\beta$. Then, by the assumptions on $\eta^{(\zeta,\ell_-)}(m; r)$, any $\text{sp}(\Gamma)$ is contained in Λ. A contour Γ is called external if $\text{sp}(\Gamma)$ is not contained in $c(\Gamma')$, for any other contour Γ'. We then apply Theorem 6.4.6.1 to each one of the "external contours"; thus there is a function $\psi \in L^\infty(\mathbb{R}^d, [-1,1])$ such that $\eta^{(\zeta,\ell_-)}(\psi; r) = 1$ on Λ and $\mathcal{F}_\Lambda(m_\Lambda|m_{\Lambda^c}) \ge \mathcal{F}_\Lambda(\psi_\Lambda|m_{\Lambda^c})$. We can now use Theorem 6.3.3.1 to modify ψ into a new function ϕ^* which coincides with ψ (and hence with m) on Λ^c being such that (see (6.3.3.2)) $|\phi^*(r) - m_\beta| \le c_\omega e^{-\omega\,\text{dist}(r,\Lambda^c_{\neq})}$, with $\mathcal{F}_\Lambda(\psi_\Lambda|\psi_{\Lambda^c}) \ge \mathcal{F}_\Lambda(\phi^*_\Lambda|\phi^*_{\Lambda^c})$. We set $\phi = \phi^*$ on Δ^c and $= m_\beta$ on Δ and by the same argument used in the proof of Theorem 6.4.6.1, see (6.4.6.7) with Δ instead of Σ_0, we get (6.4.6.10). □

6.4.7 Conditional free energy of contours

In Chap. 9 we shall study phase transitions in Ising systems with Kac potentials by using a modified version of Theorem 6.4.6.1; see Theorem 6.4.7.1 below. The problem with Theorem 6.4.6.1 is that the region where ψ and m differ may be too large. When going from spins to the present mesoscopic formulation there is an error proportional to the volume of the region involved. Such an error must be controlled by the difference of free energies of m and ψ. In Theorem 6.4.6.1 the free energy difference is proportional to the volume of $\mathrm{sp}(\Gamma)$, the region involved in the modification is instead proportional to $c(\Gamma)$. The latter includes $\mathrm{int}(\Gamma)$ whose volume may be much larger than $|\mathrm{sp}(\Gamma)|$. Theorem 6.4.7.1 below settles the problem, exploiting the symmetry of the functional under a change of sign of m.

Theorem 6.4.7.1 *In the same context as Theorem* 6.4.6.1, *there is ψ equal to m on $c(\Gamma)^c \cup \mathrm{int}^+(\Gamma)$, equal to $-m$ on $\mathrm{int}^-(\Gamma)$ and to m_β on $\mathrm{sp}(\Gamma) \setminus \delta_{\mathrm{in}}^{\ell_+}[\mathrm{sp}(\Gamma)]$. Moreover, ψ has values in $[m_\beta - \epsilon, m_\beta + \epsilon]$, $\epsilon = (1-\kappa_0)\zeta$, on B and*

$$\mathcal{F}_{\mathrm{sp}(\Gamma)}(m_{\mathrm{sp}(\Gamma)}|m_{\mathrm{sp}(\Gamma)^c}) \ge \mathcal{F}_{\mathrm{sp}(\Gamma)}(\psi_{\mathrm{sp}(\Gamma)}|\psi_{\mathrm{sp}(\Gamma)^c}) + c_1\,(\zeta^2 \ell_-^d)\,N_\Gamma, \tag{6.4.7.1}$$

$$F_{\mathrm{sp}(\Gamma)}(m_{\mathrm{sp}(\Gamma)}|m_{\mathrm{sp}(\Gamma)^c}) \ge F_{\mathrm{sp}(\Gamma)}(\psi_{\mathrm{sp}(\Gamma)}|\psi_{\mathrm{sp}(\Gamma)^c}) + c_1\,(\zeta^2 \ell_-^d)\,N_\Gamma. \tag{6.4.7.2}$$

Proof By (6.1.3.7), we can rewrite (6.4.7.2) as

$$\begin{aligned}
&\mathcal{F}_{\mathrm{sp}(\Gamma)}(m_{\mathrm{sp}(\Gamma)}|m_{\mathrm{sp}(\Gamma)^c})\\
&\quad\ge \mathcal{F}_{\mathrm{sp}(\Gamma)}(\psi_{\mathrm{sp}(\Gamma)}|\psi_{\mathrm{sp}(\Gamma)^c}) + c_1\,(\zeta^2 \ell_-^d)\,N_\Gamma\\
&\qquad + R_{\mathrm{sp}(\Gamma)}(m_{\mathrm{sp}(\Gamma)^c}) - R_{\mathrm{sp}(\Gamma)}(\psi_{\mathrm{sp}(\Gamma)^c}).
\end{aligned}$$

Since $R_{\mathrm{sp}(\Gamma)}(m_{\mathrm{sp}(\Gamma)^c})$ depends on $m_{\mathrm{sp}(\Gamma)^c}(\cdot)^2$ by the definition of ψ,

$$R_{\mathrm{sp}(\Gamma)}(m_{\mathrm{sp}(\Gamma)^c}) = R_{\mathrm{sp}(\Gamma)}(\psi_{\mathrm{sp}(\Gamma)^c}),$$

and therefore (6.4.7.2) and (6.4.7.1) are equivalent.

We proceed as in the proof of Theorem 6.4.6.1, but with $c(\Gamma)^c$ replaced by the union of $c(\Gamma)^c$ and $\bigcup_i \mathrm{int}_i^{\pm}(\Gamma)$. We thus define corridors $\Delta_j^{\pm}(i)$, $j = 1,2,3$, around $\mathrm{int}_i^{\pm}(\Gamma)$ analogous to the corridors Δ_j of Theorem 6.4.6.1 and rename $\Delta_2^{\pm}(i) \equiv \Sigma_i^{\pm}$. Then there is a function ϕ such that $\phi(r) = \pm m_\beta$ on $\Sigma_i^{\pm}$ and $= m_\beta$ on Σ_0, and $\mathcal{F}(m) \ge \mathcal{F}(\phi) - c_\omega|\mathrm{sp}(\Gamma)|e^{-\omega\ell_+/8}$. (6.4.6.9) then extends to the present case with $\Lambda_0 = \mathrm{sp}(\Gamma) \setminus \Delta$, Δ the union of all $\Delta_j^{\pm}(i)$, $j = 1,2$ and of Δ_1 and Δ_2 of Theorem 6.4.6.1. We write in (6.4.6.9) $\mathcal{F}_{\Lambda_0^c}(\phi_{\Lambda_0^c}) = \sum_i \mathcal{F}_{A_i}(\phi_{A_i})$, A_i being the maximal connected components of Λ_0^c. Exploiting the symmetry of the functional we write $\mathcal{F}_{A_i}(\phi_{A_i}) = \mathcal{F}_{A_i}(-\phi_{A_i})$ on the A_i which intersect $\mathrm{int}^-(\Gamma)$. We then define ψ to be equal to m_β on Λ_0, to ϕ on $c(\Gamma)^c$ and on the sets A_i which intersect $\mathrm{int}^+(\Gamma)$, while we set $\psi = -\phi$ on those which intersect $\mathrm{int}^-(\Gamma)$ and obtain (6.4.7.1). □

6.4.8 Scaling properties

In the applications to the Ising model, see Chap. 9, the interaction kernel $J(r,r')$ is replaced by $J_\gamma(r,r') = \gamma^d J(\gamma r, \gamma r')$ with γ typically a small quantity (theorems are proved for "γ small enough"). Denote by $\mathcal{F}_\gamma$ the functional with $J_\gamma(r,r')$, and by $\mathcal{F}$ the functional with $J(r,r')$. Then by a change of variables in the integrals,

$$\mathcal{F}_\gamma(m) = \gamma^{-d}\mathcal{F}(\tilde{m}), \quad \tilde{m}(r) = m(\gamma^{-1}r). \tag{6.4.8.1}$$

Thus the theory with γ can be obtained from the one developed so far with $J(r,r')$ by scaling by a factor γ^{-1} and the free energies by γ^{-d}. Thus contours for $\mathcal{F}_\gamma$ are defined with the parameters

$$(\zeta, \gamma^{-1}\ell_-, \gamma^{-1}\ell_+). \tag{6.4.8.2}$$

(ζ, ℓ_-, ℓ_+) is a good triple in the sense of Sect. 6.4.6. Let then Γ be a contour with parameters (6.4.8.2) and let $m \in L^\infty(\mathbb{R}^d, [-1,1])$ have Γ as a contour. Then (6.4.7.2) in Theorem 6.4.7.1 reads

$$F_{\gamma,\mathrm{sp}(\Gamma)}(m_{\mathrm{sp}(\Gamma)}|m_A) \geq F_{\gamma,\mathrm{sp}(\Gamma)}(\psi_{\mathrm{sp}(\Gamma)}|\psi_A) + \gamma^{-d} c_1\,(\zeta^2\ell_-^d)\,N_\Gamma. \tag{6.4.8.3}$$

The extra factor γ^{-d} comes from (6.4.7.2) with $\ell_- \to \gamma^{-1}\ell_-$ (according to (6.4.8.2)) or because free energies are multiplied by γ^{-d}.

6.5 Pictures and examples

To visualize the notions introduced in the last section we fix a configuration, compute the phase indicators and draw pictures of the corresponding plus/minus correct regions and contours. Let $S \subset \mathbb{R}^2$ be the coordinate square with bottom left corner the origin and side $21\ell_+$ (what we are going to say applies as well to Part III with $\ell_\pm \to \ell_{\pm,\gamma}$). In Fig. 6.1, S is partitioned into 21×21 squares of side ℓ_+, and we shall denote by C_x, $x \in \mathbb{Z}^2$ the square of side ℓ_+ with bottom left corner $x\ell_+$, x the "lattice coordinate" of $x\ell_+$. We suppose that $\theta^{(\ell_-,\ell_+,\zeta)} \equiv 1$ outside S, while the values in S are as in Fig. 6.1. $\Theta^{(\ell_-,\ell_+,\zeta)}$ is then completely determined. Its values on S are reported in Fig. 6.2, where the plus (marked by $+$), minus (marked by $-$) correct regions and the support of the contours (marked by 0) are separated by the discontinuity lines of $\Theta^{(\ell_-,\ell_+,\zeta)}$. Notice that $\Theta^{(\ell_-,\ell_+,\zeta)} \equiv 1$ on $\delta_{\mathrm{in}}^{\ell_+}[S]$, $\delta_{\mathrm{in}}^{\ell_+}[S]$. Thus, the configuration we are considering is one of those which arise when computing the plus diluted partition function in S; see Sect. 9.2.6 for the Ising case and Sect. 10.3.2 for LMP.

The configuration of Fig. 6.1 has three contours, Γ_1, Γ_2 and Γ_3: $\mathrm{sp}(\Gamma_1)$ is the region between the discontinuity lines of Fig. 6.2 which contains $C_{(17,3)}$, $\mathrm{sp}(\Gamma_2)$ the one which contains $C_{(16,17)}$ and $\mathrm{sp}(\Gamma_3)$ the one containing $C_{(5,11)}$ (i.e. the squares marked 0 in Fig. 6.2). Γ_1 and Γ_2 are both external, plus contours, Γ_3 is a minus

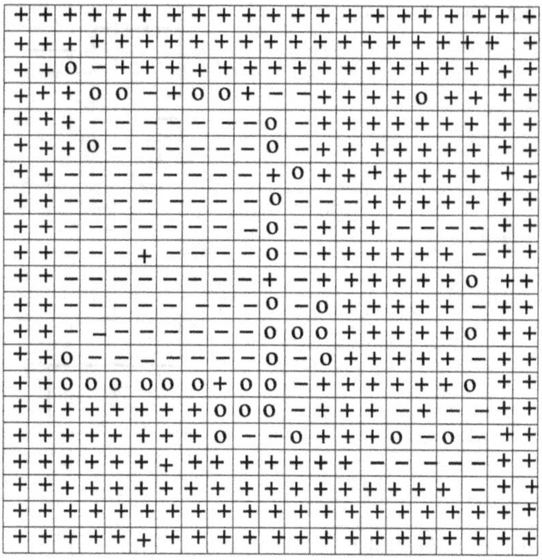

Fig. 6.1 Values of $\theta^{(\ell_-,\ell_+,\zeta)}$ on S

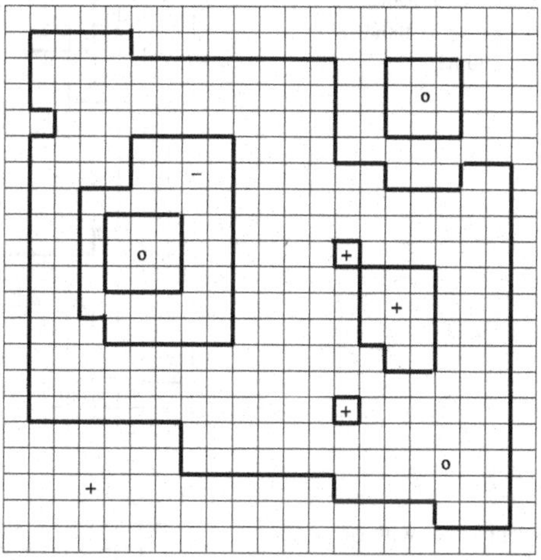

Fig. 6.2 *Discontinuity lines* of $\Theta^{(\ell_-,\ell_+,\zeta)}$ and regions where it is $+$, $-$ and 0

contour. $c(\Gamma_1)$ is drawn in Fig. 6.3. Γ_1 has three internal parts, $\mathrm{int}^-(\Gamma_1)$, $\mathrm{int}_1^+(\Gamma_1)$ and $\mathrm{int}_2^+(\Gamma_1)$, see Fig. 6.4, $\mathrm{int}^-(\Gamma_1)$ has a contour inside, Γ_3; see Fig. 6.5. The sets $B_{\mathrm{ext}}(\Gamma_1)$ and $A_{\mathrm{ext}}(\Gamma_1)$ are drawn in Fig. 6.6, the sets $B_i(\Gamma_1)$ and $A_i(\Gamma_1)$ are drawn in Fig. 6.7. Observe that $B_2(\Gamma_1) \cap B_3(\Gamma_1) \neq \emptyset$. Values of $\eta^{(\zeta,\ell_-)}$ are reported by blowing up portions of S; see Figs. 6.8 and 6.9.

In Fig. 6.10 we have drawn the corridors Δ_i, $i = 1, 2, 3$ (introduced in the proof of Theorem 6.4.6.1 and in Sect. 11.2.1) in a window made by a square of $\mathcal{D}^{(\ell_+)}$. The width of each strip is $\ell_+/8$. In Fig. 6.11 we have drawn the corridors Δ_4 and Δ_5 (see Sect. 11.2.1) as they look in the two squares $C_{(2,10)}$–$C_{(3,10)}$, respectively in $\mathrm{sp}(\Gamma_1)$ and $\mathrm{int}^-(\Gamma_1)$.

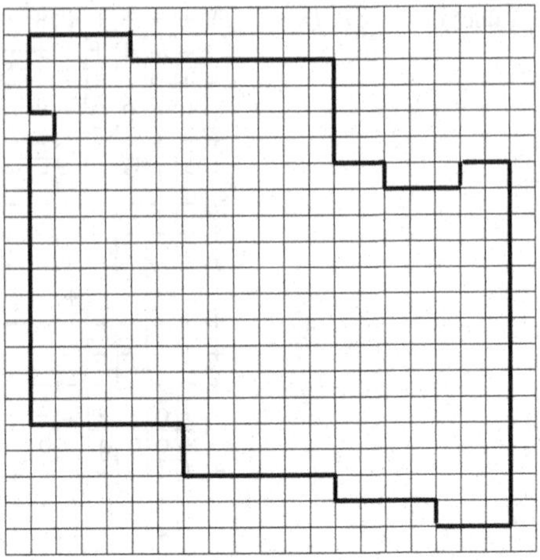

Fig. 6.3 The set $c(\Gamma_1)$

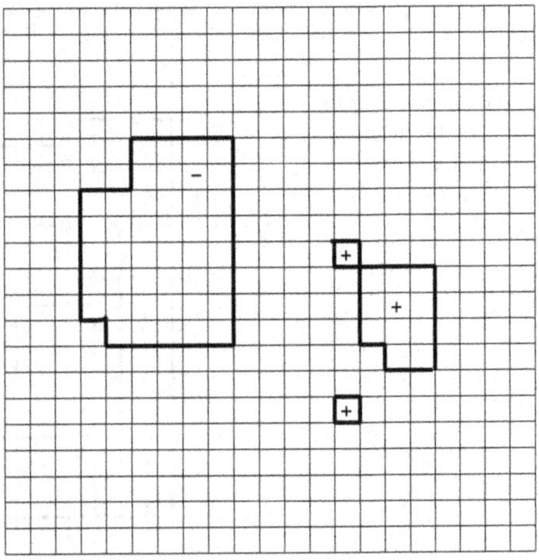

Fig. 6.4 $\mathrm{int}^-(\Gamma_1)$ is the region with $-$, $\mathrm{int}_1^+(\Gamma_1)$ is the $+$ region on *top left* and $\mathrm{int}_2^+(\Gamma_1)$ the one consisting of a single square

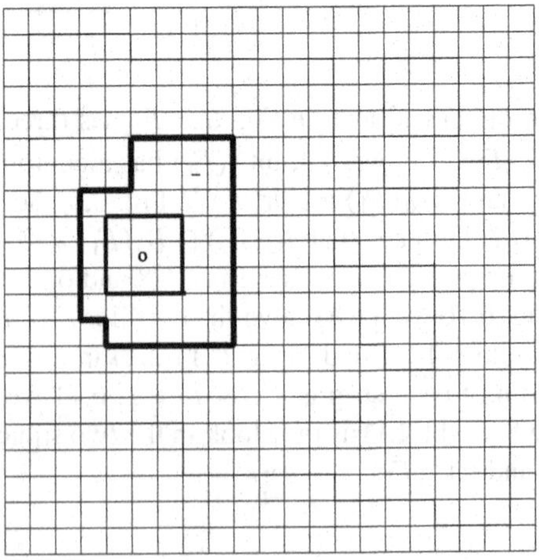

Fig. 6.5 $\mathrm{int}^-(\Gamma_1)$ with on the inside $\mathrm{sp}(\Gamma_3)$

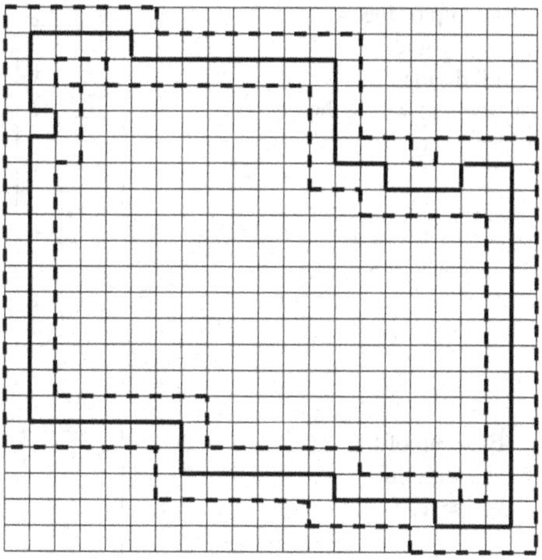

Fig. 6.6 $A_{\text{ext}}(\Gamma_1) =$ $\delta_{\text{out}}^{\ell_+}[c(\Gamma)]$ is the region between the *external dashed* and the *thick line*; $B_{\text{ext}}(\Gamma_1) =$ $\delta_{\text{in}}^{\ell_+}[c(\Gamma)]$ is the region between the *internal dashed* and the *thick line*

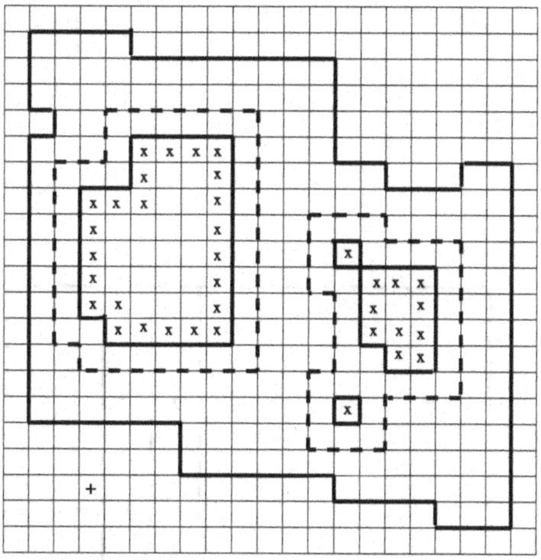

Fig. 6.7 $A_i(\Gamma_1) =$ $\delta_{\text{in}}^{\ell_+}[\text{int}_i(\Gamma_1)]$ are the regions made of the x marked squares contained in $\text{int}_i(\Gamma_1)$; $B_i(\Gamma_1) = \delta_{\text{out}}^{\ell_+}[\text{int}_i(\Gamma_1)]$ form the region between the *dashed lines* and the *thick lines* inside

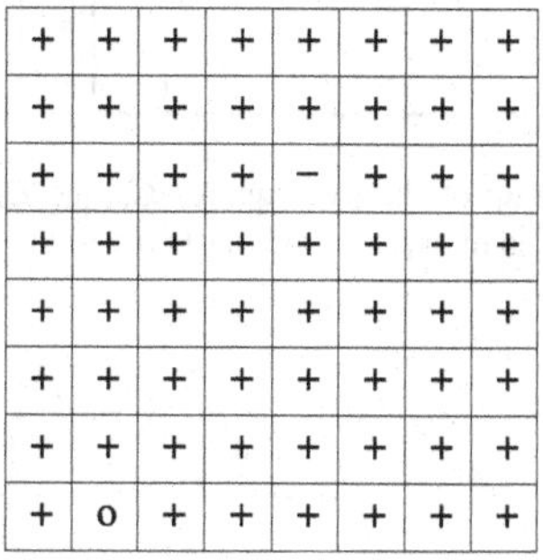

Fig. 6.8 Blow up of the squares $C_{(2,6)}$, where $\theta^{(\zeta,\ell_-,\ell_+)} = 0$ divided into squares of side ℓ_-, where the values of $\eta^{(\zeta,\ell_-)}$ are reported

+	+	+	+	+	+	+	+
+	+	+	+	+	+	+	+
+	+	+	+	−	+	+	+
+	+	+	+	+	+	+	+
+	+	+	+	+	+	+	+
+	+	+	+	+	+	+	+
+	+	+	+	+	+	+	+
+	o	+	+	+	+	+	+

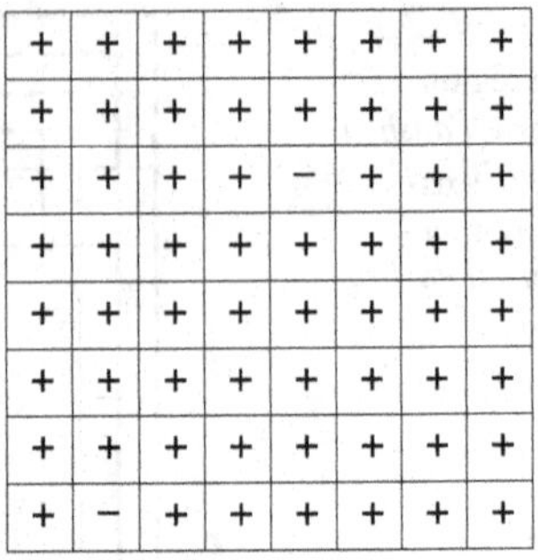

Fig. 6.9 Blow up of the squares $C_{(2,7)}$, where $\theta^{(\zeta,\ell_-,\ell_+)} = 0$ divided into squares of side ℓ_-, where the values of $\eta^{(\zeta,\ell_-)}$ are reported

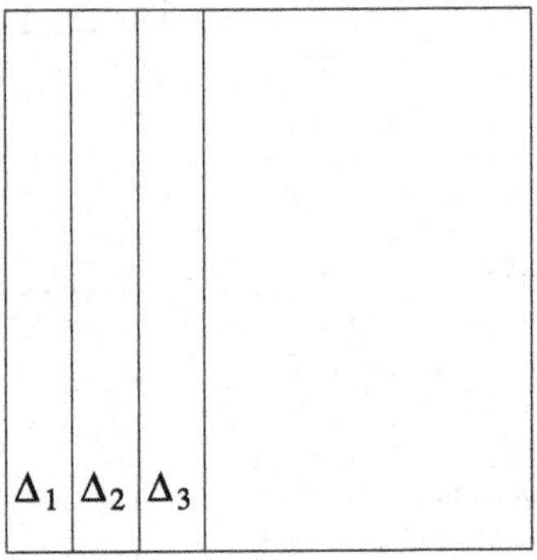

Fig. 6.10 Blow up of the squares $C_{(2,8)}$ with the corridors Δ_i, $i = 1, 2, 3$

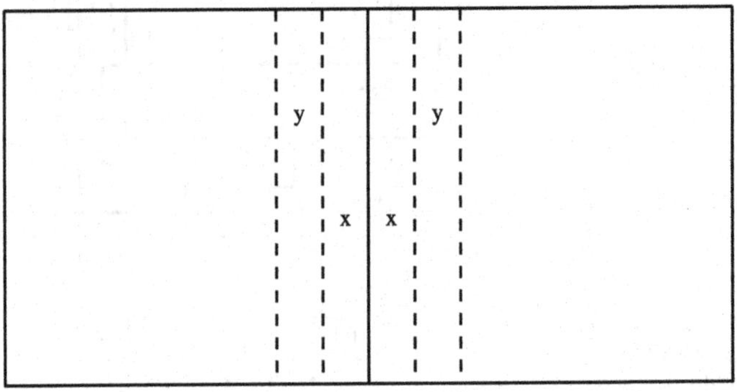

Fig. 6.11 Blow up of the two squares $C_{(3,10)}$–$C_{(4,10)}$, respectively in $\text{sp}(\Gamma_1)$ and $\text{int}^-(\Gamma_1)$; the region marked x is in Δ_4, the one marked y in Δ_5

Chapter 7
Surface tension, Gamma convergence, Wulff shape

The surface tension is the excess free energy per unit area needed to create a state with two coexisting phases. The area in the definition refers to the interface which separates the two phases, and the surface tension may depend on its orientation when the interaction is anisotropic. Phases are forced to coexist by imposing either suitable boundary conditions or constraints on the set of profiles where the free energy functional is defined, for instance the canonical constraint which fixes the total magnetization density. The main outcome of the theory, in the case of the L–P free energy functionals of Chap. 6, is that the surface tension is to a large extent independent of the way the phases are forced to coexist and it coincides with the excess free energy of a one dimensional instanton, in agreement with the predictions of the van der Waals theory.

7.1 Gamma convergence

The basic point formalized in this section is that, while interfaces are macroscopically sharp, they are instead diffuse in the mesoscopic scale with many mesoscopic profiles which correspond to a same macroscopic interface. The mesoscopic free energy of an interface is then defined by minimizing the mesoscopic excess free energy functional over all profiles corresponding to a same macroscopic interface. As we shall see, the mathematical formalization of this idea leads to a problem with Gamma convergence; hence the title of this section.

7.1.1 Finite volume corrections to the free energy

Applied to the L–P functional, Theorem 6.1.6.2 states that when $\beta > 1$ and $h = 0$, which will be tacitly assumed throughout the sequel,

$$\lim_{\Lambda \to \mathbb{R}^d} \frac{1}{|\Lambda|} \inf_{m_\Lambda \in \mathcal{X}_{\Lambda,s}} \mathcal{F}_\Lambda(m_\Lambda | m_{\Lambda^c}) = 0, \quad s \in (-m_\beta, m_\beta), \qquad (7.1.1.1)$$

where $\mathcal{X}_{\Lambda,s} = \{m_\Lambda \in L^\infty(\Lambda; [-1,1]) : \fint_\Lambda m_\Lambda = s\}$. Recall that $\mathcal{F}_\Lambda$ is the excess free energy functional defined by subtracting from $F_\Lambda(m_\Lambda | m_{\Lambda^c})$ its absolute minimum $\phi_\beta(m_\beta)|\Lambda|$, $\phi_\beta(m_\beta)$ being in fact the thermodynamical free energy density of the system when $s \in [-m_\beta, m_\beta]$. The limit in (7.1.1.1) is over all van Hove sequences $\Lambda \to \mathbb{R}^d$ but for the sequel it is convenient to restrict ourselves to the simpler case of an increasing sequence of cubes. The proof of (7.1.1.1) is obtained by exhibiting a particular m_Λ such that $\mathcal{F}_\Lambda(m_\Lambda|m_{\Lambda^c})/|\Lambda| \to 0$. A simple choice for m_Λ is to set it equal to m_β and to $-m_\beta$ in the two regions of Λ separated by an hyperplane chosen to satisfy the canonical constraint that the average magnetization is s. The free energy excess of such a function is due to the interaction between the plus and the minus regions and to the interaction with the boundaries. It thus scales as L^{d-1} (L the side of Λ) and since in (7.1.1.1) we divide by L^d, the limit as $L \to \infty$ vanishes.

The next order corrections to the free energy are thus defined by the quantity

$$\frac{1}{L^{d-1}} \inf_{m_\Lambda \in \mathcal{X}_{\Lambda,s}} \mathcal{F}_\Lambda(m_\Lambda|m_{\Lambda^c}). \tag{7.1.1.2}$$

We want to single out in (7.1.1.2) the contribution of the interface between pluses and minuses from the interaction with the boundaries. The easiest way is just to kill the latter by taking for Λ a torus:

$$\begin{aligned} \mathcal{F}_\Lambda^{\text{per}}(m_\Lambda) := &\int_\Lambda f_{\beta,h}\big(m_\Lambda(r)\big)\, dr \\ &+ \frac{1}{4}\int_\Lambda\int_\Lambda J^{\text{per}}(r,r')[m_\Lambda(r) - m_\Lambda(r')]^2\, dr'\, dr. \end{aligned} \tag{7.1.1.3}$$

Equation (7.1.1.3) differs from its analogue (6.1.3.6) by having J in the latter replaced by J^{per}, where for any r and r' in Λ,

$$J^{\text{per}}(r,r') = \sum_{n \in \mathbb{Z}^d} J(r, r' + nL), \tag{7.1.1.4}$$

with $L > 1$ the side of Λ ($L > 1$ to avoid self-interaction). $\frac{1}{L^{d-1}} \times \inf_{m_\Lambda \in \mathcal{X}_{\Lambda,s}} \mathcal{F}_\Lambda^{\text{per}}(m_\Lambda|m_{\Lambda^c})$ is then interpreted as the excess free energy density of the interface implicitly imposed on the system by the constraint $\mathcal{X}_{\Lambda,s}$. The previous bound obtained by splitting Λ into two rectangles, one where we put $m_\Lambda = -m_\beta$ and the other where $m_\Lambda = m_\beta$, can be improved in two ways. One by optimizing over the shape of the region where we separate the two phases, the other one by relaxing the condition $m_\Lambda = \pm m_\beta$ to allow for more general ways to connect the two values across the interface. As we shall see in Sect. 7.2 we can actually separate the analysis, by studying first the variational problem where we optimize the profile realizing a given interface and then by looking for the optimal shape of the interface, the Wulff problem. The former problem is studied in this section, the Wulff problem in the next one.

7.1.2 Main definitions

To implement the approach outlined in the previous subsection we need to study the excess free energy of profiles which approximate a given interface. The latter is defined by the way the torus is split into $\Lambda = \Lambda_+ \cup \Lambda_-$, $\Lambda_+ \cap \Lambda_- = \emptyset$, which are the regions where the plus and minus phases are localized. We thus want to study the excess free energy of profiles m_Λ "close" to the function equal to m_β in Λ_+ and to $-m_\beta$ in Λ_-. We have essentially two points to clarify, how does the splitting into Λ_+ and Λ_- changes when we vary Λ? What does closeness mean?

The first problem is simply solved by a scaling argument: we introduce the unit torus Ω in $\mathbb{R}^d$ and regard $\Lambda = \epsilon^{-1}\Omega$, with $\epsilon^{-1} = L$ the side of Λ. We then fix a function $u \in L^\infty(\Omega, \{-m_\beta, m_\beta\})$ which defines the way Ω splits into the plus and minus phases and, for each $\epsilon > 0$, the representative of u in $\Lambda = \epsilon^{-1}\Omega$ is the function $u_\Lambda(r) := u(\epsilon r)$ on Λ. As a rule we denote by u, $m, \dots,$ functions on Ω and add a suffix Λ to denote their images under dilation by ϵ^{-1} as functions on $\Lambda = \epsilon^{-1}\Omega$.

Next issue is to define closeness, namely which profiles m_Λ are called "close to u_Λ." Here the choice is not univocal; ours is to use the $L^1(\Omega)$ norm to weight both the volume where two profiles differ and the amount of their discrepancy which is therefore a good candidate to quantify distances among profiles. We thus measure "the distance" of m_Λ from u_Λ by

$$\begin{aligned} &\fint_\Lambda |m_\Lambda - u_\Lambda| = \|m - u\|_{L^1(\Omega)}, \\ &m(r) = m_\Lambda(\epsilon^{-1} r), \;\; u(r) = u_\Lambda(\epsilon^{-1} r). \end{aligned} \tag{7.1.2.1}$$

We are now ready for the definition of the excess free energy of an interface. Recalling the bound (7.1.1.2) for profiles with an interface, we define a functional Φ_ϵ on $L^\infty(\Omega, [-1, 1])$ by setting $\Lambda = \epsilon^{-1}\Omega$ and for any $v \in L^\infty(\Omega, [-1, 1])$

$$\Phi_\epsilon(v) = \epsilon^{d-1} \mathcal{F}^{\text{per}}_\Lambda(v_\Lambda), \quad v_\Lambda(r) = v(\epsilon r). \tag{7.1.2.2}$$

Let then $u \in L^\infty(\Omega, \{-m_\beta, m_\beta\})$ (referred to hereafter as "an interface"); then the free energy of the interface u with scaling parameter $\epsilon > 0$ and accuracy $\zeta > 0$ is

$$\Psi_{\zeta,\epsilon}(u) = \inf_{\|v-u\|_{L^1(\Omega)} \le \zeta} \Phi_\epsilon(v). \tag{7.1.2.3}$$

We then call

$$\Psi_0'(u) = \lim_{\zeta\to 0} \liminf_{\epsilon\to 0} \Psi_{\zeta,\epsilon}(u), \qquad \Psi_0''(u) = \lim_{\zeta\to 0} \limsup_{\epsilon\to 0} \Psi_{\zeta,\epsilon}(u). \tag{7.1.2.4}$$

To make the definition unambiguous we would like that

$$\Psi_0'(u) = \Psi_0''(u) =: \Psi_0(u). \tag{7.1.2.5}$$

In such a case we would then interpret $\Psi_0(u)$ as the macroscopic free energy of the interface u. For consistency with the macroscopic theory, however, we also want $\Psi_0(u)$ to have the expression (1.1.1), and in such a case we can identify the surface tension as the integrand in (1.1.1).

We have followed here the definition of free energy of an interface introduced by De Giorgi with his notion of Γ-convergence, which, in the above context, is the existence of the limit (7.1.2.5).

7.1.3 Main theorem

We state below the main theorem of this chapter, referring to Appendix B for the notion of BV spaces and a list of properties used in the sequel.

Theorem 7.1.3.1 *There is a continuous function $s_\beta(\nu)$, $\nu \in \mathbb{R}^d$, $|\nu| = 1$, so that for any u in $BV(\Omega, \{\pm m_\beta\})$*

$$\Psi_0'(u) = \Psi_0''(u) = \Psi_0(u) = \int_{\partial^* E} s_\beta(\nu(r))\, d\mu(r), \tag{7.1.3.1}$$

where $E = \{r \in \Omega : u(r) = m_\beta\}$ and if its boundary ∂E is a regular surface, then $\partial^* E = \partial E$, $\nu(r)$ its unit normal and $d\mu$ the Hausdorff measure on ∂E. For general BV sets the meaning of the above is specified in Appendix B.

In the isotropic case the "surface tension" $s_\beta(\nu) = s_\beta$ is independent of ν and

$$\Psi_0(u) = s_\beta |\partial E|, \quad |\partial E| := \int_{\partial^* E} d\mu(r); \tag{7.1.3.2}$$

we thus recover the well known law of thermodynamics that in an isotropic fluid the free energy of an interface is proportional to its area (which for BV sets is $\int_{\partial^* E} d\mu(r)$), with the free energy per unit area equal to the surface tension.

The extension in the theorem from smooth to BV functions (and BV sets) may look as a mathematical subtlety and not really relevant from a physical viewpoint. I do not share this view, clearly a regular surface is easier to visualize and to treat, but the surface tension is "the force" responsible for its regularity and the surface tension is not strong enough to make the interface C^1 as we would like. Indeed BV interfaces have a finite free energy and the regularity which the surface tension implies is exactly BV because, as we shall see in Theorem 7.2.2.1, if $u \notin BV(\mathcal{T}, \{\pm m_\beta\})$, then the Gamma-limit is equal to ∞.

Theorem 7.1.3.1 is proved in this chapter, together with some additional results, more noticeably the one regarding the Wulff shape. In the next chapter we shall derive a more explicit expression for the surface tension $s_\beta(\nu)$ in terms of the free energy of a one dimensional "instanton."

7.1.4 An alternative definition of surface tension

The proof of Theorem 7.1.3.1 uses general structure theorems in geometric measure theory (see Appendix B) to approximate BV interfaces by polygons and then by localizing the analysis which is eventually reduced to a problem with flat interfaces. This part is general; the sequel is instead definitely model dependent. We shall invert the order and start from the end, namely by considering the very special case where Ω is a unit cube, $u \in L^\infty(\Omega, \{-m_\beta, m_\beta\})$ is equal to m_β in the upper and to $-m_\beta$ in the lower half of Ω, the interface parallel to a face of Ω. To study this case it is convenient to give first an alternative definition of the surface tension of a flat interface, which in the end will be proved to coincide with the surface tension of Theorem 7.1.3.1.

Let then $R_n(L; B)$ be the infinite cylinder in $\mathbb{R}^d$ whose axis is directed along the unit vector n and whose cross section is LB, B a unit cube in a $d-1$ dimensional hyperplane normal to n. $L \geq 1$ is a scaling parameter. $R_n(L, h; B)$, $h \geq 1/2$, is a finite cylinder in $R_n(L; B)$ of height $2h > 0$ with LB in the middle, (throughout the sequel the magnetic field is 0 and h will always denote a height variable). We keep here n and B fixed and to simplify notation we do not make them explicit, writing $R(L)$ and $R(L, h)$. By default in the sequel $L \geq 1$ and $h \geq 1/2$. We also introduce coordinate axes with origin the center of $R(L, h)$, x_d axis along n and the others parallel to the sides of B, so that B is a coordinate cube. Then

$$R(L, h) = \big\{(x_1, \ldots, x_d) \in \mathbb{R}^d : |x_d| \leq h,\ |x_i| \leq L,\ i = 1, \ldots, d-1\big\}. \tag{7.1.4.1}$$

Call

$$\chi(r) = m_\beta \mathbf{1}_{x_d \geq 0} - m_\beta \mathbf{1}_{x_d < 0}, \tag{7.1.4.2}$$

and χ_Δ, $\Delta \subset \mathbb{R}^d$, its restriction to Δ. Define

$$\theta(L, h) = \frac{1}{L^{d-1}} \inf_{m \in L^\infty(R(L,h),[-1,1])} \mathcal{F}^{\text{per}}_{R(L,h)}(m|\chi_{R(L,h)^c}), \tag{7.1.4.3}$$

where the superfix "per" means here periodic only on the sides of the cylinder, namely $\mathcal{F}^{\text{per}}_{R(L,h)}(m|\chi_{R(L,h)^c})$ has the usual expression (6.1.3.6) but with J replaced by

$$J^{R(L)}(r, r') = \sum_{k \in \mathbb{Z}^{d-1}} J(r, r' + Lk), \quad r, r' \in R(L). \tag{7.1.4.4}$$

We regard $\theta(L, h)$ as a finite volume approximation of the surface tension of a flat interface with normal n.

Remarks Before studying properties of $\theta(L, h)$ and its thermodynamical limit, let us comment on the definition. It certainly presumes that the interaction with the boundaries gives a negligible contribution to the excess free energy of the state optimizing the r.h.s. of (7.1.4.3), namely that such a state for large $|x_d|$ is close to

$\pm m_\beta$, on the upper and lower halves of $R(L,h)$. A similar reasoning is used in statistical mechanics where the surface tension is usually defined by expression like

$$-\frac{1}{\beta L^{d-1}}\big(\log Z^{+,-}_{R(L,h)} - \log Z^{+,+}_{R(L,h)}\big), \tag{7.1.4.5}$$

where $+-$ means spins equal to 1 and to -1 above and below $R(L,h)$ while $++$ means spins equal to 1 both above and below. Equation (7.1.4.5) is the analogue of (7.1.4.3) under the usual correspondence between $\log Z$ and $-\beta$ times the inf of the free energy functional. Here however the issue about the contribution of the boundary conditions is more delicate. With $\log Z^{+,+}_{R(L,h)}$ we in fact subtract the bulk volume term of the equilibrium free energy, just as we do in (7.1.4.3) by considering the excess free energy. However there is a surface term contributing to $\log Z^{+,+}_{R(L,h)}$ due to the boundary conditions which is certainly not negligible, as $+$ means fixing the spins equal to 1, which is not the equilibrium value at finite temperatures. With (7.1.4.5) one, however, imagines that the interaction with the top in $\log Z^{+,+}(R(L,h))$ matches the interaction with the top in $\log Z^{+,-}_{R(L,h)}$ and also that the interaction with the bottom in $\log Z^{+,+}(R(L,h))$ is like the one in the bottom in $\log Z^{+,-}_{R(L,h)}$. Such a statement presupposes a spin flip symmetry so that the $+/+$ interaction is like the $-/-$ one. Indeed when such a symmetry is not present one uses expressions like

$$-\frac{1}{\beta L^{d-1}}\left(\log Z^{+,-}_{R(L,h)} - \frac{1}{2}\log Z^{+,+}_{R(L,h)} - \frac{1}{2}\log Z^{-,-}_{R(L,h)}\right), \tag{7.1.4.6}$$

or a more intrinsic definition as in [41].

The main result in this subsection is

Theorem 7.1.4.1 *The "surface tension" $\theta := \inf_{L\ge 1, h\ge 1/2;}\theta(L,h)$ is strictly positive and bounded. Moreover*

$$\theta = \liminf_{L\to\infty}\liminf_{h\to\infty}\theta(L,h) = \liminf_{h\to\infty}\liminf_{L\to\infty}\theta(L,h), \tag{7.1.4.7}$$

and $\theta(L,h)$ converges for any fixed h when $L\to\infty$ and for any fixed L when $h\to\infty$, so that the lim inf *in* (7.1.4.7) *are actually limits.*

We start the proof with two simple lemmas:

Lemma 7.1.4.2 *Let $m \in L^\infty(R(L,h),[-1,1])$, denote by $(m, \chi_{R(L)\setminus R(L,h)})$ the function on the whole $R(L)$ whose restrictions to $R(L,h)$ and to $R(L)\setminus R(L,h)$ are m and χ. Then for any $h \ge 1/2$,*

$$\mathcal{F}^{\rm per}_{R(L,h)}(m|\chi_{R(L)\setminus R(L,h)}) = \mathcal{F}^{\rm per}_{R(L)}((m, \chi_{R(L)\setminus R(L,h)})). \tag{7.1.4.8}$$

Proof By (6.1.3.5) which holds in the periodic case as well, $\mathcal{F}^{\text{per}}_{R(L)}((m, \chi_{R(L)\setminus R(L,h)}))$ is equal to

$$\mathcal{F}^{\text{per}}_{R(L,h)}(m|\chi_{R(L,h)^c}) + \mathcal{F}^{\text{per}}_{R(L)\setminus R(L,h)}(\chi_{R(L)\setminus R(L,h)}),$$

$\mathcal{F}^{\text{per}}_{R(L)\setminus R(L,h)}(\chi_{R(L)\setminus R(L,h)}) = 0$ because the two parts are at distance ≥ 1 and thus do not interact with each other. □

Lemma 7.1.4.3 *Let* $m \in L^\infty(R(L), [-1, 1])$, $m = \chi$ *on* $R(L) \setminus R(L, h)$, *and* $\tilde{m}$ *its periodic extension to* $R(nL)$, n *a positive integer. Then*

$$L^{-d+1}\mathcal{F}^{\text{per}}_{R(L)}(m) = (nL)^{-d+1}\mathcal{F}^{\text{per}}_{R(nL)}(\tilde{m}). \tag{7.1.4.9}$$

Proof By (7.1.4.4) $\mathcal{F}^{\text{per}}_{R(nL)}(\tilde{m}) - \int_{R(nL)} f_\beta(\tilde{m})$ is equal to

$$\begin{aligned}
&\sum_{k\in\mathbb{Z}^{d-1}} \frac{1}{4}\int_{R(nL)}\int_{R(nL)} J(r, r' + nLk)[\tilde{m}(r) - \tilde{m}(r')]^2 \\
&= \frac{1}{4}\int_{R(nL)}\int_{\mathbb{R}^d} J(r, r')[\tilde{m}(r) - \tilde{m}(r')]^2,
\end{aligned} \tag{7.1.4.10}$$

because $\tilde{m}$ is periodic. Calling $\{R_p(L)\}$ the translates by $p \in \mathbb{Z}^{d-1}$ of $R(L)$ which are contained in $R(nL)$, the r.h.s. of (7.1.4.10) becomes

$$\begin{aligned}
&\sum_p \frac{1}{4}\int_{R_p(L)}\int_{\mathbb{R}^d} J(r, r')[\tilde{m}(r) - \tilde{m}(r')]^2 \\
&= \sum_p \left\{ \sum_{k\in\mathbb{Z}^{d-1}} \frac{1}{4}\int_{R_p(L)}\int_{R_p(L)} J(r, r' + kL)[\tilde{m}(r) - \tilde{m}(r')]^2 \right\},
\end{aligned}$$

having used again that $\tilde{m}$ is periodic. After adding $\int_{R(nL)} f_\beta(\tilde{m}) = \sum_p \int_{R_p(L)} f_\beta(\tilde{m})$ we obtain $n\mathcal{F}^{\text{per}}_{R(L)}(m)$ and the lemma is proved. □

Corollary 7.1.4.4 *For any* $L \geq 1$ *and* $h \geq 1/2$,

$$\theta(L, h') \leq \theta(L, h), \quad h' \geq h; \qquad \theta(nL, h) \leq \theta(L, h), \quad n \geq 1. \tag{7.1.4.11}$$

Proof By (7.1.4.8) $\mathcal{F}^{\text{per}}_{R(L,h)}(m|\chi_{R(L)\setminus R(L,h)}) = \mathcal{F}^{\text{per}}_{R(L)}((m, \chi_{R(L)\setminus R(L,h)}))$; thus, $\theta(L, h)$ is the inf over all functions m on $R(L)$ whose restriction to $R(L) \setminus R(L, h)$ is χ. Thus, $\theta(L, h')$ is the inf over a larger set of functions and the first inequality in (7.1.4.11) follows. Analogously $\theta(L, h)$ is the inf over all functions m on $\mathbb{R}^d$ which are L-periodic in the first $d - 1$ coordinates (and whose restriction to $R(L) \setminus R(L, h)$ is χ). Thus $\theta(nL, h)$ is the inf over a larger set of functions and the second inequality in (7.1.4.11) follows as well. □

To prove Theorem 7.1.4.1 we still need: • uniform upper and lower bounds on $\theta(L,h)$ and • to prove that the variations of $\theta(L',h)$ when $L' \in [nL,(n+1)L]$ vanish as $n \to \infty$.

Lemma 7.1.4.5 *There is $a > 0$ so that for any $L = n\ell$, $n \in \mathbb{N}$, ℓ as in Theorem 6.4.2.3, and $h = n\ell \ge 1/2$, $n \in \mathbb{N}$,*

$$a \le \theta(L,h) \le 2m_\beta^2. \tag{7.1.4.12}$$

Proof We use (7.1.4.8) to write $L^{d-1}\theta(L,h) \le \mathcal{F}^{\text{per}}_{R(L)}(\chi_{R(L)})$; the latter is bounded from above by

$$L^{d-1} 2m_\beta^2 \int_{-1}^{0} \left\{ \int_{x'_d>0} J(x_d e_d, x'_1 e_1 + \cdots + x'_d e_d) dx'_1 \cdots dx'_d \right\} dx_d$$
$$\le L^{d-1} 2m_\beta^2, \tag{7.1.4.13}$$

hence the upper bound in (7.1.4.12), which is indeed valid for all $L \ge 1$ and $h \ge 1/2$.

To prove the lower bound we use Theorem 6.4.2.3 (whose validity extends to the present case with J replaced by its periodic version). We choose ζ and ℓ as in Theorem 6.4.2.3 and by the assumption on L and by translational invariance we may suppose that $R(L)$ is $\mathcal{D}^{(\ell)}$-measurable. Let $m \in L^\infty(R(L),[-1,1])$ equal to χ for $r \in R(L) \setminus R(L,h)$. Let $C \in \mathcal{D}^{(\ell)}$ contained in $R(L)$ and $\{C_i\}$ the set of all $C_i \in \mathcal{D}^{(\ell)}$ which are vertical translations of C. Going down along this vertical tube, $\eta^{(\zeta,\ell)}(m;r)$ varies from 1 to -1 so that either there is a cube where $\eta^{(\zeta,\ell)}(m;r) = 0$ or else there are two contiguous cubes with $\eta^{(\zeta,\ell)}(m;r)$ equal to ± 1 on the upper and lower ones. We then construct a sequence S whose entries are for each vertical tube as above either a single cube or a pair of cubes, as explained above. We apply Theorem 6.4.2.3 with such a sequence S so that (6.4.2.4) holds with $n+p$ equal to the number of vertical tubes contained in $R(L)$. Thus, $\ell^d(2n+p) \ge L^{d-1}$ and the lower bound in (7.1.4.13) is proved. □

Lemma 7.1.4.6 *There is $c > 0$ so that the following holds. Let m be a $[-1,1]$-valued function equal to χ on $R(L) \setminus R(L,h)$ and $\tilde m$ its periodic extension to $\mathbb{R}^d$. Let B' be any unit cube with same center as B (B the cube in the definition of $R(L,h)$) and $L' \in [nL,(n+1)L]$. Then*

$$\left| L^{-d+1} \mathcal{F}^{\text{per}}_{R(L)}(m) - (L')^{-d+1} \mathcal{F}^{\text{per}}_{R(L';B')}(\tilde m_{R(L';B')}) \right| \le \frac{ch}{n}. \tag{7.1.4.14}$$

Proof Call $R_p(L)$ the translate by pL of $R(L)$, $p \in \mathbb{Z}^d$; Λ the union of all $R_p(L) \subset R(L';B')$ such that all $R_q(L)$ connected to $R_p(L)$ are also contained in $R(L';B')$; N the number of $R_p(L)$ in Λ. Denote by $\{\xi\}$ the $d-1$ dimensional lattice containing the origin obtained by translations of B'; then analogously to (7.1.4.10),

$\mathcal{F}^{\text{per}}_{R(L';B')}(\tilde m) - \int_{R(L';B')} f_\beta(\tilde m)$ is equal to

$$\sum_\xi \frac{1}{4}\int_{R(L';B')}\int_{\mathbb{R}^d} J(r,r'+L'\xi)[\tilde m(r)-\tilde m(r'+L'\xi)]^2.$$

When $r\in\Lambda$ only $\xi=0$ contributes to the sum hence the above is equal to

$$\sum_{R_p(L)\subset\Lambda} \frac{1}{4}\int_{R_p(L)}\int_{\mathbb{R}^d} J(r,r')[\tilde m(r)-\tilde m(r')]^2$$
$$+\sum_\xi \frac{1}{4}\int_{R(L';B')\setminus\Lambda}\int_{\mathbb{R}^d} J(r,r'+L'\xi)[\tilde m(r)-\tilde m(r'+L'\xi)]^2.$$

As in the proof of Lemma 7.1.4.3, $\mathcal{F}^{\text{per}}_{R(L';B')}(\tilde m) = N\mathcal{F}^{\text{per}}_{R(L)}(m)+\mathcal{R}$,

$$\mathcal{R}=\int_{R(L';B')\setminus\Lambda} f_\beta(\tilde m)+\sum_\xi \frac{1}{4}\int_{R(L';B')\setminus\Lambda}\int_{\mathbb{R}^d} J(r,r'+L'\xi)[\tilde m(r)-\tilde m(r'+L'\xi)]^2.$$

Call $\Sigma=\{r\in R(L';B')\setminus\Lambda : x_d=0\}$; then $\mathcal{R}\le c|\Sigma|(h+1)$ because the integrand vanishes if the coordinate x_d of r is $|x_d|\ge h+1$ and it is otherwise bounded. Equation (7.1.4.14) then follows because

$$\left|\frac{NL^{d-1}}{(L')^{d-1}}-1\right|\le\frac{c}{n},\qquad \frac{|\Sigma|}{(L')^{d-1}}\le\frac{c}{n}.$$ □

Corollary 7.1.4.7 *For any $L\ge 1$, $h\ge 1/2$ and $L'\in[nL,(n+1)L]$, $n\in\mathbb{N}$,*

$$\theta(L',h)\le\theta(L,h)+\frac{ch}{n}. \tag{7.1.4.15}$$

c as in Lemma 7.1.4.6.

Proof $\theta(L',h)$ is defined by taking the inf of $\mathcal{F}^{\text{per}}_{R(L')}(m)$ over all m on $R(L')$ which are equal to χ on $R(L')\setminus R(L',h)$. We increase $\theta(L',h)$ if we restrict the inf to all functions $\tilde m$ as in Lemma 7.1.4.6, hence (7.1.4.15). □

Proof of Theorem 7.1.4.1 (7.1.4.7) clearly holds as $\le$, we thus need to prove the reverse inequality. Given any $\epsilon>0$ let $L'\ge 1, h'\ge 1/2$ be such that $\theta(L',h')<\theta+\epsilon$. By (7.1.4.11)

$$\begin{aligned}&\liminf_{n\to\infty}\liminf_{h\to\infty}\theta(nL',h)\le\theta(L',h'),\\ &\liminf_{h\to\infty}\liminf_{n\to\infty}\theta(nL',h)\le\theta(L',h').\end{aligned} \tag{7.1.4.16}$$

By (7.1.4.15) we can replace the limit of $nL'\to\infty$ by the limit $L\to\infty$ thus proving that the r.h.s. of (7.1.4.7) is $\le\theta+\epsilon$, and (7.1.4.7) is proved.

By monotonicity $\theta(L,h)$ converges as $h \to \infty$ and $\theta(nL,h)$ as $n \to \infty$ and then by (7.1.4.15) also $\theta(L',h)$ as $L' \to \infty$. □

The surface tension θ defined in Theorem 7.1.4.1 may a-priori depend on the choice of the cube B used in its definition. Using Lemma 7.1.4.6 we have (details are omitted)

Theorem 7.1.4.8 *The surface tension $\theta(B)$, is independent of the orientation of the cube B.*

7.1.5 Gamma convergence when the interface is flat

In this subsection we shall prove Theorem 7.1.3.1 in the particular case of a flat interface. The analysis is not purely exemplificative as we shall use it later as a crucial ingredient in the proof of the lower bound of Gamma convergence in the general case. To have a single flat interface we replace the torus Ω by a unit cube Ω with suitable boundary conditions and take u in Theorem 7.1.3.1 equal to χ, namely equal to m_β and $-m_\beta$ in the upper and lower halves of the cube (i.e. where the last coordinate $x_d \geq 0$ and respectively $x_d < 0$).

The crucial point will be to frame the analysis with the L^1-constraint implicit in the definition of Gamma convergence into the setup of Sect. 7.1.4. With this in mind, we regard the cube $C(L) = L\Omega$ of side L as the cylinder with cross section $L\Omega_0$ ($\Omega_0 = \Omega \cap \{x_d = 0\}$) and height L, namely $C(L) = R(L, \frac{L}{2}; \Omega_0)$ with the notation of Sect. 7.1.4. The L^1-constraint is $|m - \chi|_{L^1(C(L))} \leq \delta L^d$. Instead of putting periodic conditions on the sides and $\pm m_\beta$ on top and bottom we minimize also over the boundary conditions, thus leaving the system free to choose the most favorable ones. We then set

$$\begin{aligned} \hat{\theta}' &:= \lim_{\delta \to 0} \liminf_{L \to \infty} \inf_{\|m-\chi\|_{L^1(C(L))} \leq \delta L^d} L^{-d+1} \mathcal{F}(m_{C(L)}|m_{C(L)^c}), \\ \hat{\theta}'' &:= \lim_{\delta \to 0} \limsup_{L \to \infty} \inf_{\|m-\chi\|_{L^1(C(L))} \leq \delta L^d} L^{-d+1} \mathcal{F}(m_{C(L)}|m_{C(L)^c}), \end{aligned} \tag{7.1.5.1}$$

where the inf is over all $m \in L^\infty(\mathbb{R}^d; [-1,1])$ which satisfy the L^1 bound, as indicated, while, as usual, m_Δ denotes the restriction of m to Δ.

Theorem 7.1.5.1 *We have $\hat{\theta}' = \hat{\theta}'' = \theta$, with θ as in* (7.1.4.7) *and there is $c > 0$ so that the following holds. For any $\epsilon > 0$ and $\delta > 0$ there is $L^* > 0$ and for any $L \geq L^*$ and any $\|m - \chi\|_{L^1(C(L))} \leq \delta L^d$,*

$$\mathcal{F}_{C(L)}(m_{C(L)}|m_{C(L)^c}) \geq L^{d-1}(\theta - \epsilon - c\sqrt{\delta}). \tag{7.1.5.2}$$

Proof We start from the upper bound, namely the proof that $\hat{\theta}'' \leq \theta$, the reverse inequality, $\hat{\theta}' \geq \theta$, i.e. the lower bound in Gamma convergence, will follow from (7.1.5.2) and it will be the main point in the theorem.

Indeed the upper bound is "easy as it just requires to find a good trial function" and the right choice is dictated by Theorem 7.1.4.1 (relative to cylinders with cross section Ω_0). For any $\epsilon > 0$, there are h_0 and L_0 so that $\theta(L_0, h_0) \le \theta + \epsilon/2$ and there is $m^* \in L^\infty(C(L_0, h_0), [-1, 1])$ so that

$$\frac{1}{L_0^{d-1}} \mathcal{F}^{\text{per}}_{R(L_0,h_0)}(m^* | \chi_{R(L_0)\setminus R(L_0,h_0)}) \le \theta + \epsilon.$$

Call $\tilde{m}$ the extension of m^* to $R(L_0)$ obtained by putting $\tilde{m} = \chi$ on $R(L_0) \setminus R(L_0, h_0)$. By (7.1.4.8), $\mathcal{F}^{\text{per}}_{R(L_0,h_0)}(m^* | \chi_{R(L_0)\setminus R(L_0,h_0)}) = \mathcal{F}^{\text{per}}_{R(L_0)}(\tilde{m})$. We then extend periodically $\tilde{m}$ to a function m on the whole $\mathbb{R}^d$. For L large enough, $C(L) = R(L, L/2) \supset R(L, h_0)$ and since $m = \chi$ on $C(L) \setminus R(L, h_0)$ and $|m - \chi| \le 2$ on $R(L, h_0)$,

$$\|m - \chi\|_{L^1(C(L))} = \|m - \chi\|_{L^1(R(L,h_0))} \le 2L^{d-1} 2h_0 < \delta L^d,$$

if $4h_0 < \delta L$. Given $L > L_0$ call n the integer such that $L \in [nL_0, (n + 1)L_0]$. Since there is a constant $c' > 0$ so that

$$\mathcal{F}_{C(L)}(m_{C(L)} | m_{C(L)^c}) \le \mathcal{F}^{\text{per}}_{R(L)}(m_{R(L)}) + c' L^{d-2} h_0$$

by (7.1.4.14) with $B = B' = \Omega_0$,

$$L^{-d+1} \mathcal{F}_{C(L)}(m_{C(L)} | m_{C(L)^c}) \le \theta + \epsilon + \frac{ch_0}{n},$$

so that $\hat{\theta}'' \le \theta + \epsilon$ which, by the arbitrariness of ϵ completes the proof of the upper bound.

Strategy for the lower bound We shall prove that with a negligible increase of free energy we can modify the original function m into a new function ψ equal to χ on two "thick horizontal layers" one in $x_d > 0$ the other one symmetrically in $x_d < 0$. At this point it will not be difficult to relate the free energy of ψ to the surface tension θ of Theorem 7.1.4.1. The existence of the upper layer (same proof for the lower one) is proved by showing that the original function m is close to m_β in a region Λ_+ which has the layer "in the middle," such a property is proved using that $\|m - \chi\|_{L^1} \le \sqrt{\delta} L^d$. By minimizing the free energy in Λ_+ with the given boundary conditions we shall then find ψ, this part uses the analysis in Sect. 6.3.

The upper and lower horizontal layers We choose the size of the layers in terms of the contour parameters, contours will indeed play an important role in the proof. We thus fix a triple (ζ, ℓ_-, ℓ_+) which fulfills the requests in (6.4.6.3)–(6.4.6.4) and define the i-th layer, $i \in \mathbb{Z}$, as

$$S_i = \{x \in C(L) : (x_d - \ell_+ i) \in [-\ell_+/2, \ell_+/2)\}. \tag{7.1.5.3}$$

Let

$$N = \min\{n \in \mathbb{N} : 2n\ell_+ \ge \sqrt{\delta}\, L\}. \tag{7.1.5.4}$$

We look for the first odd integer i in $\{1,\dots,2N-1\}$ such that in each one of the layers $S_i, S_{i+1}, S_{i+2N}, S_{i+2N+1}; S_{-i}, S_{-i-1}, S_{-i-2N}, S_{-i-2N-1}, m$ is "close" in L^1 to m_β (in the first four) and to $-m_\beta$ (in the last four). As we are going to see such a value of i exists as a consequence of the L^1-constraint $\|m-\chi\|_{L^1(C(L))} \le \delta L^d$. With this in mind and supposing $\sqrt{\delta}$ small enough, we define for any odd integer i in $\{1,\dots,2N-1\}$,

$$\begin{aligned}\Sigma_i &:= S_i \cup S_{i+1} \cup S_{i+2N} \cup S_{i+1+2N},\\ \Sigma_{-i} &:= S_{-i} \cup S_{-i-1} \cup S_{-i-2N} \cup S_{-i-1-2N}.\end{aligned} \tag{7.1.5.5}$$

Notice that $\Sigma_i \cap \Sigma_j = \emptyset$, $i \ne j$, $|\Sigma_i \cup \Sigma_{-i}| = 8|S_0|$ and $|\bigcup\{\Sigma_i \cup \Sigma_{-i}\}| = 8N|S_0|$.

Selection of the good layers We claim that there is an odd integer n in $\{1,\dots,2N-1\}$ such that:

$$\frac{1}{8|S_0|}\int_{\Sigma_n\cup\Sigma_{-n}} |m-\chi|\,dr \le \sqrt{\delta}. \tag{7.1.5.6}$$

This we prove as follows. Let $a_i = \frac{1}{8|S_0|}\int_{\Sigma_i\cup\Sigma_{-i}} |m-\chi|\,dx$; then (7.1.5.6) will follow from

$$a := \min_{i\,\mathrm{odd}\in\{1,\dots,2N-1\}} a_i < \sqrt{\delta}. \tag{7.1.5.7}$$

By assumption, $\delta L^d \ge \int_{C(L)} |m-\chi|\,dr \ge 8\,|S_0|\,N\,a$. By (7.1.5.4), $N \ge \frac{\sqrt{\delta}L}{2\ell_+}$ and since $|S_0| = \ell_+ L^{d-1}$, $\delta L^d \ge 8a\ell_+ L^{d-1}\frac{L\sqrt{\delta}}{2\ell_+}$, hence (7.1.5.7).

Geometrical setup We are going to use the analysis of Sect. 6.4, with a triple (ζ,ℓ_-,ℓ_+) which fulfills the requests in (6.4.6.3)–(6.4.6.4). To match with the previous geometry we shift vertically by $\ell_+/2$ the partitions $\mathcal{D}^{(\ell_\pm)}$ of Sect. 6.4 (since $J(r,r')$ is translational invariant the results of Sect. 6.4 are unchanged) and by an abuse of notation we call the shifted partitions and all the functions related to the contours by same symbols. Moreover, we use the shorthand $\eta(\cdot;\cdot)$ for $\eta^{(\zeta,\ell_-)}(\cdot;\cdot)$ and define

- $C^0(L)$ is the union of all cubes of $\mathcal{D}^{(\ell_+)}$ which are in $C(L)$ and $R^0 = R^0(L)$ is the union of all vertical translates of $C^0(L)$.
- $\mathcal{M}_n = \mathcal{M}'_n \cup \mathcal{M}''_N$ where $\mathcal{M}''_N = \delta^{\ell_+}_{\mathrm{in}}[R^0] \cap \{\bigcup_{|j|\le 4N} S_j\}$ and

$$\mathcal{M}'_n = \{r \in \Sigma_n \cap C^0(L) : \eta(m;r) < 1\} \cup \{r \in \Sigma_{-n} \cap C^0(L) : \eta(m;r) > -1\}.$$

- $\Lambda_+ = \{R^0 \setminus \delta^{\ell_+}_{\mathrm{in}}[R^0]\} \cap \{\bigcup_{j=n+1}^{n+2N} S_j\}$, $\Lambda_- = \{R^0 \setminus \delta^{\ell_+}_{\mathrm{in}}[R^0]\} \cap \{\bigcup_{-n-2N}^{j=-n-1} S_j\}$. We then call $\Lambda = \Lambda_- \cup \Lambda_+$.

A first surgery Since $\mathcal{F}_{C(L)}(m_{C(L)}|m_{C(L)^c})$ is equal to

$$\mathcal{F}_{C^0(L)}(m_{C^0(L)}|(m_{C(L)\setminus C^0(L)}, m_{C(L)^c})) + \mathcal{F}_{C(L)\setminus C^0(L)}(m_{C(L)\setminus C^0(L)}|m_{C(L)^c}),$$

we have

$$\mathcal{F}_{C(L)}(m_{C(L)}|m_{C(L)^c}) \geq \mathcal{F}_{C^0(L)}(m_{C^0(L)}|m_{C^0(L)^c}). \tag{7.1.5.8}$$

The "surgery" in the title of the paragraph refers to the operation of replacing m in $\mathcal{M}_n$ by χ leaving elsewhere m unchanged and call ϕ the new function. We claim:

$$\mathcal{F}_{C(L)}(m_{C(L)}|m_{C(L)^c}) \geq \mathcal{F}_{C^0(L)}(\phi_{C^0(L)}|\phi_{C^0(L)^c}) - c_0\sqrt{\delta}\,L^{d-1}. \tag{7.1.5.9}$$

Proof is as follows. $\mathcal{F}_{C(L)}(m_{C(L)}|m_{C(L)^c}) \geq \mathcal{F}_{C^0(L)}(\phi_{C^0(L)}|\phi_{C^0(L)^c}) - c_0'\,|\mathcal{M}_n|$. To bound $|\mathcal{M}_n'|$ we write for any $C \in \mathcal{D}^{(\ell_-)}$

$$\int_C |m-\chi| \geq \left|\int_C [m-\chi]\right| \geq \zeta \int_C \mathbf{1}_{\eta(m;r)\neq\text{sign of } x_d}. \tag{7.1.5.10}$$

Then $\zeta|\mathcal{M}_n'| \leq 8|S_0|\sqrt{\delta}$ so that

$$|\mathcal{M}_n| \leq \frac{8|S_0|\sqrt{\delta}}{\zeta} + c' N L^{d-2} \leq c\,\sqrt{\delta}\,L^{d-1}, \tag{7.1.5.11}$$

where the term $c'NL^{d-2}$ bounds the volume of $\mathcal{M}_N''$, c' and c suitable constants. Equation (7.1.5.9) is proved with $c_0 = c_0' c$.

A second surgery We first write $\mathcal{F}_{C^0(L)}(\phi_{C^0(L)}|\phi_{C^0(L)^c})$ as

$$\mathcal{F}_{\Lambda_+}(\phi_{\Lambda_+}|\phi_{\Lambda_+^c}) + \mathcal{F}_{\Lambda_-}(\phi_{\Lambda_-}|\phi_{\Lambda_-^c}) + \mathcal{F}_{C^0(L)\setminus\Lambda}(\phi_{C^0(L)\setminus\Lambda}|\phi_{\Lambda^c}).$$

Calling $\Lambda^c_{\pm,\neq} = \{r \in \delta^{\ell_+}_{\text{out}}[\Lambda_+] : \phi(r) \neq \pm m_\beta\}$, by the definition of ϕ,

$$\eta(\phi;r) = \pm 1, \quad r \in \delta^{\ell_+}_{\text{out}}[\Lambda_\pm],\ \Lambda^c_{\pm,\neq} \subset \{S_{\pm i} \cup S_{\pm(i+1+2N)}\}. \tag{7.1.5.12}$$

We apply Corollary 6.4.6.2 once with $\Lambda = \Lambda_+$, and $\Delta = S_{i+1+N} \cap \Lambda_+$ and then with $\Lambda = \Lambda_-$, and $\Delta = S_{-(i+1+N)} \cap \Lambda_-$, noticing that $\text{dist}(\Delta_\pm, \Lambda^c_{\pm,\neq}) \geq \ell_+(N-1)$. Then there is ψ equal to ϕ outside $\Lambda_\pm$ and equal to χ on $S_{n+1+N} \cap C^0(L)$ and $S_{-n-1-N} \cap C^0(L)$, such that

$$\begin{aligned}&\mathcal{F}_{C^0(L)}(\phi_{C^0(L)}|\phi_{(C^0(L))^c})\\ &\quad\geq \mathcal{F}_{C^0(L)}(\psi_{C^0(L)}|\psi_{(C^0(L))^c}) - (2c_\omega e^\omega|S_0|)e^{-\omega\ell_+(N-1)}.\end{aligned} \tag{7.1.5.13}$$

Conclusions Call

$$U = \{R^0 \setminus \delta^{\ell_+}_{\text{in}}[R^0]\} \cap \Big\{ \bigcup_{|j|<n+1+N} S_j \Big\}, \tag{7.1.5.14}$$

$2h$ the height of U and b the side of the cross section. We get

$$\mathcal{F}_{C^0(L)}(\psi_{C^0(L)}|\psi_{(C^0(L))^c}) \geq \mathcal{F}_U(\psi_U|\psi_{U^c}). \tag{7.1.5.15}$$

Since $\psi_{U^c} = \chi$ on $\delta_{\text{out}}^{\ell+}[U]$ we can replace ψ_{U^c} by χ_{U^c} on the r.h.s. of (7.1.5.15). We have

$$\mathcal{F}_U(\psi_U|\chi_{U^c}) = \mathcal{F}_{R^0}(\psi_{R^0}) = \mathcal{F}_{R^0}^{\text{per}}(\psi_{R^0}),$$

and, by the definition of θ, see Theorem 7.1.4.1,

$$\mathcal{F}_{R^0}^{\text{per}}(\psi_{R^0}) \ge b^{d-1}\theta(b,h) \ge b^{d-1}\theta. \tag{7.1.5.16}$$

By collecting the previous inequalities which have led to $\mathcal{F}_{R^0}^{\text{per}}(\psi_{R^0})$ we then obtain (7.1.5.2). □

7.1.6 Continuity of the surface tension

In the proof of Theorem 7.1.3.1 we shall use that the surface tension θ of a flat interface Σ depends continuously on the unit normal n to Σ, a property which will be proved here. Since dependence on n is the issue we resume the notation θ_n for the surface tension.

Theorem 7.1.6.1 *θ_n is a continuous function of n, $|n| = 1$.*

Proof We will prove that $|\theta_n - \theta_{n'}| \le \epsilon$ as soon as the angle ϕ between n and n' is smaller than δ_ϵ. We fix two unit vectors n and n' with a small angle ϕ between them and want to estimate $\theta_n - \theta_{n'}$. By a suitable choice of coordinates, we may suppose n along the x_1 axis and n' in the x_1, x_2 plane. By Theorem 7.1.4.8, we may define $\theta_{n'}$ using cylinders $R_{n'}(L,h;B')$ which are rotation of those $R_n(L,h;B)$ used for θ_n. With such an understanding we shall drop B and B' from the notation.

Given n' and $\epsilon > 0$, let L' and h' be such that $\theta_{n'} \le \theta_{n'}(L',h') \le \theta_{n'} + \epsilon$, then by (7.1.4.16) for any integer k (eventually $k \to \infty$),

$$\theta_{n'} \le \theta_{n'}(kL',h') \le \theta_{n'} + \epsilon; \tag{7.1.6.1}$$

$R_{n'}(kL',h')$ is the corresponding rectangle. Let $m' \in L^\infty(R_{n'}(kL'),[-1,1])$ be such that $m' = \chi$ on $R_{n'}(kL') \setminus R_{n'}(kL',h')$ and

$$\theta_{n'}(kL',h') \le \frac{1}{(kL')^{d-1}}\mathcal{F}_{R_{n'}(kL')}^{\text{per}}(m') \le \theta_{n'}(kL',h') + \epsilon. \tag{7.1.6.2}$$

Let $R_n(L,h)$ be the minimal rectangle which contains $R_{n'}(kL',h')$, see Fig. 7.1, then

$$L = kL'\cos\phi + 2h'\sin\phi; \qquad 2h = 2h'\cos\phi + kL'\sin\phi. \tag{7.1.6.3}$$

Let $A^+ := \{r \in R_n(L,h) \setminus R_{n'}(kL',h') : r \cdot n' > 0\}$ and $A^- := \{r \in R_n(L,h) \setminus R_{n'}(kL',h') : r \cdot n' \le 0\}$. Define m on $R_n(L)$ as equal to m' on $R_{n'}(kL',h')$, to $\pm m_\beta$

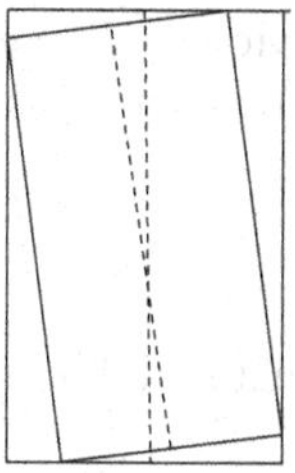

Fig. 7.1 The two *rectangles* $R_{n'}(kL', h') \subset R_n(L, h)$ in $d = 2$; *dashed lines* are their cross sections, of length kL' and L, the *angle* between the two is ϕ

on $A^\pm$ and to χ on $R_n(L) \setminus R_n(L, h)$. Then there is $c > 0$ so that

$$\mathcal{F}^{\text{per}}_{R_n(L)}(m) \le \mathcal{F}^{\text{per}}_{R_{n'}(kL')}(m') + c\big[\ell(kL')^{d-2} + L^{d-2}h\big],$$

where $\ell := h' \tan\phi$. The first term in the square bracket comes from the interaction between A^+ and A^-, the second one from the periodic boundary conditions in $R_n(L, h)$. Thus

$$\theta_n \le L^{-d+1}\mathcal{F}^{\text{per}}_{R_n(L)}(m) \le \frac{(kL')^{d-1}}{L^{d-1}}(\theta_{n'} + 2\epsilon) + c\left(h'\frac{(kL')^{d-2}}{L^{d-1}} + \frac{h}{L}\right). \qquad (7.1.6.4)$$

By letting $k \to \infty$ and using (7.1.6.3) we get $\theta_n \le (\cos\phi)^{-d+1}(\theta_{n'} + 2\epsilon) + c\tan\phi$ and by interchanging the role of n and n' we get the opposite inequality. The theorem is proved. □

7.1.7 Lower bound of Gamma convergence

In this subsection we shall prove

$$\lim_{\delta\to 0}\ \liminf_{\epsilon\to 0}\ \inf_{\|m - u^{(\epsilon)}\|_{L^1(\epsilon^{-1}\Omega)} \le \epsilon^{-d}\delta} \epsilon^{d-1}\mathcal{F}^{\text{per}}_{\epsilon^{-1}\Omega}(m) \ge \int_{\partial^\star E} \theta_{\nu(r)}\, d\mu(r), \qquad (7.1.7.1)$$

where $u \in BV(\Omega, \{\pm m_\beta\})$ and $u^{(\epsilon)}(r) = u(\epsilon r)$; $E = \{r \in \Omega : u(r) = m_\beta\}$ and $\partial^\star E$, ν, μ are respectively the boundary ∂E, its unit normal and the Hausdorff measure on ∂E if E is a regular set, for the general case we refer to Sect. B.1 of Appendix B; finally θ_ν is the surface tension θ defined in (7.1.4.7) with $R(L, h) = R_\nu(L, h)$.

The proof of (7.1.7.1) relies heavily on the following geometric measure theory properties, whose proofs are outlined in Appendix B.

Properties of BV sets which will be used in the proof of (7.1.7.1)

For any $\alpha > 0$ (I apologize, but I have run out of symbols for a small parameter!) there are n disjoint cubes $R_1, \dots, R_n$ in Ω of same side h, $\nu_1, \dots, \nu_n$ unit vectors with ν_i, $i = 1, \dots, n$, normal to a face of R_i (so that the cubes R_i are "oriented") with the following properties.

- Calling χ_i the function on R_i equal to $\pm m_\beta$ on its upper and lower half as determined by ν_i, then, for each $i = 1, \dots, n$,

$$\int_{R_i} |u - \chi_i| \le \alpha h^d. \tag{7.1.7.2}$$

- There is $c > 0$ so that for all α small enough

$$n h^{d-1} \le c \int_{\partial^* E} d\mu. \tag{7.1.7.3}$$

- There is $c' > 0$ so that calling $\Sigma_i := \partial^* E \cap R_i$, for $i = 1, \dots, n$,

$$\left| h^{d-1} \theta_{\nu_i} - \int_{\Sigma_i} \theta_{\nu(r)} \, d\mu \right| \le c' h^{d-1} \alpha, \tag{7.1.7.4}$$

and finally,

$$\lim_{\alpha \to 0} \sum_{i=1}^{n} \int_{\Sigma_i} \theta_{\nu(r)} \, d\mu = \int_{\partial^* E} \theta_{\nu(r)} \, d\mu. \tag{7.1.7.5}$$

We can now proceed with the proof of (7.1.7.1). Let $\|m - u^{(\epsilon)}\|_{L^1(\epsilon^{-1}\Omega)} \le \epsilon^{-d}\delta$ as in (7.1.7.1) and let α such that

$$\delta = \frac{\alpha h^d}{2}. \tag{7.1.7.6}$$

Then, calling $\chi_i^{(\epsilon)}(r) = \chi_i(\epsilon r)$ and using (7.1.7.2),

$$\int_{\epsilon^{-1} R_i} dr \, |m - \chi_i^{(\epsilon)}| \le \int_{\epsilon^{-1} R_i} |m - u^{(\epsilon)}| \, dr + \int_{\epsilon^{-1} R_i} |u^{(\epsilon)} - \chi_i^{(\epsilon)}| \, dr$$
$$\le \epsilon^{-d} \left\{ \delta + \int_{R_i} |u - \chi_i| \, dr \right\} =: \epsilon^{-d} 2\alpha h^d. \tag{7.1.7.7}$$

We proceed by letting first $\epsilon \to 0$ and then after $\alpha \to 0$, which, by (7.1.7.6) is the same as $\delta \to 0$. We next write

$$\Delta = \bigcup_{i=1}^{n} \epsilon^{-1} R_i. \tag{7.1.7.8}$$

For ϵ small enough the sets $\epsilon^{-1} R_i$ have mutual distance larger than the interaction range, so that (writing as usual m_Λ for the restriction of m to a region Λ)

$$\begin{aligned} \mathcal{F}^{\text{per}}_{\epsilon^{-1}\Omega}(m) &= \mathcal{F}^{\text{per}}_{\Delta^c}(m_{\Delta^c}) + \sum_{i=1}^{n} \mathcal{F}_{\epsilon^{-1} R_i}(m_{\epsilon^{-1} R_i} | m_{\Delta^c}), \\ \epsilon^{d-1} \mathcal{F}^{\text{per}}_{\epsilon^{-1}\Omega}(m) &\ge \sum_{i=1}^{n} \epsilon^{d-1} \mathcal{F}_{\epsilon^{-1} R_i}(m_{\epsilon^{-1} R_i} | m_{\Delta^c}). \end{aligned} \tag{7.1.7.9}$$

Hence

$$
\begin{aligned}
&\text{l.h.s. of (7.1.7.1)}\\
&\quad \ge \lim_{\alpha\to 0}\sum_{i=1}^{n}\Big\{\liminf_{\epsilon\to 0}\epsilon^{d-1}\\
&\quad\quad \times \inf_{\|m-\chi_i^{(\epsilon)}\|_{L^1(\epsilon^{-1}R_i)}\le \epsilon^{-d}2\alpha h^d} \mathcal{F}_{\epsilon^{-1}R_i}(m_{\epsilon^{-1}R_i}|m_{\Delta^c})\Big\}, \qquad (7.1.7.10)
\end{aligned}
$$

and the liminf over ϵ is reduced to a separate estimate over the cubes $\epsilon^{-1}R_i$.

As a consequence of Theorem 7.1.5.1, there is a constant $c>0$ so that given any $\zeta>0$ and α then for all ϵ small enough and all m such that $\|m-\chi^{(\epsilon)}\|_{L^1(\epsilon^{-1}R)}\le \epsilon^{-d}2\alpha h^d$

$$
\epsilon^{d-1}\mathcal{F}_{\epsilon^{-1}R_i}\left(m_{\epsilon^{-1}R_i}|m_{(\epsilon^{-1}R_i)^c}\right)\ge \left(\theta_{\nu_i}-\zeta-c\sqrt{\alpha}\right)h^{d-1}. \qquad (7.1.7.11)
$$

Then, going back to (7.1.7.10),

$$
\text{l.h.s. of (7.1.7.1)} \ge \liminf_{\alpha\to 0}\sum_{i=1}^{n}h^{d-1}\theta_{\nu_i}-cnh^{d-1}\sqrt{\alpha}. \qquad (7.1.7.12)
$$

By (7.1.7.3) and by (7.1.7.4) there is c'' so that

$$
\text{l.h.s. of (7.1.7.1)} \ge \lim_{\alpha\to 0}\sum_{i=1}^{n}\int_{\Sigma_i}\theta_{\nu(r)}\,d\mu - c''\sqrt{\alpha}, \qquad (7.1.7.13)
$$

which by (7.1.7.5) concludes the proof of (7.1.7.1).

7.1.8 Upper bound of Gamma convergence

In this subsection we shall prove

$$
\lim_{\delta\to 0}\limsup_{\epsilon\to 0}\inf_{\|m-u^{(\epsilon)}\|_{L^1(\epsilon^{-1}\Omega)}\le\epsilon^{-d}\delta}\epsilon^{d-1}\mathcal{F}^{\text{per}}_{\epsilon^{-1}\mathcal{T}}(m)\le\int_{\partial^\star E}\theta_{\nu(r)}\,d\mu(r), \quad (7.1.8.1)
$$

which, together with (7.1.7.1), concludes the proof of Γ-convergence, namely the validity of (7.1.3.1), with the identification $s_\beta(\nu)=\theta_\nu$.

To prove (7.1.8.1), we shall show that there exists $m^{(\epsilon)}\in L^1(\epsilon^{-1}\Omega,[-1,1])$, $\epsilon>0$, so that, writing as usual $u^{(\epsilon)}(r)=u(\epsilon r)$, $r\in\epsilon^{-1}\Omega$,

$$
\begin{aligned}
&\lim_{\epsilon\to 0}\epsilon^{d}\|m^{(\epsilon)}-u^{(\epsilon)}\|_{L^1(\epsilon^{-1}\Omega)}=0,\\
&\limsup_{\epsilon\to 0}\epsilon^{d-1}\mathcal{F}^{\text{per}}_{\epsilon^{-1}\mathcal{T}}(m^{(\epsilon)})\le\int_{\partial^\star E}\theta_{\nu(r)}\,d\mu(r),
\end{aligned} \qquad (7.1.8.2)
$$

which clearly implies (7.1.8.1). The choice of $m^{(\epsilon)}$ involves a diagonalization procedure. By Theorem B.1 of Sect. B.2 of Appendix B we can approximate in the sense of variations (i.e. so that (7.1.8.5)–(7.1.8.8) below hold), the function u by functions u_k equal to $\pm m_\beta$ inside and outside polyhedral sets E_k with boundary ∂E_k. For each k, we shall construct functions $m^{(\epsilon,L,h,k)}$ so that

$$\limsup_{h\to\infty}\limsup_{L\to\infty}\limsup_{\epsilon\to 0}\ \epsilon^d \|m^{(\epsilon,L,h,k)} - u_k^{(\epsilon)}\|_{L^1(\epsilon^{-1}\Omega)} = 0\,, \qquad (7.1.8.3)$$

$$\limsup_{h\to\infty}\limsup_{L\to\infty}\limsup_{\epsilon\to 0}\ \epsilon^{d-1}\mathcal{F}^{\text{per}}_{\epsilon^{-1}\mathcal{T}}(m^{(\epsilon,L,h,k)}) \le \int_{\partial E_k} \theta_{\nu(r)}\, d\mu_k(r). \qquad (7.1.8.4)$$

Equations (7.1.8.3) and (7.1.8.4) prove the validity of (7.1.8.1) once restricted to polyhedral sets. Since the sequence $\{u_k^{(\epsilon)}\}$ has the property (see Appendix B)

$$\epsilon^d \|u_k^{(\epsilon)} - u^{(\epsilon)}\|_{L^1(\epsilon^{-1}\Omega)} = \|u_k - u\|_{L^1(\Omega)} \to 0, \quad \text{as } k\to\infty, \qquad (7.1.8.5)$$

we then get from (7.1.8.3) and (7.1.8.4)

$$\limsup_{k\to\infty}\limsup_{h\to\infty}\limsup_{L\to\infty}\limsup_{\epsilon\to 0}\ \epsilon^d \|m^{(\epsilon,L,h,k)} - u^{(\epsilon)}\|_{L^1(\epsilon^{-1}\Omega)} = 0, \qquad (7.1.8.6)$$

$$\limsup_{k\to\infty}\limsup_{h\to\infty}\limsup_{L\to\infty}\limsup_{\epsilon\to 0}\ \epsilon^{d-1}\mathcal{F}^{\text{per}}_{\epsilon^{-1}\mathcal{T}}(m^{(\epsilon,L,h,k)}) \le \limsup_{k\to\infty}\int_{\partial E_k} \theta_{\nu(r)}\, d\mu_k(r). \qquad (7.1.8.7)$$

By Theorem B.2 of Appendix B,

$$\lim_{k\to\infty}\int_{\partial E_k} \theta_{\nu(r)}\, d\mu_k(r) = \int_{\partial^* E} \theta_{\nu(r)}\, d\mu(r), \qquad (7.1.8.8)$$

because u_k converges to u in variation and θ_ν is a bounded continuous function of the unit normal ν. By (7.1.8.6)–(7.1.8.7)–(7.1.8.8), there are $L = L(\epsilon)$, $h = h(\epsilon)$ and $k = k(\epsilon)$ so that the family $m^{(\epsilon,L(\epsilon),h(\epsilon),k(\epsilon))}$ satisfies (7.1.8.3). Thus, the proof of (7.1.8.1), follows from the existence of a family $m^{(\epsilon,L,h,k)}$ satisfying (7.1.8.3) and (7.1.8.4), which is proved below.

Notation

Here k is fixed and we shall drop it from the notation, thus writing, in the sequel, E for a polyhedral set and denoting, as usual, $u = m_\beta(\mathbf{1}_{E^c} - \mathbf{1}_E)$. The faces of E are called Σ_i, $i = 1,\dots,n$, and their normal ν_i, directed toward the plus phase. On each hyperplane which contains $\epsilon^{-1}\Sigma_i$, we introduce a partition into $d-1$ dimensional cubes of side L, the orientation of the cubes of the partition being the same for all ϵ. As already said, we shall take $L \to \infty$ after $\epsilon \to 0$, with a third parameter, $h \to \infty$ after $\epsilon \to 0$ and $L \to \infty$. We first define $m^{(\epsilon,L,h)}$ around $\epsilon^{-1}\Sigma_1$: recalling (7.1.4.3), on each rectangle $R_{\nu_1}(L,h)$ of height $2h$ and mid cross section a cube entirely contained in $\epsilon^{-1}\Sigma_1$, we choose $m^{(\epsilon,L,h)}$ so that

$$\frac{1}{L^{d-1}} \mathcal{F}_{R_{\nu_1}(L,h)}(m^{(\epsilon,L,h)}_{R_{\nu_1}(L,h)} | \chi_{R_{\nu_1}(L,h)^c}) \le \theta_{\nu_1}(L,h) + \epsilon. \tag{7.1.8.9}$$

When the mid cross section of $R_{\nu_1}(L,h)$ is not entirely contained in $\epsilon^{-1}\Sigma_1$, we set $m^{(\epsilon)} = \pm m_\beta$ in the part of $R_{\nu_1}(L,h)$ which is (vertically, w.r.t. ν_1) above and below $\epsilon^{-1}\Sigma_1 \cap R_{\nu_1}(L,h)$. We follow the same rule in the other faces, except for points (if any) where $m^{(\epsilon)}$ has already been defined. On the remaining of the space we set $m^{(\epsilon)} = u^{(\epsilon)}$.

Once h is fixed, if L is large enough, any rectangle $R_{\nu_i}(L,h)$ at distance $> L$ from the boundary of $\epsilon^{-1}\Sigma_i$ has no intersection with any of the other rectangles, then, for a suitable constant c,

$$\epsilon^{d-1} \mathcal{F}^{\text{per}}_{\epsilon^{-1}\Omega}(m^{(\epsilon)}) \le \sum_{i=1}^{n} \big((\theta_{\nu_i}(L,h) + \epsilon)|\Sigma_i| + cLh\epsilon \big). \tag{7.1.8.10}$$

Equations (7.1.8.3)–(7.1.8.4) then follow, thus completing the proof of the upper bound of Gamma convergence.

7.2 Wulff shape

Recalling from Sect. 7.1.1, the surface correction to the canonical free energy density is defined as

$$F_\epsilon(s) := \inf_{u \in L^1(\Omega;[-1,1]): \int u\, dr = s} \Phi_\epsilon(u), \tag{7.2.0.1}$$

where $s \in (-m_\beta, m_\beta)$, $\Phi_\epsilon(u) = \epsilon^{d-1} \mathcal{F}^{\text{per}}_{\epsilon^{-1}\Omega}(u(\epsilon \cdot))$; see (7.1.2.2) and Sect. 7.1.2 for notation. Since Φ_ϵ Gamma converges to Φ_0, if lim and inf commute, then $F_0(s) := \liminf_{\epsilon \to 0} F_\epsilon(s)$ would be equal to

$$W(s) := \inf_{u \in BV(\Omega;\{\pm m_\beta\}): \int u\, dr = s} \int s_\beta(\nu(r))\, d\mu(r), \tag{7.2.0.2}$$

and $F_0(s)$ given by the famous Wulff construction for the minimizer in (7.2.0.2), at least in the case of regular surface tensions. In this section we shall prove that $F_0(s) = W(s)$.

7.2.1 Coercivity

Compactness is going to play a key role. Let $u^{(\epsilon)} \in L^\infty(\Omega, [-1, 1])$, $\epsilon > 0$, be such that for some $C > 0$

$$\sup_{\epsilon>0} \Phi_\epsilon(u^{(\epsilon)}) \le C. \tag{7.2.1.1}$$

In the next subsection we shall prove that (7.2.1.1) holds for optimizing sequences. We shall prove here that:

Theorem 7.2.1.1 *If $u^{(\epsilon)}$ satisfies* (7.2.1.1), *then it converges by subsequences in $L^1(\Omega)$ and any limit point is in $BV(\Omega; \{\pm m_\beta\})$.*

Proof With $m^{(\epsilon)}(r) := u^{(\epsilon)}(\epsilon r)$, $r \in \epsilon^{-1}\Omega$, (7.2.1.1) becomes

$$\sup_{\epsilon>0} \epsilon^{d-1} \mathcal{F}^{\text{per}}_{\epsilon^{-1}\Omega}(m^{(\epsilon)}) \le C. \tag{7.2.1.2}$$

We are going to use Theorem 6.4.2.3 with $\mathcal{C}_0$ and $\mathcal{C}_{\neq}$ maximal collections relative to $m^{(\epsilon)}$, in the sense of Definition 6.4.2.1. Call $\mathcal{C}$ the union of all the cubes in $\mathcal{C}_0$ and $\mathcal{C}_{\neq}$ and p and $2n$ the number of cubes in $\mathcal{C}_0$ and respectively $\mathcal{C}_{\neq}$. Then Theorem 6.4.2.3 (whose validity extends to the case of periodic J) states that $\mathcal{F}^{\text{per}}_{\epsilon^{-1}\Omega}(m^{(\epsilon)}) \ge c_1'\zeta^2 \times \ell^d(2n+p)$, hence, by (7.2.1.2),

$$(2n+p) \le c\epsilon^{-d+1}, \quad c = \frac{C}{c_1'(\zeta^2\ell^d)}. \tag{7.2.1.3}$$

Call

$$\begin{aligned} A^{(\pm,\epsilon)} &= \big\{r \in \epsilon^{-1}\Omega : \eta^{(\zeta,\ell)}(m^{(\epsilon)}; r) = \pm 1\big\}, \\ \chi^{(\pm,\epsilon)}(r) &= \mathbf{1}_{\epsilon^{-1} r \in A^{(\pm,\epsilon)}}, \end{aligned} \tag{7.2.1.4}$$

and $|\partial A^{(\pm,\epsilon)}|$ the area of the boundary of $A^{(\pm,\epsilon)}$. $\partial A^{(+,\epsilon)}$ is union of faces of cubes common to a pair of cubes where either $\eta^{(\zeta,\ell)}(m^{(\epsilon)}; r)$ is $+1$ in one and -1 in the other or else, it is $+1$ in one and 0 in the other. In both case at least one of the two cubes is contained in $\mathcal{C}$, by maximality. Same argument applies in the minus case and since each cube has $2d$ faces, we conclude that

$$|\partial A^{(\pm,\epsilon)}| \le 2d(2n+p) \le 2dc\epsilon^{-d+1}. \tag{7.2.1.5}$$

Thus, the characteristic functions $\chi^{(\pm,\epsilon)}(r)$ on Ω have total variation bounded by $2dc$ independently of ϵ. Then, [9], there exists a subsequence ϵ_n and $\chi^{(\pm)} \in BV(\Omega; \{0, 1\})$ so that

$$\lim_{n\to\infty} \|\chi^{(\pm,\epsilon_n)} - \chi^{(\pm)}\|_{L^1(\Omega)} = 0. \tag{7.2.1.6}$$

We claim that

$$\int_{\Omega} \{\chi^{(+)} + \chi^{(-)}\} = 1. \tag{7.2.1.7}$$

Indeed $A^{(+,\epsilon)} \cup A^{(-,\epsilon)}$ covers $\epsilon^{-1}\Omega$ except for $\{r \in \epsilon^{-1}\Omega : \eta^{(\zeta,\ell)}(m^{(\epsilon)}; r) = 0\}$. By maximality the cubes where $\eta^{(\zeta,\ell)}(m^{(\epsilon)}; r) = 0$ are in $\mathcal{C}_0$; then by (7.2.1.3),

$$1 \geq \int_{\Omega} \{\chi^{(+,\epsilon)} + \chi^{(-,\epsilon)}\} \geq 1 - c\epsilon, \tag{7.2.1.8}$$

and (7.2.1.7) is proved.

Let $u^{\epsilon,\pm} = \chi^{(\pm,\epsilon)} m_\beta$, $u^\epsilon = u^{\epsilon,+} - u^{\epsilon,-}$, $u^\pm = \chi^{(\pm)} m_\beta$, $u = u^+ - u^-$. We shall conclude the proof of the theorem by showing that

$$\lim_{n\to\infty} \| u^{(\epsilon_n,\pm)} - u^{\pm} \|_{L^1(\Omega)} = 0. \tag{7.2.1.9}$$

We restrict ourselves to the plus case, the argument for the minus case being similar. Let $\delta_n \to 0$ in such a way that $\epsilon_n/\delta_n^2 \to 0$. (7.2.1.9) (with the plus) will follow from

$$\lim_{n\to\infty} \| \chi^{(+,\epsilon_n)} \mathbf{1}_{|u^{(\epsilon_n,+)} - m_\beta| \geq \delta_n} \|_{L^1(\Omega)} = 0, \tag{7.2.1.10}$$

which, going back to (7.2.1.5), reads as

$$\lim_{n\to\infty} \epsilon_n^d \big| A^{(+,\epsilon_n)} \cap \{|m^{(\epsilon_n)} - m_\beta| \geq \delta_n\}\big| = 0. \tag{7.2.1.11}$$

We decompose

$$\{|m^{(\epsilon_n)} - m_\beta| \geq \delta_n\} = \big\{\big|\, |m^{(\epsilon_n)}| - m_\beta \,\big| \geq \delta_n\big\} \cup \{|m^{(\epsilon_n)} + m_\beta| \leq \delta_n\}. \tag{7.2.1.12}$$

There is a positive constant $b > 0$ such that

$$\inf_{|\,|s| - m_\beta\,| \geq \delta} f_\beta(s) \geq b\delta^2, \tag{7.2.1.13}$$

hence $\epsilon_n^d |A^{(+,\epsilon_n)} \cap \{|\, |m^{(\epsilon_n)}| - m_\beta| \geq \delta_n\,|\}|$ is bounded by

$$\frac{\epsilon_n^d}{b\delta_n^2} \int f_\beta(m^{(\epsilon_n)})\, dr \leq \frac{\epsilon_n^d}{b\delta_n^2} \mathcal{F}^{\text{per}}_{\epsilon^{-1}\Omega}(m^{(\epsilon_n)}),$$

which, by (7.2.1.2), vanishes as $n \to \infty$ because $\epsilon_n/\delta_n^2 \to 0$. It thus remains to show that

$$\lim_{n\to\infty} \epsilon_n^d \big| A^{(+,\epsilon_n)} \cap \{|m^{(\epsilon_n)} + m_\beta| \leq \delta_n\}\big| = 0. \tag{7.2.1.14}$$

Call $B^{(+,\epsilon_n)} = \{r \in A^{(+,\epsilon_n)} : \eta^{(\zeta,\ell)}(m^{(\epsilon_n)}; r') = 1, |r' - r| \leq 1\}$. We shall first prove (7.2.1.14) with $B^{(+,\epsilon_n)}$ replacing $A^{(+,\epsilon_n)}$ and then prove that

$$\lim_{n\to\infty} \epsilon_n^d |A^{(+,\epsilon_n)} \setminus B^{(+,\epsilon_n)}| = 0, \tag{7.2.1.15}$$

thus proving (7.2.1.14) as well. With $\epsilon_n \equiv \epsilon$, $\mathcal{F}^{\text{per}}_{\epsilon^{-1}\Omega}(m^{(\epsilon)})$ is

$$\geq \frac{1}{4}\int_{B^{(\epsilon,+)}\cap\{|m^{(\epsilon)}+m_\beta|\leq\delta_n\}} \left(\int J(r,r')[m^{(\epsilon)}(r)-m^{(\epsilon)}(r')]^2 dr'\right)dr$$

$$\geq -\frac{1}{2}\int_{B^{(\epsilon,+)}\cap\{|m^{(\epsilon)}+m_\beta|\leq\delta_n\}} m^{(\epsilon)}(r)\left(\int J(r,r')m^{(\epsilon)}(r')dr'\right)dr.$$

Therefore, by (6.3.2.6), for ℓ and ζ small enough,

$$\mathcal{F}^{\text{per}}_{\epsilon_n^{-1}\Omega}\left(m^{(\epsilon_n)}\right) \geq \frac{1}{2}\left|B^{(\epsilon_n,+)}\cap\{|m^{(\epsilon_n)}+m_\beta|\leq\delta_n\}\right|(m_\beta-\delta_n)(m_\beta-2\zeta),$$

hence (7.2.1.14) with $B^{(\epsilon_n,+)}$ follows from $\epsilon_n^d|B^{(\epsilon_n,+)}\cap\{|m^{(\epsilon_n)}+m_\beta|\leq\delta_n\}|\leq[(m_\beta-\delta_n)(m_\beta-2\zeta)]^{-1}2C\epsilon_n$.

To prove (7.2.1.15) let $\mathcal{C}$ be the set associated to $m^{(\epsilon_n)}$ via Definition 6.4.2.1 and notice that $A^{(+,\epsilon_n)}\setminus B^{(+,\epsilon_n)}\subset\bigcup_{C\subset\mathcal{C}}\{r:\text{dist}(r,C)\leq 2\}$, then, for ℓ small enough,

$$|A^{(+,\epsilon_n)}\setminus B^{(+,\epsilon_n)}|\leq 8^d(2n+p)\leq 8^d c\epsilon^{-d+1}, \quad \text{see (7.2.1.3)},$$

and (7.2.1.15) is proved. □

7.2.2 Interfaces which are not in BV

As a first application of the above theorem, we complete the analysis of Γ-convergence of Theorem 7.1.3.1, by proving:

Theorem 7.2.2.1 *If $u\in L^1(\Omega,\{\pm m_\beta\})$ is not in $BV(\Omega,\{\pm m_\beta\})$, then $\Phi'_0(u)=+\infty$, $\Phi'_0(u)$ being defined in* (7.1.2.4).

Proof Suppose by contradiction that $\Phi'_0(u)<\infty$ which means that there is a sequence $u^{(\epsilon)}\in L^\infty(\Omega,[-1,1])$, so that $u^{(\epsilon)}\to u$ in $L^1(\Omega)$ and $\Phi_\epsilon(u^{(\epsilon)})\to\Phi'_0(u)$. Then there is C so that $\Phi_\epsilon(u^{(\epsilon)})<C$ and a subsequence converging in $L^1(\Omega)$ to a function in $BV(\Omega,\{\pm m_\beta\})$, against the assumption that u is not in BV. The corollary is proved. □

7.2.3 The Wulff problem

Using Theorem 7.2.1.1 together with the Gamma convergence result of Theorem 7.1.3.1, we have

Theorem 7.2.3.1 *Let $s \in (-m_\beta, m_\beta)$; then*

$$\lim_{\epsilon\to 0} \inf_{u\in L^1(\Omega;[-1,1]):\int_\Omega u=s} \Phi_\epsilon(u) = W(s), \tag{7.2.3.1}$$

where Φ_ϵ is defined in (7.1.2.2) *and $W(s)$ in* (7.2.0.1).

Proof We shall first prove that $W(s)$ is a lower bound for the l.h.s. of (7.2.3.1) and then that it is also an upper bound, thus proving (7.2.3.1).

Lower bound Let Δ be a cube in Ω such that the function $\hat{u}$ equal to $\pm m_\beta$ inside and outside Δ has integral equal to s. Then there is $C > 0$ so that $\Phi_\epsilon(\hat{u}) \le C$ for all $\epsilon > 0$ and $F_\epsilon(s) \le C$, $F_\epsilon(s)$ the inf in (7.2.3.1), as defined in (7.2.0.1). Let $u^{(\epsilon)} \in L^1(\Omega, [-1, 1])$ be such that

$$\int_\Omega u^{(\epsilon)}\, dr = s, \qquad \lim_{\epsilon\to 0} \left|\Phi_\epsilon(u^{(\epsilon)}) - F_\epsilon(s)\right| = 0. \tag{7.2.3.2}$$

Then $u^{(\epsilon)}$ satisfies (7.2.1.1) and by Theorem 7.2.1.1 there is a subsequence ϵ_n such that $u^{(\epsilon_n)}$ converges in $L^1(\Omega)$ to a limit point $u \in BV(\Omega; \{\pm m_\beta\})$ and, by possibly taking a subsequence,

$$\liminf_{\epsilon\to 0} F_\epsilon(s) = \lim_{n\to\infty} F_{\epsilon_n}(s). \tag{7.2.3.3}$$

By the definition of Gamma convergence,

$$\liminf_{n\to\infty} \Phi_{\epsilon_n}(u^{(\epsilon_n)}) \ge \Phi_0(u),$$

and, by (7.2.3.2), $\liminf_{\epsilon\to 0} F_\epsilon(s) \ge \Phi_0(u)$. Moreover, $\Phi_0(u) \ge W(s)$ because $\int_\Omega u\, dr = s$, as u is the L^1 limit of functions which satisfy the same condition. The lower bound is proved.

Upper bound We shall next show that $W(s) \ge \liminf_{\epsilon\to 0} F_\epsilon(s)$ which, together with the lower bound will prove (7.2.3.1). We proceed by contradiction and suppose that there is $v \in BV(\Omega, \pm m_\beta)$, $\int_\Omega v = s$, such that

$$\Phi_0(v) < \liminf_{\epsilon\to 0} F_\epsilon(s). \tag{7.2.3.4}$$

Then, by Gamma convergence, there is a sequence $v^{(\epsilon)} \in L^1(\Omega, [-1, 1])$ which converges to v in L^1 and such that

$$\Phi_0(v) = \lim_{\epsilon\to 0} \Phi_\epsilon(v^{(\epsilon)}). \tag{7.2.3.5}$$

This would be a contradiction if $\int_\Omega v^{(\epsilon)} = s$, because then $\Phi_\epsilon(v^{(\epsilon)}) \ge F_\epsilon(s)$ and $\Phi_0(v) \ge \liminf F_\epsilon(s)$, against (7.2.3.4). To conclude the argument, it suffices to show that we can modify $v^{(\epsilon)}$ in such a way that the constraint is verified for all ϵ and, at the same time, the possible additional free energy cost due to the modification of $v^{(\epsilon)}$ is negligible. Partition Ω into equal cubes of side $\delta > 0$; then supposing for instance $\int_\Omega v^{(\epsilon)} > s$, there exists a cube Δ_+ where

$$\int_{\Delta_+} v^{(\epsilon)} \geq s\delta^d. \tag{7.2.3.6}$$

Call $d_\epsilon = \int_\Omega v^{(\epsilon)} - s > 0$ and $u^{(\epsilon)}$ the function obtained from $v^{(\epsilon)}$ by replacing $v^{(\epsilon)}(r)$, $r \in \Delta_+$, by $-m_\beta$. Then $\int_\Omega u^{(\epsilon)} \leq s + d_\epsilon - s\delta^d - m_\beta \delta^d$. Since $s > -m_\beta$ and $d_\epsilon \to 0$ as $\epsilon \to 0$, the r.h.s. becomes smaller than s for ϵ small. By continuity for each $\epsilon > 0$ small enough there is a cube Δ' contained in Δ_+ such that if we do the replacement only in Δ', then the modified function has integral equal to s. Calling $u^{(\epsilon)}$ the modified function, we have $\Phi_\epsilon(v^{(\epsilon)}) \geq \Phi_\epsilon(u^{(\epsilon)}) - c\delta^{d-1}$, $c > 0$ a constant and by taking δ small enough we reach a contradiction. □

Chapter 8
One dimensional interfaces

In this chapter we shall complete the analysis of the coexistence of phases in systems described by the L–P free energy functional (6.1.3.1). We shall first assume a planar symmetry, thus reducing it to a one dimensional problem whose analysis will take most of the chapter. We shall prove the existence and uniqueness modulo translations of "the instanton." This is a special function, which minimizes the one dimensional L–P free energy functional in the class of functions which are asymptotically definitively positive and negative, respectively, when $x \to \pm\infty$. We shall prove that the instanton is monotonically increasing and converges exponentially as $x \to \pm\infty$ to the equilibrium values $\pm m_\beta$ (we suppose throughout the chapter that $h = 0$ and $\beta > 1$). We shall also prove the stability properties of the instanton with respect to the time evolution, in particular, the existence of a spectral gap for the evolution linearized around the instanton, and then we use it to prove first a "local" and then a "global" stability result. Finally, in Sect. 8.6 we shall prove the validity of the planar symmetry assumption showing that the free energy of the instanton is equal to the surface tension computed in Chap. 7.

8.1 Instantons

8.1.1 van der Waals theory of surface tension

The van der Waals theory of the liquid–vapor phase coexistence starts with the assumption that molecules interact with each other by short range repulsive and long range attractive forces. van der Waals then conjectures that the system can be studied in terms of a free energy functional which is structurally similar to our L–P functional of (6.1.3.1) with $h = 0$ and $\beta > 1$. Indeed, in such a case we have proved that the system has a phase transition with two pure phases, described respectively by profiles constantly equal to $+m_\beta$ and to $-m_\beta$, where m_β is the positive solution of the mean field equation $m_\beta = \tanh\{\beta m_\beta\}$. Coexistence between the two phases is then determined by the surface tension, and the van der Waals theory gives a precise prescription for computing the surface tension summarized in the following three points.

- With the underlying assumption of planar symmetry, we reduce the system to one dimensional interactions. Let e be a unit vector in $\mathbb{R}^d$, x and x' in $\mathbb{R}$. Define

$$J^{(e)}(x,x') = \int_{(y,e)=0} J(xe, x'e + y)\,dy, \tag{8.1.1.1}$$

where dy is the Lebesgue measure on the $(d-1)$-dimensional plane $(y, e) = 0$, orthogonal to e and passing through the origin. $J^{(e)}(x, x')$ is the interaction energy density between two planes orthogonal to e passing through xe and $x'e$, and it describes the interaction energy of a profile $m(r)$ which is constant on planes orthogonal to e. $J^{(e)}(x, x')$ is a one dimensional interaction which satisfies the conditions stated in Sect. 6.1.2.

• Denote by $\mathcal{F}^{(e)}(m)$ the functional (6.1.3.1) in $d = 1$ with interaction $J^{(e)}(x, x')$ (as already mentioned, we have set $h = 0$ and $\beta > 1$). The van der Waals surface tension in the direction e is then defined by

$$S_\beta(e) = \inf_{m \in \mathcal{N}} \mathcal{F}^{(e)}(m), \tag{8.1.1.2}$$

where

$$\mathcal{N} = \left\{ m \in L^\infty(\mathbb{R}, [-1, 1]) : \limsup_{x \to -\infty} m(x) < 0; \ \liminf_{x \to \infty} m(x) > 0 \right\}. \tag{8.1.1.3}$$

• The final assumption in the van der Waals theory is that the inf in (8.1.1.2) is actually attained.

Definition 8.1.1.1 A function $\bar{m}^{(e)}(x) \in L^\infty(\mathbb{R}; [-1, 1])$ is an *instanton* if

$$S_\beta(e) = \mathcal{F}^{(e)}(\bar{m}^{(e)}), \tag{8.1.1.4}$$

and moreover: $\lim_{x \to \pm\infty} \bar{m}^{(e)}(x) = \pm m_\beta$ and for all x,

$$\bar{m}^{(e)}(x) = \tanh\{\beta J^{(e)} \star \bar{m}^{(e)}(x)\}. \tag{8.1.1.5}$$

We shall prove that for the L–P functional an instanton actually exists, that it is unique (modulo translations), and finally we shall show that the planar symmetry assumption holds (thus, it is no longer an assumption). We shall then complete the analysis of Chap. 7 by showing that $S_\beta(e)$ defined in (8.1.1.2) or (8.1.1.4) is the same as the surface tension $s_\beta(e)$ defined in Theorem 7.1.3.1.

8.1.2 Instantons: main theorem

As the direction e is generic but fixed, we hereafter drop it from the notation, writing J and $\mathcal{F}$ for the one dimensional interaction and the free energy functional.

Theorem 8.1.2.1 *For any $\beta > 1$, the following holds.*

(i) *The variational problem* (8.1.1.2) *has a minimizer which is unique up to translations.*
(ii) *The mean field equation* (8.1.1.5) *has a solution in $\mathcal{N}$ which is unique in $\mathcal{N}$ up to translations.*

(iii) *There is* $\bar{m} \in \mathcal{N}$ *with the following properties. For any* $m \in \mathcal{N}$, *there is* $\xi \in \mathbb{R}$ *so that* $\lim_{t\to\infty} \|T_t(m) - \bar{m}_\xi\|_\infty = 0$, *where* $\bar{m}_\xi(x) = \bar{m}(x-\xi)$ *and* $T_t(m)$ *is the solution of* (6.2.2.7) *which starts from* m. $\bar{m}$ *is a strictly increasing, anti-symmetric function which converges exponentially fast to* $\pm m_\beta$ *as* $x \to \pm\infty$.
(iv) *The function* $\bar{m}$ *defined in* (iii) *minimizes* (8.1.1.2) *and it is therefore an instanton.*

Theorem 8.1.2.1 characterizes the instanton in three ways: as the minimizer of the free energy, as the solution of the mean field equation in $\mathcal{N}$ and as the limit of orbits starting in $\mathcal{N}$. It is not at all surprising that such apparently different procedures give the same result, as (8.1.1.5) is the Euler–Lagrange equation $D\mathcal{F} = 0$, and since the time evolution is defined by (6.2.2.4), $D\mathcal{F} = 0$ is also the equation satisfied by profiles which are stationary in time. The various statements in Theorem 8.1.2.1 are in fact strongly correlated.

Theorem 8.1.2.2 *If* (iii) *in Theorem* 8.1.2.1 *holds, then* (i), (ii) *and* (iv) *hold as well.*

Proof Let $\bar{m}$ be the function whose existence is claimed in (iii). The proof of (ii) goes as follows. Let $m \in \mathcal{N}$. Since $\lim_{t\to\infty} T_t(m) = \bar{m}_\xi$, for any $s > 0$ $\bar{m}_\xi = \lim_{t\to\infty} T_{t+s}(m) = T_s(\lim_{t\to\infty} T_t(m)) = T_s(\bar{m}_\xi)$, thus $\bar{m}_\xi$ is time invariant, hence it solves the mean field equation (8.1.1.5). On the other hand if $m \in \mathcal{N}$ solves (8.1.1.5), then $m = T_t(m)$ and by (iii) there is ξ so that $m = \lim_{t\to\infty} T_t(m) = \bar{m}_\xi$. For the proof of (i), let $m_n \in \mathcal{N}$ be any minimizing sequence for (8.1.1.2):

$$\lim_{n\to\infty} \mathcal{F}(m_n) = \inf_{m\in\mathcal{N}} \mathcal{F}(m).$$

Let $T_t(m_n)$ be the solution of (6.2.2.7) starting from m_n; by (iii) there is $\xi_n \in \mathbb{R}$ such that $T_t(m_n) \to \bar{m}_{\xi_n}$ as $t \to \infty$. By Theorem 6.2.9.1, $\mathcal{F}(m_n) \geq \mathcal{F}(\bar{m}_{\xi_n}) = \mathcal{F}(\bar{m})$ (the last equality holds because $\mathcal{F}$ is invariant by translations), thus

$$\mathcal{F}(\bar{m}) = \inf_{m\in\mathcal{N}} \mathcal{F}(m).$$

Uniqueness (modulo translations) of the minimizer of (8.1.1.2): let $\tilde{m} \in \mathcal{N}$ be a minimizer for (8.1.1.2). Using the Comparison Theorem (Theorem 6.2.10.2) and the Barrier Lemma (Theorem 6.2.7.1), it can be seen (the proof is omitted) that $T_t(\tilde{m}) \in \mathcal{N}$ for all t. By Theorem 6.2.8.1, for $t > 0$: $\mathcal{F}(T_t(\tilde{m})) < \mathcal{F}(\tilde{m})$ unless $T_t(\tilde{m}) = \tilde{m}$; hence, $T_t(\tilde{m}) = \tilde{m}$ and therefore $\tilde{m}$ solves the mean field equation (8.1.1.5). Thus $\tilde{m} = \bar{m}_\xi$ for some $\xi \in \mathbb{R}$, as we have already proved that (iii) implies (ii).

The proof of (iv) is contained in the above proof of (i) and (ii) and therefore Theorem 8.1.2.2 is also proved. □

We have thus reduced the proof of Theorem 8.1.2.1 to the proof of its statement (iii).

8.1.3 Existence of instantons

The first step in the proof of Theorem 8.1.2.2 is to prove the existence of instantons. Existence and uniqueness are trivial for the Ginzburg–Landau functional, as in one dimension the stationary solutions of (6.2.2.6) satisfy

$$\frac{d^2m(x)}{dx^2} = w'(m(x)), \tag{8.1.3.1}$$

which is the equation of motion of a point on the line with mass 1 under a force of potential $-w(m)$, m its position, x the time. Since $w(m)$ is a double well potential with equal minima at $\pm m_\beta$, $w(\pm m_\beta) = 0$ (same setting as Theorem 8.1.2.1 above), the instanton is the orbit of 0 energy and positive velocity, which connects the two "positions" $-m_\beta$ and $+m_\beta$ as "time" runs from $-\infty$ to ∞. Motions with higher energies are unbounded; those with lower energies are oscillatory without reaching $\pm m_\beta$; hence, we conclude to the existence and uniqueness (modulo translations, i.e. time shift in the above picture) of the instanton.

Theorem 8.1.3.1 *For any $\beta > 1$, there exists a continuous, antisymmetric, monotonically increasing instanton.*

Proof The proof of Theorem 8.1.3.1 uses dynamics. Let $m(x)$ be a continuous, antisymmetric function, monotonically increasing in [0, 1] from 0 to m_β and thereafter constantly equal to m_β.

- For any $t \geq 0$, $T_t(m)$ is a continuous (by Theorem 6.2.3.3) antisymmetric, non-decreasing function with values in $[-m_\beta, m_\beta]$ (the proof of the latter statements is postponed).
- Let u be a limit point of $T_t(m)$; the existence of limit points is proved in Theorem 6.2.9.1 and by the same theorem u solves (8.1.1.5) and $\mathcal{F}(u) \leq \mathcal{F}(m) < \infty$.
- u is a non-decreasing, antisymmetric function of x, being the limit of such functions. In particular, therefore, $u(x)$ has limits, say $u_\pm$ as $x \to \pm\infty$; by the antisymmetry of u, $u_+ = -u_-$.
- $\mathcal{F}(u)$ is bounded from below by the first integral in (6.1.3.1) which is indeed infinite, unless $|u_\pm| = m_\beta$. Thus, $u_\pm = \pm m_\beta$ and $u \in \mathcal{N}$; hence we see the existence of an instanton.

The above proof uses properties of $T_t(m)$ which are established after the following preliminary lemma, where $A(m) := -m + \tanh\{\beta J * m\}$.

Lemma *Let $\mathcal{P}$ be a linear map from $L^\infty(\mathbb{R})$ into itself such that $\mathcal{P}(A(m)) = A(\mathcal{P}(m))$. Then $\mathcal{P}(T_t(m)) = T_t(\mathcal{P}(m))$.*

Proof Since $\frac{d}{dt}\mathcal{P}(T_t(m)) = \mathcal{P}(A(T_t(m))) = A(\mathcal{P}(T_t(m)))$, $\mathcal{P}(T_t(m))$ solves the equation of motion with initial datum $\mathcal{P}(m)$. It thus follows that, by uniqueness, $\mathcal{P}(T_t(m)) = T_t(\mathcal{P}(m))$, and the lemma is proved. □

We can now prove the properties of $T_t(m)$ used above.

• First, we prove that if m is non-decreasing then also $T_t(m)$ is non-decreasing. Call D_a, $a \geq 0$, the shift operator on $L^\infty(\mathbb{R})$, i.e. $D_a u(x) = u(x+a)$. D_a satisfies the assumptions of the lemma, because $J(x+a, y+a) = J(x, y)$. Let $n_a(t) := D_a T_t(m)$; then $n_a(t)$ solves (6.2.2.7) and $n_a(t) = T_t(D_a m)$; since $D_a m \geq m$ (by the choice of m see the beginning of the proof) then, by the Comparison Theorem, Theorem 6.2.10.3, $T_t(D_a m) \geq T_t(m)$. Thus, $D_a T_t(m) \geq T_t(m)$, i.e. $T_t(m)(x+a) \geq T_t(m)(x)$ for all x and all $a \geq 0$.

• Proof that $\|T_t(m)\|_\infty \leq m_\beta$. By the Comparison Theorem, $-m_\beta \leq T_t(m) \leq m_\beta$ because this holds at $t = 0$ and $T_t(\pm m_\beta) = \pm m_\beta$.

• Proof that $T_t(m)(x) = -T_t(m)(-x)$. Denote by R the reflection operator on $L^\infty(\mathbb{R})$, i.e. $Ru(x) = u(-x)$; R satisfies the assumptions of the Lemma because $J(x, y) = J(-x, -y)$, see Sect. 4.2.1. Then $RT_t(m) = T_t(Rm)$. Denote by S the change of sign operator: $Su = -u$, S satisfies the assumptions of the Lemma because $A(m) := -m + \tanh\{\beta J * m\}$ is odd, i.e. $SA(m) = -A(m) = A(-m) = A(Sm)$. Then $ST_t(m) = T_t(Sm)$. Since m is antisymmetric, $Sm = Rm$ and therefore $RT_t(m) = ST_t(m)$, i.e. $T_t(m)(x) = -T_t(m)(-x)$. □

8.2 Shape of instantons

In this section we shall prove that the instanton $\bar{m}(x)$ defined in Theorem 8.1.3.1 is a strictly increasing, regular function of x which converges exponentially fast to $\pm m_\beta$ as $x \to \pm\infty$. We shall also determine the asymptotic behavior of its derivatives, which is again exponential. The main results are presented in Theorem 8.2.1.2 below. We shall, however, omit for brevity the proof of Theorem 8.2.1.2 and prove a weaker version, Theorem 8.2.1.3, which is nonetheless sufficient to prove (iii) of Theorem 8.1.2.1 and hence, by Theorem 8.1.2.2, also the whole Theorem 8.1.2.1.

8.2.1 A heuristic argument

All results here are based on the analysis of the identities

$$\bar{m}'(x) = p(x)J' * \bar{m}(x) = p(x)J * \bar{m}'(x), \quad p(x) := \beta[1 - \bar{m}(x)^2], \qquad (8.2.1.1)$$

obtained by differentiating (8.1.1.5) with respect to x; the first equality proves the regularity of $\bar{m}'$ which is then used in the integration by parts which yields the second one.

If $\bar{m}'(x) \approx e^{-\alpha x}$ as $x \to \infty$, then we can identify the value of α. Suppose in fact that to leading orders as $x \to \infty$, we can replace in $\bar{m}'(x) = p(x)J * \bar{m}'(x)$, i.e. the second equality in (8.2.1.1), $\bar{m}'(x) \to e^{-\alpha x}$ and $p(x) \to p_\infty$, where

$$p_\infty = \beta(1 - m_\beta^2) = \lim_{x \to \infty} p(x), \quad p_\infty < 1; \qquad (8.2.1.2)$$

then

$$p_\infty \int J(0,x)e^{-\alpha x}\,dx = 1. \tag{8.2.1.3}$$

To prove $p_\infty < 1$ in (8.2.1.2), we write $\beta(1-m_\beta^2) = \frac{d\tanh(\beta m)}{dm}|_{m=m_\beta}$, the latter being smaller than 1 because $\tanh\{\beta m\} > m$ for $0 < m < m_\beta$, with equality at the endpoints. We omit the proof that

Lemma 8.2.1.1 *The integral*

$$f(\alpha) := \int J(0,x)e^{-\alpha x}\,dx \tag{8.2.1.4}$$

is an even function of α; $f(0)=1$, $f(\infty)=\infty$ and $f'(\alpha)>0$ for $\alpha>0$. As a consequence, for any $p\in(0,1)$ there is a unique positive solution α_p of

$$pf(\alpha_p) = 1. \tag{8.2.1.5}$$

α_p is a strictly decreasing function of p with limits ∞ and 1 as $p\to 0$, respectively $p\to 1$.

Thus (8.2.1.3) has a unique positive solution $\alpha = \alpha_{p_\infty}$, leading to the conjecture that $\bar m'(x) \approx e^{-\alpha x}$, which is indeed correct:

Theorem 8.2.1.2 *The instanton $\bar m$ defined in Theorem 8.1.3.1 is in $C^\infty(\mathbb{R})$, its derivative $\bar m'$ is strictly positive and there are $a>0$, $\alpha_0>\alpha$ and $c>0$ so that for $x\ge 0$*

$$\begin{aligned}&\left|\bar m(x) - (m_\beta - ae^{-\alpha x})\right| + \left|\bar m'(x) - a\alpha e^{-\alpha x}\right| + \left|\bar m''(x) + a\alpha^2 e^{-\alpha x}\right|\\ &\quad \le ce^{-\alpha_0 x}.\end{aligned} \tag{8.2.1.6}$$

We omit the proof of Theorem 8.2.1.2 and prove a weaker version which gives however all the ingredients for the proof of uniqueness and stability of the instantons:

Theorem 8.2.1.3 *The instanton $\bar m$ is in $C^\infty(\mathbb{R})$, $\bar m'(x)>0$ for all x and for any $\theta\in(0,\alpha)$ there is s such that*

$$\bar m'(x) \le 2\beta m_\beta \|J\|_\infty e^{-\theta(x-s)}, \quad \textit{for all } |x|\ge s, \tag{8.2.1.7}$$

$$\bar m(x) \ge m_\beta - 2\beta m_\beta \|J\|_\infty \theta^{-1} e^{-\theta(x-s)}, \quad \textit{for all } |x|\ge s. \tag{8.2.1.8}$$

Proof In this subsection we shall prove the statements about regularity and strict monotonicity, the proof of (8.2.1.7) and (8.2.1.8) is given in the next subsection. By the second equality in (8.2.1.1) we get

$$\|\bar m'\|_\infty \le \beta\|J\|_\infty 2m_\beta, \tag{8.2.1.9}$$

and by differentiating the first equality in (8.2.1.1), recalling that J is in C^2, we have $\bar{m}'' = p'J' * \bar{m} + pJ' * \bar{m}'$; since $p' = -2\bar{m}\bar{m}'$, $\bar{m}''$ indeed exists and is bounded. By iteration, $\bar{m}$ is in C^∞.

To prove that $\bar{m}' > 0$ we suppose by contradiction that there is x_0 such that $\bar{m}'(x_0) = 0$. Set $x = x_0$ in (8.2.1.1). Since $\bar{m}'(y)$ and $J(x, y)$ are both non-negative, by the continuity of $\bar{m}'(y)$, we have that $\bar{m}'(y) = 0$ for all y such that $J(x_0, y) > 0$. By repeating the argument starting from any such y, we conclude that $\bar{m}'(y) = 0$ whenever $J^2(x_0, y) > 0$ and by successive iterations whenever $J^n(x_0, y) > 0$ for any positive integer n. Since $J(x, y)$ is symmetric and not identically 0, the above set of y covers the whole line, thus concluding that $\bar{m}' \equiv 0$. This contradicts the fact that $\bar{m}(x) \to \pm m_\beta$ as $x \to \pm\infty$. □

8.2.2 Exponential decay

In this subsection we shall prove that $\bar{m}'$ decays to 0 as $|x| \to \infty$ exponentially with any exponent $\theta < \alpha$.

Lemma 8.2.2.1 *For any $p \in (p_\infty, 1)$ there is $s > 0$ such that $p(s) = p$. Let $\theta = -\log p > 0$; then,*

$$\bar{m}'(x) \le 2m_\beta \|J\|_\infty e^{-\theta(x-s)}, \quad \textit{for all } |x| \ge s. \tag{8.2.2.1}$$

Proof By symmetry we may just consider the case $x \ge 0$. Since $p(x)$ is a continuous function which decreases from $p(0) = \beta > 1$ to $p_\infty < 1$ as $x \to +\infty$, see (8.2.1.3), for any $p \in (p_\infty, 1)$ there is $s > 0$ so that $p(s) = p$ and

$$p(x) \le p < 1, \quad \text{for all } x \ge s. \tag{8.2.2.2}$$

From (8.2.1.1) we deduce that, for $x \ge s$, $\bar{m}'(x) \le p\,\|J\|_\infty 2m_\beta$.

If $x > s + 1$, the terms $\bar{m}'(y)$ in the convolution $J \star \bar{m}'(x)$ have $y > s$. Then $\bar{m}'(x) \le p^2 \|J\|_\infty 2m_\beta$. Analogously, writing $\theta = -\log p$ and $[x - s]$ for the integer part of $x - s$,

$$\bar{m}'(x) \le p^{[x-s]+1} \|J\|_\infty 2m_\beta \le (2m_\beta \|J\|_\infty) e^{-\theta(x-s)},$$

thus proving the lemma. □

The above result is not yet what we are aiming at, because the sup of the values of θ is $-\log p_\infty$ which is smaller than α, as given by (8.2.1.3). To improve, we need a more accurate analysis of the equation (8.2.1.1) in the interval $x \ge s$, s as in (8.2.2.2), which is what we shall do next. Let $x \ge s$ and

$$R_s(x, y) = p(x) J(x, y) \mathbf{1}_{y \ge s}, \qquad R(x, y) = p(x) J(x, y). \tag{8.2.2.3}$$

Lemma 8.2.2.2 (Green functions) *Let s be as in* (8.2.2.2); *then, for any* $x \geq s$

$$\bar{m}'(x) = \int_{s-1}^{s} G_s(x, y)\bar{m}'(y)\, dy, \tag{8.2.2.4}$$

where the Green function $G_s(x, y)$, $x \geq s$, $y \in [s-1, s)$, *is defined as*

$$G_s(x, y) := \sum_{n=0}^{\infty} \int_s^{\infty} R_s^n(x, z) R(z, y)\, dz, \tag{8.2.2.5}$$

the series converging uniformly in x and y.

Proof By (8.2.1.1), for any positive integer N

$$\bar{m}'(x) = \sum_{n=0}^{N-1} \int_s^{\infty} R_s^n(x, z)\, dz \int_{s-1}^{s} R(z, y)\bar{m}'(y)\, dy$$
$$+ \int_s^{\infty} R_s^N(x, y)\bar{m}'(y)\, dy. \tag{8.2.2.6}$$

Recalling (8.2.1.9), the n-th term of the sum in (8.2.2.6) is bounded by $p(s)^n C$, and the last term by $p(s)^N C$. Then the series in (8.2.2.5) converges by (8.2.2.2) so that (8.2.2.4) follows by letting $N \to \infty$ in (8.2.2.6). □

Proof of (8.2.1.7) and (8.2.1.8) The proof of (8.2.1.8) is obtained from (8.2.1.7) by integrating the latter between x and ∞; thus we only need to prove (8.2.1.7).

Let $\theta \in (0, \alpha)$ and s solve $p(s) f(\theta) = 1$ (see (8.2.1.4) for notation), which indeed has solution because $f(\theta) < f(\alpha) = p_\infty^{-1}$. Calling $p \equiv p(s)$, we have from (8.2.2.3), for $x \geq s$,

$$R_s(x, y) \leq pJ(x, y)\mathbf{1}_{y \geq s}, \qquad R(x, y) \leq pJ(x, y). \tag{8.2.2.7}$$

We then write

$$K_s(x, y) := pJ(x, y)e^{\theta(x-y)}\mathbf{1}_{y \geq s}, \qquad K(x, y) := pJ(x, y)e^{\theta(x-y)} \tag{8.2.2.8}$$

and get from (8.2.2.4)

$$\bar{m}'(x) \leq e^{-\theta x} \sum_{n=0}^{\infty} \int_s^{\infty} K_s^n(x, z) \left\{ \int_{s-1}^{s} K(z, y)e^{\theta y}\bar{m}'(y)\, dy \right\} dz.$$

By (8.2.1.9),

$$\bar{m}'(x) \leq e^{-\theta(x-s)}[2\beta m_\beta \|J\|_\infty]$$
$$\times \sum_{n=0}^{\infty} \int_s^{\infty} K_s^n(x, z) \left\{ \int_{s-1}^{s} K(z, y)\, dy \right\} dz. \tag{8.2.2.9}$$

To conclude the proof of the theorem we need to show that the series is bounded by 1.

The kernel $K(x, y)$ is a probability kernel and the n-th term of the series in (8.2.2.9) denotes the exit probability from the interval $[s, \infty)$ exactly in $n+1$ steps. Thus the series is bounded by 1 (it is actually equal to 1). □

8.3 Spectral gap

8.3.1 Spectral gap in Hilbert space

The evolution (6.2.2.7) linearized around $\bar{m}$ is described by the equation

$$\frac{\partial v}{\partial t} = Lv = -v + pJ \star v, \quad p = \beta(1 - \bar{m}^2), \tag{8.3.1.1}$$

namely (8.3.1.1) is obtained by writing $m = \bar{m} + \epsilon v$ in (6.2.2.7), expanding in ϵ and retaining only the first order, the zero-th order drops because $\bar{m}$ is a stationary solution.

The analysis depends non-trivially on the context in which the equation (8.3.1.1) is studied. In this subsection (8.3.1.1) is regarded as an evolution equation in the Hilbert space $L^2(\mathbb{R})$ while later it will be studied as an equation in the Banach space $L^\infty(\mathbb{R})$. Correspondingly L is first an operator on $L^2(\mathbb{R})$ and then on $L^\infty(\mathbb{R})$. In the former case it is actually more convenient to work in the weighted space $L^2(\mathbb{R}, p^{-1}dx)$ as L becomes then self-adjoint, indeed calling $(\cdot, \cdot)_0$ the scalar product in $L^2(\mathbb{R}, p^{-1}dx)$

$$(v, Lw)_0 = -\int v(x)w(x)p^{-1}dx + \int\int J(x, y)v(x)w(y)\,dxdy, \tag{8.3.1.2}$$

which depends symmetrically on v and w. Despite the different signs on the right hand side of (8.3.1.2), the spectrum of L lies entirely on the non-positive axis:

Theorem 8.3.1.1 *L is a bounded, self-adjoint operator, with spectrum in $\mathbb{R}_-$; 0 is a simple eigenvalue with eigenvector $\bar{m}'$, then there is a spectral gap $\omega > 0$, namely*

$$(v, Lv)_0 \le -\omega(v, v)_0, \quad \text{for all } v \text{ such that } (v, \bar{m}')_0 = 0. \tag{8.3.1.3}$$

Proof L is clearly a bounded self-adjoint operator with eigenvalue 0 and eigenvector $\bar{m}'$ (as proved by differentiating (8.1.1.5) with respect to x). To locate the rest of the spectrum, we rewrite the quadratic form as

$$(v, Lv)_0 = -\frac{1}{2}\int\int J(x, y)\bar{m}'(x)\bar{m}'(y)\left[\frac{v(x)}{\bar{m}'(x)} - \frac{v(y)}{\bar{m}'(y)}\right]^2 \le 0, \tag{8.3.1.4}$$

which proves that 0 is a simple eigenvalue and that the spectrum lies in $\mathbb{R}_-$.

To prove the gap property we write $L = L_0 + K$, where:

$$L_0 v(x) = -v(x) + \beta[1 - m_\beta{}^2](J \star v)(x),$$
$$K v(x) = \beta[m_\beta{}^2 - \bar{m}^2(x)](J \star v)(x). \tag{8.3.1.5}$$

The above decomposition of L foresees the use of Weyl's theorem. We shall indeed first determine, using Fourier analysis, the spectrum of L_0, and then prove that K is compact.

We claim that the spectrum of L_0 lies in the interval: $[-1 - \beta(1 - \bar{m}_\beta^2), -1 + \beta(1 - \bar{m}_\beta^2)]$ and it is therefore strictly contained in $(-\infty, 0)$ because $\beta(1 - \bar{m}_\beta^2) < 1$. This would be obvious if L_0 was an operator on $L^2(\mathbb{R}, dx)$ because in such a space the norm of J is bounded by 1 (which can be proved by a Fourier transform, see below). The statement, however, is true also in $L^2(\mathbb{R}, p^{-1}dx)$, as we are going to prove. The resolvent equation for J is

$$J\phi - \lambda\phi = f, \quad \lambda \in \mathbb{C}, \; f \in L^2(\mathbb{R}, p^{-1}dx). \tag{8.3.1.6}$$

Denoting by $\hat{g}(\cdot)$ the Fourier transform of g, $|\hat{J}(k)| < \hat{J}(0) = 1$ for $k \neq 0$. For λ in the resolvent set of J, we get $\hat{\phi}(k) = \frac{\hat{f}(k)}{\hat{J}(k) - \lambda}$. Thus (8.3.1.6) has a solution if λ is such that the denominator is different from zero for all k. This proves that the norm of J is bounded by 1.

The bounded operator K is compact because it maps the bounded sets of $L^2(\mathbb{R}, p^{-1}dx)$ into relatively compact sets in the same space. In fact, by the regularity of the convolution term and using that $(m_\beta{}^2 - \bar{m}^2(\cdot)) \to 0$ exponentially fast, it follows that uniformly in $\|\phi\| \le 1$:

- For any $\epsilon > 0$ there is $s_\epsilon > 0$ so that

$$\int_{|x|>s} |K\phi|^2 p^{-1} dx < \epsilon, \quad \text{if } s \ge s_\epsilon.$$

- For any $\epsilon > 0$ there is $h_\epsilon > 0$ such that

$$\int |K\phi(x+h) - K\phi(x)|^2 p^{-1} dx < \epsilon, \qquad \text{for all } h < h_\epsilon.$$

By Weyl theorem [150], the essential spectrum is invariant under compact perturbations. Then the full operator L has the same essential spectrum as L_0 so that its eigenvalue 0 cannot be a cluster point of the spectrum. □

8.3.2 *Spectral gap with sup norms*

The above proof uses extensively the Hilbert structure of the space and its extension to L^∞ is not straightforward. L^∞ on the other hand is a more natural and convenient space for the analysis of stability of the instantons, as we shall see in Sect. 8.5. We shall prove a spectral gap for e^{Lt} regarded as an operator on spaces with sup and weighted-sup norms. The analysis is constructive and applies to a more general class of systems.

For any positive function w we define the "weighted space"

$$X_w = \{u : |u|_w = \|w^{-1}u\|_\infty < \infty\}. \tag{8.3.2.1}$$

In particular we shall consider the weights $w(x) = e_\delta(x)$, $e_\delta(x) = e^{\delta|x|}$, with $|\delta| < \alpha$ so that by Theorem 8.2.1.3 for any $v \in X_{e_\delta}$, the L^2-orthogonal projection of v along $\bar{m}'$

$$N_v = \frac{(v, \bar{m}')_0}{(\bar{m}', \bar{m}')_0}, \tag{8.3.2.2}$$

is well defined.

Theorem 8.3.2.1 (Spectral gap with sup norms) *For any $|\delta| < \alpha$, there are ω and c both positive so that for any $v \in X_\delta$*

$$\left|e^{Lt}\tilde{v}\right|_{e_\delta} \le ce^{-\omega t}\left|\tilde{v}\right|_{e_\delta}, \quad \tilde{v} := v - N_v \bar{m}'. \tag{8.3.2.3}$$

The proof of the theorem is to a large extent constructive, its main point is a reformulation in terms of Markov chains, see the next subsection. The problem is well studied in the literature, see for instance Meyn and Tweedie [171]. Our proof underlines methods and techniques of statistical mechanics, in particular the Dobrushin theory of Sect. 3.2.

8.3.3 Reduction to Markov chains

The operator $M = 1 + L$, L as in (8.3.1.1), is an integral operator with kernel $M(x.y) = p(x)J(x, y) \ge 0$. M has an eigenvalue 1 with eigenvector $\bar{m}'$ which is a strictly positive function. Then

$$K(x, y) = \frac{M(x, y)\bar{m}'(y)}{\bar{m}'(x)} \tag{8.3.3.1}$$

is a transition probability kernel; hence, the idea to study M (or equivalently L) in terms of properties of K and/or of the Markov chain generated by K. The extension to more general kernels M is discussed in Sect. 8.4.7.

To estimate the sup norm of $M^n u$ in terms of K we need to do the following operations. First we define $\phi = u/\bar{m}'$; then, compute $K^n\phi$ and finally estimate the sup norm of $\bar{m}'K^n\phi$. Notice that we want a bound in terms of the sup norm of u and not of ϕ, the sup norm of u being the sup norm of $\bar{m}'\phi$. Thus when working with K the relevant norm is $|\cdot|_w$, $w^{-1} = \bar{m}'$, recall (8.3.2.1), and this is why in the following analysis of Markov chains we are going to use weighted norms. We shall relate the w-weighted sup norms to the Vaserstein distance with a metric which grows at infinity as w and estimate the Vaserstein distance using Dobrushin theory.

8.4 Dobrushin theory. II Applications to Markov chains

The Dobrushin's uniqueness theory includes both Gibbs and Markov processes, we have seen in Part I its applications to statistical mechanics here we shall discuss Markov chains. Using the Dobrushin approach we shall prove a spectral gap theorem for a class of Markov chains in $\mathbb{R}$ characterized by smooth kernels which asymptotically have a drift toward the origin. We shall then show in Sect. 8.4.7 that a class of operators (of Perron–Frobenius type) which includes $M = 1 + L$, can be reduced to the above Markov chains and from this Theorem 8.3.2.1 will then follow, see Sect. 8.4.8.

8.4.1 Spectral gap for Markov chains

We denote by $K(x, y)$ a transition probability kernel on $\mathbb{R}$, i.e. $K(\cdot, \cdot) \geq 0$ and $\int K(\cdot, y)\, dy = 1$, and by $K^n(x, y)$ the kernel of its n-th power,

$$K^n(x, y) = \int K(x, z_1) \cdots K(z_{n-1}, y) dz_1 \cdots dz_{n-1}.$$

A probability μ on $\mathbb{R}$ is invariant under K if

$$\mu(Ku) = \mu(u), \qquad \text{for all } u \in L^\infty(\mathbb{R}), \tag{8.4.1.1}$$

$$Ku = \int K(\cdot, y)u(y)\, dy, \qquad \mu(u) = \int u(x)\, \mu(dx). \tag{8.4.1.2}$$

The main assumptions on K concern its behavior at infinity where we demand a drift toward the origin (such a condition is implied as we shall see by Assumption K1 below) and smoothness around the origin (Assumption K2 below). The two conditions at infinity and around the origin must match and the Assumptions K1–K2 make this quantitative (the assumptions could be greatly relaxed and are stated in view of the applications to the instantons theory). There are a few parameters whose collection will be denoted by π which enter in the assumptions, the behavior of the chain will be uniform independently of the particular kernel $K(x, y)$ for same values of the parameters π.

Assumptions on K

- K1. $K(x, y)$ is supported by $|x - y| \leq 1$. Moreover there is a measurable, positive, divergent at ∞ function $w(x)$, $x \in \mathbb{R}$, and coefficients $r_0 < 1$, r_1, s_0, w_-, w_+, α_-

and α_+ all positive, so that

$$w_- e^{\alpha_- |x|} \le w(x) \le w_+ e^{\alpha_+ |x|}, \tag{8.4.1.3}$$

$$K w(x) \le r_0 w(x) + a, \quad a := r_1 w_+ e^{\alpha_+ s_0}. \tag{8.4.1.4}$$

Let then $s_1 > 0$ be such that

$$w_- e^{\alpha_- s_1} \frac{1-r_0}{4a} \ge 1, \qquad \frac{1-r_0}{2as_1} \le 1. \tag{8.4.1.5}$$

- K2. There are $b_1 \in (0, 1/8)$ and a positive integer k_1 so that $K^{k_1}(x,y) \ge 4b_1$ for any $|x| \le s_1$ and $|y| \le 1/2$.

We shorthand $\pi = (r_0, r_1, s_0, s_1, \alpha_\pm, w_\pm)$ and say that (K, w) satisfies K1 and K2 with π if the above holds. By K2 we can construct a coupling $Q^{(k_1)}_{x_0,x_1}$ of $K^{k_1}(x_0, y)\,dy$ and $K^{k_1}(x_1, y)\,dy$ so that

$$Q^{(k_1)}_{x_0,x_1}(\{x = x'\}) \ge 4b_1, \quad \text{for any } |x_0| \le s_1,\ |x_1| \le s_1. \tag{8.4.1.6}$$

Indeed, $Q^{(k_1)}_{x_0,x_1}(dydy') = 4b_1 \delta(y-y') \mathbf{1}_{|y|\le 1/2} + P(y,y')dydy'$, where

$$P(y,y') = \frac{1}{1-4b_1}\big(K^{k_1}(x_0,y) - 4b_1 \mathbf{1}_{|y|\le 1/2}\big)\big(K^{k_1}(x_1,y') - 4b_1 \mathbf{1}_{|y'|\le 1/2}\big).$$

Notice that $Q^{(k_1)}_{x_0,x_1}$ depends in a measurable way on x_0 and x_1.

In the next theorem we shall see that the spectral gap is determined by the quantity

$$\theta := \frac{4-2b_1}{4-b_1}, \qquad \text{and} \quad \text{since } b_1 \in \left(0, \frac{1}{8}\right); \text{ then } \quad \theta \in \left(\frac{30}{31}, 1\right). \tag{8.4.1.7}$$

Theorem 8.4.1.1 *If (K, w) satisfies K1 and K2 there is a unique K-invariant measure μ such that $X_w \subset L^1(\mathbb{R}, d\mu)$ (X_w as in (8.3.2.1)) and for any $u \in X_w$*

$$\left|K^n(u - \mu(u))\right|_w \le c e^{-\lambda n} \left|u - \mu(u)\right|_w, \quad e^{-\lambda} = \theta^{1/(k_1+\kappa)}, \tag{8.4.1.8}$$

with θ as in (8.4.1.7), and where c and κ are positive coefficients which depend only on the parameter π associated to (K, w).

The proof is given in Sects. 8.4.3–8.4.6. In (8.4.7.5) we shall identify the spectral gap ω of Theorem 8.3.2.1 as

$$\omega = 1 - e^{-\lambda}. \tag{8.4.1.9}$$

8.4.2 Invariant measure

We shall prove here the existence of an invariant measure for the Markov chain with a transition probability kernel K which satisfies the Assumption K1 of Sect. 8.4.1. The proof is a corollary of the following inequality which will be repeatedly used in the sequel. Recalling that $r_0 < 1$,

$$\int K^n(x,y)w(y)\,dy \le r_0^n w(x) + \frac{a}{1-r_0}. \tag{8.4.2.1}$$

Equation (8.4.2.1) is simply obtained by iterating (8.4.1.4) n times.

By (8.4.2.1) for any $N > 1$ the probability

$$\mu_N(dx) := \frac{1}{N}\sum_{n=0}^{N-1} K^n(0,x)dx, \tag{8.4.2.2}$$

satisfies the bound

$$\int w(x)\mu_N(dx) \le \frac{w(0)/N + a}{1-r_0}.$$

By (8.4.1.3), $\lim_{|x|\to\infty} w(x) = \infty$; then, by the Prokhorov theorem, see for instance Theorem 6.7 in Chap. II of [179], $\{\mu_N\}$ is tight and converges weakly by subsequences to a K-invariant probability μ for which

$$\mu(w) \le \frac{a}{1-r_0}. \tag{8.4.2.3}$$

Thus $w \in L^1(\mathbb{R}, \mu(dx))$ and $X_w \subset L^1(\mathbb{R}, \mu(dx))$.

8.4.3 Couplings and Vaserstein distance

We reformulate here (8.4.1.8) in terms of the Vaserstein distance of powers of the transition probabilities. We rewrite the term $K^n[u-\mu(u)]$ in (8.4.1.8) first as $K^n u - \mu(K^n u)$ and then as

$$K^n[u-\mu(u)](x) = \int_{\mathbb{R}}\int_{\mathbb{R}^2} \big(u(x') - u(y')\big) Q^{(n)}_{x,y}(dx'dy')\mu(dy), \tag{8.4.3.1}$$

where $Q^{(n)}_{x,y}(dx'dy')$ is any joint representation of $K^n(x,dx')$ and $K^n(y,dy')$ which depends on x, y in a measurable way.

We define a distance on $\mathbb{R}$ by setting

$$D(x,y) = \mathbf{1}_{x\ne y}\big\{w(x) + w(y)\big\}. \tag{8.4.3.2}$$

We have $u(x) - u(y) = \frac{\tilde{u}(x)}{w(x)} w(x) - \frac{\tilde{u}(y)}{w(y)} w(y)$, $\tilde{u} = u - \mu(u)$, so that

$$\big|u(x) - u(y)\big| \le D(x, y)|u - \mu(u)|_w. \tag{8.4.3.3}$$

We shall prove in the next subsections that there are $C, \lambda > 0$ and for any n a joint representation $Q^{(n)}_{x,y}(dx'dy')$ which depends on x, y in a measurable way, so that

$$\int_{\mathbb{R}^2} D(x', y')\, Q^{(n)}_{x,y}(dx'dy') \le Ce^{-\lambda n} D(x, y). \tag{8.4.3.4}$$

We then get from (8.4.3.1)

$$\big|K^n[u - \mu(u)](x)\big| \le Ce^{-\lambda n}|u - \mu(u)|_w \int_{\mathbb{R}} D(x, y)\, \mu(dy). \tag{8.4.3.5}$$

By (8.4.2.3) $w \in L^1(\mathbb{R}; \mu(dx))$, then, recalling that $\inf_x w(x) \ge w_- > 0$,

$$\int_{\mathbb{R}} D(x, y)\, \mu(dy) \le w(x) + \mu(w) \le C' w(x), \quad C' = 1 + \frac{\mu(w)}{w_-},$$

and

$$\big|K^n[u - \mu(u)](x)\big| \le CC'e^{-\lambda n}|u - \mu(u)|_w w(x), \tag{8.4.3.6}$$

which proves (8.4.1.8) with $c = CC'$. The proof of (8.4.3.4) is quite involved. We shall prove that there is $s \ge s_1$ (s_1 as in (8.4.1.5) and s only determined by the parameters π defined by the Assumptions K1–K2), so that

$$\int_{\mathbb{R}^2} Q^{(n)}_{x,y}(dx'dy') D_s(x', y') \le C_s e^{-\lambda n} D_s(x, y), \tag{8.4.3.7}$$

$$D_s(x, y) = \mathbf{1}_{x \ne y}\big\{w_s(x) + w_s(y)\big\}, \tag{8.4.3.8}$$

$$w_s(x) = \begin{cases} w(x)/w(s) & \text{if } |x| \ge s, \\ 1 & \text{otherwise.} \end{cases} \tag{8.4.3.9}$$

By (8.4.1.3), $w_- \le w(x) \le w_+ e^{\alpha + s}$, $|x| \le s$, (8.4.3.4) then follows with

$$C = \frac{w_+}{w_-} e^{\alpha+s}\, C_s, \tag{8.4.3.10}$$

so that, in conclusion,

$$\text{(8.4.1.8) follows from (8.4.3.7) if } C_s \le c',\ c' = c'(\pi), \tag{8.4.3.11}$$

namely if C_s is bounded uniformly for all chains with same parameters π.

8.4.4 Choice of parameters

We start by defining two functions R_s and n_s of s for $s \ge s_1$. Recall that b_1, s_1, k_1, $w_\pm$, $\alpha_\pm$, a and r_0 are all parameters which enter in the Assumptions K1–K2 of Sect. 8.4.1.

- R_s is such that $R_s > s$ and $\frac{w_- e^{\alpha_- R_s}}{w_+ e^{\alpha_+ s}} \ge \frac{sa}{1-r_0}$; the latter implies

$$w_s(x) \ge \frac{sa}{1-r_0}, \qquad \text{for all } |x| \ge R_s. \tag{8.4.4.1}$$

- $n_s = k_1 + k$, where $k = k(s)$ is a positive integer larger than s and such that

$$r_0^k \sup_{|x|\le R_s} w(x) \le \frac{1}{2s}. \tag{8.4.4.2}$$

In the next lemma we shall denote by $P^{(n_s)}_{x_0,x_1}(dydy')$, $s > s_1$, the following coupling of $K^{n_s}(x_0, y)dy$ and $K^{n_s}(x_1, y')dy'$, with $n_s = k + k_1$ and $Q^{(k_1)}_{y_0,y_1}(dydy')$ the coupling in (8.4.1.6) (which depends on y_0, y_1 in a measurable way):

$$\begin{aligned}\int_{(y_0,y_1)\in\mathbb{R}^2} & K^k(x_0, y_0)K^k(x_1, y_1)\big\{\mathbf{1}_{|y_0|\le s_1,|y_1|\le s_1}\, Q^{(k_1)}_{y_0,y_1}(dydy') \\ &+ [1-\mathbf{1}_{|y_0|\le s_1,|y_1|\le s_1}]\, K^{k_1}(y_0, y)K^{k_1}(y_1, y')dydy'\big\}dy_0dy_1.\end{aligned} \tag{8.4.4.3}$$

Evidently $P^{(n_s)}_{x_0,x_1}(dydy')$ depends in a measurable way on x_0 and x_1 and moreover:

Lemma *For all $s > s_1$ the following three inequalities hold:*

$$\inf_{|x_0|\le R_s,|x_1|\le R_s} P^{(n_s)}_{x_0,x_1}(y = y') \ge b_1, \tag{8.4.4.4}$$

$$\int_{|y|\ge s} dy\, K^{n_s}(x, y)w_s(y) \le \left(\frac{1}{2s} + \frac{a}{1-r_0}\right)\frac{1}{w(s)}, \quad \textit{for all } |x| \le R_s, \tag{8.4.4.5}$$

$$\int dy\, K^{n_s}(x, y)w_s(y) \le \left(\frac{1}{2s} + \frac{1}{sw(s)} + \frac{1-r_0}{as}\right) w_s(x), \quad \textit{for all } |x| \ge R_s. \tag{8.4.4.6}$$

Proof *Proof of* (8.4.4.4). Let $|x_0| \le R_s$ and $|x_1| \le R_s$; then

$$P^{(n_s)}_{x_0,x_1}(y = y') \ge \left\{\inf_{|x|\le R_s}\int_{|y|\le s_1} K^k(x, y)\right\}^2 \inf_{|y_0|\le s_1,|y_1|\le s_1} Q^{(k_1)}_{y_0,y_1}(y = y').$$

By (8.4.1.6) the last inf is bounded from below by $4b_1$, we shall prove below that the curly bracket is bounded from below by $1/2$, hence (8.4.4.4). Indeed, by (8.4.1.4) and (8.4.1.5), $\int_{|y|>s_1} dy\, K^k(x,y)$ is bounded by

$$\frac{1-r_0}{4a}\int_{|y|>s_1} dy\, K^k(x,y)w(y) \le \frac{1-r_0}{4a}\left(r_0^k w(x) + \frac{a}{1-r_0}\right),$$

so that, by (8.4.4.2) and (8.4.1.5), $\int_{|y|>s_1} dy\, K^k(x,y) \le \frac{1-r_0}{8as} + \frac{1}{4} \le \frac{1}{2}$ for $|x| \le R_s$, thus concluding the proof of (8.4.4.4).

Proof of (8.4.4.5). Recalling (8.4.1.4) we have

$$\int_{|y|\ge s} dy\, K^{n_s}(x,y)w_s(y) \le \frac{1}{w(s)}\left(r_0^{n_s} w(x) + \frac{a}{1-r_0}\right). \tag{8.4.4.7}$$

Since $|x| \le R_s$, $w(x) \le \sup_{|y|\le R_s} w(y)$, then, by (8.4.4.2), the r.h.s. is bounded by

$$\frac{1}{w(s)}\left(\frac{1}{2s} + \frac{a}{1-r_0}\right).$$

Proof of (8.4.4.6). For $|x| > R_s$ the l.h.s. of (8.4.4.6) is bounded by

$$\frac{1}{w(s)}\left(r_0^{n_s} w(x) + \frac{a}{(1-r_0)}\right) + \int_{|y|\le s} dy\, K^{n_s}(x,y),$$

which is bounded by $(r_0^{n_s} + \frac{a}{(1-r_0)w(x)} + \frac{1}{w_s(x)})w_s(x)$, hence (8.4.4.6) because $r_0^{n_s} \le \frac{1}{2s}$ and by (8.4.4.1), $w(x) \ge \frac{sa}{1-r_0}\, w(s)$. □

The parameter *s*

Let b_1 be as in (8.4.1.6) and (8.4.4.4), call

$$\epsilon = \frac{b_1}{8}, \tag{8.4.4.8}$$

and choose $s \ge s_1$ so large that the following three inequalities are satisfied:

$$\frac{w_- e^{\alpha_- R_s}}{w_+ e^{\alpha_+ s}} \ge 8, \qquad \inf_{|x|\ge R_s} w_s(x) \ge 8, \tag{8.4.4.9}$$

$$\int w_s(y)K^{n_s}(x,y)\,dy \le \epsilon w_s(x), \qquad \text{for all } |x| \ge R_s, \tag{8.4.4.10}$$

$$\sup_{|x|\le R_s} \int_{|y|\ge s} w_s(y)K^{n_s}(x,y)\,dy \le \epsilon. \tag{8.4.4.11}$$

The first inequality in (8.4.4.9) is satisfied for s large because its l.h.s. is $\ge \frac{sa}{1-r_0}$; the second one follows using (8.4.4.1). The conditions (8.4.4.10) and (8.4.4.11) can be enforced by choosing s large enough because of (8.4.4.6) and respectively (8.4.4.5) and therefore the definition of s is well posed.

8.4.5 Construction of a joint representation

In this subsection we will construct a joint representation $Q^{(n)}_{x,x'}(dy\,dy')$ of $K^n(x,y)dy$ and $K^n(x',y)dy$ and then show that it satisfies the bound (8.4.3.4). We start from the special value $n = n_s$, the same s chosen at the end of Sect. 8.4.4. Of course we set for $x = x'$

$$Q^{(n_s)}_{x,x}(dy\,dy') = K^{n_s}(x,y)dy\delta(y-y')dy'. \tag{8.4.5.1}$$

If $x \neq x'$ and either $|x| \geq R_s$ or $|x'| \geq R_s$ we just take the product:

$$Q^{(n_s)}_{x,x'}(dy\,dy') := \{K^{n_s}(x,y)dy\}\,\{K^{n_s}(x',y')dy'\}. \tag{8.4.5.2}$$

If instead both $|x| \leq R_s$ and $|x'| \leq R_s$, we set

$$Q^{(n_s)}_{x,x'}(dy\,dy') = P^{(n_s)}_{x,x'}(dy\,dy'), \tag{8.4.5.3}$$

with $P^{(n_s)}_{x,x'}(dy\,dy')$ defined in (8.4.4.3). Notice that $Q^{(n_s)}_{x,x'}(dy\,dy')$ depends in a measurable way on x, x'. To complete the definition of the joint representation of $K^n(x,y)dy$ and $K^n(x',y)dy$ for general values of n, we write $n = mn_s + q$, with m and q non-negative integers, $q < n_s$. We then set

$$Q^{(n)}_{x,x'}(dy_m\,dy'_m) = \int dy_0 dy'_0 K^q(x,y_0)K^q(x',y'_0) \times \int_{y_1,\dots,y_{m-1}} \prod_{i=1}^{m} Q^{(n_s)}_{y_{i-1}\,y'_{i-1}}(dy_i\,dy'_i). \tag{8.4.5.4}$$

To prove that this is a joint representation, we integrate a bounded measurable function $f(y_m)$; $Q^{(n_s)}(dy_m\,dy'_m|y_{m-1}\,y'_{m-1})$ is a joint representation of $K^{(n_s)}(y_{m-1},y_m)\,dy_m$ and of $K^{(n_s)}(y'_{m-1},y'_m)\,dy'_m$; then

$$\int dy_0 dy'_0 K^q(x,y_0)K^q(x',y'_0)\int_{y_1,\dots,y_{m-1}} \prod_{i=1}^{m-1} Q^{(n_s)}(dy_i\,dy'_i|y_{i-1}\,dy'_{i-1})g(y_{m-1}),$$

where $g(x) = \int dy\,K^{n_s}(x,y)\,f(y)$. Iterating we get

$$\int dy dy'\,Q^{(n)}(dy\,dy'|x,x')f(y) = \int dy\,K^n(x,y)f(y).$$

The analogous result is obtained when integrating $f(y')$; hence (8.4.5.4) is a joint representation.

8.4.6 Bounds on the Vaserstein distance

We now have the required joint representation, and we need to show that it satisfies the bound in (8.4.3.4). We prove it iteratively, and to this end we begin with the special case $n = n_s$.

Lemma *For any x and x'*

$$\int D_s(y, y')\, Q^{(n_s)}_{x\,x'}(dy\, dy') \le \theta D_s(x, x'), \tag{8.4.6.1}$$

where θ is defined in (8.4.1.7).

Proof By (8.4.5.1), (8.4.6.1) obviously holds for $x = x'$. Let us take $x \neq x'$ and observe preliminarily that by (8.4.1.7) $\theta > 1/2$ and $\epsilon := b_1/8 < 1/8$. Let $x \neq x'$. Suppose first that $|x| \ge R_s$, $|x'| \ge R_s$. Then by (8.4.5.2) and (8.4.4.10)

$$\int D_s(y, y')\, Q^{(n_s)}_{x\,x'}(dy\, dy') \le \epsilon D_s(x, x') \le \theta D_s(x, x').$$

Let $|x| \ge R_s$, $|x'| < R_s$; then, $Q^{(n_s)}_{x\,x'}(dy\, dy')$ is given by (8.4.5.2). Recalling (8.4.3.9), using (8.4.4.10) and (8.4.4.11) we get

$$\int D_s(y, y')\, Q^{(n_s)}_{x\,x'}(dy\, dy'|x\, x') \le \epsilon w_s(x) + 1 + \epsilon.$$

The last expression is bounded by

$$\theta w_s(x) + \theta w_s(x') + \big(-[\theta - \epsilon] w_s(x) + 1 + \epsilon\big) \le \theta D_s(x, x'), \tag{8.4.6.2}$$

because the bracket term is negative: in fact, $\theta - \epsilon > \frac{1}{2} - \frac{1}{8} > 1/4$ and $w_s(x) \ge 8$ by (8.4.4.9); hence

$$(\theta - \epsilon) w_s(x) \ge \frac{w_s(x)}{4} \ge 2 > 1 + \epsilon.$$

Finally, let $|x| < R_s$, $|x'| < R_s$, $x \neq x'$; then, $Q^{(n_s)}_{x\,x'}(dy\, dy')$ is given by (8.4.5.3). We write

$$\int D_s(y, y') Q^{(n_s)}_{x\,x'}(dy\, dy') \le \int_{|y| \ge s} w_s(y) \{K^{(n_s)}(x, y) + K^{(n_s)}(x', y)\} dy$$
$$+ 2 \int_{y \neq y', |y| < s, |y'| < s} Q^{(n_s)}_{x\,x'}(dy\, dy'),$$

because $D_s(y, y') = 2$ when both $|y| \le s$ and $|y'| \le s$. Using (8.4.4.11) and (8.4.4.4) and bounding $\epsilon = b_1/8 < b_1/2$, we then see (using that θ, which is defined in

(8.4.1.7), obeys $\theta > 1 - b_1/2$),

$$\int D_s(y, y') Q^{(n_s)}_{x\,x'}(dy\,dy')$$
$$\le 2[1 + \epsilon - b_1] < 2\left[1 - \frac{b_1}{2}\right] \le 2\theta = \theta D_s(x, x'). \qquad \square$$

Let us write, as before, $n = mn_s + q$ and use (8.4.5.4) for $Q^{(n)}_{x\,x'}(dy\,dy')$. By (8.4.6.1) we have

$$\int D_s(y, y') Q^{(n)}_{x\,x'}(dy\,dy')$$
$$\le \theta^m \int \{w_s(y) + w_s(y')\} K^{(q)}(x, y) K^{(q)}(x, y') dy\,dy'. \qquad (8.4.6.3)$$

By (8.4.2.1) the r.h.s. is bounded by

$$\theta^m \left\{ r_0^q [w_s(x) + w_s(x')] + \frac{2a}{(1 - r_0) w(s)} + 2 \right\},$$

with the last term due to the region $\{|y| \le s\} \cup \{|y'| \le s\}$. Thus the l.h.s. of (8.4.6.3) is bounded by

$$\left\{ \left(r_0^q + \frac{2a}{(1 - r_0)(w_s(x) + w_s(x'))} + \frac{2}{w_s(x) + w_s(x')} \right) \theta^{-q} \right\}$$
$$\times \theta^{n/n_s} D_s(x, x'), \qquad (8.4.6.4)$$

which using (8.4.4.1) proves (8.4.3.7) with $C_s \le (1 + \frac{1}{s} + \frac{1 - r_0}{a})(\frac{1}{2})^{-n_s}$, $e^{-\lambda} = \theta^{1/(k_1 + \kappa)}$ and $\kappa = n_s - k_1$, in agreement with (8.4.1.8).

By (8.4.3.11) the proof of Theorem 8.4.1.1 is complete.

8.4.7 Perron–Frobenius kernels and Markov chains

Theorem 8.4.1.1 applies to transition probability kernels and to Markov chains. In many cases, however, as in Theorem 8.3.2.1, we are not given a probability kernel, and thus we have to do some preliminary work to reduce to such a case as where we can apply Theorem 8.4.1.1. We call $M(x, y)$ a Perron–Frobenius kernel if $M(x, y) \in L^\infty(\mathbb{R} \times \mathbb{R}, \mathbb{R}_+)$ and if there are a bounded, positive function $\psi(x)$ and a positive number λ_0, so that

$$\int M(x, y) \psi(y)\, dy = \lambda_0 \psi(x). \qquad (8.4.7.1)$$

We shall later give examples of Perron–Frobenius kernels; here we proceed by assuming that M is a Perron–Frobenius kernel with λ_0 and ψ as above. Then

$$K(x,y) = \frac{M(x,y)\psi(y)}{\lambda_0\psi(x)}, \tag{8.4.7.2}$$

is a transition probability kernel. Set

$$e_\delta(x) = e^{\delta|x|}, \qquad w(x) = \big(\psi(x)e^{-\delta|x|}\big)^{-1}, \qquad w = \psi^{-1}e_\delta, \tag{8.4.7.3}$$

and call

$$L = -1 + M. \tag{8.4.7.4}$$

Theorem 8.4.7.1 *Let K as in* (8.4.7.2) *and w as in* (8.4.7.3); *suppose that the pair (K,w) satisfies the Assumptions K1 and K2 stated in Sect.* 8.4.1. *Let c, λ and μ as be in Theorem* 8.4.1.1, *λ_0 and ψ as in* (8.4.7.2), *L as in* (8.4.7.4), *and let us write $\tilde v := v - \psi\mu(v/\psi)$. Then we have*

$$\big|e^{Lt}\tilde v\big|_{e_\delta} \le ce^{-\gamma t}\big|\tilde v\big|_{e_\delta}, \quad \gamma = 1 - \lambda_0 e^{-\lambda}, \tag{8.4.7.5}$$

with $|\cdot|_{e_\delta}$ defined in (8.3.2.1). *Observe that the r.h.s. vanishes exponentially as $t\to\infty$ if $\gamma > 0$.*

Proof Let U: $Uv = \frac{v}{\psi}$; then

$$U(L+1)v = \lambda_0 KUv, \tag{8.4.7.6}$$

and we have

$$U\tilde v = Uv - \mu(Uv). \tag{8.4.7.7}$$

Write

$$e^{Lt}\tilde v = e^{-t}e^{(L+1)t}\tilde v = \sum_{n=0}^\infty e^{-t}\frac{t^n}{n!}(L+1)^n\tilde v. \tag{8.4.7.8}$$

By (8.4.7.6), $(L+1) = \lambda_0 U^{-1}KU$; then, after telescopic cancelations,

$$e^{Lt}\tilde v = \sum_{n=0}^\infty e^{-t}\frac{(\lambda_0 t)^n}{n!}U^{-1}K^nU\tilde v \tag{8.4.7.9}$$

and

$$\big|e^{Lt}\tilde v\big|_{e_\delta} \le \sum_{n=0}^\infty e^{-t}\frac{(\lambda_0 t)^n}{n!}\big|U^{-1}K^nU\tilde v\big|_{e_\delta}. \tag{8.4.7.10}$$

Since U^{-1} is multiplication by ψ, $|u|_w = \|w^{-1}u\|_\infty$ and $w^{-1} = \psi e_{-\delta}$,

$$\big|U^{-1}K^nU\tilde v\big|_{e_\delta} \le \big|K^nU\tilde v\big|_w. \tag{8.4.7.11}$$

Call $u := Uv$; then, by (8.4.7.7), $U\tilde{v} = u - \mu(u)$. We then apply (8.4.1.8) and get

$$\left|K^n U\tilde{v}\right|_w \le ce^{-\lambda n}\left|U\tilde{v}\right|_w = ce^{-\lambda n}\left|\tilde{v}\right|_{e_\delta}. \tag{8.4.7.12}$$

Going back to (8.4.7.10), we get

$$\left|e^{Lt}\tilde{v}\right|_{e_\delta} \le c\sum_{n=0}^{\infty} e^{-t}\frac{(\lambda_0 t)^n}{n!}e^{-\lambda n}\left|\tilde{v}\right|_{e_\delta}.$$

□

8.4.8 Proof of Theorem 8.3.2.1

In this subsection we shall prove Theorem 8.3.2.1, which follows from Theorem 8.4.7.1 applied to

$$M(x,x') = p(x)J(x,x'), \quad p(x) = \beta\big(1 - \bar{m}(x)^2\big), \tag{8.4.8.1}$$

once we prove that such a M is a Perron–Frobenius kernel which satisfies K1–K2. This will be done in Lemma 8.4.8.1 below for a slightly larger class of kernels: those which satisfy the following.

Conditions verified by M

- M1. There are $\psi(x)$ strictly positive and bounded and $\lambda_0 > 0$, so that

$$M\psi = \lambda_0\psi. \tag{8.4.8.2}$$

  In the specific case of (8.4.8.1), $\psi = \bar{m}'$ and $\lambda_0 = 1$.
- M2. $p(x)$ is a continuous function; there are $s_0 > 0$, p_- and p_+, so that

$$0 < p_- \le p(x) \le p_+ < 1, \quad \text{for all } |x| \ge s_0, \tag{8.4.8.3}$$

  and $p_+ < \lambda_0$.
  (In our case this is a consequence of $\lim_{|x|\to\infty} p(x) = \beta(1 - m_\beta^2) < 1$.)
- M3. Given any s_1, there are $\nu > 0$ and n_0, so that for any x and y in $[-s_1, s_1]$,

$$\inf_{|x'-y|\le 1} M^{n_0}(x,x') \ge e^{-\nu n_0} \tag{8.4.8.4}$$

  which easily follows from (8.4.8.1).

In our case (8.4.7.3) becomes

$$w(x) = \big(\bar{m}'(x)e^{-\delta|x|}\big)^{-1}, \qquad e_\delta(x) = e^{\delta|x|}, \tag{8.4.8.5}$$

and (8.4.7.2)

$$K(x,y) := \frac{p(x)}{\bar{m}'(x)} J(x,y)\bar{m}'(y). \tag{8.4.8.6}$$

Notice that the invariant measure μ, whose existence is proved in general in Theorem 8.4.1.1, obeys the explicit expression

$$\mu(dx) = Cp(x)^{-1}\bar{m}'(x)^2 dx \tag{8.4.8.7}$$

(in general $\mu(dx) = C\psi(x)\phi(x)\,dx$ if ψ and ϕ are right and left positive eigenvectors of M with positive eigenvalue λ_0; in our case $\phi = p^{-1}\bar{m}'$).

Let s_0 and $p_\pm$ as in M2 above, define $\alpha_+^0 > \alpha_-^0 > 0$ so that

$$\frac{p_\pm}{\lambda_0}\int J(0,x)e^{\alpha_\mp^0 x} = 1, \tag{8.4.8.8}$$

then

Lemma 8.4.8.1 *The pair (K,w), K as in* (8.4.8.6) *and w as in* (8.4.8.5) *with $|\delta| < \alpha_-^0$, the latter as in* (8.4.8.8), *satisfies the Assumptions K1–K2 of Sect.* 8.4.1, *with the same s_0, $\alpha_\pm = \alpha_\pm^0 - \delta$ and with $w_\pm$ determined by s_0, $p_\pm$, λ_0, δ, n_0, v and $\|M\|_\infty$.*

Proof We start from K1. Since $\bar{m}'$ is continuous it has a minimum and a maximum in $[-s_0, s_0]$, say at x' and x'' respectively. To have a bound for their ratio which depends only on the parameters in M1–M3, we write $\psi(x') = \lambda_0^{-n_0}\int M^{n_0}(x',x)\psi(x)dx$. By (8.4.8.4),

$$\psi(x') \ge \lambda_0^{-n_0} e^{-vn_0}\int_{|y-x''|\le 1}\psi(y) \ge e^{-vn_0}\lambda_0^{-n_0}\int \frac{M(x'',y)}{\|M\|_\infty}\psi(y)$$
$$\ge e^{-vn_0}\frac{\lambda_0^{1-n_0}}{\|M\|_\infty}\psi(x''),$$

so that

$$e^{-vn_0}\frac{\lambda_0^{1-n_0}}{\|M\|_\infty} \le \frac{\psi(x')}{\psi(x'')} \le 1. \tag{8.4.8.9}$$

Let $x > s_0$. Then (the following bound, stronger than needed here, will be used in the proof of Lemma 8.4.9.1)

$$\psi(x) \ge \sum_k \int_{s_0 \le x_i \le x, i \le k-1; x_k < s_0} \left(\frac{p_-}{\lambda_0}\right)^k J(x,x_1)\cdots J(x_{k-1},x_k)\psi(x').$$

Calling $\pi(x,y) = (p_-/\lambda_0)e^{-\alpha_+^0(y-x)}$ and c_+ the probability that the chain $\{x_n\}_{n\ge 0}$ with transition probability π starting from 0 is never positive and goes to $-\infty$

($c_+ > 0$ because the chain has a negative drift) we have

$$\psi(x) \geq c_+ e^{-\alpha_+^0(x-s_0+1)}\psi(x'). \tag{8.4.8.10}$$

By the same argument, $\psi(x) \geq c_+ e^{-\alpha_+^0(|x|-s_0+1)}\psi(x')$ when $x < -s_0$.

For the upper bound we have, supposing $x > s_0$,

$$\psi(x) \leq \sum_k \int_{s_0 \leq x_i, i \leq k-1; x_k < s_0} \left(\frac{p_+}{\lambda_0}\right)^k J(x, x_1)\cdots J(x_{k-1}, x_k)\psi(x''). \tag{8.4.8.11}$$

Call now $\pi(x, y) = (p_+/\lambda_0)e^{-\alpha_-^0(y-x)}$; then

$$\psi(x) \leq e^{-\alpha_-^0(|x|-s_0)}\psi(x''), \tag{8.4.8.12}$$

because the chain diverges to $-\infty$ with probability 1, as it has a negative drift.

By (8.4.8.10), $w := \psi^{-1}e^{\delta|x|}$ is bounded for $|x| > s_0$ by

$$w(x) \leq [c_+\psi(x')e^{\alpha_+^0(s_0-1)}]^{-1}e^{(\alpha_+^0+\delta)|x|},$$

while, for $|x| \leq s_0$, $w(x) \leq \psi(x')^{-1}e^{(\alpha_+^0+\delta)|x|}$. Hence

$$w_+ = \psi(x')^{-1}\max\{[c_+e^{\alpha_+^0(s_0-1)}]^{-1}, 1\}.$$

Analogously, by (8.4.8.12) $w(x) \geq [c_-\psi(x'')e^{\alpha_-^0 s_0)}]^{-1}e^{(\alpha_-^0+\delta)|x|}$ for $|x| > s_0$, while, for $|x| \leq s_0$, we have $w(x) \geq e^{-\alpha_-^0 s_0}\psi(x'')^{-1}e^{(\alpha_-^0+\delta)|x|}$. Hence,

$$w_- = [\psi(x'')e^{\alpha_-^0 s_0}]^{-1}\min\{[c_-e^{\alpha_-^0}]^{-1}, 1\}.$$

To prove (8.4.1.4) we write for $|x| \geq s_0$

$$Kw(x) = \int \frac{M(x, y)}{\lambda_0\psi(x)}e^{\delta y} \leq \left\{\frac{p_+}{\lambda_0}\int J(0, y)e^{\delta y}\right\}w(x) =: r_0 w(x), \tag{8.4.8.13}$$

and since $|\delta| < \alpha_-^0$, $r_0 < 1$. For $|x| < s_0$,

$$Kw(x) \leq \frac{e^{|\delta|}\|M\|_\infty}{\lambda_0}w(x), \tag{8.4.8.14}$$

so that $r_1 = \|M\|_\infty \frac{e^{|\delta|}}{\lambda_0}$. The remaining properties follow from M3, i.e. from the regularity of the kernel. Details are omitted. □

8.4.9 Finite volume instantons

Another class of kernels in which the above theory has been applied are $d = 1$ kernels,

$$M(x, x') = p^*(x)J^{\text{neum}}(x, x'), \tag{8.4.9.1}$$

where x and x' vary in a "large" interval $[-\ell, \ell]$, $\ell \gg 1$, with reflections at the boundaries:

$$J^{\text{neum}}(x,x') = \{J(x,x') + J(x,\ell + (\ell - x'))\} \\ + \{J(x,x') + J(x,-\ell - (x' + \ell))\}.$$

We have $p^*(x) = \beta \cosh^{-2}(\beta J^{\text{neum}} * m)$ with m "close" to an instanton. The kernel (8.4.9.1) arises in the study of the finite volume analogue of the instanton with $J^{\text{neum}}(x,x')$, which simulates Neumann boundary conditions [100]. The aim of the analysis is to determine the stability properties of a one dimensional manifold which connects the finite volume analogue of the instanton to the equilibrium states $m \equiv m_\beta$ and $m \equiv -m_\beta$.

In [99] it is proved that if ℓ is large enough and m is sufficiently close in sup norm to $\bar{m}$, then the conditions M1–M3 stated in Sect. 8.4.8 are verified. In particular, it is proved that M1 holds and λ_0 can be made as close to 1 as desired by a suitable choice of the parameters. Let $\alpha_\pm^0 > 0$ as in (8.4.8.8), so that $\alpha_+^0 > \alpha_-^0 > \alpha^* := -\log(p_+/\lambda_0)$.

Lemma 8.4.9.1 *Let* $K(x,y) := \frac{M(x,x')\psi_0(x')}{\lambda_0\psi(x)}$, w *as in* (8.4.7.3) *with* $|\delta| < \alpha^*$; *then* (K,w) *satisfies the Assumptions K1–K2 of Sect.* 8.4.1, *with the same* s_0, $\alpha_\pm = \alpha_\pm^0 - \delta$ *and with* $w_\pm$ *determined by* s_0, $p_\pm$, λ_0, δ, n_0, ν *and* $\|M\|_\infty$.

Proof Since M is regular, (8.4.8.2) shows that ψ is continuous; it has therefore a minimum and a maximum in $[-s_0, s_0]$, say at x' and x'', respectively, and the same argument as used in the proof of Lemma 8.4.8.1 shows that (8.4.8.9) holds.

Let x be such that $|x| > s_0$. The same argument as in the proof of Lemma 8.4.8.1 shows that (8.4.8.10) remains valid and, for the upper bound, (8.4.8.11) also holds. The analysis of (8.4.8.11) requires instead some changes due to the fact that $J(x,y) \neq J^{\text{neum}}(x,y)$, when $|x| \geq \ell - 1$. We observe that for any k in (8.4.8.11) and any $x_1, \ldots, x_k$, there is $i < k$ such that $x_{i+1}, \ldots, x_k$ are all $< |x|$ and $|x_i| \leq |x| + 1$. Then $\psi(x)$ is bounded by $c_- := \frac{1}{1 - p_+/\lambda_0}$ times the sup over $|x_0| \in [|x|, |x|+1]$ of

$$\sum_k \int_{s_0 \leq |x_i| \leq |x|, i \leq k-1; |x_k| < s_0} \left(\frac{p_+}{\lambda_0}\right)^k J(x_0,x_1) \cdots J(x_{k-1},x_k)\psi(x''). \qquad (8.4.9.2)$$

c_- bounds the initial excursions past $|x|$. We then have

$$\psi(x) \leq c_- e^{-\alpha_-^0(|x| - s_0 - 1)}\psi(x''). \qquad (8.4.9.3)$$

By (8.4.8.10), $w = \psi^{-1}e^{-\delta|x|}$ is bounded for $|x| > s_0$ by

$$w(x) \leq [c_+\psi(x')e^{\alpha_+^0 s_0}]^{-1} e^{(\alpha_+^0 - \delta)|x|},$$

while, for $|x| \leq s_0$, $w(x) \leq \psi(x')^{-1}e^{(\alpha_+^0 - \delta)|x|}$. Hence

$$w_+ = \psi(x')^{-1}\max\{[c_+e^{\alpha_+^0 s_0}]^{-1}, 1\}.$$

Analogously, $w(x) \geq [c_-\psi(x'')e^{\alpha_-^0(s_0+1)}]^{-1}e^{(\alpha_-^0-\delta)|x|}$, $|x| > s_0$, by (8.4.8.12), while for $|x| \leq s_0$, $w(x) \geq e^{-\alpha_-^0 s_0}\psi(x'')^{-1}e^{(\alpha_-^0-\delta)|x|}$; hence,

$$w_- = [\psi(x'')e^{\alpha_-^0 s_0}]^{-1}\min\{[c_- e^{\alpha_-^0}]^{-1}, 1\}.$$

To prove (8.4.1.4), we write for $|x| \geq s_0$

$$Kw(x) = \int \frac{M(x,y)}{\lambda_0\psi(x)}e^{-\delta y} \leq \frac{e^{|\delta|}p_+}{\lambda_0}w(x), \tag{8.4.9.4}$$

thus identifying $r_0 := \frac{e^{|\delta|}p_+}{\lambda_0} < 1$ because $|\delta| < \alpha^*$. Notice that for $|x| < \ell - 1$ we could use the same argument as in (8.4.8.13), which gives $r_0 < 1$ for $|\delta| < \alpha_-^0$. For $|x| < s_0$ the bound is as in (8.4.8.14) and $r_1 = \|M\|_\infty \frac{e^{|\delta|}}{\lambda_0}$. We have thus proved K1. K2 follows from M3 as in the proof of Lemma 8.4.8.1, but details are omitted. □

8.5 Global stability and uniqueness

The stability in the heading refers to the "manifold" $\mathcal{M}$ of all translates of the instanton $\bar{m}$: we shall prove that $\mathcal{M}$ attracts (exponentially fast) the whole set $\mathcal{N}$ defined in (8.1.1.3). Notice that such a statement does not at all mean stability of a single instanton; in fact, even if $m \in \mathcal{N}$ is a small perturbation of the instanton $\bar{m}$; yet, $T_t(m)$ may not converge to $\bar{m}$ as $t \to \infty$, it will, however, converge to a translate of $\bar{m}$ (which depends on m).

8.5.1 The center of a function

The shifted instantons are defined as

$$\bar{m}_\xi(x) = \bar{m}(x - \xi), \quad \xi \in \mathbb{R}. \tag{8.5.1.1}$$

$\bar{m}_\xi$ solves (8.1.1.5) and ξ is its "center," 0 being therefore the center of $\bar{m}$; $\mathcal{M} := \{\bar{m}_\xi, \xi \in \mathbb{R}\}$ is the "instanton manifold" and, for any $\delta > 0$, we denote by

$$\mathcal{M}_\delta = \bigcup_{\xi \in \mathbb{R}} \{m \in L^\infty(\mathbb{R}, [-1,1]) : \|m - \bar{m}_\xi\|_\infty < \delta\} \tag{8.5.1.2}$$

its "δ-neighborhood."

We define a center also for profiles which are not instantons: $\xi \in \mathbb{R}$ is a center of m if

$$\left(m - \bar{m}_\xi, \bar{m}'_\xi\right)_\xi = 0, \tag{8.5.1.3}$$

where $(\cdot,\cdot)_\xi$ denotes the scalar product in $L_2(\mathbb{R}, d\nu_\xi)$, and

$$d\nu_\xi(x) = \{\beta[1-\bar{m}_\xi(x)^2]\}^{-1}\,dx.$$

When $\xi = 0$, the scalar product $(\cdot,\cdot)_\xi$ becomes the one defined in Sect. 8.3.1 when studying the spectral gap. Thus, the center ξ of m specifies an element $\bar{m}_\xi \in \mathcal{M}$ such that the two directions, one pointing from $\bar{m}_\xi$ to m and the other one tangent to $\mathcal{M}$ at $\bar{m}_\xi$ are mutually orthogonal (with respect to the ξ-dependent scalar product). The next theorem proves that any $m \in \mathcal{M}_\delta$, with δ small enough, has a unique center. We shall also see that the position of the center of $m = \bar{m}_\xi + v$ deviates from ξ to first order in v by $-N_{v,\xi}$, where

$$N_{v,\xi} = \frac{(v,\bar{m}'_\xi)_\xi}{(\bar{m}'_\xi,\bar{m}'_\xi)_\xi}. \tag{8.5.1.4}$$

When $\xi = 0$, $N_{v,0} = N_v$, the latter having been defined in (8.3.2.2).

Theorem 8.5.1.1 *Any $m \in \mathcal{N}$ has a center ($\mathcal{N}$ is defined in (8.1.1.3)) and there is $\delta > 0$, so that any $m \in \mathcal{M}_\delta$ has a unique center, $\xi(m)$. There is $c > 0$, so that if $m \in \mathcal{M}_{\delta/2}$ and $\|m-n\|_\infty < \delta/2$, then $n \in \mathcal{M}_\delta$ and*

$$|\xi(m) - \xi(n)| \le c\|m-n\|_\infty. \tag{8.5.1.5}$$

Finally, for any $\|v\|_\infty < \delta$ and any $\xi_0 \in \mathbb{R}$ and N_{v,ξ_0} as in (8.5.1.4), calling $m = \bar{m}_{\xi_0} + v$,

$$\big|\xi(m) - (\xi_0 - N_{v,\xi_0})\big| \le c\|v\|_\infty^2, \qquad |N_{v,\xi_0}| \le c\|v\|_\infty. \tag{8.5.1.6}$$

Proof Since both $\bar{m}'$ and the density $(\beta[1-\bar{m}^2])^{-1}$ of $d\nu$ are symmetric, while $\bar{m}$ is antisymmetric, $(\bar{m}_\xi, \bar{m}'_\xi)_\xi = 0$, and (8.5.1.3) becomes

$$\left(m,\bar{m}'_\xi\right)_\xi = 0. \tag{8.5.1.7}$$

By (8.1.1.3), and recalling that $\bar{m}'_\xi$ vanishes exponentially fast as $|x| \to \infty$,

$$\limsup_{\xi\to-\infty}\left(m,\bar{m}'_\xi\right)_\xi < 0, \qquad \liminf_{\xi\to\infty}\left(m,\bar{m}'_\xi\right)_\xi > 0,$$

so that the l.h.s. of (8.5.1.7) is negative as $\xi \to -\infty$ and positive as $\xi \to +\infty$. Being a continuous function, it must vanish for some value of ξ; hence, m has a center.

We write the elements m in the δ-ball around $\bar{m}_{\xi_0}$ as

$$m = \bar{m}_{\xi_0} + \epsilon u, \quad 0 \le \epsilon < \delta,\ \|u\|_\infty = 1; \tag{8.5.1.8}$$

our estimates will only depend on ϵ and will be uniform on u. We are going to prove that any such m has a unique center. Without loss of generality we may suppose $\xi_0 = 0$; we shall write g' for the derivative of g with respect to x.

With this notation, (8.5.1.7) becomes

$$\big(\bar{m}, \bar{m}'_\xi\big)_\xi = -\epsilon\big(u, \bar{m}'_\xi\big)_\xi. \tag{8.5.1.9}$$

Since $\|u\|_\infty = 1$,

$$\epsilon\big|\big(u, \bar{m}'_\xi\big)_\xi\big| \le \epsilon \frac{2m_\beta}{\beta(1-m_\beta^2)}, \quad \text{for any } \xi \in \mathbb{R}. \tag{8.5.1.10}$$

For any ξ the l.h.s. of (8.5.1.9) can be written as

$$\begin{aligned}\big(\bar{m}, \bar{m}'_\xi\big)_\xi &= \beta^{-1}\int [1-\bar{m}(x-\xi)^2]^{-1}\bar{m}(x)\bar{m}'(x-\xi)\,dx \\ &= \beta^{-1}\int [1-\bar{m}(y)^2]^{-1}\bar{m}(y+\xi)\bar{m}'(y)\,dy.\end{aligned} \tag{8.5.1.11}$$

Then

$$\frac{d}{d\xi}\big(\bar{m}, \bar{m}'_\xi\big)_\xi = \beta^{-1}\int [1-\bar{m}(y)^2]^{-1}\bar{m}'(y+\xi)\bar{m}'(y)\,dy > 0$$

and

$$\alpha_0 := \inf_{|\xi|\le 1} \frac{d}{d\xi}\big(\bar{m}, \bar{m}'_\xi\big)_\xi > 0. \tag{8.5.1.12}$$

Since $(\bar{m}, \bar{m}'_\xi)_\xi = 0$ at $\xi = 0$, then $|(\bar{m}, \bar{m}'_\xi)_\xi| \ge \alpha_0$ for all $|\xi| \ge 1$.

Our first condition on $\delta > 0$ is that it should be so small that

$$\delta \frac{2m_\beta}{\beta(1-m_\beta^2)} < \alpha_0. \tag{8.5.1.13}$$

Then there is no solution to (8.5.1.9) when $|\xi| \ge 1$ and $\epsilon < \delta$, because $|(\bar{m}, \bar{m}'_\xi)_\xi| \ge \alpha_0$, while, by (8.5.1.10)–(8.5.1.13), $\epsilon|(u, \bar{m}'_\xi)_\xi| < \alpha_0$.

We shall next show that there is $K > 0$ so that, if $\|u\|_\infty \le 1$, then

$$\left|\frac{d}{d\xi}\big(u, \bar{m}'_\xi\big)_\xi\right| \le K. \tag{8.5.1.14}$$

To this end, we compute the derivative

$$\frac{d}{d\xi}\frac{\bar{m}'_\xi}{1-\bar{m}_\xi^2} = -2\left(\frac{\bar{m}'_\xi}{1-\bar{m}_\xi^2}\right)^2 - \frac{\bar{m}''_\xi}{1-\bar{m}_\xi^2}.$$

By differentiating $\bar{m}'_\xi = \beta[1-\bar{m}_\xi^2]J \star \bar{m}'_\xi$, we get

$$|\bar{m}''_\xi| \le \beta|J'| \star \bar{m}'_\xi + 2\beta\bar{m}'_\xi J \star \bar{m}'_\xi,$$

and (8.5.1.14) follows from the bound $\bar{m}'(x) \le ce^{-\omega|x|}$, c and ω being positive constants. We do not need here the sharp estimate of (8.2.1.6).

We choose $\delta > 0$, so that both (8.5.1.13) and (8.5.1.15) below hold, where

$$\delta K < \frac{\alpha_0}{2}, \tag{8.5.1.15}$$

and K as in (8.5.1.14). Thus, by (8.5.1.12), for $\epsilon < \delta$ and $|\xi| \le 1$

$$\frac{d}{d\xi}\{(\bar{m}, \bar{m}'_\xi)_\xi + \epsilon(u, \bar{m}'_\xi)_\xi\} \ge \frac{\alpha_0}{2} > 0, \tag{8.5.1.16}$$

which proves that the center is unique in $|\xi| \le 1$. Since we have already proved that no $|\xi| \ge 1$ is a center of m, we have proved the existence and global uniqueness of the center, so that the center $\xi(m)$ is well defined in $\mathcal{M}_\delta$ with $\delta > 0$ satisfying (8.5.1.13) and (8.5.1.15).

The proof of (8.5.1.5) is as follows. Without loss of generality, we may suppose that $m = \bar{m} + \epsilon u$, $\epsilon < \delta/2$ and $\|u\|_\infty = 1$. Let $\lambda v = n - m$, $|\lambda| < \delta/2$, $\|v\|_\infty = 1$, so that

$$n = \bar{m} + \epsilon u + \lambda v, \qquad \|\epsilon u + \lambda v\|_\infty < \delta. \tag{8.5.1.17}$$

We can use for both m and n the previous analysis. $\xi(n)$ is the zero of

$$g(\xi) := (\bar{m}, \bar{m}'_\xi)_\xi + \epsilon(u, \bar{m}'_\xi)_\xi + \lambda(v, \bar{m}'_\xi)_\xi, \tag{8.5.1.18}$$

and $\xi(m)$ of the function with $\lambda = 0$. Write $g(\xi(n)) = g(\xi(m)) + \int_{\xi(m)}^{\xi(n)} g'(\xi)\, d\xi$,

$$0 = g(\xi(n)) = \lambda(v, \bar{m}'_{\xi(m)})_{\xi(m)} + \int_{\xi(m)}^{\xi(n)} g'(\xi)\, d\xi. \tag{8.5.1.19}$$

We have already seen that $|\xi(n)| \le 1$ and $|\xi(m)| \le 1$; thus, all ξ in the integral have $|\xi| \le 1$, and, by (8.5.1.16), $g'(\xi) \ge \alpha_0/2$, so that

$$\left|\int_{\xi(m)}^{\xi(n)} g'(\xi)\, d\xi\right| \ge \frac{\alpha_0}{2}|\xi(n) - \xi(m)|,$$

$$|\xi(n) - \xi(m)| \le \left\{\frac{2}{\alpha_0}\left|(\lambda v, \bar{m}'_{\xi(m)})_{\xi(m)}\right|\right\}.$$

Recalling that $\lambda v = n - m$ and using (8.5.1.10),

$$|\xi(n) - \xi(m)| \le \frac{4m_\beta}{\alpha_0\beta(1 - m_\beta^2)}\|n - m\|_\infty, \tag{8.5.1.20}$$

and we thus conclude the proof of (8.5.1.5).

The proof of (8.5.1.6) is as follows. By (8.5.1.20) with $n = \bar{m}$

$$|\xi(m)| \le \epsilon\kappa, \qquad \epsilon = \|m - \bar{m}\|_\infty, \qquad \kappa = \frac{4m_\beta}{\alpha_0\beta(1 - m_\beta^2)}. \tag{8.5.1.21}$$

Recalling (8.5.1.6) and the fact that $\xi_0 = 0$, let $\epsilon u = m - \bar{m}$, $\|u\|_\infty = 1$,

$$\epsilon R_\epsilon := \xi(m) - (-\epsilon N_u) = \xi(m) + \epsilon N_u.$$

By (8.5.1.21), since $|N_u| \le c\|u\|_\infty \le c$, $|R_\epsilon| \le \kappa + c =: C$ for all ϵ small enough. To prove (8.5.1.6) we must improve the bound showing that $|R_\epsilon| \le c'\epsilon$, c' a constant. By a Taylor expansion on the l.h.s. of (8.5.1.9) rewritten as in (8.5.1.11), we get

$$\begin{aligned}(\bar{m}, \bar{m}'_\xi)_\xi &= \int \frac{\bar{m}'(y)}{\beta[1-\bar{m}(y)^2]} \\ &\quad \times \left\{\bar{m}(y) + \epsilon(R_\epsilon - N_u)\bar{m}'(y) + \frac{\epsilon^2(R_\epsilon - N_u)^2}{2}\bar{m}''_{\xi_1}(y)\right\}, \qquad (8.5.1.22)\end{aligned}$$

with $\xi_1 = \xi_1(y)$, $|\xi_1| \le \epsilon\kappa$; see (8.5.1.21).

Analogously, the r.h.s. of (8.5.1.9) can be written as

$$-\frac{\epsilon}{\beta}\int u(y)\left\{\frac{\bar{m}'(y)}{[1-\bar{m}(y)^2]} - \epsilon(R_\epsilon - N_u)\left(\frac{\bar{m}''_{\xi_2}(y)}{[1-\bar{m}_{\xi_2}(y)^2]} + \frac{2\bar{m}'_{\xi_2}(y)^2\bar{m}_{\xi_2}(y)}{[1-\bar{m}_{\xi_2}(y)^2]^2}\right)\right\}. \qquad (8.5.1.23)$$

The term without ϵ in (8.5.1.22) vanishes as the integrand is odd; the term with ϵN_u in (8.5.1.22) cancels with the first term in (8.5.1.23), by the definition of N_u. The integrands decay exponentially as they contain $\bar{m}'$ or $\bar{m}''$ as a factor; $\|u\|_\infty \le 1$, by assumption, and R_ϵ and N_u are also bounded, as already noticed. We then have, after equating (8.5.1.22) and (8.5.1.23),

$$|R_\epsilon|(\bar{m}', \bar{m}')_0 \le \epsilon c'.$$

□

8.5.2 *Local stability*

In this subsection we shall prove:

Theorem 8.5.2.1 (Local stability) *There exist c^*, δ^* and α all positive, such that, if $m \in L^\infty(\mathbb{R}, [-1,1])$ and $\|m - \bar{m}_{\xi_0}\|_\infty < \delta^*$, with ξ_0 the center of m, then there is ξ such that*

$$\|T_t(m) - \bar{m}_\xi\|_\infty \le c^* e^{-\alpha t}, \qquad (8.5.2.1)$$

$$|\xi - \xi_0| \le c^*\|m - \bar{m}_{\xi_0}\|_\infty^2. \qquad (8.5.2.2)$$

Proof By translational invariance we may assume without loss of generality that the center ξ_0 of m is 0. Set $v(\cdot, t) = T_t(m) - \bar{m}$. Call

$$M(x,y) = \beta[1-\bar{m}(x)^2]J(x,y), \qquad M\,v(x) = \int dy\, M(x,y)v(y), \qquad (8.5.2.3)$$

then $L - M = -1$ and

$$\frac{\partial v}{\partial t} = Lv + \big\{\tanh(\beta J \star [\bar{m} + v]) - \tanh(\beta J \star \bar{m}) - M\, v\big\}, \tag{8.5.2.4}$$

which we rewrite in integral form as $v(\cdot, t) = \Psi(v)(\cdot, t)$, where $\Psi(u)(\cdot, t)$ is the sum of $e^{Lt} v(\cdot, 0)$ and

$$\int_0^t e^{L(t-s)} \big\{\tanh(\beta J \star [\bar{m} + u(\cdot, s)]) - \tanh(\beta J \star \bar{m}) - Mu(\cdot, s)\big\}.$$

Ψ is regarded as a map from X into itself, where X is the space of measurable functions $u(r, t)$, $r \in \mathbb{R}$, $t \in \mathbb{R}_+$, which are continuous as functions of t with values in $L^\infty(\mathbb{R})$, indeed we know that $v(r, t)$ is in such a space.

By Theorem 8.3.2.1 (on the spectral gap with sup norms) we get from (8.3.2.3)

$$\|\Psi(u)(\cdot, t)\|_\infty \le a e^{-\omega t} \|v(\cdot, 0)\|_\infty + b \int_0^t ds\, \|u(\cdot, s)\|_\infty^2, \tag{8.5.2.5}$$

where ω, a and b are positive constants. We have used that the center of m is the origin so that $v(x, 0)$ has no component along $\bar{m}'$. We shall use the following corollary of Theorem 8.5.1.1. There are δ and c positive so that if $m = \bar{m} + v$, $\|v\|_\infty < \delta$, then m has a unique center ξ and

$$m = \bar{m} + v = \bar{m}_\xi + w, \quad \|w\|_\infty \le c\|v\|_\infty. \tag{8.5.2.6}$$

To prove the inequality in (8.5.2.6) we write $\|w\|_\infty \le \|v\|_\infty + \|m - \bar{m}_\xi\|_\infty \le \|v\|_\infty + \|\bar{m}'_\xi\|_\infty |\xi|$, hence (8.5.2.6); we have used (8.5.1.5) with $n = \bar{m}$ and $\xi(n) = 0$.

Having fixed a, b and c as above, we claim that

$$\|v(\cdot, t)\|_\infty \le 2a e^{-\omega t} \|v(\cdot, 0)\|_\infty, \quad \text{for all } t \le T(\|v(\cdot, 0)\|_\infty), \tag{8.5.2.7}$$

where $T(x)$ is defined for $x > 0$ small enough by

$$a\, e^{-\omega T(x)} = b \frac{4a^2}{2\omega} x. \tag{8.5.2.8}$$

We shall choose δ^* so that $T(\|v(\cdot, 0)\|_\infty)$ is well defined (and large as required below).

Proof of (8.5.2.7). By (8.5.2.5) the inequality in (8.5.2.7) is satisfied for $t > 0$ small enough. We need to prove that its validity extends up to $T \equiv T(\|v(\cdot, 0)\|_\infty)$. Arguing by contradiction, assume that it holds only up to some time $S < T \equiv T(\|v(\cdot, 0)\|_\infty)$, namely that $[0, S]$ is the largest interval from 0 where the inequality in (8.5.2.7) holds. By (8.5.2.5) for any $t \le S$

$$\|v(\cdot, t)\|_\infty \le a e^{-\omega t} \|v(\cdot, 0)\|_\infty + b \int_0^t ds\, \big(2a e^{-\omega s} \|v(\cdot, 0)\|_\infty\big)^2, \tag{8.5.2.9}$$

hence by (8.5.2.8), and since $t \le S < T$, we have

$$\|v(\cdot,t)\|_\infty \le ae^{-\omega t}\|v(\cdot,0)\|_\infty + b\frac{4a^2}{2\omega}\|v(\cdot,0)\|_\infty^2, \qquad b\frac{4a^2}{2\omega}\|v(\cdot,0)\|_\infty < ae^{-\omega t}.$$

Thus $\|v(\cdot,t)\|_\infty < 2ae^{-\omega t}\|v(\cdot,0)\|_\infty$ for all $t \le S$. Since $S < T$, by continuity it continues to hold up to some $S' \in (S, T]$, thus leading to the desired contradiction. In conclusion, $S = T$ and (8.5.2.7) is proved.

We have thus proved (8.5.2.7) with T as in (8.5.2.8) and $x = \|v(\cdot,0)\|_\infty$. Let c be as in (8.5.2.6) assuming without loss of generality that $c \ge 1$.

$$\text{Let } S: \quad c2ae^{-\omega S} < \frac{1}{2} \quad \text{and} \quad \delta^* < \delta, \qquad T(\delta^*) \ge S. \tag{8.5.2.10}$$

δ is as in Theorem 8.5.1.1. Then, by (8.5.2.6) with $\|v(\cdot,0)\|_\infty \le \delta^*$,

$$T_S(m) = \bar m + v^{(1)} = \bar m_{\xi(S)} + w^{(1)}, \tag{8.5.2.11}$$

with $\|w^{(1)}\|_\infty \le \frac{\|v(\cdot,0)\|_\infty}{2}$, $\|v^{(1)}\|_\infty \le \frac{\|v(\cdot,0)\|_\infty}{2}$ and $\xi(S)$ the center of $T_S(m)$. Since $\|w^{(1)}\|_\infty < \delta^*$ we can repeat the argument starting from $T_S(m)$ as, by translational invariance, the whole argument is independent of the center of the initial instanton, which for $T_S(m)$ is $\xi(S)$. We then have, after n iterations,

$$T_{nS}(m) = \bar m_{\xi((n-1)S)} + v^{(n)} = \bar m_{\xi(nS)} + w^{(n)},$$
$$\|w^{(n)}\|_\infty \le \frac{\|v(\cdot,0)\|_\infty}{2^n}, \qquad \|v^{(n)}\|_\infty \le \frac{\|v(\cdot,0)\|_\infty}{2^n}. \tag{8.5.2.12}$$

We can also bound the increments of the centers: $\xi(nS)$, the center of $T_{nS}(m)$, is in fact the center of $\bar m_{\xi((n-1)S)} + v^{(n)}$; therefore, by (8.5.1.5)

$$\big|\xi(nS) - \xi([n-1]S)\big| \le c\|v^{(n)}\|_\infty \le c2^{-n}\|v(\cdot,0)\|_\infty.$$

$\xi(nS)$ then converges as $n \to \infty$ to a limit ξ with $|\xi| \le 2c\|v(\cdot,0)\|_\infty$ and

$$\big|\xi(nS) - \xi\big| \le 2c2^{-n}\|v(\cdot,0)\|_\infty.$$

As a consequence (since $\bar m'$ is bounded)

$$\|\bar m_{\xi(nS)} - \bar m_\xi\|_\infty \le c'2^{-n}\|v(\cdot,0)\|_\infty,$$

and by (8.5.2.12)

$$\|T_{nS}(m) - \bar m_\xi\|_\infty \le (c'+1)2^{-n}\|v(\cdot,0)\|_\infty. \tag{8.5.2.13}$$

By using (8.5.2.7) for the number of times in the interior of the intervals $[nS, (n+1)S]$, we get (8.5.2.1). To prove the [quadratic] bound (8.5.2.2) we choose the length of the first time interval t_0 in a way which depends on the initial condition; namely, we set $t_0 := T(\|v(\cdot,0)\|_\infty)$. The other time intervals are all equal to S, as before. Set again for simplicity $\xi_0 = 0$. By (8.5.2.7) with $t = t_0$ and (8.5.2.8) with $T = t_0$, we get

$$\|v(\cdot,t_0)\|_\infty \le 2b\frac{4a^2}{2\omega}\|v(\cdot,0)\|_\infty^2, \tag{8.5.2.14}$$

hence, as in the proof of (8.5.2.11), denoting by ξ_0 the center of $T_{t_0}(m)$,

$$T_{t_0}(m) = \bar{m} + v(\cdot,t_0) = \bar{m}_{\xi_0} + w. \tag{8.5.2.15}$$

We have $\|w\|_\infty \le c[2b\frac{4a^2}{2\omega}]\|v(\cdot,0)\|_\infty^2$, $|\xi_0| \le C\|w\|_\infty$. Calling $\psi = T_{t_0}(m)$ and using (8.5.2.13) with $T_{nS}(\psi)$, $\psi = \bar{m}_{\xi_0} + w$, ξ_0 the center of ψ, we get, by (8.5.2.15),

$$\lim_{t\to\infty} T_t(m) = \lim_{t\to\infty} T_t(\psi) = \bar{m}_\xi,$$

$$|\xi| \le |\xi_0| + C|w| \le 2Cc\left[2b\frac{4a^2}{2\omega}\right]\|v(\cdot,0)\|_\infty^2.$$

□

8.5.3 Trapping orbits between instantons

By clever use of super- and subsolutions, Fife and Mc Leod [121], have proved for the Allen–Cahn equation that any orbit starting from a datum $m \in \mathcal{N}$, see (8.1.1.3), is eventually trapped between two instantons. Their proof applies also to the non-local evolution equation (6.2.2.7), as proved in [92]; but with the background properties already established, it is more convenient to present an alternative proof.

Theorem 8.5.3.1 *Let $m \in \mathcal{N}$, see* (8.1.1.3)*; then there are $\xi_\pm$ and $\delta(t)$, so that*

$$\bar{m}_{\xi_-} - \delta(t) \le T_t(m) \le \bar{m}_{\xi_+} + \delta(t), \quad \text{for all } t, \tag{8.5.3.1}$$

with $\delta(t) \to 0$ exponentially fast as $t \to \infty$.

We postpone the proof of the theorem to first deal with the following lemma:

Lemma 8.5.3.2 *Let $m \in \mathcal{N}$, $m(x,t) = T_t(m)(x)$. Then, for any $\epsilon > 0$ there are t_0 and ℓ, so that*

$$\begin{aligned} &m(x,t_0) \le m_\beta + \epsilon \quad \text{for all } x \quad \text{and} \\ &m(x,t_0) \ge m_\beta - \epsilon \quad \text{for all } x \ge \ell, \end{aligned} \tag{8.5.3.2}$$

$$\begin{aligned} &m(x,t_0) \ge -m_\beta - \epsilon \quad \text{for all } x \quad \text{and} \\ &m(x,t_0) \le -m_\beta + \epsilon \quad \text{for all } x \le -\ell. \end{aligned} \tag{8.5.3.3}$$

Proof Denote by $T_t(u)$ the solution at time t of the Cauchy problem (6.2.2.7) with initial datum u. Then, $T_t(m) \le T_t(1)$ (1 the function constantly equal to 1) and $\|T_t(1) - m_\beta\|_\infty \le ce^{-t}$. Indeed, if the initial datum is a constant, (6.2.2.7) becomes a simple ordinary differential equation. We choose $t_0 \ge s$ with $ce^{-s} = \epsilon$, so that $m(x,t_0) \le m_\beta + \epsilon$ for all x. To prove (8.5.3.2) we still need to prove that $m(x,t_0) \ge m_\beta - \epsilon$ for all $x \ge \ell$. By (8.1.1.3), there are x_0 and $\zeta > 0$ so that $m(x,0) \ge \zeta$ for all $x \ge x_0$. Let t_0 be so large that $T_{t_0}(\zeta) \ge m_\beta - \epsilon/2$. With reference to Theorem 6.2.7.1, let $V \ge \beta e^2$ be such that

$$C \exp\left\{-t_0 V \log \frac{V}{\beta e}\right\} \le \frac{\epsilon}{2}. \tag{8.5.3.4}$$

Set $u(x) = m(x,0)$ for $x \ge x_0$ and $u(x) = m(x_0,0)$ for $x \le x_0$, so that $u(x) \ge \zeta$ everywhere. Then $T_{t_0}(u) \ge m_\beta - \epsilon/2$, everywhere, and by Theorem 6.2.7.1,

$$\left|T_{t_0}(u) - T_{t_0}(m)\right| \le C \exp\left\{-t_0 V \log \frac{V}{\beta e}\right\} \le \frac{\epsilon}{2}, \quad \text{on } \{x \ge x_0 + V t_0\}.$$

Hence

$$m(x,t_0) \ge u(x,t_0) - \frac{\epsilon}{2} \ge m_\beta - \epsilon, \quad \text{for all } x \ge \ell := x_0 + V t_0,$$

thus proving (8.5.3.2). The proof of (8.5.3.3) is analogous and we omit it. □

Proof of Theorem 8.5.3.1 We only prove the lower bound in (8.5.3.1), the proof of the upper bound is analogous and omitted. Let $a > 0$ be such that $\bar{m}(x) \le -m_\beta + \epsilon$ for all $x \le -a$. Then

$$\bar{m}_{\ell+a}(x) = \bar{m}(x - [\ell + a]) \le -m_\beta + \epsilon, \quad \text{for all } x \le \ell, \tag{8.5.3.5}$$

so that, by the first inequality in (8.5.3.3), $m(x,t_0) \ge \bar{m}_{\ell+a}(x) - 2\epsilon$ for all $x \le \ell$. By the second inequality in (8.5.3.2), $m(x,t_0) \ge \bar{m}_{\ell+a}(x) - 2\epsilon$ for all $x \ge \ell$; hence

$$m(x,t_0) \ge \bar{m}_{\ell+a}(x) - 2\epsilon, \quad \text{for all } x. \tag{8.5.3.6}$$

Then, for any $s \ge 0$ and any ξ,

$$m(\cdot, t_0 + s) \ge T_s(\bar{m}_{\ell+a} - 2\epsilon) = \bar{m}_\xi(\cdot) + \{T_s(\bar{m}_{\ell+a} - 2\epsilon) - \bar{m}_\xi(\cdot)\}. \tag{8.5.3.7}$$

For ϵ small enough by Theorem 8.5.2.1, there is ξ so that with such a ξ, the curly bracket's sup norm vanishes exponentially fast. We set $\delta(t) = 1$ for $t \le t_0$, and we have

$$\delta(t) = \left\|T_{t-t_0}\left(\bar{m}_{\ell+a} - 2\epsilon\right) - \bar{m}_\xi\right\|_\infty, \quad \text{for } t > t_0, \tag{8.5.3.8}$$

so that (8.5.3.7) gives the lower bound in (8.5.3.1). The proof of the upper bound is analogous and we omit it. □

8.5.4 Uniqueness

The main result in this subsection is that the only stationary solutions in $\mathcal{N}$ are translates of the instanton:

Theorem 8.5.4.1 (Uniqueness) *Let $m^\star \in C(\mathbb{R}, [-1, 1])$ be a solution of the mean field equation* (8.1.1.5) *which is in $\mathcal{N}$, i.e. it satisfies* (8.1.1.3). *Then $m^\star \in \mathcal{M}$; i.e., there is ξ so that $m^\star = \bar{m}_\xi$.*

Proof Since $m^\star$ solves (8.1.1.5), it is a stationary solution of (6.2.2.7), $m^\star(x,t) = m^\star(x,0) = m^\star(x)$. By taking the limit as $t \to \infty$ in (8.5.3.1) with $m(x,t)$ replaced by $m^\star(x,t)$, we deduce the existence of $\xi_- \geq \xi_+$, so that

$$\bar{m}_{\xi_-} \leq m^\star \leq \bar{m}_{\xi_+}. \tag{8.5.4.1}$$

We shall assume that ξ_- and ξ_+ are extremal with such a property.

Let $\epsilon > 0$, $a = \xi_+ + \epsilon$ (so that $\bar{m}_a < \bar{m}_{\xi_+}$), and

$$m(x) := \max\{\bar{m}_a, m^\star(x)\}, \quad \bar{m}_a \leq m \leq \bar{m}_{\xi_+}. \tag{8.5.4.2}$$

If ϵ is small enough, we can use Theorem 8.5.2.1, to conclude that there is ξ so that $T_t(m) \to \bar{m}_\xi$ as $t \to \infty$. Since $m^\star \leq m$ and $T_t(m^\star) = m^\star$, $m^\star \leq \bar{m}_\xi$. In the lemma below we shall show that (for ϵ small enough) $\xi > \xi_+$, unless $m^\star = \bar{m}_{\xi_+}$; by the maximality of ξ_+, this implies $m^\star = \bar{m}_{\xi_+}$ and the theorem is proved. □

Lemma 8.5.4.2 *For all $\epsilon > 0$ small enough, the following holds. Let m be a continuous function and $\bar{m}_\epsilon \leq m \leq \bar{m}$. Then, if $m \neq \bar{m}$,*

$$\lim_{t\to\infty} T_t(m) = \bar{m}_\xi, \quad \xi > 0. \tag{8.5.4.3}$$

Proof Let $\lambda \in [0, 1]$ and

$$u_\lambda := \bar{m} + \lambda[m - \bar{m}]. \tag{8.5.4.4}$$

If ϵ is small enough, by Theorem 8.5.2.1, for any $\lambda \in [0, 1]$, $T_t(u_\lambda)$ converges to an instanton, whose center will be denoted by ξ_λ^f. By (8.5.4.4), $\partial u_\lambda / \partial\lambda = m - \bar{m} \leq 0$. Thus, u_λ is a non-increasing function of λ, and, by monotonicity, also the function

$$\lambda \to \bar{m}_{\xi_\lambda^f} = \lim_{t\to\infty} T_t(u_\lambda)$$

is non-increasing. We then conclude that ξ_λ^f is a non-decreasing function of λ. Since $u_1 = m$, the lemma will follow from proving that, if $m \neq \bar{m}$, then for $\lambda > 0$ small enough, $\xi_\lambda^f > 0$. This we shall prove next.

If ϵ is small enough, u_λ has a center ξ_λ^0, which, by (8.5.1.6) satisfies

$$\left| \xi_\lambda^0 + \lambda \frac{(m - \bar{m}, \bar{m}')}{(\bar{m}', \bar{m}')} \right| \leq c\lambda^2 \|m - \bar{m}\|_\infty^2. \tag{8.5.4.5}$$

By (8.5.2.4),

$$|\xi_\lambda^f - \xi_\lambda^0| \le c\|u_\lambda - \bar{m}_{\xi_\lambda^0}\|_\infty^2. \tag{8.5.4.6}$$

By (8.5.4.4),

$$\|u_\lambda - \bar{m}_{\xi_\lambda^0}\|_\infty^2 \le 2\lambda^2\|m - \bar{m}\|_\infty^2 + 2\|\bar{m} - \bar{m}_{\xi_\lambda^0}\|_\infty^2, \tag{8.5.4.7}$$

and using (8.5.4.5),

$$\|\bar{m} - \bar{m}_{\xi_\lambda^0}\|_\infty^2 \le \|\bar{m}'\|_\infty^2 (\xi_\lambda^0)^2 \le C\lambda^2. \tag{8.5.4.8}$$

Thus, by (8.5.4.6), for a suitable constant c', $|\xi_\lambda^f - \xi_\lambda^0| \le c'\lambda^2$; hence, by (8.5.4.5),

$$\xi_\lambda^f \ge -\lambda\frac{(m - \bar{m}, \bar{m}')}{(\bar{m}', \bar{m}')} - \lambda^2\left(c\|m - \bar{m}\|_\infty^2 + c'\right), \tag{8.5.4.9}$$

which shows that for $\lambda > 0$ small enough, $\xi_\lambda^f > 0$, unless $m = \bar{m}$. □

8.5.5 Global stability

In this subsection we conclude the stability analysis of the instanton manifold by proving:

Theorem 8.5.5.1 (Global stability) *Let $m \in \mathcal{N}$; then, there is ξ so that*

$$\lim_{t\to\infty} \|T_t(m) - \bar{m}_\xi\|_\infty = 0. \tag{8.5.5.1}$$

Proof By (6.2.6.1), there are $u \in C^1(\mathbb{R}, [-1, 1])$ and a sequence $t_n \to \infty$, such that, uniformly in the compacts,

$$\lim_{t_n\to\infty} T_{t_n}(m) = u. \tag{8.5.5.2}$$

By Theorem 8.5.3.1 the convergence is in fact in sup norm, and there are $\xi_\pm$, so that

$$\bar{m}_{\xi_-} \le u \le \bar{m}_{\xi_+}. \tag{8.5.5.3}$$

Then $\mathcal{F}(u) < \infty$ and by Theorem 6.2.9.1 any limit point v of $T_t(u)$ satisfies (6.2.4.1). By (8.5.5.3) and the Comparison Theorem, Theorem 6.2.10.2, $v \in \mathcal{N}$; then, by uniqueness, Theorem 8.5.4.1, $v = \bar{m}_\xi$, for some ξ. Since $T_t(u)$ is trapped between two instantons, uniform convergence on the compacts becomes convergence in sup norms; hence, for any $\epsilon > 0$, there is s_k so that

$$\|T_{s_k}(u) - \bar{m}_\xi\|_\infty \le \epsilon. \tag{8.5.5.4}$$

We have seen that $T_{t_n}(m) \to u$ in sup norm; then, by Theorem 6.2.3.4,

$$\lim_{t_n \to \infty} T_{s_k}\big(T_{t_n}(m)\big) = T_{s_k}(u). \tag{8.5.5.5}$$

Hence, there is t_n such that

$$\|T_{s_k+t_n}(m) - T_{s_k}(u)\|_\infty \le \epsilon. \tag{8.5.5.6}$$

Then $\|T_{s_k+t_n}(m) - \bar m_\xi\|_\infty \le 2\epsilon$, and, if ϵ is small enough, by Theorem 8.5.2.1, $T_{s+s_k+t_n}(m) \to \bar m_\xi$, as $s \to \infty$, exponentially fast and uniformly. □

8.6 Surface tension and the instanton free energy

In this section, we shall prove that the free energy $\mathcal{F}(\bar m)$ of the one dimensional instanton is equal to the surface tension s_β defined in Sect. 7.1.2. We shall first prove in Sect. 8.6.1 that $\mathcal{F}(\bar m) \ge s_\beta$ and then in Sect. 8.6.2 the opposite inequality.

8.6.1 The lower bound

The surface tension is the excess free energy per unit area of an interface separating two equilibrium phases. As argued in Sect. 7.1.2 its value for the L–P functional is the quantity $s_\beta(e)$, defined in Theorem 7.1.3.1 by (7.1.3.1). With Theorem 7.1.3 we have proved that $s_\beta(e) = \theta_e$ (dependence on β is not made explicit in the latter), where θ_e is defined in Sect. 7.1.4 starting from the expression

$$\theta_e(L,h) = \frac{1}{L^{d-1}} \inf_m \mathcal{F}^{\text{per}}_{R(L,h)}(m|\chi_{R(L)\setminus R(L,h)}). \tag{8.6.1.1}$$

The inf is over all m in $L^\infty(R(L,h),[-1,1])$, $R(L,h)$ being the d-dimensional channel with height $2h$ along the direction e and as cross section a $d-1$ dimensional cube of side L; for ease of reference we take the x_d axis along the direction e. The interaction is made periodic in the coordinates x_i, $i \ne d$, and χ is the function equal to m_β when $x_d \ge h$, and to $-m_\beta$ when $x_d \le -h$. The values of χ in the strip $|x_d| \le h$ are not relevant, because the interaction and hence $R(L,h)$ are made periodic orthogonally to e. θ_e is then defined from $\theta_e(L,h)$ by taking (for instance) first its limit as $h \to \infty$ and then as $L \to \infty$.

Let

$$\theta'_e(L,h) = \frac{1}{L^{d-1}} \inf_{\{m \text{ symm}\}} \mathcal{F}^{\text{per}}_{R(L,h)}(m|\chi_{R(L)\setminus R(L,h)}), \tag{8.6.1.2}$$

where the inf is now over the subset of functions $m(r)$ in the previous class which depend only on the coordinate x_d. Then $\theta'_e(L,h) \ge \theta_e(L,h)$, and

$$\theta'_e(L,h) = \inf_{m \in L^\infty([-h,h],[-1,1])} \mathcal{F}^{(e)}_{[-h,h]}(m|\chi_{[-h,h]^c}), \tag{8.6.1.3}$$

where $\mathcal{F}^e$ is the one dimensional functional with interaction $J^{(e)}$ as in (8.1.1.1), and $\chi_{[-h,h]^c}$ is the function equal to $\pm m_\beta$ when $x_d > h$, respectively $x_d < -h$. Observe that $\theta'_e(L,h)$ does no longer depend on L, so that we can drop L from the notation and simply write $\theta'_e(h)$. Then, if $h > 1/2$,

$$\theta'_e(h) = \inf_{m=\chi \text{ on } [-h,h]^c} \mathcal{F}^{(e)}(m); \tag{8.6.1.4}$$

hence, recalling the definition of $\mathcal{N}$ in (8.1.1.3),

$$\theta'_e(h) \geq \inf_{m\in\mathcal{N}} \mathcal{F}^{(e)}(m) = S_\beta(e) \tag{8.6.1.5}$$

and

$$\theta'_e = \lim_{h\to\infty} \theta'_e(h) \geq S_\beta(e). \tag{8.6.1.6}$$

On the other hand, $S_\beta(e) = \mathcal{F}^{(e)}(\bar{m})$, $\bar{m}$ the instanton associated to the interaction $J^{(e)}$, and since $\bar{m}$ converges exponentially to its asymptotes, for any ϵ, there is h_ϵ so that, denoting by $\bar{m}_\epsilon$ the function equal to $\bar{m}$ on $[-h_\epsilon, h_\epsilon]$ and to χ on its complement,

$$\left|\mathcal{F}^{(e)}(\bar{m}) - \mathcal{F}^{(e)}(\bar{m}_\epsilon)\right| \leq \epsilon. \tag{8.6.1.7}$$

Thus, $\theta'_e \leq S_\beta(e) + \epsilon$, which, together with (8.6.1.6) shows that

$$\theta'_e = S_\beta(e), \qquad \theta_e \leq S_\beta(e). \tag{8.6.1.8}$$

8.6.2 Equality among surface tensions

We shall prove here the reverse inequality $\theta_e \geq S_\beta(e)$, and hence that $\theta_e = S_\beta(e)$ and since $\theta_e = s_\beta(e)$, that the van der Waals surface tension $S_\beta(e)$ has the correct value, $s_\beta(e)$. In other words, the planar symmetry which is an assumption in the van der Waals theory, is automatically satisfied in our model. The main feature responsible for such a property will be the non-negativity of the interaction.

Theorem 8.6.2.1 $S_\beta(e) \leq s_\beta(e)$ *and by* (8.6.1.8), $S_\beta(e) = s_\beta(e)$.

Proof By the same argument as used in Sect. 8.6.1 for $\theta'_e(h)$, we see that $\theta_e(L,h)$, defined in (8.6.1.1), is bounded by

$$\theta_e(L,h) \geq \inf_{m\in\mathcal{N}_L} L^{-d+1} \mathcal{F}^{\text{per}}_{R(L,\infty)}(m), \tag{8.6.2.1}$$

where

$$\mathcal{N}_L = \Big\{m \in L^\infty(R(L,\infty); [-1,1]) : \limsup_{x_d\to-\infty} \sup_{x_1,\dots,x_{d-1}} m(x_1,\dots,x_d) < 0;$$

$$\liminf_{x_d\to\infty} \inf_{x_1,\dots,x_{d-1}} m(x_1,\dots,x_d) > 0\Big\}. \tag{8.6.2.2}$$

We are going to prove that

$$\inf_{m \in \mathcal{N}_L} L^{-d+1} \mathcal{F}^{\text{per}}_{R(L,\infty)}(m) = S_\beta(e). \tag{8.6.2.3}$$

Then

$$\theta_e = \lim_{L\to\infty} \lim_{h\to\infty} \theta_e(L,h) \geq S_\beta(e), \tag{8.6.2.4}$$

thus concluding the proof of the theorem.

Proof of (8.6.2.4). The proof is essentially the same as in the $d=1$ case. Calling $x = x_d$ and $y = (x_1, \ldots, x_{d-1})$,

$$\bar{m}_L(x,y) := \bar{m}^{(e)}(x) \mathbf{1}_{(x,y)\in R(L,\infty)} \tag{8.6.2.5}$$

is an instanton solving

$$\bar{m}_L(r) = \tanh\{\beta J^{\text{per}} * \bar{m}_L(r)\}, \quad r \in R(L,\infty), \tag{8.6.2.6}$$

J^{per} being the periodic extension of J on $R(L,\infty)$. Postponing the proof that $\bar{m}_L$ is the minimizer in (8.6.2.3), we have

$$\mathcal{F}^{\text{per}}_{R(L,\infty)}(m) \geq \mathcal{F}^{\text{per}}_{R(L,\infty)}(\bar{m}_L) = L^{d-1}\mathcal{F}^{(e)}(\bar{m}^{(e)}) = L^{d-1} S_\beta(e),$$

thus proving (8.6.2.4).

As in the $d=1$ case, we prove that $\bar{m}_L$ is the minimizer of $\mathcal{F}^{\text{per}}_{R(L,\infty)}$ in $\mathcal{N}_L$ by showing that any orbit starting from $\mathcal{N}_L$ is attracted by a x_1-translate of $\bar{m}_L$. The first point is to prove a spectral gap for the linearized operator, traditionally denoted by L. To keep such a notation we rename the size L of the cube: $L = \epsilon^{-1}$. With $r = (x,y)$ below, we have

$$\begin{aligned} L_\epsilon &= -1 + M_\epsilon, \qquad M_\epsilon(r,r') = p_\epsilon(r) J^{\text{per}}(r,r'), \\ p_\epsilon(r) &= \beta(1 - \bar{m}_L(x)^2). \end{aligned} \tag{8.6.2.7}$$

We then have to consider the transition probability kernel

$$K_\epsilon(r,r') := \frac{p_\epsilon(r)}{\bar{m}'_\epsilon(r)} J^{\text{per}}(r,r') \bar{m}'_\epsilon(r). \tag{8.6.2.8}$$

The kernel defined in (8.6.2.8) satisfies K1–K2 relative to a weight $w = \bar{m}'_\epsilon(x)^{-1} \times e_\delta(x)^{-1}$, $|\delta| < \alpha^0_+$. After the spectral gap, the rest of the proof is quite model independent; we prove first local and then global stability of the manifold of x_d-translates of the instantons in $\mathcal{N}_{\epsilon^{-1}}$ just as in the $d=1$ case, and we omit the details. $\square$

8.7 Notes and references to Part II

I went back after a long time to Kac potentials from dynamics. I was trying with Lebowitz and Orlandi to make mathematically rigorous an observation of Rothman and Zaleski [196]: that the speed of phase segregation (in their numerical simulations) improves dramatically by increasing the length of the interaction. To simulate this we considered in [163] a simple exclusion process with an additional small drift for jumps in the direction of the more densely occupied regions. We half accomplished our task, as we proved that on a suitable scaling limit the evolution is indeed ruled by a diffusion equation with a density dependent diffusion coefficient which becomes negative when the density is in a certain interval (ρ', ρ''). Unfortunately, we could only analyze initial data with values either in $[0, \rho')$ or $(\rho'', 1]$, leaving out the most interesting cases. Dal Passo and De Mottoni in an unpublished paper showed us that in $d = 1$, there is a space dependent (on a mesoscopic scale) stationary solution which, however, does not connect ρ' to ρ'', as we naively thought but two other values, $\rho_0 < \rho'$ and $\rho_1 > \rho''$. Herbert Spohn then explained to us that the whole issue has a clear statistical mechanics meaning in terms of Kac potentials and thermodynamics (of phase transitions).

I discovered to my great surprise that not much had been done after the pioneering papers of Kac, Lebowitz and Penrose in the sixties and seventies, and that many important issues of Kac potentials had in fact been left open. I felt that before dynamics the "easier equilibrium questions" had to be settled first. There is a huge literature on non-local equations and functionals coming from physics but also from biology, population dynamics... and also, more recently, on the web, internet, ... I am not able to make an even short survey of all that and just mention a few papers more closely related to the topics of this book. While absolute minimizers are easy to find in the Ginzburg–Landau and L–P like functionals, due to the ferromagnetic nature of their energies, the problem becomes hard if the interaction is also repulsive. Such a case is studied in [86] for a system in a bounded interval with a qualitative analysis of the bifurcations which occur when the temperature is varied. Non-local interactions have been considered in [191–193]. The existence, uniqueness and stability of instanton solutions for L–P functionals are proved in [92, 94], and the relations with Glauber dynamics are established in detail in [93, 97, 98]. In [4], there is a simplified proof of the existence and uniqueness of the instanton for monotonic interactions using rearrangement inequalities. Rearrangements are also used in [65, 67], to study interfaces in systems described by non-local free energy functionals with two species. The functional arising from the LMP model is not ferromagnetic as the L–P functional; it, however, turns out that in an interval (β_c, β^*), $\beta^* \in (\beta_c, \beta_0)$, the function $K_{\beta,\lambda(\beta)}(\rho)$ in $(\rho_{\beta,-}, \rho_{\beta,+})$, see Sect. 10.2.2, is strictly increasing and the analysis of Chap. 6 can be reproduced without problems. However, when $\beta > \beta^*$, $K_{\beta,\lambda(\beta)}(\rho)$ is decreasing at $\rho_{\beta,+}$; see Fig. 10.2. In [134] it has been proved that the instanton still exists and that it is unique in $(\beta^*, \beta^* + \epsilon)$, $\epsilon > 0$ small, but it is no longer monotonous. It develops oscillations when approaching $\rho_{\beta,+}$. The question of oscillations is investigated theoretically and numerically in [88] in more dimensions.

In the presence of a magnetic field "the instanton moves," i.e. there exist traveling fronts connecting stable and metastable states as proved for the L–P functional in [96, 178]; traveling fronts for non-local equations are also discussed in [7, 15, 18, 81, 83, 115, 121]. The existence of instantons is an interesting open question when the order parameter is not a scalar and there are several equilibrium states. Which are the pairs connected by an instanton? The issue is discussed in [8]. From a statistical mechanics perspective an interesting example comes from the Potts model (with $Q \geq 3$ states) at the critical point where there are Q "ordered states" and a disordered one. In the n.n. case it has been rigorously proved in [159, 170] that there is wetting; namely, there appears a layer of the disordered phase at the interface between two distinct ordered phases. This may imply that there is not an instanton solution connecting two distinct ordered states, or maybe that there is such a solution, but it is not a minimizer.

From a mathematical point of view it looks quite reductive to assume that the local term $f_\beta(m)$ in the L–P functional has the form $-\frac{m^2}{2} + a(m)$, with $a(m)$ a convex function. Indeed, more general convex functions have a spin interpretation as the free energy of some "reference interaction" as discussed in Sect. 6.1.4. The case when $a(m)$ is, for instance, a double well cannot have such an interpretation; instantons may become discontinuous, and we may have stationary solutions also in the presence of an external field [16, 17].

As we have seen the analysis of interfaces is closely related to Γ-convergence and geometric measure theory. As general references for Gamma convergence see the books by Dal Maso [87], and Braides [49], for geometric measure theory and BV spaces the books by Evans and Gariepy [116], and by Ambrosio, Fusco and Pallara [9]. I have studied the Wulff problem and the Wulff construction starting from statistical mechanics in the pioneering paper of Dobrushin, Kotecký and Shlosman [112]. The first proof of Gamma-convergence for the L–P functional is in [6]; the Wulff problem is considered in [19]. A generalization of the analysis to non-local functionals with more general interactions (including the anisotropic case) is due to Alberti and Bellettini [4, 5]; I have followed their approach in Chap. 7. The survival of the Wulff shape when the fixed magnetization is volume dependent, and the difference with one equilibrium value vanishes is studied for non-local functionals in [66].

The gradient flow of the perimeter functional (1.1.1) is motion by curvature. The question of deriving motion by curvature from particle/spin systems has been much studied. A first derivation is in Bonaventura [42], and then in Katsoulakis and Souganidis [151], for a system of particles which in the hydrodynamic limit are described by a reaction–diffusion equation. In [91], it is proved that in a parabolic scaling the non-local evolution equation (6.2.2.7) converges to the case of motion by curvature. The analogous result for the Glauber dynamics in Ising spins with Kac potentials is proved in [93] and in [58, 153] and [152] (for anisotropic interactions) where convergence (to viscosity solutions in [153] and [152]) is extended past the appearance of singularities in the limit motion. For the notion of viscosity solutions of motion by curvature, see [117]. In [14] Barles and Souganidis have proved criteria for the convergence to generalized motion by curvature, which essentially reduce the analysis to the case of classical solutions. The theory includes anisotropic interactions and uses ideas related to the De Giorgi frontiers.

The mobility coefficient in the motion by curvature is related to the surface tension by an Einstein relation, as clarified by Spohn [203]. Its validity in the Ising case with Kac potential has been proved in [57], and in [20] in the anisotropic case, where also strict positivity of the stiffness tensor is proved.

All the above results on convergence to motion by curvature have been obtained by suitably tuning the parabolic space–time scaling parameter (say ϵ) with the Kac scaling parameter γ or other parameters which enter in the definition of the microscopic evolution. The choice is made in such a way that the microscopic evolution is close to the mesoscopic evolution for a time sufficiently long for the latter to be well approximated by the motion by curvature. A direct link between microscopic evolution and motion by curvature has only been proved, as far as I know, by Spohn [203] in the n.n. Ising model at 0 temperature and with some assumptions of monotonicity on the profiles, and by Funaki and Spohn for SOS models [127]. More is known about the derivation of motion of interfaces in the context of the "Stefan problem" [27, 190]. See also [105], where the Stefan problem is derived from a non-local version of the phase field equations.

Stationary solutions (bumps) on the line with an external magnetic field have been studied in [99, 100] and in bounded domains with Neumann conditions (with and without magnetic field) in [21, 62].

Part III
Phase transitions in systems with Kac potentials

Chapter 9
Ising systems with Kac potentials

Part I of this book has been devoted to statistical mechanics, Part II to the mesoscopic theory of free energy functionals; in Part III the two issues come together in the analysis of phase transitions in systems with Kac potentials. Phase transitions and minimization problems are indeed closely related in statistical mechanics since Peierls presented his pioneering ideas on how to prove that the ground states of the Ising model survive at non-zero temperature. Pirogov and Sinai have next developed a theory which extends the validity of the approach to a large class of lattice systems.

The mean field idea is an apparently different road to phase transitions, also often present in the statistical mechanics literature since its early stages. The mean field model is clearly an approximation and the van der Waals theory of liquid–vapor phase transitions is maybe the first quantitative step to apply the mean field ideas to physical systems. A deeper understanding came with the rigorous derivation of the van der Waals theory from statistical mechanics due to Kac and Lebowitz–Penrose in the sixties. The basic question in "limit theories" as the above is always the same: does a phase transition present in the limit persist in true systems before the limit is taken? Since the mean field is approximated by systems with long range interactions, the question is whether models with long but finite range interactions behave as a mean field does, at least to the extent that they share the type of phase transitions. This is analogous to the classical problem of the persistence of ground states at small positive temperatures, which is the object of the original work of Pirogov and Sinai. A small temperature corresponds to a small inverse interaction range, and the hope is to prove the validity of the Peierls argument and of the Pirogov–Sinai theory in such a new context, where the mean field minimizers play the role of the ground states and the inverse interaction range becomes the small parameter of the theory. One of the main advantages of this approach is that it is not necessarily confined to lattice systems; a perturbation of the mean field may also include continuum particle systems and in this way the theory of phase transitions has indeed been extended to the continuum.

As is made clear by the work of Lebowitz–Penrose and Gates–Penrose the relevant quantity to look at rather than being the mean field free energy is the corresponding free energy functional. The meta-theorem we hope for would say that if the free energy functional has distinct minimizers, then its Kac approximation has distinct DLR measures. In this chapter we will prove that this is indeed correct for the Ising model with ferromagnetic Kac interactions, and in the next chapters that the same holds for the LMP model. The latter is a particle model introduced by Lebowitz, Mazel and Presutti to study phase transitions in $\mathbb{R}^d$. Unlike Ising, in the LMP model there is no symmetry between the plus and minus phases and the whole machinery of the Pirogov–Sinai theory is needed.

Errico Presutti, *Scaling Limits in Statistical Mechanics and Microstructures in Continuum Mechanics*, © Springer 2009

Even though our analysis is restricted to the above two models, its applicability seems much wider. There are a few stability conditions on the free energy functionals which must be verified and then the procedure seems to be rather model independent.

9.1 Main results

The phase diagram of an Ising system with Kac potential and the phase diagram of the corresponding mean field model are close to each other when the scaling parameter γ of the Kac potential is sufficiently small. More precisely, we shall prove that for Ising systems with ferromagnetic Kac potentials at zero magnetic field and in $d \geq 2$ dimensions, the inverse critical temperature $\beta_c(\gamma)$ is close for small γ to its mean field value which, in the models we consider, is equal to 1. It follows indeed from the Dobrushin's high temperature uniqueness theorem, see Corollary 3.1.3.3, that for any $\beta < 1$ and γ correspondingly small, there is only one DLR measure. We shall prove here the converse statement, which is the main result of the chapter: for any $\beta > 1$, there is $\gamma(\beta) > 0$, so that if $\gamma \leq \gamma(\beta)$ the set of DLR measures is not a singleton. It then follows from the Griffiths inequalities (see Sect. 9.1) that $\beta_c(\gamma) \leq \beta$ for $\gamma \leq \gamma(\beta)$. The upper and the lower bounds together prove that $|\beta_c(\gamma) - 1|$ vanishes as $\gamma \to 0$. The analogous statement on the existence of two distinct DLR measures has been proved in Theorem 3.1.2.1 for the n.n. ferromagnetic Ising model at low temperatures. Our analysis will indeed show that Ising with Kac interactions at $\beta > 1$ behaves, after suitable coarse graining, as n.n. Ising at temperature $\gamma^{-d}\beta$, if γ is sufficiently small.

We shall also state without proof a stronger theorem, which says that there are two and only two ergodic (with respect to space translations) DLR measures and that their σ-algebra at infinity is trivial. The analogous statement will be proved in the next chapters for the LMP model, and the proof could be easily adapted to the simpler Ising case.

9.1.1 The model

In this chapter we shall study the Ising spin system in $\mathbb{Z}^d$, $d \geq 2$, with zero magnetic field, $h = 0$, and with Kac potential $J_\gamma(x, y) = \gamma^d J(\gamma x, \gamma y)$. Referring to Sect. 2.1 for the general notation and definitions, we just recall here that $\gamma > 0$ is the Kac "scaling parameter"; that the interaction energy of two spins $\sigma(x)$ and $\sigma(y)$ is $-J_\gamma(x, y)\sigma(x)\sigma(y)$; and that $J(r, r')$ is a smooth, symmetric, translational invariant probability kernel supported by $|r - r'| \leq 1$ (the precise assumptions on J_γ are stated in Sect. 4.2.1). We shall refer to this hereafter for brevity as the Ising model with Kac potential.

9.1.2 Bounds on the inverse critical temperature

For any $\gamma > 0$, the model is a finite range, ferromagnetic Ising system to which the general considerations of Sect. 3.1 apply. Thus, we already know from the general theory that, if $h \neq 0$, then there is a unique DLR measure, while, if $h = 0$, then there is a critical inverse temperature $\beta_c(\gamma) > 0$, so that for all $\beta < \beta_c(\gamma)$ there is only one DLR measure, while for all $\beta > \beta_c(\gamma)$ there are at least two distinct DLR measures $\mu^{\pm}_{\beta,\gamma}$. They can be obtained by taking the thermodynamic limit of finite volume Gibbs measures with, respectively, + and − boundary conditions.

The specific structure of the model allows for more detailed information, in particular, about the behavior of $\beta_c(\gamma)$ for small γ. A lower bound on $\beta_c(\gamma)$ follows from the Dobrushin uniqueness condition (3.1.3.4) and an upper bound from the Peierls argument of Sect. 3.1.2. The issue is discussed below.

9.1.3 Lower bounds on the inverse critical temperature

The quantity $j_\gamma := \sum_{x\neq 0} J_\gamma(0,x)$ is called the "total interaction strength." As $\gamma \to 0$, j_γ converges to 1, as j_γ is the Riemann sum of the integral $\int J(0,r)\,dr = 1$. Due to the smoothness of J, there is a constant $c > 0$, so that

$$|j_\gamma - 1| \leq c\gamma. \tag{9.1.3.1}$$

Theorem 9.1.3.1 *The inverse critical temperature $\beta_c(\gamma)$ is bounded from below:*

$$\beta_c(\gamma) \geq j_\gamma^{-1} \geq \frac{1}{1+c\gamma}, \tag{9.1.3.2}$$

with c as in (9.1.3.1).

Proof The proof is contained in the Dobrushin uniqueness theorem. In fact, by Corollary 3.1.3.3 the uniqueness statement follows from the condition (3.1.3.4), which in our case, using translational invariance, reads $\beta j_\gamma < 1$; hence (9.1.3.2). □

Since $1 = \beta_c^{\mathrm{mf}}$ is the inverse mean field critical temperature, β_c^{mf} is a lower bound (in the limit $\gamma \to 0$) for the true inverse critical temperature—a statement which remains valid in a large variety of models.

9.1.4 Upper bounds on the inverse critical temperature

The proof that there is a critical temperature, namely that $\beta_c(\gamma) < \infty$, is already contained in the n.n. results of Chap. 3, at least if we suppose that $J_\gamma(x,y) \geq$

$b(\gamma) > 0$ when $|x - y| = 1$. In fact, by the Griffiths inequalities the average value of the spin at 0 in the plus Gibbs state in a region $\Lambda \ni 0$ is bounded from below by $G_{\gamma,\Lambda}(\sigma(0)|\mathbf{1}_{\Lambda^c}) \geq G^{\text{n.n.}}_{\gamma,\Lambda}(\sigma(0)|\mathbf{1}_{\Lambda^c})$, where the r.h.s. is the Gibbs measure with n.n. interaction of strength $b(\gamma)$. Theorem 3.1.2.1 shows that $G^{\text{n.n.}}_{\gamma,\Lambda}(\sigma(0)|\mathbf{1}_{\Lambda^c}) \geq \alpha > 0$ uniformly in Λ if β is large enough; hence, a fortiori $G_{\gamma,\Lambda}(\sigma(0)|\mathbf{1}_{\Lambda^c}) \geq \alpha$. Thus, there are two distinct DLR measures (produced by taking thermodynamic limits with all $+1$ and all -1 boundary conditions, respectively) and $\beta_c(\gamma) < \infty$. However, the bound obtained in this way for $\beta_c(\gamma)$ is far from optimal (as for γ small, $b(\gamma) \approx \gamma^d$, as most of the interaction has been dropped), and the main result in this chapter will be to show that the Peierls argument of Sect. 3.1.2 can be reproduced without cutting the interaction to the n.n. case, so that the interaction is totally active. Indeed, in the next section we shall prove the following.

Theorem 9.1.4.1 *In the Ising model with Kac potential (of Sect.* 9.1.1*) with* $d \geq 2$ *the following holds. Given any* $\delta > 0$*, there is* $\gamma(\delta) > 0$*, so that* $\beta_c(\gamma) < 1 + \delta$ *for any* $\gamma < \gamma(\delta)$*. In particular, given any* $\beta > 1$ *for any* γ *small enough, there are two DLR measures* $\mu^{\pm}_{\beta,\gamma}$*, such that* $0 < \mu^{+}_{\beta,\gamma}(\sigma(x)) = -\mu^{-}_{\beta,\gamma}(\sigma(x))$.

In Chap. 12 we shall prove a theorem for the LMP particle model whose proof can be easily extended to cover the following.

Theorem 9.1.4.2 *In the same context as of Theorem* 9.1.4.1*, the two DLR measures* $\mu^{\pm}_{\beta,\gamma}$ *have trivial* σ*-algebra at infinity, they are translational invariant and any other translational invariant DLR measure is a convex combination of* $\mu^{\pm}_{\beta,\gamma}$ *(namely, they are the only ergodic DLR measures).*

9.2 Existence of two distinct DLR measures

In this section we shall prove Theorem 9.1.4.1 by "constructing" two distinct DLR measures obtained as limits of Gibbs measures with two different classes of boundary conditions, related to each other by a spin flip. We shall prove that this flipping gives rise to a change in the distribution of the spins which propagates throughout the whole system, and which persists into the thermodynamic limit.

9.2.1 Scheme of proof

The proof of Theorem 9.1.4.1 will exploit the scaling properties of the interaction to relate Gibbs probabilities and non-local free energy functionals; this part uses the coarse graining estimates proved in Sect. 4.2, when studying the Lebowitz–Penrose limit. In this way the Peierls estimates established in Sect. 6.4 for the free energy functional extend to the Ising spins, proving Peierls bounds on the probability of contours. The occurrence of a phase transition then follows from classical arguments analogous to those outlined in Sect. 3.1.2 in the n.n. case. In the sequel, we shall extensively use the notation, definitions and theorems of Sects. 6.3–6.4, which, together with Sect. 4.2, provide the background for this chapter.

9.2.2 A heuristic argument

To describe the ideas behind the proof of Theorem 9.1.4.1 we consider a simplified model with a new, no longer translational invariant interaction $J^*_\gamma(x,y)$ which is constant on the cubes of side $\ell_{-,\gamma}$, $\ell_{-,\gamma}=\gamma^{-1+\alpha}$, $\alpha>0$ being a small parameter, and for simplicity we assume $\ell_{-,\gamma}\in\{2^n$ and $n\in\mathbb{N}\}$. As we shall see, J_γ and J^*_γ are very close to each other when γ is small. Let $\mathcal{D}^{(\ell_{-,\gamma})}$ be the partition of $\mathbb{R}^d$ into cubes of side $\ell_{-,\gamma}$ as in Sect. 4.2.2, call $C_i=C_i^{(\ell_{-,\gamma})}$, $i\in\ell_{-,\gamma}\mathbb{Z}^d$, the cube of $\mathcal{D}^{(\ell_{-,\gamma})}$ which contains i; and $\ell_{-,\gamma}\mathbb{Z}^d$ the lattice with mesh $\ell_{-,\gamma}$. With such a notation we define $J^*_\gamma(x,y)=J_\gamma(i,j)$, $x\in C_i$, $y\in C_j$. To exploit the constancy of J^*_γ we introduce the block spins $m(i)$, where $m(i)$ is the magnetization density in the cube C_i: $m(i)=\frac{1}{\ell^d_{-,\gamma}}\sum_{x\in C_i}\sigma(x)$. Let Λ be a bounded $\mathcal{D}^{(\ell_{-,\gamma})}$-measurable region, $\Lambda^*:=\ell_{-,\gamma}\mathbb{Z}^d\cap\Lambda$; then the energy $H^*_{\gamma,\Lambda}(\sigma_\Lambda)$ with couplings J^*_γ is $H^*_{\gamma,\Lambda}(\sigma)=\ell^d_{-,\gamma}K^*_{\gamma,\Lambda^*}(m_{\Lambda^*})+\text{const}$, where

$$K^*_{\gamma,\Lambda^*}(m_{\Lambda^*})=-\frac{1}{2}\sum_{i,j\in\Lambda^*}[\ell^d_{-,\gamma}J^*_\gamma(i,j)]m_{\Lambda^*}(i)m_{\Lambda^*}(j).$$

Notice that the new pair interaction $[\ell^d_{-,\gamma}J^*_\gamma(i,j)]$ has strength $\sum_j\ell^d_{-,\gamma}J^*_\gamma(i,j)$, which converges to 1 as $\gamma\to 0$. The "const" above takes into account the self-interaction of the spins, which are present in $K^*_\gamma(m)$ and absent in $H^*_\gamma(\sigma)$. The original partition function $Z^*_{\gamma,\Lambda}$ becomes, after dropping the "const"

$$Z^*_{\gamma,\Lambda^*}=\sum_{\{m_{\Lambda^*}\}}e^{-(\beta\ell^d_{-,\gamma})[K_{\gamma,\Lambda^*}(m_{\Lambda^*})-\beta^{-1}I_{\Lambda^*}(m_{\Lambda^*})]},\tag{9.2.2.1}$$

where $I_{\Lambda^*}(m_{\Lambda^*})=\sum_{i\in\Lambda^*}I(m(i))$ and $I(m(i))$ is the entropy

$$I(m(i))=\frac{1}{\ell^d_{-,\gamma}}\log\Big\{\#\text{ of }\sigma_{C_i}\text{ such that }\sum_{x\in C_i}\sigma(x)=m(i)|C_i|\Big\}.\tag{9.2.2.2}$$

Equation (9.2.2.1) is the usual Gibbs partition function in Λ^* for block spin variables m_{Λ^*} with hamiltonian $H_{\text{new}}:=[K_{\gamma,\Lambda^*}(m_{\Lambda^*})-\beta^{-1}I_{\Lambda^*}(m_{\Lambda^*})]$ and inverse temperature $\beta\ell^d_{-,\gamma}$. Since $\beta\ell^d_{-,\gamma}\to\infty$ as $\gamma\to 0$, we are in the low temperature regime, where well developed techniques are available for studying phase transitions.

The actual proof of Theorem 9.1.4.1 requires one to deal with the true interaction J_γ, rather than the simplified one of the above heuristics; and to adapt the low temperature analysis in Sect. 3.1.2 of the n.n. interaction to the effective hamiltonian obtained after the block spin transformation. The effective hamiltonian will be related to the non-local free energy functional of Chap. 6 and the proof of the Peierls bounds will come from the solution of a variational problem for a free energy functional, which has been studied in Chap. 6. As we shall see we can avoid an exact computation of the effective hamiltonian which describes the interactions among block spins by using a coarse graining analysis à la Lebowitz–Penrose.

9.2.3 Coarse graining and block spins

To implement the program outlined at the end of the previous subsection we do a coarse graining transformation which uses three basic parameters, two lengths $\ell_\pm$ and an accuracy $\zeta > 0$, which all depend on γ. There are wide margins for their choices, but, for the sake of definiteness, we fix them once and for all by

$$\ell_{\pm,\gamma} = \gamma^{-(1\pm\alpha)}, \qquad \zeta = \gamma^a, \quad 1 \gg \alpha \gg a > 0. \tag{9.2.3.1}$$

Precise bounds can be evinced from the proofs, but since there is no critical dependence, we shall not enter into more details.

For notational simplicity, we suppose that $\ell_{\pm,\gamma} \in \{2^n, n \in \mathbb{N}_+\}$; this is a restriction on γ and α which could be removed by asking for proportionality rather than equality in (9.2.3.1), and choosing the constants in such a way that $\mathcal{D}^{(\ell_{+,\gamma})}$ is coarser than $\mathcal{D}^{(\gamma^{-1})}$, which is coarser than $\mathcal{D}^{(\ell_{-,\gamma})}$; see Sect. 4.2.2 for notation and definitions (the condition is automatically satisfied if $\ell_{\pm,\gamma}, \gamma^{-1} \in \{2^n, n \in \mathbb{N}_+\}$).

The basic objects are three phase indicators designed to define the presence or absence of a phase at a local level. We need notions progressively stronger; hence the use of several phase indicators. They involve block spin averages:

$$\sigma^{(\ell)}(r) := \frac{1}{|C_r^{(\ell)}|} \sum_{x \in C_r^{(\ell)} \cap \mathbb{Z}^d} \sigma(x), \quad r \in \mathbb{R}^d. \tag{9.2.3.2}$$

Our three phase indicators are denoted by

$$\eta^{(\zeta,\ell_{-,\gamma})}(\sigma; r), \qquad \theta^{(\zeta,\ell_{-,\gamma},\ell_{+,\gamma})}(\sigma; r), \qquad \Theta^{(\zeta,\ell_{-,\gamma},\ell_{+,\gamma})}(\sigma; r); \quad \sigma \in \mathcal{X}, r \in \mathbb{R}^d.$$

As ζ, $\ell_{-,\gamma}$ and $\ell_{+,\gamma}$ have been fixed, see (9.2.3.1), we simply write in the sequel η, θ and Θ unless we want to underline their dependence on the parameters.

- $\eta(\sigma; r) = \pm 1$ if $|\sigma^{(\ell_{-,\gamma})}(r) \mp m_\beta| \le \zeta$, and $= 0$ otherwise.
- $\theta(\sigma; r) = 1$ $[= -1]$, if $\eta(\sigma; r') = 1$ $[= -1]$, for all $r' \in C_r^{(\ell_{+,\gamma})}$, and $= 0$ otherwise.
- $\Theta(\sigma; r) = 1$ $[= -1]$, if $\theta(\sigma; r) = 1$ $[= -1]$, on $C_r^{(\ell_{+,\gamma})} \cup \delta_{\text{out}}^{\ell_{+,\gamma}}[C_r^{(\ell_{+,\gamma})}]$ and $= 0$ otherwise.

This is exactly the same definition as given in Sect. 6.4.4 once we identify $\sigma^{(\ell)}(r)$ with $\fint_{C_r^{(\ell)}} m(r')dr'$; $\ell_{\pm,\gamma}$ with $\ell_\pm$ and the interaction range γ^{-1} with 1. We use the same notation, and refer to Sect. 6.4.4 for the definition of the "± correct regions," plus/minus contours and their geometry.

9.2.4 *Weight of a contour*

Call $G_{\gamma,\Lambda,\sigma} = G_{\gamma,\Lambda}(\cdot|\sigma_{\Lambda^c})$ the Gibbs conditional probability in Λ, given σ_{Λ^c}, (σ_{Λ^c} denoting the restriction of σ to Λ^c).

Definition The weight $W_\gamma^+(\Gamma;\sigma)$ of a plus contour $\Gamma = (\text{sp}(\Gamma), \eta_\Gamma)$ is defined for any σ, such that $\eta(\sigma;\cdot) = 1$ on $\delta_{\text{out}}^{\gamma^{-1}}[c(\Gamma)]$ ($c(\Gamma)$ as in (6.4.4.1)) by

$$\frac{G_{\gamma,c(\Gamma),\sigma}(\{\eta(\sigma'_{c(\Gamma)};r) = \eta_\Gamma(r), r \in \text{sp}(\Gamma);\ \Theta(\sigma'_{c(\Gamma)};r) = \pm 1, r \in A^\pm(\Gamma)\})}{G_{\gamma,c(\Gamma),\sigma}(\{\Theta(\sigma'_{c(\Gamma)};r) = 1, r \in \text{sp}(\Gamma) \cup A(\Gamma)\})}. \tag{9.2.4.1}$$

$A^\pm(\Gamma)$ is as in (6.4.4.4), and we have $A(\Gamma) = A^+(\Gamma) \cup A^-(\Gamma)$. $W_\gamma^-(\Gamma;\sigma)$ is defined by interchanging plus and minus.

9.2.5 *Peierls bounds*

The key estimate in the proof of Theorem 9.1.4.1 is as follows.

Theorem 9.2.5.1 (Peierls bounds) *There is $c_1 > 0$ so that for any ± contour Γ and any σ such that $\eta(\sigma;\cdot) = \pm 1$ on $\delta_{\text{out}}^{\gamma^{-1}}[c(\Gamma)]$,*

$$W_\gamma^\pm(\Gamma;\sigma) \le \exp\left\{ -\beta\frac{c_1}{2}(\zeta^2\ell_{-,\gamma}^d) N_\Gamma \right\}, \quad N_\Gamma = \frac{|\text{sp}(\Gamma)|}{\ell_{+,\gamma}^d}. \tag{9.2.5.1}$$

The proof of the theorem is postponed to Sect. 9.3. By comparison with the energy of the contours in the n.n. Ising model of Sect. 3.1.2, and recalling that N_Γ is the number of block spins present in the contour Γ, we observe that the system of block spins defined by Θ resembles an Ising model at inverse effective temperature proportional to $(\zeta^2\ell_{-,\gamma}^d) = \gamma^{-d+2a+\alpha d}$. The coefficient diverges as $\gamma \to 0$ ("almost like" γ^{-d}, because $2a + \alpha d \ll 1$), and the proof then proceeds as in the n.n. case at low temperatures showing (see Corollary 9.2.8.2) that plus and minus boundary conditions produce distinct measures in the thermodynamic limit. The notion of plus and minus boundary conditions is, however, a delicate issue, which we address in the next subsection.

9.2.6 Diluted Gibbs measures

We here introduce the important notion of "diluted Gibbs measures"—diluted because there are less contours than normally. The motivation is to define finite volume Gibbs measures which look like the measures in the internal part of a contour, clearly having in mind an iteration scheme. Each internal part of a contour is characterized by Θ being constantly equal to a non-zero value on the boundary, Theorem 6.4.4.1. With this in mind, denote by σ^+ any configuration such that $\eta(\sigma^+; r) \equiv 1$. Then the plus diluted Gibbs measure in a bounded $\mathcal{D}^{(\ell_{+,\gamma})}$ region Λ (by default, Λ in the sequel will always denote any such region) with boundary condition σ^+ is

$$G^+_{\gamma,\Lambda,\sigma^+}(\cdot) := G_{\gamma,\Lambda,\sigma^+}\big(\cdot \,|\{\Theta((\sigma_\Lambda, \sigma^+_{\Lambda^c}); r) = 1,\ r \in \delta_{\text{in}}^{\ell_{+,\gamma}}[\Lambda]\}\big), \tag{9.2.6.1}$$

where $((\sigma_\Lambda, \sigma^+_{\Lambda^c}))$ is equal to σ_Λ and $\sigma^+_{\Lambda^c}$ on Λ and Λ^c and

$$G^+_{\gamma,\Lambda,\sigma^+}(\sigma_\Lambda) := \frac{e^{-\beta H_\gamma(\sigma_\Lambda|\sigma^+_{\Lambda^c})}}{Z^+_{\gamma,\Lambda,\sigma^+_{\Lambda^c}}} \mathbf{1}_{\Theta((\sigma_\Lambda,\sigma^+_{\Lambda^c});r)=1\ r\in\delta_{\text{in}}^{\ell_{+,\gamma}}[\Lambda]}, \tag{9.2.6.2}$$

$$Z^+_{\gamma,\Lambda,\sigma^+_{\Lambda^c}} = \sum_{\sigma_\Lambda:\Theta((\sigma_\Lambda,\sigma^+_{\Lambda^c});r)=1\ r\in\delta_{\text{in}}^{\ell_{+,\gamma}}[\Lambda]} e^{-\beta H_\gamma(\sigma_\Lambda|\sigma^+_{\Lambda^c})}, \tag{9.2.6.3}$$

see Figs. 6.1 and 6.2 in Sect. 6.5 for an example.

The minus diluted Gibbs measures are defined by interchanging the role of plus and minus. Notice that $G^\pm_{\gamma,\Lambda,\sigma^\pm}(\cdot)$ does not depend on the values of $\sigma^\pm$ outside $\delta_{\text{out}}^{\gamma^{-1}}[\Lambda]$; thus, the condition that $\eta(\sigma^\pm; r) = \pm 1$ for all r is stronger than needed. It just follows from the definition, and the reader may check its validity in the example of Fig. 6.1,

Theorem 9.2.6.1 *Let $\Lambda_0 = \Lambda \setminus \delta_{\text{in}}^{\ell_{+,\gamma}}[\Lambda]$; then, with $G^+_{\gamma,\Lambda,\sigma^+}(\cdot)$ probability 1, any contour Γ of a configuration σ_Λ is such that*

$$c(\Gamma) \subset \Lambda_0 \quad \text{and} \quad \delta_{\text{in}}^{\ell_{+,\gamma}}[\Lambda_0] \cap c(\Gamma) = \delta_{\text{in}}^{\ell_{+,\gamma}}[\Lambda_0] \cap \text{sp}(\Gamma), \tag{9.2.6.4}$$

with $\eta_\Gamma(r) = 1$ for all $r \in \delta_{\text{in}}^{\ell_{+,\gamma}}[\Lambda_0] \cap \text{sp}(\Gamma)$. In particular, if Γ is a minus contour, then $c(\Gamma) \subset \Lambda_0 \setminus \delta_{\text{in}}^{\ell_{+,\gamma}}[\Lambda_0]$. Analogous statements hold for $G^-_{\gamma,\Lambda,\sigma^-}(\cdot)$.

9.2.7 Probability of a contour

In the sequel, we further restrict the regions Λ by requiring them to be simply connected. This is essential in the proofs, because the main trick in proving the Peierls bounds is a spin flip in the whole internal part of a contour, and this is not possible if the interior of the contour intersects Λ^c, where the boundary conditions are fixed and cannot be flipped.

For simplicity, we refer below to the plus diluted Gibbs measure, but the same results hold for the minus one. We denote by $\mathcal{X}(\Gamma)$ the set of all configurations which have Γ as one of their contours.

Theorem 9.2.7.1 *If Λ is simply connected,*

$$G^{\pm}_{\gamma,\Lambda,\sigma^{\pm}}\big(\mathcal{X}(\Gamma)\big) \le \|W_\gamma(\Gamma;\cdot)\|_\infty. \tag{9.2.7.1}$$

Proof We shall prove the theorem for $G^{+}_{\gamma,\Lambda,\sigma^{+}}$ with Γ a plus contour, as this is what is needed in the proof of the Peierls bounds. The geometry below can be visualized in the example of Figs. 6.1 and 6.2 in Sect. 6.5. By Theorem 9.2.6.1 $c(\Gamma) \subset \Lambda_0$; then, using the DLR property and writing $A^{\pm}$ and A_{ext} for $A^{\pm}(\Gamma)$ and $A_{\text{ext}}(\Gamma)$ (see (6.4.4.2) and (6.4.4.5)), we get by the DLR property,

$$G^{+}_{\gamma,\Lambda,\sigma^{+}}\big(\mathcal{X}(\Gamma)\big) = \frac{G_{\gamma,\Lambda,\sigma^{+}}(f(\sigma'_{\Lambda\setminus c(\Gamma)}))}{G_{\gamma,\Lambda,\sigma^{+}}(\{\Theta((\sigma'_\Lambda,\sigma^{+}_{\Lambda^c});r)=1, r\in\delta_{\text{in}}^{\ell_+,\gamma}[\Lambda]\})}, \tag{9.2.7.2}$$

where $f(\sigma'_{\Lambda\setminus c(\Gamma)})$ is equal to the $G_{\gamma,c(\Gamma),\sigma'_\Lambda}$ probability of all $\sigma_{c(\Gamma)}$ in

$$\big\{(\sigma_{c(\Gamma)},\sigma'_{\Lambda\setminus c(\Gamma)},\sigma^{+}_{\Lambda^c})\in\mathcal{X}(\Gamma);\,\Theta(\sigma_{c(\Gamma)},\sigma'_{\Lambda\setminus c(\Gamma)},\sigma^{+}_{\Lambda^c};\cdot)=1,\ \text{on}\,\delta_{\text{in}}^{\ell_+,\gamma}[\Lambda]\big\}.$$

Then $f(\sigma'_{\Lambda\setminus c(\Gamma)})$ is bounded by

$$G_{\gamma,c(\Gamma),\sigma'_\Lambda}\big(\mathcal{X}_0(\Gamma)\big)\mathbf{1}_{\Theta((\sigma'_{\Lambda\setminus c(\Gamma)},\sigma^{+}_{\Lambda^c\cup c(\Gamma)});r)=1,r\in\delta_{\text{in}}^{\ell_+,\gamma}[\Lambda]\cup A_{\text{ext}}},$$

where $\mathcal{X}_0(\Gamma)$ is the set

$$\big\{\sigma_{c(\Gamma)}:\eta(\sigma_{c(\Gamma)};r)=\eta_\Gamma(r), r\in\text{sp}(\Gamma);\,\Theta(\sigma_{c(\Gamma)};r)=\pm1, r\in A^{\pm}(\Gamma)\big\}.$$

By (9.2.4.1), $G_{\gamma,c(\Gamma),\sigma'_\Lambda}(\mathcal{X}_0(\Gamma))$ is bounded by

$$\|W^{+}_\gamma(\Gamma;\cdot)\|_\infty G_{\gamma,c(\Gamma),\sigma'_\Lambda}\big(\Theta(\sigma_{c(\Gamma)};r)=1, r\in\text{sp}(\Gamma)\cup A^{\pm}(\Gamma)\big).$$

Thus, using backwards the DLR property, the numerator in (9.2.7.2) is bounded by $\|W^{+}_\gamma(\Gamma;\cdot)\|_\infty$ times

$$G_{\gamma,\Lambda,\sigma^{+}}\big(\{\Theta((\sigma'_{\Lambda\setminus c(\Gamma)},\sigma^{+}_{\Lambda^c\cup c(\Gamma)});r)=1, r\in\delta_{\text{in}}^{\ell_+,\gamma}[\Lambda]\cup A_{\text{ext}}\}\cap\{\Theta(\sigma'_\Lambda;r)=1,$$
$$r\in\text{sp}(\Gamma)\cup A^{\pm}(\Gamma)\}\big).$$

If σ'_Λ is in the intersection of the two sets, then $\Theta((\sigma'_\Lambda,\sigma^{+}_{\Lambda^c});r)=1, r\in\delta_{\text{in}}^{\ell_+,\gamma}[\Lambda]$, because, by (9.2.6.4), the only part of $c(\Gamma)$ connected to $\delta_{\text{in}}^{\ell_+,\gamma}[\Lambda]$ is contained in $\text{sp}(\Gamma)$, and $\eta(\sigma'_\Lambda;r')=1, r'\in\text{sp}(\Gamma)$. Hence the above is bounded by $\|W^+_\gamma(\Gamma;\cdot)\|_\infty \cdot G_{\gamma,\Lambda,\sigma^+}(\{\Theta((\sigma'_\Lambda,\sigma^+_{\Lambda^c});r)=1, r\in\delta^{\ell_+,\gamma}_{\text{in}}[\Lambda]\})$ and (9.2.7.2) is bounded by $\|W^+_\gamma(\Gamma;\cdot)\|_\infty$. □

By an inductive procedure in the numbers of contours, Theorem 9.2.7.1 can be extended as follows (as we shall not use such stronger statement, we omit the (simple) proof):

Theorem 9.2.7.2 *If Λ is simply connected, for any family $(\Gamma_1,\dots,\Gamma_n)$ of contours*

$$G^+_{\gamma,\Lambda,\sigma^+}\big(\mathcal{X}(\Gamma_1,\dots,\Gamma_n)\big) \le \prod_{i=1}^n \|W_\gamma(\Gamma_i;\cdot)\|_\infty. \tag{9.2.7.3}$$

9.2.8 Counting contours

In this subsection we shall show the persistence of the memory of the boundary conditions; a result which will imply the existence of two distinct DLR measures.

Theorem 9.2.8.1 *Suppose that the Peierls bounds in* (9.2.5.1) *are satisfied for all contours Γ, and that γ is small enough. Then for any Λ, $\sigma^\pm$ and $G^\pm_{\gamma,\Lambda,\sigma^\pm}$ as in Theorem* 9.2.6 *and any $r\in\Lambda$*

$$G^\pm_{\gamma,\Lambda,\sigma^\pm}(\{\Theta(\sigma;r)=\pm1\}) \ge 1 - \exp\left\{-\beta\frac{c_1}{4}\,(\zeta^2\ell^d_{-,\gamma})\right\}. \tag{9.2.8.1}$$

Proof For notational simplicity, we restrict ourselves to the plus case. We claim that, calling the collection of all plus contours $\{\Gamma\}^+$,

$$\{\Theta(\sigma;r)<1\}\subset \bigcup_{\Gamma\in\{\Gamma\}^+: r\in c(\Gamma)} \mathcal{X}(\Gamma). \tag{9.2.8.2}$$

The proof goes as follows. There must be a contour Γ for which $c(\Gamma)\ni r$; otherwise r would be connected to Λ^c by a path which does not intersect contours and then $\Theta(\sigma:r)=1$. By the definition of contours, if also Γ' is such that $c(\Gamma')\ni r$, then either $c(\Gamma)\subset c(\Gamma')$ or vice versa; thus, there is a largest $c(\Gamma)$, and such a Γ is a plus contour, because $\Theta=1$ on Λ^c.

By (9.2.8.2), (9.2.7.1) and (9.2.5.1) we have

$$G^\pm_{\gamma,\Lambda,\sigma^\pm}(\{\Theta(\sigma;r)<1\}) \le \sum_{\Gamma: c(\Gamma)\ni r} e^{-\beta(c_1/2)\,(\zeta^2\ell^d_{-,\gamma})N_\Gamma}. \tag{9.2.8.3}$$

We shall prove that, for γ small enough, we have

$$\sum_{\Gamma:c(\Gamma)\ni r} \exp\left\{-\beta\frac{c_1}{4}\,(\zeta^2\ell^d_{-,\gamma})N_\Gamma\right\} \le 1, \tag{9.2.8.4}$$

which will then conclude the proof of (9.2.8.3), because $N_\Gamma \geq 3^d$.

Proof of (9.2.8.4) We first observe that there are at most $3^{(\ell_{+,\gamma}/\ell_{-,\gamma})^d N_\Gamma}$ contours with the same $\mathrm{sp}(\Gamma)$, as the above is the number of $\{0, \pm 1\}$-valued, $\mathcal{D}^{(\ell_{-,\gamma})}$-measurable functions on $\mathrm{sp}(\Gamma)$. Thus, using (9.2.5.1) and shrinking sets by a factor $\ell_{+,\gamma}$, so that the partition $\mathcal{D}^{(\ell_{+,\gamma})}$ becomes $\mathcal{D}^{(1)}$,

$$\text{l.h.s. of (9.2.8.4)} \leq \sum_{D:c(D)\ni r} \exp\left\{\left(\gamma^{-2\alpha d}\log 3 - \beta\frac{c_1}{4}(\zeta^2\ell_{-,\gamma}^d)\right)|D|\right\}, \qquad (9.2.8.5)$$

where D denotes a bounded, connected, $\mathcal{D}^{(1)}$-measurable set, and $c(D) = D \cup \mathrm{int}(D)$, $|D|$ the volume of D.

Without loss of generality, we may assume $r = 0$. We decompose the set $\{D : c(D) \ni 0\}$ into equivalence classes, calling D and D' equivalent if D' is a shift of D along the first coordinate direction. The following three properties hold: all elements in the same equivalence class give the same contribution to the sum on the r.h.s. of (9.2.8.5); there is at least one element D in each class such that $D \ni 0$; and there are at most $|D|$ elements in each class. Thus,

$$\text{l.h.s. of (9.2.8.4)} \leq \sum_{D:0\in D} |D| \exp\left\{\left(\gamma^{-2\alpha d}\log 3 - \beta\frac{c_1}{4}(\zeta^2\ell_{-,\gamma}^d)\right)|D|\right\}. \qquad (9.2.8.6)$$

Since $|D| \geq 1$, we have $|D| \leq e^{|D|}$, so that the l.h.s. of (9.2.8.4) is bounded by

$$\sum_{D:0\in D} \exp\left\{\left(\gamma^{-2\alpha d}\log 3 + 1 - \beta\frac{c_1}{4}(\zeta^2\ell_{-,\gamma}^d)\right)|D|\right\}. \qquad (9.2.8.7)$$

The coefficient multiplying $|D|$ in the last term diverges to $-\infty$ as $\gamma \to 0$ and (9.2.8.4) follows from Lemma 3.1.2.4 for γ small enough. □

□

Corollary 9.2.8.2 *For any γ so small that the r.h.s. of (9.2.8.1) is $> 1/2$, there is an increasing sequence Λ_n of bounded $\mathcal{D}^{(\ell_{+,\gamma})}$-measurable regions which invades the whole space, and a configuration σ, $\eta(\sigma;\cdot) \equiv 1$, so that*

$$\lim_{n\to\infty} G^{\pm}_{\gamma,\Lambda_n,\pm\sigma} = \mu^{\pm}_\gamma, \quad \textit{weakly}, \qquad (9.2.8.8)$$

where $\mu^+_\gamma \neq \mu^-_\gamma$ are DLR measures.

Proof Let $\{C_n\}$ be an increasing sequence of $\mathcal{D}^{(\ell_{+,\gamma})}$-measurable cubes which invades the whole space. By compactness, there is a subsequence $\{\Lambda_n\}$ of $\{C_n\}$ so that $G^+_{\gamma,\Lambda_n,\sigma}$ converges weakly. Call μ^+_γ the limit. By spin flip symmetry, $G^-_{\gamma,\Lambda_n,-\sigma}$

converges to a measure μ_γ^- which is the spin flip image of μ_γ^+. We shall first prove that $\mu_\gamma^+ \neq \mu_\gamma^-$. As soon as $\Lambda_n \ni 0$, by Theorem 9.2.8.1,

$$G^+_{\gamma,\Lambda,\sigma}(\{\Theta(\sigma;0)=1\}) \geq b := 1 - \exp\{-\beta(c_1/4)(\zeta^2\ell^d_{-,\gamma})\} > 1/2,$$

the last equality for γ small enough. Therefore,

$$\mu_\gamma^+(\{\Theta(\sigma;0)=1\}) > \frac{1}{2}, \qquad \mu_\gamma^-(\{\Theta(\sigma;0)=-1\}) > \frac{1}{2},$$

the second inequality holds because μ_γ^- is the spin flip image of μ_γ^+.

Thus, $\mu_\gamma^-(\{\Theta(\sigma;0)=1\}) < 1/2$; hence $\mu_\gamma^- \neq \mu_\gamma^+$. We shall next show that μ_γ^+ is DLR; this implies that also its spin flip image, μ_γ^-, is DLR, because the hamiltonian is invariant under spin flip. Given any bounded set Δ, $G^+_{\gamma,\Lambda_n,\sigma}$ is in $\mathcal{G}_\Delta$ (the set of all Gibbs measures in Δ; see Sect. 2.1.6) as soon as $\Delta \subset \Lambda_n \setminus \delta_{\text{in}}^{\ell_{+,\gamma}}[\Lambda_n]$. Since $\mathcal{G}_\Delta$ is weakly closed, μ_γ^+ is in $\mathcal{G}_\Delta$, and, by the arbitrariness of Δ, it follows that μ_γ^+ is DLR. □

9.3 Proof of the Peierls bounds

In this section we shall prove Theorem 9.2.5.1. By symmetry we may restrict ourselves to plus contours. The fraction on the r.h.s. of (9.2.4.1) is the ratio of two partition functions, $W_\gamma^+(\Gamma;\sigma) = \frac{\mathcal{N}(\sigma_{A_{\text{ext}}})}{\mathcal{D}(\sigma_{A_{\text{ext}}})}$, where, calling

$$D^\pm = \delta_{\text{in}}^{\gamma^{-1}}[\text{int}^\pm(\Gamma)], \qquad D = D^+ \cup D^-, \qquad \text{int}_0^\pm(\Gamma) = \text{int}^\pm(\Gamma) \setminus D \quad (9.3.0.1)$$

(see Sect. 6.4.4 for the notation), and writing $\sigma_D = (\sigma_{D^-}, \sigma_{D^+})$, the numerator $\mathcal{N}(\sigma_{A_{\text{ext}}})$ is equal to

$$\sum_{\sigma_D:\eta(\sigma_{D^\pm};r)=\pm 1, r\in D^\pm} e^{-\beta H_{\gamma,D}(\sigma_D)} Z^-_{\gamma,\text{int}_0^-(\Gamma),\sigma_{D^-}} Z^+_{\gamma,\text{int}_0^+(\Gamma),\sigma_{D^+}}$$
$$\times \sum_{\sigma_{\text{sp}(\Gamma)}:\eta(\sigma_{\text{sp}(\Gamma)};r)=\eta_\Gamma(r), r\in\text{sp}(\Gamma)} e^{-\beta H_{\gamma,\text{sp}(\Gamma)}(\sigma_{\text{sp}(\Gamma)}|\sigma_D,\sigma_{A_{\text{ext}}})}, \quad (9.3.0.2)$$

while the denominator $\mathcal{D}(\sigma_{A_{\text{ext}}})$ is

$$\sum_{\sigma_D:\eta(\sigma_D;r)=1, r\in D} e^{-\beta H_\gamma(\sigma_D)} Z^+_{\gamma,\text{int}_0^-(\Gamma),\sigma_{D^-}} Z^+_{\gamma,\text{int}_0^+(\Gamma),\sigma_{D^+}}$$
$$\times \sum_{\sigma_{\text{sp}(\Gamma)}:\eta(\sigma_{\text{sp}(\Gamma)};r)=1, r\in\text{sp}(\Gamma)} e^{-\beta H_{\gamma,\text{sp}(\Gamma)}(\sigma_{\text{sp}(\Gamma)}|\sigma_D,\sigma_{A_{\text{ext}}})}. \quad (9.3.0.3)$$

We shall proceed by bounding one by one the terms appearing on the r.h.s. of (9.3.0.2), obtaining in the end a product of two factors: one is the Peierls bound (that we need to prove), the other one is the partition function $\mathcal{D}(\sigma_{A_{\rm ext}})$, which simplifies with the denominator and concludes the proof of Theorem 9.2.5.1.

9.3.1 Spin flip symmetry

The proof of the Peierls bounds is simple in our Ising model because of spin flip symmetry; in the LMP particle model (studied in the next chapter) the symmetry is missing and the analysis is much more complex. By spin flip symmetry, with σ_{D^-} as in (9.3.0.2),

$$e^{-\beta H_{\gamma,D^-}(\sigma_{D^-})} = e^{-\beta H_{\gamma,D^-}(-\sigma_{D^-})}, \qquad Z^-_{\gamma,{\rm int}_0^-(\Gamma),\sigma_{D^-}} = Z^+_{\gamma,{\rm int}_0^-(\Gamma),-\sigma_{D^-}},$$

so that $\mathcal{N}(\sigma_{A_{\rm ext}})$ becomes equal to

$$\sum_{\sigma_D:\eta(\sigma_D;r)=1,r\in D} e^{-\beta H_{\gamma,D}(\sigma_D)}\, Z^+_{\gamma,{\rm int}_0^-(\Gamma),\sigma_{D^-}}\, Z^+_{\gamma,{\rm int}_0^+(\Gamma),\sigma_{D^+}}$$
$$\times \sum_{\sigma_{{\rm sp}(\Gamma)}:\eta(\sigma_{{\rm sp}(\Gamma)};r)=\eta_\Gamma(r),r\in{\rm sp}(\Gamma)} e^{-\beta H_{\gamma,{\rm sp}(\Gamma)}(\sigma_{{\rm sp}(\Gamma)}|(-\sigma_{D^-},\sigma_{D^+},\sigma_{A_{\rm ext}})}. \qquad (9.3.1.1)$$

In (9.3.1.1) we have a plus diluted partition function also in ${\rm int}_0^-(\Gamma)$ which lets us foresee a cancelation with the one appearing in $\mathcal{D}(\sigma_{A_{\rm ext}})$.

9.3.2 Coarse-graining, reduction to a variational problem

The "gain term" $\exp\{-\beta\frac{c_1}{2}(\zeta^2\ell^d_{-,\gamma})N_\Gamma\}$ in (9.2.5.1) comes from the last factor in (9.3.1.1) (and the corresponding one in $\mathcal{D}(\sigma_{A_{\rm ext}})$). The analogue in the nearest neighbor case is just the energy difference between plus–minus and plus–plus interactions. Here the analysis is not as simple, because we have a full (constrained) partition function to estimate. The key point is that we can use the Lebowitz–Penrose approach to reduce it to a variational problem. Using (4.2.2.12) with $\mathcal{A}$ the whole space, and since for γ small $\epsilon(\gamma)\le 2\gamma^{1/2}$ we obtain by (4.2.2.10) the following bound for the last term in (9.3.1.1):

$$\exp\Big\{-\beta\Big(\inf_{\eta(m_{{\rm sp}(\Gamma)};r)=\eta_\Gamma(r),r\in{\rm sp}(\Gamma)} K(m_{{\rm sp}(\Gamma)}) - c\gamma^{1/2}|{\rm sp}(\Gamma)|\Big)\Big\}, \qquad (9.3.2.1)$$

$$K(m_{{\rm sp}(\Gamma)}) := F_{\gamma,{\rm sp}(\Gamma)}\big(m_{{\rm sp}(\Gamma)}|(-\sigma^{(\gamma^{-1/2})}_{D^-},\sigma^{(\gamma^{-1/2})}_{D^+},\sigma^{(\gamma^{-1/2})}_{A_{\rm ext}})\big),$$

where $F_{\gamma,{\rm sp}(\Gamma)^-}(\cdot|\cdot)$ is the L–P free energy functional defined in (4.2.2.11). The "error" $\beta c\gamma^{1/2}|{\rm sp}(\Gamma)| = \beta c\gamma^{1/2}|\ell^d_{+,\gamma}N_\Gamma$ is harmless, because it is much smaller (for γ small) than the gain term we shall obtain. Observe that we cannot afford an analogous procedure in $\text{int}^-(\Gamma)$, as the error $\beta c\gamma^{1/2}|\text{int}^-(\Gamma)|$ may be much larger than the gain term. For $\text{int}^-(\Gamma)$ transition to continuum is not an option we can choose and we thus need other strategies. In the present case this is the spin flip symmetry used earlier.

9.3.3 The energy estimate

The energy estimate refers to a lower bound for the minimization problem in (9.3.2.2). Here we face right away a new difficulty: the interaction energy with the boundary in $F_{\gamma,\text{sp}(\Gamma)^-}(\cdot|\cdot)$ has order $\exp(\beta c \frac{|\delta_{\text{out}}^{\ell_{+,\gamma}}[\text{int}(\Gamma)]|}{\ell_{+,\gamma}^d} \ell_{+,\gamma}^{d-1}\gamma^{-1})$, which, for γ small, may be much larger than the gain term $\beta\frac{c_1}{2}(\zeta^2\ell_{-,\gamma}^d)N_\Gamma$. What helps us here is the definition of contours which has been conceived just for this purpose; namely, with the property that there are corridors of size $\ell_{+,\gamma}$ at the boundaries of $\text{sp}(\Gamma)$, where η is constantly equal to 1 or to -1. The purpose of such "safety zones" is explained in detail in Sect. 6.4, to which the reader is referred; here, we just quote the results.

By (6.4.8.3), there is a function $\psi_{\text{sp}(\Gamma)}(r)$, $r \in \text{sp}(\Gamma)$ with values in $(m_\beta - (1-\kappa_0)\zeta$,
$m_\beta + (1-\kappa_0)\zeta)$, $\kappa_0 > 0$, such that

$$F_{\gamma,\text{sp}(\Gamma)}\big(m_{\text{sp}(\Gamma)}|(-\sigma_{D^-}^{(\gamma^{-1/2})}, \sigma_{D^+}^{(\gamma^{-1/2})}, \sigma_{A_{\text{ext}}}^{(\gamma^{-1/2})})\big)$$
$$\geq c_1(\zeta^2\ell_{-,\gamma}^d)N_\Gamma + F_{\gamma,\text{sp}(\Gamma)}\big(\psi_{\text{sp}(\Gamma)}|(\sigma_{D^-}^{(\gamma^{-1/2})}, \sigma_{D^+}^{(\gamma^{-1/2})}, \sigma_{A_{\text{ext}}}^{(\gamma^{-1/2})})\big). \quad (9.3.3.1)$$

Note that the spins $\sigma_{D^-}^{(\gamma^{-1/2})}$ have changed sign.

9.3.4 Reconstruction of the partition function

With $[\cdot]_\gamma$ the integer part term defined just before (4.2.3.6), by (4.2.2.13) with $\mathcal{A}$ the set $\{\sigma_{\text{sp}(\Gamma)} : \sigma_{\text{sp}(\Gamma)}^{(\gamma^{-1/2})} = [\psi_{\text{sp}(\Gamma)^-}]_\gamma\}$, and using again that in (4.2.2.13) we have $\epsilon(\gamma) \leq 2\gamma^{1/2}$,

$$e^{-\beta F_{\gamma,\text{sp}(\Gamma)}(\psi_{\text{sp}(\Gamma)}|(\sigma_D^{(\gamma^{-1/2})}, \sigma_{A_{\text{ext}}}^{(\gamma^{-1/2})}))}$$
$$\leq e^{\beta c\gamma^{1/2}|\text{sp}(\Gamma)|}$$
$$\times \Big\{ \sum_{\sigma_{\text{sp}(\Gamma)}:\sigma_{\text{sp}(\Gamma)}^{(\gamma^{-1/2})}(;r)=[\psi]_\gamma(r), r\in\text{sp}(\Gamma)} e^{-\beta H_{\gamma,\text{sp}(\Gamma)}(\sigma_{\text{sp}(\Gamma)}|(\sigma_D,\sigma_{A_{\text{ext}}}))} \Big\}. \quad (9.3.4.1)$$

9.3.5 Conclusion

It follows from the definition of $[\cdot]_\gamma$ that $|[\psi_{\mathrm{sp}(\Gamma)}]_\gamma - \psi_{\mathrm{sp}(\Gamma)}| \le 2\gamma^{d/2}$ for γ small enough; then

$$[\psi_{\mathrm{sp}(\Gamma)}]_\gamma \in \left(m_\beta - (1-\kappa_0)\zeta - 2\gamma^{d/2}, m_\beta + (1-\kappa_0)\zeta + 2\gamma^{d/2}\right).$$

Since $\zeta = \gamma^a$ and $a < d/2$, for γ small enough, $\eta([\psi_{\mathrm{sp}(\Gamma)}]_\gamma; r) = 1$ on $\mathrm{sp}(\Gamma)$. The l.h.s. of (9.3.4.1) is then bounded by

$$e^{\beta c \gamma^{1/2} |\mathrm{sp}(\Gamma)|} \left\{ \sum_{\sigma_{\mathrm{sp}(\Gamma)}:\eta(\sigma_{\mathrm{sp}(\Gamma)};r)=1, r\in \mathrm{sp}(\Gamma)} e^{-\beta H_{\gamma,\mathrm{sp}(\Gamma)}(\sigma_{\mathrm{sp}(\Gamma)}|(\sigma_D,\sigma_{A_{\mathrm{ext}}}))} \right\}.$$

Collecting all the above bounds we then get from (9.3.1.1) that $\mathcal{N}(\sigma_{A_{\mathrm{ext}}})$ is bounded by

$$\sum_{\sigma_D:\eta(\sigma_D;r)=1, r\in D} e^{-\beta H_{\gamma,D}(\sigma_D)} Z^+_{\gamma,\mathrm{int}_0^-(\Gamma),\sigma_{D^-}} Z^+_{\gamma,\mathrm{int}_0^+(\Gamma),\sigma_{D^+}}$$
$$\times e^{2\beta c\gamma^{1/2}|\mathrm{sp}(\Gamma)| - \beta c_1 (\zeta^2 (\ell_{-,\gamma})^d) N_\Gamma}$$
$$\times \sum_{\sigma_{\mathrm{sp}(\Gamma)}:\eta(\sigma_{\mathrm{sp}(\Gamma)};r)=1, r\in \mathrm{sp}(\Gamma)} e^{-\beta H_{\gamma,\mathrm{sp}(\Gamma)}(\sigma_{\mathrm{sp}(\Gamma)}|(\sigma_D,\sigma_{A_{\mathrm{ext}}}))}.$$

For γ small enough, we have $2c\gamma^{1/2}|\mathrm{sp}(\Gamma)| \le \frac{1}{2}\{c_1 (\zeta^2(\ell_{-,\gamma})^d) N_\Gamma\}$ and the above is bounded by $e^{-\beta(c_1/2)(\zeta^2(\ell_{-,\gamma})^d) N_\Gamma} \mathcal{D}(\sigma_{A_{\mathrm{ext}}})$ thus concluding the proof of Theorem 9.2.5.1.

Chapter 10
The LMP model and the Pirogov–Sinai strategy

In this and in the next chapter we shall study systems of particles in the continuum. The ultimate goal is the derivation of a phase diagram like the one on the left in Fig. 10.1, typical of single-component fluids, where the P–T quadrant (pressure versus temperature) is divided into regions where the system is respectively in the gaseous, liquid and solid phases. Compared to the simple Ising phase diagram on the right of Fig. 10.1, it becomes evident that the structures we want to investigate are much richer and complex. Moreover, the hamiltonians of particle models are typically invariant under the full Euclidean group, and one of the issues is the spontaneous breaking of the Euclidean symmetry with the appearance of crystalline structures in the ground states and at positive temperatures, to be compared with the simpler breaking of the spin flip symmetry in the Ising model. Elastic and non-elastic behaviors, spatial patterns, microstructures, ... are among the many issues which then naturally arise. Compared to the lattice spins studied so far, the analysis is thus infinitely more complex but also correspondingly intriguing.

From a mathematically rigorous point of view, the theory is still in a preliminary stage, by no means comparable to the state of the lattice theories. While the gaseous phase is fairly well understood (absence of phase transitions at high temperatures and/or low densities, validity of the cluster expansion, ...), almost nothing is known rigorously about phase transitions, except for the Widom–Rowlinson model of two component fluids [198], some long range, one dimensional systems [146], and a class of models with Kac potentials, LMP models, introduced recently by Lebowitz, Mazel and Presutti [164], for which techniques similar to those used in Chap. 9 apply. LMP models are the object of these last three chapters. They should be regarded as a compromise between realistic models of fluids and mathematically treatable systems.

Intermolecular forces are often described by Lennard–Jones potentials

$$V(r) = ar^{-12} - br^{-6}, \quad a, b > 0,$$

where r is the intermolecular distance, molecules being represented here by points of $\mathbb{R}^d$ (their internal structure is completely neglected). The ground states of Lennard–Jones-like hamiltonians have been recently characterized [204] as configurations of points on a triangular lattice. Persistence of the picture at non-zero temperatures is instead a completely open problem. Translational invariance is certainly restored in $d = 2$ at positive temperatures (for smooth potentials) as implied by a general theorem by Frölich and Pfister [125].

The LMP model mimics the Lennard–Jones (L–J) hamiltonian. Observe that the "typical" L–J energy density $e(\rho)$, ρ being the "particle density," has a negative minimum at the ground state density of the L–J hamiltonian; $e(\rho)$ then increases

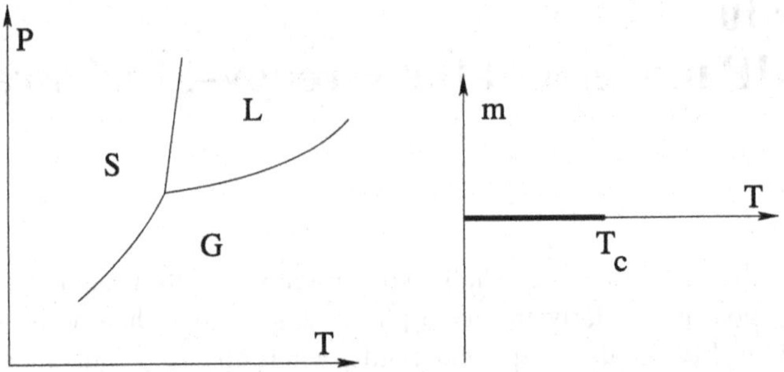

Fig. 10.1 On the *left* the phase diagram of a fluid. *S*, *L* and *G* stand for solid, liquid and gas; on the *right* the phase diagram of the Ising model. A *thick line* means that phases coexist

to 0 as ρ decreases to 0 (corresponding to large inter-particles distances) and to ∞ as $\rho \to \infty$ (at small inter-particles distances the L–J potential diverges). The LMP model is defined in terms of a function $e(\rho)$ with the same features described above. We preliminarily need to specify the particle density $\rho(\cdot)$ of a particle configuration q. Postponing to Sect. 10.2.4 the general definition, here we discuss a particular choice. Let q be a particle configuration, that is, a finite collection of points in $\mathbb{R}^d$, and let $r \in \mathbb{R}^d$. Then the local density $\rho(r)$ at r (in the configuration q) is $\frac{|q \cap B_R(r)|}{|B_R(r)|}$, where $B_R(r) \subset \mathbb{R}^d$ is the sphere of radius R and center r, $|q \cap \Lambda|$ the number of points of q in $\Lambda \subset \mathbb{R}^d$. The LMP energy $H(q)$ is then defined as

$$H(q) = \int_{\mathbb{R}^d} e\left(\frac{|q \cap B_R(r)|}{|B_R(r)|}\right), \quad e(\rho) = \frac{\rho^4}{4!} - \frac{\rho^2}{2}.$$

In this example $e(\rho)$ has a negative minimum at $\rho = 1$; it diverges as $\rho \to \infty$ and vanishes as $\rho \to 0$.

R will play the role of the inverse Kac parameter, $R \leftrightarrow \gamma^{-1}$, and choosing R large enough we shall be able to extend to the LMP model the analysis of Chap. 9 on Ising systems with Kac potentials. To introduce the argument we start in the next section from a mean field.

10.1 Mean field models

In this section we shall study mean field interactions. Already at the mean field level, however, there are problems, due to the large accumulation of particles and we start by discussing "stability of matter" for short range interactions.

10.1.1 Stable interactions

A system of identical point particles in $\mathbb{R}^d$ is conceptually analogous to an Ising configuration. To implement the analogy let us divide $\mathbb{R}^d$ into small cubic cells of "ultra-microscopic side a." If the system is "physically reasonable" most of the cells will be empty; negligibly many will have multiple occupancy, and the dependence on the position of the particle inside an occupied cell will also be negligible. Under such an approximation, the states become elements of the space $\{0, 1\}^{a\mathbb{Z}^d}$. Such a system (isomorphic to the Ising model) is called "a lattice gas" and in the course of the proofs we shall actually use in Sect. 11.4.1 lattice gas approximations on $a\mathbb{Z}^d$ of our continuum model.

We have thus transformed a particle configuration into a lattice gas configuration, but the resulting system has an un-physical topological structure, and there are problems with "stability of matter" as we are going to discuss. The natural topology of the lattice description is not physically correct when applied to the particles. Consider for instance particle configurations consisting of a single point. If the position of the point moves continuously in $\mathbb{R}^d$ (the topology must be such that) the particle configuration consisting of that point also changes continuously. Its image on the lattice $a\mathbb{Z}^d$ is instead not continuous, as the configuration jumps when the particle changes cell.

The second issue, maybe more serious, is related to the assumption of single occupancy of cells. If in analogy with the Ising ferromagnetic interaction we suppose that there is an attractive interaction among the particles, then each pair of particles in the same site contributes to the energy by a term $-U$, $U > 0$. This term favors the presence of more particles in the same cell. This effect is in contrast to the entropy, which wants to spread particles to increase the phase space volume. The entropy depresses the presence of n particles in the same cell by a factor $\frac{\epsilon^n}{n!}$, ϵ being the volume of the cell. The Gibbs factor goes like $e^{\beta U n(n-1)/2}$: the energy wins and $\lim_{n\to\infty} \frac{\epsilon^n e^{\beta U n(n-1)/2}}{n!} = \infty$. This proves that an Ising-like, attractive interaction is incompatible with the lattice approximation of particle systems. The problem is real. It is a fact that matter is not statistically stable under interactions which at small distances are negative (attractive): in the thermodynamic limit there would be a collapse of matter with infinitely many particles in bounded regions. Statistical mechanics thus requires the addition of stabilizing, repulsive forces at short distances strong enough to win against the attractive, Ising-like, interaction. Repulsive forces in the spin language are anti-ferromagnetic interactions and this already hints at the difficulties we are going to face when dealing with continuous particle systems. In the next subsection we shall rephrase in a more formal way the above considerations in the context of mean field interactions.

10.1.2 Pure mean field

The particles analogue of the mean field Ising model of Sect. 4.1 is a system of n particles in a bounded Borel measurable region Λ, where the energy of a configuration

$q = (r_1, \dots, r_n)$ is

$$H_\Lambda^{\mathrm{mf}}(r_1, \dots, r_n) = -\frac{n^2}{2|\Lambda|} - \lambda n; \tag{10.1.2.1}$$

λ the chemical potential. It would have been more correct to write $n(n-1)$ instead of n^2, in order to interpret the first term as the sum of all pair interactions, when the coupling constant is identically $|\Lambda|^{-1}$ (as in the Ising case). However, the difference can be absorbed in the chemical potential and (10.1.2.1) has a more convenient form for the sequel. The canonical partition function is

$$Z_{n,\Lambda}^{\mathrm{mf}} = \frac{1}{n!}\int_{\Lambda^n} e^{-\beta H_\Lambda^{\mathrm{mf}}(r_1,\dots,r_n)} dr_1 \cdots dr_n = e^{-\beta(-\frac{n^2}{2|\Lambda|} - \lambda n - \frac{1}{\beta}\log\frac{|\Lambda|^n}{n!})}. \tag{10.1.2.2}$$

By the Stirling formula (A.2.1),

$$\lim_{n,|\Lambda|\to\infty: n/|\Lambda|\to\rho} -\frac{1}{\beta|\Lambda|}\log Z_{n,\Lambda}^{\mathrm{mf}} = -\frac{\rho^2}{2} - \lambda\rho + \frac{1}{\beta}\rho(\log\rho - 1). \tag{10.1.2.3}$$

The thermodynamical pressure is the Legendre transform of the free energy, see (4.1.4.2) for the Ising case, and it is thus given by

$$\sup_{\rho\ge 0}\left\{\lambda\rho + \frac{\rho^2}{2} - \frac{1}{\beta}\rho(\log\rho - 1)\right\} = +\infty.$$

In contrast to the Ising case, in fact, the canonical free energy density, i.e. the r.h.s. of (10.1.2.3) as a function of the density $\rho \in \mathbb{R}_+$ is unbounded from below. The problem is absent in the Ising model because the magnetization density (which plays the role of ρ) is a priori bounded (as it ranges in the interval $[-1, 1]$). In (10.1.2.3), the energy decreases as $-\rho^2$ (due to a negative pair interaction), while the entropy increases only as $\rho\log\rho$ and it is therefore not capable to contrast the energy divergence. As a result, when particles are exchanged with a reservoir with chemical potential λ, equilibrium is reached when the free energy is minimized. This occurs with the collapse of matter because of a too strong attractive interaction.

If we change the sign of the interaction, making it repulsive, then the problem disappears, as the analogue of (10.1.2.3) has now $+\rho^2/2$ and it is therefore bounded from below. But the limit free energy is now a convex function of ρ, so that together with the divergencies we have also lost the phase transitions. The lesson is that we must introduce repulsive forces to stabilize the matter, but, at the same time, we must also keep the attractive interactions to have a phase transition.

10.1.3 Mean field with hard cores

As we want a negative interaction term (attractive forces), we must find a way to avoid the occurrence of large densities. This can easily be achieved by restricting the phase space by force to

$$\mathcal{X}_\Lambda^a = \bigcup_n \mathcal{X}_{n,\Lambda}^a, \quad \mathcal{X}_{n,\Lambda}^a = \Big\{(r_1, \ldots, r_n) \in \Lambda^n : \min_{i \neq j} |r_i - r_j| > a\Big\}, \tag{10.1.3.1}$$

where $a > 0$ is called the "hard core length"; the number of particles in finite volumes is then bounded. The constraint in (10.1.3.1) may be interpreted as due to a pair interaction which is equal to $+\infty$ whenever the two particles are at distance $\leq a$. This is not completely un-physical, the hard core plus the mean field attractive interaction in fact mimics the L–J potential with its strong short range repulsive interaction and long range attractive tail.

The canonical partition function in this model is

$$\begin{aligned} Z_{n,\Lambda}^{\mathrm{mf};a} &= \frac{1}{n!} \int_{\mathcal{X}_{n,\Lambda}^a} e^{-\beta H_\Lambda^{\mathrm{mf}}(r_1,\ldots,r_n)}\, dr_1 \cdots dr_n \\ &= \exp\left\{-\beta\left(-\frac{n^2}{2|\Lambda|} - \lambda n + \frac{|\Lambda|}{\beta} f_{n/|\Lambda|,\Lambda}^a\right)\right\}, \end{aligned} \tag{10.1.3.2}$$

where $f_{n/|\Lambda|,\Lambda}^a$ is the pure hard core free energy in Λ with density $n/|\Lambda|$:

$$f_{n/|\Lambda|,\Lambda}^a = -\frac{1}{|\Lambda|} \log Z_{n,\Lambda}^a, \quad Z_{n,\Lambda}^a = \frac{1}{n!} \int_{\mathcal{X}_{n,\Lambda}^a} dr_1 \cdots dr_n. \tag{10.1.3.3}$$

As in the lattice case, it is possible to prove convergence in the thermodynamic limit, namely

$$\lim_{n,|\Lambda| \to \infty : n/|\Lambda| \to \rho} f_{n/|\Lambda|,\Lambda}^a = f_\rho^a, \tag{10.1.3.4}$$

with f_ρ^a a convex function which becomes $+\infty$ when the density ρ is larger than the close packing density ρ_{cp}^a. It then follows that the canonical mean field free energy is

$$\phi_{\beta,\lambda}^{\mathrm{mf};a}(\rho) := \lim_{n,|\Lambda| \to \infty : n/|\Lambda| \to \rho} \frac{-1}{\beta |\Lambda|} \log Z_{n,\Lambda}^{\mathrm{mf};a} = -\frac{\rho^2}{2|\Lambda|} - \lambda\rho + \frac{1}{\beta} f_\rho^a. \tag{10.1.3.5}$$

The mean field pressure $P_{\beta,\lambda}^{\mathrm{mf};a}$ is then

$$P_{\beta,\lambda}^{\mathrm{mf};a} := -\inf_{\rho \geq 0} \phi_{\beta,\lambda}^{\mathrm{mf};a}(\rho). \tag{10.1.3.6}$$

Phase transitions in this picture appear as a loss of convexity for $\phi_{\beta,0}^{\mathrm{mf};a}(\rho)$, and we shall see below that such a phenomenon indeed occurs in a certain range of temperatures and densities. By letting $\beta \to \infty$ in (10.1.3.5) with ρ smaller than the close packing density, $\phi_{\beta,\lambda}^{\mathrm{mf};a}(\rho) \to -\frac{\rho^2}{2|\Lambda|} - \lambda\rho$, which is strictly concave. Then, by continuity, $\phi_{\beta,\lambda}^{\mathrm{mf};a}(\rho)$ is not convex for large β, and the grand canonical free energy

$f^{\text{mf};a}_{\beta,\rho}$, which is the convex envelope of $\phi^{\text{mf};a}_{\beta,0}(\rho)$, has a graph with a flat part. We have thus reproduced the scenario we had in the mean field Ising model, with a phase transition at small temperatures, but the present model has a much richer structure. The hard core free energy f^a_ρ may in fact itself have a phase transition. In $d=1$ the hard core system is trivial: by describing the particle positions in terms of variables defined by the distances between successive hard rods, the system becomes in fact an ideal point particle gas. On the contrary in $d \ge 2$ there should be a phase transition, with the appearance of "ordered states" at densities near close-packing. Such states would mimic crystalline structures and the phase transition would be into a "solid phase." While there is numerical evidence of this phenomenon, a rigorous proof is still missing, as far as I know.

10.1.4 The Lebowitz–Penrose limit

The (statistical mechanics) correct version of the mean field approach involves Kac potentials, and the mean field hamiltonian is then replaced by

$$H_\gamma(r_1,\dots,r_n) = -\lambda n - \frac{1}{2}\sum_{i\neq j} J_\gamma(r_i,r_j). \tag{10.1.4.1}$$

$(r_1,\dots,r_n)$ is a configuration with n particles at $(r_1,\dots,r_n)$ and $J_\gamma(r,r') = \gamma^d J(\gamma r, \gamma r')$, with $J(r,r')$ the same kernel as in Sect. 4.2.1; in particular, $J(r,r') \ge 0$ is a smooth symmetric probability kernel with compact support which depends on $r'-r$. Due to the minus sign in (10.1.4.1) and the positivity of J, the interaction is attractive, and stability is enforced by restricting the phase space to $\mathcal{X}^a_\Lambda$ or in other words by adding a hard core interaction with radius $a>0$. The grand canonical partition function is then

$$Z^a_{\gamma,\Lambda} = 1 + \sum_{n\ge 1}\frac{1}{n!}\int_{\mathcal{X}^a_{n,\Lambda}} e^{-\beta H_\gamma(r_1,\dots,r_n)}\, dr_1\cdots dr_n, \tag{10.1.4.2}$$

and we have the analogue of [the Lebowitz–Penrose] Theorem 4.2.1.1 (the proof is omitted).

Theorem 10.1.4.1 *For any β and λ and any increasing van Hove sequence of bounded, Borel measurable, regions Λ which invades $\mathbb{R}^d$*

$$\lim_{\gamma\to 0}\lim_{\Lambda\to\mathbb{R}^d} \frac{\log Z^a_{\gamma,\Lambda}}{\beta|\Lambda|} = P^{\text{mf};a}_{\beta,\lambda}. \tag{10.1.4.3}$$

Thus the graph of the pressure $P^a_{\gamma,\beta,\lambda}$ (defined by the l.h.s. of (10.1.4.3), without taking the first limit $\gamma\to 0$) as a function of λ with β fixed, converges to the graph of $P^{\text{mf};a}_{\beta,\lambda}$ as $\gamma\to 0$. The latter has a cusp, which indicates the occurrence of a phase transition but as already discussed in Chap. 4 $P^a_{\gamma,\beta,\lambda}$ may very well be smooth for all γ with the cusp appearing only in the limit (as indeed happens in $d = 1$). While Theorem 10.1.4.1 does not imply that there are phase transitions at $\gamma > 0$, this nevertheless may very well happen as we have seen in Chap. 9 for the Ising model with Kac potentials and the hope is to extend such an analysis to the present case. As we are going to explain, there are problems in this program that we do not know how to solve, but fortunately we can modify the model in such a way that a complete analysis can be carried through.

The analysis in Chap. 9 is based on a perturbative study of the case $\gamma = 0$, formally described by a non-local, free energy functional. In the present case the latter is

$$\mathcal{F}(\rho) = \int f_{\beta,\lambda}(\rho)dr - \frac{1}{2}\int\int J(r,r')\big(\rho(r)-\rho(r')\big)^2 dr\,dr',$$

$$f_{\beta,\lambda}(\rho) = \phi^{\mathrm{mf};a}_{\beta,\lambda}(\rho) - \min_{\rho\ge 0}\phi^{\mathrm{mf};a}_{\beta,\lambda}(\rho), \tag{10.1.4.4}$$

where the density profile $\rho \in L^\infty(\mathbb{R}^d, [0, \rho^a_{cp}))$, ρ^a_{cp} the close packing density. For β large there is $\lambda = \lambda(\beta)$ for which $f_{\beta,\lambda(\beta)}(\rho)$ has two minimizers, $\rho_{\beta,\pm}$ and so far there is no big difference with Ising. The idea in Chap. 9 was then to use the functional to prove Peierls estimates, which then imply the occurrence of a phase transition. Contours are defined just as in Chap. 9 and the problem is to estimate/bound their weights. Recall from Sect. 9.2.4 that the weight of a contour involves ratios of partition functions both defined on the spatial support of the contour: the one in the numerator has the constraint that the configurations should be compatible with the contour, the one in the denominator instead is over configurations which are close to $\rho_{\beta,+}$ (or to $\rho_{\beta,-}$, depending on the contour we are considering). This part of the analysis can be carried through also here by using the Lebowitz–Penrose approach to estimate the partition functions in terms of the functional (10.1.4.4) and then studying the corresponding minimization problems. So far everything goes; the problem arises because in the computation of the weight of a contour there are also other factors which can again be expressed as ratios of partition functions. These are partition functions in the interior parts of the contour, and we need to take the ratio with different boundary conditions, one "close" to $\rho_{\beta,+}$ the other to $\rho_{\beta,-}$. In the Ising case things could be arranged in such a way that the boundary conditions are obtained one from the other by spin flip; then by the spin flip symmetry the two partition functions are equal and their ratio is 1. In general the ratio of two partition functions goes like the surface, but for our applications we need to prove that the proportionality coefficient is suitably small. The analysis of the surface correction to the pressure, i.e. of the dependence on the shape of the region and boundary conditions, is usually carried out using cluster expansion techniques. Even if the Kac interaction was missing, a cluster expansion for pure hard cores is known to be valid only for small densities, and we do not know whether the densities we have to deal with meet such requirements. Moreover, we cannot expect this to work unless we suitably tune the chemical potential λ in terms of γ; at least this is what we have to do in the LMP model, where all this can be carried through.

10.2 The LMP model

We shall first describe the men field version of the model and then the actual particle system.

10.2.1 The mean field LMP model

The problems outlined in the last subsection arise from a poor control of the hard core interaction, which has been added to stabilize the long attractive tail of the forces. Potentials which are sufficiently positive at the origin, like the superstable interactions, do stabilize the system as well and can replace the hard cores, but the situation does not improve at all and the analysis may even be harder. Another possibility, which is the one pursued in LMP, is to use Kac potentials also for the repulsive forces. We first describe the LMP model in its mean field version.

The LMP mean field hamiltonian $H_\Lambda(q)$ (in a bounded region Λ) has the form $H_{\lambda,\Lambda}(q) = |\Lambda| e_\lambda(\rho),\ \ \rho = \frac{|q|}{|\Lambda|}$ where $e_\lambda(\cdot)$ has the meaning of an energy density. The assumption here is that such an energy depends only on the total particle density $\rho = |q|/|\Lambda|$. It remains to choose the energy density $e_\lambda(\cdot)$. λ is interpreted as a chemical potential, so that $e_\lambda(\rho) = -\lambda\rho + e_0(\rho)$. As already argued in the beginning of this chapter, to mimic a fluid, $e_0(\rho)$ should be a decreasing function of the density ρ from 0 till when the energy reaches its minimum, which corresponds to an optimal disposition of the molecules. A further increase of ρ causes an increase of the energy, which we suppose to diverge to $+\infty$ as $\rho \to \infty$. The LMP choice which fulfills such requirements is

$$e_\lambda(\rho) = -\lambda\rho - \frac{\rho^2}{2} + \frac{\rho^4}{4!}. \tag{10.2.1.1}$$

Indeed $e_0(\rho)$ decreases from 0 to its negative minimum when ρ varies in $[0,1]$ and then increases to $+\infty$ as $\rho \in [1,\infty)$. The reader must not be fooled by the double well shape of (10.2.1.1), which is actually not there, because ρ, being a particles density, must be ≥ 0 and indeed phase transitions come, as they should, as a result of the competition of energy versus entropy. As already remarked, the first term, $-\lambda\rho$, is the energy of the chemical potential λ, the term $-\rho^2/2$ comes from the attractive part of the interaction, and, as discussed earlier, it can be ascribed to pair interactions; the stabilizing term $\rho^4/4!$ may be interpreted as due to four body repulsive interactions, as we shall see in the sequel.

The LMP mean field canonical partition function is

$$Z^{\text{mf}}_{n,\Lambda} = \frac{1}{n!}\int_{\Lambda^n} e^{-\beta|\Lambda| e_\lambda(n/|\Lambda|)}, \tag{10.2.1.2}$$

while the grand canonical one is $Z_\Lambda^{\mathrm{mf}} = \sum_{n=0}^{\infty} Z_{n,\Lambda}^{\mathrm{mf}}$. We omit the [elementary] proof that in the thermodynamic limit $|\Lambda| \to \infty$, the canonical mean field free energy is

$$\phi_{\beta,\lambda}(\rho) = \lim_{|\Lambda|,n\to\infty:\frac{n}{|\Lambda|}\to\rho} \frac{-1}{\beta|\Lambda|} \log Z_{n,\Lambda}^{\mathrm{mf}} = e_\lambda(\rho) + \frac{1}{\beta}\rho(\log\rho - 1), \quad (10.2.1.3)$$

while the grand canonical pressure is the Legendre transform of $\phi_{\beta,0}(\rho)$:

$$\begin{aligned} p_{\beta,\lambda} &= \lim_{|\Lambda|\to\infty} \frac{1}{\beta|\Lambda|} \log Z_\Lambda^{\mathrm{mf}} \\ &= \sup_{\rho\geq 0}\left\{\lambda\rho - \left[-\frac{\rho^2}{2} + \frac{\rho^4}{4!} + \frac{1}{\beta}\rho(\log\rho - 1)\right]\right\}. \end{aligned} \quad (10.2.1.4)$$

The repulsive interaction with the quartic term $\frac{\rho^4}{4!}$ makes the sup finite dominate, for large ρ, the negative quadratic interaction.

10.2.2 *Thermodynamics of the mean field LMP model*

In this subsection we state the main properties of $\phi_{\beta,\lambda}(\rho)$ as a function of β, λ and ρ, which will be used in the sequel. Since the proofs are just computational they are omitted. The critical points of $\phi_{\beta,\lambda}(\rho)$ as a function of ρ, namely the solutions of the mean field equation

$$\frac{d}{d\rho}\left\{e_\lambda(\rho) + \frac{1}{\beta}\rho(\log\rho - 1)\right\} = 0 \quad (10.2.2.1)$$

have the form (see Fig. 10.2)

$$\rho = \exp\{-\beta e_\lambda'(\rho)\} =: K_{\beta,\lambda}(\rho). \quad (10.2.2.2)$$

We start by studying the convexity properties of $\phi_{\beta,\lambda}(\rho)$ as a function of ρ for fixed β and λ; see Fig. 10.2. Since $\phi_{\beta,\lambda}(\rho) = -\lambda\rho + \phi_{\beta,0}(\rho)$ they are independent of λ and in the next proposition we set $\lambda = 0$.

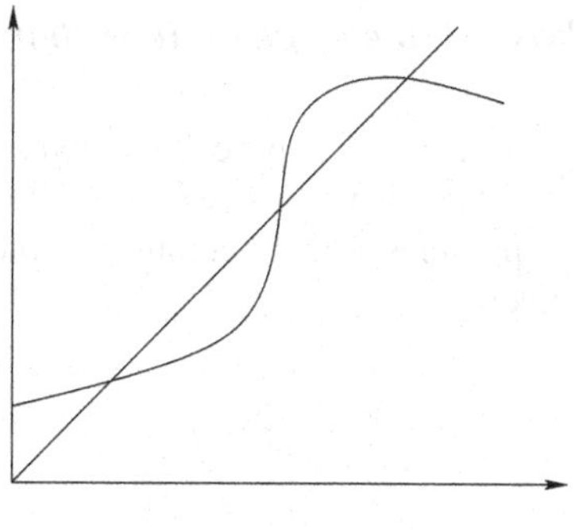

Fig. 10.2 Graph of $K_{\beta,\lambda(\beta)}(\rho)$ with $\beta > \beta_c$ and "close" to β_0

Proposition 10.2.2.1 (Convexity properties of $\phi_{\beta,0}(\cdot)$) *There is a critical inverse temperature $\beta_c = (3/2)^{3/2}$ such that $\phi_{\beta,0}(\rho)$ is convex for $\beta \le \beta_c$, while for $\beta > \beta_c$ it has two inflection points $0 < s_-(\beta) < s_+(\beta)$, being concave for $\rho \in (s_-(\beta), s_+(\beta))$ and convex for $\rho \notin (s_-(\beta), s_+(\beta))$.*

A mean field phase transition corresponds to a multiplicity of minimizers of $\phi_{\beta,\lambda}(\cdot)$. In Fig. 10.2 there are three critical points.

Proposition 10.2.2.2 (Phase transitions) *For any $\beta > \beta_c$ there is $\lambda(\beta)$ so that $\phi_{\beta,\lambda(\beta)}(\cdot)$ has two global minimizers, $\rho_{\beta,-} < \rho_{\beta,+}$ (and a local maximum at $\rho_{\beta,0}$). Moreover, there is c_1' so that for any $\zeta > 0$ small enough*

$$\phi_{\beta,\lambda(\beta)}(\rho) \ge \phi_{\beta,\lambda(\beta)}(\rho_{\beta,\pm}) + c_1'\zeta^2, \quad \text{if } \min\{|\rho - \rho_{\beta,+}|, |\rho - \rho_{\beta,}|\} \ge \zeta. \quad (10.2.2.3)$$

We shall need the properties of $\phi_{\beta,\lambda}(\cdot)$, of the map $K_{\beta,\lambda}(\rho)$ of (10.2.2.2) and of its critical points for λ in a neighborhood of $\lambda(\beta)$. Fig. 10.2 corresponds to $\lambda = \lambda(\beta)$, $\beta > \beta_c$. $\rho_{\beta,-}$ and $\rho_{\beta,+}$ are the abscissas of the first and third intersections with the diagonal. The new feature with respect to the mean field Ising case is that the analogue of $K_{\beta,\lambda}(\rho)$ in the latter is monotonically increasing; see Fig. 4.2 in Sect. 4.1.2. Non-monotonicity of $K_{\beta,\lambda}(\rho)$ has important consequences on the structure of the instanton, which for β sufficiently large (as in Fig. 10.2) has oscillations [134].

Proposition 10.2.2.3 (Critical points) *For any $\beta > \beta_c$ there is an interval $(\lambda_-(\beta), \lambda_+(\beta))$ containing $\lambda(\beta)$ and for any λ in the interval $\phi_{\beta,\lambda}(\cdot)$ it has two local minima $\rho_{\beta,\lambda,\pm}$ which are differentiable functions of λ and $\frac{d}{d\lambda}(\phi_{\beta,\lambda}(\rho_{\beta,\lambda,+}) - \phi_{\beta,\lambda}(\rho_{\beta,\lambda,-})) = \rho_{\beta,\lambda,-} - \rho_{\beta,\lambda,+} < 0$. For all $\beta > \beta_c$,*

$$\frac{d}{d\rho} K_{\beta,\lambda(\beta)}(\rho)\Big|_{\rho=\rho_{\beta,\pm}} \equiv K'_{\beta,\lambda(\beta)}(\rho_{\beta,\pm}) < 1, \quad (10.2.2.4)$$

the condition (10.2.2.4) *being equivalent to $\phi''_{\beta,\lambda(\beta)}(\rho_{\beta,\pm}) > 0$. Moreover, there exists $\beta_0 > \beta_c$ such that*

$$K'_{\beta,\lambda(\beta)}(\rho_{\beta,\pm}) > -1 \quad \text{for all } \beta \in (\beta_c, \beta_0). \quad (10.2.2.5)$$

10.2.3 Phase space of particle systems

The definition of the phase space for systems of particles in the continuum is more delicate than in lattice systems. We have already introduced phase space and configurations of particles without comments when studying the mean field case; they were respectively

$$\mathcal{X}_\Lambda^{\mathrm{lbl}} = \bigcup_{n=0}^{\infty} \Lambda^n, \qquad q = (r_1, \ldots, r_n) \in \mathcal{X}_\Lambda^{\mathrm{lbl}}. \quad (10.2.3.1)$$

This is the physically correct definition when the particles are distinguishable, and for this reason we have added the superscript "lbl," which stands for labelled. We are instead interested in the unlabelled case; namely, we want to consider as identical all configurations $(r_{\pi(1)},\dots,r_{\pi(n)})$ obtained from $(r_1,\dots,r_n)$ by a permutation $(\pi(1),\dots,\pi(n))$ of $(1,\dots,n)$. We call $\mathcal{X}_\Lambda$ the phase space obtained from $\mathcal{X}_\Lambda^{\rm lbl}$ under such an identification. $\mathcal{X}_\Lambda$ can be described as the space of finite subsets of Λ with an integer for each point of the subset which counts how many particles are present in that point. The case of multiple occupancy, however, is measure theoretically negligible, as we are going to see. The analogue of the counting measure in the Ising model is in fact in continuum systems the Lebesgue measure, usually called "the free measure." The free measure on $\mathcal{X}_\Lambda$, denoted by $d\nu_\Lambda(q_\Lambda)$, $q_\Lambda\in\mathcal{X}_\Lambda$, is defined so that its restriction to the sub-space of $\mathcal{X}_\Lambda$ with n particles has support on elements $(r_1,\dots,r_n)$ where all first coordinates (denoted by x_i) of r_i are distinct and it is then given by $\mathbf{1}_{x_1<\dots<x_n}dr_1\cdots dr_n$. We complete the definition by setting $\nu_\Lambda(\emptyset)=1$, $\emptyset$ being the configuration with no particles.

Any function f on $\mathcal{X}_\Lambda$ can be identified with a function on $\mathcal{X}_\Lambda^{\rm lbl}$ constantly equal to $f(q)$ on all elements of $\mathcal{X}_\Lambda^{\rm lbl}$ which correspond to q, and such functions on $\mathcal{X}_\Lambda^{\rm lbl}$ are called "symmetric." We can correspondingly identify the free measure on $\mathcal{X}_\Lambda$ with a measure on $\mathcal{X}_\Lambda^{\rm lbl}$ by requiring that the integral of any symmetric function on $\mathcal{X}_\Lambda^{\rm lbl}$ should be equal to the integral over $d\nu_\Lambda(q_\Lambda)$ of the corresponding function on $\mathcal{X}_\Lambda$. The "canonical choice" is

$$\int_{\mathcal{X}_\Lambda} f(q)d\nu_\Lambda(q)=\sum_{n=0}^{\infty}\frac{1}{n!}\int_{\Lambda^n} f(r_1,\dots,r_n)dr_1\cdots dr_n. \qquad (10.2.3.2)$$

The measure on the r.h.s. will be called the free measure on $\mathcal{X}_\Lambda^{\rm lbl}$ and for brevity and by abuse of notation, the free measure. Notice the factor $1/n!$ in agreement with (10.1.2.2).

$\mathcal{X}_\Lambda$ is naturally imbedded in $\mathcal{X}_{\Lambda'}$ with $\Lambda'\supset\Lambda$. We can thus write q without specifying to which $\mathcal{X}_\Lambda$ it belongs, and for the moment we may restrict ourselves to such configurations. In Chap. 12 we shall need to extend the notion to particle configurations in the whole space and with infinitely many particles. In this and in the next chapter we shall always work with systems in bounded regions and it is convenient to consider $\mathcal{X}_\Lambda$ equipped with the topology inherited from $\mathcal{X}_\Lambda^{\rm lbl}$. In Chap. 12 we shall come back on the issue.

We conclude this discussion by observing that there is a natural notion of addition of two particle configurations, $q+q'$, being the configuration which collects all the particles of q and q'; the notion is well defined because particles are undistinguishable.

10.2.4 The LMP hamiltonian

The local version of the mean field LMP energy is an example of many body Kac potentials. With in mind the energy (10.2.1.1), for each point $r\in\mathbb{R}^d$ we introduce a "window" through which we can only see a "portion $B_R(r)$ of the whole space," recall $B_R(r) = \{r' \in \mathbb{R}^d : |r' - r| \le R\}$. In this window, the density of a particle configuration q is $\rho(q;r) = |q \cap B_R(r)|/|B_R(r)|$, $|q \cap B_R(r)|$ the number of particles of q in $B_R(r)$, $|B_R(r)|$ the volume of $B_R(r)$. We can then define a local energy density which we set equal to $e_\lambda(\rho(q;r))$, e_λ as in (10.2.1.1). The integral of $e_\lambda(\rho(q;r))$ over $r \in \mathbb{R}^d$ finally defines the total energy of the configuration q.

It is convenient to generalize a bit the procedure by introducing more general convolution kernels (more general, that is, than the characteristic functions of the set $B_R(r)$ considered above). The density at r of a configuration $q = (r_1, \dots, r_n, \dots)$ is then

$$J_\gamma * q(r) := \sum_i J_\gamma(r, r_i), \tag{10.2.4.1}$$

supposing that there are finitely many r_i in each compact of $\mathbb{R}^d$. The notation $J_\gamma * q(r)$ reminds one of convolutions and indeed the r.h.s. of (10.2.4.1) may be interpreted as the convolution of J_γ with the generalized function $\sum \delta(r - r_i)$ which is the sum of Dirac deltas at the positions r_i of the particles in the configuration q; hence the notation $J_\gamma * q(r)$ used in (10.2.4.1). In agreement with such notation we shall also write

$$J_\gamma * (q + \bar{q})(r) := \sum_i J_\gamma(r, r_i) + \sum_i J_\gamma(r, \bar{r}_i), \tag{10.2.4.2}$$

where $\bar{q} = (\bar{r}_1, \dots, \bar{r}_n, \dots)$ and if f is a non-negative bounded measurable function

$$J_\gamma * (q + f)(r) := \sum_i J_\gamma(r, r_i) + \int J_\gamma(r, r') f(r'). \tag{10.2.4.3}$$

For notational simplicity we take the same probability kernel $J_\gamma(r, r')$ as introduced in Sect. 4.2.1 and define (for configurations with finitely many particles) the LMP hamiltonian as

$$H_{\gamma,\lambda}(q) = \int_{\mathbb{R}^d} e_\lambda(J_\gamma * q(r)). \tag{10.2.4.4}$$

We can write more explicitly (10.2.4.4) as

$$H_{\gamma,\lambda}(q) = -\lambda n - \frac{1}{2} \sum_{i,j} V_\gamma^{(2)}(r_i, r_j) + \frac{1}{4!} \sum_{i_1,\dots,i_4} V_\gamma^{(4)}(r_{i_1}, \dots, r_{i_4}), \tag{10.2.4.5}$$

where $V_\gamma^{(2)}(r_1, r_2) = \int J_\gamma(r, r_1) J_\gamma(r, r_2)\, dr$ and

$$V_\gamma^{(4)}(r_1, \dots, r_4) = \int J_\gamma(r, r_1) \cdots J_\gamma(r, r_4) dr.$$

(10.2.4.5) is the energy of the two and four body Kac potential (except for the fact that we are including the self-interaction, i.e. the terms with equal indices in the sums in (10.2.4.5)). Notice that while $J_\gamma(r,r')$ is supported by $|r-r'|\le\gamma^{-1}$, the range of the interaction is actually $2\gamma^{-1}$. The two and four body potentials which arise from our construction are not the most general ones as they appear as a convolution product of a same kernel and thus have positivity properties with important implications, as we shall see.

The energy of a configuration q, given another configuration $\bar q$, is

$$H_{\gamma,\lambda}(q|\bar q)=\int_{\mathbb{R}^d}\{e_\lambda(J_\gamma*(q+\bar q)(r))-e_\lambda(J_\gamma*\bar q(r))\}, \tag{10.2.4.6}$$

see (10.2.4.2), and if f is a non-negative bounded measurable function, using (10.2.4.3),

$$H_{\gamma,\lambda}(q|\bar q+f)$$
$$=\int_{\mathbb{R}^d}\{e_\lambda(J_\gamma*(q+\bar q+f)(r))-e_\lambda(J_\gamma*(\bar q+f)(r))\}. \tag{10.2.4.7}$$

10.2.5 The LMP finite volume Gibbs measures

The LMP Gibbs measure in a bounded measurable region Λ with boundary conditions $\bar q$ is the probability measure on $\mathcal{X}_\Lambda$ given by

$$G_{\gamma,\lambda,\Lambda,\bar q}(dq_\Lambda)=Z^{-1}_{\gamma,\lambda,\Lambda,\bar q}e^{-\beta H_{\gamma,\lambda,\Lambda}(q_\Lambda|\bar q)}d\nu_\Lambda(q_\Lambda), \tag{10.2.5.1}$$

where the partition function $Z_{\gamma,\lambda,\Lambda,\bar q}$ is the normalization factor and $d\nu_\Lambda(q)$ "the free measure" on $\mathcal{X}_\Lambda$ defined in Sect. 10.2.3. As β is fixed, it is dropped from the notation.

Usually in statistical mechanics the boundary conditions are given by a particle configuration $\bar q$. It is, however, technically convenient in the proofs to include the possibility of a continuous density, so that $\bar q$ in (10.2.5.1) may either be a particle configuration or a continuous density ρ or even the sum of the two. The corresponding energy has been defined in Sect. 10.2.4.

10.2.6 The LMP free energy functional

The mesoscopic free energy functional for the LMP system is described here; its derivation in the limit as $\gamma\to 0$ is contained in the proofs in the next chapter. Mesoscopic states are non-negative Borel measurable functions $\rho(r)$, $r\in\mathbb{R}^d$, which are interpreted as a particle density. The "mesoscopic" free energy is then a functional defined on the space of density functions which thus assigns to each possible state its free energy. We start by defining the free energy functional $F_{\gamma,\lambda,\Lambda}(\rho_\Lambda|\rho_{\Lambda^c})$ in a bounded region Λ with $\rho_\Lambda\in L^\infty(\Lambda;\mathbb{R}_+)$ and with $\rho_{\Lambda^c}\in L^\infty(\Lambda^c;\mathbb{R}_+)$ acting as a boundary condition: it is simply given by the LMP energy (extended from particles to continuous states) minus β^{-1} times the entropy:

$$F_{\gamma,\lambda,\Lambda}(\rho_\Lambda|\rho_{\Lambda^c}) = \int_{\mathbb{R}^d} \{e_\lambda(J_\gamma * (\rho_\Lambda + \rho_{\Lambda^c})) - e_\lambda(J_\gamma * \rho_{\Lambda^c})\} - \int_\Lambda \frac{S(\rho_\Lambda)}{\beta},$$

$$S(\rho) = -\rho(\log\rho - 1).$$

Here $J_\gamma * f$ is the usual convolution of the kernel J_γ with the function f. We shall simply write $F_{\gamma,\lambda,\Lambda}(\rho_\Lambda)$ when $\rho_{\Lambda^c} \equiv 0$; the dependence on β is at the moment not made explicit. If necessary we add β as a subscript. The dependence on γ of the functional can be scaled out. Denote

$$\rho^*(r) = \rho(\gamma r), \qquad \Lambda^* = \gamma\Lambda, \qquad F_{\lambda,\Lambda}(\cdot|\cdot) = F_{1,\lambda,\Lambda}(\cdot|\cdot); \tag{10.2.6.1}$$

then

$$F_{\gamma,\lambda,\Lambda}(\rho_\Lambda|\rho_{\Lambda^c}) = \gamma^{-d} F_{\lambda,\Lambda^*}(\rho^*_{\Lambda_\gamma}|\rho^*_{\Lambda^{*c}}). \tag{10.2.6.2}$$

Thus once expressed in mesoscopic units, the free energy inherits from the change of variables a factor γ^{-d} which will make the effective inverse temperature diverge proportionally to γ^{-d}. As in the Ising case with Kac potentials, this is the key factor in the proof of phase transitions. Hereafter in this subsection we restrict ourselves to $\gamma = 1$, dropping the superscript $*$. Observe that if ρ has compact support, then $F_{\lambda,\Lambda}(\rho)$ is independent of Λ once Λ is so large as to contain the support of ρ, so that $F_\lambda(\rho) = \lim_{\Lambda \nearrow \mathbb{R}^d} F_{\lambda,\Lambda}(\rho)$ defines a functional on $L_0^\infty(\mathbb{R}^d, \mathbb{R}_+)$, the space of bounded, measurable, non-negative functions with compact support. $F_\lambda(\rho)$ can be written as

$$F_\lambda(\rho) = \int_{\mathbb{R}^d} \left\{ e_\lambda(J * \rho) - J * \frac{S(\rho)}{\beta} \right\},$$

because, for $\rho \in L_0^\infty(\mathbb{R}^d, \mathbb{R}_+)$, $\int_{\mathbb{R}^d} J * S(\rho) = \int_{\mathbb{R}^d} S(\rho)$. The expression with $J * S(\rho)$ is more convenient because it has a definite sign. Indeed, recalling that $\phi_{\beta,\lambda}(\cdot)$ has been defined in (10.2.1.3),

$$\left\{ e_\lambda(J * \rho) - J * \frac{S(\rho)}{\beta} - \phi_{\beta,\lambda}(\rho_{\beta,\lambda}) \right\} \ge 0, \tag{10.2.6.3}$$

where $\rho_{\beta,\lambda}$ is a minimizer of $\phi_{\beta,\lambda}(\rho)$.

Proof of (10.2.6.3): by adding and subtracting $\frac{S(J*\rho)}{\beta}$

$$\{\phi_{\beta,\lambda}(J * \rho) - \phi_{\beta,\lambda}(\rho_{\beta,\lambda})\} + \frac{1}{\beta}\{S(J * \rho) - J * S(\rho)\} \ge 0, \tag{10.2.6.4}$$

because the two curly brackets are both non-negative: the first one by definition, the second one by convexity.

In analogy with the functional (6.1.3.1) of Chap. 6, we can then define the excess free energy functional $\mathcal{F}_\lambda$ as a $[0, \infty]$-valued, extremes included, functional on any measurable $\rho \geq 0$, by the formula

$$\mathcal{F}_\lambda(\rho) = \int_{\mathbb{R}^d} \Big\{ e_\lambda(J * \rho) - J * \frac{S(\rho)}{\beta} - \phi_{\beta,\lambda}(\rho_{\beta,\lambda}) \Big\}, \tag{10.2.6.5}$$

which can be rewritten as

$$\mathcal{F}_\lambda(\rho) = \int_{\mathbb{R}^d} \{\phi_{\beta,\lambda}(J * \rho) - \phi_{\beta,\lambda}(\rho_{\beta,\lambda})\} + \frac{1}{\beta}\{S(J * \rho) - J * S(\rho)\}.$$

It then immediately follows that:

Theorem 10.2.6.1 *The minimum of $\mathcal{F}_\lambda$ is 0, the minimizers are functions constantly equal to a minimizer of $\phi_{\beta,\lambda}$, and any other function has strictly positive free energy (possibly infinite).*

Thus $\mathcal{F}_{\lambda(\beta)}$, $\beta > \beta_c$, has two minimizers, the functions constantly equal to $\rho_{\beta,-}$ and to $\rho_{\beta,+}$. As we shall see, a similar result holds for the LMP particles system as well. We shall in fact prove that there are two phases with densities close to $\rho_{\beta,\pm}$; however, the chemical potential λ is not precisely equal to $\lambda(\beta)$, but only close to $\lambda(\beta)$ (to which it converges as $\gamma \to 0$).

10.3 Contours and Peierls bounds

Theorem 10.2.6.1 shows that at $\lambda = \lambda(\beta)$, $\lambda(\beta)$ as in Proposition 10.2.2.2, the free energy functional $\mathcal{F}_{\lambda(\beta)}$ has two and only two minimizers which are the functions constantly equal to $\rho_{\beta,+}$ and to $\rho_{\beta,-}$. We shall prove the existence of a phase transition in the LMP system by a perturbative argument which shows that the typical particle configurations in the Gibbs measures after a suitable coarse graining look like the minimizers of $\mathcal{F}_{\lambda(\beta)}$.

10.3.1 Phase indicators and contours

The scaling parameters are the same as in the Ising case:

$$\ell_{\pm,\gamma} = \gamma^{-(1\pm\alpha)}, \qquad \zeta = \gamma^a, \tag{10.3.1.1}$$

with $1 \gg \alpha \gg a > 0$ and, for simplicity, we suppose that $\ell_{\pm,\gamma}, \gamma^{-1} \in \{2^n, n \in \mathbb{N}_+\}$. For γ small, which is the regime we are interested in, $1 \ll \ell_{-,\gamma} \ll \gamma^{-1} \ll \ell_{+,\gamma}$, thus in the cubes of $\mathcal{D}^{(\ell_{-,\gamma})}$ there are typically many particles to make statistics reliable; yet the cubes are so small that the interaction felt by particles in the same cube is approximately the same. Local closeness (of the particle density in cubes of $\mathcal{D}^{(\ell_{-,\gamma})}$) to $\rho_{\beta,+}$ (or to $\rho_{\beta,-}$) with accuracy ζ must extend to regions of size $\ell_{+,\gamma}$ to call the configuration in the plus (or minus) equilibrium.

To quantify the above considerations we introduce three phase indicators: $\eta^{(\zeta,\ell_{-,\gamma})}(q;r)$, $\theta^{(\zeta,\ell_{-,\gamma},\ell_{+,\gamma})}(q;r)$ and $\Theta^{(\zeta,\ell_{-,\gamma},\ell_{+,\gamma})}(q;r)$. Since in the whole sequel ζ and $\ell_{\pm,\gamma}$ are fixed as in (10.3.1.1) we drop them from the notation, writing simply η, θ and Θ. $\eta(q;r)=\pm 1$ if $|\mathrm{Av}^{(\ell_{-,\gamma})}(q;r)-\rho_{\beta,\pm}|\le\zeta$,

$$\mathrm{Av}^{(\ell_{-,\gamma})}(q;r)=\frac{|q\cap C_r^{(\ell_{-,\gamma})}|}{\ell_{-,\gamma}^d}; \tag{10.3.1.2}$$

otherwise $\eta(q;r)=0$. θ and Θ are then defined in terms of η as explained in Sect. 6.4.4, reading the definitions of Sect. 6.4.4 with $\ell_\pm\leftrightarrow\ell_{\pm,\gamma}$.

Recall that the contours of a configuration q are the pairs $\Gamma=(\mathrm{sp}(\Gamma),\eta_\Gamma)$, where $\mathrm{sp}(\Gamma)$, the spatial support of Γ, is a maximal connected component of $\{r:\Theta(q;r)=0\}$, while $\eta_\Gamma(r)$, $r\in\mathrm{sp}(\Gamma)$, its specification, is the restriction of $\eta(q;r)$ to $\mathrm{sp}(\Gamma)$. An abstract contour Γ is a pair produced by some configuration q. We shall tacitly suppose that all contours Γ have bounded spatial support. We refer again to Sect. 6.4.4 for a geometrical notion and properties of contours and to Sect. 6.5 for examples and pictures.

10.3.2 Diluted Gibbs measures

We shall prove phase transitions by introducing two classes of boundary conditions and showing that they give rise in the thermodynamic limit to two distinct measures. As in the Ising case, we consider "diluted Gibbs measures"; their definition involves special boundary conditions as well as constraints on the structure of the configurations close to the boundaries. The constraints are such that contours cannot reach the boundaries, and for this reason the partition functions are called diluted.

Definition 1 A configuration $\bar q$ is a *plus boundary condition* relative to a bounded $\mathcal{D}^{(\ell_{+,\gamma})}$-measurable region Λ if there is a configuration $q^+\in\mathcal{X}^+$, i.e. such that $\eta(q^+;r)=1$ for all $r\in\mathbb{R}^d$ and $\bar q$ and q^+ are the same in the region $\{r\in\Lambda^c:\mathrm{dist}(r,\Lambda)\le 2\gamma^{-1}\}$.

Definition 2 The *plus diluted Gibbs measure* in a bounded $\mathcal{D}^{(\ell_{+,\gamma})}$-measurable region Λ with plus boundary conditions $\bar q$ is

$$G^+_{\gamma,\lambda,\Lambda,\bar q}(dq_\Lambda):=\frac{1}{Z^+_{\gamma,\lambda,\Lambda,\bar q}}e^{-\beta H_{\gamma,\lambda,\Lambda}(q_\Lambda|\bar q_{\Lambda^c})}\\ \times\mathbf{1}_{\Theta((q_\Lambda+q^+_{\Lambda^c});r)=1\, r\in\delta_{\mathrm{in}}^{\ell_{+,\gamma}}[\Lambda]}\nu_\Lambda(dq_\Lambda), \tag{10.3.2.1}$$

where $q^+\in\mathcal{X}^+$, ν_Λ is defined in Sect. 10.2.3 and

$$Z^+_{\gamma,\lambda,\Lambda,\bar q}=\int_{\mathcal{X}_\Lambda}e^{-\beta H_{\gamma,\lambda,\Lambda}(q_\Lambda|\bar q_{\Lambda^c})}\mathbf{1}_{\Theta((q_\Lambda+q^+_{\Lambda^c});r)=1\, r\in\delta_{\mathrm{in}}^{\ell_{+,\gamma}}[\Lambda]}\nu_\Lambda(dq_\Lambda) \tag{10.3.2.2}$$

is the *plus diluted partition function.*

Definition 3 *Minus boundary conditions, minus diluted Gibbs measures* and *minus diluted partition functions* are defined analogously.

10.3.3 Weight of a contour

We refer again to Sect. 6.4.4 for notation and properties of contours, in particular for the definition of the sets $\mathrm{sp}(\Gamma)$, $c(\Gamma)$, $\mathrm{int}(\Gamma)$, $A_{\mathrm{ext}}(\Gamma)$, $A^{\pm}(\Gamma)$ and $\mathrm{int}^{\pm}(\Gamma)$. All these sets, in agreement with the notation, are independent of which configuration q gives rise to the contour Γ, namely which $q \in \mathcal{X}(\Gamma)$, where

$$\mathcal{X}(\Gamma) = \{q : \Gamma \text{ is a contour for } q\}. \tag{10.3.3.1}$$

Recall finally that Γ is a plus or minus contour if $\Theta(q;r) = \pm 1$ on $A_{\mathrm{ext}}(\Gamma) = \delta_{\mathrm{out}}^{\ell_{+,\gamma}}[c(\Gamma)]$. Given a plus contour Γ and a plus boundary condition q^+ for $c(\Gamma)$, we define the weight $W_\gamma^+(\Gamma;\bar{q})$ of Γ to be equal to

$$\frac{G_{\gamma,\lambda,c(\Gamma),q^+}(\eta(q_{c(\Gamma)};r) = \eta_\Gamma(r), r \in \mathrm{sp}(\Gamma);\ \Theta(q_{c(\Gamma)};r) = \pm 1, r \in A^{\pm}(\Gamma))}{G_{\gamma,\lambda,c(\Gamma),q^+}(\eta(q_{c(\Gamma)};r) = 1, r \in \mathrm{sp}(\Gamma);\ \Theta(q_{c(\Gamma)};r) = 1, r \in A^{\pm}(\Gamma))}. \tag{10.3.3.2}$$

The weight $W_{\gamma,\lambda}^-(\Gamma;q^-)$ of a minus contour Γ is defined analogously. Since the range of the interaction is $2\gamma^{-1}$, we obviously have the following.

Lemma 10.3.3.1 *The weights $W_{\gamma,\lambda}^{\pm}(\Gamma;q^{\pm})$ depend only on the restriction of $q^{\pm}$ to $\{r \in c(\Gamma)^c : \mathrm{dist}(r, c(\Gamma)) \le 2\gamma^{-1}\}$.*

10.3.4 Peierls bounds

The Peierls bounds hold with constant c if

$$W_{\gamma,\lambda}^{\pm}(\Gamma;q^{\pm}) \le \exp\Big\{-\frac{c}{2}(\zeta^2\ell_{-,\gamma}^d)\,N_\Gamma\Big\}, \quad N_\Gamma = \frac{|\mathrm{sp}(\Gamma)|}{\ell_{+,\gamma}^d},$$

for all bounded plus and minus contours Γ and all plus, minus boundary conditions $q^{\pm}$. The bound is exactly the same as in (9.2.5.1) for Ising. Notice that N_Γ is the number of cubes of the partition $\mathcal{D}^{(\ell_{+,\gamma})}$ contained in $\mathrm{sp}(\Gamma)$.

Theorem 10.3.4.1 *For any $\beta \in (\beta_c, \beta_0)$ there are c_1 (whose explicit expression is given in* (10.3.4.3) *below), $\gamma_\beta > 0$ and $\lambda_{\beta,\gamma}$, $\gamma \le \gamma_\beta$, so that for any $\pm$ contour Γ and any $\pm$ boundary condition $q^{\pm}$ relative to $c(\Gamma)$,*

$$W_{\gamma,\lambda}^{\pm}(\Gamma;q^{\pm}) \le \exp\Big\{-\beta\frac{c_1}{2}(\zeta^2\ell_{-,\gamma}^d)\,N_\Gamma\Big\}. \tag{10.3.4.1}$$

As a corollary of Theorem 10.3.4.1 we have:

Theorem 10.3.4.2 *For any $\beta \in (\beta_c, \beta_0)$ let c_1, γ_β, γ and $\lambda_{\beta,\gamma}$ as in Theorem* 10.3.4.1; *then for any bounded, simply connected, $\mathcal{D}^{(\ell_{+,\gamma})}$ measurable region Λ, any plus/minus boundary condition $q^\pm$ and any $r \in \Lambda$,*

$$G^{\pm}_{\gamma,\lambda_{\beta,\gamma},\Lambda,q^\pm}(\{\Theta(q;r) = \pm 1\}) \geq 1 - \exp\left\{-\beta\frac{c_1}{4}(\zeta^2\ell^d_{-,\gamma})\right\}. \qquad (10.3.4.2)$$

Proof The proof is exactly as in Chap. 9 for the Ising case, so that its details are omitted and we shall only sketch the main points. For the sake of definiteness consider plus boundary conditions. The proof starts by showing that the probability of having configurations with a contour Γ is bounded by the r.h.s. of (10.3.4.1). Here we use that Λ is simply connected. The proof is the same (changing spins into particles configurations) as that of Theorem 9.2.7.1. Since the event $\{\Theta(q;r) < 1\}$ is contained in the set of configurations such that the point r is "surrounded by contours," (10.3.4.2) follows from a counting (of contours) argument, just as in Sect. 9.2.8. □

Theorem 10.3.4.2 implies that for γ small enough the difference between the diluted Gibbs measures $G^{+}_{\gamma,\lambda_{\beta,\gamma},\Lambda,q^+}(dq)$ and $G^{-}_{\gamma,\lambda_{\beta,\gamma},\Lambda,q^-}(dq)$ survives in the thermodynamic limit $\Lambda \nearrow \mathbb{R}^d$ and a phase transition occurs. The implications of Theorem 10.3.4.2 on the structure of the DLR measures will be discussed in Chap. 12, where it is proved that indeed there are a plus and a minus DLR measure (these are distinct), that they are translational invariant and that any translational invariant measure is a convex combination of the two.

The constant c_1

In the actual proof of Theorem 10.3.4.1 we establish the validity of the Peierls bounds for all constant c small enough: in the course of the proofs several constraints on the smallness of the constant are successively required which in the end determine c_1. Its specific value is not particularly important, because the bounds are not optimal; nonetheless the choice of many parameters in the proof depends on c_1 and there is a real danger that the reader may feel that the proof is based on a circular argument. For this reason we specify here c_1 selecting the constraints in the proof which determine its value. At this stage they look unmotivated, but as said before it is important for the sequel to consider c_1 as fixed, namely as a true constant. The mean field candidate for c_1 is the constant c_1' which appears in (10.2.2.3), but we shall only have $c_1 < c_1'$:

$$c_1 = \frac{1}{4 \cdot 3^d \cdot 32\beta\rho_{\max}}, \qquad (10.3.4.3)$$

with $\rho_{\max}$ a "density cutoff" which is any positive number $> 2\rho_{\beta,+}$ which satisfies the three inequalities, (10.3.4.4)–(10.3.4.5)–(10.3.4.6) below:

$$\frac{c_1'}{16} > \frac{1}{32\beta\rho_{\max}} \tag{10.3.4.4}$$

(hence $c_1 < c_1'$); moreover, calling $b_0 > 0$ a constant such that $e_\lambda(\xi) \geq -b_0\xi$ for all $\xi \geq 0$ and all $|\lambda - \lambda(\beta)| \leq 1$, see (11.1.2.1), we need

$$\log \rho_{\max} \geq \beta[1 + b_0], \tag{10.3.4.5}$$

and we need that for all $K \geq 1$

$$1 \leq -\frac{1}{4K\rho_{\max}} \log \sum_{n \geq \rho_{\max} K} \frac{(Ke^{\beta b_0})^n}{n!}. \tag{10.3.4.6}$$

The right hand side diverges as $\log \rho_{\max}$ as $\rho_{\max} \to \infty$ (as can be proved using the Stirling formula (A.2.1)); hence the existence of a number $\rho_{\max}$ which satisfies (10.3.4.6).

10.4 The naive scheme of proof

In this section we outline "a natural scheme of proof" of Theorem 10.3.4.1, which is based on the proof of the Peierls bounds given in Sect. 9.3 for the Ising model with Kac potentials. Due to the absence in LMP of an analogue of the spin flip symmetry, which was an essential ingredient in the Ising case, the analysis cannot be carried through to the end, and a change of strategy will be needed. Nonetheless, it gives a good motivation for our successive analysis based on the Pirogov–Sinai theory. We fix $\beta \in (\beta_c, \beta_0)$ and drop it from the notation when confusion may not arise.

10.4.1 Absence of symmetry

As in Sect. 9.3 we start by writing the ratio (10.3.3.2) of probabilities in the definition of the weight of a contour as the ratio of two partition functions. The formulas are the same after we change σ into q and replace sums by integrals (with respect to the free measure); indeed referring for the sake of definiteness to a plus contour Γ, analogously to (9.3.0.2) we have

$$W^+_{\gamma,\lambda}(\Gamma; q^+) = \frac{\mathcal{N}^+_{\gamma,\lambda}(\Gamma, q^+)}{\mathcal{D}^+_{\gamma,\lambda}(\Gamma, q^+)}, \tag{10.4.1.1}$$

where calling $D^\pm = \delta_{\text{out}}^{2\gamma^{-1}}[\text{int}^\pm(\Gamma)]$ (in LMP the range of the interaction is $2\gamma^{-1}$), $D = D^+ \cup D^-$, $A_{\text{ext}} = \delta_{\text{out}}^{\ell_{+,\gamma}}[c(\Gamma)]$ and writing $\text{sp}(\Gamma)^-$ for $\text{sp}(\Gamma) \setminus D$, the numerator $\mathcal{N}^+_{\gamma,\lambda}(\Gamma, q^+)$ is equal to

$$\int_{q_D:\eta(q_D;r)=\pm 1, r\in D^\pm} e^{-\beta H_{\gamma,\lambda,D}(q_D)} Z^-_{\gamma,\lambda,\text{int}^-(\Gamma),q_{D^-}} Z^+_{\gamma,\lambda,\text{int}^+(\Gamma),q_{D^+}}$$

$$\times \int_{q_{\text{sp}(\Gamma)^-}:\eta(q_{\text{sp}(\Gamma)^-};r)=\eta_\Gamma(r),r\in\text{sp}(\Gamma)^-} e^{-\beta H_{\gamma,\lambda,\text{sp}(\Gamma)^-}(q_{\text{sp}(\Gamma)^-}|q_D+q^+_{A_{\text{ext}}})}, \qquad (10.4.1.2)$$

while the denominator $\mathcal{D}^+_{\gamma,\lambda}(\Gamma, q^+)$ is

$$\int_{q_D:\eta(q_D;r)=1, r\in D} e^{-\beta H_{\gamma,\lambda,D}(q_D)} Z^+_{\gamma,\lambda,\text{int}^-(\Gamma),q_{D^-}} Z^+_{\gamma,\lambda,\text{int}^+(\Gamma),q_{D^+}}$$

$$\times \int_{q_{\text{sp}(\Gamma)^-}:\eta(q_{\text{sp}(\Gamma)^-};r)=1,r\in\text{sp}(\Gamma)^-} e^{-\beta H_{\gamma,\lambda,\text{sp}(\Gamma)^-}(q_{\text{sp}(\Gamma)^-}|q_D+q^+_{A_{\text{ext}}})}. \qquad (10.4.1.3)$$

The integrals above are with respect to the measures $\nu_D(dq_D)$ and $\nu_{\text{sp}(\Gamma)^-}(dq_{\text{sp}(\Gamma)^-})$: by default and unless otherwise indicated, integrals are always relative to the free measure which will be omitted (as above) when it is clear from the context.

The strategy in the Ising model was to successively manipulate $\mathcal{N}^+_{\gamma,\lambda}(\Gamma, q^+)$, finding at the end an upper bound given by the product of $\mathcal{D}^+_{\gamma,\lambda}(\Gamma, q^+)$ times a factor identified with the Peierls gain, thus concluding the proof of the Peierls bound. The manipulations can be divided into those acting on the spatial support of the contour, $\text{sp}(\Gamma)$, and those on the internal parts of the contour, $\text{int}^\pm(\Gamma)$. The former use a Lebowitz–Penrose coarse graining argument and optimization analysis on the L–P free energy functional, the latter exploit the spin flip symmetry of the interaction.

The manipulations on the spatial support of the contour can be repeated in our LMP model (under the assumption that $\beta \in (\beta_c, \beta_0)$). They lead to the following bound:

$$\mathcal{N}^+_{\gamma,\lambda}(\Gamma, q^+) \le e^{-\beta c_1 \zeta^2 \ell^d_{-,\gamma} N_\Gamma + \beta c \gamma^{1/2}|\text{sp}(\Gamma)|}$$

$$\times \int_{\eta(q_{\text{sp}(\Gamma)};\cdot)=1} e^{-\beta H_{\gamma,\text{sp}(\Gamma)}(q_{\text{sp}(\Gamma)}|q^+_{A_{\text{ext}}})}$$

$$\times \mathbf{1}_{|\text{Av}^{(\gamma^{-1/2})}(q_{\text{sp}(\Gamma)};r)-\rho_{\beta,+}|\le\gamma^{1/2}, r\in D}$$

$$\times \left\{ e^{\beta[I^-_{\gamma,\lambda(\beta)}(\text{int}^-(\Gamma))-I^+_{\gamma,\lambda(\beta)}(\text{int}^-(\Gamma))]} Z^-_{\gamma,\lambda,\text{int}^-(\Gamma),\chi^-_{D^-}} \right\}$$

$$\times Z^+_{\gamma,\lambda,\text{int}^-(\Gamma),\chi^+_{D^+}}, \qquad (10.4.1.4)$$

where $\mathrm{Av}^{(\ell)}(q;r) = |q \cap C_r^{(\ell)}|/\ell^d$, while $I^{\pm}_{\gamma,\lambda(\beta)}(\Lambda)$ is a surface term:

$$I^{\pm}_{\gamma,\lambda(\beta)}(\Lambda) = \int_{\Lambda^c} \{e_{\lambda(\beta)}(\rho_{\beta,\pm}) - e_{\lambda(\beta)}(J_\gamma * \rho_{\beta,\pm}\mathbf{1}_{\Lambda^c})\} - \int_{\Lambda} e_{\lambda(\beta)}(J_\gamma * \rho_{\beta,\pm}\mathbf{1}_{\Lambda^c}). \tag{10.4.1.5}$$

We shall actually prove only an analogue of (10.4.1.4) and for a particular value $\lambda = \lambda_{\beta,\gamma}$, but the proof extends to all $|\lambda - \lambda(\beta)| < c\gamma^{1/2}$ with γ small enough and it can be adapted to cover (10.4.1.4).

Observe that the factor $\beta c \gamma^{1/2}|\mathrm{sp}(\Gamma)|$ in the exponent of the first term on the r.h.s. of (10.4.1.4), being equal to $\beta c \gamma^{1/2}\ell^d_{+,\gamma} N_\Gamma$, is an infinitesimal fraction (as $\gamma \to 0$) of the other term in the exponent, so that the crucial term becomes $\{e^{\beta[I^-_{\gamma,\lambda(\beta)}(\mathrm{int}^-(\Gamma)) - I^+_{\gamma,\lambda(\beta)}(\mathrm{int}^-(\Gamma))]} Z^-_{\gamma,\lambda,\chi^-_{D^-}}(\mathrm{int}^-(\Gamma))\}$. To reconstruct $\mathcal{D}^+_{\gamma,\lambda}(\Gamma, q^+)$, we need to change

$$\{e^{\beta[I^-_{\gamma,\lambda(\beta)}(\mathrm{int}^-(\Gamma)) - I^+_{\gamma,\lambda(\beta)}(\mathrm{int}^-(\Gamma))]} Z^-_{\gamma,\lambda,\mathrm{int}^-(\Gamma),\chi^-_{D^-}}\} \to Z^+_{\gamma,\lambda,\mathrm{int}^-(\Gamma),\chi^+_{D^-}}.$$

This is the only symmetry we actually need between the minus and the plus phases, yet even in such a weak form equality does not seem to be true. However (and this is the way out), we can afford an error as large as $e^{\beta c \gamma^\kappa N_\Gamma}$, $N_\Gamma := |\mathrm{sp}(\Gamma)|/\ell^d_{+,\gamma}$, provided $\kappa > 2a - (1-\alpha)d$. Recall that in fact the Peierls gain is proportional to $\zeta^2 \ell^d_{-,\gamma} N_\Gamma$, $\zeta = \gamma^a$, $\ell_{-,\gamma} = \gamma^{-(1-\alpha)}$.

The leading term in the partition function is $e^{\beta P_{\gamma,\lambda}|\mathrm{int}^-(\Gamma)|}$, which is therefore the same for the two partition functions, and thus the whole point is to estimate the next term, i.e. the surface corrections to the pressure, proving that they are as small as $e^{\beta c \gamma^\kappa N_\Gamma}$ (at least when the boundary conditions "are perfect," i.e. given by $\chi^\pm$) provided $\kappa > 2a - (1-\alpha)d$. This is the hardest part of the whole proof and the validity of the estimate depends critically on λ, which so far was arbitrary (in the interval $|\lambda - \lambda(\beta)| < c\gamma^{1/2}$).

10.4.2 Surface corrections to the pressure

In Sects. 11.4, 11.5 and 11.6 of the next chapter we shall prove the following theorem:

Theorem SCP (Surface corrections to the pressure) *For any $\beta \in (\beta_c, \beta_0)$, there are $c > 0$, $\gamma_\beta > 0$ and $\lambda_{\beta,\gamma}$, $\gamma \in (0, \gamma_\beta)$, $|\lambda(\beta) - \lambda_{\beta,\gamma}| \le c\gamma^{1/2}$, so that, for any bounded $\mathcal{D}^{(\ell_{+,\gamma})}$-measurable region Λ,*

$$\frac{e^{\beta I^-_{\gamma,\lambda(\beta)}(\Lambda)} Z^-_{\gamma,\lambda_{\beta,\gamma},\Lambda,\chi^-_{\Lambda^c}}}{e^{\beta I^+_{\gamma,\lambda(\beta)}(\Lambda)} Z^+_{\gamma,\lambda_{\beta,\gamma},\Lambda,\chi^+_{\Lambda^c}}} \le e^{c\gamma^{1/2}|\delta^{\ell_{+,\gamma}}_{\mathrm{out}}[\Lambda]|}. \tag{10.4.2.1}$$

Moreover, since in (10.4.1.4) $|\text{Av}^{(\gamma^{-1/2})}(q_D;r) - \rho_{\beta,+}| \le \gamma^{1/2}, r \in D$, there is $c > 0$ so that for any such q_D

$$\frac{Z^+_{\gamma,\lambda_{\beta,\gamma},\text{int}^-(\Gamma),\chi^+_{D^-}}}{Z^+_{\gamma,\lambda_{\beta,\gamma},\text{int}^-(\Gamma),q_{D^-}}} \le e^{c\gamma^{1/2}|\delta^{\ell_+,\gamma}_{\text{out}}[\text{int}^-(\Gamma)]|}. \tag{10.4.2.2}$$

This is easy; it is proved, with other simple energy bounds, in Sect. 11.1.2. After using (10.4.2.1) and (10.4.2.2), we get from (10.4.1.4) for γ small enough

$$\mathcal{N}^+_{\gamma,\lambda}(\Gamma,q^+) \le e^{-\beta(c_1/2)\zeta^2\ell^d_{-,\gamma}N_\Gamma}\mathcal{D}^+_{\gamma,\lambda}(\Gamma,q^+), \tag{10.4.2.3}$$

which completes the proof of the Peierls bounds.

The crucial step of the whole proof is thus the bound (10.4.2.1). The analysis of the surface corrections to the pressure is generally based on cluster expansion techniques whose applicability is proved as a consequence of the Peierls bounds. We are thus caught in a circular argument with the flavor of the catch 22 paradox: to prove the Peierls bounds we need Theorem SCP, but to prove Theorem SCP we need the Peierls bounds. The situation, however, is not as bad as it looks. The key point is that the proof of Theorem SCP needs Peierls bounds only for contours whose spatial supports are contained in the interior parts of the original region $c(\Gamma)$, which suggests an inductive argument on the volume as originally proposed by Pirogov–Sinai. Knowing the validity of the Peierls bounds in regions of a given volume we prove their validity in regions whose volume is "one step" larger. The inductive procedure is complicated by the necessity of adjusting at each step the chemical potential, whose precise value is only fixed in the limit. The most convenient way to proceed, I believe, is to follow the version of the Pirogov–Sinai theory proposed by Zahradnik [209, 210]. This will involve a repetition of the scheme outlined in this section for a new class of systems, where the contours weights are modified, their values depending on some "cutoff" parameter. For each system in the class we shall prove the validity of the Peierls bounds; we shall then recognize that for a particular value of the parameter in this class the system is equal to the original one, thus concluding that the Peierls bounds are verified. The ideas which motivate the approach and the actual scheme are presented in the next section. Its implementation will then be carried out in the next chapter.

10.5 The Pirogov–Sinai scheme

The physical picture to have in mind (and which will eventually be established rigorously) is the following one. There is a special value $\lambda_{\beta,\gamma}$ of the chemical potential where the plus and the minus phases coexist. If we vary λ away from $\lambda_{\beta,\gamma}$ at constant β, we run into the one phase regions: one of the two phases present at $\lambda_{\beta,\gamma}$ persists, the other one becomes "metastable" and disappears. A way to describe the metastable phase has been proposed by Lebowitz and Penrose. As argued at length in this book, a stable phase should be regarded as a homogeneous sea with rare and small islands of the other phase, separated from the sea by contours. At phase coexistence the phase is selected by the boundary conditions. If we vary the chemical potential "opposite to" the boundary conditions, so that the boundary conditions favor the no longer stable phase, then we see the formation of a huge island of the truly stable phase which occupies most of the space, as the large cost of the contour surrounding such an island (which grows as its surface) is overcome by the volume gain of having the correct phase in the bulk. Following Lebowitz and Penrose, let us then forbid by brute force "large" islands; for small ones the cost of a contour wins over the volume gain and islands are again rare, just as at the phase coexistence. Lebowitz and Penrose used these ideas to define metastable states; Pirogov and Sinai used them to implement a proof of the existence of phase transitions. The point in the Pirogov–Sinai approach is that the metastable phase defined with the constraint that contours (and islands) cannot be too large has a free energy larger than in the truly stable phase: thus stable and metastable phases are recognized by their free energies and the right chemical potential is selected by requiring equality of the two and phase coexistence then follows. The beauty of the approach is that it requires to compute free energies of states where contours are rare: they are rare in the stable phase just because they are improbable, while they are rare in the metastable phase by definition of the latter. Thus a real computation along these lines looks feasible and indeed this is the way we shall prove Theorems 10.3.4.1–10.3.4.2. We shall follow the approach proposed by Milos Zahradnik, which uses the notion of cutoff weights. The point is that a constraint which literally forbids contours larger than some given value modifies also the stable phase, as any contour, no matter how large it is, has anyway a positive probability to occur. It is thus convenient to allow for all contours, but to give them a weight which is the true one only if the true weight is small, as in the Peierls bounds; otherwise it is given a fictitiously small value. In the stable phase the cutoff (if properly chosen) is not reached and the state is not modified by this procedure. Thus, the right chemical potential is selected as that value for which the weights of both the plus and the minus contours do not reach the cutoff. Then in both phases the weights of the contours are the true ones and they are small, as they are smaller than the cutoff. Thus the Peierls bounds hold and we have a phase transition.

10.5.1 Cutoff weights

While the meaning of forbidding large contours is clear and easily implementable, less obvious is it what "changing the weight of a contour" means. We shall discuss the plus case; the minus one is completely analogous and not considered explicitly. The starting point is to write the plus diluted partition functions as partition functions on the restricted ensembles $\mathcal{X}_\Lambda^+ = \{q : \eta(q;r) = 1, r \in \Lambda\}$.

The set of all contours as a graph

• We denote by $\{\Gamma\}^+$ the collection of all plus contours with bounded spatial support. Recalling that two sets are connected if their closures have non-empty intersection, we draw edges joining any two elements Γ and Γ' of $\{\Gamma\}^+$ if $\mathrm{sp}(\Gamma)$ is connected to $\mathrm{sp}(\Gamma')$, thus equipping $\{\Gamma\}^+$ with a graph structure. (Notice that two connected contours cannot appear in a configuration q.)

• $\mathcal{B}^+$ is the space of all finite subsets of $\{\Gamma\}^+$ made of elements which are mutually disconnected. Elements of $\mathcal{B}^+$ are denoted by $\underline{\Gamma}$.

• $\mathcal{B}^{+,\mathrm{ext}}$ is the subset of $\mathcal{B}^+$ made of mutually "external" contours: namely if $\underline{\Gamma} \in \mathcal{B}^{+,\mathrm{ext}}$ for any $\Gamma \neq \Gamma'$ in $\underline{\Gamma}$, $c(\Gamma) \cap c(\Gamma') = \emptyset$.

• $\{\Gamma\}^+_\Lambda$, $\mathcal{B}^+_\Lambda$ and $\mathcal{B}^{+,\mathrm{ext}}_\Lambda$ (by default Λ denotes a bounded $\mathcal{D}^{(\ell_+,\gamma)}$-measurable region) denote the previous quantities with the additional restriction that all contours have spatial support in $\Lambda \setminus \delta_{\mathrm{in}}^{\ell_+,\gamma}[\Lambda]$, i.e. [the spatial support of the] contours is not connected to Λ^c.

• Let $W^+_{\gamma,\lambda}(\emptyset, q^+) = 1$ and, recalling (10.3.4.1), let us write

$$W^+_{\gamma,\lambda}(\underline{\Gamma}, q^+) = \prod_{\Gamma \in \underline{\Gamma}} W^+_{\gamma,\lambda}(\underline{\Gamma}, q^+), \quad \underline{\Gamma} \in \mathcal{B}^+,\ q^+ \in \mathcal{X}^+. \tag{10.5.1.1}$$

• If $\underline{\Gamma} \in \mathcal{B}^{+,\mathrm{ext}}$ we write

$$\mathrm{int}(\underline{\Gamma}) = \bigcup_{\Gamma \in \underline{\Gamma}} \mathrm{int}(\Gamma), \qquad \mathrm{ext}(\underline{\Gamma}) = \bigcap_{\Gamma \in \underline{\Gamma}} \mathrm{ext}(\Gamma). \tag{10.5.1.2}$$

Partition functions in restricted ensembles

The diluted partition function in a region Λ can be written as a partition function in $\mathcal{X}^+_\Lambda = \{q \in \mathcal{X}_\Lambda : \eta(q,r) = 1, r \in \Lambda\}$:

Theorem 10.5.1.1 *For any bounded $\mathcal{D}^{(\ell_+,\gamma)}$-measurable region Λ and any plus b.c. q^+,*

$$Z^+_{\gamma,\lambda,\Lambda,q^+} = \sum_{\underline{\Gamma} \in \mathcal{B}^+_\Lambda} \int_{q_\Lambda \in \mathcal{X}^+_\Lambda} W^+_{\gamma,\lambda}(\underline{\Gamma}, q_\Lambda)\, e^{-\beta H_{\gamma,\lambda,\Lambda}(q_\Lambda | q^+_{\Lambda^c})}, \tag{10.5.1.3}$$

where $q^+_{\Lambda^c}$ is made of all the particles of q^+ which are in Λ^c.

Proof Recalling the definition in (10.4.1.1) of $\mathcal{N}^+_{\gamma,\lambda}(\Gamma, q)$, we have

$$Z^+_{\gamma,\lambda,\Lambda,q^+} = \sum_{\underline{\Gamma} \in \mathcal{B}^{+,\mathrm{ext}}_\Lambda} \int_{\mathcal{X}^+_{\Lambda \cap \mathrm{ext}(\underline{\Gamma})}} \nu_{\Lambda \cap \mathrm{ext}(\underline{\Gamma})}(dq)$$
$$\times e^{-\beta H_{\gamma,\lambda,\Lambda \cap \mathrm{ext}(\underline{\Gamma})}(q | q^+ \cap \Lambda^c)} \prod_{\Gamma \in \underline{\Gamma}} \mathcal{N}^+_{\gamma,\lambda}(\Gamma, q).$$

Rewriting $\mathcal{N}^+_{\gamma,\lambda}(\Gamma,q)$ in terms of $\mathcal{D}^+_{\gamma,\lambda}(\Gamma,q)$ and $W^+_{\gamma,\lambda}(\underline{\Gamma},q)$ we get

$$Z^+_{\gamma,\lambda,\Lambda,q^+} = \sum_{\underline{\Gamma}\in\mathcal{B}^{+,\mathrm{ext}}_\Lambda} \int_{\mathcal{X}^+_{\Lambda\cap\{\mathrm{sp}(\underline{\Gamma})\cup\mathrm{ext}(\underline{\Gamma})\}}} \nu_{\Lambda\cap\{\mathrm{sp}(\underline{\Gamma})\cup\mathrm{ext}(\underline{\Gamma})\}}(dq)$$
$$\times\, e^{-\beta H_{\gamma,\lambda,\Lambda\cap\{\mathrm{sp}(\underline{\Gamma})\cup\mathrm{ext}(\underline{\Gamma})\}}(q|q^+\cap\Lambda^c)}\, W^+_{\gamma,\lambda}(\underline{\Gamma},q)\, Z^+_{\gamma,\lambda,\mathrm{int}(\underline{\Gamma}),q}. \qquad (10.5.1.4)$$

By using again (10.5.1.4) to rewrite $Z^+_{\gamma,\mathrm{int}(\underline{\Gamma}),q}$ and then iterating this procedure, we get (10.5.1.3). □

Fictitious weights

Throughout the sequel $\hat{W}^\pm_{\gamma,\lambda}(\Gamma;q^\pm)$, $\Gamma\in\{\Gamma\}^+$, $q^\pm$ $\pm$ boundary conditions for $c(\Gamma)$, denote strictly positive numbers which, like the true weights, see Lemma 10.3.3.1, depend only on the restriction of $q^\pm$ to $\{r\in c(\Gamma)^c : \mathrm{dist}(r,c(\Gamma))\le 2\gamma^{-1}\}$. Recalling (10.5.1.3), we define for any bounded, simply connected $\mathcal{D}^{(\ell_+,\gamma)}$-measurable region Λ and any $\pm$ b.c. $q^\pm$ for $c(\Gamma)$

$$\hat{Z}^\pm_{\gamma,\lambda,\Lambda,q^\pm} = \sum_{\underline{\Gamma}\in\mathcal{B}^\pm_\Lambda} \int_{\mathcal{X}^+_\Lambda} \hat{W}^\pm_{\gamma,\lambda}(\underline{\Gamma},q_\Lambda)\, e^{-\beta H_{\gamma,\lambda,\Lambda}(q_\Lambda|q^\pm_{\Lambda^c})} \qquad (10.5.1.5)$$

(the integral over the free measure $\nu_\Lambda(dq_\Lambda)$), where, if $\underline{\Gamma}=(\Gamma_1,\dots,\Gamma_n)$,

$$\hat{W}^\pm_{\gamma,\lambda}(\underline{\Gamma},q) = \prod_{i=1}^n \hat{W}^\pm_{\gamma,\lambda}(\Gamma_i,q), \quad \underline{\Gamma}=(\Gamma_1,\dots,\Gamma_n). \qquad (10.5.1.6)$$

By (10.5.1.3) if the weights of the contours are the true ones, then $\hat{Z}^\pm_{\gamma,\lambda,\Lambda,q}$ are the plus and minus diluted partition functions. We should regard (10.5.1.5) as the definition of $\hat{Z}^\pm_{\gamma,\lambda,\Lambda,q}$ as a function of the variables $\{\hat{W}^\pm_{\gamma,\lambda}(\Gamma;q)$, Γ a plus, respectively, minus contour, whose spatial support is in $\Lambda\setminus\delta^{\ell_+,\gamma}_{\mathrm{in}}[\Lambda]\}$. We next define $\hat{N}^+_{\gamma,\lambda}(\Gamma,q^+)$ and $\hat{D}^+_{\gamma,\lambda}(\Gamma,q^+)$ as functions of the weights $\hat{W}^\pm_{\gamma,\lambda}(\cdot;\cdot)$ as follows:

$$\hat{N}^+_{\gamma,\lambda}(\Gamma,q^+) = \int_{q_{\mathrm{sp}(\Gamma)}:\eta(q_{\mathrm{sp}(\Gamma)};r)=\eta_\Gamma(r),r\in\mathrm{sp}(\Gamma)} e^{-\beta H_{\gamma,\lambda,\mathrm{sp}(\Gamma)}(q_{\mathrm{sp}(\Gamma)}|q^+_{A_{\mathrm{ext}}})}$$
$$\times\, \hat{Z}^-_{\gamma,\lambda,\mathrm{int}^-(\Gamma),q_{\mathrm{sp}(\Gamma)}}\, \hat{Z}^+_{\gamma,\lambda,\mathrm{int}^+(\Gamma),q_{\mathrm{sp}(\Gamma)}}, \qquad (10.5.1.7)$$

$$\hat{D}^+_{\gamma,\lambda}(\Gamma,q^+) = \int_{q_{\mathrm{sp}(\Gamma)}:\eta(q_{\mathrm{sp}(\Gamma)};r)=1,r\in\mathrm{sp}(\Gamma)} e^{-\beta H_{\gamma,\lambda,\mathrm{sp}(\Gamma)}(q_{\mathrm{sp}(\Gamma)}|q^+_{A_{\mathrm{ext}}})}$$
$$\times\, \hat{Z}^+_{\gamma,\lambda,\mathrm{int}^-(\Gamma),q_{\mathrm{sp}(\Gamma)}}\, \hat{Z}^+_{\gamma,\lambda,\mathrm{int}^+(\Gamma),q_{\mathrm{sp}(\Gamma)}}. \qquad (10.5.1.8)$$

$\hat{N}^{-}_{\gamma,\lambda}(\Gamma, q^{-+})$ and $\hat{D}^{-}_{\gamma,\lambda}(\Gamma, q^{-})$ are defined analogously. As said before, they are regarded as functions of the weights $\hat{W}^{\pm}_{\gamma,\lambda}(\cdot;\cdot)$ and like the true weights, also $\hat{N}^{+}_{\gamma,\lambda}(\Gamma, q^{+})$ and $\hat{D}^{+}_{\gamma,\lambda}(\Gamma, q^{+})$ depend only on the restriction of $q^{\pm}$ to $\{r \in c(\Gamma)^c : \mathrm{dist}(r, c(\Gamma)) \le 2\gamma^{-1}\}$.

The cutoff weights

We are now ready for the final step, the choice of the weights. With c_1 as in (10.3.4.3) we have:

Theorem 10.5.1.2 *There is a unique choice of $\hat{W}^{\pm}_{\gamma,\lambda}(\Gamma; q^{\pm})$ such that for any $\pm$ contour Γ and any $q^{\pm}$,*

$$\hat{W}^{\pm}_{\gamma,\lambda}(\Gamma; q^{\pm}) = \min\left\{ \frac{\hat{N}^{\pm}_{\gamma,\lambda}(\Gamma, q)}{\hat{D}^{\pm}_{\gamma,\lambda}(\Gamma, q)}, e^{-\beta \frac{c_1}{100}(\zeta^2 \ell^d_{-,\gamma}) N_\Gamma} \right\}, \tag{10.5.1.9}$$

with $\hat{N}^{\pm}_{\gamma,\lambda}(\Gamma, q)$ and $\hat{D}^{\pm}_{\gamma,\lambda}(\Gamma, q)$ as in (10.5.1.6)–(10.5.1.7) *and* (10.5.1.5) *and hence dependent on the weights $\hat{W}^{\pm}_{\gamma,\lambda}(\cdot;\cdot)$. Moreover the weights $\hat{W}^{\pm}_{\gamma,\lambda}(\Gamma; q^{\pm})$ depend only on the restriction of $q^{\pm}$ to $\{r \in c(\Gamma)^c : \mathrm{dist}(r, c(\Gamma)) \le 2\gamma^{-1}\}$.*

Proof The proof is by induction on the size of the spatial support of the contours. To this end we introduce the notion of

Ordering regions and contours A bounded, $\mathcal{D}^{(\ell_+,\gamma)}$-measurable region Λ has "order" k if k equals the maximal integer n of elements $(\Gamma_1, \ldots, \Gamma_n) \in \mathcal{B}^+_\Lambda$ such that $\mathrm{sp}(\Gamma_{i+1})$ is contained in $\mathrm{c}(\Gamma_i)$, for all $i = 1, \ldots, n-1$. Notice that it is immaterial whether we use $\mathcal{B}^+_\Lambda$ or $\mathcal{B}^-_\Lambda$.

A contour Γ has order k if $k-1$ is the maximal order of $\mathrm{int}^{\mp}_i(\Gamma)$; namely, there is a connected component of $\mathrm{int}(\Gamma)$ which has order $k-1$ and all the others have order $\le k-1$.

Weights of contours of "order 1" If a plus [minus] contour Γ has order 1, this means that there is no other contour whose spatial support is contained in $\mathrm{int}^{\mp}(\Gamma)$. Then the formulas (10.5.1.7) and (10.5.1.8) define univocally $\hat{N}^{\pm}_{\gamma,\lambda}(\Gamma, q^{\pm})$ and $\hat{D}^{\pm}_{\gamma,\lambda}(\Gamma, q^{\pm})$, because the functions $Z^{\mp}_{\gamma,\lambda,\mathrm{int}^{\mp}(\Gamma),q_D}$ which appear in their definitions do not depend on the weights of the contours, as on the r.h.s. of (10.5.1.5) the sum over $\underline{\Gamma}$ is empty. Thus $\hat{N}^{\pm}_{\gamma,\lambda}(\Gamma, q^{\pm})$ and $\hat{D}^{\pm}_{\gamma,\lambda}(\Gamma, q^{\pm})$ are explicitly known, and the r.h.s. of (10.5.1.9) defines unambiguously the weight of all contours of order 1, and they depend only on $q^{\pm}$ restricted to $\{r \in c(\Gamma)^c : \mathrm{dist}(r, c(\Gamma)) \le 2\gamma^{-1}\}$.

Inductive definition of the contours weights Having defined the weights of contours of order 1, we can use (10.5.1.7) and (10.5.1.8) for contours of order 2, because on the r.h.s. of (10.5.1.5) written for $Z^{\mp}_{\gamma,\lambda,\mathrm{int}^{\mp}(\Gamma),q_D}$ the sum is over contours of order 1.

(10.5.1.9) thus defines the weights of contours of order 2 which again depend only on $q^\pm$ restricted to $\{r \in c(\Gamma)^c : \mathrm{dist}(r, c(\Gamma)) \le 2\gamma^{-1}\}$. Proceeding iteratively, we then define the weight of any contour Γ. The last statement follows from (10.5.1.9) and the remark after (10.5.1.8). □

10.5.2 Recovering the true weights

As anticipated in the introduction to the section, an important point of the scheme is that within the context of "the fictitious cutoff weights model" we can decide whether it is the true one.

Theorem 10.5.2.1 *Assume that for any plus or minus contour Γ*

$$\hat{W}^{\pm}_{\gamma,\lambda}(\Gamma; q^\pm) < e^{-\beta \frac{c_1}{100}(\zeta^2 \ell^d_{-,\gamma}) N_\Gamma}; \tag{10.5.2.1}$$

then

$$\hat{W}^{\pm}_{\gamma,\lambda}(\Gamma; q^\pm) = W^{\pm}_{\gamma,\lambda}(\Gamma; q^\pm), \qquad \hat{Z}^{\pm}_{\gamma,\lambda,\Lambda,q^\pm} = Z^{\pm}_{\gamma,\lambda,\Lambda,q^\pm}, \tag{10.5.2.2}$$

where $W^{\pm}_{\gamma,\lambda}(\Gamma; q_{\Lambda^\pm})$ are the true weights, defined in (10.4.1.1) *and $Z^{\pm}_{\gamma,\lambda,\Lambda,q^\pm}$ the plus minus diluted partition functions.*

Proof The proof is again by induction. Assume that the first equality in (10.5.2.2) holds for contours of order $\le n$; then, we claim, the second equality in (10.5.2.2) holds for regions Λ of order $\le n$. By Theorem 10.5.1.1 we write $Z^{+}_{\gamma,\lambda,\Lambda,q^+}$ as in (10.5.1.3). Since Λ has order $\le n$, all Γ_i in $\underline{\Gamma}$ have order $\le n$, and hence $W_{\gamma,\lambda}(\underline{\Gamma}, q_\Lambda) = \hat{W}_{\gamma,\lambda}(\underline{\Gamma}, q_\Lambda)$. By comparing (10.5.1.3) and (10.5.1.5), we get $\hat{Z}^{+}_{\gamma,\lambda,\Lambda,q^+} = Z^{+}_{\gamma,\lambda,\Lambda,q^+}$. The same proof applies to minus diluted partition functions.

Let us now prove the first equality in (10.5.2.2) for a contour Γ of order $n+1$. By the assumption (10.5.2.1) we get from (10.5.1.9)

$$\hat{W}^{\pm}_{\gamma,\lambda}(\Gamma; q^\pm) = \frac{\hat{N}^{\pm}_{\gamma,\lambda}(\Gamma, q^\pm)}{\hat{D}^{\pm}_{\gamma,\lambda}(\Gamma, q^\pm)}. \tag{10.5.2.3}$$

The functions $\hat{Z}^{\pm}_{\gamma,\lambda,\mathrm{int}^\pm(\Gamma),q^+}$ appearing on the r.h.s. of (10.5.1.7)–(10.5.1.8) are equal to the corresponding plus and minus diluted partition functions, because the regions involved have order n, and this is because Γ has order $n+1$. Thus $\hat{N}^{\pm}_{\gamma,\lambda}(\Gamma, q^\pm) = \mathcal{N}^{\pm}_{\gamma,\lambda}(\Gamma, q^\pm)$ and $\hat{D}^{\pm}_{\gamma,\lambda}(\Gamma, q^\pm) = \mathcal{D}^{\pm}_{\gamma,\lambda}(\Gamma, q^\pm)$, and by comparing (10.5.2.3) and (10.4.1.1) we conclude the proof of the first equality in (10.5.2.2) when Γ has order $n+1$.

To complete the proof by induction we need to prove the first equality in (10.5.2.2) when Γ has order 1. In such a case, by definition the connected components of the interior of Γ are regions of order 0 where no contour can be contained so that the previous argument can be applied. □

Corollary 10.5.2.2 *To prove Theorem* 10.3.4.1 *it suffices to show that for γ small enough for all Γ*

$$\hat{W}^{\pm}_{\gamma,\lambda}(\Gamma;q^{\pm}) = \frac{\hat{N}^{\pm}_{\gamma,\lambda}(\Gamma,q^{\pm})}{\hat{D}^{\pm}_{\gamma,\lambda}(\Gamma,q^{\pm})} \le e^{-\beta\frac{c_1}{2}(\zeta^2\ell^d_{-,\gamma})N_\Gamma}. \tag{10.5.2.4}$$

Proof By Theorem 10.5.2.1 if (10.5.2.4) holds, then

$$W^{\pm}_{\gamma,\lambda}(\Gamma;q^{\pm}) = \hat{W}^{\pm}_{\gamma,\lambda}(\Gamma;q^{\pm}) \le e^{-\beta\frac{c_1}{2}(\zeta^2\ell^d_{-,\gamma})N_\Gamma}. \tag{10.5.2.5}$$

□

Chapter 11
Phase transitions in the LMP model

In this chapter we shall prove the Peierls bounds stated in Theorem 10.3.4.1. We fix the inverse temperature β in the open interval (β_c, β_0) (and since it is fixed we shall often drop it from the notation). By Corollary 10.5.2.2 it suffices to prove that for γ small enough (10.5.2.4) holds for all contours Γ. The analysis leading to (10.5.2.4) has been rather general and it has not used in any essential way the specific features of the LMP model, which instead enter massively in its proof, which is the object of the present chapter. The proof of (10.5.2.4) immediately follows from the three basic estimates below, which involve quantities whose definition is recalled here for the readers' convenience:

Notation $\alpha > 0$ and $\ell_{\pm,\gamma}$ are defined in (10.3.1.1); $\hat{Z}^{\pm}_{\gamma,\lambda,\Lambda,q^{\pm}}$ in (10.5.1.5); $\hat{N}^{\pm}_{\gamma,\lambda}$ and $\hat{D}^{\pm}_{\gamma,\lambda}$ are equal to the expressions (10.5.1.7) and (10.5.1.8), respectively; the constant c_1 has been chosen in Sect. 10.3.4. We also use the shorthand notation:

$$\chi^{\pm}_{\Delta}(r) = \rho_{\beta,\pm}\mathbf{1}_{r\in\Delta}, \qquad \chi^{\pm} = \chi^{\pm}_{\mathbb{R}^d}. \tag{11.0.2.1}$$

Equality of pressures There are functions $P^{\pm}_{\gamma,\lambda}$ which depend continuously on $\lambda \in [\lambda(\beta) - 1, \lambda(\beta) + 1]$ such that for any van Hove sequence of $\mathcal{D}^{(\ell_{+,\gamma})}$-measurable regions Λ_n and any $\pm$ Λ_n-boundary conditions $q^{\pm}_n$ we have

$$\lim_{n\to\infty} \frac{\log \hat{Z}^{\pm}_{\gamma,\lambda,\Lambda_n,q^{\pm}_n}}{\beta|\Lambda_n|} = P^{\pm}_{\gamma,\lambda}. \tag{11.0.2.2}$$

Moreover, there are $c'' > 0$ and $\lambda_{\beta,\gamma} \in [\lambda(\beta) - c''\gamma^{1/2}, \lambda(\beta) + c''\gamma^{1/2}]$ so that

$$P^{+}_{\gamma,\lambda_{\beta,\gamma}} = P^{-}_{\gamma,\lambda_{\beta,\gamma}} \quad \text{for all } \gamma \text{ small enough.} \tag{11.0.2.3}$$

The energy estimate There is $c' > 0$ so that given any γ small enough, for all plus contours Γ

$$\frac{\hat{N}^{+}_{\gamma,\lambda_{\beta,\gamma}}(\Gamma, q^{+})}{\hat{D}^{+}_{\gamma,\lambda_{\beta,\gamma}}(\Gamma, q^{+})} \le e^{-\beta(c_1\zeta^2 - c'\gamma^{1/2-2\alpha d})\ell^d_{-,\gamma} N_\Gamma} \frac{e^{\beta I^{-}_{\gamma,\lambda(\beta)}(\text{int}^{-}(\Gamma))} \hat{Z}^{-}_{\gamma,\lambda_{\beta,\gamma},\text{int}^{-}(\Gamma),\chi^{-}}}{e^{\beta I^{+}_{\gamma,\lambda(\beta)}(\text{int}^{-}(\Gamma))} \hat{Z}^{+}_{\gamma,\lambda_{\beta,\gamma},\text{int}^{-}(\Gamma),\chi^{+}}}, \tag{11.0.2.4}$$

with $I^{\pm}_{\gamma,\lambda(\beta)}$ as in (10.4.1.5). An analogous bound holds for the minus contours.

Surface correction to the pressure There is $c' > 0$ so that for any γ small enough the following holds: for all plus contours Γ

$$\left| \log \left\{ \frac{e^{\beta I^{\pm}_{\gamma,\lambda(\beta)}(\mathrm{int}^-(\Gamma))} \hat{Z}^{\pm}_{\gamma,\lambda_{\beta,\gamma},\mathrm{int}^-(\Gamma),\chi^{\pm}}}{e^{\beta|\mathrm{int}^-(\Gamma)| P^{\pm}_{\gamma,\lambda_{\beta,\gamma}}}} \right\} \right| \le c' \gamma^{1/2} \ell^d_{+,\gamma} N_\Gamma. \qquad (11.0.2.5)$$

An analogous bound (with $\mathrm{int}^+(\Gamma)$ instead of $\mathrm{int}^-(\Gamma)$) holds for minus contours.

Equation (10.5.2.4) with $\lambda = \lambda_{\beta,\gamma}$ follows immediately, for all γ small enough, from (11.0.2.4) and (11.0.2.5); thus Theorem 10.3.4.1 is reduced to the proof of (11.0.2.4) and (11.0.2.5), which will take the whole chapter. We give below a short outline.

Equality of pressures In Sect. 11.1 and in a complementary section, Sect. 11.7, we shall prove (11.0.2.2) and (11.0.2.3). Existence of the thermodynamic limit for the two pressures $P^{\pm}_{\gamma,\lambda}$ is standard and we report it mainly for the sake of completeness. It differs from the proof in the Ising case (Sect. 2.3.3), because variables are here unbounded and there is an additional term in the partition function which takes into account the contribution of the weights of the contours. The first problem is not too hard because of the special structure of the LMP hamiltonian. The latter can be handled because the Peierls bounds satisfied by the cutoff weights imply that typically contours are rare and small. The equality of pressures follows from a coarse graining argument by which we shall prove closeness of the pressures to their mean field values. Then, since the difference of the plus and minus mean field pressures changes sign when λ varies in an interval around $\lambda(\beta)$, the same happens to the difference $P^{+}_{\gamma,\lambda} - P^{-}_{\gamma,\lambda}$. Continuity of $P^{+}_{\gamma,\lambda} - P^{-}_{\gamma,\lambda}$ in λ then implies the existence of a zero. Its uniqueness is a much deeper question, which is, however, here unessential as we just want to prove the existence of phase transitions.

The energy estimate The validity of (11.0.2.4) is not restricted to the special value $\lambda = \lambda_{\beta,\gamma}$, but it holds as well for all $\lambda \in [\lambda(\beta) - c\gamma^{1/2}, \lambda(\beta) + c\gamma^{1/2}]$ (c a positive constant and γ small enough): however, since we do not need such a stronger statement we shall only prove (11.0.2.4). Its proof, which is considerably more involved than the previous equality of pressures, is divided into two steps, respectively in Sects. 11.2 and 11.3, and some more technical parts are reported in Sect. 11.8. In the first step we prove that we can factorize with "negligible error" the estimate in $\mathrm{int}(\Gamma)$ from the one in $\mathrm{sp}(\Gamma)$. In this step we bound the l.h.s. of (11.0.2.4) by a product of three factors: one is the fraction appearing on the r.h.s. of (11.0.2.4); another factor is a constrained partition function in $\mathrm{sp}(\Gamma)$ and finally the last one is "the negligible error" $e^{\beta c \gamma^{1/2-2\alpha d} \ell^d_{-,\gamma} N_\Gamma}$, $c > 0$ a constant. The second step in the proof of (11.0.2.4) involves a bound on the above constrained partition function in $\mathrm{sp}(\Gamma)$ which yields the gain factor $e^{-\beta(c_1\zeta^2 - c\gamma^{1/2-2\alpha d})\ell^d_{-,\gamma} N_\Gamma}$, $c > 0$ another constant. Combining the two we then get (11.0.2.4).

In both steps we use coarse graining arguments. The error in coarse graining is bounded by $e^{\beta c \gamma^{1/2}|\mathrm{sp}(\Gamma)|} = e^{\beta c \gamma^{1/2-2\alpha d}\ell_{-,\gamma}^{d} N_\Gamma}$, which is the "negligible factor" mentioned above. Just as in the Ising model the key point is that we can afford to make errors of this size, because it is a small fraction (as $\gamma \to 0$) of the gain term in the Peierls bounds. Thus, in both steps we have a reduction, after coarse graining, to variational problems with the LMP free energy functional. They involve two different regions, one is at the boundary between interior and spatial support of Γ, the other is in the bulk of the spatial support. In the former, we exploit the definition of contours which implies that the boundary of $\mathrm{int}^{\pm}(\Gamma)$ is in the middle of a "large region" (whose size has the order of $\ell_{+,\gamma}$), where $\eta(\cdot;\cdot)$ is identically equal to ± 1, respectively. By the strong stability properties of the LMP free energy functional, the minimizers are then proved to converge exponentially to $\rho_{\beta,\pm}$ with the distance from the boundaries. Here we use the assumption that $\beta \in (\beta_c, \beta_0)$, i.e. the mean field operator $K_{\beta,\lambda(\beta)}$ is a contraction; see Proposition 10.2.2.3. We then conclude that with a negligible error we have "thick corridors" where the minimizers are equal to $\rho_{\beta,\pm}$, thus separating the regions outside and inside the corridors.

After this step we have plus/minus partition functions in $\mathrm{int}^{\pm}(\Gamma)$ with boundary conditions $\rho_{\beta,\pm}$ and there still is a variational problem in the region $\mathrm{sp}(\Gamma)$ with the constraint that profiles should be compatible with the presence of the contour Γ. The analysis (far from trivial) of such a minimization problem leads to the gain factor in the Peierls bound. The argument is a refinement of the one used in the proof of Theorem 10.2.6.1.

Surface correction to the pressure The proof of (11.0.2.5) is definitely the most delicate point in the whole scheme and it is carried out in Sects. 11.4, 11.5, 11.6; there is also a complementary section, Sect. 11.9, where we report (without proof) some theorems on cluster expansions which play an important role in the proofs. In classical Pirogov–Sinai models where we perturb the ground states, the partition functions $\hat{Z}^{\pm}_{\mathrm{int}^{\pm}(\Gamma)}$ have only the contribution of the contour weights (the restricted ensembles $\mathcal{X}^{\pm}$ are in fact singletons consisting each one of a ground state). The Peierls bounds which again hold by definition (because the weights are cutoff weights!) prove the validity of a cluster expansion, from which (11.0.2.5) then follows. A natural requirement for the extension of the theory to the continuum where $\mathcal{X}^{\pm}$ has a non-trivial structure is the one proposed in [54, 55], which requires the validity of a cluster expansion. We shall prove instead that a generalized version of the Dobrushin uniqueness theorem holds and show that this implies (11.0.2.5). Exponential decay of the correlations in the restricted ensembles is the key property here, and its validity may well go beyond the region of validity of a cluster expansion and Dobrushin uniqueness, as argued by Dobrushin and Shlosman [110]. An example (in the context of Kac potentials) is in the recent analysis of the Potts model in the continuum [102, 103], and, presumably, in the same LMP model for $\beta \geq \beta_0$.

11.1 Choice of the chemical potential

In this section we shall prove (11.0.2.2) and (11.0.2.3) and thus select the value $\lambda = \lambda_{\beta,\gamma}$ of the chemical potential for which a phase transition will be proved to occur. As mentioned before, the argument proves existence but not uniqueness. We restrict λ to the interval $[\lambda(\beta)-1, \lambda(\beta)+1]$ and we shall prove that $\lambda_{\beta,\gamma}$ lies in such an interval, actually the much stronger bound $\lambda_{\beta,\gamma} \in [\lambda(\beta)-c''\gamma^{1/2}, \lambda(\beta)+c''\gamma^{1/2}]$ (c'' a positive constant) will be proved to hold for all γ small enough.

Notational remarks (for this section) By default all sets are bounded, $\mathcal{D}^{(\ell_{+},\gamma)}$-measurable regions of $\mathbb{R}^d$. Plus/minus boundary conditions are in general denoted by $q^{\pm}$, where $q^{\pm}$ may stand either for a particle configuration or a density function. The plus/minus restricted ensembles are denoted by

$$\mathcal{X}_{\Lambda}^{\pm} = \{q_{\Lambda} \in \mathcal{X}_{\Lambda} : \eta(q_{\Lambda}; r) = \pm 1, r \in \Lambda\}; \qquad \mathcal{X}^{\pm} = \mathcal{X}_{\mathbb{R}^d}^{\pm}, \tag{11.1.0.1}$$

and the plus/minus partition functions with plus/minus b.c. $q^{\pm}$ are

$$\hat{Z}^{\pm}_{\gamma,\lambda,\Lambda,q^{\pm}} = \int_{\mathcal{X}_{\Lambda}^{\pm}} e^{-\beta H_{\gamma,\lambda}(q_{\Lambda}|q^{\pm}_{\Lambda^c})} X^{\pm}_{\gamma,\lambda,q_{\Lambda}}(\Lambda), \tag{11.1.0.2}$$

where $q^{\pm}_{\Lambda^c} = q^{\pm} \cap \Lambda^c$ and $X^{\pm}_{\gamma,\lambda,q_{\Lambda}}(\Lambda) = \sum_{\underline{\Gamma} \in \mathcal{B}^{+}_{\Lambda}} \hat{W}^{\pm}_{\gamma,\lambda}(\underline{\Gamma}, q_{\Lambda})$. Recall that the spatial supports of the contours in an element of $\mathcal{B}^{+}_{\Lambda}$ are contained in $\Lambda_0 := \Lambda \setminus \delta_{\text{in}}^{\ell_{+},\gamma}[\Lambda]$, so that the weights depend only on the configuration in $\Lambda_0 \cup \delta_{\text{out}}^{2\gamma^{-1}}[\Lambda_0]$ (we shall use this remark in the proof of Lemma 11.2.2.3). The main result in this section is the following.

Theorem 11.1.0.3 *For any $\lambda \in [\lambda(\beta)-1, \lambda(\beta)+1]$ the following holds.*

- *For any van Hove sequence (see Sect. 2.3.2) Λ_n of $\mathcal{D}^{(\ell_{+},\gamma)}$-measurable regions and any sequence $q_n^{\pm}$ of $\pm$ b.c. the following limits exist:*

$$\lim_{n\to\infty} \frac{1}{\beta|\Lambda_n|} \log \hat{Z}^{\pm}_{\gamma,\lambda,\Lambda_n,q_n^{\pm}} = P^{\pm}_{\gamma,\lambda}, \tag{11.1.0.3}$$

and they are independent of the sequence $\{\Lambda_n, q_n^{\pm}\}$.
- *$P^{\pm}_{\gamma,\lambda}$ depend continuously on λ and there is a constant $c > 0$ such that*

$$P^{+}_{\gamma,\lambda} - P^{-}_{\gamma,\lambda} \gtrless 0 \quad \text{if } \lambda = \lambda(\beta) \pm c\gamma^{1/2}. \tag{11.1.0.4}$$

As an immediate corollary of Theorem 11.1.0.3 we have:

Theorem 11.1.0.4 *For all γ small enough there is $\lambda_{\beta,\gamma}$ such that*

$$P^{+}_{\gamma,\lambda_{\beta,\gamma}} = P^{-}_{\gamma,\lambda_{\beta,\gamma}}, \qquad |\lambda_{\beta,\gamma} - \lambda(\beta)| \le c\gamma^{1/2}, \quad c \text{ as in (11.1.0.4)}. \tag{11.1.0.5}$$

Remarks The two partition functions $\hat{Z}^{\pm}_{\gamma,\lambda,\Lambda,q^{\pm}}$ in (11.1.0.3) do not differ only by the boundary conditions; they are in fact defined on different phase spaces, i.e. the restricted ensembles $\mathcal{X}^{\pm}_{\Lambda}$ of (11.1.0.1), and the contour weights entering in their definition are also different: there is therefore no apparent reason why the corresponding pressures may coincide. However, if the weights are the true weights, then by Theorem 10.5.1.1 the two partition functions are equal to the original diluted partition functions, and thus they differ from each other only by a surface term. Thus, the fact that the two pressures are equal is already an indication that at $\lambda_{\beta,\gamma}$ true and cutoff weights may coincide and hence that the Peierls bound holds. The main steps in the proof of Theorem 11.1.0.3 are:

- $P^{\pm}_{\gamma,\lambda}$ are continuous functions of λ for γ small enough and with the notation of Proposition 10.2.2.3, $\lim_{\gamma\to 0} P^{\pm}_{\gamma,\lambda} = p^{\pm;\mathrm{mf}}_{\lambda} := -\phi_{\beta,\lambda}(\rho_{\beta,\lambda,\pm})$ uniformly in λ in a neighborhood of $\lambda(\beta)$;
- $p^{+;\mathrm{mf}}_{\lambda(\beta)} = p^{-;\mathrm{mf}}_{\lambda(\beta)}$ (as stated in Proposition 10.2.2.2);
- $\frac{d}{d\lambda}\{p^{+;\mathrm{mf}}_{\lambda} - p^{-;\mathrm{mf}}_{\lambda}\}|_{\lambda=\lambda(\beta)} > 0$ (as stated in Proposition 10.2.2.3).

By combining these ingredients and restricting to γ small enough, we shall prove Theorem 11.1.0.3 and as a consequence Theorem 11.1.0.4 as well.

11.1.1 Contribution of the contours to the partition function

The first statement (11.1.0.3) of Theorem 11.1.0.3 is about the existence of the pressure in the thermodynamical limit. There are two main differences with the Ising case: the energy estimates are more complex because the local number of particles and the corresponding energy are unbounded, and there is a new contribution to the partition function, the factor $X^{\pm}_{\gamma,\lambda,q}(\Lambda)$ which depends on the contour weights. We shall reduce this to the Ising case using upper and lower bounds which are proved in this and in the next subsection, respectively for $X^{\pm}_{\gamma,\lambda,q}(\Lambda)$ and for the energy. The proofs are simple but technical and, in a first reading, they may be skipped.

By default in the sequel, Λ, Λ', $\Delta, \ldots$ are bounded $\mathcal{D}^{(\ell_{+,\gamma})}$ measurable regions in $\mathbb{R}^d$. It is convenient to extend the definition of $X^{\pm}_{\gamma,\lambda,q}(\Lambda)$ by setting for $\Lambda \subseteq \Lambda'$ and $N \in \mathbb{N} \cup \infty$,

$$X^{+,N}_{\gamma,\lambda,q}(\Lambda;\Lambda') = \sum_{\underline{\Gamma}\in\mathcal{B}^{+}_{\Lambda'}} \prod_{\Gamma\in\underline{\Gamma}} \{\mathbf{1}_{N_\Gamma \le N,\, \mathrm{sp}(\Gamma)\cap\Lambda\neq\emptyset} \hat{W}^{+}_{\gamma,\lambda}(\Gamma,q)\},$$
$$X^{+}_{\gamma,\lambda,q}(\Lambda) = X^{+,\infty}_{\gamma,\lambda,q}(\Lambda;\Lambda), \tag{11.1.1.6}$$

observing that

$$\mathrm{sp}(\Gamma)\cap\Lambda \neq \emptyset \quad \Leftrightarrow \quad \mathrm{sp}(\Gamma)\cap\delta^{\ell_{+,\gamma}}_{\mathrm{in}}[\Lambda]\neq\emptyset, \tag{11.1.1.7}$$

because Λ and $\mathrm{sp}(\Gamma)$ are $\mathcal{D}^{(\ell_{+,\gamma})}$ measurable. To simplify the notation in this subsection we shall often drop γ, λ and q from $X^{\pm}_{\gamma,\lambda,q}(\Lambda)$.

Lemma 11.1.1.1 (Lower bounds) *For any* $N \ge 0$

$$X^+(\Lambda) \ge X^{+,N}(\Lambda), \qquad X^{+,0}(\Lambda) = 1, \tag{11.1.1.8}$$

$$X^+(\Lambda, \Lambda') \ge X^+(\Delta, \Lambda'), \quad \Delta \subset \Lambda \subseteq \Lambda', \tag{11.1.1.9}$$

$$X^+(\Lambda) \ge X^+(\Lambda \setminus \Delta) X^+(\Delta). \tag{11.1.1.10}$$

Proof The inequality in (11.1.1.8) holds, because in $X^+(\Lambda)$ there are more terms than in $X^{+,N}(\Lambda)$ and they are all positive. $X^{+,0}(\Lambda)$ has only the term with no contours, whose weight is 1; hence $X^{+,0}(\Lambda) = 1$. Equation (11.1.1.9) follows because if $\underline{\Gamma} \in \mathcal{B}^+_{\Lambda'}$ contributes to $X^+(\Delta, \Lambda')$; then each $\Gamma \in \underline{\Gamma}$ is such that $\mathrm{sp}(\Gamma) \cap \Delta \neq \emptyset$. As a consequence, $\mathrm{sp}(\Gamma) \cap \Lambda \neq \emptyset$ as well and therefore $\underline{\Gamma}$ contributes to $X^+(\Lambda, \Lambda')$. An analogous argument proves (11.1.1.10). Indeed, if $\underline{\Gamma}$ in $\mathcal{B}^+_{\Lambda \setminus \Delta}$ and $\underline{\Gamma}'$ in $\mathcal{B}^+_{\Delta}$ then $\underline{\Gamma} \cup \underline{\Gamma}' \in \mathcal{B}^+_{\Lambda}$, recall in fact that the spatial supports of the contours in an element of $\mathcal{B}^+_{\Lambda}$ are contained in $\Lambda \setminus \delta_{\mathrm{in}}^{\ell_{+,\gamma}}[\Lambda]$. □

Lemma 11.1.1.2 (Upper bounds) $X^+(\Lambda; \Lambda')$, $\Lambda' \supset \Lambda$, *is a non-decreasing function of* Λ *and* Λ' *(see* (11.1.1.15) *below) and for any* $\Delta \subset \Lambda \subseteq \Lambda'$

$$X^+(\Lambda; \Lambda') \le X^+(\Lambda \setminus \Delta; \Lambda')\, X^+(\Delta; \Lambda'). \tag{11.1.1.11}$$

Moreover, there is a constant $b > 0$ *so that*

$$X^+(\Lambda; \mathbb{R}^d) \le \Big(1 + [e^b e^{-\beta \frac{c_1}{100}(\zeta^2 \ell^d_{-,\gamma})}]^{3^d}\Big)^{|\Lambda|/\ell^d_{+,\gamma}}, \tag{11.1.1.12}$$

$$X^+(\Lambda; \mathbb{R}^d) \le X^+(\Lambda)\Big(1 + [e^b e^{-\beta \frac{c_1}{100}(\zeta^2 \ell^d_{-,\gamma})}]^{3^d}\Big)^{|\delta_{\mathrm{in}}^{\ell_{+,\gamma}}[\Lambda]|/\ell^d_{+,\gamma}}, \tag{11.1.1.13}$$

$$X^+(\Lambda) \le X^{+,N}(\Lambda)\Big(1 + [e^b e^{-\beta \frac{c_1}{100}(\zeta^2 \ell^d_{-,\gamma})}]^{N}\Big)^{|\Lambda|/\ell^d_{+,\gamma}}. \tag{11.1.1.14}$$

Proof Resuming the full notation, we shall prove that

$$X^+_{\gamma,\lambda,q_{\Lambda'}}(\Lambda; \Lambda') \le X^+_{\gamma,\lambda,q_{\Lambda''}}(\Lambda; \Lambda''),$$
$$\Lambda \subset \Lambda' \subset \Lambda'', \; q_{\Lambda''} \cap \Lambda' = q_{\Lambda'}. \tag{11.1.1.15}$$

Indeed, if $\Lambda'' \supset \Lambda'$ each $\underline{\Gamma}$ in the sum defining $X^+_{\gamma,\lambda,q_{\Lambda'}}(\Lambda; \Lambda')$ is also in $X^+_{\gamma,\lambda,q_{\Lambda''}}(\Lambda; \Lambda'')$; hence (11.1.1.15). The proof that $X^+(\Lambda_1; \Lambda') \ge X^+(\Lambda; \Lambda')$, $\Lambda \subset \Lambda_1 \subseteq \Lambda'$, is analogous and we omit it.

Proof of (11.1.1.11). If $\underline{\Gamma}$ contributes to $X^+(\Lambda; \Lambda')$, then $\underline{\Gamma} = \underline{\Gamma}' \cup \underline{\Gamma}''$, where $\underline{\Gamma}'$ is the collection of all $\Gamma \in \underline{\Gamma}$ such that $\mathrm{sp}(\Gamma) \cap \Delta \neq \emptyset$; $\underline{\Gamma}''$ is then the collection of all $\Gamma \in \underline{\Gamma}$ such that $\mathrm{sp}(\Gamma) \cap \Delta = \emptyset$. Hence $\mathrm{sp}(\Gamma) \cap (\Lambda \setminus \Delta) \neq \emptyset$. Then $\underline{\Gamma}'$ contributes to $X^+(\Delta; \Lambda')$ and $\underline{\Gamma}''$ to $X^+(\Lambda \setminus \Delta; \Lambda')$; hence (11.1.1.11).

Claim *Let $\{C_i,\ i=1,\dots,N\}$ be the cubes in $\mathcal{D}^{(\ell_{+,\gamma})}$ contained in Λ and $\{C'_i,\ i=1,\dots,M\}$ those contained in $\delta_{\text{in}}^{\ell_{+,\gamma}}[\Lambda]$; then*

$$X^+(\Lambda;\mathbb{R}^d)\le \prod_{i=1}^{N}\Big\{1+\sum_{\Gamma:\text{sp}(\Gamma)\supset C_i}\hat W^+_{\gamma,\lambda}(\Gamma)\Big\}, \tag{11.1.1.16}$$

$$X^+(\Lambda;\mathbb{R}^d)\le X^+(\Lambda)\prod_{i=1}^{M}\Big\{1+\sum_{\Gamma:\text{sp}(\Gamma)\supset C'_i}\hat W^+_{\gamma,\lambda}(\Gamma)\Big\}. \tag{11.1.1.17}$$

Proof of (11.1.1.16) Any $\underline{\Gamma}=(\Gamma_1,\dots,\Gamma_k)$ which contributes to $X^+(\Lambda;\mathbb{R}^d)$ is made of contours Γ_i each one having $\text{sp}(\Gamma_i)\subset\Lambda$, and hence containing a cube C_j, $j\in\{1,\dots,N\}$. Then $\hat W^+_{\gamma,\lambda}(\underline{\Gamma})$ is one of the terms which arise when expanding the r.h.s. of (11.1.1.16), which is therefore proved. □

Proof of (11.1.1.17) Any $\underline{\Gamma}$ which contributes to $X^+(\Lambda;\mathbb{R}^d)$ can be written as $\underline{\Gamma}=\underline{\Gamma}'\cup\underline{\Gamma}''$ with $\underline{\Gamma}'$ the maximal subset of $\underline{\Gamma}$ which is in $\mathcal{B}^+_\Lambda$. Then $\underline{\Gamma}''$ is the collection of all $\Gamma\in\underline{\Gamma}$ such that $\text{sp}(\Gamma)\cap\delta_{\text{in}}^{\ell_{+,\gamma}}[\Lambda]\ne\emptyset$. In fact, $\text{sp}(\Gamma)\cap\Lambda\ne\emptyset$ and $\text{sp}(\Gamma)\not\subset\Lambda^0$, $\Lambda^0=\Lambda\setminus\delta_{\text{in}}^{\ell_{+,\gamma}}[\Lambda]$ (otherwise $\Gamma\in\underline{\Gamma}'$) and the claim follows from (11.1.1.7). Thus, $\underline{\Gamma}''$ contributes to $X^+(\delta_{\text{in}}^{\ell_{+,\gamma}}[\Lambda];\mathbb{R}^d)$, and therefore $X^+(\Lambda;\mathbb{R}^d)\le X^+(\Lambda)X^+(\delta_{\text{in}}^{\ell_{+,\gamma}}[\Lambda];\mathbb{R}^d)$. Equation (11.1.1.17) then follows from (11.1.1.16). The claim is proved. □

Proof of (11.1.1.12)–(11.1.1.13). By (11.1.1.16) each factor in the product on the r.h.s. of (11.1.1.16) is bounded by

$$1+\sum_{\text{sp}(\Gamma)\supset C_i} e^{-bN_\Gamma}\{e^{bN_\Gamma}\hat W^+_{\gamma,\lambda}(\Gamma,q)\}\le 1+e^{b\,3^d}e^{-\beta\frac{c_1}{100}(\zeta^2\ell^d_{-,\gamma})\,3^d}, \tag{11.1.1.18}$$

where b is such that $\sum_{\text{sp}(\Gamma)\supset C_i}e^{-bN_\Gamma}\le 1$. Its existence is proved in Lemma 3.1.2.4; the last inequality in (11.1.1.18) holds because for γ small enough the curly bracket is a decreasing function of N_Γ whose minimal value is 3^d. (11.1.1.12) then follows from (11.1.1.16) and (11.1.1.18). By the same argument (11.1.1.13) follows from (11.1.1.17) and (11.1.1.18).

Proof of (11.1.1.14). Any $\underline{\Gamma}\in\mathcal{B}^+_\Lambda$ can be written as $\underline{\Gamma}=\underline{\Gamma}'\cup\underline{\Gamma}''$, with $\underline{\Gamma}'$ consisting of all $\Gamma\in\underline{\Gamma}$ with $N_\Gamma\le N$; thus if $\Gamma\in\underline{\Gamma}''$ then $N_\Gamma>N$. Hence

$$X^+(\Lambda)\le X^{+,N}(\Lambda)X''(\Lambda),\quad X''(\Lambda)=1+\sum_{\underline{\Gamma}''\ne\emptyset}\prod_{\Gamma\in\underline{\Gamma}''}\mathbf{1}_{N_\Gamma>N}\hat W^+_{\gamma,\lambda}(\Gamma,q).$$

$X''(\Lambda)$ has the same structure as $X^+(\Lambda)$, but with weights identical to the old ones when $N_\Gamma>N$ and $N_\Gamma=0$, otherwise they are equal to 0. Then $X''(\Lambda)$ is bounded by the r.h.s. of (11.1.1.18) with 3^d replaced by N (as in $X''(\Lambda)$, if $\text{sp}(\Gamma)\supset C_i$ then $N_\Gamma>N$) and (11.1.1.14) follows. □

11.1.2 Energy bounds (for the LMP hamiltonian)

In this subsection we shall prove bounds on the LMP interaction energy among particles. In general, this may be very hard because the number of particles and the energy are unbounded, but the particular form of the LMP hamiltonian and the fact that we are in the restricted ensemble where the particles density is bounded makes life easy. We obviously have the following.

Lemma 11.1.2.1 *The derivative* $e'_\lambda(\xi) = -\lambda - \xi + \frac{\xi^3}{6}$ *is bounded in the compacts and there is* b_0 *so that for any* $\lambda \in [\lambda(\beta)-1, \lambda(\beta)+1]$ $e'_\lambda(\cdot) \ge -b_0$. *As a consequence for any* $\lambda \in [\lambda(\beta)-1, \lambda(\beta)+1]$ *and for any* ξ_0 *and* ξ *positive*

$$e_\lambda(\xi_0+\xi) - e_\lambda(\xi_0) \ge -b_0\xi, \qquad e_\lambda(\xi) \ge -b_0\xi. \tag{11.1.2.1}$$

Write

$$\mathrm{Av}^{(\ell)}(q;r) = \frac{|q \cap C_r^{(\ell)}|}{\ell^d}, \qquad \mathrm{Av}^{(\ell)}(\rho;r) = \fint_{C_r^{(\ell)}} \rho(r')dr'. \tag{11.1.2.2}$$

Lemma 11.1.2.2 *Assume* $\mathrm{Av}^{(\gamma^{-1})}(q;r') \le \rho_{\max}$ *(see Sect.* 10.3.4*) for all* r': $\mathrm{dist}(r, C_{r'}^{(\gamma^{-1})}) \le \gamma^{-1}$; *then*

$$J_\gamma * q(r) \le 3^d \|J\|_\infty \rho_{\max}. \tag{11.1.2.3}$$

In particular, if $\mathrm{Av}^{(\ell_{-,\gamma})}(q;\cdot) \le \rho_{\beta,+} + \zeta$ *(which is verified if* $q \in \mathcal{X}^+$*), then*

$$J_\gamma * q \le 3^d \|J\|_\infty (\rho_\beta^+ + \zeta). \tag{11.1.2.4}$$

Analogous statements hold if instead of q *there is a density* ρ.

Proof (11.1.2.3) follows from the bound

$$J_\gamma(r,r') \le \|J\|_\infty \gamma^d \mathbf{1}_{\mathrm{dist}(r, C_{r'}^{(\gamma^{-1})}) \le \gamma^{-1}}, \tag{11.1.2.5}$$

as there are at most 3^d cubes in $\mathcal{D}^{(\gamma^{-1})}$ at distance $\le \gamma^{-1}$ from r. (11.1.2.4) then follows from (11.1.2.3) because $\mathrm{Av}^{(\gamma^{-1})}(q;r) \le \sup_{r'} \mathrm{Av}^{(\ell_{-,\gamma})}(q;r')$ (we are assuming that γ^{-1} is an integer multiple of $\ell_{-,\gamma}$ and that $\mathcal{D}^{(\gamma^{-1})}$ is a partition coarser than $\mathcal{D}^{(\ell_{-,\gamma})}$). □

Lemma 11.1.2.3 *There is* c' *so that for any* q_Λ *such that* $\mathrm{Av}^{(\ell_{-,\gamma})}(q_\Lambda;r) \le \rho_{\max}$, $r \in \Lambda$, *and for any particle configuration or density function* q^+ *such that* $\mathrm{Av}^{(\ell_{-,\gamma})}(q^+;r) \le \rho_{\max}$, $r \in \Lambda^c$ *(in particular if* $q_\Lambda \in \mathcal{X}_\Lambda^+$ *and* $q^+ \in \mathcal{X}^+$*),*

$$|H_{\gamma,\lambda,\Lambda}(q_\Lambda|q^+_{\Lambda^c})| \le c'|\Lambda|, \quad \textit{for all } |\lambda - \lambda(\beta)| \le 1. \tag{11.1.2.6}$$

If also q'' *is such that* $\mathrm{Av}^{(\ell_-,\gamma)}(q'';r) \le \rho_{\max}$, $r \in \Lambda^c$, *then*

$$|H_{\gamma,\lambda,\Lambda}(q_\Lambda|q^+_{\Lambda^c}) - H_{\gamma,\lambda,\Lambda}(q_\Lambda|q''_{\Lambda^c})| \le c'|\partial\Lambda|\gamma^{-1}, \quad \text{for all } \lambda. \tag{11.1.2.7}$$

Finally, for all q_Λ, q_{Λ^c} *and all* $|\lambda - \lambda(\beta)| \le 1$

$$H_{\gamma,\lambda,\Lambda}(q_\Lambda|q_{\Lambda^c}) \ge -b_0|q_\Lambda|, \tag{11.1.2.8}$$

with b_0 *as in Lemma* 11.1.2.1.

Proof To prove (11.1.2.6), we recall that

$$H_{\gamma,\lambda,\Lambda}(q_\Lambda|q^+_{\Lambda^c}) = \int_{\Lambda\cup\delta^{\gamma^{-1}}_{\text{out}}[\Lambda]} \{e_\lambda(J_\gamma * (q_\Lambda + q^+_{\Lambda^c})) - e_\lambda(J_\gamma * (q^+_{\Lambda^c}))\}.$$

By (11.1.2.3) the arguments of e_λ are $\le 3^d\|J\|_\infty\rho_{\max}$, so that by Lemma 11.1.2.1,

$$|\{e_\lambda(J_\gamma * (q_\Lambda + q^+_{\Lambda^c})) - e_\lambda(J_\gamma * (q^+_{\Lambda^c}))\}| \le cJ_\gamma * q_\Lambda \le c3^d\|J\|_\infty\rho_{\max};$$

hence we have (11.1.2.6). Analogously

$$\text{l.h.s. of (11.1.2.7)} \le \int_{\delta^{\gamma^{-1}}_{\text{in}}[\Lambda]\cup\delta^{\gamma^{-1}}_{\text{out}}[\Lambda]} \Big|e_\lambda(J_\gamma * (q_\Lambda + q^+_{\Lambda^c})) - e_\lambda(J_\gamma * q^+_{\Lambda^c}) - e_\lambda(J_\gamma * (q_\Lambda + q''_{\Lambda^c})) + e_\lambda(J_\gamma * q''_{\Lambda^c})\Big|;$$

hence (11.1.2.7), because the arguments in the energy are bounded as before. (11.1.2.8) follows because, by (11.1.2.1), $H_{\gamma,\lambda}(q_\Lambda|q_{\Lambda^c}) \ge -b_0\int J_\gamma * q_\Lambda$ and, calling $q_\Lambda = (r_1,\dots,r_N)$, $\int J_\gamma * q_\Lambda \le \int\sum_{i=1}^N J_\gamma(r,r_i) = N = |q_\Lambda|$. □

11.1.3 Coarse graining

With the estimates of the two previous subsections we have all the ingredients to reproduce here in LMP the proofs of Sect. 2.3 about the existence in the Ising model of the thermodynamic limit for the pressure, thus proving (11.1.0.3) and proving that both $P^\pm_{\gamma,\lambda}$ are continuous functions of $\lambda \in [\lambda(\beta)-1, \lambda(\beta)+1]$. The analysis is simple if one has familiarity with these techniques but may be not totally straightforward for unexperienced readers. The compromise is to report it in all detail, but in a complementary section, Sect. 11.7. There are thus two alternatives: either to read first Sect. 11.7 and then come back here, or to skip Sect. 11.7 and continue reading from this point with the above statements accepted as proven.

To prove that $P^+_{\gamma,\lambda} - P^-_{\gamma,\lambda}$ changes sign as λ varies we shall use closeness to the mean field values and for this we shall use a Lebowitz–Penrose theorem with coarse graining techniques. The main result in this subsection is an estimate on "constrained partition functions" in terms of the LMP free energy functional which will be used often in the sequel. We set:

Definition 11.1.3.1 Let $\tilde{\mathcal{M}}_\Lambda$, Λ a $\mathcal{D}^{(\gamma^{-1/2})}$-measurable region, be the *space of non-negative*, $\mathcal{D}^{(\gamma^{-1/2})}$-measurable functions on Λ.

With $\mathrm{Av}^{(\ell)}(q;\cdot)$ as in (11.1.2.2), for any "constraint" $\mathcal{B} \subset \tilde{\mathcal{M}}_\Lambda$ we define

$$Z_{\gamma,\lambda,\Lambda,\bar{q}}(\mathcal{B}) = \int_{q_\Lambda:\mathrm{Av}^{(\gamma^{-1/2})}(q_\Lambda;\cdot)\in\mathcal{B}} e^{-\beta H_\gamma(q_\Lambda|\bar{q}_{\Lambda^c})}, \tag{11.1.3.1}$$

where Λ hereafter denotes a bounded $\mathcal{D}^{(\ell_{-,\gamma})}$-measurable region; $\bar{q}$ is a boundary condition, either denoting a particle configuration or a density function or a sum of the two. We shall suppose throughout the subsection that

$$\mathrm{Av}^{(\ell_{-,\gamma})}(\bar{q};r) \le \rho_{\max}, \quad r \in \Lambda^c, \tag{11.1.3.2}$$

$\rho_{\max}$ being defined in the paragraph "The constant c_1" in Sect. 10.3.4. Observe that (11.1.3.2) is satisfied (for γ small enough) if $\eta(\bar{q};r) = 1$ for $r \in \Lambda^c$. We also restrict the constraints $\mathcal{B}$ as follows.

Definition 11.1.3.2 (Assumptions on $\mathcal{B}$) We suppose that

$$\mathcal{B} \subset \left\{ \rho_\Lambda \in \tilde{\mathcal{M}}_\Lambda : \fint_{C_r^{(\ell_{-\gamma})}} \rho_\Lambda \le \rho_{\max}, r \in \Lambda \right\} \tag{11.1.3.3}$$

and that there is c so that for any $\rho_\Lambda \in \mathcal{B}$ there is q_Λ so that

$$\mathrm{Av}^{(\gamma^{-1/2})}(q_\Lambda;\cdot) \in \mathcal{B} \quad \text{and} \quad |\mathrm{Av}^{(\gamma^{-1/2})}(q_\Lambda;\cdot) - \rho_\Lambda| \le c\gamma^{d/2}. \tag{11.1.3.4}$$

Coarse grained variables

We shall coarse grain with mesh $\gamma^{-1/2}$. We write

$$\tilde{J}_\gamma(r,r') = \fint_{C_r^{(\gamma^{-1/2})}} \fint_{C_{r'}^{(\gamma^{-1/2})}} J_\gamma(r_1,r_1') \tag{11.1.3.5}$$

and add a tilde to denote quantities which are computed with $\tilde{J}_\gamma$ instead of J_γ; in particular, $\tilde{H}_\gamma$ and $\tilde{Z}$ are the energy and partition function with $\tilde{J}_\gamma$. We next define the "tilde" mesoscopic free energy functional, setting for $\rho_\Lambda \in \tilde{\mathcal{M}}_\Lambda$

$$\tilde{F}_{\gamma,\lambda,\Lambda}(\rho_\Lambda|\bar{q}) = \tilde{H}_{\gamma,\lambda,\Lambda}(\rho_\Lambda|\bar{q}_{\Lambda^c}) - \frac{1}{\beta}\int_\Lambda S(\rho_\Lambda),$$
$$S(t) = -t(\log t - 1). \tag{11.1.3.6}$$

Our main result here is:

Theorem 11.1.3.3 *There is $c > 0$ so that for any $\bar q$ satisfying* (11.1.3.2) *and any $\mathcal{B}$ as in Definition* 11.1.3.2:

$$\Big|\log Z_{\gamma,\lambda,\Lambda,\bar q}(\mathcal{B}) + \beta \inf_{\rho_\Lambda \in \mathcal{B}} \tilde F_{\gamma,\lambda,\Lambda}(\rho_\Lambda|\bar q)\Big| \le c\gamma^{1/2}|\Lambda|. \tag{11.1.3.7}$$

We postpone the proof of Theorem 11.1.3.3 to Lemma 11.1.3.4, which will be used to replace the partition function $Z_{\gamma,\lambda,\Lambda,\bar q}(\mathcal{B})$ in (11.1.3.7) by its tilde analogue.

Lemma 11.1.3.4 *There is $c > 0$ so that for all $\bar q$ satisfying* (11.1.3.2),

$$e^{-c\gamma^{1/2}|\Lambda|} \le \frac{Z_{\gamma,\lambda,\Lambda,\bar q}(\mathcal{B})}{\tilde Z_{\gamma,\lambda,\Lambda,\bar q}(\mathcal{B})} \le e^{c\gamma^{1/2}|\Lambda|}. \tag{11.1.3.8}$$

Proof We write the shorthand $q = q_\Lambda + \bar q_{\Lambda^c}$, i.e. the configuration equal to q_Λ in Λ and to $\bar q$ in Λ^c; then

$$|\tilde H_{\gamma,\Lambda}(q_\Lambda|\bar q_{\Lambda^c}) - H_{\gamma,\Lambda}(q_\Lambda|\bar q_{\Lambda^c})|$$
$$\le \int_{\Lambda \cup \delta_{\text{out}}^{\gamma^{-1}}[\Lambda]} \{[e_\lambda(\tilde J_\gamma * q) - e_\lambda(J_\gamma * q)] - [e_\lambda(\tilde J_\gamma * \bar q_{\Lambda^c}) - e_\lambda(J_\gamma * \bar q_{\Lambda^c})]\}.$$

By (11.1.2.3) $\tilde J_\gamma * q \le a\rho_{\max}$ and $J_\gamma * q \le a\rho_{\max}$, $a = 3^d\|J\|_\infty$. Then by Lemma 11.1.2.1 there is a constant c so that

$$|\tilde H_{\gamma,\Lambda}(q_\Lambda|\bar q_{\Lambda^c}) - H_{\gamma,\Lambda}(q_\Lambda|\bar q_{\Lambda^c})|$$
$$\le \int_{\Lambda \cup \delta_{\text{out}}^{\gamma^{-1}}[\Lambda]} c\{|\tilde J_\gamma * q - J_\gamma * q| + |\tilde J_\gamma * \bar q_{\Lambda^c} - J_\gamma * \bar q_{\Lambda^c}|\}. \tag{11.1.3.9}$$

Since the gradient of J is uniformly bounded

$$|\tilde J_\gamma(r,r') - J_\gamma(r,r')| \le c\gamma^{d+1/2}\mathbf{1}_{\text{dist}(C_r^{(\gamma^{-1/2})},C_{r'}^{(\gamma^{-1/2})}) \le \gamma^{-1}}, \tag{11.1.3.10}$$

so that the l.h.s. of (11.1.3.9) is bounded by $\le c'|\Lambda|\gamma^{1/2}$. □

Proof of Theorem 11.1.3.3

By Lemma 11.1.3.4 we can replace $Z_{\gamma,\lambda,\Lambda,\bar q}(\mathcal{B})$ by its tilde analogue with an error $c\gamma^{1/2}|\Lambda|$; it is therefore sufficient to prove (11.1.3.7) with the tilde partition function. The energy $\tilde H_{\gamma,\Lambda}(q_\Lambda|\bar q_{\Lambda^c})$ depends only on the number of particles $n_x = |q_\Lambda \cap C_x^{(\gamma^{-1/2})}|$ in the cubes $C_x^{(\gamma^{-1/2})}$, $x \in \gamma^{-1/2}\mathbb{Z}^d$, contained in Λ. We shall thus write, by abuse of notation, $\tilde H_{\gamma,\Lambda}(\{n_x\}|\bar q_{\Lambda^c})$ for the energy of any configuration whose particles numbers are $\{n_x\}$. Denoting by $\rho(\{n_x\})$ the function in $\tilde{\mathcal{M}}_\Lambda$ defined as $\rho(\{n_x\})(r) = \gamma^{d/2} n_x$ if $r \in C_x^{(\gamma^{-1/2})}$,

$$\tilde{Z}_{\gamma,\lambda,\Lambda,\bar{q}}(\mathcal{B}) = \sum_{\{n_x\}:\rho(\{n_x\})\in\mathcal{B}} \left\{\prod_x \frac{(\gamma^{-d/2})^{n_x}}{n_x!}\right\} e^{-\beta \tilde{H}_{\gamma,\Lambda}(\{n_x\}|\bar{q}_{\Lambda^c})}. \tag{11.1.3.11}$$

We use the Stirling formula in (11.1.3.11) to replace $\log n_x!$ by the entropy. The error for each x is bounded proportionally to $\log\gamma^{-1}$, as $n_x \le \rho_{\max}\ell^d_{-,\gamma}$. Since there are $|\Lambda|\gamma^{d/2}$ values of x,

$$\log \tilde{Z}_{\gamma,\lambda,\Lambda,\bar{q}}(\mathcal{B}) \ge -c(\gamma^{d/2}\log\gamma^{-1})|\Lambda| - \beta \inf_{\{n_x\}:\rho(\{n_x\}\in\mathcal{B}} \tilde{F}_{\gamma,\lambda,\Lambda}(\rho(\{n_x\})|\bar{q}).$$

Let ρ be the minimizer of $\tilde{F}_{\gamma,\lambda,\Lambda}(\cdot)|\bar{q})$ on $\mathcal{B}$. By (11.1.3.4) there are $\{n_x\}$ and c so that $\rho(\{n_x\}) \in \mathcal{B}$ and $|\rho(\{n_x\}) - \rho| \le c\gamma^{d/2}$ on Λ. The lower bound (11.1.3.7) then follows using that $|S(\gamma^{d/2}(n+t)) - S(\gamma^{d/2}n)| \le c'\gamma^{d/2}\log\gamma^{-1}$ for all $|t| \le c$ and $n \le \rho_{\max}\ell^d_{-,\gamma}$ (the $\log\gamma^{-1}$ term missing in the analogous bound for the energy which is instead differentiable).

To prove the upper bound in (11.1.3.7) we observe that due to the constraint in (11.1.3.11), $n_x \le \rho_{\max}\ell^d_{-,\gamma}$, so that n_x may take at most $\rho_{\max}\ell^d_{-,\gamma}$ values; hence the sum over $\{n_x\}$ has at most $(\rho_{\max}\ell^d_{-,\gamma})^{|\Lambda|\gamma^{d/2}}$ terms and

$$\log \tilde{Z}_{\gamma,\lambda,\Lambda,\bar{q}}(\mathcal{B}) \le -\beta \inf_{\rho_\Lambda\in\mathcal{B}} \tilde{F}_{\gamma,\lambda,\Lambda}(\rho_\Lambda|\bar{q}) + c(\gamma^{d/2}\log\gamma^{-1})|\Lambda|.$$

11.1.4 A variational problem with constraints

Observe that given any $c > 0$ for all γ small enough,

$$|\rho_{\beta,\lambda,\pm} - \rho_{\beta,\pm}| < \zeta \quad \text{for } |\lambda - \lambda(\beta)| \le c\gamma^{1/2}. \tag{11.1.4.1}$$

Proposition 11.1.4.1 *Let λ be such that $|\rho_{\beta,\lambda,\pm} - \rho_{\beta,\pm}| < \zeta$; then*

$$\inf_{\rho_\Lambda:\eta(\rho_\Lambda;\cdot)\equiv\pm1} F_{\gamma,\lambda,\Lambda}(\rho_\Lambda|\rho_{\beta,\lambda,\pm}\mathbf{1}_{\Lambda^c}) = F_{\gamma,\lambda,\Lambda}(\rho_{\beta,\lambda,\pm}\mathbf{1}_\Lambda|\rho_{\beta,\lambda,\pm}\mathbf{1}_{\Lambda^c})$$
$$= \phi_{\beta,\lambda}(\rho_{\beta,\lambda,\pm})|\Lambda| + I^{\pm}_{\gamma,\lambda}(\Lambda), \tag{11.1.4.2}$$

where $I^{\pm}_{\gamma,\lambda}(\Lambda)$ is equal to

$$\int_{\Lambda^c} \{e_\lambda(\rho_{\beta,\lambda,\pm}) - e_\lambda(\rho_{\beta,\lambda,\pm}J_\gamma * \mathbf{1}_{\Lambda^c})\} - \int_\Lambda e_\lambda(\rho_{\beta,\lambda,\pm}J_\gamma * \mathbf{1}_{\Lambda^c}). \tag{11.1.4.3}$$

The same results hold for $\tilde{F}_{\gamma,\lambda,\Lambda}$ with J_γ replaced by $\tilde{J}_\gamma$.

Proof The proof is similar to the proof of Theorem 10.2.6.1, where Λ is a torus, and it is based on a smart way to rewrite of $F_{\gamma,\lambda,\Lambda}(\rho_\Lambda|\rho_{\Lambda^c})$. Denote by ρ the function equal to ρ_Λ on Λ and to ρ_{Λ^c} on Λ^c. Then regarding $\rho_{\Lambda^c}=0$ on Λ and recalling that $S(0)=0$, so that $S(\rho_{\Lambda^c})=S(\rho_{\Lambda^c})\mathbf{1}_{\Lambda^c}$,

$$F_{\gamma,\lambda,\Lambda}(\rho_\Lambda|\rho_{\Lambda^c}) = \int_{\mathbb{R}^d}\{\phi_{\beta,\lambda}(J_\gamma*\rho)-\phi_{\beta,\lambda}(J_\gamma*\rho_{\Lambda^c})\}+\frac{1}{\beta}\{S(J_\gamma*\rho)-S(\rho)\}$$
$$-\frac{1}{\beta}\{S(J_\gamma*\rho_{\Lambda^c})-S(\rho_{\Lambda^c})\}. \tag{11.1.4.4}$$

In our case $\rho_{\Lambda^c}=\rho_{\beta,\lambda,\pm}\mathbf{1}_{\Lambda^c}$ so that we can write the integral of the sum as the sum of the integrals, and in the integral with $\{S(J_\gamma*\rho)-S(\rho)\}$ we can replace $S(\rho)$ by $J_\gamma*S(\rho)$. Then $F_{\gamma,\lambda,\Lambda}(\rho_\Lambda|\rho_{\Lambda^c})$ becomes

$$\int_{\mathbb{R}^d}\{\phi_{\beta,\lambda}(J_\gamma*\rho)-\phi_{\beta,\lambda}(J_\gamma*\rho_{\Lambda^c})\}+\frac{1}{\beta}\int_{\mathbb{R}^d}\{S(J_\gamma*\rho)-J_\gamma*S(\rho)\}$$
$$-\frac{1}{\beta}\int_{\mathbb{R}^d}\{S(J_\gamma*\rho_{\Lambda^c})-S(\rho_{\Lambda^c})\}. \tag{11.1.4.5}$$

Since $\rho_{\Lambda^c}=\rho_{\beta,\lambda,\pm}\mathbf{1}_{\Lambda^c}$ and $\eta(\rho_\Lambda;\cdot)\equiv\pm1$, for all γ small enough the first curly bracket is minimized by setting $\rho_\Lambda=\rho_{\beta,\lambda,\pm}$; the second curly bracket by convexity is non-negative and vanishes when $\rho_\Lambda=\rho_{\beta,\lambda,\pm}\mathbf{1}_\Lambda$; the third one is independent of ρ_Λ and the first equality in (11.1.4.2) is proved. To prove the second one it is better to look at the original expression for $F_{\gamma,\lambda,\Lambda}(\rho_{\beta,\lambda,\pm}\mathbf{1}_\Lambda|\rho_{\beta,\lambda,\pm}\mathbf{1}_{\Lambda^c})$ and set $\rho_\Lambda=\rho_{\beta,\lambda,\pm}$. Recalling the definition of $I^\pm_{\gamma,\lambda}(\Lambda)$ in (11.1.4.3), we then get

$$F_{\gamma,\lambda,\Lambda}(\rho_{\beta,\lambda,\pm}\mathbf{1}_\Lambda|\rho_{\beta,\lambda,\pm}\mathbf{1}_{\Lambda^c})=\int_\Lambda\Big\{e_\lambda(\rho_{\beta,\lambda,\pm})-\frac{1}{\beta}S(\rho_{\beta,\lambda,\pm})\Big\}+I^\pm_{\gamma,\lambda}(\Lambda).$$

The curly brackets are equal to $\phi_{\beta,\lambda}(\rho_{\beta,\lambda,\pm})$. The same proof works for $\tilde J_\gamma$ and $\tilde F_{\gamma,\lambda}$. □

11.1.5 Equality of the plus and minus pressures

In this subsection we shall complete the proof of (11.1.0.4).

Lemma 11.1.5.1 *There is a constant c' so that for any λ such that $|\rho_{\beta,\lambda,\pm}-\rho_{\beta,\pm}|<\zeta$,*

$$|P^\pm_{\gamma,\lambda}-p^{\pm;\mathrm{mf}}_\lambda|\le c'\gamma^{1/2}. \tag{11.1.5.1}$$

Proof Let Δ_n be an increasing sequence of $\mathcal{D}^{(\ell_{+,\gamma})}$-measurable cubes which invades $\mathbb{R}^d$; then (by the existence of thermodynamic limit, see (11.7.1.3))

$$P^{\pm}_{\gamma,\lambda} = \lim_{n\to\infty} \frac{1}{\beta|\Delta_n|} \log \hat{Z}^{\pm}_{\gamma,\lambda,\Delta_n,\rho_{\beta,\lambda,\pm}}.$$

Upper bound. By (11.1.1.12)

$$\hat{Z}^{\pm}_{\gamma,\lambda,\Delta_n,\rho_{\beta,\lambda,\pm}} \le 2^{|\Delta_n|/\ell^d_{+,\gamma}} Z_{\gamma,\lambda,\Delta_n,\rho_{\beta,\lambda,\pm}}(\{\eta(\cdot;\cdot)\equiv\pm 1\}), \tag{11.1.5.2}$$

and the upper bound follows from Lemmas 11.1.3.4, Theorem 11.1.3.3 and Proposition 11.1.4.1, because $\lim_{n\to\infty}\frac{I^{\pm}_{\gamma,\Delta_n}}{|\Delta_n|}=0$.

Lower bound. By Lemma 11.1.1.1 $\hat{Z}^{\pm}_{\gamma,\lambda,\Delta_n,\rho_{\beta,\lambda,\pm}} \ge Z_{\gamma,\lambda,\Delta_n,\rho_{\beta,\lambda,\pm}}(\{\eta(\cdot;\cdot)\equiv\pm 1\})$ and then the proof proceeds as in the upper bound. □

Lemma 11.1.5.2 *There are δ and a both positive so that*

$$\frac{d}{d\lambda}\big(p^{+;\mathrm{mf}}_{\lambda} - p^{-;\mathrm{mf}}_{\lambda}\big) > a, \qquad |\lambda-\lambda(\beta)|\le\delta. \tag{11.1.5.3}$$

Proof $\frac{d}{d\lambda}(p^{+;\mathrm{mf}}_{\lambda} - p^{-;\mathrm{mf}}_{\lambda})|_{\lambda=\lambda(\beta)} = \rho_{\beta,+}-\rho_{\beta,-}>0.$ □

Proof of (11.1.0.4)

We are going to show that:

Claim *Let c be such that $ca>2c'$, a as in* (11.1.5.3) *and c' as in* (11.1.5.1), *then for all γ small enough (in particular such that* (11.1.4.1) *holds)*

$$\begin{aligned} P^{+}_{\gamma,\lambda(\beta)+c\gamma^{1/2}} - P^{-}_{\gamma,\lambda(\beta)+c\gamma^{1/2}} &> 0,\\ P^{+}_{\gamma,\lambda(\beta)-c\gamma^{1/2}} - P^{-}_{\gamma,\lambda(\beta)-c\gamma^{1/2}} &< 0. \end{aligned} \tag{11.1.5.4}$$

Proof By Lemmas 11.1.5.2 and 11.1.5.1, writing $\lambda'':=\lambda(\beta)+c\gamma^{1/2}$, for all γ small enough

$$P^{+}_{\gamma,\lambda''} - P^{-}_{\gamma,\lambda''} \ge p^{+;\mathrm{mf}}_{\lambda''} - p^{-;\mathrm{mf}}_{\lambda''} - 2c'\gamma^{1/2} \ge a(\lambda''-\lambda(\beta)) - 2c'\gamma^{1/2} > 0,$$

having used that $p^{+;\mathrm{mf}}_{\lambda(\beta)} - p^{+;\mathrm{mf}}_{\lambda(\beta)} = 0$. Analogously, if $\lambda'=\lambda(\beta)-c\gamma^{1/2}$,

$$P^{+}_{\gamma,\lambda'} - P^{-}_{\gamma,\lambda'} \le -a(\lambda(\beta)-\lambda') + 2c'\gamma^{1/2} < 0.$$ □

11.2 Reduction to perfect boundary conditions

As an intermediate step in the proof of (11.0.2.4), we shall prove in this section that we can separate the estimate in $\text{int}(\Gamma)$ from the one in $\text{sp}(\Gamma)$. We shall define a set B_0 in $\delta_{\text{in}}^{\ell_{+,\gamma}}[\text{sp}(\Gamma)]$ made of several corridors which separate $\text{int}(\Gamma)$ from the rest of $\text{sp}(\Gamma)$, showing that with "negligible error" the interaction with B_0 can be replaced by an interaction with perfect boundary conditions, i.e. the density constantly equal in each connected component either to $\rho_{\beta,+}$ or to $\rho_{\beta,-}$. The main result is stated in Theorem 11.2.2.1, after setting some geometrical notation and definitions which will be used throughout this and the next section.

11.2.1 Regions at the boundaries of sp(Γ)

Without loss of generality (proofs are identical) we restrict our treatment to plus contours Γ, set $\ell := \ell_{+,\gamma}/8$ and (see Sect. 6.5, in particular Figs. 6.10, 6.11, recalling that $\ell_\pm$ should be replaced by $\ell_{\pm,\gamma}$)

$$\Delta_1 := \delta_{\text{out}}^{\ell}[c(\Gamma)^c], \qquad \Delta_2 := \delta_{\text{out}}^{\ell}[c(\Gamma)^c \cup \Delta_1], \qquad \Delta_3 := \delta_{\text{out}}^{\ell}[c(\Gamma)^c \cup \Delta_1 \cup \Delta_2].$$

These are successive corridors that we meet when we move from $c(\Gamma)^c$ into $\text{sp}(\Gamma)$. In all of them $\eta = 1$, while $\{r : \eta(\cdot; r) < 1\}$ is far away i.e. at least $\ell_{+,\gamma} - (3/8)\ell_{+,\gamma}$. When approaching $\text{sp}(\Gamma)$ from $\text{int}(\Gamma)$ we see that

$$\Delta_4^{\pm} := \delta_{\text{out}}^{\ell}[\text{int}^{\pm}(\Gamma)] \cup \delta_{\text{in}}^{\ell}[\text{int}^{\pm}(\Gamma)], \qquad \Delta_5^{\pm} := \delta_{\text{out}}^{\ell}[\Delta_4^{\pm}]$$

(see Fig. 6.11). By the definition of contours the above plus [minus] corridors are in a region where $\eta = 1$ [$\eta = -1$] and the distance from where $\eta < 1$ [$\eta > -1$] is $\ell_{+,\gamma} - (2/8)\ell_{+,\gamma}$. We then write $B = B^+ \cup B^-$, $B_0 = B_0^+ \cup B_0^-$

$$B^- = \Delta_4^- \cup \Delta_5^-, \qquad B^+ = \Delta_4^+ \cup \Delta_5^+ \cup \Delta_1 \cup \Delta_2 \cup \Delta_3, \tag{11.2.1.1}$$

$$B_0^- = \Delta_4^- \cap \text{sp}(\Gamma), \qquad B_0^+ = \Delta_2 \cup (\Delta_4^+ \cap \text{sp}(\Gamma)), \tag{11.2.1.2}$$

and finally define

$$\begin{aligned} \Lambda &= \text{sp}(\Gamma) \setminus \big(\Delta_1 \cup \Delta_2 \cup (\Delta_4 \cap \text{sp}(\Gamma))\big), \\ \Lambda' &= \Lambda \setminus \big(\Delta_3 \cup (\Delta_5 \cap \text{sp}(\Gamma))\big), \end{aligned} \tag{11.2.1.3}$$

$\Delta_4 = \Delta_4^+ \cup \Delta_4^-$, $\Delta_5 = \Delta_5^+ \cup \Delta_5^-$.

11.2.2 Main results

Let $\tilde{\mathcal{F}}$ be as in (11.1.3.6), $\chi^{\pm}$ as in (11.0.2.1) and let us use the shorthand $\chi_{B_0} = \chi^+_{B_0^+} + \chi^-_{B_0^-}$. With such a notation we can state the main result in this section:

Theorem 11.2.2.1 *There is $c > 0$ so that for all γ small enough,*

$$\hat{N}^{+}_{\gamma,\lambda_{\beta,\gamma}}(\Gamma, q^{+}) \le e^{-\beta \tilde{\mathcal{F}}_{\gamma,\lambda(\beta),B_0}(\chi_{B_0}) + c\gamma^{1/2}|B|} \hat{Z}^{-}_{\gamma,\lambda_{\beta,\gamma},\mathrm{int}^{-}(\Gamma),\chi^{-}_{B_0^{-}}}$$
$$\times \hat{Z}^{+}_{\gamma,\lambda_{\beta,\gamma},\mathrm{int}^{+}(\Gamma),\chi^{+}_{B_0^{+}}} Z_{\gamma,\lambda_{\beta,\gamma},\Delta_1,\chi^{+}_{B_0^{+}}+q^{+}_{c(\Gamma)^c}}(\{\eta = 1 \text{ on } \Delta_1\})$$
$$\times Z_{\gamma,\lambda_{\beta,\gamma},\Lambda,\chi_{B_0}}(\{\eta = \eta_\Gamma \text{ on } \Lambda\}). \qquad (11.2.2.1)$$

The proof of Theorem 11.2.2.1 uses coarse graining in B, reducing the case to a minimization problem with the LMP free energy functional. The results are anticipated in Theorem 11.2.2.4 (which is proved in the next section) and used to complete the proof of Theorem 11.2.2.1. To perform coarse graining we shall use Theorem 11.1.3.3, and thus we need to check that the constraints satisfy the assumptions in Definition 11.1.3.2. To this end we shall in the sequel use:

Lemma 11.2.2.2 *Let A be a bounded $\mathcal{D}^{(\ell_{-,\gamma})}$-measurable region, ϕ a $\mathcal{D}^{(\ell_{-,\gamma})}$-measurable function on A with values in $\{0, \pm 1\}$; then for γ small enough*

$$\mathcal{B} = \{\rho_A \in \tilde{\mathcal{M}}_A : \eta(\rho_A; \cdot) = \phi, \mathrm{Av}^{(\ell_{-,\gamma})}(\rho_A; \cdot) \le \rho_{\max}\} \qquad (11.2.2.2)$$

satisfies the assumptions in Definition 11.1.3.2 *with constant* $c = 1$.

Proof Assume for instance [the other cases are similar and omitted] that on a cube $C \in \mathcal{D}^{(\ell_{-,\gamma})}$ the average $\mathrm{Av}^{(\ell_{-,\gamma})}(\rho_A; \cdot) \in [\rho_{\beta,+}, \rho_{\beta,+} + \zeta]$. Then on each cube $C_x \in \mathcal{D}^{(\gamma^{-1/2})}$ contained in C we call n_x the integer part of $\gamma^{-d/2}\rho_A(x)$ (recall that by definition ρ_A is constant on C_x). Let q be any configuration with n_x particles in C_x for all x as above. Then $0 \le \rho_A - \mathrm{Av}^{(\gamma^{-1/2})}(q; \cdot) \le \gamma^{d/2}$, so that $|\mathrm{Av}^{(\ell_{-,\gamma})}(q; \cdot) - \mathrm{Av}^{(\ell_{-,\gamma})}(\rho_A; \cdot)| \le \gamma^{d/2}$ and $\eta(q; \cdot) = 1$ on C if $\gamma^{d/2} < \zeta$. □

Lemma 11.2.2.3 *There is a non-negative measure $d\mu_{c(\Gamma)\setminus B}$ absolutely continuous with respect to the free measure which has support on $\{\eta(q_{c(\Gamma)\setminus B}; r) = \pm 1$ for all $r \in c(\Gamma) \setminus B$ at distance $\le 4\gamma^{-1}$ from $B^{\pm}\}$, and it is such that*

$$\hat{N}^{+}_{\gamma,\lambda_{\beta,\gamma}}(\Gamma, q^{+}) = \int Z_{\gamma,\lambda_{\beta,\gamma},B,\bar{q}}(\{\eta = \pm 1 \text{ on } B^{\pm}\}) d\mu_{c(\Gamma)\setminus B}, \qquad (11.2.2.3)$$

where $\bar{q} := q_{c(\Gamma)\setminus B} + q^{+}_{c(\Gamma)^c}$ and $Z_{\gamma,\lambda,\Lambda,\bar{q}}(\mathcal{B})$ is the partition function with the constraint $\mathcal{B}$ defined in (11.1.3.1).

Proof Recalling the remark before Theorem 11.1.0.3, we can write the partition functions in the definition of $\hat{N}^{+}_{\gamma,\lambda_{\beta,\gamma}}(\Gamma, q^{+})$, see (10.5.1.7), as

$$\int_{\mathcal{X}^{+}_{\mathrm{int}^{+}(\Gamma)}} e^{-\beta H_{\gamma,\lambda_{\beta,\gamma},\mathrm{int}^{+}(\Gamma)}(q_{\mathrm{int}^{+}(\Gamma)}|q_{\mathrm{sp}(\Gamma)})} X^{+}_{\gamma,\lambda_{\beta,\gamma},q_{\mathrm{int}^{+}(\Gamma)\setminus B}}(\mathrm{int}^{+}(\Gamma)).$$

The partition function in $\mathrm{int}^-(\Gamma)$ has an analogous expression. Define

$$\phi_+(q_{\mathrm{int}^+(\Gamma)\setminus B}) = \mathbf{1}_{q_{\mathrm{int}^+(\Gamma)\setminus B}\in \mathcal{X}^+_{\mathrm{int}^+(\Gamma)\setminus B}} e^{-\beta H_{\gamma,\lambda_{\beta,\gamma},\mathrm{int}^+(\Gamma)\setminus B}(q_{\mathrm{int}^+(\Gamma)\setminus B})} \times X^+_{\gamma,\lambda_{\beta,\gamma},q_{\mathrm{int}^+(\Gamma)\setminus B}}(\mathrm{int}^+(\Gamma)),$$

and ϕ_- by the analogous expression. Let

$$\phi_0(q_{\mathrm{sp}(\Gamma)\setminus B}) = \mathbf{1}_{\eta(q_{\mathrm{sp}(\Gamma)\setminus B};r)=\eta_\Gamma(r),r\in \mathrm{sp}(\Gamma)\setminus B} e^{-\beta H_{\gamma,\lambda_{\beta,\gamma},\mathrm{sp}(\Gamma)\setminus B}(q_{\mathrm{sp}(\Gamma)\setminus B})}.$$

By setting $\phi = \phi_0\phi_+\phi_-$, $d\mu_{c(\Gamma)\setminus B} := \phi d\nu_{c(\Gamma)\setminus B}$, $d\nu_{c(\Gamma)\setminus B}$ the free measure, (11.2.2.3) becomes an identity. □

Observe that the partition function in (11.2.2.3) is equal to the product

$$Z_{\gamma,\lambda_{\beta,\gamma},B^+,\bar q}(\{\eta=1 \text{ on } B^+\}) Z_{\gamma,\lambda_{\beta,\gamma},B^-,q_{c(\Gamma)\setminus B}}(\{\eta=-1 \text{ on } B^-\}),$$

so that we shall bound separately the contribution of B^+ and B^-. For γ small enough the range of $\tilde J_\gamma$ is $\le 4\gamma^{-1}$; moreover, $\eta(\bar q; r) = 1$ when $\mathrm{dist}(r, B^+) \le 4\gamma^{-1}$ and $\eta(q_{c(\Gamma)\setminus B}; r) = -1$ when $\mathrm{dist}(r, B^-) \le 4\gamma^{-1}$; thus the boundary conditions have $\eta(\cdot;\cdot) = \pm 1$. Then by Theorem 11.1.3.3 (to which we also refer for notation) there is a constant c so that for all γ small enough (recall that $\bar q$ is defined in (11.2.2.3), $\tilde{\mathcal F}$ in (11.1.3.6) and $\tilde{\mathcal M}_\Lambda$ in Definition 11.1.3.1)

$$\begin{aligned}\log Z_{\gamma,\lambda_{\beta,\gamma},B,\bar q}(\{\eta=\pm 1 \text{ on } B^\pm\}) &\le c\gamma^{1/2}|B| - \beta \inf_{\rho_{B^+}\in\tilde{\mathcal M}_{B^+}:\eta(\rho_{B^+};\cdot)\equiv 1} \tilde{\mathcal F}_{\gamma,\lambda_{\beta,\gamma},B^+}(\rho_{B^+}|\bar q)\\ &\quad - \beta \inf_{\rho_{B^-}\in\tilde{\mathcal M}_{B^-}:\eta(\rho_{B^-};\cdot)\equiv -1} \tilde{\mathcal F}_{\gamma,\lambda_{\beta,\gamma},B^-}(\rho_{B^-}|q_{c(\Gamma)\setminus B}).\end{aligned} \tag{11.2.2.4}$$

The term in $\tilde{\mathcal F}$ with the chemical potential is $-\lambda_{\beta,\gamma}\int_{B^\pm}\rho_{B^\pm}(r)dr$; then, recalling (11.1.0.5),

$$|\tilde{\mathcal F}_{\gamma,\lambda_{\beta,\gamma},B^\pm}(\rho_{B^\pm}|q) - \tilde{\mathcal F}_{\gamma,\lambda(\beta),B^\pm}(\rho_{B^\pm}|q)| \le c|B|\gamma^{1/2}. \tag{11.2.2.5}$$

Theorem 11.2.2.4 *There are positive constants c and ω and positive functions $\rho^*_{B^\pm}\in\tilde{\mathcal M}_{B^\pm}$ ($\tilde{\mathcal M}$ as in Definition 11.1.3.1), $\eta(\rho^*_{B^\pm}; r) = \pm 1$, $r\in B^\pm$, so that $\rho^*_{B^\pm}(r) = \rho_{\beta,\pm}$, $r\in B_0^\pm$ and for all $\rho_{B^\pm}\in\tilde{\mathcal M}_{B^\pm} : \eta(\rho_{B^\pm};\cdot)\equiv \pm 1$,*

$$\tilde{\mathcal F}_{\gamma,\lambda(\beta),B^\pm}(\rho_{B^\pm}|\bar q) \ge \tilde{\mathcal F}_{\gamma,\lambda(\beta),B^\pm}(\rho^*_{B^\pm}|\bar q) - ce^{-\omega\gamma\ell_{+,\gamma}}. \tag{11.2.2.6}$$

Postponing to Sect. 11.2.4 the proof of Theorem 11.2.2.4, we proceed with the proof of Theorem 11.2.2.1. Using (11.2.2.5), we get from (11.2.2.4) the result that there is $c > 0$ so that

$$\begin{aligned}\log Z_{\gamma,\lambda_{\beta,\gamma},B,\bar{q}}(\{\eta = \pm 1 \text{ on } B^\pm\}) \\ \le -\beta \tilde{\mathcal{F}}_{\gamma,\lambda(\beta),B^+}(\rho^*_{B^+}|\bar{q}) \\ -\beta \tilde{\mathcal{F}}_{\gamma,\lambda(\beta),B^-}(\rho^*_{B^-}|q_{c(\Gamma)\setminus B}) + c\gamma^{1/2}|B|. \end{aligned} \tag{11.2.2.7}$$

Using the formula

$$\tilde{\mathcal{F}}_{\gamma,\lambda,A\cup B}(\rho_{A\cup B}|\rho_{(A\cup B)^c}) = \tilde{\mathcal{F}}_{\gamma,\lambda,A}(\rho_A|\rho_{(A\cup B)^c}) + \tilde{\mathcal{F}}_{\gamma,\lambda,B}(\rho_B|\rho_{(A\cup B)^c} + \rho_A)$$

and setting

$$B_1^\pm := (B^\pm \setminus B_0^\pm) \cap \mathrm{sp}(\Gamma), \qquad B_2^\pm := B^\pm \cap \mathrm{int}^\pm(\Gamma), \qquad B_0 = B_0^+ \cup B_0^-,$$

we have

$$\begin{aligned}\tilde{\mathcal{F}}_{\gamma,\lambda(\beta),B^-}(\rho^*_{B^-}|q_{c(\Gamma)\setminus B}) \\ = \tilde{\mathcal{F}}_{\gamma,\lambda(\beta),B_1^-}(\rho^*_{B_1^-}|\chi^-_{B_0^-} + q_{\mathrm{sp}(\Gamma)\setminus B}) \\ + \tilde{\mathcal{F}}_{\gamma,\lambda(\beta),B_2^-}(\rho^*_{B_2^-}|\chi^-_{B_0^-} + q_{\mathrm{int}^-(\Gamma)\setminus B}) + \tilde{\mathcal{F}}_{\gamma,\lambda(\beta),B_0^-}(\chi^-_{B_0^-}).\end{aligned}$$

By using (11.2.2.5) we replace $\lambda(\beta)$ by $\lambda_{\beta,\gamma}$ in the first two terms with an error bounded by $c\gamma^{1/2}|B|$. Since $\rho^*_{B^-} \in \mathcal{B} := \{\eta = -1 \text{ on } B_1^-\}$ so that, using Lemma 11.2.2.2 and Theorem 11.1.3.3 "backwards" to reconstruct the partition function, we have

$$\begin{aligned}-\beta \tilde{\mathcal{F}}_{\gamma,\lambda(\beta),B^-}(\rho^*_{B^-}|q_{c(\Gamma)\setminus B}) \\ \le c\gamma^{1/2}|B| - \beta \tilde{\mathcal{F}}_{\gamma,\lambda(\beta),B_0^-}(\chi^-_{B_0^-}) \\ + \log Z_{\gamma,\lambda_{\beta,\gamma},B_1^-,\chi^-_{B_0^-}+q_{\mathrm{sp}(\Gamma)\setminus B}}(\eta = -1 \text{ on } B_1^-) \\ + \log Z_{\gamma,\lambda_{\beta,\gamma},B_2^-,\chi^-_{B_0^-}+q_{\mathrm{int}^-(\Gamma)\setminus B}}(\eta = -1 \text{ on } B_2^-).\end{aligned} \tag{11.2.2.8}$$

Since an analogous bound holds for $-\beta \tilde{\mathcal{F}}_{\gamma,\lambda(\beta),B^+}(\hat{\rho}_{B^+}|\bar{q})$, we get from (11.2.2.3) that $\hat{N}^+_{\gamma,\lambda_{\beta,\gamma}}(\Gamma, q^+)$ is bounded by the integral over $d\mu_{c(\Gamma)\setminus B}(q_{c(\Gamma)\setminus B})$ of $e^{-\beta \tilde{\mathcal{F}}_{\gamma,\lambda(\beta),B_0}(\chi_{B_0}) + c\gamma^{1/2}|B|}$ times

$$\begin{aligned}Z_{\gamma,\lambda_{\beta,\gamma},B_2^-,\chi^-_{B_0^-}+q_{\mathrm{int}^-(\Gamma)\setminus B}}(\eta = -1 \text{ on } B_2^-) \\ \times Z_{\gamma,\lambda_{\beta,\gamma},B_2^+,\chi^+_{B_0^+}+q_{\mathrm{int}^+(\Gamma)\setminus B}}(\eta = 1 \text{ on } B_2^+) \\ \times Z_{\gamma,\lambda_{\beta,\gamma},B_1^-\cup B_1^+,q_{\mathrm{sp}(\Gamma)\setminus B}+\chi_{B_0}+q^+_{c(\Gamma)^c}}(\{\eta = \pm 1 \text{ on } B_1^\pm\}).\end{aligned}$$

Recalling from the proof of Lemma 11.2.2.3 that

$$d\mu_{c(\Gamma)\setminus B}(q_{c(\Gamma)\setminus B}) = \phi(q_{c(\Gamma)\setminus B}) d\nu_{c(\Gamma)\setminus B}(q_{c(\Gamma)\setminus B}),$$

and using the expression for the density $\phi(q_{c(\Gamma)\setminus B})$ we get (11.2.2.1), thus proving Theorem 11.2.2.1 (pending the validity of Theorem 11.2.2.4, which will be proved in Sect. 11.2.4). Thus we achieve the goal set for this section, namely to separate the estimate in int(Γ) from that in sp(Γ).

11.2.3 Minimization in a neighborhood of equilibrium

For the first time we shall use the assumption that $\beta \in (\beta_c, \beta_0)$, which enters essentially in the proof of Proposition 11.2.3.1 below. The analysis here parallels the one in Sect. 6.3 for the L–P functional, Proposition 11.2.3.1 corresponding to Theorem 6.3.3.1. To have a simpler notation we consider only the plus case and call Λ a bounded $\mathcal{D}^{(\ell_-,\gamma)}$-measurable region, $\bar\rho$ a non-negative function with support on Λ^c where $\eta(\bar\rho;\cdot) = 1$ (in the setup of the previous subsection $\Lambda = B^+$).

Proposition 11.2.3.1 *There is $\kappa > 0$ and a positive function $\hat\rho_\Lambda \in \tilde{\mathcal{M}}_\Lambda$ ($\tilde{\mathcal{M}}_\Lambda$ as in Definition* 11.1.3.1) *so that*

$$\inf_{\rho_\Lambda \in \tilde{\mathcal{M}}_\Lambda : \eta(\rho_\Lambda;\cdot)\equiv 1} \tilde{\mathcal{F}}_{\gamma,\lambda(\beta),\Lambda}(\rho_\Lambda|\bar\rho) = \tilde{\mathcal{F}}_{\gamma,\lambda(\beta),\Lambda}(\hat\rho_\Lambda|\bar\rho). \tag{11.2.3.1}$$

Moreover, for all $r \in \Lambda$,

$$|\hat\rho_\Lambda(r) - \rho_{\beta,+}| \le (1-\kappa)\zeta, \quad \hat\rho_\Lambda(r) = e^{-\beta \tilde J_\gamma * e'_{\lambda(\beta)}(\tilde J_\gamma * [\rho_\Lambda + \bar\rho])}, \quad r \in \Lambda. \tag{11.2.3.2}$$

Proof Since $\rho_\Lambda \in \tilde{\mathcal{M}}_\Lambda$, ρ_Λ is a function of n real variables, $n = |\Lambda|/\gamma^{-d/2}$, which vary in a compact set specified by the constraint $\eta(\rho_\Lambda;\cdot) \equiv 1$ (which indeed, by (10.3.1.2), defines a compact set in $\mathbb{R}^n$). Since $\tilde{\mathcal{F}}_{\gamma,\lambda(\beta),\Lambda}(\rho_\Lambda|\bar\rho)$ is continuous it has a minimizer $\hat\rho_\Lambda \in \tilde{\mathcal{M}}_\Lambda$.

A local mean field equation Even though $\hat\rho_\Lambda$ is a minimizer of a variational problem with constraints, yet $\hat\rho_\Lambda$ satisfies the Euler–Lagrange equations as if the constraints were absent. As we shall see, the Euler–Lagrange equations are the same as the local mean field equation in (11.2.3.2); hence the title of the paragraph (the analogous property for the L–P functional is proved in Theorem 6.3.2.2).

Denote by $\hat\rho$ the function equal to $\hat\rho_\Lambda$ on Λ and to $\bar\rho$ on Λ^c and, given any $C_x = C_x^{(\ell_-,\gamma)} \subset \Lambda$, consider the variational problem

$$\left\{\tilde{\mathcal{F}}_{\gamma,\lambda(\beta),C_x}(\rho_{C_x}|\hat\rho_{C_x^c}),\ \rho_{C_x} \in \tilde{\mathcal{M}}_{C_x}\right\} \longrightarrow \text{minimum}. \tag{11.2.3.3}$$

The constraint $\eta(\rho_{C_x};\cdot)=1$ has been dropped! Since $\tilde{\mathcal{F}}_{\gamma,\lambda(\beta),C_x}(\rho_{C_x}|\hat\rho_{C_x^c})$ is smooth and diverges at ∞, (11.2.3.3) has at least one solution $\tilde\rho_{C_x}$, and hence

$$\tilde{\mathcal{F}}_{\gamma,\lambda(\beta),C_x}(\hat\rho_{C_x}|\hat\rho_{C_x^c})\ge \tilde{\mathcal{F}}_{\gamma,\lambda(\beta),C_x}(\tilde\rho_{C_x}|\hat\rho_{C_x^c}). \tag{11.2.3.4}$$

Since $\tilde\rho_{C_x}$ is a minimizer of a problem with no constraint, it satisfies the Euler–Lagrange equations $D\tilde F_{\gamma,\lambda(\beta),C_x}(\rho_{C_x}|\hat\rho_{C_x^c})=0$ which (after an explicit computation whose details are omitted) satisfies the equation:

$$\rho_{C_x}=e^{-\beta\tilde J_\gamma * e'_{\lambda(\beta)}(\tilde J_\gamma*[\rho_{C_x}+\hat\rho_{C_x^c}])}. \tag{11.2.3.5}$$

Uniqueness of solutions of (11.2.3.5) To prove that (11.2.3.5) has a unique solution, we introduce the map

$$\rho_{C_x}\longrightarrow T(\rho_{C_x}):=e^{-\beta\tilde J_\gamma * e'_{\lambda(\beta)}(\tilde J_\gamma*[\rho_{C_x}+\hat\rho_{C_x^c}])} \tag{11.2.3.6}$$

on the space $\tilde{\mathcal{M}}_{C_x}$ (equipped with sup norm). T has range on the functions ρ_{C_x} with values in a bounded interval $(0,e^{\beta b_0}]$, $e'_{\lambda(\beta)}(\cdot)\ge -b_0$; see Lemma 11.1.2.1. Thus the fixed points of T are functions $\le e^{\beta b_0}$ and we may thus restrict the case to functions $\rho_{C_x}\le e^{\beta b_0}$. The validity of (11.1.2.3) extends to $\tilde J_\gamma$ provided we replace 3^d by 5^d (as $\tilde J_\gamma$ has range $\le 2\gamma^{-1}$). Then recalling that $\eta(\hat\rho_{C_x^c};r)=1$ and supposing $\rho_{C_x}\le e^{\beta b_0}$,

$$\tilde J_\gamma*[\rho_{C_x}+\hat\rho_{C_x^c}]\le 5^d\|J_\gamma\|_\infty(e^{\beta b_0}+\rho_\beta^+ +\zeta)=:R.$$

Let $\psi,\phi\in\tilde{\mathcal{M}}_{C_x}$; $\|\psi\|_\infty, \|\phi\|_\infty\le e^{\beta b_0}$, $s\in[0,1]$; then

$$\begin{aligned}|T(\psi)-T(\phi)|&=\left|\int_0^1\frac{d}{ds}T\big(s\psi+(1-s)\phi\big)\,ds\right|\\&\le e^{\beta b_0}\beta\Big\{\max_{0\le\xi\le R}|e''_{\lambda(\beta)}(\xi)|\Big\}\|\tilde J_\gamma*(\psi-\phi)\|_\infty\\&\le e^{\beta b_0}\beta\Big\{\max_{0\le\xi\le R}|e''_{\lambda(\beta)}(\xi)|\Big\}\|J_\gamma\|_\infty(\gamma\ell_{-,\gamma})^d\|\psi-\phi\|_\infty.\end{aligned}$$

Thus, if γ is small enough the map T is a contraction and it has therefore a unique fixed point which satisfies (11.2.3.5). Hence we see the uniqueness of solutions of (11.2.3.5) and of the minimizer in (11.2.3.3).

Bounds on the solution $\tilde\rho_{C_x}$ of (11.2.3.5) The goal in this paragraph is to prove that for γ small enough,

$$|\tilde\rho_{C_x}-\rho_{\beta,+}|<(1-\kappa)\zeta,\quad \kappa>0. \tag{11.2.3.7}$$

Claim *With χ^+ the function constantly equal to $\rho_{\beta,+}$, for γ small enough*

$$|\tilde{J}_\gamma * [\tilde{\rho}_{C_x} + \hat{\rho}_{C_x^c} - \chi^+]| \le \|J_\gamma\|_\infty (\gamma \ell_{-,\gamma})^d e^{\beta b_0} + c'\gamma\ell_{-,\gamma} + \zeta$$
$$\le \zeta + c''\gamma^\alpha \le 2\zeta. \qquad (11.2.3.8)$$

Proof The first term on the r.h.s. of (11.2.3.8) bounds $|\tilde{J} * [\tilde{\rho}_{C_x} - \chi^+_{C_x}]|$; recall that $\tilde{\rho}_{C_x}$ and χ^+ are bounded by $e^{\beta b_0}$. The second term arises when we change $\tilde{J}_\gamma$ into $\bar{J}_\gamma$, where

$$\bar{J}_\gamma(r,r') = \fint_{C_r^{(\ell_{-,\gamma})}} \fint_{C_{r'}^{(\ell_{-,\gamma})}} J_\gamma(r_1, r_1'). \qquad (11.2.3.9)$$

Finally, ζ bounds $\bar{J}_\gamma * [\hat{\rho}_{C_x^c} - \chi^+_{C_x^c}]$ because $\eta(\hat{\rho};\cdot) = 1$. The second inequality in (11.2.3.8) is derived recalling that $\ell_{-,\gamma} = \gamma^{-(1-\alpha)}$, and the last one follows for γ small enough because $\zeta = \gamma^a$, $\alpha \gg a$. The proof of (11.2.3.8) is complete.

Write $A(\rho) := e^{-\beta \tilde{J}_\gamma * e'_{\lambda(\beta)}(\tilde{J}_\gamma * \rho)}$ with $\rho = \tilde{\rho}_{C_x} + \hat{\rho}_{C_x^c}$. Denote by $e'_{\lambda(\beta)}(\xi_1)$ the max of $e'_{\lambda(\beta)}(\xi)$ over $\xi \in [\rho_{\beta,+} - 2\zeta, \rho_{\beta,+} - 2\zeta]$ and by $|e''_{\lambda(\beta)}(\xi_2)|$ the max of $|e''_{\lambda(\beta)}(\xi)|$ over $\xi \in [\rho_{\beta,+} - 2\zeta, \rho_{\beta,+} - 2\zeta]$. Since $\rho_{\beta,+} = e^{-\beta e'_{\lambda(\beta)}(\rho_{\beta,+})}$,

$$|\tilde{\rho}_{C_x} - \rho_{\beta,+}| = \left| \int_0^1 \frac{d}{ds} A(s\rho + (1-s)\chi^+)\, ds \right|$$
$$\le e^{-\beta e'_{\lambda(\beta)}(\xi_1)} \beta |e''_{\lambda(\beta)}(\xi_2)|\, |\tilde{J} * \tilde{J} * (\tilde{\rho}_{C_x} + \hat{\rho}_{C_x^c} - \rho^+_\beta)|$$
$$\le e^{-\beta e'_{\lambda(\beta)}(\xi_1)} \beta |e''_{\lambda(\beta)}(\xi_2)| (\zeta + c''\gamma^\alpha), \qquad (11.2.3.10)$$

because $\{s\rho + (1-s)\rho_{\beta,+}\} \in [\rho_{\beta,+} - 2\zeta, \rho_{\beta,+} - 2\zeta]$ for all s by (11.2.3.8). Finally, the last inequality in (11.2.3.10) follows from (11.2.3.8).

By Proposition 10.2.2.3, $|e^{-\beta e'_{\lambda(\beta)}(\rho_{\beta,+})} \beta e''_{\lambda(\beta)}(\rho_{\beta,+})| < 1$; hence (11.2.3.7) follows from (11.2.3.10) for γ small enough.

Conclusions By (11.2.3.7) $\eta(\tilde{\rho}_{C_x}; r) = 1$, $r \in C_x$, so that

$$\tilde{\mathcal{F}}_{\gamma,\lambda(\beta),C_x}(\hat{\rho}_{C_x}|\hat{\rho}_{C_x^c}) \le \tilde{\mathcal{F}}_{\gamma,\lambda(\beta),C_x}(\tilde{\rho}_{C_x}|\hat{\rho}_{C_x^c}),$$

which proves that $\hat{\rho}_{C_x^c}$ is a minimizer of (11.2.3.3), and since we have proved uniqueness $\hat{\rho}_{C_x^c} = \tilde{\rho}_{C_x^c}$ and the last inequality in (11.2.3.1) follows from (11.2.3.7). Proposition 11.2.3.1 is proved. □

The characterization of $\hat{\rho}_\Lambda$ as the solution of the mean field equation (11.2.3.1) is the key ingredient in the proof of the following theorem, where ω_β is defined by

$$e^{-\omega_\beta} = \max\left\{ |e^{-\beta e'_{\lambda(\beta)}(\rho_{\beta,+})} \beta e''_{\lambda(\beta)}(\rho_{\beta,+})|, |e^{-\beta e'_{\lambda(\beta)}(\rho_{\beta,-})} \beta e''_{\lambda(\beta)}(\rho_{\beta,-})| \right\}. \qquad (11.2.3.11)$$

By Proposition 10.2.2.3 the r.h.s. of (11.2.3.11) is < 1, so that $\omega_\beta > 0$.

Proposition 11.2.3.2 *Given any ω in $(0, \omega_\beta)$ there is $c_0 > 0$ so that for all γ small enough*

$$|\hat\rho_\Lambda(r) - \rho_{\beta,+}| \le c_0 e^{-\omega\,\mathrm{dist}(r,\Lambda^c)/(2\gamma^{-1})}, \quad r \in \Lambda. \tag{11.2.3.12}$$

Denote by ρ^*_Λ the function obtained by changing $\hat\rho_\Lambda$ into $\rho_{\beta,+}$ on

$$\Lambda' := \bigcup_{C_x \in \mathcal{D}^{(\ell_{-,\gamma})}:\mathrm{dist}(C_x,\Lambda^c)\ge \ell_{+,\gamma}/20} C_x, \tag{11.2.3.13}$$

while elsewhere $\rho^*_\Lambda = \hat\rho_\Lambda$. Then

$$\tilde{\mathcal{F}}_{\gamma,\lambda(\beta),\Lambda}(\hat\rho_\Lambda|\bar\rho) \ge \tilde{\mathcal{F}}_{\gamma,\lambda(\beta),\Lambda}(\rho^*_\Lambda|\bar\rho) - c_0|\Lambda'|e^{-\omega\gamma^{-\alpha}/40}. \tag{11.2.3.14}$$

Proof Denoting by $\hat\rho$ the function equal to $\hat\rho_\Lambda$ on Λ and to $\bar\rho$ on Λ^c, by (11.2.3.10) (recall $\tilde\rho_{C_x} = \hat\rho_{C_x}$) we have for all $r \in \Lambda$

$$\begin{aligned} |\hat\rho_\Lambda(r) - \rho_{\beta,+}| &\le (\beta e^{-\beta e'_{\lambda(\beta)}(\xi_1)}|e''_{\lambda(\beta)}(\xi_2)|)\tilde J * \tilde J * |\hat\rho_\Lambda - \rho_{\beta,+}| \\ &\le e^{-\omega}\tilde J * \tilde J * |\hat\rho - \rho_{\beta,+}| \end{aligned} \tag{11.2.3.15}$$

for γ small enough. If $r \in \Lambda'$ we can iterate (11.2.3.15) a number of times which is the integer part of the ratio of $\ell_{+,\gamma}/8$ with $4\gamma^{-1}$ (which bounds the range of $\tilde J * \tilde J *$). After such a number of iterations we bound $|\hat\rho - \rho_{\beta,+}| \le (1-\kappa)\zeta$, thus proving (11.2.3.12).

Proof of (11.2.3.14). $|\tilde{\mathcal{F}}_{\gamma,\lambda(\beta),\Lambda}(\hat\rho_\Lambda|\bar\rho) - \tilde{\mathcal{F}}_{\gamma,\lambda(\beta),\Lambda}(\rho^*_\Lambda|\bar\rho)|$ is bounded by

$$\int \Big\{ e'_{\lambda(\beta)}(\xi_1)\tilde J_\gamma * |\hat\rho_\Lambda - \rho^*_\Lambda| + \frac{1}{\beta}|S(\hat\rho_\Lambda) - S(\rho^*_\Lambda)| \Big\},$$

and since $\hat\rho_\Lambda$ and ρ^*_Λ are bounded away from 0 (and $S(\cdot)$ is differentiable away from 0) (11.2.3.14) follows from (11.2.3.13). □

11.2.4 Proof of Theorem 11.2.2.4

Theorem 11.2.2.4 follows from Propositions 11.2.3.1 and 11.2.3.2 (which hold as well for the minus case) with ω in Theorem 11.2.2.4 proportional to ω_β, ω_β as in (11.2.3.11); for instance $\omega = \omega_\beta/40$, ω_β as in (11.2.3.11).

11.3 The energy estimate

In this section we shall prove (11.0.2.4) starting from (11.2.2.1). The main step is the following bound of its last factor:

$$\log Z_{\gamma,\lambda_{\beta,\gamma},\Lambda,\chi_{\Lambda^c}}(\{\eta=\eta_\Gamma \text{ on } \Lambda\})$$
$$\le -\beta\big(\phi|\Lambda| + I_{\gamma,\lambda(\beta)}(\chi^+_\Lambda;\chi^+_{B^+_0}) + I(\chi^-_\Lambda;\chi^-_{B^-_0}) + c_1 N_\Gamma \zeta^2 \ell^d_{-,\gamma}\big)$$
$$+ c\gamma^{1/2}|\operatorname{sp}(\Gamma)|. \qquad (11.3.0.1)$$

We refer to Sect. 11.2.1 for the definitions of the various sets, in particular $\Lambda \subset \operatorname{sp}(\Gamma)$ is the set defined in (11.2.1.3); $\chi_{\Lambda^c} = \rho_{\beta,\pm}$ on $B^\pm_0$, and the values outside $B^\pm_0$ are irrelevant, because points of Λ only interact with points of $B^\pm_0$;

$$\phi := \phi_{\beta,\lambda(\beta)}(\rho_{\beta,+}) = \phi_{\beta,\lambda(\beta)}(\rho_{\beta,-}), \qquad (11.3.0.2)$$

and $I_{\gamma,\lambda(\beta)}(\chi^\pm_\Lambda;\chi^\pm_{B^\pm_0})$ is defined in (11.3.1.1). They are closely related to the expression (10.4.1.5).

The proof of (11.3.0.1) uses coarse graining. Reduction to the free energy functional is not a straightforward application of Sect. 11.1.3, because the number of particles in the present context is not bounded. The particular structure of the LMP hamiltonian greatly simplifies the problem which otherwise would be far from trivial. The analysis is technical and is reported in full detail in Sect. 11.8. There are again two alternatives: either to read first Sect. 11.8 and then go directly to the next subsection, or just read the text as it is where we summarize the results of Sect. 11.8 in Proposition 11.3.0.1 below, stated after some definitions.

Let $\tilde{\mathcal{M}}_\Lambda$ be as in Definition 11.1.3.1. We "localize" the excess free energy stored in ρ_Λ in terms of a penalty for (i) deviations from $\rho_{\beta,+}$ and $\rho_{\beta,-}$ and (ii) jumps from $\rho_{\beta,+}$ and $\rho_{\beta,-}$. With this in mind, we call $\mathcal{C}'_0$ the collection of all cubes $C^0_i \in \mathcal{D}^{(\ell_{-,\gamma})}$ contained in Λ', see (11.2.1.3), such that $|\fint_{C^0_i}\rho_\Lambda - \rho_{\beta,\pm}| > \frac{\zeta}{2}$. We also call $\mathcal{C}'_{\neq}$ a collection of pairs (C'_i, C''_i) of cubes in $\mathcal{D}^{(\ell_{-,\gamma})}$ both contained in Λ' which are contiguous and such that $|\fint_{C'_i}\rho_\Lambda - \rho_{\beta,+}| \le \frac{\zeta}{2}$ and $|\fint_{C''_i}\rho_\Lambda - \rho_{\beta,-}| \le \frac{\zeta}{2}$. $\mathcal{C}'_{\neq}$ should be "maximal," namely any pair (C', C'') as above must have at least one among C' and C'' already appearing in the other pairs. Denoting by N'_0 and $N'_{\neq}$ the total number of cubes in $\mathcal{C}'_0$ and $\mathcal{C}'_{\neq}$, respectively, we shall prove in Sect. 11.8:

Proposition 11.3.0.1 *There is $c > 0$, so that $\log Z_{\gamma,\lambda_{\beta,\gamma},\Lambda,\chi_{\Lambda^c}}(\{\eta=\eta_\Gamma$ on $\Lambda\})$ is bounded for all γ small enough by*

$$-\min_{m\ge 0}\Big\{m\ell^d_{-,\gamma}\frac{\zeta}{2} + \beta \inf_{\rho_\Lambda\in\mathcal{G}_{2m}} \tilde{\mathcal{F}}_{\gamma,\lambda(\beta),\Lambda}(\rho_\Lambda|\chi_{\Lambda^c})\Big\} + c\gamma^{1/2}|\Lambda|, \qquad (11.3.0.3)$$

where

$$\mathcal{G}_m := \Big\{\rho_\Lambda \in \tilde{\mathcal{M}}_\Lambda : \rho_\Lambda \le \rho_{\max},\ N'_0(\rho_\Lambda) + N'_{\neq}(\rho_\Lambda) \ge \frac{3^{-d}N_\Gamma - m}{2}\Big\}. \qquad (11.3.0.4)$$

11.3.1 The interaction terms $I^{\pm}_{\gamma,\lambda(\beta)}$

In this subsection we shall split "in a smart way" the interaction between two regions into a sum of two terms, and in this way we shall explain in (11.3.1.3) and (11.3.1.6) below the origin of the interaction terms $I^{\pm}_{\gamma,\lambda(\beta)}(\Lambda)$ defined in (10.4.1.5). Let Δ and Λ be possibly unbounded regions, ρ_Δ and ρ_Λ densities in Δ and Λ, $\rho = \rho_\Delta + \rho_\Lambda$. We then set

$$I_{\gamma,\lambda}(\rho_\Delta;\rho_\Lambda) := \int_\Lambda \{e_\lambda(J_\gamma * \rho) - e_\lambda(J_\gamma * \rho_\Lambda)\} - \int_\Delta e_\lambda(J_\gamma * \rho_\Lambda), \qquad (11.3.1.1)$$

observing that $I_{\gamma,\lambda(\beta)}(\chi^{\pm}_\Lambda;\chi^{\pm}_{\Lambda^c}) = I^{\pm}_{\gamma,\lambda(\beta)}(\Lambda)$, the latter as in (10.4.1.5). Since $H_{\gamma,\lambda,\Delta}(\rho_\Delta|\rho_{\Delta^c}) = \int\{e_\lambda(J_\gamma * \rho) - e_\lambda(J_\gamma * \rho_{\Delta^c})\}$, by (11.3.1.1)

$$H_{\gamma,\lambda,\Delta}(\rho_\Delta|\rho_{\Delta^c}) = \int_\Delta e_\lambda(J_\gamma * \rho) + I_{\gamma,\lambda}(\rho_\Delta;\rho_{\Delta^c}), \qquad (11.3.1.2)$$

so that

$$H_{\gamma,\lambda(\beta),\Delta}(\chi^{\pm}_\Delta|\chi^{\pm}_{\Delta^c}) = |\Delta| e_{\lambda(\beta)}(\rho_{\beta,\pm}) + I_{\gamma,\lambda(\beta)}(\chi^{\pm}_\Delta;\chi^{\pm}_{\Delta^c}). \qquad (11.3.1.3)$$

Thus $I^{\pm}_{\gamma,\lambda(\beta)}(\Lambda) = I_{\gamma,\lambda(\beta)}(\chi^{\pm}_\Delta;\chi^{\pm}_{\Delta^c})$ is the difference between the energy in Δ of the equilibrium profile $\chi^{\pm}_\Delta$ with equilibrium b.c. $\chi^{\pm}_{\Delta^c}$ and the mean field equilibrium energy in Δ, which is $|\Delta| e_{\lambda(\beta)}(\rho_{\beta,\pm})$.

We can extend the above relations as follows. Since the interaction energy between Δ and Δ^c is $U_{\gamma,\lambda}(\rho_\Delta|\rho_{\Delta^c}) = H_{\gamma,\lambda,\Delta}(\rho_\Delta|\rho_{\Delta^c}) - H_{\gamma,\lambda,\Delta}(\rho_\Delta)$,

$$U_{\gamma,\lambda}(\rho_\Delta|\rho_{\Delta^c}) = \int \{e_\lambda(J_\gamma * \rho) - e_\lambda(J_\gamma * \rho_{\Delta^c}) - e_\lambda(J_\gamma * \rho_\Delta)\}; \qquad (11.3.1.4)$$

hence $U_{\gamma,\lambda}(\rho_\Delta|\rho_{\Delta^c}) = I_{\gamma,\lambda}(\rho_\Delta;\rho_{\Delta^c}) + I_{\gamma,\lambda}(\rho_{\Delta^c};\rho_\Delta)$, showing that the terms $I_{\gamma,\lambda}$ arise from splitting the interaction into two terms. By (11.3.1.1)

$$H_{\gamma,\lambda,\Delta}(\rho_\Delta) = \int_\Delta e_\lambda(J_\gamma * \rho) - I_{\gamma,\lambda}(\rho_{\Delta^c};\rho_\Delta), \qquad (11.3.1.5)$$

which yields

$$H_{\gamma,\lambda(\beta),\Delta}(\chi^{\pm}_\Delta) = |\Delta| e_{\lambda(\beta)}(\rho_{\beta,\pm}) - I_{\gamma,\lambda(\beta)}(\chi^{\pm}_{\Delta^c};\chi^{\pm}_\Delta), \qquad (11.3.1.6)$$

providing another interpretation for $I_{\gamma,\lambda(\beta)}(\chi^{\pm}_{\Delta^c};\chi^{\pm}_\Delta)$, alternative to (11.3.1.3).

11.3.2 Free energy cost of contours

Theorem 11.3.2.1 below quantifies the free energy cost of density profiles $\rho \le \rho_{\max}$. Recall that in Proposition 11.3.0.1 we have made a reduction to such a case. Recall that $\chi_{\Lambda^c} = \rho_{\beta,\pm}$ on $B_0^\pm$ and that Λ ($\Lambda \subset \text{sp}(\Gamma)$ as in (11.2.1.3)) interacts only with $B_0^\pm$ so that the values of χ_{Λ^c} elsewhere are unimportant. Recall also that $I^\pm_{\gamma,\lambda(\beta)}$ are defined in (10.4.1.5) and N_0' and $N_{\neq}'$ in Proposition 11.3.0.1.

Theorem 11.3.2.1 *For all γ small enough, if $\rho_\Lambda \le \rho_{\max}$ then*

$$\tilde{\mathcal{F}}_{\gamma,\lambda(\beta),\Lambda}(\rho_\Lambda|\chi_{\Lambda^c}) \ge \phi|\Lambda| + I_{\gamma,\lambda(\beta)}(\chi_\Lambda^+; \chi_{B_0^+}^+) + I_{\gamma,\lambda(\beta)}(\chi_\Lambda^-; \chi_{B_0^-}^-) + 2\cdot 3^d c_1(N_0' + N_{\neq}')\zeta^2 \ell^d_{-,\gamma}, \tag{11.3.2.1}$$

with c_1 as in the paragraph "The constant c_1" in Sect. 10.2.6.

Proof We here introduce the main notation used in the proof:

Notation Λ and Λ' are as in (11.2.1.3); by their definition, $\text{dist}(\Lambda', \Lambda^c) > 2\gamma^{-1}$. We write $\mathcal{C}_0' = \{C_i^0\}$, $C_i^0 \in \mathcal{D}^{(\ell_{-,\gamma})}$, and, by the definition of $\mathcal{C}_0'$:

$$\left| \fint_{C_i^0} \rho - \rho_{\beta,-} \right| > \frac{\zeta}{2}, \qquad \left| \fint_{C_i^0} \rho - \rho_{\beta,+} \right| > \frac{\zeta}{2}. \tag{11.3.2.2}$$

$\mathcal{C}_{\neq}' = \{(C_i', C_i''\}$, with C_i', C_i'' in $\mathcal{D}^{(\ell_{-,\gamma})}$ mutually connected and such that

$$\left| \fint_{C_i'} \rho - \rho_{\beta,+} \right| \le \frac{\zeta}{2}, \qquad \left| \fint_{C_i''} \rho - \rho_{\beta,-} \right| \le \frac{\zeta}{2}. \tag{11.3.2.3}$$

We write $C_i^{\neq} := C_i' \cup C_i''$ and use the following notation:

$$\mathcal{R} = \tilde{J}_\gamma * (\rho_\Lambda + \chi_{\Lambda^c}), \tag{11.3.2.4}$$

$$D := \Lambda \setminus D', \quad D' = \left\{ r \in \Lambda : |\mathcal{R} - \rho_{\beta,s}| \le \frac{\zeta}{4},\ |s| = 1 \right\}. \tag{11.3.2.5}$$

Claim $\tilde{\mathcal{F}}_{\gamma,\lambda(\beta),\Lambda}(\rho_\Lambda|\chi_{\Lambda^c}) = F_1 + F_2 + F_3$, *where*

$$F_1 = \int_\Lambda \phi_{\beta,\lambda(\beta)}(\mathcal{R}), \qquad F_2 = \frac{1}{\beta}\int_\Lambda \{S(\mathcal{R}) - \tilde{J}_\gamma * S(\rho_\Lambda + \chi_{\Lambda^c})\},$$

$$\begin{aligned} F_3 = &\int_{\Lambda^c} \{\phi_{\beta,\lambda(\beta)}(\mathcal{R}) - \phi_{\beta,\lambda(\beta)}(\tilde{J}_\gamma * \chi_{\Lambda^c})\} \\ &+ \frac{1}{\beta}\int_{\Lambda^c} \{S(\mathcal{R}) - \tilde{J}_\gamma * S(\rho_\Lambda + \chi_{\Lambda^c})\} \\ &- \int_\Lambda \phi_{\beta,\lambda(\beta)}(\tilde{J}_\gamma * \chi_{\Lambda^c}) - \frac{1}{\beta}\int \{S(\tilde{J}_\gamma * \chi_{\Lambda^c}) - \tilde{J}_\gamma * S(\chi_{\Lambda^c})\}. \end{aligned}$$

Proof Split $\phi_{\beta,\lambda(\beta)}(\cdot) = e_{\beta,\lambda(\beta)}(\cdot) - \frac{S(\cdot)}{\beta}$. The sum of the energy terms in F_1, F_2 and F_3 reconstructs the energy term in $\tilde{\mathcal{F}}_{\gamma,\lambda(\beta),\Lambda}$. The term with $S(\mathcal{R})$ in F_1 cancels with the one in F_2, those in F_3 cancel with each other. When summing $F_2 + F_3$ we can replace $\int_\Lambda \tilde{J}_\gamma * S(\rho_\Lambda + \chi_{\Lambda^c}) + \int_{\Lambda^c} \tilde{J}_\gamma * S(\rho_\Lambda + \chi_{\Lambda^c})$ by $\{\int_\Lambda S(\rho_\Lambda + \chi_{\Lambda^c}) + \int_{\Lambda^c} S(\rho_\Lambda + \chi_{\Lambda^c})\}$, hence by $\{\int_\Lambda S(\rho_\Lambda) + \int_{\Lambda^c} S(\chi_{\Lambda^c})\}$. The last term $\int_{\Lambda^c} S(\chi_{\Lambda^c})$ (which actually appears with a minus sign) cancels with the last term in F_3, and thus only $\int_\Lambda S(\rho_\Lambda)$ remains, which is the entropy term appearing in $\tilde{\mathcal{F}}_{\gamma,\lambda(\beta),\Lambda}$. We are left with the entropy terms of the form $S(\tilde{J}_\gamma * \chi_{\Lambda^c})$ which appear in the first, third and fourth integrals in F_3: they all cancel with each other, and therefore $\tilde{\mathcal{F}}_{\gamma,\lambda(\beta),\Lambda}(\rho_\Lambda|\chi_{\Lambda^c}) = F_1 + F_2 + F_3$. The claim is proved. □

The three bounds We shall prove that for all γ small enough,

$$F_1 \geq \phi|\Lambda| + c_1' \frac{\zeta^2}{16}|D|, \tag{11.3.2.6}$$

$$F_2 \geq 2 \cdot 3^d c_1 \zeta^2 N_0' \ell_{-,\gamma}^d - \frac{c_1'}{16}\zeta^2|D| + c\zeta N_{\neq}' \ell_{-,\gamma}^d, \tag{11.3.2.7}$$

$$F_3 \geq I_{\gamma,\lambda(\beta)}(\chi_\Lambda^+; \chi_{B_0^+}^+) + I_{\gamma,\lambda(\beta)}(\chi_\Lambda^-; \chi_{B_0^-}^-). \tag{11.3.2.8}$$

Theorem 11.3.2.1 follows immediately from these bounds, which will be proved hereafter.

Proof of (11.3.2.6) and of (11.3.2.8) Equation (11.3.2.6) follows from (10.2.2.3). To prove (11.3.2.8) we first observe: • the second integral in F_3 is non-negative; • $\phi_{\beta,\lambda(\beta)}(\mathcal{R}) \geq \phi$; • $f(\tilde{J}_\gamma * \chi_{\Lambda^c}) = f(\tilde{J}_\gamma * \chi^+) + f(\tilde{J}_\gamma * \chi^-)$ on $\Lambda \cup \delta_{\text{out}}^{4\gamma^{-1}}[\Lambda]$. Then

$$\begin{aligned} F_3 \geq & \int_{B_0^+} \{\phi - \phi_{\beta,\lambda(\beta)}(\tilde{J}_\gamma * \chi^+)\} + \int_{B_0^-} \{\phi - \phi_{\beta,\lambda(\beta)}(\tilde{J}_\gamma * \chi^-)\} \\ & - \int_\Lambda \phi_{\beta,\lambda(\beta)}(\tilde{J}_\gamma * \chi^+) - \int_\Lambda \phi_{\beta,\lambda(\beta)}(\tilde{J}_\gamma * \chi^-) \\ & - \frac{1}{\beta} \int \{S(\tilde{J}_\gamma * \chi^+) + S(\tilde{J}_\gamma * \chi^-) - \tilde{J}_\gamma * S(\chi^+) - \tilde{J}_\gamma * S(\chi^-)\}. \end{aligned}$$

The last integral is equal to $\{\int S(\tilde{J}_\gamma * \chi^+) + S(\tilde{J}_\gamma * \chi^-) - S(\chi^+) - S(\chi^-)\}$ and $\int\{-S(\chi^+) - S(\chi^-)\} = -\int_{B_0^+} S(\chi^+) - \int_{B_0^-} S(\chi^-)$, which cancels with the entropy term in $\int_{B_0^\pm} \phi$. The terms with $S(\tilde{J}_\gamma * \chi^\pm)$ cancel with each other and with those in $\phi_{\beta,\lambda(\beta)}(\tilde{J}_\gamma * \chi^\pm)$, and (11.3.2.8) follows from (11.3.1.3).

The proof of Theorem 11.3.2.1 is now reduced to the proof of (11.3.2.7). The bound will come from the following entropy inequality:

Lemma 11.3.2.2 *Let $0 \le \rho \le \rho_{\max}$, write the shorthand $v(r) = (\tilde{J}_\gamma * \rho)(r)$; then*

$$S\big(v(r)\big) - (\tilde{J}_\gamma * S(\rho))(r) \ge \frac{1}{2\rho_{\max}} \int \tilde{J}_\gamma(r,r')\big(\rho(r') - v(r)\big)^2. \tag{11.3.2.9}$$

Proof Writing $u = \rho(r')$,

$$S(u) = S(v) + S'(v)(u-v) + \int_v^u S''(z)(u-z)dz.$$

By assumption, $u \le \rho_{\max}$ and $v \le \rho_{\max}$; then if $u > v$:

$$\int_v^u S''(z)(u-z)dz \le -\frac{1}{\rho_{\max}} \int_v^u (u-z)dz = -\frac{(u-v)^2}{2\rho_{\max}}.$$

If $v > u$, $\int_v^u S''(z)(u-z)dz = \int_u^v S''(u+v-y)(v-y)dy$, and we get the same bound as when $u > v$. Thus

$$S(u) \le S(v) + S'(v)(u-v) - \frac{(u-v)^2}{2\rho_{\max}}.$$

By integrating over $\tilde{J}_\gamma(r,r')dr'$ the linear term vanishes, and we get (11.3.2.9). □

Proof of (11.3.2.7) Let C_i^0 and $C_i^{\neq}$ be as in the beginning of the proof; $\rho = \rho_\Lambda + \chi_{\Lambda^c}$ so that $\mathcal{R} = \tilde{J}_\gamma * \rho$. Let $V(A) = \frac{1}{2\beta\rho_{\max}} \int_{r\in A}\int_{r'\in A} \tilde{J}_\gamma(r,r')(\rho(r') - \mathcal{R}(r))^2$; then by Lemma 11.3.2.2

$$F_2 \ge \sum V(C_i^0) + \sum V(C_i^{\neq}). \tag{11.3.2.10}$$

Let $J_\gamma^*(r,r') = \fint_{C_{r'}^{(\ell_{-,\gamma})}} \tilde{J}_\gamma(r,r'')\,dr''$, $M_\gamma(r,r') := \tilde{J}_\gamma(r,r') - J_\gamma^*(r,r')$. There is $c_3 > 0$ so that $\sup_{r'} \int |M(r,r')| \le c_3\gamma^\alpha$. Since $(\rho - \mathcal{R})^2 \le \rho_{\max}^2$, denoting by r_i a point in C_i^0,

$$V(C_i^0) \ge \frac{1}{2\beta\rho_{\max}} \int_{r\in\Lambda} J_\gamma^*(r,r_i) \int_{r'\in C_i^0} \big(\rho(r') - \mathcal{R}(r)\big)^2 - \frac{c_3\gamma^\alpha\rho_{\max}^2\ell_{-,\gamma}^d}{2\beta\rho_{\max}}.$$

By Cauchy–Schwartz,

$$\ell_{-,\gamma}^d \int_{r'\in C_i^0} \big(\rho(r') - \mathcal{R}(r)\big)^2 \ge \left(\int_{r'\in C_i^0} [\rho(r') - \mathcal{R}(r)]\right)^2. \tag{11.3.2.11}$$

If $r \in D'$, $(\int_{r'\in C_i^0}[\rho(r') - \mathcal{R}(r)])^2 \ge (\ell_{-,\gamma}^d \frac{\zeta}{4})^2$, and hence

$$V(C_i^0) \ge \frac{\zeta^2\ell_{-,\gamma}^d}{32\beta\rho_{\max}} \int_{r\in D'} J_\gamma^*(r,r_i) - c_3\frac{\rho_{\max}\gamma^\alpha\ell_{-,\gamma}^d}{2\beta}. \tag{11.3.2.12}$$

We write

$$\int_{r\in D'} J_\gamma^*(r,r_i) = \int_{r\in\Lambda} J_\gamma^*(r,r_i) - \int_{r\in D} J_\gamma^*(r,r_i), \quad \int_{r\in\Lambda} J_\gamma^*(r,r_i) = 1,$$

because $\mathrm{dist}(r_i,\Lambda^c) > 4\gamma^{-1}$. Since $\frac{c_1'}{16} > \frac{1}{32\beta\rho_{\max}}$, see (10.3.4.4), and $4c_1 3^d = (32\beta\rho_{\max})^{-1}$, see (10.3.4.3),

$$\sum_i V(C_i^0) \ge \left(4c_1 3^d \zeta^2 - c_3 \frac{\rho_{\max}\gamma^\alpha}{2\beta}\right) N_0' \ell_{-,\gamma}^d - \frac{c_1'}{16}\zeta^2 |D|, \tag{11.3.2.13}$$

and for γ small enough

$$\sum_i V(C_i^0) \ge 2\cdot 3^d c_1 \zeta^2 N_0' \ell_{-,\gamma}^d - \frac{c_1'}{16}\zeta^2|D|. \tag{11.3.2.14}$$

The bound of $V(C_i^{\neq})$ is similar. Write $p_i(r) := \fint_{C_i^{\neq}} \tilde{J}_\gamma(r,r')\,dr'$; then since $\mathrm{dist}(C_i^{\neq},\Lambda^c) > 4\gamma^{-1}$, $\int_\Lambda p_i(r)\,dr = 1$. There is c_4 so that

$$\sup_{r'\in C_i\neq} |p_i(r) - \tilde{J}_\gamma(r,r')| \le c_4\gamma^\alpha;$$

then

$$V(C_i^{\neq}) \ge \frac{1}{2\beta\rho_{\max}} \int_{r\in\Lambda}\int_{r'\in C_i^{\neq}} p_i(r)\big(\rho_\Lambda(r') - \mathcal{R}(r)\big)^2 - c_4 \frac{\rho_{\max}\gamma^\alpha 2\ell_{-,\gamma}^d}{2\beta}.$$

Calling the two cubes whose union is $C_i^{\neq}$ C_i' and C_i'', and ξ being a vector such that $r'+\xi$ runs over all points of C_i'' when r' varies in C_i', then for $r\in\Lambda$,

$$\begin{aligned}
&\int_{r'\in C_i'\cup C_i''} \big(\rho_\Lambda(r') - \mathcal{R}(r)\big)^2 \\
&= \int_{r'\in C_i'} \Big\{\big(\rho_\Lambda(r') - \mathcal{R}(r)\big)^2 + \big(\rho_\Lambda(r'+\xi) - \mathcal{R}(r)\big)^2\Big\} \\
&\ge \int_{r'\in C_i'} \frac{1}{2}\big(\rho_\Lambda(r') - \rho_\Lambda(r'+\xi)\big)^2 \\
&\ge \frac{\ell_{-,\gamma}^{-d}}{2}\left(\int_{r'\in C_i'} [\rho_\Lambda(r') - \rho_\Lambda(r'+\xi)]\right)^2,
\end{aligned}$$

having used Cauchy–Schwartz as in (11.3.2.11). We have

$$\left|\int_{r'\in C_i'} [\rho(r') - \rho(r'+\xi)]\right| \ge \ell_{-,\gamma}^d\big(|\rho_{\beta,+} - \rho_{\beta,-}| - 2\zeta\big).$$

For γ small enough $(|\rho_{\beta,+}-\rho_{\beta,-}|-2\zeta)\geq|\rho_{\beta,+}-\rho_{\beta,-}|/2$; hence

$$V(C_i^{\neq})\geq\frac{|\rho_{\beta,+}-\rho_{\beta,-}|^2}{16\beta\rho_{\max}}\ell_{-,\gamma}^d-c_4\frac{\rho_{\max}\gamma^{\alpha}2\ell_{-,\gamma}^d}{2\beta},$$

and (11.3.2.7) is proved for γ small enough. □

11.3.3 Proof of the energy estimate

We conclude here the proof of (11.0.2.4). By (11.8.1.4), (11.3.0.3) and (11.3.2.1) there is c so that $\log Z_{\gamma,\lambda_{\beta,\gamma},\Lambda,\chi_{\Lambda^c}}(\{\eta=\eta_\Gamma \text{ on } \Lambda\})$ is bounded by

$$-\beta\Bigg(\phi|\Lambda|+I_{\gamma,\lambda(\beta)}(\chi_\Lambda^+;\chi_{B_0^+}^+)+I(\chi_\Lambda^-;\chi_{B_0^-}^-)$$
$$+\min_{m\geq 0}\left\{m\ell_{-,\gamma}^d\frac{\zeta}{2}+2\cdot 3^d c_1\frac{3^{-d}N_\Gamma-2m}{2}\zeta^2\ell_{-,\gamma}^d\right\}\Bigg)+c\gamma^{1/2}|\mathrm{sp}(\Gamma)|.$$

For all γ small enough the min is achieved at $m=0$ and (11.3.0.1) is proved. Using the bound (11.3.0.1) in (11.2.2.1), $\hat{N}^+_{\gamma,\lambda_{\beta,\gamma}}(\Gamma,q^+)$ is then bounded by

$$e^{-\beta c_1 N_\Gamma\zeta^2\ell_{-,\gamma}^d+c\gamma^{1/2}|\mathrm{sp}(\Gamma)|}\hat{Z}^-_{\gamma,\lambda_{\beta,\gamma},\mathrm{int}^-(\Gamma),\chi^-_{B_0^-}}\hat{Z}^+_{\gamma,\lambda_{\beta,\gamma},\mathrm{int}^+(\Gamma),\chi^+_{B_0^+}}$$
$$\times Z_{\gamma,\lambda_{\beta,\gamma},\Delta_1,\chi^+_{B_0^+}+q^+_{c(\Gamma)^c}}(\{\eta=1 \text{ on } \Delta_1\})$$
$$\times e^{-\beta(\tilde{\mathcal{F}}_{\gamma,\lambda(\beta),B_0}(\chi_{B_0})+\phi|\Lambda|+I_{\gamma,\lambda(\beta)}(\chi^+_\Lambda;\chi^+_{B_0^+})+I_{\gamma,\lambda(\beta)}(\chi^-_\Lambda;\chi^-_{B_0^-}))}. \qquad (11.3.3.1)$$

Since $\tilde{\mathcal{F}}_{\gamma,\lambda(\beta),B_0}(\chi_{B_0})=\tilde{\mathcal{F}}_{\gamma,\lambda(\beta),B_0^+}(\chi^+_{B_0^-})+\tilde{\mathcal{F}}_{\gamma,\lambda(\beta),B_0^-}(\chi_{B_0^-})$, by (11.3.1.6) and the definition of ϕ in (11.3.0.2),

$$\tilde{\mathcal{F}}_{\gamma,\lambda(\beta),B_0}(\chi_{B_0})$$
$$=\phi|B_0|-I_{\gamma,\lambda(\beta)}(\chi^+_\Lambda;\chi^+_{B_0^+})-I_{\gamma,\lambda(\beta)}(\chi^-_\Lambda;\chi^-_{B_0^-})$$
$$-I_{\gamma,\lambda(\beta)}(\chi^+_{\Delta_1};\chi^+_{B_0^+})-I_{\gamma,\lambda(\beta)}(\chi^+_{\mathrm{int}^+(\Gamma)};\chi^+_{B_0^+})-I_{\gamma,\lambda(\beta)}(\chi^-_{\mathrm{int}^-(\Gamma)};\chi^-_{B_0^-});$$

therefore, the exponent in the last term of (11.3.3.1) becomes $-\beta$ times

$$\phi|\Lambda\cup B_0|-I_{\gamma,\lambda(\beta)}(\chi^+_{\Delta_1};\chi^+_{B_0^+})-I_{\gamma,\lambda(\beta)}(\chi^+_{\mathrm{int}^+(\Gamma)};\chi^+_{B_0^+})-I_{\gamma,\lambda(\beta)}(\chi^-_{\mathrm{int}^-(\Gamma)};\chi^-_{B_0^-}).$$

Going backwards we replace

$$\phi|\Lambda\cup B_0|=\tilde{\mathcal{F}}_{\gamma,\lambda(\beta),\Lambda\cup B_0}(\chi^+_{\Lambda\cup B_0})+I_{\gamma,\lambda(\beta)}(\chi^+_{\Delta_1};\chi^+_{B_0^+})+I_{\gamma,\lambda(\beta)}(\chi^+_{\mathrm{int}(\Gamma)};\chi^+_{B_0}).$$

The last term is $I_{\gamma,\lambda(\beta)}(\chi^{+}_{\mathrm{int}^+(\Gamma)};\chi^{+}_{B_0})+I_{\gamma,\lambda(\beta)}(\chi^{+}_{\mathrm{int}^-(\Gamma)};\chi^{+}_{B_0})$. Then, after some cancelations, by (11.3.3.1) $\hat{N}^{+}_{\gamma,\lambda_{\beta,\gamma}}(\Gamma,q^+)$ is bounded by

$$K e^{-\beta c_1 N_\Gamma \zeta^2 \ell^d_{-,\gamma}+c\gamma^{1/2}|\mathrm{sp}(\Gamma)|}\frac{\hat{Z}^{-}_{\gamma,\lambda_{\beta,\gamma},\mathrm{int}^-(\Gamma),\chi^{-}_{B_0^-}}\, e^{\beta I_{\gamma,\lambda(\beta)}(\chi^{-}_{\mathrm{int}^-(\Gamma)};\chi^{-}_{B_0^-})}}{\hat{Z}^{+}_{\gamma,\lambda_{\beta,\gamma},\mathrm{int}^-(\Gamma),\chi^{+}_{B_0^-}}\, e^{\beta I_{\gamma,\lambda(\beta)}(\chi^{+}_{\mathrm{int}^-(\Gamma)};\chi^{+}_{B_0^-})}},$$

where K is a shorthand for

$$\hat{Z}^{+}_{\gamma,\lambda_{\beta,\gamma},\mathrm{int}^+(\Gamma),\chi^{+}_{B_0^+}}\,\hat{Z}^{+}_{\gamma,\lambda_{\beta,\gamma},\mathrm{int}^-(\Gamma),\chi^{+}_{B_0^-}}\,Z_{\gamma,\lambda_{\beta,\gamma},\Delta_1,\chi^{+}_{B_0^+}+q^{+}_{c(\Gamma)^c}}(\{\eta=1 \text{ on } \Delta_1\})$$
$$\times\, e^{-\beta(\tilde{\mathcal{F}}_{\gamma,\lambda(\beta),\Lambda\cup B_0}(\chi^{+}_{\Lambda\cup B_0}))}, \qquad (11.3.3.2)$$

and the proof of (11.0.2.4) will follow from $K \le e^{c\gamma^{1/2}|\mathrm{sp}(\Gamma)|}\hat{D}^{+}_{\gamma,\lambda_{\beta,\gamma}}(\Gamma,q^+)$. Let $\mathcal{B}$ be the subset of $\tilde{\mathcal{M}}_{\Lambda\cup B}$ consisting of the single configuration $\rho=[\rho_{\beta,+}]$, $[\rho_{\beta,+}]$ being the closest element of $\{\gamma^{d/2}n, n\in\mathbb{N}\}$ to $\rho_{\beta,+}$. Then

$$e^{-\beta\tilde{\mathcal{F}}_{\gamma,\lambda(\beta),\Lambda\cup B_0}(\chi^{+}_{\Lambda\cup B_0})} \le Z_{\gamma,\lambda_{\beta,\gamma},\Lambda\cup B_0}(\mathcal{B})e^{c\gamma^{1/2}|\Lambda\cup B_0|}.$$

Let $\bar{q}$ be any configuration in $\Lambda\cup B_0$ such that $\mathrm{Av}^{(\gamma^{-1/2})}(\bar{q};\cdot)\in\mathcal{B}$; then

$$\hat{Z}^{+}_{\gamma,\lambda_{\beta,\gamma},\mathrm{int}^+(\Gamma),\chi^{+}_{B_0^+}} \le \hat{Z}^{+}_{\gamma,\lambda_{\beta,\gamma},\mathrm{int}^+(\Gamma),\bar{q}}\, e^{c\gamma^{d/2}|B_0^+|}.$$

Analogous bounds hold for the other partition functions in (11.3.3.2) so that K is bounded by

$$e^{c\gamma^{1/2}|\mathrm{sp}(\Gamma)|}\int_{\mathrm{Av}^{(\gamma^{-1/2})}(\bar{q};\cdot)\in\mathcal{B}} e^{-\beta H_{\gamma,\lambda_{\beta,\gamma},\Lambda\cup B_0}(\bar{q})}\,\hat{Z}^{+}_{\gamma,\lambda_{\beta,\gamma},\mathrm{int}^+(\Gamma),\bar{q}}$$
$$\times\,\hat{Z}^{+}_{\gamma,\lambda_{\beta,\gamma},\mathrm{int}^-(\Gamma),\bar{q}}\,Z_{\gamma,\lambda_{\beta,\gamma},\Delta_1,\bar{q}+q^{+}_{c(\Gamma)^c}}(\{\eta=1 \text{ on } \Delta_1\})d\nu_{\Lambda\cup B_0}(\bar{q});$$

hence $K \le e^{c\gamma^{1/2}|\mathrm{sp}(\Gamma)|}\hat{D}^{+}_{\gamma,\lambda_{\beta,\gamma}}(\Gamma,q^+)$.

11.4 Surface corrections to the pressure

In this section we shall prove (11.0.2.5) and as explained in the beginning of the chapter, together with (11.0.2.4) and (11.0.2.3) (which have already been proved) this will complete the proof of Theorem 10.3.4.1. (11.0.2.5) is undoubtedly the hardest point of the whole proof. We distinguish five main steps:

Reduction to a lattice gas

We shall approximate our system by a lattice gas in which particles can only stay with single occupancy on a lattice $a\mathbb{Z}^d$, of mesh $a > 0$. By fixing a region Λ and choosing suitably the chemical potential of the lattice gas we shall prove that the lattice gas partition function and the Gibbs measure converge to the true quantities as $a \to 0$. The advantage of working with a lattice gas will become clear when studying couplings of Gibbs measures, as it automatically settles some delicate measurability problems.

An effective hamiltonian for the contour weights

In this step we shall write the contour contribution to the partition function $X^{\pm}_{\gamma,\lambda_{\beta,\gamma},q}(\Lambda)$, see (11.1.1.6), as equal to $e^{-\beta K_{\gamma,\Lambda}(q)}$, $X^{\pm}$. Since $X^{\pm}_{\gamma,\lambda_{\beta,\gamma},q}(\Lambda)$ is positive this is obviously possible, but the key point is that $K_{\gamma,\Lambda}(q)$ has all the good properties of a hamiltonian. This is shown by using well established theorems on a "cluster expansion," whose applicability follows from the smallness of the weights, a property which is automatically satisfied because they are cutoff weights. This part of the analysis is standard and common to all classical Pirogov–Sinai models where a ground state is perturbed: in such cases $\mathcal{X}^{\pm}$ are singletons and the only contribution to the partition function is due to the weights of the contours. By cluster expansion methods it is then proved that the effective hamiltonian is "nice" which directly implies control of the surface corrections to the pressure. In our case this is still far, because the $\mathcal{X}^{\pm}$ have a non-trivial structure.

An interpolation formula for the partition function

The main point of the whole analysis will be to prove that the correlations in the lattice gas with mesh a decay exponentially, uniformly as $a \to 0$. In this third step we shall prove that if such a decay property holds, then the surface corrections to the pressure are as small as needed. The idea that decay of correlations and surface corrections may be related is natural but a rigorous treatment of the property goes back, I believe, to Dobrushin and Shlosman in [110], where they proved such a connection in the case of large temperatures by interpolating between the actual value of β and $\beta = 0$ where the system is described by a Bernoulli (independent) process. We use the same approach interpolating between an independent measure (corresponding formally to the mean field limit $\gamma = 0$) and the actual value of γ. We shall prove in this way that the surface corrections to the pressure are given by a sum of difference of expectations of local functions. The expectations are with respect to the Gibbs measure with the interpolating hamiltonian in Λ (where we are computing the partition function) and then in the infinite volume. The estimate then splits into two: one is when the local function (of which expectations are taken) is localized away from Λ^c and we can then exploit the exponential decay of the correlations; the other case is when the local function is localized close to Λ^c. We shall prove closeness of both expectations to the mean field values for γ small. The two estimates are stated in Theorem 11.4.5.1 and Theorem 11.4.5.2. We shall prove that, pending the validity of such theorems, (11.0.2.5) is correct.

A generalized Dobrushin uniqueness theorem

The beauty and the strength of the high temperature Dobrushin uniqueness theorem, see Sect. 3.2, is that the uniqueness condition is a simple criterion which involves only the Vaserstein distance between one site conditional probabilities. Such a condition is obviously not satisfied by the lattice gas approximation of the LMP model, which in fact has a phase transition. However if we do the computations restricting the boundary conditions of the one site conditional probability to configurations which on average have a density close to mean field equilibrium, we find indeed that the Vaserstein distance between conditional probabilities is as small as required by the Dobrushin criterion. This seems to settle the whole matter, because we are working in the restricted ensemble where configurations are by definition close to the mean field case. There is a subtle point though, which causes a lot of trouble, namely if we first take, as we should, the Gibbs measure on the restricted ensemble and then compute its one site conditional probability, we do not always find the same result as when we first compute the one site conditional Gibbs measure and then restrict the boundary conditions (to be in the restricted ensemble). The two differ only for very special configurations (those which are "at the boundaries of the constraint" which defines the restricted ensemble). For such very special configurations the Dobrushin condition is not satisfied. On the other hand, the Dobrushin theorem requires uniformity on the conditioning and we cannot use it as it is. We shall solve the problem by showing first that bad configurations have small probability and by next extending the Dobrushin theorem to such a setup. This is the content of Sect. 11.5, where we develop the theory in a more general context for possible applications to other systems with long range interactions, in particular the Ising model with Kac potentials of Chap. 9 in restricted ensembles.

Conclusions

This is the last, concluding step in which we prove that the conditions for uniqueness established in the previous Dobrushin theory are actually verified by our interpolating hamiltonian. We shall then check that Theorems 11.4.5.1 and 11.4.5.2 hold, thus concluding the proof of (11.0.2.5).

Notational remarks

Since the analysis of the plus and of the minus case are similar, we shall mainly consider the former and to simplify notation we rename in this section

$$\Lambda := \text{int}^-(\Gamma), \qquad \chi^+_{\Lambda^c} = \rho_{\beta,+}\mathbf{1}_{\Lambda^c}, \tag{11.4.0.1}$$

$\chi^+_{\Lambda^c}$ acting as boundary condition outside Λ. Recall also that the chemical potential is fixed at the value $\lambda_{\beta,\gamma}$ for which the pressures in the plus and the minus restricted ensembles coincide, $P^{\pm}_{\gamma,\lambda_{\beta,\gamma}} =: P_{\gamma,\lambda_{\beta,\gamma}}$.

11.4.1 Lattice gas

We start with some notation and definitions. The phase space of the lattice gas is $\{0,1\}^{a\mathbb{Z}^d}$, $a\mathbb{Z}^d$ the lattice, and we suppose $a \in \{2^{-n}, n \in \mathbb{N}\}$. As in the Ising case we denote the configurations by $\sigma = \sigma(x), x \in a\mathbb{Z}^d$. Here, however, $\sigma(x)$ has the values 0 and 1. $\{0,1\}^{a\mathbb{Z}^d}$ is naturally embedded in $\mathcal{X}$ by associating to each $\sigma \in \{0,1\}^{a\mathbb{Z}^d}$ the particle configuration $q \in \mathcal{X}$ which has one particle at each $x \in a\mathbb{Z}^d$ where (and only there) $\sigma(x) = 1$. By abuse of notation we shall often denote by the same σ its corresponding element in $\mathcal{X}$, in particular when writing $J_\gamma * \sigma$. By default in the sequel all regions are $\mathcal{D}^{(a)}$-measurable and eventually $\mathcal{D}^{(\ell_+,\gamma)}$-measurable. If Λ is such a region, σ_Λ denotes a configuration in $\Lambda \cap a\mathbb{Z}^d$, and we shall write

$$\mathcal{X}^+_{\Lambda,a} := \{\sigma_\Lambda : \eta(\sigma_\Lambda; r) = 1, r \in \Lambda\}, \tag{11.4.1.1}$$

where

$$\eta(\sigma_\Lambda; r) = 1 : \frac{1}{\ell^d_{-,\gamma}} \sum_{x \in C_r^{(\ell_-,\gamma)} \cap a\mathbb{Z}^d} \sigma_\Lambda(x) \in [\rho_{\beta,+} - \zeta, \rho_{\beta,+} - \zeta]. \tag{11.4.1.2}$$

Setting

$$J_\gamma * \sigma(r) = \sum_{x \in a\mathbb{Z}^d} J_\gamma(r,x)\sigma(x), \tag{11.4.1.3}$$

we define the hamiltonian by

$$H_{\gamma,\Lambda}(\sigma_\Lambda|\sigma^c_\Lambda) = \int e_{\lambda_{\beta,\gamma}}(J_\gamma * \sigma_\Lambda + J_\gamma * \sigma_{\Lambda^c}) - e_{\lambda_{\beta,\gamma}}(J_\gamma * \sigma_{\Lambda^c}), \tag{11.4.1.4}$$

with an analogous formula if σ_{Λ^c} is replaced by a continuous density. Notice that in the hamiltonian we have dropped the chemical potential from the suffix, as it will be constantly held equal to $\lambda_{\beta,\gamma}$, but we keep it in $e_{\lambda_{\beta,\gamma}}$.

Unlike the Ising case, here the free measure is not trivial as it gives weight 1 to the state $\sigma(x) = 0$ and weight a^d to the state $\sigma(x) = 1$. Writing $|\sigma_\Lambda| = \sum_{x\in\Lambda} \sigma_\Lambda(x)$, the lattice gas partition function is then

$$Z^+_{\gamma,a,\Lambda,\bar{q}} = \sum_{\sigma_\Lambda \in \mathcal{X}^+_{\Lambda,a}} (a^d)^{|\sigma_\Lambda|} e^{-\beta H_{\gamma,\Lambda}(\sigma_\Lambda|\bar{q}_{\Lambda^c})} X^+_{\gamma,\sigma_\Lambda}(\Lambda). \tag{11.4.1.5}$$

$X^+_{\gamma,\sigma_\Lambda}(\Lambda)$ is as in (11.1.1.6) (with σ_Λ regarded as an element of $\mathcal{X}_\Lambda$). The boundary conditions $\bar{q}_{\Lambda^c}$ will eventually be specified as the perfect boundary conditions $\chi^+_{\Lambda^c}$.

We shall denote by $\mu^+_{\gamma,a,\Lambda,\bar q}$ the Gibbs probability on $\mathcal{X}^+_{\Lambda,a} \times \mathcal{B}^+_\Lambda$ defined by

$$\mu^+_{\gamma,a,\Lambda,\bar q}(\sigma_\Lambda, \underline{\Gamma}) := \frac{(a^d)^{|\sigma_\Lambda|} e^{-\beta H_{\gamma,\Lambda}(\sigma_\Lambda|\bar q_{\Lambda^c})} \hat W^+_\gamma(\underline{\Gamma}, \sigma_\Lambda)}{Z^+_{\gamma,a,\Lambda,\bar q}}. \tag{11.4.1.6}$$

11.4.2 Lattice gas approximations

Given a bounded measurable function f on $\mathcal{X}_\Lambda \times \mathcal{B}^+_\Lambda$ let $f^{(a)}$ be the function on the same space $\mathcal{X}_\Lambda \times \mathcal{B}^+_\Lambda$ defined by

$$f^{(a)}(r_1, \dots, r_n; \underline{\Gamma}) := \oint_{C^{(a)}_{r_1}} \cdots \oint_{C^{(a)}_{r_n}} f(r'_1, \dots, r'_n; \underline{\Gamma}), \tag{11.4.2.1}$$

where $(r_1, \dots, r_n)$ is the generic element of $\mathcal{X}_\Lambda$ and $C^{(a)}_{r_i}$ the element of $\mathcal{D}^{(a)}$ which contains r_i. By restricting $q = (r_1, \dots, r_n)$, $f^{(a)}$ becomes a function on $\mathcal{X}^+_{\Lambda,a} \times \mathcal{B}^+_\Lambda$ denoted by the same symbol. Calling the Gibbs measure on $\mathcal{X}^+_\Lambda \times \mathcal{B}^+_\Lambda$ $d\mu^+_{\gamma,\Lambda,\bar q}(q_\Lambda, \underline{\Gamma})$, we have:

Theorem 11.4.2.1 *With the above notation and assumptions, for any $\bar q \in \mathcal{X}^+$, $\lim_{a\to 0} Z^+_{\gamma,a,\Lambda,\bar q} = \hat Z^+_{\gamma,\Lambda,\bar q}$ while $\lim_{a\to 0} \sum_{\underline{\Gamma}\in\mathcal{B}^+_\Lambda} \sum_{\sigma_\Lambda \in \mathcal{X}^+_{\Lambda,a}} f^{(a)}(\sigma_\Lambda, \underline{\Gamma}) \times \mu^+_{\gamma,a,\Lambda,\bar q}(\sigma_\Lambda, \underline{\Gamma})$ is equal to $\sum_{\underline{\Gamma}\in\mathcal{B}^+_\Lambda} \int_{\mathcal{X}^+_\Lambda} f(q_\Lambda, \underline{\Gamma}) d\mu^+_{\gamma,\Lambda,\bar q}(q_\Lambda, \underline{\Gamma})$.*

A stronger statement actually holds: there are constants c and c' so that calling $c_\gamma = e^{c\ell^d_{-,\gamma}}$, uniformly in Λ

$$e^{-(c'a + c_\gamma a^d)|\Lambda|} \le \frac{Z^+_{\gamma,a,\Lambda,\chi^+_{\Lambda^c}}}{\hat Z^+_{\gamma,\chi^+_{\Lambda^c}}(\Lambda)} \le e^{(c'a + c_\gamma a^d)|\Lambda|}. \tag{11.4.2.2}$$

The proof of (11.4.2.2) is not difficult but lengthy so that we shall modify some of the proofs in this section in order to avoid its use: (11.4.2.2) is thus stated without proof. We postpone the proof of Theorem 11.4.2.1 to after some preliminary lemmas.

Given a mesh a and a particle configuration q we denote by $q^{(a)}$ another particle configuration whose particles are only at the sites $a\mathbb{Z}^d$, the number of particles of $q^{(a)}$ at $x \in a\mathbb{Z}^d$ being $|q \cap C^{(a)}_x|$; $q^{(a)}$ does not have to be a lattice gas configuration because of multiple occupancy.

Lemma 11.4.2.2 *There is $c > 0$ so that for all plus contours Γ and all $\bar q \in \mathcal{X}^+$,*

$$e^{-c\gamma a|\mathrm{sp}(\Gamma)|} \le \frac{\hat W^+_\gamma(\underline{\Gamma}, \bar q^{(a)})}{\hat W^+_\gamma(\underline{\Gamma}, \bar q)} \le e^{c\gamma a|\mathrm{sp}(\Gamma)|}, \tag{11.4.2.3}$$

the chemical potential $\lambda = \lambda_{\beta,\gamma}$ *being dropped from the notation.*

Proof We have

$$\left| \log \frac{\hat{W}^+_\gamma(\Gamma, \bar{q}^{(a)})}{\hat{W}^+_\gamma(\Gamma, \bar{q})} \right| \le \left| \log \left\{ \frac{\hat{N}^+_\gamma(\Gamma, \bar{q}^{(a)})}{\hat{D}^+_\gamma(\Gamma, \bar{q}^{(a)})} \frac{\hat{D}^+_\gamma(\Gamma, \bar{q})}{\hat{N}^+_\gamma(\Gamma, \bar{q})} \right\} \right|;$$

$\log\{\frac{\hat{N}^+_\gamma(\Gamma,\bar{q}^{(a)})}{\hat{N}^+_\gamma(\Gamma,\bar{q})}\}$ is bounded by

$$\beta \sup_{q_{c(\Gamma)}:\eta(q_{c(\Gamma)};r)=1, r\in\delta^{\ell_{+,\gamma}}_{\text{in}}[c(\Gamma)]} |H_{\gamma,c(\Gamma)}(q_{c(\Gamma)}|\bar{q}_{c(\Gamma)^c}) - H_{\gamma,c(\Gamma)}(q_{c(\Gamma)}|\bar{q}^{(a)}_{c(\Gamma)^c})|.$$

By (11.1.2.7) the above energy difference is bounded by $c'\gamma a|\operatorname{sp}(\Gamma)|$. The same argument applied to the ratio of $\hat{D}^+_\gamma(\Gamma, q^{(a)})$ and $\hat{D}^+_\gamma(\Gamma, q)$ concludes the proof. □

Lemma 11.4.2.3 *There is* $c' > 0$ *so that for any* $\underline{\Gamma} \in \mathcal{B}^+_\Lambda$*, any* $q_\Lambda \in \mathcal{X}^+_\Lambda$ *and* $\bar{q} \in \mathcal{X}^+$*,*

$$e^{-c'\gamma a|\Lambda|} \le \frac{\hat{W}^+_\gamma(\underline{\Gamma}, q^{(a)}_\Lambda)\, e^{-\beta H_{\gamma,\Lambda}(q^{(a)}_\Lambda|\bar{q}_{\Lambda^c})}}{\hat{W}^+_\gamma(\underline{\Gamma}, q_\Lambda)\, e^{-\beta H_{\gamma,\Lambda}(q_\Lambda|\bar{q}_{\Lambda^c})}} \le e^{c'\gamma a|\Lambda|}. \tag{11.4.2.4}$$

Proof As in the proof of Lemma 11.4.2.2, there is a constant $c > 0$ so that

$$\left|H_{\gamma,\Lambda}(q_\Lambda|\bar{q}_{\Lambda^c}) - H_{\gamma,\Lambda}(q^{(a)}_\Lambda|\bar{q}_{\Lambda^c})\right| \le c|q_\Lambda|\gamma a$$

where $|q_\Lambda|$, the number of particles in q_Λ, is bounded by $(\rho_{\beta,+} + \zeta)|\Lambda|$. Then (11.4.2.4) follows from Lemma 11.4.2.2, because $\bigcup_{\Gamma\in\underline{\Gamma}} \operatorname{sp}(\Gamma) \subset \Lambda$. □

Proof of Theorem 11.4.2.1 Write $\sum_{\underline{\Gamma}\in\mathcal{B}^+_\Lambda} \int_{\mathcal{X}^+_\Lambda} f(q_\Lambda, \underline{\Gamma}) d\mu^+_{\gamma,\Lambda,\bar{q}}(q_\Lambda, \underline{\Gamma})$ as

$$\frac{1}{\hat{Z}^+_{\gamma,\bar{q}_{\Lambda^c}}(\Lambda)} \sum_{\underline{\Gamma}\in\mathcal{B}^+_\Lambda} \int \cdots \int \mathbf{1}_{\eta(q_\Lambda;\cdot)\equiv 1} f(q_\Lambda, \underline{\Gamma})$$

$$\times\, e^{-\beta H_{\gamma,\Lambda}(q_\Lambda|\bar{q}_{\Lambda^c})} \hat{W}^+_\gamma(\underline{\Gamma}, q_\Lambda) \prod_{i\in a\mathbb{Z}^d\cap\Lambda} d\nu_{C^{(a)}_i}(q_{C^{(a)}_i}), \tag{11.4.2.5}$$

where q_Λ is the collection of all $q_{C^{(a)}_i}$. Let A' be the set of configurations such that $|q_{C^{(a)}_i}| \le 1$ for all $i \in a\mathbb{Z}^d \cap \Lambda$, and let A'' be its complement. For γ small enough the contribution of A'' is bounded after using (11.1.1.12) and (11.1.2.8) by

$$\frac{\|f\|_\infty}{\hat{Z}^+_{\gamma,\bar{q}_{\Lambda^c}}(\Lambda)} 2^{|\Lambda|/\ell^d_{+,\gamma}} \frac{|\Lambda|}{a^d} e^{(|\Lambda|/a^d - 1)a^d e^{\beta b_0}} \sum_{n\ge 2} \frac{(a^d e^{\beta b_0})^n}{n!} \le \|f\|_\infty a^d |\Lambda| e^{c|\Lambda|}.$$

Indeed, the factor $|\Lambda|/a^d$ counts the number of choices for a cube $C_i^{(a)}$ with ≥ 2 particles and the last sum its contribution. The contribution of the other cubes where the sum is not constrained (we have dropped from the beginning the $\eta = 1$ constraint) is bounded by $e^{(|\Lambda|/a^d-1)a^d e^{\beta b_0}}$. We have also bounded $\hat{Z}^+_{\gamma,\bar{q}_{\Lambda^c}}(\Lambda) \geq e^{-c'|\Lambda|}$, which follows after using (11.1.2.6) by restricting to configurations for which in each cube $C_i^{(\ell_-,\gamma)}$ in Λ there are an integer part of $\rho_{\beta,+}\ell^d_{-,\gamma}$ particles, and then using the Stirling formula. In conclusion therefore, the contribution of A'' vanishes in the limit $a \to 0$ (recall that Λ is fixed). In the integral over A' we use Lemma 11.4.2.3 to replace $\hat{W}^+_\gamma(\underline{\Gamma}, q_\Lambda)\, e^{-\beta H_{\gamma,\Lambda}(q_\Lambda|\bar{q}^+_{\Lambda^c})}$ by $\hat{W}^+_\gamma(\underline{\Gamma}, q^{(a)}_\Lambda)\, e^{-\beta H_{\gamma,\Lambda}(q^{(a)}_\Lambda|\bar{q}_{\Lambda^c})}$. By an argument similar to the previous one, the error is bounded by $[e^{c'\gamma a|\Lambda|} - 1]\|f\|_\infty e^{c|\lambda|}$, which also vanishes as $a \to 0$. Thus modulo terms which vanish as $a \to 0$, (11.4.2.5) becomes

$$\frac{Z^+_{\gamma,a,\Lambda,\bar{q}}}{\hat{Z}^+_{\gamma,\bar{q}}(\Lambda)} \sum_{\underline{\Gamma}\in\mathcal{B}^+_\Lambda} \sum_{\sigma_\Lambda\in\mathcal{X}^+_{\Lambda,a}} f^{(a)}(\sigma_\Lambda, \underline{\Gamma})\mu^+_{\gamma,a,\Lambda,\bar{q}}(\sigma_\Lambda, \underline{\Gamma}).$$

The same argument used so far with $f \equiv 1$ shows that the first ratio converges to 1 as $a \to 0$, thus concluding the proof of Theorem 11.4.2.1. □

11.4.3 Cluster expansion, effective hamiltonian

Since $X^+_{\gamma,\sigma_\Lambda}(\Lambda) = \sum_{\underline{\Gamma}\in\mathcal{B}^+_\Lambda} W^+_\gamma(\underline{\Gamma}, \sigma_\Lambda)$ is positive it can be written as

$$X^+_{\gamma,\sigma_\Lambda}(\Lambda) = e^{-\beta K^+_{\gamma,\Lambda}(\sigma_\Lambda)}, \tag{11.4.3.1}$$

so that

$$Z^+_{\gamma,a,\Lambda,\sigma^+_{\Lambda^c}} = \sum_{\sigma_\Lambda\in\mathcal{X}^+_{\Lambda,a}} e^{-\beta[H_{\gamma,\Lambda}(\sigma_\Lambda|\sigma^+_{\Lambda^c})+K^+_{\gamma,\Lambda}(\sigma_\Lambda)]}. \tag{11.4.3.2}$$

By an inductive procedure on the volume we can always write a hamiltonian in terms of potentials. In our case

$$K^+_{\gamma,\Lambda}(\sigma_\Lambda) = \sum_{\Delta\subseteq\Lambda} U^+_{\gamma,\Delta}(\sigma), \tag{11.4.3.3}$$

where the sum is over bounded $\mathcal{D}^{(\ell_+,\gamma)}$-measurable regions Δ and $U^+_{\gamma,\Delta}(\sigma) \equiv U^+_{\gamma,\Delta}(\sigma_\Delta)$. The energy in Λ plus the interaction with Λ', $\Lambda' \cap \Lambda = \emptyset$, is then

$$K^+_{\gamma,\Lambda,\Lambda'}(\sigma_\Lambda|\sigma_{\Lambda'}) = \sum_{\Delta\subseteq(\Lambda\cup\Lambda'):\Delta\cap\Lambda\neq\emptyset} U^+_{\gamma,\Delta}(\sigma_{\Lambda\cup\Lambda'}). \tag{11.4.3.4}$$

When $\Lambda' = \Lambda^c$, it is omitted from the suffix and we write

$$K^+_{\gamma,\Lambda}(\sigma_\Lambda|\sigma_{\Lambda^c}) = K^+_{\gamma,\Lambda,\Lambda^c}(\sigma_\Lambda|\sigma_{\Lambda^c}). \tag{11.4.3.5}$$

So far it was only a matter of notation, but the representation (11.4.3.2) becomes effective once we prove that K^+_γ is a "nice" hamiltonian; namely: the many body potentials $U^+_{\gamma,\Delta}$ are suitably small as $|\Delta|$ increases. The bounds are given in Theorem 11.4.3.1 below, where we use the following notation:

- $\{D\}$ is the family of all $\mathcal{D}^{(\ell_{+,\gamma})}$-measurable sets which are bounded, connected and such that $N_D \ge 3$, N_D being the number of $\mathcal{D}^{(\ell_{+,\gamma})}$ cubes in D.
- The $\{D\}$-distance of r from a $\mathcal{D}^{(\ell_{+,\gamma})}$ measurable set A is (calling C_r the element of $\mathcal{D}^{(\ell_{+,\gamma})}$ which contains r)

$$d_{\{D\}}(r,A) := \min\big\{N_D | D \in \{D\},\ D \cap \big(C_r \cup \delta_{\text{out}}^{\ell_{+,\gamma}}[C_r]\big) \ne \emptyset,$$
$$D \cap \big(A \cup \delta_{\text{out}}^{\ell_{+,\gamma}}[A]\big) \ne \emptyset\big\}. \tag{11.4.3.6}$$

We shall often use in the sequel the following lower bound: there is a constant $\kappa > 0$ so that

$$d_{\{D\}}(r,A) \ge \max\left\{3^d, \kappa \frac{\text{dist}(r,A)}{\ell_{+,\gamma}}\right\}. \tag{11.4.3.7}$$

Theorem 11.4.3.1 *$U^+_{\gamma,\Delta} \equiv 0$ unless $\Delta \in \{D\}$ and there is $c_{\text{pol}} > 0$ so that the following holds. For any $b > 0$ and all γ correspondingly small,*

$$\beta \sum_{\Delta \ni r} e^{bN_\Delta} \|U^+_{\gamma,\Delta}\|_\infty \le e^{-\beta c_{\text{pol}}(\zeta^2 \ell^d_{-,\gamma})}, \quad r \in \Lambda. \tag{11.4.3.8}$$

Moreover, for any $r \in \Lambda$ and any set A,

$$\beta \sum_{\Delta \ni r, \Delta\cap A \ne \emptyset} |U^+_{\gamma,\Delta}(\sigma_\Delta)| \le 3^d e^{-\beta c_{\text{pol}}(\zeta^2 \ell^d_{-,\gamma}) d_{\{D\}}(r,A)}. \tag{11.4.3.9}$$

The proof of Theorem 11.4.3.1 is reported in Sect. 11.9 as a corollary of well known theorems on a cluster expansion quoted from the literature without proofs. Even though we shall refer to the LMP model, the validity is much more general: the only relevant quantities are the weights $W^+_\gamma(\Gamma,\sigma)$ regarded as functions of Γ, the variables σ (which are model dependent) enter only as parameters and the space that they belong to is immaterial, provided the weights are small enough. In particular, what follows applies to the Ising model of Chap. 9. Indeed the analogue of Theorem 10.5.1.1 holds for the Ising case, and the diluted partition function $Z^+_{\gamma,\Lambda,\sigma^+_{\Lambda^c}}$ defined in (9.2.6.3) can be written as

$$Z^+_{\gamma,\Lambda,\sigma^+_{\Lambda^c}} = \sum_{\underline{\Gamma}\in\mathcal{B}^+_\Lambda} \sum_{\sigma_\Lambda \in \mathcal{X}^+_\Lambda} W^+_\gamma(\underline{\Gamma},\sigma_\Lambda)\, e^{-\beta H_{\gamma,\Lambda}(\sigma_\Lambda|\sigma^+_{\Lambda^c})},$$

thus identifying $X^+_{\gamma,\sigma_\Lambda}(\Lambda) = \sum_{\underline{\Gamma}\in\mathcal{B}^+_\Lambda} W^+_\gamma(\underline{\Gamma},\sigma_\Lambda)$.

11.4.4 An interpolation formula

After the lattice gas approximation and the introduction of the effective hamiltonian $K^+_{\gamma,\Lambda}(\sigma_\Lambda)$, we are left with the analysis of the partition function

$$Z^+_{\gamma,a,\Lambda;1} := \sum_{\sigma_\Lambda\in\mathcal{X}^+_{\Lambda,a}} (a^d)^{|\sigma_\Lambda|} e^{-\beta h^+_{\gamma,a,\Lambda;1}(\sigma_\Lambda|\chi^+_{\Lambda^c})}, \tag{11.4.4.1}$$

where $h^+_{\gamma,a,\Lambda;1}(\sigma_\Lambda|\chi^+_{\Lambda^c}) = H_{\gamma,\Lambda}(\sigma_\Lambda|\chi^+_{\Lambda^c}) + K^+_{\gamma,\Lambda}(\sigma_\Lambda)$, Λ being a bounded $\mathcal{D}^{(\ell_{+,\gamma})}$-measurable region, eventually set equal to $\text{int}^-(\Gamma)$. The rules of the game are "*all estimates must be uniform as* $a \to 0$." We interpolate between (11.4.4.1) and the reference partition function defined by the same expression (11.4.4.1), but with $h^+_{\gamma,a,\Lambda;1}(\sigma_\Lambda|\chi^+_{\Lambda^c})$ replaced by

$$h^+_{\gamma,a,\Lambda;0}(\sigma_\Lambda) = \sum_{x\in\Lambda} e'_{\lambda(\beta)}(\rho_{\beta,+})(\sigma_\Lambda - a^d\rho_{\beta,+}). \tag{11.4.4.2}$$

Namely, we write

$$\log\frac{Z^+_{\gamma,a,\Lambda;1}}{Z^+_{\gamma,a,\Lambda;0}} = -\beta\int_0^1 \Big\langle [H_{\gamma,\Lambda}(\sigma_\Lambda|\chi^+_{\Lambda^c}) + K^+_{\gamma,\Lambda}(\sigma_\Lambda) - h^+_{\gamma,a,\Lambda;0}(\sigma_\Lambda)]\Big\rangle_{\gamma,a,\Lambda;u} du, \tag{11.4.4.3}$$

where $\langle\cdot\rangle_{\gamma,a,\Lambda;u}$ denotes expectation with respect to the Gibbs measure on $\mathcal{X}^+_{\Lambda,a}$ and

$$m^+_{\gamma,a,\Lambda;u}(\sigma_\Lambda) = \frac{e^{-\beta h^+_{\gamma,a,\Lambda,;u}(\sigma_\Lambda|\chi^+_{\Lambda^c})}}{Z^+_{\gamma,a,\Lambda,\chi^+;u}}. \tag{11.4.4.4}$$

$h^+_{\gamma,a,\Lambda;u}(\sigma_\Lambda|\chi^+_{\Lambda^c})$, $u\in[0,1]$, is the interpolating hamiltonian

$$h^+_{\gamma,a,\Lambda;u}(\sigma_\Lambda|\chi^+_{\Lambda^c}) = u\{H_{\gamma,\Lambda}(\sigma_\Lambda|\chi^+_{\Lambda^c}) + K^+_{\gamma,\Lambda}(\sigma_\Lambda)\} + (1-u)h^\pm_{\gamma,a;0}(\sigma_\Lambda), \tag{11.4.4.5}$$

and we have $Z^+_{\gamma,a,\Lambda;u} = \sum_{\sigma_\Lambda\in\mathcal{X}^+_{\Lambda,a}} (a^d)^{|\sigma_\Lambda|} e^{-\beta h^+_{\gamma,a,\Lambda;u}(\sigma_\Lambda|\chi^+_{\Lambda^c})}$.

By translation invariance under multiples of $\ell_{+,\gamma}$, for any bounded $\mathcal{D}^{(\ell_{+,\gamma})}$-measurable region Λ and denoting by C a cube in $\mathcal{D}^{(\ell_{+,\gamma})}$,

$$\log Z^+_{\gamma,a,\Lambda;0} = \frac{|\Lambda|}{|C|}\log Z^+_{\gamma,a,C;0} =: |\Lambda|p^+_{\gamma,a;0}, \tag{11.4.4.6}$$

$p^+_{\gamma,a;0}$ being the pressure of the reference system. (11.4.4.3) then becomes

$$\frac{1}{\beta}\log Z^+_{\gamma,a,\Lambda} = p^+_{\gamma,a;0}|\Lambda| - \int_0^1 \Big\langle [H_{\gamma,\Lambda}(\sigma_\Lambda|\chi^+_{\Lambda^c}) + K^+_{\gamma,\Lambda}(\sigma_\Lambda) - h^+_{\gamma,a,\Lambda;0}(\sigma_\Lambda)]\Big\rangle_{\gamma,a,\Lambda;u}. \tag{11.4.4.7}$$

The choice of the reference hamiltonian was unimportant so far, but it becomes essential in what follows. The choice (11.4.4.2) has two advantages: it is simple, entailing just one body interactions; the corresponding average particle density is the mean field equilibrium density $\rho_{\beta,+}$, to leading orders as $\gamma \to 0$ and $a \to 0$. We give a heuristic proof of the statement by considering $\langle\sigma(x)\rangle^0_{\gamma,a,\Lambda;0}$ the mean local density for the free system without the constraint to be in $\mathcal{X}^+_{\Lambda,a}$:

$$\langle\sigma(x)\rangle^0_{\gamma,a,\Lambda;0} = \frac{1}{Z_{\gamma,a,\Lambda;0}} \sum_{\sigma_\Lambda\in\mathcal{X}_{\Lambda,a}} (a^d)^{|\sigma_\Lambda|}\sigma_\Lambda(x)e^{-\beta e'_{\lambda(\beta)}(\rho_{\beta,+})[|\sigma_\Lambda|-\rho_{\beta,+}|\Lambda|]},$$

which is equal to $\frac{a^d e^{-\beta e'_{\lambda(\beta)}(\rho_{\beta,+})}}{1+a^d e^{-\beta e'_{\lambda(\beta)}(\rho_{\beta,+})}}$ and $\approx a^d e^{-\beta e'_{\lambda(\beta)}(\rho_{\beta,+})} = a^d\rho_{\beta,+}$, having neglected terms with a^{2d}. By the law of large numbers, deviations from $\rho_{\beta,+}$ are small for small γ, which indicates that the constraint of being in $\mathcal{X}^+_{\Lambda,a}$ in the true reference Gibbs measure must not be important and thus $\langle\sigma(x)\rangle^0_{\gamma,a,\Lambda;0}$ must also be close to $\rho_{\beta,+}$ as claimed.

11.4.5 The main estimates

The proof of (11.0.2.5) is a consequence of the two theorems below, as we shall show in Sect. 11.4.6. By default, Λ and Δ in the sequel denote bounded $\mathcal{D}^{(\ell_{+,\gamma})}$-measurable sets, C, C_i and C_j being elements of $\mathcal{D}^{(\ell_{-,\gamma})}$; $J_\gamma * \sigma$ is the convolution defined in (11.4.1.3) and

$$\Lambda_0 = \Lambda \setminus \delta^{\ell_{+,\gamma}}_{\text{in}}[\Lambda], \qquad \Lambda_{00} = \Lambda_0 \setminus \delta^{\ell_{+,\gamma}}_{\text{in}}[\Lambda_0]. \tag{11.4.5.1}$$

Theorem 11.4.5.1 (Exponential decay) *There are c and $\omega > 0$ so that for all $u \in [0,1]$, for all γ and $a > 0$ small enough, there is a probability measure $m^+_{\gamma,a;u}$ on $\mathcal{X}^+_{\Lambda,a}$ (whose expectation is denoted by $\langle\cdot\rangle_{\gamma,a;u}$) which is invariant under translations in $\ell_{+,\gamma}\mathbb{Z}^d$, and it is such that for any bounded $\mathcal{D}^{(\ell_{+,\gamma})}$-measurable set Λ and any $r \in \Lambda_{00}$:*

$$\Big|\langle e_{\lambda_{\beta,\gamma}}(J_\gamma * \sigma_\Lambda(r))\rangle_{\gamma,a,\Lambda;u} - \langle e_{\lambda_{\beta,\gamma}}(J_\gamma * \sigma_\Lambda(r))\rangle_{\gamma,a;u}\Big| \le c\gamma^{-d} \sum_{C\subset\delta^{2\gamma^{-1}}_{\text{in}}[\Lambda]} e^{-\omega\gamma\,\text{dist}(r,C)}. \tag{11.4.5.2}$$

For any $\Delta \subset \Lambda$,

$$\left|\langle U^+_{\gamma,\Delta}(\sigma_\Delta)\rangle_{\gamma,a,\Lambda;u} - \langle U^+_{\gamma,\Delta}(\sigma_\Delta)\rangle_{\gamma,a;u}\right| \le c|\Delta|\|U^+_{\gamma,\Delta}(\cdot)\|_\infty \sum_{C\subset\delta_{\text{in}}^{2\gamma^{-1}}[\Lambda]} e^{-\omega\gamma\,\text{dist}(C,\Delta)}, \tag{11.4.5.3}$$

and, finally, for any $C \subset \Lambda_{00}$,

$$\left|\left\langle \sum_{y\in C\cap a\mathbb{Z}^d} \sigma_\Lambda(y)\right\rangle_{\gamma,a,\Lambda;u} - \left\langle \sum_{y\in C\cap a\mathbb{Z}^d} \sigma_\Lambda(y)\right\rangle_{\gamma,a;u}\right| \le c\ell^d_{-,\gamma} \sum_{C'\subset\delta_{\text{in}}^{2\gamma^{-1}}[\Lambda]} e^{-\omega\gamma\,\text{dist}(C',C)}. \tag{11.4.5.4}$$

Theorem 11.4.5.1 as well as Theorem 11.4.5.2 below will be proved in a more general form in Sect. 11.6. The bounds (11.4.5.2), (11.4.5.3) and (11.4.5.4) show exponential decay from the boundaries which in (11.4.5.3) yields uniform smallness because the potentials $U^+_{\gamma,\Delta}$ are small. In the other cases smallness is obtained by restricting the case to Λ_{00}, and hence staying far from Λ^c. Near the boundaries we shall instead use closeness to the mean field, which indeed holds in the whole Λ:

Theorem 11.4.5.2 (Closeness to mean field) *There is* $c > 0$ *so that for all* γ *small enough, all* $u \in [0, 1]$, *all* Λ *as in Theorem* 11.4.5.1 *and all* $r \in \Lambda$

$$\left\langle |J_\gamma * [\sigma_\Lambda - \chi^+_\Lambda](r)|\right\rangle_{\gamma,a,\Lambda;u} \le c\gamma^{1/2}. \tag{11.4.5.5}$$

Moreover, for any cube $C \in \mathcal{D}^{(\ell_{-,\gamma})}$ *contained in* Λ,

$$\left|\left\langle \frac{1}{|C|} \sum_{y\in C\cap a\mathbb{Z}^d} \sigma_\Lambda(y)\right\rangle_{\gamma,a,\Lambda;u} - \rho_{\beta,+}\right| \le c\gamma^{1/2}. \tag{11.4.5.6}$$

The inequalities (11.4.5.5) *and* (11.4.5.6) *hold as well for the* $\langle\cdot\rangle_{\gamma,a;u}$ *expectation.*

11.4.6 Corollaries

A consequence of Theorems 11.4.5.1 and 11.4.5.2 we shall prove here:

Corollary 11.4.6.1 *There is* c *so that for all* γ *small enough the following holds: for any bounded* $\mathcal{D}^{(\ell_{+,\gamma})}$*-measurable region* Λ *and all* a *small enough*

$$\left|\frac{1}{\beta}\log Z^+_{\gamma,a,\Lambda} - P^+_{\gamma,\lambda_{\beta,\gamma}}|\Lambda| - I^+_{\gamma,\lambda(\beta)}(\Lambda)\right| \le c|\delta_{\text{out}}^{\ell_{+,\gamma}}[\Lambda]|\gamma^{1/2}. \tag{11.4.6.1}$$

By applying Corollary 11.4.6.1 and Theorem 11.4.2.1 with $\Lambda = \text{int}^-(\Gamma)$ we get (11.0.2.5), which is thus proved, pending the validity of "the main estimates" of Sect. 11.4.5. Their proofs are reported in Sect. 11.6 as an application of the Dobrushin theory of the next section.

In the sequel of the subsection we shall prove Corollary 11.4.6.1. Let C be the $\mathcal{D}^{(\ell_{+,\gamma})}$ cube which contains the origin; define

$$p^+_{\gamma,a} := p^+_{\gamma,a;0} - \Big\langle \frac{1}{|C|}\sum_{y\in C} e_{\lambda_{\beta,\gamma}}(J_\gamma * \sigma(y)) + \sum_{\Delta\ni 0} \frac{U^+_{\gamma,\Delta}(\sigma_\Delta)}{|\Delta|}$$
$$- e'_{\lambda(\beta)}(\rho_{\beta,+})\Big\{\frac{1}{|C|}\sum_{y\in C}\sigma_\Lambda(y) - \rho_{\beta,+}\Big\}\Big\rangle_{\gamma,a;u}.$$

Thus $p^+_{\gamma,a;0} = p^+_{\gamma,a} + \langle\cdot\rangle_{a;u}$, which, inserted in (11.4.4.7), yields (since the limit measure $m^+_{\gamma,a;u}$ is invariant under translations in $\ell_{+,\gamma}\mathbb{Z}^d$)

$$\frac{1}{\beta}\log Z^+_{\gamma,a,\Lambda} = p^+_{\gamma,a}|\Lambda| - \mathcal{I} - R_U - R_\sigma, \tag{11.4.6.2}$$

where (the integrals below being over $u \in [0, 1]$)

$$\mathcal{I} = \int_0^1 \Big\{\big\langle H_{\gamma,\Lambda}(\sigma_\Lambda|\chi^+_{\Lambda^c})\big\rangle_{\gamma,a,\Lambda;u}$$
$$- \frac{|\Lambda|}{|C|}\int_C \big\langle e_{\lambda_{\beta,\gamma}}(J_\gamma * \sigma(r))\big\rangle_{\gamma,a;u} dr\Big\}, \tag{11.4.6.3}$$

$$R_U = \int_0^1 \Big\{\big\langle K^+_{\gamma,\Lambda}(\sigma_\Lambda)\big\rangle_{\gamma,a,\Lambda;u} - |\Lambda|\Big\langle\sum_{\Delta\ni 0}\frac{U^+_{\gamma,\Delta}(\sigma_\Delta)}{|\Delta|}\Big\rangle_{\gamma,a;u}\Big\}, \tag{11.4.6.4}$$

$$R_\sigma = \int_0^1 e'_{\lambda(\beta)}(\rho_{\beta,+})\sum_{x\in\Lambda}\{\langle\sigma_\Lambda(x)\rangle_{a,\Lambda;u} - \langle\sigma_\Lambda(x)\rangle_{\gamma,a;u}\} \tag{11.4.6.5}$$

(having used again invariance of the limit measure under translations by $\ell_{+,\gamma}$).

Estimates on $\mathcal{I}$

We shall prove that there is $c > 0$ so that

$$|\mathcal{I} - I^+_{\gamma,\lambda(\beta)}(\Lambda)| \le c\gamma^{1/2}|\delta^{\ell_{+,\gamma}}_{\text{out}}[\Lambda]|. \tag{11.4.6.6}$$

Write $H_{\gamma,\Lambda}(\sigma_\Lambda|\chi^+_{\Lambda^c}) = \int_\Lambda e_{\lambda_{\beta,\gamma}}(J_\gamma * (\sigma_\Lambda + \chi^+_{\Lambda^c})) + I_{\gamma,\lambda_{\beta,\gamma}}(J_\gamma * \sigma_\Lambda; \chi^+_{\Lambda^c})$, see (11.3.1.2), $I_{\gamma,\lambda_{\beta,\gamma}}(J_\gamma * \sigma_\Lambda; \chi^+_{\Lambda^c})$ being defined in (11.3.1.1). Split the integral over

Λ into the integral over Λ_{00}, see (11.4.5.1), and over $\Lambda \setminus \Lambda_{00}$. By using translational invariance of $\langle\cdot\rangle_{\gamma,a;u}$ we then get $\mathcal{I} = I^+_{\gamma,\lambda(\beta)}(\Lambda) + \mathcal{I}_1 + \mathcal{I}_2 + \mathcal{I}_3$:

$$\mathcal{I}_1 := \int_0^1 \int_{\Lambda_{00}} \{\langle e_{\lambda_{\beta,\gamma}}(J_\gamma * \sigma_\Lambda(r))\rangle_{\gamma,a,\Lambda;u} - \langle e_{\lambda_{\beta,\gamma}}(J_\gamma * \sigma_\Lambda(r))\rangle_{\gamma,a;u}\}$$

$$\mathcal{I}_2 := \int_0^1 \langle I_{\gamma,\lambda_{\beta,\gamma}}(J_\gamma * \sigma_\Lambda; \chi^+_{\Lambda^c})\rangle_{\gamma,a,\Lambda;u} dr - I^+_{\gamma,\lambda(\beta)}(\Lambda)$$

$$\mathcal{I}_3 := \int_0^1 \int_{\Lambda\setminus\Lambda_{00}} (\{\langle e_{\lambda_{\beta,\gamma}}(J_\gamma * [\sigma_\Lambda + \chi^+_{\Lambda^c}](r))\rangle_{\gamma,a,\Lambda;u} - e_{\lambda(\beta)}(\rho_{\beta,+})\}$$
$$- \{\langle e_{\lambda_{\beta,\gamma}}(J_\gamma * \sigma_\Lambda(r))\rangle_{\gamma,a;u} - e_{\lambda(\beta)}(\rho_{\beta,+})\}).$$

The number of cubes $C \in \mathcal{D}^{(\ell_{-,\gamma})}$ which are in $\delta_{\text{in}}^{2\gamma^{-1}}[\Lambda]$ is bounded by

$$\frac{|\delta_{\text{in}}^{\ell_{+,\gamma}}[\Lambda]|}{\ell^d_{+,\gamma}} \frac{\ell^{d-1}_{+,\gamma}}{(2\gamma^{-1})^{d-1}} \frac{(2\gamma^{-1})^d}{\ell^d_{-,\gamma}} \le 3^d |\delta_{\text{out}}^{\ell_{+,\gamma}}[\Lambda]| \gamma^{(1-\alpha)d+\alpha}.$$

Then, by (11.4.5.2), $|\mathcal{I}_1|$ is bounded by

$$c\gamma^{-d} |\delta_{\text{out}}^{\ell_{+,\gamma}}[\Lambda]| \gamma^{(1-\alpha)d+\alpha} \int_{|r|\ge \ell_{+,\gamma}} e^{-\gamma\omega|r|} \le c'\gamma^{-\alpha(d-1)} e^{-\omega\gamma^{-\alpha}} |\delta_{\text{out}}^{\ell_{+,\gamma}}[\Lambda]|.$$

As observed after (11.3.1.3), $I^+_{\gamma,\lambda(\beta)}(\chi^+_\Lambda; \chi^+_{\Lambda^c}) = I^+_{\gamma,\lambda(\beta)}(\Lambda)$ with the latter equal to (10.4.1.5), so that recalling (11.3.1.3), $\mathcal{I}_2$ is equal to the integral over u of the $\langle\cdot\rangle_{\gamma,a,\Lambda;u}$ expectation of

$$\int_{\delta_{\text{out}}^{\ell_{+,\gamma}}[\Lambda]} \{e_{\lambda_{\beta,\gamma}}(J_\gamma * [\sigma_\Lambda + \chi^+_{\Lambda^c}]) - e_{\lambda(\beta)}(\rho_{\beta,+})\} + \{e_{\lambda_{\beta,\gamma}}(J_\gamma * \sigma_\Lambda)$$
$$- e_{\lambda(\beta)}(J_\gamma * \chi^+_\Lambda)\} - \int_{\delta_{\text{in}}^{\ell_{+,\gamma}}[\Lambda]} \{e_{\lambda_{\beta,\gamma}}(J_\gamma * \sigma_\Lambda) - e_{\lambda(\beta)}(J_\gamma * \chi^+_\Lambda)\}.$$

Since the arguments of $e_{\lambda_{\beta,\gamma}}(\cdot)$ are bounded (because of the support properties of $\langle\cdot\rangle_{\gamma,a,\Lambda;u}$), and since $|\lambda_{\beta,\gamma} - \lambda(\beta)| \le c\gamma^{1/2}$, we get

$$|\mathcal{I}_2| \le c \int_0^1 \int_{\delta_{\text{in}}^{\ell_{+,\gamma}}[\Lambda] \cup \delta_{\text{out}}^{\ell_{+,\gamma}}[\Lambda]} (\langle |J_\gamma * [\sigma(r) - \chi_\Lambda](r)|\rangle_{\gamma,a,\Lambda;u} + \gamma^{1/2}),$$

and by (11.4.5.5) $|\mathcal{I}_2| \le c|\delta_{\text{out}}^{\ell_{+,\gamma}}[\Lambda]|\gamma^{1/2}$. The same bound by analogous arguments holds for $|\mathcal{I}_3|$; hence (11.4.6.6).

Bounds on R_U

We write $|\Lambda| \sum_{\Delta \ni 0} \langle \frac{U^+_{\gamma,\Delta}(\sigma_\Delta)}{|\Delta|} \rangle_{\gamma,a;u} = \int_\Lambda \sum_{\Delta \ni r} \langle \frac{U^+_{\gamma,\Delta}(\sigma_\Delta)}{|\Delta|} \rangle_{\gamma,a;u} dr$ and for each $r \in \Lambda_{00}$ we distinguish the sets Δ contained in Λ_{00} from the others. Denoting by C^+ the cubes of $\mathcal{D}^{(\ell_{+,\gamma})}$,

$$\left| \int_\Lambda \sum_{\Delta \ni r} \left\langle \frac{U^+_{\gamma,\Delta}(\sigma_\Delta)}{|\Delta|} \right\rangle_{\gamma,a;u} dr - \sum_{\Delta \subset \Lambda_{00}} \langle U^+_{\gamma,\Delta}(\sigma_\Delta) \rangle_{\gamma,a;u} \right|$$

$$\le \int_\Lambda \sum_{\Delta \ni r, \Delta \cap \Lambda^c_{00} \neq \emptyset} \frac{\| U^+_{\gamma,\Delta} \|_\infty}{|\Delta|} \le \sum_{C^+ \subset \Lambda \setminus \Lambda_{00}} \sum_{\Delta \supset C^+} \| U^+_{\gamma,\Delta} \|_\infty .$$

Then

$$|R_U| \le \sum_{\Delta \subset \Lambda_{00}} \int_0^1 \left| \langle U^+_{\gamma,\Delta}(\sigma_\Delta) \rangle_{\gamma,a,\Lambda;u} - \langle U^+_{\gamma,\Delta}(\sigma_\Delta) \rangle_{\gamma,a;u} \right| du$$
$$+ 2 \sum_{C^+ \subset \Lambda \setminus \Lambda_{00}} \sum_{\Delta \supset C^+} \| U^+_{\gamma,\Delta}(\cdot) \|_\infty . \qquad (11.4.6.7)$$

Fix a cube $C \in \mathcal{D}^{(\ell_{-,\gamma})}$ which is in $\delta^{2\gamma^{-1}}_{\text{in}}[\Lambda]$. For any $\Delta \in \Lambda_{00}$ the set of $\{r \in \Delta : \text{dist}(r, C) \le 2\,\text{dist}(\Delta, C)\}$ has volume larger than 1. Then by (11.4.5.3)

$$\left| \langle U^+_{\gamma,\Delta}(\sigma_\Delta) \rangle_{\gamma,a,\Lambda;u} - \langle U^+_{\gamma,\Delta}(\sigma_\Delta) \rangle_{\gamma,a;u} \right|$$
$$\le c|\Delta| \| U^+_{\gamma,\Delta}(\cdot) \|_\infty \sum_{C \in \delta^{2\gamma^{-1}}_{\text{in}}[\Lambda]} \int_\Delta e^{-\gamma\omega\,\text{dist}(r,C)/2} .$$

Calling the first and second term on the r.h.s. of (11.4.6.7) R'_U and R''_U, we have

$$R'_U \le c \sum_{\Delta \ni 0} |\Delta| \| U^+_{\gamma,\Delta}(\cdot) \|_\infty \sum_{C \in \delta^{2\gamma^{-1}}_{\text{in}}[\Lambda]} \int_{\mathbb{R}^d} e^{-\gamma\omega\,\text{dist}(r,C)/2} .$$

Since $|\Delta| = \ell^d_{+,\gamma} N_\Delta \le \ell^d_{+,\gamma} e^{N_\Delta}$, we have by (11.4.3.8) (with a new constant c)

$$R'_U \le c \ell^d_{+,\gamma} e^{-\beta c_{\text{pol}}(\zeta^2 \ell^d_{-,\gamma})} |\delta^{2\gamma^{-1}}_{\text{in}}[\Lambda]| \ell^{-d}_{-,\gamma} \gamma^{-d} \int_{\mathbb{R}^d} e^{-\omega|r|/2} .$$

Analogous arguments are used to bound R''_U obtaining

$$|R_U| \le c |\delta^{\ell_{+,\gamma}}_{\text{in}}[\Lambda]| e^{-\beta c_{\text{pol}}(\zeta^2 \ell^d_{-,\gamma})/2} . \qquad (11.4.6.8)$$

Bounds on R_σ

We split the sum over x in (11.4.6.5) over $x \in \Lambda_{00}$ and $x \in \Lambda \setminus \Lambda_{00}$. In the former we sum over $C_i \subset \Lambda_{00}$ and $x \in C_i$ and use (11.4.5.4); for the latter we use (11.4.5.6) after adding and subtracting $\rho_{\beta,+}|\Lambda \setminus \Lambda_{00}|$. We then get for γ small enough

$$|R_\sigma| \le c\ell^d_{-,\gamma} \sum_{C_j \subset \delta_{\text{in}}^{2\gamma^{-1}}[\Lambda]} \sum_{C_i:\text{dist}(C_i,C_j)\ge \ell_{+,\gamma}} e^{-\omega\gamma\,\text{dist}(C_i,C_j)} + 2c\gamma^{1/2}|\Lambda \setminus \Lambda_{00}|$$
$$\le c'|\delta_{\text{in}}^{\ell_{+,\gamma}}[\Lambda]|\gamma^{1/2}. \tag{11.4.6.9}$$

Conclusions

Using the above estimates, (11.4.6.2) becomes

$$\left| \frac{1}{\beta} \log Z^+_{\gamma,a,\Lambda} - p^+_{\gamma,a}|\Lambda| - I^+_{\gamma,\lambda(\beta)}(\Lambda) \right| \le c|\delta_{\text{in}}^{\ell_{+,\gamma}}[\Lambda]|\gamma^{1/2}. \tag{11.4.6.10}$$

More precisely, there are $\gamma^* > 0$ and $a^* > 0$ so that for all $\gamma \le \gamma^*$, all bounded $\mathcal{D}^{(\ell_{+,\gamma})}$-measurable regions Λ and all $a \le a^*$, (11.4.6.10) is verified. We claim that

$$\lim_{a\to 0} p^+_{\gamma,a} = P^+_{\gamma,\lambda_{\beta,\gamma}}. \tag{11.4.6.11}$$

Proof For any $\epsilon > 0$ and all cubes Λ large enough, $|\frac{\log Z^+_{\gamma,\Lambda}}{\beta|\Lambda|} - P^+_{\gamma,\lambda_{\beta,\gamma}}| \le \frac{\epsilon}{3}$. By (11.4.6.10), for Λ large also $I^+_{\gamma,\lambda(\beta)}(\Lambda) + c|\delta_{\text{in}}^{\ell_{+,\gamma}}[\Lambda]|\gamma^{1/2} \le \frac{\epsilon}{3}|\Lambda|$, and given any such Λ let $a > 0$ be so small that $|\frac{\log Z^+_{\gamma,\Lambda}}{\beta|\Lambda|} - \frac{\log Z^+_{\gamma,a,\Lambda}}{\beta|\Lambda|}| \le \frac{\epsilon}{3}$ (see Theorem 11.4.2.1). The three inequalities yield $|p^+_{\gamma,a} - P^+_{\gamma,\lambda_{\beta,\gamma}}| \le \epsilon$; (11.4.6.1) as well as Corollary 11.4.6.1 are then proved. □

11.5 Dobrushin theory. III Relativized uniqueness

In this section we extend the Dobrushin uniqueness theory of Sect. 3.2 to cases where the uniqueness condition (3.1.3.17) is not satisfied uniformly on the conditioning spins, namely (3.1.3.17) holds with parameters $r(x, y)$ defined with the sup replaced by a sup over a reduced set of conditioning configurations. We shall then talk of a "relativized Dobrushin condition." We shall prove that if the complement of "the good set of conditioning configurations" has a suitably small measure, then uniqueness still holds.

Unfortunately the result is not as general as the above hints, as several more specific assumptions are needed; however, the theory includes both the Ising models with Kac potentials and the LMP particle system both in the restricted ensembles.

To give an idea we begin by showing why in LMP a relativized Dobrushin condition may hold. In Sect. 11.5.2 we shall define the class of models included in our theory; assumptions are stated in a successive subsection, and the results are in Sect. 11.5.4. Proofs start in Sect. 11.5.5.

11.5.1 Revisiting the Dobrushin uniqueness condition

We shall study here the LMP model in its lattice approximation, see Sect. 11.4.2, calling a the lattice mesh. We first consider the unrestricted phase space $\mathcal{X}_a$. The one-site Gibbs conditional probability $G_{\gamma,a,x;u}(\{\sigma(x)=1\}|\sigma_{x^c})$ is

$$G_{\gamma,a,x;u}\big(\{\sigma(x)=1\}|\sigma_{x^c}\big) = \frac{(a^d e^{-\beta(1-u)e'_{\lambda(\beta)}(\rho_{\beta,+})})e^{-\beta u E_x(\sigma_{x^c})}}{1+(a^d e^{-\beta(1-u)e'_{\lambda(\beta)}})e^{-\beta u E_x(\sigma_{x^c})}}, \qquad (11.5.1.1)$$

$$E_x(\sigma_{x^c}) = \int e_{\lambda_{\beta,\gamma}}(t_{x,r}+J_\gamma(r,x)) - e_{\lambda_{\beta,\gamma}}(t_{x,r}), \quad t_{x,r} = \sum_{y\neq x} J_\gamma(r,y)\sigma_{x^c}(y).$$

Write $R \equiv R(G_{\gamma,a,x;u}(\cdot|\sigma_{x^c}), G_{\gamma,a,x}(\cdot|\sigma'_{x^c}))$, the Vaserstein distance (see Sect. 3.2.1 for notation and definitions), and let

$$\theta := \beta e^{-\beta e'_{\lambda(\beta)}(\rho_{\beta,+})}|e''_{\lambda(\beta)}(\rho_{\beta,+})| < 1, \quad \beta\in(\beta_c,\beta_0). \qquad (11.5.1.2)$$

Lemma 11.5.1.1 *For any $\epsilon>0$ there is $\gamma^*>0$ so that for all $\gamma\le\gamma^*$, all $a>0$ small enough, all $u\in[0,1]$ and all configurations σ and σ' in $\mathcal{X}_a^+$,*

$$R \le (\theta+\epsilon)a^d \sum_{y\neq x}\left\{\int J_\gamma(r,y)J_\gamma(x,r)\right\}|\sigma'_{x^c}(y)-\sigma_{x^c}(y)|. \qquad (11.5.1.3)$$

Proof We have $R = |G_{\gamma,a,x;u}(\{\sigma(x)=1\}|\sigma_{x^c}) - G_{\gamma,a,x;u}(\{\sigma(x)=1\}|\sigma'_{x^c})|$; see the proof of Theorem 3.1.3.6. By interpolating,

$$R = \left|\int_0^1 \frac{d}{ds}\left(\frac{(a^d e^{-\beta(1-u)e'_{\lambda(\beta)}(\rho_{\beta,+})})e^{-\beta u E_x^*(s)}}{1+(a^d e^{-\beta(1-u)e'_{\lambda(\beta)}(\rho_{\beta,+})})e^{-\beta u E_x^*(s)}}\right)\right|,$$

$$E_x^*(s) := \int e_{\lambda_{\beta,\gamma}}([st_{x,r}+(1-s)t'_{x,r}]+J_\gamma(r,x)) - e_{\lambda_{\beta,\gamma}}([st_{x,r}+(1-s)t'_{x,r}]).$$

Then R is bounded by

$$\int_0^1 \left|a^d e^{-\beta[(1-u)e'_{\lambda(\beta)}(\rho_{\beta,+})+uE_x^*(s)]}\beta\sum_{y\neq x}\kappa(x,y,s)(\sigma_{x^c}(y)-\sigma'_{x^c}(y))\right|ds + ca^{2d},$$

where ca^{2d} bounds the term obtained by differentiating the denominator, while the coefficients $\kappa(x,y,s)$ have the expression

$$\int \Big(e'_{\lambda_{\beta,\gamma}}([st_{x,r}+(1-s)t'_{x,r}]+J_\gamma(r,x))-e'_{\lambda_{\beta,\gamma}}([st_{x,r}+(1-s)t'_{x,r}])\Big)J_\gamma(r,y);$$

$|E^*_x(s)-e'_{\lambda_{\beta,\gamma}}(\rho_{\beta,+})|\le c'\zeta$ because $|t_{x,r}-\rho_{\beta,+}|\le c\zeta$, $|t'_{x,r}-\rho_{\beta,+}|\le c\zeta$, and since $J(x,r)\le c\gamma^d$,

$$\left|\kappa(x,y,s)-e''_{\lambda_{\beta,\gamma}}(\rho_{\beta,+})\int J_\gamma(r,y)J_\gamma(x,r)\right|\le c\zeta\mathbf{1}_{|x-y|\le 2\gamma^{-1}};$$

(11.5.1.3) then follows because $|\lambda_{\beta,\gamma}-\lambda(\beta)|\le c\gamma^{1/2}$. □

We thus recognize $r(x,y)=(\theta+\epsilon)a^d\int J_\gamma(r,y)J_\gamma(x,r)$ and recover the uniqueness condition $\sup_x\sum_{y\ne x}r(x,y)<1$ for a and γ small enough. The problem arises because the one site conditional probability of the Gibbs measure

$$\pi_\Lambda(\sigma_\Lambda):=Z^{-1}(a^d e^{-\beta(1-u)e'_{\lambda(\beta)}(\rho_{\beta,+})})^{|\sigma_\Lambda|}e^{-\beta u H_{\gamma,\lambda_{\beta,\gamma},\sigma_\Lambda}(\sigma_\Lambda|\sigma_{\Lambda^c})}$$

regarded as a probability on $\mathcal{X}^+_{\Lambda,a}$ is not given by (11.5.1.1)!

Let $N_\Delta(\sigma)=\sum_{y\in\Delta\cap a\mathbb{Z}^d}\sigma(y)$, $C\in\mathcal{D}^{(\ell_-,\gamma)}$, $x\in C$, and

$$n_-=\min_{\sigma:\eta(\sigma;x)=1}N_C(\sigma),\qquad n_+=\max_{\sigma_C:\eta(\sigma_C;x)=1}N_C(\sigma),$$

$$\mathcal{B}^{(1)}_x=\{\sigma\in\mathcal{X}^+_a:N_{C\setminus x}(\sigma)=n_--1\},\tag{11.5.1.4}$$
$$\mathcal{B}^{(2)}_x=\{\sigma\in\mathcal{X}^+_a:N_{C\setminus x}(\sigma)=n_+\}.$$

Writing $\pi_{x,\Lambda}(\cdot\,|\,\sigma_{x^c})$ for the conditional probability of π_Λ given $\sigma_{\Lambda\setminus x}$, then $\pi_{x,\Lambda}(\cdot\,|\,\sigma_{x^c})=G_{\gamma,a,x;u}(\{\sigma(x)=1\}|\sigma_{x^c})$ if σ is in $\mathcal{B}^c_x$, $\mathcal{B}_x=\mathcal{B}^{(1)}_x\cup\mathcal{B}^{(2)}_x$ (notice that $\mathcal{B}_x$ does not depend on the value $\sigma(x)$), while

$$\pi_{x,\Lambda}\big(\sigma(x)=1\,|\,\sigma_{x^c}\big)=\begin{cases}1 & \text{if } \sigma_{x^c}\in\mathcal{B}^{(1)}_x,\\ 0 & \text{if } \sigma_{x^c}\in\mathcal{B}^{(2)}_x.\end{cases}\tag{11.5.1.5}$$

Thus, $R\equiv R(\pi_{x,\Lambda}(\cdot|\sigma_{x^c}),\pi_{x,\Lambda}(\cdot|\sigma'_{x^c}))$ is bounded as in (11.5.1.3) if both σ and σ' are in $\mathcal{B}^c_x$ but not when they are not both in $\mathcal{B}_x$. The theory in this section proves uniqueness and exponential decay of correlations under the assumption that the probability of $\mathcal{B}_x$ is suitably small. The proof that our LMP model satisfies such a smallness assumption will be given in Lemma 11.6.1.1. The smallness assumption will distinguish between $\mathcal{B}^{(1)}_x$ and $\mathcal{B}^{(2)}_x$; notice in fact that if for instance $\sigma\in\mathcal{B}^{(1)}_x$ and $\sigma'\in\mathcal{B}^c_x$, then

$$R\le\pi_{x,\Lambda}\big(\sigma(x)=1\,|\,\sigma'_{\Lambda\setminus x}\big)\le 2a^d,\tag{11.5.1.6}$$

for a small enough.

11.5.2 The setup

• *Phase space.* The "unrestricted" phase space of our models is $S^{a\mathbb{Z}^d}$, with S a finite set and $a\mathbb{Z}^d$ the lattice with mesh $a \in \{2^{-n} n \in \mathbb{N}\}$. The elements of $S^{a\mathbb{Z}^d}$ are called spin configurations and are denoted by σ: $\sigma(x) \in S$, $x \in a\mathbb{Z}^d$, the spin at the lattice point x. By default in the sequel, $x, y, \dots$ indicate elements of $a\mathbb{Z}^d$, regions $\Lambda \subset \mathbb{R}^d$ are always $\mathcal{D}^{(a)}$-measurable and identified with $\Lambda \cap a\mathbb{Z}^d$, so that $\sum_{x\in\Lambda}(\cdot) \equiv \sum_{x\in\Lambda\cap a\mathbb{Z}^d}(\cdot)$.

S is equipped with a distance $d(s,s')$, $d(s,s') \ge 1$ if $s \ne s'$. The diameter of S is $d(S) = \max_{s,s'\in S} d(s,s')$, and for any two probabilities ν and ν' on $S^\Delta := S^{\Delta\cap a\mathbb{Z}^d}$ we denote

$$d_\Delta(\sigma,\sigma') = \sum_{x\in\Delta} d(\sigma(x),\sigma'(x)), \qquad R_\Delta(\nu,\nu') = \inf_Q E_Q(d_\Delta(\sigma,\sigma')),$$

where Q ranges over all couplings of ν and ν'; E_Q means expectation with respect to Q. $R_\Delta(\cdot,\cdot)$ is the Vaserstein distance between probabilities on S^Δ. If ν and ν' are probabilities on S^Λ, $\Lambda \supset \Delta$, by abuse of notation $R_\Delta(\nu,\nu')$ denotes the Vaserstein distance of the marginal distributions of ν and ν' on S^Δ.

Remarks We have in mind two specific examples: the Ising model with $S = \{-1,1\}$ and the lattice gas where $S = \{0,1\}$. In the LMP model we need estimates uniform as $a \to 0$, and thus a plays a fundamental role, in the Ising and in genuine lattice systems a is unessential and fixed equal to 1.

We are going to compare two probabilities μ_Λ and μ'_Λ on S^Λ which are supposed to be absolutely continuous with respect to two "reference probabilities" π_Λ and π'_Λ. In LMP the former are the Gibbs measures with (the interpolating) hamiltonian $h^+_{\gamma,a,\Lambda;u}$ (see (11.4.4.5)) and conditioned to $\mathcal{X}^+_\Lambda$; the reference measures are the same but without the hamiltonian K_γ (determined by the contours weights and with infinite range). In the Ising case we may avoid the interpolating hamiltonian unless we want to study the surface corrections to the pressure.

• *The reference measures π_Λ and π'_Λ.* We suppose that there is $\ell_0 \in \{2^n, n \in \mathbb{N}\}$ and a family of sets $\{\mathcal{X}_{C_i} \in S^\Lambda, i \in \ell_0\mathbb{Z}^d\}$ ($C_i = C_i^{(\ell_0)}$ the cube of $\mathcal{D}^{(\ell_0)}$ containing i) so that the following holds. Restricting ourselves in the sequel to bounded $\mathcal{D}^{(\ell_0)}$-measurable regions, given any such region Λ, we write $I := \{i \in \ell_0\mathbb{Z}^d : C_i \subset \Lambda\}$ and suppose that for any $i \in I$ there is a set $\mathcal{X}_{C_i} \subset S^{C_i}$ such that

$$\pi_\Lambda(\mathcal{X}_{C_i}) = \pi'_\Lambda(\mathcal{X}_{C_i}) = 1, \qquad \max_{\sigma_{C_i},\sigma'_{C_i}\in\mathcal{X}_{C_i}} d_{C_i}(\sigma_{C_i},\sigma'_{C_i}) \le c_0\ell_0^d. \tag{11.5.2.1}$$

We conclude the paragraph by setting the following notation: if $\Delta \subset \Lambda$, $\pi_{\Delta,\Lambda}(\cdot|\sigma_{\Lambda\setminus\Delta})$ and $\pi'_{\Delta,\Lambda}(\cdot|\sigma_{\Lambda\setminus\Delta})$ denote the conditional probabilities of π_Λ and π'_Λ on S^Δ, $\Delta \subset \Lambda$, given $\sigma_{\Lambda\setminus\Delta}$.

Remarks In the Ising and in the LMP model $\ell_0 = \ell_{-,\gamma}$, $\mathcal{X}_{C_i} = \{\sigma : \eta(\sigma; i) = 1\}$. Observe that $\max_{\sigma_{C_i},\sigma'_{C_i} \in S^{C_i}} d_{C_i}(\sigma_{C_i}, \sigma'_{C_i}) = d(S)a^{-d}\ell_0^d$ so that the second inequality in (11.5.2.1) is not at all innocent when a is small. In the LMP case $\mathcal{X}_{C_i}$ is a constraint which imposes the number of particles in C_i to be bounded proportionally to $|C_i|$ independently of a. Since the maximal distance of two configurations with $\le N$ particles is bounded by $2N$, the constraint $\mathcal{X}_{C_i}$ greatly reduces the distance among configurations.

- *The "Gibbs measures" μ_Λ and μ'_Λ.* There is $\ell_1 \in \{2^n, n \in \mathbb{N}\}$, $\ell_1 > \ell_0$, and functions $U_\Delta(\sigma_{\Delta\cap\Lambda})$, $U'_\Delta(\sigma_{\Delta\cap\Lambda})$, Δ varying in the set of bounded $\mathcal{D}^{(\ell_1)}$-measurable regions, so that

$$\frac{d\mu_\Lambda}{d\pi_\Lambda}(\sigma_\Lambda) = \frac{1}{Z}\ \exp\Big\{-\beta \sum_{\Delta:\Delta\cap\Lambda\neq\emptyset} U_\Delta(\sigma_{\Delta\cap\Lambda})\Big\},$$
$$\frac{d\mu_\Lambda}{d\pi'_\Lambda}(\sigma_\Lambda) = \frac{1}{Z'}\ \exp\Big\{-\beta \sum_{\Delta:\Delta\cap\Lambda\neq\emptyset} U'_\Delta(\sigma_{\Delta\cap\Lambda})\Big\}, \tag{11.5.2.2}$$

and

$$U_\Delta(\sigma_{\Lambda\cap\Delta}) = U'_\Delta(\sigma_{\Lambda\cap\Delta}) \quad \text{if } \Delta \subseteq \Lambda. \tag{11.5.2.3}$$

Remarks We have added β in the definition of U_Δ and U'_Δ to agree with the notation in LMP. In the Ising case and LMP $\ell_1 = \ell_{+,\gamma}$.

11.5.3 The assumptions

In the assumptions on μ_Λ and μ'_Λ are distinguished assumptions on the pairs $\pi_\Lambda, \pi'_\Lambda$ and U_Δ, U'_Δ.

Assumptions on the conditional probabilities of π_Λ and π'_Λ

- *Finite range.* There is $R \in \{2^n, n \in \mathbb{N}\}$, $R \in (\ell_0, \ell_1)$, so that for any $\Delta \subset \Lambda$: $\text{dist}(\Delta, \Lambda^c) > R$, $\pi_{\Delta,\Lambda}(\sigma_\Delta|\sigma_{\Lambda\setminus\Delta}) = \pi_{\Delta,\Lambda}(\sigma_\Delta|\sigma'_{\Lambda\setminus\Delta})$ if $\sigma_{\Lambda\setminus\Delta}(x) = \sigma'_{\Lambda\setminus\Delta}(x)$ for all $x \in \Lambda\setminus\Delta$ with $\text{dist}(x, \Delta) \le R$; the same holds for π'_Λ. (In the Ising case $R = \gamma^{-1}$, in LMP $R = 2\gamma^{-1}$.)
- *One site conditional probabilities.* There are $r(x, y) \ge 0$

$$r(x, y) = 0, \quad \text{for } |x - y| \ge R, \tag{11.5.3.1}$$

and there is a set $\Lambda^c_{\neq} \subseteq \Lambda^c$ so that the following holds. Writing $C_i = C_i^{(\ell_0)}$ and

$$I = \{i \in \ell_0\mathbb{Z}^d : C_i \subset \Lambda\}, \qquad I^0 = \{i \in I : \text{dist}(C_i, \Lambda^c_{\neq}) > R\}, \tag{11.5.3.2}$$

then for any $i \in I^0$ and $x \in C_i$ there are two sets $\mathcal{B}_x^{(1)}$, $\mathcal{B}_x^{(2)}$ in $\mathcal{X}_{C_i}$ which depend on $\sigma_{C_i \setminus x}$, so that with $\mathcal{B}_x = \mathcal{B}_x^{(1)} \cup \mathcal{B}_x^{(2)}$

$$R_x\big(\pi_{x,\Lambda}(\cdot|\sigma_{\Lambda\setminus x}), \pi'_{x,\Lambda}(\cdot|\sigma'_{\Lambda\setminus x})\big) \le \sum_{y\neq x} r(x,y)\, d_y(\sigma_\Lambda, \sigma'_\Lambda), \quad \sigma_\Lambda, \sigma'_\Lambda \notin \mathcal{B}_x \tag{11.5.3.3}$$

(further assumptions on $r(x,y)$ will be stated in the paragraph *Assumptions on the parameters*). Moreover, there is $c_2 > 0$ (the constant c_1 is the one appearing in the Peierls bound and we have thus jumped from c_0 in (11.5.2.1) to c_2) so that if $\sigma'_\Lambda \in \mathcal{B}_x^{(2)}, \sigma_\Lambda \notin \mathcal{B}_x$ or vice versa

$$R_x\big(\pi_{x,\Lambda}(\cdot|\sigma_{\Lambda\setminus x}), \pi'_{x,\Lambda}(\cdot|\sigma'_{\Lambda\setminus x})\big) \le c_2 a^d. \tag{11.5.3.4}$$

Remarks In LMP $\mathcal{B}_x$, $\mathcal{B}_x^{(1)}$ and $\mathcal{B}_x^{(2)}$ are specified in Sect. 11.5.1; in particular, see Lemma 11.5.1.1 for (11.5.3.3) and (11.5.1.6) for (11.5.3.4).

• *One block conditional probabilities.* There is ϵ_2 so that for all $i \in I_0$, all $x \in C_i$ and for all $\sigma_{\Lambda\setminus C_i}$,

$$\begin{aligned} &\pi_{C_i,\Lambda}\big(\mathcal{B}_x^{(1)}|\sigma_{\Lambda\setminus C_i}\big) \le \epsilon_2 a^d, \qquad \pi'_{C_i,\Lambda}\big(\mathcal{B}_x^{(1)}|\sigma_{\Lambda\setminus C_i}\big) \le \epsilon_2 a^d,\\ &\pi_{C_i,\Lambda}\big(\mathcal{B}_x^{(2)}|\sigma_{\Lambda\setminus C_i}\big) \le \epsilon_2, \qquad \pi'_{C_i,\Lambda}\big(\mathcal{B}_x^{(2)}|\sigma_{\Lambda\setminus C_i}\big) \le \epsilon_2. \end{aligned} \tag{11.5.3.5}$$

Assumptions on U_Δ and U'_Δ

There are ϵ_3 and $\omega_3 > 0$ so that for any $i, j \in I$,

$$\beta \sum_{\Delta\supset C_i, \Delta\supset C_j} \|U_\Delta\|_\infty \le \epsilon_3\, e^{-\omega_3 |i-j|}, \tag{11.5.3.6}$$

and the same holds for U'_Δ. Moreover, (11.5.2.3) is extended to

$$U_\Delta(\sigma_{\Lambda\cap\Delta}) = U'_\Delta(\sigma_{\Lambda\cap\Delta}) \quad \text{if } \Delta \cap \Lambda^c_{\neq} = \emptyset. \tag{11.5.3.7}$$

Assumptions on the parameters

With I and I_0 as in (11.5.3.2), call

$$\begin{aligned} &c_3 = \sum_{i\in\ell_0\mathbb{Z}^d} e^{-\omega_3|i|}, \qquad r^*(i) = \max_{x\in C_i} \sum_{y\in C_i} r(x,y), \quad i \in I_0,\\ &r^*(i,j) = a^{-d}|C_i| \max_{x\in C_i, y\in C_j} r(x,y), \quad i \in I_0, j \in I \end{aligned} \tag{11.5.3.8}$$

(recall $|C_i| = \ell_0^d$). Assume $r^*(i) < 1$, $c' = c_2 + d(S)$, for $i \in I_0$ and $j \in I$ and

$$\theta(i,j) := \begin{cases} \dfrac{1}{1-r^*(i)}\{r^*(i,j) + 2\epsilon_2 c'\ell_0^d\} + 4c_0\ell_0^d\epsilon_3 c_3 \\ \quad \text{if } \operatorname{dist}(C_i, C_j) \le R, \\ 4c_0\ell_0^d\epsilon_3 e^{-\omega_3|i-j|} \\ \quad \text{if } \operatorname{dist}(C_i, C_j) > R. \end{cases} \tag{11.5.3.9}$$

Assume finally that there is $r^* < 1$ so that for all $i \in I$

$$\sum_{j \neq i} \theta(i,j) \le r^* < 1. \tag{11.5.3.10}$$

11.5.4 Main results

Under the above assumptions the following theorem holds:

Theorem 11.5.4.1 *Let $\delta \in (0, \omega_3)$ and $r \in (r^*, 1)$ be such that for any $i \in I_0$, I and I_0 are defined in* (11.5.3.2), *$\theta(i,j)$ in* (11.5.3.9),

$$\sum_{j \neq i} \theta(i,j) e^{\delta|i-j|} \le r < 1 \tag{11.5.4.1}$$

(*the existence of δ and r follows from* (11.5.3.10)–(11.5.3.9)). *Then there is $c > c_0$ and a coupling Q of μ_Λ and μ'_Λ so that for any $i \in I_0$*

$$E_Q(d_{C_i}) \le c\ell_0^d \sum_{j \in I \setminus I_0} e^{-\delta|i-j|}. \tag{11.5.4.2}$$

Corollary 11.5.4.2 *Let $f(\sigma) = f(\sigma_\Delta)$, $\Delta \subseteq \bigcup_{i=1}^n C_i$, $C_i \in \mathcal{D}^{(\ell_0)}$; then*

$$|\mu_\Lambda(f) - \mu'_\Lambda(f)| \le c\ell_0^d 2\|f\|_\infty \sum_{i=1}^n \sum_{j \in I \setminus I_0} e^{-\delta|i-j|}. \tag{11.5.4.3}$$

Proof (11.5.4.3) follows from (11.5.4.2) after writing

$$|f(\sigma_\Delta) - f(\sigma'_\Delta)| \le 2\|f\|_\infty \mathbf{1}_{\sigma_\Delta \neq \sigma'_\Delta} \le 2\|f\|_\infty \sum_{i=1}^n d_{C_i}(\sigma_\Lambda, \sigma'_\Lambda).$$ □

Before starting the proof of Theorem 11.5.4.1, we state some rather elementary bounds on the Vaserstein distance which will be used in the proofs. For future applications we state the results in a more general setting than what is needed here.

11.5.5 Two simple bounds on the Vaserstein distance

Let Ω be a complete, separable metric space with distance $d(\omega,\omega')$ and, after fixing arbitrarily an element $\omega_0 \in \Omega$, write $|\omega| = d(\omega,\omega_0)$. Let ν be a positive measure on Ω; if Ω is countable, ν is the counting measure.

Theorem 11.5.5.1 *Let h and v be such that for all $t \in [0,1]$,*

$$Z_t = \int e^{-[h(\omega)+tv(\omega)]}\nu(d\omega) < \infty. \tag{11.5.5.1}$$

Set

$$m_t(\omega) = Z_t^{-1}e^{-[h(\omega)+tv(\omega)]}, \qquad \mu_t(d\omega) = m_t(\omega)\nu(d\omega). \tag{11.5.5.2}$$

Then

$$R(\mu_1,\mu_0) \le \sup_{0\le t\le 1}\big(\mu_t(|\omega|\,|v|) + \mu_t(|\omega|)\mu_t(|v|)\big). \tag{11.5.5.3}$$

In particular, supposing $|\omega|$ bounded in Ω,

$$R(\mu_1,\mu_0) \le 2\Big(\sup_{\omega\in\Omega}|\omega|\Big)\sup\big(|v(\omega)|\big). \tag{11.5.5.4}$$

Proof Repeating the argument in the proof of Theorem 3.2.3.1, let

$$m(\omega) = \min\{m_1(\omega), m_0(\omega)\}, \qquad C = 1 - \int m(\omega)\nu(d\omega),$$

$$P(d\omega d\omega') = \Big\{m(\omega)\delta_{\omega-\omega'} + \frac{1}{C}[m_1(\omega)-m(\omega)][m_0(\omega')-m(\omega')]\Big\} \times \nu(d\omega)\nu(d\omega').$$

P is a coupling of μ_1 and μ_0, and therefore

$$\begin{aligned} R(\mu_1,\mu_0) &\le \int_{\Omega\times\Omega} d(\omega,\omega')P(d\omega d\omega') \\ &\le \int_\Omega |\omega|\big([m_1(\omega)-m(\omega)] + [m_0(\omega)-m(\omega)]\big)\nu(d\omega) \\ &= \int_\Omega |\omega|\,|m_1(\omega)-m_0(\omega)|\nu(d\omega), \end{aligned}$$

having bounded $d(\omega,\omega') \le |\omega| + |\omega'|$ and having integrated over the missing variable.

Equation (11.5.5.3) is then obtained by writing $m_1(\omega) - m_0(\omega) = \int_0^1 \frac{d}{dt}m_t(\omega)$. □

In the previous theorem we have changed the measure by varying the hamiltonian; in the next one the change is due to taking the conditional probability on some event. It is shown that if the conditioning event has a large probability, then the distance between the original measure and the conditioned one is small.

Theorem 11.5.5.2 *Let $A \subset \Omega$ be a measurable set, μ a probability on Ω and μ_A the probability μ conditioned to A. Then*

$$R(\mu, \mu_A) \le \left\{2 \sup_{\omega \in \Omega} |\omega|\right\} \mu(A^c). \tag{11.5.5.5}$$

Proof Let

$$Q(d\omega, d\omega') = \mathbf{1}_{\omega \in A} \mu(d\omega)\delta_\omega(d\omega') + \mathbf{1}_{\omega \in A^c} \mu(d\omega)\mu_A(d\omega'),$$

where $\delta_\omega(d\omega')$ is the probability supported by ω. Let f be any bounded, measurable function on Ω; then

$$\int f(\omega) Q(d\omega, d\omega') = \int_A f(\omega)\mu(d\omega) + \int_{A^c} f(\omega)\mu(d\omega)\mu_A(d\omega') = \mu(f),$$

$$\begin{aligned} \int f(\omega') Q(d\omega, d\omega') &= \int_A f(\omega)\mu(d\omega) + \mu(A^c) \int f(\omega')\mu_A(d\omega') \\ &= \mu_A(f)\mu(A) + \mu_A(f)\mu(A^c) = \mu_A(f). \end{aligned}$$

Hence Q is a coupling and

$$R(\mu, \mu_A) \le \int d(\omega, \omega') Q(d\omega, d\omega') \le \int (|\omega| + |\omega'|) \mathbf{1}_{\omega \in A^c} \mu(d\omega)\mu_A(d\omega'),$$

which proves (11.5.5.5). Actually the Vaserstein distance is bounded by the average

$$R(\mu, \mu_A) \le \int_{\omega \in A^c} \int_{\omega' \in A} d(\omega, \omega')\mu(d\omega)\mu_A(d\omega'). \tag{11.5.5.6}$$

□

11.5.6 *Proof of Theorem 11.5.4.1*

We start with a short outline of the proof:

Scheme of the proof

We regard our system as a process $\{\xi_i,\ i \in I\}$, with values in $\mathcal{X}_{C_i}$, recalling that $C_i \in \mathcal{D}^{(\ell_0)}$, $i \in \Lambda \cap \ell_0 \mathbb{Z}^d$ (the collection of all such i being called I). Thus $\xi_i = \sigma_{C_i}$

is the collection of all spins $\{\sigma_\Lambda(x), x \in C_i\}$. Recall that the distance between two elements $\xi_i = \sigma_{C_i}$ and $\xi_i' = \sigma'_{C_i}$ is $\sum_{x\in C_i} |\sigma_{C_i}(x) - \sigma'_{C_i}(x)|$, which, by (11.5.2.1), is bounded by $c_0\ell_0^d$. We shall show that the ξ_i-process satisfies the Dobrushin uniqueness condition (3.1.3.17) (relative to the distance above), so that Theorem 11.5.4.1 will follow from Theorem 3.1.3.5. To verify the Dobrushin uniqueness condition we need to estimate the Vaserstein distance between the conditional probabilities of ξ_i for different values of the conditioning variables ξ_j, $j \neq i$.

A preliminary definition of $\theta(i, j)$

In the sequel, unless otherwise stated, $i \in I^0$ (see (11.5.3.2)). As a first application of Theorem 11.5.5.1 we shall bound the influence on the interaction of the potentials U_Δ with $\Delta \cap \Lambda^c \neq \emptyset$. Denote by $\mu_{C_i,\Lambda}(\cdot|\sigma_{\Lambda\setminus C_i})$ the probability μ_Λ conditioned to $\sigma_{\Lambda\setminus C_i}$ and define $\nu_{C_i,\Lambda}(\cdot|\sigma_{\Lambda\setminus C_i})$ as

$$\frac{d\nu_{C_i,\Lambda}(\cdot|\sigma_{\Lambda\setminus C_i})}{d\mu_{C_i,\Lambda}(\cdot|\sigma_{\Lambda\setminus C_i})} = \exp\left\{\beta \sum_{\Delta\supset C_i, \Delta\cap\Lambda^c\neq\emptyset} U_\Delta\right\}. \tag{11.5.6.1}$$

Notice the plus sign in the exponent which cancels exactly the corresponding terms present in (11.5.2.2). An analogous expression defines $\nu'_{C_i,\Lambda}(\cdot|\sigma_{\Lambda\setminus C_i})$ in terms of $\mu'_{C_i,\Lambda}(\cdot|\sigma_{\Lambda\setminus C_i})$.

By Theorem 11.5.5.1, by (11.5.3.6) and recalling that the sets Δ are connected so that if $\Delta \ni i$ and $\Delta \cap \Lambda^c \neq \emptyset$, there is $j \in I \setminus I_0$ so that $\Delta \ni j$, we have

$$R_{C_i}\big(\mu_{C_i,\Lambda}(\cdot|\sigma_{\Lambda\setminus C_i}), \nu_{C_i,\Lambda}(\cdot|\sigma_{\Lambda\setminus C_i})\big) \le 2[c_0\ell_0^d]\epsilon_3 \sum_{j\in I\setminus I_0} e^{-\omega_3|i-j|},$$

and using again the triangular inequality,

$$\begin{aligned} &R_{C_i}\big(\mu_{C_i,\Lambda}(\cdot|\sigma_{\Lambda\setminus C_i}), \mu'_{C_i,\Lambda}(\cdot|\sigma'_{\Lambda\setminus C_i})\big)\\ &\quad\le R_{C_i}\big(\nu_{C_i,\Lambda}(\cdot|\sigma_{\Lambda\setminus C_i}), \nu'_{C_i,\Lambda}(\cdot|\sigma'_{\Lambda\setminus C_i})\big) + 4[c_0\ell_0^d]\epsilon_3 \sum_{j\in I\setminus I_0} e^{-\omega_3|i-j|}. \end{aligned}$$

By Theorem 3.1.3.4 we conclude

$$\begin{aligned} &R_{C_i}\big(\mu_{C_i,\Lambda}(\cdot|\sigma_{\Lambda\setminus C_i}), \mu'_{C_i,\Lambda}(\cdot|\sigma'_{\Lambda\setminus C_i})\big)\\ &\quad\le 4[c_0\ell_0^d]\epsilon_3 \sum_{j\in I\setminus I_0} e^{-\omega_3|i-j|} + \sum_{j\neq i, j\in I} \theta^*(i,j) d_{C_j}(\sigma_{C_j}, \sigma'_{C_j}), \end{aligned} \tag{11.5.6.2}$$

where

$$\theta^*(i,j) = \sup_{\sigma_\Lambda=\sigma'_\Lambda \text{on } \Lambda\setminus C_j} \frac{R_{C_i}(\nu_{C_i,\Lambda}(\cdot|\sigma_{\Lambda\setminus C_i}), \nu'_{C_i,\Lambda}(\cdot|\sigma'_{\Lambda\setminus C_i}))}{d_{C_j}(\sigma_{C_j}, \sigma'_{C_j})}, \tag{11.5.6.3}$$

with the understanding that the r.h.s. is equal to 0 when $\sigma_{C_j} = \sigma'_{C_j}$.

We shall prove Theorem 11.5.4.1 by showing that $\theta^*(i,j) \le \theta(i,j)$, the latter as in (11.5.3.9).

The case $i \in I^0$ and $\{j : \mathrm{dist}(C_i, C_j) > R\}$

By the finite range assumption, $\pi_{C_i,\Lambda}(\cdot|\sigma_{\Lambda\setminus C_i}) = \pi'_{C_i,\Lambda}(\cdot|\sigma'_{\Lambda\setminus C_i})$, and the contribution to $\theta^*(i,j)$ in this case is only due to the hamiltonian difference

$$\sum_{\Delta\subset\Lambda:\Delta\supset C_i,\Delta\supset C_j} [U_\Delta(\sigma_\Delta) - U_\Delta(\sigma'_\Delta)]. \tag{11.5.6.4}$$

By Theorem 11.5.5.1 and by (11.5.3.6)

$$\theta^*(i,j) \le 2[c_0\ell_0^d]2\epsilon_3 e^{-\omega_3|i-j|}, \qquad \mathrm{dist}(C_i, C_j) > R, \tag{11.5.6.5}$$

which proves (11.5.3.9) for $i \in I^0$ and $\{j : \mathrm{dist}(C_i, C_j) > R\}$.

The case $i \in I^0$ and $\{j : \mathrm{dist}(C_i, C_j) \le R\}$

By the triangular inequality and using Theorem 11.5.5.1 and (11.5.3.6) with c_3 as in (11.5.3.8),

$$\theta^*(i,j) \le 2[c_0\ell_0^d]2\epsilon_3 c_3 + \sup_{\sigma_\Lambda=\sigma'_\Lambda \text{ on } \Lambda\setminus C_j} \frac{R_{C_i}(\pi_{C_i,\Lambda}(\cdot|\sigma_{\Lambda\setminus C_i}), \pi'_{C_i,\Lambda}(\cdot|\sigma'_{\Lambda\setminus C_i}))}{d_{C_j}(\sigma_{C_j}, \sigma'_{C_j})}. \tag{11.5.6.6}$$

We shall next use Theorem 3.2.2.1 to bound $R_{C_i}(\pi_{C_i,\Lambda}(\cdot|\sigma_{\Lambda\setminus C_i}), \pi'_{C_i,\Lambda}(\cdot|\sigma'_{\Lambda\setminus C_i}))$ in terms of the Vaserstein distances $R_x(\pi_{x,\Lambda}(\cdot|\sigma_{\Lambda\setminus x}), \pi'_{x,\Lambda}(\cdot|\sigma'_{\Lambda\setminus x}))$.

We first exclude the case where $\sigma_{\Lambda\setminus x}$ and $\sigma'_{\Lambda\setminus x}$ are both in $\mathcal{B}_x$. Then, by (11.5.3.3)–(11.5.3.4),

$$\begin{aligned} &R_x\big(\pi_{x,\Lambda}(\cdot|\sigma_{\Lambda\setminus x}), \pi'_{x,\Lambda}(\cdot|\sigma'_{\Lambda\setminus x})\big) \\ &\quad \le \sum_{y\ne x} r(x,y) d_y(\sigma_\Lambda, \sigma'_\Lambda) + d(S)\{\mathbf{1}_{\sigma_\Lambda\in\mathcal{B}_x^{(1)}} + \mathbf{1}_{\sigma'_\Lambda\in\mathcal{B}_x^{(1)}}\} \\ &\quad + c_2 a^d\{\mathbf{1}_{\sigma_\Lambda\in\mathcal{B}_x^{(2)}} + \mathbf{1}_{\sigma'_\Lambda\in\mathcal{B}_x^{(2)}}\}. \end{aligned} \tag{11.5.6.7}$$

The remaining case, $\sigma_{\Lambda\setminus x}$ and $\sigma'_{\Lambda\setminus x}$ both in $\mathcal{B}_x$, is bounded by introducing a third configuration $\sigma''_{\Lambda\setminus x} \notin \mathcal{B}_x$ and using the triangular inequality. We then obtain the same bound (11.5.6.7) which thus holds in all cases.

Then by Theorem 3.2.2.1 there is a coupling Q of $\pi_{C_i,\Lambda}(\cdot|\sigma_{\Lambda\setminus C_i})$ and $\pi'_{C_i,\Lambda}(\cdot|\sigma'_{\Lambda\setminus C_i})$ such that by (11.5.3.5), writing $c' = c_2 + d(S)$,

$$E_Q(d_x) \le \sum_{y\neq x, y\in C_i} r(x,y)E_Q(d_y) + 2\epsilon_2 a^d c' + \sum_{y\in C_j} r(x,y)d_y(\sigma_\Lambda, \sigma'_\Lambda). \tag{11.5.6.8}$$

By Theorem 3.2.4.1, using the notation in (11.5.3.8)

$$E_Q(d_x) \le \frac{a^d}{1-r^*(i)}\left\{2\epsilon_2 c' + \frac{1}{|C_i|} r^*(i,j) d_{C_j}(\sigma_\Lambda, \sigma'_\Lambda)\right\}; \tag{11.5.6.9}$$

hence

$$E_Q(d_{C_i}) \le \frac{1}{1-r^*(i)}\left\{2\epsilon_2 \ell_0^d c' + r^*(i,j) d_{C_j}(\sigma_\Lambda, \sigma'_\Lambda)\right\}. \tag{11.5.6.10}$$

Since $d_{C_j}(\sigma_\Lambda, \sigma'_\Lambda) \neq 0$ then $d_{C_j}(\sigma_\Lambda, \sigma'_\Lambda) \ge 1$, and thus we conclude from (11.5.6.6) that

$$\theta^*(i,j) \le 4[c_0\ell_0^d]\epsilon_3 c_3 + \frac{1}{1-r^*(i)}\{r^*(i,j) + 2\epsilon_2 \ell_0^d c'\}, \qquad \text{dist}(C_i, C_j) \le R, \tag{11.5.6.11}$$

thus completing the proof that $\theta^*(i,j) \le \theta(i,j)$, the latter as in (11.5.3.9).

The case $i \in I \setminus I^0$

In such a case we use the trivial bound

$$R_{C_i}\big(\mu_{C_i,\Lambda}(\cdot|\sigma_{\Lambda\setminus C_i}), \mu'_{C_i,\Lambda}(\cdot|\sigma'_{\Lambda\setminus C_i})\big) \le c_0\ell_0^d, \quad i \in I \setminus I^0, \tag{11.5.6.12}$$

as $c_0\ell_0^d$ is the max of d_{C_i}.

Conclusions

We have proved so far that (11.5.6.2) holds for all $i \in I_0$ and with $\theta^*(i,j) \le \theta(i,j)$, while the bound (11.5.6.12) holds when $i \in I \setminus I^0$. Then by Theorem 3.1.3.4, there is a coupling Q of μ_Λ and μ'_Λ such that

$$E_Q(d_{C_i}) \le \begin{cases} 4[c_0\ell_0^d]\epsilon_3 \sum_{j\in I\setminus I_0} e^{-\omega_3|i-j|} + \sum_{j\neq i, j\in I} \theta(i,j) E_Q(d_{C_j}), & i \in I_0, \\ c_0\ell_0^d, & i \in I \setminus I_0. \end{cases}$$

Writing $v(i) := E_Q(d_{C_i})$, $i \in I_0$, $v(i) \le \sum_{j\in I_0} \theta(i,j)v(j) + \alpha(i)$ and

$$\alpha(i) = 4[c_0\ell_0^d]\epsilon_3 \sum_{j\in I\setminus I_0} e^{-\omega_3|i-j|} + c_0\ell_0^d \sum_{j\in I\setminus I_0} \theta(i,j).$$

By (11.5.3.10) the assumptions in Theorem 3.2.4.1 are satisfied and

$$E_Q(d_{C_i}) = v(i) \le u(i) := \sum_{j\in I_0} g_{I_0}(i,j)\alpha(j),$$

where $g_{I_0}(i,j)$ is the Green function associated to the kernel $\theta(i,j)$. By (11.5.4.1) and Theorem 3.2.5.3, we finally get

$$E_Q(d_{C_i}) \le \frac{1}{1-r} \max_{k\in I_0}\{e^{-\delta|i-k|}\alpha(k)\},$$

and (11.5.4.2)–(11.5.4.3) follow.

11.5.7 Bounds on the correlation functions

This subsection is, finally, devoted to the analysis of the LMP model: it will be used in the proof of Theorem 11.4.5.2. The results therefore are less general and sharp than what a more accurate analysis would yield. By default, in the sequel the spins take values 0 and 1; the one and two body correlation functions of a measure μ_Λ on $\{0,1\}^{\Lambda\cap a\mathbb{Z}^d}$ are defined for $x \ne y \in \Lambda \cap a\mathbb{Z}^d$, as

$$\begin{aligned} \rho_\Lambda(x) &:= a^{-d}\mu_\Lambda\big(\sigma_\Lambda(x) = 1\big), \\ \rho_\Lambda(x,y) &:= a^{-2d}\mu_\Lambda\big(\sigma_\Lambda(x)\sigma_\Lambda(y) = 1\big), \end{aligned} \tag{11.5.7.1}$$

while the two body truncated correlation function is

$$\rho_\Lambda^T(x,y) := \rho_\Lambda(x,y) - \rho_\Lambda(x)\rho_\Lambda(y). \tag{11.5.7.2}$$

These are "density correlations" because $\sigma_\Lambda(x)/a^d$ is the number of particles in the cell $C_x^{(a)} \in \mathcal{D}_x^{(a)}$ divided by the volume of the cell. We shall prove that the one and two body correlations are bounded (uniformly in a) and that the truncated correlation functions are bounded by an expression which in the LMP model vanishes as $\gamma \to 0$ uniformly in a. We need a few extra assumptions.

Setup and extra assumptions

The setup is the same as in Sect. 11.5.2 with $\mu'_\Lambda = \mu_\Lambda$ and $\Lambda^c_{\neq} = \emptyset$, see (11.5.3.2), so that $I_0 = I$ in the present case. Besides the assumption stated in Sect. 11.5.3 we make the following assumptions.

- There are constants c_4 and c_5 so that for any $x \in \Lambda$ and any $y \neq x$ in Λ

$$\sup_{\sigma_{\Lambda\setminus x}\notin \mathcal{B}_x^{(1)}} \pi_{x,\Lambda}\big(\sigma(x)=1\,|\,\sigma_{\Lambda\setminus x}\big) \le c_4 a^d, \qquad r(x,y) \le c_5 a^d R^{-d}, \quad (11.5.7.3)$$

$r(x,y)$ being as in (11.5.3.3). By (11.5.3.10)

$$\sup_{x\in\Lambda\cap a\mathbb{Z}^d} \sum_{y\neq x} r(x,y) \le r' \le r^* < 1. \qquad (11.5.7.4)$$

- For any $C_i \in \mathcal{D}^{(\ell_0)}$ in Λ and for any pair $x \neq x^*$ in C_i, calling $C' = C_i \setminus x^*$,

$$\begin{aligned} &\pi_{C',\Lambda}\big(\{\sigma_{C'} : (\sigma_{C'}, \sigma_{\Lambda\setminus C'}(x)) \in \mathcal{B}_x^{(1)}\}\,|\,\sigma_{\Lambda\setminus C'}\big) \le \epsilon_2 a^d, \\ &\pi_{C',\Lambda}\big(\{\sigma_{C'} : (\sigma_{C'}, \sigma_{\Lambda\setminus C'}(x)) \in \mathcal{B}_x^{(2)}\}\,|\,\sigma_{\Lambda\setminus C'}\big) \le \epsilon_2. \end{aligned} \qquad (11.5.7.5)$$

- The constants c_3 and ϵ_3 in (11.5.3.6) and (11.5.3.8) are such that

$$\epsilon_4 := \epsilon_3 c_3 : \quad e^{2\epsilon_4} \le 2. \qquad (11.5.7.6)$$

Lemma 11.5.7.1 *Let μ_Λ satisfy the above assumptions; then for any set $C' \subset C$, $C \in \mathcal{D}^{(\ell_0)}$ in Λ,*

$$e^{-2\epsilon_4} \le \frac{\mu_{C',\Lambda}(\sigma_{C'}\,|\,\sigma_{\Lambda\setminus C'})}{\pi_{C',\Lambda}(\sigma_{C'}\,|\,\sigma_{\Lambda\setminus C'})} \le e^{2\epsilon_4}. \qquad (11.5.7.7)$$

Proof By definition $\mu_{C',\Lambda}(\sigma_{C'}\,|\,\sigma_{\Lambda\setminus C'})$ is equal to

$$Z^{-1}\pi_{C',\Lambda}(\sigma_{C'}\,|\,\sigma_{\Lambda\setminus C'}) \exp\Big\{-\beta \sum_{\Delta\supset C} U_\Delta(\sigma_{\Delta\cap\Lambda})\Big\},$$

where $\sigma_{\Delta\cap\Lambda}$ is the restriction to $\Delta\cap\Lambda$ of the configuration $\sigma_\Lambda = (\sigma_{C'}, \sigma_{\Lambda\setminus C'})$; Z is the normalization factor, namely the $\pi_{C',\Lambda}(\cdot\,|\,\sigma_{\Lambda\setminus C'})$ expectation of

$$\exp\Big\{-\beta \sum_{\Delta\supset C} U_\Delta\big(\sigma'_{C'\cap(\Delta\cap\Lambda)}, \sigma_{(\Delta\cap\Lambda)\setminus C'}\big)\Big\}.$$

The upper bound in (11.5.7.7) then follows from (11.5.3.6), the definition of c_3 in (11.5.3.8) and on recalling that $\epsilon_4 = \epsilon_3 c_3$. The lower bound is proved similarly. □

We are now ready for the main result in this subsection:

Theorem 11.5.7.2 *There is c independent of a so that for any $x \in \Lambda$*

$$\rho_\Lambda(x) \le c, \qquad (11.5.7.8)$$

and for any pair $x \neq x^*$ in Λ ($d(S) = 1$ and r' as in (11.5.7.4) below), we have

$$\rho_\Lambda^T(x,x^*) \le \frac{c_4}{1-r'}\big([e^{2\epsilon_4}-1]c_4 + 2\epsilon_2 a^d (d(S)+c_2) + 2c_5 R^{-d}\big). \qquad (11.5.7.9)$$

Proof Before starting the proof, notice that in the applications to LMP the r.h.s. of (11.5.7.9) is bounded by $c\gamma^d$, $c>0$ being a constant, as the coefficients ϵ_i are exponentially small in γ^{-1} and R is proportional to γ^{-1}.

Proof of (11.5.7.8) Recall that when $S=\{0,1\}$, $\sigma_\Lambda \in B_x^{(1)} \Rightarrow \sigma_\Lambda(x)=1$ and $\sigma_\Lambda \in B_x^{(2)} \Rightarrow \sigma_\Lambda(x)=0$. Then, calling C_i the cube containing x, we have

$$
\begin{aligned}
a^d \rho_\Lambda(x) &= \mu_\Lambda(\mathcal{B}_x^{(1)}) + \mu_\Lambda\big((\mathcal{B}_x^{(1)})^c \cap \{\sigma_\Lambda(x)=1\}\big)\\
&\le E_{\mu_\Lambda}\big(\mu_{C_i,\Lambda}(\mathcal{B}_x^{(1)}\,|\sigma_{\Lambda\setminus C_i})\big) + E_{\mu_\Lambda}\big(\mathbf{1}_{\sigma_{\Lambda\setminus x}\notin \mathcal{B}_x^{(1)}}\mu_{x,\Lambda}\big(\sigma_\Lambda(x)=1\,|\,\sigma_{\Lambda\setminus x}\big)\big).
\end{aligned}
$$

By Lemma 11.5.7.1 $\mu_{C_i,\Lambda}(\mathcal{B}_x^{(1)}\,|\sigma_{\Lambda\setminus C_i}) \le 2\pi_{C_i,\Lambda}(\mathcal{B}_x^{(1)}\,|\sigma_{\Lambda\setminus C_i}) \le 2\epsilon_2 a^d$, the last inequality by (11.5.3.5). Analogously, $\mu_{x,\Lambda}(\sigma_\Lambda(x)=1\,|\,\sigma_{\Lambda\setminus x}) \le 2\pi_{x,\Lambda}(\sigma_\Lambda(x)=1\,|\,\sigma_{\Lambda\setminus x}) \le 2c_4 a^d$, by (11.5.7.3) and because $\sigma_{\Lambda\setminus x}\notin \mathcal{B}_x^{(1)}$. Thus, $a^d C_\Lambda(x) \le 2\epsilon_2 a^d + 2c_4 a^d$; hence (11.5.7.8).

Proof of (11.5.7.9) Let μ_s, $s\in S=\{0,1\}$, be the measure on $S^{a\mathbb{Z}^d\cap[\Lambda\setminus x^*]}$ defined by

$$
\mu_s(\sigma_{\Lambda\setminus x^*}) := \mu_{\Lambda\setminus x^*,\Lambda}\big(\sigma_{\Lambda\setminus x^*}\ \big|\ \{\sigma(x^*)=s\}\big); \tag{11.5.7.10}
$$

$\rho_\Lambda^T(x,x^*)$ can then be written in terms of couplings $Q_{s,s'}$ of μ_s and $\mu_{s'}$:

$$
\begin{aligned}
\rho_\Lambda^T(x,x^*) &= a^{-2d}\sum_{s,s'} s\,\mu_\Lambda\big(\sigma_\Lambda(x^*)=s\big)\mu_\Lambda\big(\sigma_\Lambda(x^*)=s'\big)\\
&\quad\times \sum_{\sigma_{\Lambda\setminus x^*},\sigma'_{\Lambda\setminus x^*}} \{\sigma_{\Lambda\setminus x^*}(x)-\sigma'_{\Lambda\setminus x^*}(x)\}\\
&\quad\times Q_{s,s'}(\sigma_{\Lambda\setminus x^*},\sigma'_{\Lambda\setminus x^*}).
\end{aligned} \tag{11.5.7.11}
$$

Claim *There is a coupling $Q_{s,s'}$ of μ_s and $\mu_{s'}$ so that*

$$
\begin{aligned}
&Q_{s,s'}\big(d_x(\sigma_\Lambda,\sigma'_\Lambda)\big)\\
&\quad\le \frac{a^d}{1-r'}\big([e^{2\epsilon_4}-1]c_4 + 2\epsilon_2 a^d(d(S)+c_2)+2c_5 R^{-d}\big).
\end{aligned} \tag{11.5.7.12}
$$

Proof of the claim Let $\Lambda' = \Lambda\setminus x^*$. By Lemma 11.5.7.1

$$
\begin{aligned}
&R_x\big(\mu_{x,\Lambda}\big(\cdot\,|\{\sigma_{\Lambda'\setminus x},\sigma(x^*)=s\}\big),\mu_{x,\Lambda}\big(\cdot|\{\sigma'_{\Lambda'\setminus x},\sigma(x^*)=s'\}\big)\big)\\
&\quad= \big|\mu_{x,\Lambda}\big(\sigma(x)=1|\{\sigma_{\Lambda'\setminus x},\sigma(x^*)=s\}\big)\\
&\qquad - \mu_{x,\Lambda}\big(\sigma(x)=1|\{\sigma'_{\Lambda'\setminus x},\sigma(x^*)=s'\}\big)\big|\\
&\quad\le [e^{2\epsilon_4}-1]2c_4a^d + \big|\pi_{x,\Lambda}\big(\sigma(x)=1|\{\sigma_{\Lambda'\setminus x},\sigma(x^*)=s\}\big)\\
&\qquad - \pi_{x,\Lambda}\big(\sigma(x)=1|\{\sigma'_{\Lambda'\setminus x},\sigma(x^*)=s'\}\big)\big|.
\end{aligned}
$$

The last term is $R_x(\pi_{x,\Lambda}(\cdot|\{\sigma_{\Lambda'\setminus x}, \sigma(x^*) = s\}), \pi_{x,\Lambda}(\cdot|\{\sigma'_{\Lambda'\setminus x}, \sigma(x^*) = s'\}))$. Then by (11.5.3.3), (11.5.3.4), (11.5.7.3) and Theorem 3.2.2.1 there is a coupling $Q_{s,s'}$ so that

$$\begin{aligned} Q_{s,s'}\big(d_x(\sigma_\Lambda, \sigma'_\Lambda)\big) &\le [e^{2\epsilon_4} - 1]2c_4 a^d + 2\epsilon_2 a^d (d(S) + c_2) \\ &\quad + r(x, x^*)d(s, s') + \sum_{y \ne x, x^*} r(x, y) Q_{s,s'}\big(d_y(\sigma_\Lambda, \sigma'_\Lambda)\big); \end{aligned} \tag{11.5.7.13}$$

and hence (11.5.7.12). The proof of the claim is complete.

Equation (11.5.7.9) follows directly from (11.5.7.12). In fact the sum over s, s' in (11.5.7.11) is actually restricted to $s = 1$ and $s' = 0$. By (11.5.7.8), $\mu_\Lambda(\sigma_\Lambda(x^*) = 1) \le c a^d$ and then (11.5.7.12), (11.5.7.11) yield (11.5.7.9). □

11.6 Surface corrections to the pressure, conclusion

We shall prove here Theorems 11.4.5.1 and 11.4.5.2 (see Sects. 11.6.3 and 11.6.4) and thus conclude the proof of (11.0.2.5) and of Theorem 10.3.4.1. With $m^+_{\gamma,a,\Lambda;u}(\sigma_\Lambda)$ defined in (11.4.4.4), we use the shorthand

$$\mu_\Lambda(\sigma_\Lambda) = m^+_{\gamma,a,\Lambda;u}(\sigma_\Lambda); \qquad \mu'_\Lambda(\sigma_\Lambda) = m^+_{\gamma,a,\Lambda';u}(\sigma_\Lambda|\sigma_{\Lambda'\setminus\Lambda}), \tag{11.6.0.1}$$

where Λ and $\Lambda' \supset \Lambda$ are bounded $\mathcal{D}^{(\ell_{+,\gamma})}$-measurable regions and $\sigma_{\Lambda'\setminus\Lambda}$ is any fixed configuration in $\mathcal{X}^+_{\Lambda'\setminus\Lambda,a}$. To compare μ_Λ and μ'_Λ we shall use the theory of Sect. 11.5: we shall first identify the various quantities of Sect. 11.5 in our LMP model and then check that all assumptions are satisfied.

The reference measures π_Λ and π'_Λ

We take $S = \{0, 1\}$, $\ell_0 = \ell_{-,\gamma}$ and $\ell_1 = \ell_{+,\gamma}$, μ_Λ and μ'_Λ as in (11.6.0.1). The reference measures π_Λ and π'_Λ of Sect. 11.5 are

$$\begin{aligned} \pi_\Lambda(\sigma_\Lambda) &= \frac{1}{Z}\,(a^d e^{-\beta(1-u)e'_{\lambda(\beta)}(\rho_{\beta,+})})^{|\sigma_\Lambda|} e^{-\beta u H_{\gamma,\Lambda}(\sigma_\Lambda|\chi^+_{\Lambda^c})}, \\ \pi'_\Lambda(\sigma_\Lambda) &= \frac{1}{Z'}(a^d e^{-\beta(1-u)e'_{\lambda(\beta)}(\rho_{\beta,+})})^{|\sigma_\Lambda|} e^{-\beta u H_{\gamma,\Lambda}(\sigma_\Lambda|\sigma'_{\Lambda'\setminus\Lambda}, \chi^+_{(\Lambda')^c})}. \end{aligned} \tag{11.6.0.2}$$

They are measures on $\mathcal{X}^+_{\Lambda,a}$ which is the product of $\mathcal{X}^+_{C_i,a}$, $i \in I = \ell_{-\gamma}\mathbb{Z}^d \cap \Lambda$. The coefficient c_0 in (11.5.2.1) is then $c_0 = 2(\rho_{\beta,+} + \zeta)$ and the interaction range R is $R = 2\gamma^{-1}$. The sets $\mathcal{B}^{(1)}_x$ and $\mathcal{B}^{(2)}_x$ are those defined in (11.5.1.4).

The potentials U_Δ and U'_Δ

We write $\frac{d\mu_\Lambda(\sigma_\Lambda)}{d\pi_\Lambda(\sigma_\Lambda)} = Z^{-1}\exp\{-\beta\sum_{\Delta\subset\Lambda} U^+_{\gamma,\Delta}(\sigma_\Delta)\}$,

$$\frac{d\mu'_\Lambda(\sigma_\Lambda)}{d\pi'_\Lambda(\sigma_\Lambda)} = Z^{-1}\exp\Big\{-\beta\sum_{\Delta\cap\Lambda\neq\emptyset,\Delta\subset\Lambda'} U^+_{\gamma,\Delta}(\sigma^*_\Delta)\Big\},\quad \sigma^* = (\sigma_\Lambda,\sigma'_{\Lambda'\setminus\Lambda}).$$

U_Δ and U'_Δ of Sect. 11.5 are then $U_\Delta(\sigma_\Delta) = U^+_{\gamma,\Delta}(\sigma_\Delta)\mathbf{1}_{\Delta\subset\Lambda}$ and

$$U'_\Delta(\sigma_\Delta) = U^+_{\gamma,\Delta}(\sigma_\Delta,\sigma'_{\Delta\cap\{\Lambda'\setminus\Lambda\}})\mathbf{1}_{\Delta\cap\Lambda\neq\emptyset,\Delta\subset\Lambda'}. \tag{11.6.0.3}$$

11.6.1 *Compatibility of the parameters*

Having completed the identification of the quantities in Sect. 11.5, we now need to check that the various coefficients satisfy the condition for the validity of Theorem 11.5.4.1, Corollary 11.5.4.2 and Theorem 11.5.7.2.

The coefficients ϵ_3, ω_3 and c_3

Let C_i, C_j be cubes of $\mathcal{D}^{(\ell_{-,\gamma})}$ contained in Λ; then by (11.4.3.9)–(11.4.3.7), $\beta\sum_{\Delta\supset C_i,\Delta\supset C_j}\|U_\Delta\|_\infty$ is bounded by

$$\{3^d e^{-\beta c_{\rm pol}(\zeta^2\ell^d_{-,\gamma})3^d/2}\}\{e^{-\beta c_{\rm pol}(\zeta^2\ell^d_{-,\gamma})\kappa\,{\rm dist}(C_i,C_j)/(2\ell_{+,\gamma})}\}. \tag{11.6.1.1}$$

Hence (11.5.3.6) holds with

$$\epsilon_3 = 3^d e^{-\beta c_{\rm pol}(\zeta^2\ell^d_{-,\gamma})3^d/2},\qquad \omega_3 = \beta c_{\rm pol}\kappa\frac{\zeta^2\ell^d_{-,\gamma}}{2\ell_{+,\gamma}}. \tag{11.6.1.2}$$

The coefficient c_3 in (11.5.3.8) is by (11.6.1.2)

$$c_3 = \sum_{i\in\ell_0\mathbb{Z}^d} e^{-\beta c_{\rm pol}\kappa\zeta^2\ell^d_{-,\gamma}(2\ell_{+,\gamma})^{-1}|i|} \le c\exp\{-c'\gamma^{-(d-1-(\alpha+1)d-2a)}\}, \tag{11.6.1.3}$$

c and c' being positive constants.

The parameters $r(x,y)$, $r^*(i)$, $r^*(i,j)$; the coefficients c_4 and c_5

The one site conditional probability $\pi_{x,\Lambda}(\cdot|\sigma_{\Lambda\setminus x})$ is

$$\pi_{x,\Lambda}\big(\sigma(x)=1|\sigma_{\Lambda\setminus x}\big) = \begin{cases}1 & \text{if } \sigma_{\Lambda\setminus x}\in\mathcal{B}^{(1)}_x,\\ 0 & \text{if } \sigma_{\Lambda\setminus x}\in\mathcal{B}^{(2)}_x,\end{cases} \tag{11.6.1.4}$$

while, when $\sigma_{\Lambda\setminus x} \notin \mathcal{B}_x$, it is given by the r.h.s. of (11.5.1.1). Analogous expressions hold for π'_Λ. Then by (11.1.2.8), the coefficient c_4 in (11.5.7.3) is $c_4 = e^{\beta b_0}$. Call the Vaserstein distance $R_x(\sigma_{\Lambda\setminus x}, \sigma'_{\Lambda\setminus x})$,

$$\inf_{Q} \sum_{\sigma(x),\sigma'(x)} |\sigma(x) - \sigma'(x)| Q_{\sigma_{\Lambda\setminus x},\sigma'_{\Lambda\setminus x}}(\sigma(x), \sigma'(x)), \tag{11.6.1.5}$$

the inf being over all the joint representations of $\pi_{x,\Lambda}(\cdot|\sigma_{\Lambda\setminus x})$ and $\pi_{x,\Lambda}(\cdot|\sigma'_{\Lambda\setminus x})$. If σ and σ' are both in $\mathcal{B}^c_x$, then we can use (11.5.1.3) to conclude that the parameters $r(x, y)$ in (11.5.3.3) are (with $\theta < 1$ below as in (11.5.1.2))

$$r(x, y) := (\theta + \epsilon)a^d \left\{ \int J_\gamma(r, y) J_\gamma(x, r) \right\}, \tag{11.6.1.6}$$

where $\epsilon > 0$ can be made as small as desired by restricting to small γ. Then

$$r(x, y) \le (\theta + \epsilon)a^d \|J\|_\infty 2^d (2\gamma^{-1})^{-d} \mathbf{1}_{|x-y|\le 2\gamma^{-1}}, \qquad c_5 = \|J\|_\infty 2^d$$

with c_5 as in (11.5.7.3) (having used that $(\theta + \epsilon) < 1$).

The coefficient $r^*(i)$ in (11.5.3.8) is bounded by

$$r^*(i) \le c\gamma^d \ell^d_{-,\gamma} = c\gamma^{\alpha d}, \tag{11.6.1.7}$$

and the coefficient $r^*(i, j)$ in (11.5.3.8) is then identified with

$$r^*(i, j) = \ell^d_{-,\gamma}(\theta + \epsilon) \max_{x\in C_i, y\in C_j} \left\{ \int J_\gamma(r, y) J_\gamma(x, r) \right\}. \tag{11.6.1.8}$$

The coefficient ϵ_2

We next check the condition (11.5.3.5) on the "one block conditional probabilities" which determines the parameter ϵ_2. The lemma below, which concludes this subsection, covers also the case considered in (11.5.7.5) and is used as one of the extra assumptions needed to bound the correlation functions.

Lemma 11.6.1.1 *There is $c > 0$ so that for all γ and a small enough the following holds. Let $C \in \mathcal{D}^{(\ell_{-,\gamma})}$, $\mathrm{dist}(C, \Lambda^c) > 2\gamma^{-1}$, $x \in C$, $\sigma_{\Lambda\setminus C} \in \mathcal{X}^+_{\Lambda\setminus C,a}$; then*

$$\pi_{C,\Lambda}\big(\mathcal{B}^{(1)}_x | \sigma_{\Lambda\setminus C}\big) \le a^d e^{-c\zeta^2 \ell^d_{-,\gamma}}, \qquad \pi_{C,\Lambda}\big(\mathcal{B}^{(2)}_x | \sigma_{\Lambda\setminus C}\big) \le e^{-c\zeta^2 \ell^d_{-,\gamma}}. \tag{11.6.1.9}$$

The same bound holds if C is replaced by $C' = C \setminus x^*$, $x^* \neq x$ in C for the conditional probability given $\sigma_{\Lambda\setminus C'}$. Equation (11.6.1.9) identifies the coefficient ϵ_2 as

$$\epsilon_2 = e^{-c\zeta^2 \ell^d_{-,\gamma}}. \tag{11.6.1.10}$$

Proof We shall only prove (11.6.1.9), as its analogue with C' can easily be recovered from the arguments below. Before starting the proof we write more explicitly the reference measure. Recalling that points in Λ^c do not interact with points in C, let

$$p_C(\sigma_C|\sigma_{\Lambda\setminus C}) = Z^{-1}\big(a^d e^{-\beta(1-u)e'_{\lambda(\beta)}(\rho_{\beta,+})}\big)^{|\sigma_C|} e^{-\beta u H_{\gamma,\lambda_{\beta,\gamma},C}(\sigma_C|\sigma_{\Lambda\setminus C})} \tag{11.6.1.11}$$

be a probability on $\{0,1\}^C$ (which is a shorthand for $C\cap a\mathbb{Z}^d$; Z is the normalization factor), and there are no constraints so far. Writing

$$N_C=\sum_{y\in C}\sigma_C(y),\qquad N_{C\setminus x}=\sum_{y\in C\setminus x}\sigma_C(y), \tag{11.6.1.12}$$

the constraint is $N_C\in[n_-,n_+]$, see (11.5.1.4), so that

$$\pi_{C,\Lambda}(\sigma'_C|\sigma_{\Lambda\setminus C})=p_C\big(\sigma'_C\,\big|\,\sigma_{\Lambda\setminus C},\{N_C\in[n_-,n_+]\}\big) \tag{11.6.1.13}$$

and

$$\begin{aligned}\pi_{C,\Lambda}\big(\mathcal{B}^{(1)}_x|\sigma_{\Lambda\setminus C}\big)&=\frac{p_C(\{N_{C\setminus x}=n_--1;\sigma_C(x)=1\}\,|\,\sigma_{\Lambda\setminus C})}{p_C(\{N_C\in[n_-,n_+]\}\,|\,\sigma_{\Lambda\setminus C})},\\ \pi_{C,\Lambda}\big(\mathcal{B}^{(2)}_x\,|\,\sigma_{\Lambda\setminus C}\big)&=\frac{p_C(\{N_{C\setminus x}=n_+;\sigma_C(x)=0\}\,|\,\sigma_{\Lambda\setminus C})}{p_C(\{N_C\in[n_-,n_+]\}\,|\,\sigma_{\Lambda\setminus C})}.\end{aligned} \tag{11.6.1.14}$$

We are now ready for the proof of (11.6.1.9) which proceeds by bounding the probabilities in the numerators of (11.6.1.14). We obviously have $p_C(\{N_{C\setminus x}=n_+;\sigma_C(x)=0\}\,|\,\sigma_{\Lambda\setminus C})\le p_C(\{N_{C\setminus x}=n_+\,|\,\sigma_{\Lambda\setminus C})$. We also have

$$\begin{aligned}&p_C\big(\{N_{C\setminus x}=n_--1;\sigma_C(x)=1\}\,\big|\,\sigma_{\Lambda\setminus C}\big)\\&\quad\le ca^d p_C\big(\{N_{C\setminus x}=n_--1\}\,\big|\,\sigma_{\Lambda\setminus C}\big),\end{aligned} \tag{11.6.1.15}$$

which follows by conditioning on $\sigma_{\{x\}^c}$. The conditional measure is then given by the r.h.s. of (11.5.1.1), and (11.6.1.15) follows because $E_x(\cdot)$ is bounded.

We thus need to bound $p_C(\{N_{C\setminus x}=n_+\}\,|\,\sigma_{\Lambda\setminus C})$ and $p_C(\{N_{C\setminus x}=n_--1\}\,|\,\sigma_{\Lambda\setminus C})$; as the two are similar we shall only consider the former. We have

$$p_C\big(\{N_{C\setminus x}=n_+\}\,\big|\,\sigma_{\Lambda\setminus C}\big)\le p_C\big(\{N_C\in\{n_+,n_++1\}\}\,\big|\,\sigma_{\Lambda\setminus C}\big). \tag{11.6.1.16}$$

Going back to (11.6.1.11),

$$\left|H_{\gamma,\lambda_{\beta,\gamma},C}(\sigma_C|\sigma_{\Lambda\setminus C})-\int_{\mathbb{R}^d}e'_{\lambda_{\beta,\gamma}}(J_\gamma*\sigma_{\Lambda\setminus C})J_\gamma*\sigma_C\right|\le c\gamma^d\ell^{2d}_{-,\gamma}=c\gamma^{\alpha d}\ell^d_{-,\gamma}.$$

Denoting by $\bar J_\gamma(r,r')=\fint_{C_r^{(\ell_{-,\gamma})}}\fint_{C_{r'}^{(\ell_{-,\gamma})}}J_\gamma(r_1,r_1')$, by the smoothness of J_γ,

$$\left|H_{\gamma,\lambda_{\beta,\gamma},C}(\sigma_C|\sigma_{\Lambda\setminus C})-\int_{\mathbb{R}^d}e'_{\lambda_{\beta,\gamma}}(J_\gamma*\sigma_{\Lambda\setminus C})\bar J_\gamma*\sigma_C\right|\le c'\gamma^{\alpha}\ell^d_{-,\gamma}.$$

Let y be any point in C,

$$\lambda_{\text{eff}} = -(1-u)e'_{\lambda(\beta)}(\rho_{\beta,+}) - u \int \bar{J}_\gamma(r,y)e'_{\lambda_{\beta,\gamma}}(J_\gamma * \sigma_{\Lambda\setminus C}(r)), \tag{11.6.1.17}$$

and call P the probability on $\{\mathbb{N}\cap[0,M]\}$, $M = \frac{\ell^d_{-,\gamma}}{a^d}$, defined as

$$P(n) = Z^{-1}(a^d e^{\beta\lambda_{\text{eff}}})^n \binom{M}{n}. \tag{11.6.1.18}$$

Z^{-1} is the normalization. Then by (11.6.1.16)

$$p_C\big(\{N_{C\setminus x} = n_+\}\,\big|\,\sigma_{\Lambda\setminus C}\big) \le e^{2c'\gamma^\alpha \ell^d_{-,\gamma}}\, P\big(\{n = n_+, n_+ + 1\}\big). \tag{11.6.1.19}$$

Analogously

$$p_C\big(\{N_C \in [n_-, n_+]\}\,\big|\,\sigma_{\Lambda\setminus C}\big) \ge e^{-2c'\gamma^\alpha \ell^d_{-,\gamma}}\, P\big(\{n \in [n_-, n_+]\}\big). \tag{11.6.1.20}$$

What is left in the proof is just elementary; we only need the Stirling formula, see (A.2.3), and a Taylor expansion, plus some properties of the mean field free energy. By (11.6.1.18) for any $n \in [n_-, n_+]$,

$$\left| P(n) - \frac{e^{-\ell^d_{-,\gamma}\{\rho(\log\rho - 1) + \beta\lambda_{\text{eff}}\rho\}}}{(2\pi\ell^d_{-,\gamma}\rho)^{1/2}} \right| \le c a^d \ell^{2d}_{-,\gamma}, \quad \rho := \frac{n}{\ell^d_{-,\gamma}}. \tag{11.6.1.21}$$

The minimum in the exponent is reached at $\rho^* := e^{\beta\lambda_{\text{eff}}}$. Recalling (11.6.1.17) and using that $|e^{\beta e'_{\lambda(\beta)}(\rho_{\beta,+})}\beta e''_{\lambda(\beta)}(\rho_{\beta,+})| < 1$, there is $\kappa_0 > 0$ so that for all γ and a small enough,

$$|\rho^* - \rho_{\beta,+}| \le (1-\kappa_0)\zeta. \tag{11.6.1.22}$$

Thus, after a second-order Taylor expansion, for all γ and a small enough,

$$P\big(\{n = n_+, n_+ + 1\}\big) \le 2e^{-\ell^d_{-,\gamma}(\kappa_0\zeta)^2/(4\rho_{\beta,+})}\, \frac{e^{-\ell^d_{-,\gamma}\{\rho^*(\log\rho^* - 1) + \beta\lambda_{\text{eff}}\rho^*\}}}{(2\pi\ell^d_{-,\gamma}\rho^*)^{1/2}},$$

while, calling n the integer part of $\ell_{-,\gamma,d}\rho^*$ and $\rho = n/\ell_{-,\gamma^d}$,

$$P\big(\{n \in [n_-, n_+]\}\big) \ge \frac{e^{-\ell^d_{-,\gamma}\{\rho(\log\rho - 1) + \beta\lambda_{\text{eff}}\rho\}}}{(2\pi\ell^d_{-,\gamma}\rho)^{1/2}},$$

thus obtaining (11.6.1.9) (we omit the details). □

11.6.2 Main implications

We shall check here the validity of (11.5.4.1) determining the decay parameter δ and then state Corollary 11.5.4.2 and Theorem 11.5.7.2 in the present context. Recalling that $\theta(i,j)$ is defined in (11.5.3.9), by the bounds proved so far we have that if $\mathrm{dist}(C_i,C_j)\le 2\gamma^{-1}$ then for any $\epsilon'>0$ and correspondingly for all $\gamma>0$ small enough:

$$\theta(i,j)\le \ell^d_{-,\gamma}(\theta+\epsilon')J^2_\gamma(i,j),\quad J^2_\gamma(i,j):=\int J_\gamma(r,j)J_\gamma(r,i)\,dr. \tag{11.6.2.1}$$

When instead $\mathrm{dist}(C_i,C_j)>2\gamma^{-1}$, there is a constant $c>0$ so that

$$\theta(i,j)\le c\ell^d_{-,\gamma}\epsilon_3 e^{-\omega_3|i-j|},\quad \epsilon_3 \text{ and } \omega_3 \text{ as in (11.6.1.2).} \tag{11.6.2.2}$$

Since ω_3 diverges as $\gamma\to 0$, for any $\epsilon''>0$ and all γ small enough

$$\sum_{j:\mathrm{dist}(C_i,C_j)>2\gamma^{-1}} \theta(i,j)e^{|i-j|}\le\epsilon'',\qquad \sum_{j:\mathrm{dist}(C_i,C_j)\le 2\gamma^{-1}} \ell^d_{-,\gamma}J^2_\gamma(i,j)\le 1+\epsilon''.$$

Let $\omega>0$ be such that $\theta e^{2\omega}<1$, recalling that $\theta:=\beta e^{-\beta e'_{\lambda(\beta)}(\rho_{\beta,+})}\times |e''_{\lambda(\beta)}(\rho_{\beta,+})|<1$; then for all γ small enough

$$\sum_{j\ne i}\theta(i,j)e^{\gamma\omega|i-j|}\le\epsilon''+\epsilon' e^{2\omega}(1+\epsilon'')(\theta+\epsilon')=r<1. \tag{11.6.2.3}$$

For all γ small enough $1-r\ge(1-\theta e^{2\omega})/2$; hence (11.5.4.1) with

$$\delta=\gamma\omega,\qquad r\le 1-\frac{1-\theta e^{2\omega}}{2}. \tag{11.6.2.4}$$

Then Proposition 11.6.2.1 below (which will be used in the next chapter) follows directly from Theorem 11.5.4.1, Proposition 11.6.2.2 from Corollary 11.5.4.2 and Proposition 11.6.2.3 from Theorem 11.5.7.2:

Proposition 11.6.2.1 *There is c so that for all $\gamma\le\gamma^*$ the following holds uniformly in a. Let $u=1$, μ'_Λ, μ''_Λ as in (11.6.0.1) with boundary conditions $\sigma'_{\Lambda'\setminus\Lambda}$, $\sigma''_{\Lambda''\setminus\Lambda}$. Then there is a coupling Q of μ'_Λ and μ''_Λ so that for all $C\in\mathcal{D}^{(\ell_{-,\gamma})}$ contained in $\Lambda\setminus\delta^{2\gamma^{-1}}_{\mathrm{in}}[\Lambda]$*

$$E_Q(d_C)\le c\ell^d_{-,\gamma}\sum_{C_j\in\delta^{2\gamma^{-1}}_{\mathrm{in}}[\Lambda]} e^{-\omega\gamma\,\mathrm{dist}(C,C_j)}, \tag{11.6.2.5}$$

with $\omega>0$ as in (11.6.2.3).

Proposition 11.6.2.2 *There is c so that for all $\gamma \le \gamma^*$ the following holds uniformly in a, u and all measures μ_Λ and μ'_Λ as above. Let $f(\sigma) = f(\sigma_\Delta)$, $\Delta \subset \Lambda_{00}$ (Λ_{00} as in (11.4.5.1)) a $\mathcal{D}^{(\ell_{-,\gamma})}$-measurable set; then ($C_i$ and C_j below both in $\mathcal{D}^{(\ell_{-,\gamma})}$)*

$$|\mu_\Lambda(f) - \mu'_\Lambda(f)| \le c\ell^d_{-,\gamma} \|f\|_\infty \sum_{C_i \in \Delta} \sum_{C_j \in \delta_{\text{in}}^{2\gamma^{-1}}[\Lambda]} e^{-\omega\gamma \operatorname{dist}(C_i, C_j)}, \tag{11.6.2.6}$$

with $\omega > 0$ as in (11.6.2.3).

Proposition 11.6.2.3 *Denote $\rho(x) := a^{-d}\mu_\Lambda(\{\sigma_\Lambda(x) = 1\})$, μ_Λ as in (11.6.0.1) (the dependence on Λ, γ, a and u is not made explicit). Then there is c so that for any $u \in [0, 1]$, any bounded $\mathcal{D}^{(\ell_{+,\gamma})}$-measurable region Λ, any $x \in \Lambda$*

$$\rho(x) \le c, \tag{11.6.2.7}$$

and for any $y \in \Lambda$, $y \ne x$,

$$a^{-2d}\left|\mu_\Lambda\big([\sigma(x) - a^d\rho(x)][\sigma(y) - a^d\rho(y)]\big)\right| \le c\gamma^d. \tag{11.6.2.8}$$

11.6.3 Proof of Theorem 11.4.5.1

As we shall see, Theorem 11.4.5.1 is a corollary of Proposition 11.6.2.2. Fix an increasing sequence of bounded $\mathcal{D}^{(\ell_{+,\gamma})}$-measurable regions Λ_n. With μ_{Λ_n} as in (11.6.0.1) we have:

Lemma 11.6.3.1 *μ_{Λ_n} converges weakly to a measure $m^+_{\gamma,a,u}$ invariant under translations of $\ell_{+,\gamma}$ and independent of the sequence Λ_n.*

Proof By (11.6.2.6), $\{\mu_{\Lambda_n}(f)\}$, f being a cylindrical function, is a Cauchy sequence and the limits $m^+_{\gamma,a;u}(f)$ define a measure $m^+_{\gamma,a;u}$ on $\mathcal{X}^+_{\gamma,a}$. Using again (11.6.2.6) we have independence of the sequence $\{\Lambda_n\}$; hence $m^+_{\gamma,a;u}$ is invariant under translations of $\ell_{+,\gamma}$. □

By Proposition 11.6.2.2 and the uniformity on the boundary conditions defining μ'_Λ, it follows that for $\Lambda_n \supset \Lambda$ (Λ_n an element of the sequence defining $m^+_{\gamma,a;u}$)

$$|\mu_\Lambda(f) - \mu_{\Lambda_n}(f)| \le c\ell^d_{-,\gamma} \|f\|_\infty \sum_{C_i \in \Delta} \sum_{C_j \in \delta_{\text{in}}^{2\gamma^{-1}}[\Lambda]} e^{-\omega\gamma \operatorname{dist}(C_i, C_j)}. \tag{11.6.3.1}$$

By Lemma 11.6.3.1 and since (11.6.3.1) holds for all Λ_n,

$$|\mu_\Lambda(f) - m^+_{\gamma,a;u}(f)| \le c\ell^d_{-,\gamma} \|f\|_\infty \sum_{C_i \in \Delta} \sum_{C_j \in \delta_{\text{in}}^{2\gamma^{-1}}[\Lambda]} e^{-\omega\gamma \operatorname{dist}(C_i, C_j)}, \tag{11.6.3.2}$$

and by choosing f suitably we prove (11.4.5.2), (11.4.5.3) and (11.4.5.4).

11.6.4 Proof of Theorem 11.4.5.2

We have completed in the previous subsection the proof of Theorem 11.4.5.1 and we start now the proof of Theorem 11.4.5.2, which by (11.4.6.1) will then complete the proof of the Peierls bounds. We postpone the proof of:

Proposition 11.6.4.1 *There is c so that for any $u \in [0,1]$, any bounded $\mathcal{D}^{(\ell_+,\gamma)}$-measurable region Λ, any $x \in \Lambda$ and with $\rho(x)$ as in Proposition* 11.6.2.3

$$|\rho(x) - \rho_{\beta,+}| \le c\sqrt{\gamma}. \tag{11.6.4.1}$$

Equation (11.4.5.6) is a straightforward consequence of (11.6.4.1); in fact by (11.6.4.1) the proof of (11.4.5.5) is reduced to (expectations are with respect to μ_Λ as in (11.6.0.1))

$$\left\langle \left(\sum_y J_\gamma(r,y)[\sigma_\Lambda(y) - a^d\rho(y)] \right)^2 \right\rangle \le c\gamma, \tag{11.6.4.2}$$

for all a small enough. By Proposition 11.6.2.3, see (11.6.2.8),

$$\begin{aligned} &\left\langle \left(\sum_y J_\gamma(r,y)[\sigma_\Lambda(y) - a^d\rho(y)] \right)^2 \right\rangle \\ &\quad \le c\gamma^d a^{2d} \sum_{y\ne y'} J_\gamma(r,y)J_\gamma(r,y') \\ &\qquad + \sum_y J_\gamma(r,y)^2[a^d\rho(y) - (a^d\rho(y))^2]; \end{aligned} \tag{11.6.4.3}$$

hence (11.4.5.5). We are thus left with the proof of Proposition 11.6.4.1.

11.6.5 Proof of Proposition 11.6.4.1

Recall that $\rho(x) = \mu_\Lambda(\sigma_\Lambda(x) = 1)$, $x \in \Lambda$, μ_Λ as in (11.6.0.1), and the dependence on γ, a, Λ and u is not made explicit. We shall prove (11.6.4.1) by proving that $\rho(\cdot)$ satisfies an equation close to the mean field equation for $\rho_{\beta,+}$. Let $x \in \Lambda$, ψ being a function on a set containing $a\mathbb{Z}^d$, $J_\gamma^{(x)} * \psi(r) = \sum_{y\in a\mathbb{Z}^d, y\ne x} J_\gamma(r,y)\psi(y)$ and

$$\lambda_x(\psi) = -\int J_\gamma(x,r) e'_{\lambda_{\beta,\gamma}}\big(J_\gamma^{(x)} * \psi(r)\big). \tag{11.6.5.1}$$

We extend σ_Λ and $\chi^+_{\Lambda^c}$ as equal to 0 outside Λ, respectively, in Λ.

Lemma 11.6.5.1 *There is a constant $c > 0$ so that for all $x \in \Lambda$, all $\sigma_\Lambda \in \mathcal{X}^+_{a,\Lambda}$ and all a and γ small enough*

$$|J_\gamma^{(x)} * (\sigma_\Lambda + a^d \chi^+_{\Lambda^c}) - \rho_{\beta,+}| \le c\gamma\ell_{-,\gamma} + \zeta,$$
$$|J_\gamma^{(x)} * (a^d \rho + a^d \chi^+_{\Lambda^c}) - \rho_{\beta,+}| \le c\gamma\ell_{-,\gamma} + \zeta. \tag{11.6.5.2}$$

Proof Denote by $J_\gamma^{(\ell_{-,\gamma})}(r,r')$ the average $\fint_{C_{r'}^{(\ell_{-,\gamma})}} J_\gamma(r,r'')dr''$. By the smoothness of J_γ, $|J_\gamma^{(\ell_{-,\gamma})}(r,r') - J_\gamma(r,r')| \le c\gamma^d(\gamma\ell_{-,\gamma})\mathbf{1}_{\mathrm{dist}(r,C_{r'}^{(\ell_{-,\gamma})})}$, so that

$$\Big|\sum_{y\in\Lambda:y\neq x}\{J_\gamma(x,y) - J_\gamma^{(\ell_{-,\gamma})}(x,y)\}\sigma_\Lambda(y)\Big| \le c\gamma\ell_{-,\gamma}. \tag{11.6.5.3}$$

Since $\sigma_\Lambda \in \mathcal{X}^+_{a,\Lambda}$, $|\sum_{y\in\Lambda:y\neq x} J_\gamma^{(\ell_{-,\gamma})}(x,y)\{\sigma_\Lambda(y) - a^d\rho_{\beta,+}\}| \le \zeta$ and therefore

$$|J_\gamma^{(x)} * \sigma_\Lambda + J_\gamma * a^d\chi^+_{\Lambda^c} - \rho_{\beta,+}| \le c\gamma\ell_{-,\gamma} + \zeta + |J_\gamma^{(x)} * a^d\chi^+ - \rho_{\beta,+}|.$$

The last term is bounded by $c\gamma^d$ (for all a and γ small enough), hence the first inequality in (11.6.5.2). The second inequality in (11.6.5.2) follows from the first one because $a^d\rho(x)$ is the expectation of $\sigma(x)$. □

Lemma 11.6.5.2 *There is a constant c so that for all $x \in \Lambda$ and a small enough,*

$$\Big|\rho(x) - \mu_\Lambda\Big(e^{-\beta[(1-u)e'_{\lambda(\beta)}(\rho_{\beta,+}) + u\lambda_x(\sigma_\Lambda + a^d\chi^+_{\Lambda^c})]}\Big)\Big| \le c\gamma^d. \tag{11.6.5.4}$$

Proof We write $a^d\rho(x)$, $x \in \Lambda$, as $\mu_\Lambda(\mu_\Lambda(\sigma(x) = 1|\sigma_{\Lambda\setminus x}))$ which is

$$\mu_\Lambda\big(\mathbf{1}_{B_x^c}\,\mu_\Lambda(\sigma(x) = 1\big|\sigma_{\Lambda\setminus x})\big) + \mu_\Lambda\big(\mathbf{1}_{B_x}\,\mu_\Lambda(\sigma(x) = 1\big|\sigma_{\Lambda\setminus x})\big).$$

By (11.5.1.5) $\mu_\Lambda(\mathbf{1}_{B_x}\,\mu_\Lambda(\sigma(x) = 1|\sigma_{\Lambda\setminus x})) = \mu_\Lambda(\mathbf{1}_{B_x^{(1)}}\,\mu_\Lambda(\sigma(x) = 1|\sigma_{\Lambda\setminus x}))$; then by (11.5.7.7), (11.5.7.6) and (11.6.1.9)

$$\Big|\rho(x) - a^{-d}\mu_\Lambda\big(\mathbf{1}_{B_x^c}\,\mu_\Lambda(\sigma(x) = 1\big|\sigma_{\Lambda\setminus x})\big)\Big| \le 2e^{-c\zeta^2\ell^d_{-,\gamma}}. \tag{11.6.5.5}$$

Set

$$v_x = \int [e_{\lambda_{\beta,\gamma}}(J_\gamma^{(x)} * [\sigma_\Lambda + a^d\chi^+_{\Lambda^c}] + J_\gamma(r,x))$$
$$- e_{\lambda_{\beta,\gamma}}(J_\gamma^{(x)} * [\sigma_\Lambda + a^d\chi^+_{\Lambda^c}])]\,dr + \sum_{\Delta\ni x,\Delta\subset\Lambda} U^+_{\gamma,\Delta}(\sigma_{\Delta\setminus x}, 1_x);$$

then $\mu_\Lambda(\sigma(x) = 1|\sigma_{\Lambda\setminus x}) = Z^{-1}(a^d e^{-\beta(1-u)e'_{\lambda(\beta)}(\rho_{\beta,+})})e^{-\beta u v_x}$. There are constants c and c', so that $|Z^{-1} - 1| \le ca^d$ and

$$\Big|\mu_\Lambda\big(\sigma(x) = 1\big|\sigma_{\Lambda\setminus x}\big) - (a^d e^{-\beta(1-u)e'_{\lambda(\beta)}(\rho_{\beta,+})})e^{-\beta u v_x}\Big| \le c'a^{2d}. \tag{11.6.5.6}$$

$|\mu_\Lambda(\mathbf{1}_{\mathcal{B}_x^c} a^d e^{-\beta[(1-u)e'_{\lambda(\beta)}(\rho_{\beta,+})+u v_x]}) - \mu_\Lambda(a^d e^{-\beta[(1-u)e'_{\lambda(\beta)}(\rho_{\beta,+})+u v_x]})|$ is then bounded by $ca^d\mu_\Lambda(\mathcal{B}_x)$ and $\mu_\Lambda(\mathcal{B}_x) \le 2\pi_\Lambda(\mathcal{B}_x) \le 2(1+a^d)e^{-c\zeta^2\ell^d_{-,\gamma}}$ (by using (11.5.7.7), (11.5.7.6) and (11.6.1.9)). Then (11.6.5.4) is a consequence of

$$|v_x - \lambda_x(\sigma_\Lambda + a^d\chi^+_{\Lambda^c})| \le c\{\gamma^d + e^{-\beta c_{\text{pol}}(\zeta^2\ell^d_{-,\gamma})}\}$$

(since, by Lemma 11.6.5.1, $e'_{\lambda_{\beta,\gamma}}(\cdot)$ is bounded). □

Lemma 11.6.5.3 *There is a positive constant c so that*

$$|\rho(x) - e^{-\beta[(1-u)e'_{\lambda(\beta)}(\rho_{\beta,+})+u\lambda_x(a^d\rho+a^d\chi^+_{\Lambda^c})]}| \le c\gamma^{d/2}. \tag{11.6.5.7}$$

Proof Let $F(x,t) = e^{-\beta[(1-u)e'_{\lambda(\beta)}(\rho_{\beta,+})+tu\lambda_x(a^d\rho+a^d\chi^+_{\Lambda^c})+(1-t)u\lambda_x(a^d\sigma_\Lambda+\chi^+_{\Lambda^c})]}$. By Lemma 11.6.5.2 the l.h.s. of (11.6.5.7) is bounded by $|\mu_\Lambda(F(x,1)-F(x,0))| + c\gamma^d$; $|\mu_\Lambda(F(x,1)-F(x,0))|$ is bounded by

$$\int_0^1 \left|\mu_\Lambda\big(F(x,t)\beta\{\lambda_x(a^d\rho + a^d\chi^+_{\Lambda^c}) - \lambda_x(a^d\sigma_\Lambda + \chi^+_{\Lambda^c})\}\big)\right| dt.$$

Thus, there is a constant c so that $\mu_\Lambda(|F(x,1)-F(x,0)|) \le c\mu_\Lambda(|J^{(x)}_\gamma * \sigma_\Lambda - J^{(x)}_\gamma * a^d\rho|)$. By Cauchy–Schwartz

$$\begin{aligned}
&\big[\mu_\Lambda\big(|F(x,1)-F(x,0)|\big)\big]^2 \\
&\quad\le c^2\Big\{\sum_{y\in\Lambda} J_\gamma(x,y)^2 a^d\rho(y) \\
&\qquad + \sum_{y\neq y'\in\Lambda} J_\gamma(x,y)J_\gamma(x,y')\big|\mu_\Lambda\big([\sigma(y)-a^d\rho(y)][\sigma(y')-a^d\rho(y')]\big)\big|\Big\}.
\end{aligned}$$

By (11.6.2.8) the r.h.s. is bounded proportionally to γ^d and (11.6.5.7) follows. □

Proof of Proposition 11.6.4.1

We claim that, similarly to (11.6.5.7), there is c so that

$$|\rho_{\beta,+} - e^{-\beta[(1-u)e'_{\lambda(\beta)}(\rho_{\beta,+})+u\lambda_x(a^d\rho_{\beta,+})]}| \le c\gamma^{1/2}. \tag{11.6.5.8}$$

The proof follows using $\rho_{\beta,+} = e^{-\beta e'_{\lambda(\beta)}(\rho_{\beta,+})}$, $|\lambda_{\beta,\gamma} - \lambda(\beta)| \le c\gamma^{1/2}$ and, for all a small enough, $|a^d\sum_{y\in a\mathbb{Z}^d\setminus 0} J_\gamma(r,y) - 1| \le c\gamma^d$. Then, writing $F(x,t) = e^{-\beta[(1-u)e'_{\lambda(\beta)}(\rho_{\beta,+})+tu\lambda_x(a^d\rho(\cdot)+a^d\chi^+_{\Lambda^c})+(1-t)u\lambda_x(a^d\rho_{\beta,+})]}$, $t\in[0,1]$,

$$|\rho(x) - \rho_{\beta,+}| \le \int_0^1 F(x,t)\beta u|\lambda_x(a^d\rho + a^d\chi^+_{\Lambda^c}) - \lambda_x(a^d\rho^*_{a,u}(\cdot))|\,dt + c\gamma^{1/2};$$

c a constant. $|\lambda_x(a^d\rho + a^d\chi^+_{\Lambda^c})) - \lambda_x(a^d\rho_{\beta,+})|$ is bounded by

$$e''_{\lambda_{\beta,\gamma}}(\xi)\int J_\gamma(x,r)\Big\{a^d\sum_{y\neq x, y\in\Lambda} J_\gamma(r,y)|\rho(y)-\rho_{\beta,+}|\Big\},$$

where $|\xi - \rho_{\beta,+}| \le c\zeta$, c a constant. Since $\beta e^{-\beta e'_{\lambda(\beta)}(\rho_{\beta,+})} e''_{\lambda(\beta)}(\rho_{\beta,+}) < 1$, for γ small enough $\beta F(x,t)|e''_{\lambda_{\beta,\gamma}}(\xi)| \le \kappa < 1$ uniformly. Then

$$|\rho(x)-\rho_{\beta,+}| \le \kappa a^d \sum_{y\neq x, y\in\Lambda}\Big\{\int J_\gamma(x,r)J_\gamma(r,y)\Big\}|\rho(y)-\rho_{\beta,+}| + c\sqrt{\gamma}.$$

For all a and γ small enough $\kappa a^d\sum_y\{\int J_\gamma(0,r)J_\gamma(r,y)\} \le \kappa' < 1$, and hence

$$|\rho(x)-\rho_{\beta,+}| \le c\sqrt{\gamma} + \kappa' \max_{y\in\Lambda}|\rho(y)-\rho_{\beta,+}|.$$

Thus, $|\rho(x)-\rho_{\beta,+}| \le c(1-\kappa')^{-1}\sqrt{\gamma}$.

11.7 Complements to Section 11.1

Throughout this section the chemical potential λ belongs to the interval $[\lambda(\beta)-1, \lambda(\beta)+1]$, and $\gamma > 0$ is as small as required in Sects. 11.1.1 and 11.1.2.

11.7.1 Convergence by subsequences

Convergence by subsequences to $P^+_{\gamma,\lambda}$ is proved in Corollary 11.7.1.3.

Lemma 11.7.1.1 *There is $c > 0$ so that if $C \in \mathcal{D}^{(\ell_{-},\gamma)}$ and $r \in C$,*

$$e^{|C|} \ge \int_{\eta(q_C;r)=\pm 1} \nu_C(dq_C) \ge e^{-c|C|}. \tag{11.7.1.1}$$

Proof The upper bound follows by dropping the constraint $\eta(q_C;r)=\pm 1$. For the lower bound we write

$$\int_{\eta(q_C;r)=\pm 1} \nu_\Lambda(dq_C) \ge \frac{|C|^n}{n!},$$

with n the integer part of $|C|\rho_{\beta,\pm}$. (11.7.1.1) then follows using the Stirling formula. □

Lemma 11.7.1.2 *There is $c > 0$ so that for any $q^+ \in \mathcal{X}^+$, or $q^+ = \chi^+$,*

$$-c \le \frac{1}{\beta|\Lambda|} \log \hat{Z}^+_{\gamma,\lambda,\Lambda,q^+} \le c, \quad |\lambda - \lambda(\beta)| \le 1. \tag{11.7.1.2}$$

Proof *Lower bound.* By (11.1.1.8), $X^+(\Lambda) \ge 1$. Then using (11.1.2.6)

$$\hat{Z}^+_{\gamma,\lambda,\Lambda,q^+} \ge \int_{\mathcal{X}^+_\Lambda} e^{-\beta H_{\gamma,\lambda}(q_\Lambda|q^+_{\Lambda^c})} \nu_\Lambda(dq_\Lambda) \ge e^{-\beta c'|\Lambda|} \int_{\mathcal{X}^+_\Lambda} \nu_\Lambda(dq_\Lambda),$$

which by (11.7.1.1) proves the lower bound in (11.7.1.2).

Upper bound. By (11.1.1.12), $X^+(\Lambda) \le X^+(\Lambda;\mathbb{R}^d) \le 2^{|\Lambda|/\ell^d_{+,\gamma}}$; then by (11.1.2.8) and Lemma 11.7.1.1

$$\hat{Z}^+_{\gamma,\lambda,\Lambda,q^+} \le e^{|\Lambda|/\ell^d_{+,\gamma} + \beta b_0 |\Lambda|} \int_{\mathcal{X}_\Lambda} \nu_\Lambda(dq_\Lambda) \le e^{(\ell^{-d}_{+,\gamma} + \beta b_0 + 1)|\Lambda|}. \qquad \square$$

Corollary 11.7.1.3 *For any γ small enough and any λ such that $|\lambda - \lambda(\beta)| \le 1$ there is $P^+_{\gamma,\lambda}$ and an increasing sequence $\{\Delta_k\}$ of $\mathcal{D}^{(\ell_{+,\gamma})}$-measurable cubes which invades $\mathbb{R}^d$ such that*

$$\lim_{k\to\infty} \frac{1}{\beta|\Delta_k|} \log \hat{Z}^+_{\gamma,\lambda,\Delta_k,\chi^+} = P^+_{\gamma,\lambda}. \tag{11.7.1.3}$$

11.7.2 Existence of the thermodynamic limit

Using Sects. 11.1.1 and 11.1.2 we shall extend the convergence by subsequences proved in Corollary 11.7.1.3 to any van Hove sequence (see Sect. 2.3.2 for the definition of van Hove sequences). The argument is essentially the same as in the Ising case.

Proposition 11.7.2.1 *For any van Hove sequence $\{\Lambda_n\}$ and any sequence of plus b.c. $\{q^+_n\}$ (or $q^+_n = \chi^+$)*

$$\lim_{n\to\infty} \frac{1}{\beta|\Lambda_n|} \log \hat{Z}^+_{\gamma,\lambda,\Lambda_n,q^+_n} = P^+_{\gamma,\lambda}, \tag{11.7.2.1}$$

with $P^+_{\gamma,\lambda}$ as in (11.7.1.3).

Proof Let Δ_k be an element of the converging sequence defined in Corollary 11.7.1.3 and $\{\Delta_k(i),\ 1 \le i \le N_k\}$ the elements of the partition of $\mathbb{R}^d$ into Δ_k and its translates which are contained in Λ; call their union Λ^0_k.

Lower bound. Write $D = \Delta_k(i)$ and ℓ_k the length of its side, and write

$$H_{\gamma,\lambda,\Lambda}(q_\Lambda|q^+_{\Lambda^c}) = H_{\gamma,\lambda,D}(q_D|q_{\Lambda\setminus D}, q^+_{\Lambda^c}) + H_{\gamma,\lambda,\Lambda\setminus D}(q_{\Lambda\setminus D}|q^+_{\Lambda^c}).$$

By (11.1.2.7) $|H_{\gamma,\lambda,D}(q_D|q_{\Lambda\setminus D},q^+_{\Lambda^c})-H_{\gamma,\lambda,D}(q_D|\chi^+_{D^c})|\le c'\ell_k^{d-1}\gamma^{-1}$. Iterate to all cubes $\Delta_k(\cdot)$, use (11.1.2.6) to bound the energy $H_{\gamma,\lambda,\Lambda\setminus\Lambda_k^0}(q_{\Lambda\setminus\Lambda_k^0}|q^+_{\Lambda^c})$; then

$$\left|H_{\gamma,\lambda,\Lambda}(q_\Lambda|q^+_{\Lambda^c})-\sum_{i=1}^{N_k}H_{\gamma,\lambda,\Delta_k(i)}(q_{\Delta_k(i)}|\chi^+_{\Delta_k(i)^c})\right|$$
$$\le c'\{\ell_k^{d-1}\gamma^{-1}N_k+|\Lambda\setminus\Lambda_k^0|\}. \qquad (11.7.2.2)$$

By (11.1.1.10), $X^+(\Lambda)\ge\prod_{i=1}^{N_k}X^+(\Delta_k(i))$ so that, by (11.7.2.2),

$$\frac{1}{\beta|\Lambda|}\log\hat Z^+_{\gamma,\lambda,\Lambda,q^+_{\Lambda^c}}\ge\frac{|\Lambda_k^0|}{|\Lambda|}\frac{1}{\beta|\Delta_k|}\log\hat Z^+_{\gamma,\lambda,\Delta_k,\chi^+}$$
$$-\frac{c'}{|\Lambda|}\left(\ell_k^{d-1}\gamma^{-1}\frac{|\Lambda_k^0|}{\ell_k^d}+|\Lambda\setminus\Lambda_k^0|\right). \qquad (11.7.2.3)$$

Set $\Lambda=\Lambda_n$ and let $n\to\infty$; then $|(\Lambda_n)_k^0|/|\Lambda_n|\to1$, by the definition of van Hove sequences, so that

$$\liminf_{n\to\infty}\frac{1}{\beta|\Lambda_n|}\log\hat Z^+_{\gamma,\lambda,\Lambda_n,q^+_{\Lambda_n^c}}\ge\frac{1}{\beta|\Delta_k|}\log\hat Z^+_{\gamma,\lambda,\Delta_k,\chi^+}-c'\ell_k^{-1}\gamma^{-1}.$$

Letting $k\to\infty$, $\liminf_{n\to\infty}\frac{1}{\beta|\Lambda_n|}\log\hat Z^+_{\gamma,\lambda,\Lambda_n,q^\pm_{\Lambda_n^c}}\ge P^+_{\gamma,\lambda}$, thus completing the proof of the lower bound.

Upper bound. We bound $X^+_{\gamma,\lambda,q_\Lambda}(\Lambda)\le X^+_{\gamma,\lambda,q}(\Lambda;\mathbb{R}^d)$, $q\in\mathcal{X}^+$, $q\cap\Lambda=q_\Lambda$. By repeatedly applying (11.1.1.11)

$$X^+_{\gamma,\lambda,q_\Lambda}(\Lambda)\le X^+_{\gamma,\lambda,q}(\Lambda\setminus\Lambda_k^0;\mathbb{R}^d)\prod_{i=1}^{N_k}X^+_{\gamma,\lambda,q}(\Delta_k(i);\mathbb{R}^d).$$

By (11.1.1.12) $X^+(\Lambda\setminus\Lambda_k^0;\mathbb{R}^d)\le2^{|\Lambda\setminus\Lambda_k^0|/\ell^d_{+,\gamma}}$ while, by (11.1.1.13),

$$X^+_{\gamma,\lambda,q}(\Delta_k(i);\mathbb{R}^d)\le X^+_{\gamma,\lambda,q_{\Delta_k(i)}}(\Delta_k(i))2^{|\delta_{\rm in}^{\ell_{+,\gamma}}[\Delta_k(i)]|/\ell^d_{+,\gamma}},$$

with $|\delta_{\rm in}^{\ell_{+,\gamma}}[\Delta_k(i)]|/\ell^d_{+,\gamma}\le2d(\ell_k/\ell_{+,\gamma})^{d-1}$. Then

$$X^+_{\gamma,\lambda,q_\Lambda}(\Lambda)\le2^{|\Lambda\setminus\Lambda_k^0|/\ell^d_{+,\gamma}}2^{N_k2d(\ell_k/\ell_{+,\gamma})^{d-1}}\prod_{i=1}^{N_k}X^+_{\gamma,\lambda,q_{\Delta_k(i)}}(\Delta_k(i)).$$

By (11.7.2.2) $\frac{1}{\beta|\Lambda|}\log\hat Z^+_{\gamma,\lambda,\Lambda,q^+_{\Lambda^c}}$ is bounded by

$$\frac{|\Lambda_k^0|}{|\Lambda|}\frac{1}{\beta|\Delta_k|}\log\hat Z^+_{\gamma,\lambda,\Delta_k,\rho_{\beta,+}}+\frac{c'}{|\Lambda|}\left(\ell_k^{d-1}\gamma^{-1}\frac{|\Lambda_k^0|}{|\Delta_k|}+|\Lambda\setminus\Lambda_k^0|\right)$$

$$+\frac{|\Lambda \setminus \Lambda_k^0|}{\beta|\Lambda|\ell_{+,\gamma}^d}+\frac{|\Lambda_k^0|2d\ell_k^{d-1}}{\beta|\Lambda|\ell_k^d\ell_{+,\gamma}^{d-1}}. \tag{11.7.2.4}$$

Setting $\Lambda = \Lambda_n$ let $k \to \infty$ after $n \to \infty$, then $\limsup_{n\to\infty} \frac{1}{\beta|\Lambda_n|} \log \hat{Z}^{\pm}_{\gamma,\lambda,\Lambda_n,q^{\pm}_{\Lambda_n^c}} \le P^{+}_{\gamma,\lambda}$, which completes the proof of the upper bound. □

Corollary 11.7.2.2 *Let Δ_k be a $\mathcal{D}^{(\ell_{+,\gamma})}$-measurable cube of side $\ell_k = \ell_{+,\gamma}2^k$; then for any $\lambda \in [\lambda(\beta)-1, \lambda(\beta)+1]$,*

$$\left|P^{+}_{\gamma,\lambda}-\frac{1}{\beta|\Delta_k|}\log \hat{Z}^{+}_{\gamma,\lambda,\Delta_k,\chi^+}\right| \le \frac{1}{\ell_k}\left(c'\gamma^{-1}+\frac{2d}{\beta\ell_{+,\gamma}^{d-1}}\right), \tag{11.7.2.5}$$

c' being as in Lemma 11.1.2.3.

Proof It follows directly from (11.7.2.3) and (11.7.2.4) because the translates of Δ_k cover exactly Δ_n. □

11.7.3 Continuous dependence on the chemical potential

In this subsection we shall prove that the $P^{\pm}_{\gamma,\lambda}$ are continuous functions of λ. This is not a consequence of the pressure being a convex function of λ because in our case λ appears also in the cutoff weights $\hat{W}^{\pm}_{\gamma,\lambda}(\Gamma,q)$ and we thus need to prove uniform continuity of the latter as well. We shall indeed prove that $\log \hat{N}^{\pm}_{\gamma,\lambda}(\Gamma,q)$ and $\log \hat{D}^{\pm}_{\gamma,\lambda}(\Gamma,q)$ are for each Γ both uniformly continuous, which, recalling (10.5.1.9), implies the uniform continuity of $\hat{W}^{\pm}_{\gamma,\lambda}(\Gamma,q)$. The basic quantity to estimate has the following structure (referring for simplicity to the plus case):

$$Z_{\lambda,\mu,q_{\Lambda^c};i} := \int_{\mathcal{B}_i} e^{-\beta H_{\gamma,\lambda,\Lambda}(q_\Lambda|q_{\Lambda^c})} f_i(\mu;q_\Lambda), \quad i=1,2,3. \tag{11.7.3.1}$$

When $i=1$, $\mathcal{B}_1 = \mathcal{X}^{+}_{\Lambda}$, $f_1(\mu;q_\Lambda) = X^{+}_{\gamma,\mu,q_\Lambda}(\Lambda)$, so that $Z_{\lambda,\lambda;1} = Z^{+}_{\gamma,\lambda,\Lambda,q}$; when $i=2$, $\Lambda = c(\Gamma)$, $\mathcal{B}_2 = \{q_{c(\Gamma)} \in \mathcal{X}_{c(\Gamma)} : \eta(q_{c(\Gamma)};r) = \eta_\Gamma(r), r \in \mathrm{sp}(\Gamma);\ \Theta(q_{c(\Gamma)};r) = \pm 1, r \in \mathrm{int}^{\pm}(\Gamma)\}$ and $f_2(\mu;q_\Lambda) = X^{-}_{\gamma,\mu,q_{\mathrm{int}^-(\Gamma)}}(\mathrm{int}^-(\Gamma)) \times X^{+}_{\gamma,\mu,q_{\mathrm{int}^+(\Gamma)}}(\mathrm{int}^+(\Gamma))$, so that $Z_{\lambda,\lambda;2} = \hat{N}^{+}_{\gamma,\lambda}(\Gamma,q)$. Finally, when $i=3$, $\Lambda = c(\Gamma)$, $\mathcal{B}_3 = \mathcal{X}^{+}_{c(\Gamma)}$ and $f_3(\mu;q_\Lambda) = X^{-}_{\gamma,\mu,q_{\mathrm{int}^-(\Gamma)}}(\mathrm{int}^-(\Gamma))X^{+}_{\gamma,\mu,q_{\mathrm{int}^+(\Gamma)}}(\mathrm{int}^+(\Gamma))$, so that $Z_{\lambda,\lambda;3} = \hat{D}^{+}_{\gamma,\lambda}(\Gamma,q)$.

In all cases we eventually set $\mu = \lambda$, but it is convenient to keep them distinct and prove uniform continuity in both. We shall use the following bounds:

$$1 \le f_i(\mu;q_\Lambda) \le 2^{|\Lambda)|/\ell_{+,\gamma}^d}, \quad i=1,2,3. \tag{11.7.3.2}$$

the lower bound being obtained by using (11.1.1.8), the upper bound by (11.1.1.12). The proof is by induction on the volume and the basic step is:

Lemma 11.7.3.1 *Assume $f_i(\mu; q_\Lambda)$ is a continuous function of $\mu \in [\lambda(\beta) - 1, \lambda(\beta) + 1]$, uniformly in $q_\Lambda \in \mathcal{B}_i$, $i = 1, 2, 3$. Then $\log Z_{\lambda,\mu,q_{\Lambda^c};i}$ is a continuous function of λ and μ in $|\lambda - \lambda(\beta)| \le 1$ uniformly in q_{Λ^c}, $i = 1, 2, 3$.*

Proof Uniform continuity in μ. We write

$$\log \frac{Z_{\lambda,\mu,q_{\Lambda^c};i}}{Z_{\lambda,\mu',q_{\Lambda^c};i}}$$

$$= \log\left(1 + \frac{\int_{\mathcal{B}_i} e^{-\beta H_{\gamma,\lambda,\Lambda}(q_\Lambda|q_{\Lambda^c})}[f_i(\mu; q_\Lambda) - f_i(\mu'; q_\Lambda)]}{\int_{\mathcal{B}_i} e^{-\beta H_{\gamma,\lambda,\Lambda}(q_\Lambda|q_{\Lambda^c})} f_i(\mu'; q_\Lambda)}\right). \tag{11.7.3.3}$$

By assumption for any $\epsilon > 0$ there is $\delta > 0$ so that if $|\mu - \mu'| \le \delta$ then $|f_i(\mu; q_\Lambda) - f_i(\mu'; q_\Lambda)| < \epsilon$, hence by (11.7.3.2), $|\log \frac{Z_{\lambda,\mu,q_{\Lambda^c};i}}{Z_{\lambda,\mu',q_{\Lambda^c};i}}| \le \max|\log(1 \pm \epsilon)|$.

Uniform continuity in λ. We have

$$\frac{d}{d\lambda} \log Z_{\lambda,\mu,q_{\Lambda^c};i} = \frac{\int_{\mathcal{B}_i} \beta|q_\Lambda|\, e^{-\beta H_{\gamma,\lambda,\Lambda}(q_\Lambda|q_{\Lambda^c})} f_i(\mu; q_\Lambda)}{\int_{\mathcal{B}_i} e^{-\beta H_{\gamma,\lambda,\Lambda}(q_\Lambda|q_{\Lambda^c})} f_i(\mu; q_\Lambda)}. \tag{11.7.3.4}$$

For $i = 1, 3$, $|q_\Lambda| \le (\rho_{\beta,+} + \zeta)|\Lambda|$ so that the l.h.s. of (11.7.3.4) is bounded by the same quantity. When $i = 2$, calling $B = \text{int}(\Gamma) + \delta_{\text{out}}^{\ell_{+,\gamma}}[\text{int}(\Gamma)]$, we claim that

$$\left|\frac{d}{d\lambda} \log Z_{\lambda,\mu,q_{\Lambda^c};i}\right| \le \beta 2^{|c(\Gamma)|/\ell_{+,\gamma}^d} \frac{\int |q_{c(\Gamma)}|\, e^{\beta b_0 |q_{c(\Gamma)}|} d\nu_{c(\Gamma)}}{\int_{\mathcal{X}_B^+} e^{-\beta c'|q_B|} d\nu_B}.$$

We bound the numerator on the r.h.s. of (11.7.3.4) by dropping the constraint $\mathcal{B}_2$, using (11.1.2.8) to bound the energy and (11.7.3.2) to bound the term f_2; the denominator is bounded by imposing $q_{c(\gamma)} \cap \{c(\Gamma) \setminus B\} = \emptyset$, bounding the energy using (11.1.2.6) and using the lower bound in (11.7.3.2). The integral in the numerator is equal to $z|c(\Gamma)|e^{z|c(\Gamma)|}$ with $z = e^{\beta b_0}$; the one in the denominator is bounded from below by $e^{-(c'(\rho_{\beta,+} - \zeta) + c)|B|}$, having used (11.7.1.1). □

Lemma 11.7.3.2 *For each plus/minus contour Γ, $\hat{W}^{\pm}_{\gamma,\lambda}(\Gamma, q)$ is a continuous function of λ uniformly in $|\lambda - \lambda(\beta)| \le 1$ and in q.*

Proof We shall prove uniform continuity of $\log \hat{N}^{\pm}_{\gamma,\lambda,\mu}(\Gamma, q)$ and $\log \hat{D}^{\pm}_{\gamma,\lambda,\mu}(\Gamma, q)$ which are respectively equal to $\log Z^{\pm}_{\lambda,\mu,q_{\Lambda^c};i}$, $i = 2, 3$, the latter defined as in (11.7.3.1) for the plus case and by the analogue of (11.7.3.1) in the minus case.

The proof is by induction on the volume $|c(\Gamma)|$ of the contours Γ. By the induction hypothesis we thus suppose uniform continuity of $\log \hat{N}^{\pm}_{\gamma,\lambda,\mu}(\Gamma', q)$ and $\log \hat{D}^{\pm}_{\gamma,\lambda,\mu}(\Gamma', q)$ for all Γ' such that $|c(\Gamma')| < |c(\Gamma)|$. Then the weights

$\hat{W}^{\pm}_{\gamma,\lambda}(\Gamma',q)$ are uniformly continuous and so are the terms $X^{\pm}_{\gamma,\mu,q_{\mathrm{int}^{\pm}(\Gamma)}}(\mathrm{int}^{\pm}(\Gamma))$; the assumptions in Lemma 11.7.3.1 are then verified and the induction step is proved. To complete the induction proof we need to analyze the first step, but in the smallest Γ there is no interior part and the $X^{\pm}$ terms missing, i.e. $f_i(\mu;q_\Lambda)=1$ for which Lemma 11.7.3.1 remains obviously valid. □

Proposition 11.7.3.3 *$P^{\pm}_{\gamma,\lambda}$ depend continuously on λ in $[\lambda(\beta)-1,\lambda(\beta)+1]$.*

Proof We just consider the plus case. By Corollary 11.7.2.2, for any $\epsilon>0$ there is n so that for any $\lambda\in[\lambda(\beta)-1,\lambda(\beta)+1]$

$$\left|P^{+}_{\gamma,\lambda}-\frac{1}{\beta|\Delta_n|}\log\hat{Z}^{+}_{\gamma,\lambda,\Delta_n,\chi^+}\right|\le\epsilon. \tag{11.7.3.5}$$

By Lemma 11.7.3.1 (with $i=1$) $\log\hat{Z}^{+}_{\gamma,\lambda,\Delta_n,\chi^+}$ is uniformly continuous in λ, because by Lemma 11.7.3.2 $X^{+}_{\gamma,\mu,q}(\Lambda)$ is uniformly continuous. □

11.8 Complements to Section 11.3

Here is a short outline of the proof of Proposition 11.3.0.1. In Sect. 11.8.1 we shall localize in a collection of cubes of $\mathcal{D}^{(\ell_-,\gamma)}$ the region where large deviations from equilibrium occur due to the constraint $\{\eta(q_\Lambda;\cdot)=\eta_\Gamma(\cdot)\}$. They are of various types: there are those for which the particle density exceeds some threshold value, those for which it does not but not being close to $\rho_{\beta,\pm}$ and finally those for which it is close to $\rho_{\beta,+}$ but in a neighboring cube being close to $\rho_{\beta,-}$. In Sect. 11.8.2 we shall prove that we can reduce to the case where the threshold value is never exceeded. The other cases, considered in the successive subsections, are studied using the coarse graining of Sect. 11.1.3 and with reduction to minimization problems for the free energy functional $\tilde{\mathcal{F}}_\gamma$.

11.8.1 Localization of the excess energy

The analysis here parallels that in Sect. 6.4.2 for the L–P functional. Let Λ and Λ' be as in (11.2.1.3), $\rho_{\max}>\rho^{+}_{\beta}+\zeta$ as in the paragraph *The constant* c_1 of Sect. 10.2.6. In analogy with Definition 6.4.2.1, given a configuration q_Λ we define two collections of cubes of $\mathcal{D}^{(\ell_-,\gamma)}$:

$$\begin{aligned}\mathcal{C}_0&=\left\{C^{(\ell_-,\gamma)}_x\subset\Lambda':\eta(q_\Lambda;x)=0,\ \mathrm{Av}^{(\ell_-,\gamma)}(q_\Lambda;x)\le\rho_{\max}\right\},\\ \mathcal{C}_>&=\left\{C^{(\ell_-,\gamma)}_x\subset\Lambda':\mathrm{Av}^{(\ell_-,\gamma)}(q_\Lambda;x)>\rho_{\max}\right\}.\end{aligned} \tag{11.8.1.1}$$

and a collection $\mathcal{C}_{\neq}$ of pairs of cubes $\{(C_{x_i'}^{(\ell_{-,\gamma})}, C_{x_i''}^{(\ell_{-,\gamma})})\}$ with the following properties: • $1 = \eta(q_\Lambda; x_i') = -\eta(q_\Lambda; x_i'')$; • all cubes are in Λ' and no cube appears twice; • $\mathcal{C}_{\neq}$ is maximal, namely: if there is another pair (C', C'') which satisfies the first two properties then necessarily at least one among C' and C'' is a cube already present among the cubes of $\mathcal{C}_{\neq}$. Finally, let us denote

$$N_0, N_>, N_{\neq} \text{ the number of all the cubes in } \mathcal{C}_0, \mathcal{C}_>, \mathcal{C}_{\neq}. \tag{11.8.1.2}$$

Lemma 11.8.1.1 *With the above notation,*

$$\{\eta(q_\Lambda; \cdot) = \eta_\Gamma(\cdot)\} \subseteq \{N_0 + N_> + N_{\neq} \geq 3^{-d} N_\Gamma\}, \tag{11.8.1.3}$$

$$Z_{\gamma,\lambda_{\beta,\gamma},\Lambda,\chi_{\Lambda^c}}(\{\eta(q_\Lambda; \cdot) = \eta_\Gamma(\cdot)\}) \leq \mathcal{I} = \int_{N_0+N_>+N_{\neq}\geq \frac{N_\Gamma}{3^d}} e^{-\beta H_{\gamma,\lambda_{\beta,\gamma},\Lambda}(q_\Lambda|\chi_{\Lambda^c})}. \tag{11.8.1.4}$$

Proof (11.8.1.4) is an immediate consequence of (11.8.1.3), which will be proved next. If q_Λ is such that $\eta(q_\Lambda; \cdot) = \eta_\Gamma(\cdot)$ on Λ, then $\eta(q_\Lambda; r) = 0$ implies $r \in \mathrm{sp}(\Gamma) \setminus \delta_{\mathrm{in}}^{\ell_{+,\gamma}}[\mathrm{sp}(\Gamma)]$ which is "well" inside Λ'. If $\eta(q_\Lambda; x) = -\eta(q_\Lambda; y) = 1$ with $C_x^{(\ell_{-,\gamma})}$ connected to $C_y^{(\ell_{-,\gamma})}$ then both cubes are in Λ', because η is constant on the connected components of $\delta_{\mathrm{in}}^{\ell_{+,\gamma}}[\mathrm{sp}(\Gamma)]$. Thus the condition of being in Λ' in the definition of $\mathcal{C}_0$, $\mathcal{C}_>$ and $\mathcal{C}_{\neq}$ is automatically satisfied if $\eta(q_\Lambda; \cdot) = \eta_\Gamma(\cdot)$. This will play an important role in the proof which is an adaptation to the present context of the proof of Theorem 6.4.5.1. With $C_x^{(\ell_{+,\gamma})} \in \mathcal{D}^{(\ell_{+,\gamma})}$, denote

$$\mathcal{C}^+ = \mathcal{C}^+(q_\Lambda) = \{C_x^{(\ell_{+,\gamma})}, x \in \ell_{+,\gamma}\mathbb{Z}^d : C_x^{(\ell_{+,\gamma})} \ni y, C_y^{(\ell_{-,\gamma})} \in \mathcal{C}_0 \cup \mathcal{C}_> \cup \mathcal{C}_{\neq}\},$$

where, by abuse of notation, $\mathcal{C}_0 \cup \mathcal{C}_> \cup \mathcal{C}_{\neq}$ denotes the collection of all the cubes in $\mathcal{C}_0$, in $\mathcal{C}_>$ and in $\mathcal{C}_{\neq}$. Denoting by N the number of cubes in $\mathcal{C}^+$, $N \leq N_0 + N_> + N_{\neq}$, so that it will suffice to prove that $N \geq 3^{-d} N_\Gamma$. We postpone the proof that

$$\bigcup_{C \in \mathcal{C}(q_\Lambda)} (C \cup \delta_{\mathrm{out}}^{\ell_{+,\gamma}}[C]) \supseteq \mathrm{sp}(\Gamma). \tag{11.8.1.5}$$

By (11.8.1.5), $\ell_{+,\gamma}^d N_\Gamma$ (which is the volume of $\mathrm{sp}(\Gamma)$) is $\leq N 3^d \ell_{+,\gamma}^d$ (because the number of cubes connected to a given one is $3^d - 1$); then $N \geq 3^{-d} N_\Gamma$ and the lemma is proved, pending (11.8.1.5). Let $C_x^{(\ell_{+,\gamma})} \subset \mathrm{sp}(\Gamma)$, then:

Case 1: $\theta(q_\Lambda; x) = 0$ so that, see above, $C_x^{(\ell_{+,\gamma})} \subset \mathrm{sp}(\Gamma) \setminus \delta_{\mathrm{in}}^{\ell_{+,\gamma}}[\mathrm{sp}(\Gamma)]$ and Case 1a: there is $y \in \ell_{-,\gamma}\mathbb{Z}^d \cap C_x^{(\ell_{+,\gamma})}$ such that $\eta(q_\Lambda; y) = 0$; if no such y exists, Case 1b, there are $z, z' \in \ell_{-,\gamma}\mathbb{Z}^d \cap C_x^{(\ell_{+,\gamma})}$ such that $C_z^{(\ell_{-,\gamma})}$ and $C_{z'}^{(\ell_{-,\gamma})}$ are connected and $\eta(q_\Lambda; z) = -\eta(q_\Lambda; z') \neq 0$. In Case 1a $C_y^{(\ell_{-,\gamma})} \in \mathcal{C}_0 \cup \mathcal{C}_>$; in Case 1b at least one of the two cubes $C_z^{(\ell_-,\gamma)}$ and $C_{z'}^{(\ell_-,\gamma)}$ is in $\mathcal{C}_{\neq}$ by the maximality of $\mathcal{C}_{\neq}$. Thus in Case 1, $C_x^{(\ell_+,\gamma)}$ is contained in the set on the l.h.s. of (11.8.1.5).

Case 2: $\theta(q_\Lambda;x)=1$. Case 2a: there is $y\in\ell_{+,\gamma}\mathbb{Z}^d\cap\delta_{\text{out}}^{\ell_+,\gamma}[C_x^{(\ell_+,\gamma)}]$ such that $\theta(q_\Lambda;y)=0$, then by the analysis of Case 1, $C_y^{(\ell_+,\gamma)}\in\mathcal{C}^+$ so that $C_x^{(\ell_+,\gamma)}$ is contained in the set on the l.h.s. of (11.8.1.5). Case 2b: if Case 2a does not hold, by the definition of contours there is $y\in\ell_{+,\gamma}\mathbb{Z}^d\cap\delta_{\text{out}}^{\ell_+,\gamma}[C_x^{(\ell_+,\gamma)}]$ such that $\theta(q_\Lambda;y)=-\theta(q_\Lambda;x)$ and at least one among $C_x^{(\ell_+,\gamma)}$ and $C_y^{(\ell_+,\gamma)}$ is in $\text{sp}(\Gamma)\setminus\delta_{\text{in}}^{\ell_+,\gamma}[\text{sp}(\Gamma)]$. Then there are $z\in\ell_{-,\gamma}\mathbb{Z}^d\cap C_x^{(\ell_+,\gamma)}$ and $z'\in\ell_{-,\gamma}\mathbb{Z}^d\cap C_y^{(\ell_+,\gamma)}$ such that $\eta(q_\Lambda;z)=-\eta(q_\Lambda;z')\neq 0$ and $C_z^{(\ell_-,\gamma)}$ and $C_{z'}^{(\ell_-,\gamma)}$ are connected. Since, say, $z\in\text{sp}(\Gamma)\setminus\delta_{\text{in}}^{\ell_+,\gamma}[\text{sp}(\Gamma)]$, both z and z' are in Λ' and the same argument for the Case 1b shows that at least one among $C_z^{(\ell_-,\gamma)}$ and $C_{z'}^{(\ell_-,\gamma)}$ is in $\mathcal{C}_{\neq}$; hence at least one among $C_x^{(\ell_+,\gamma)}$ and $C_y^{(\ell_+,\gamma)}$ is in $\mathcal{C}$ so that, again, $C_x^{(\ell_+,\gamma)}$ is contained in the set on the l.h.s. of (11.8.1.5).

Case 3: $\theta(q_\Lambda;x)=-1$ is just as Case 2 and the proof is complete. □

11.8.2 Large density fluctuations

The goal for this subsection is to reduce the minimization problem in (11.8.2.7) below by exploiting the special structure of the LMP energy.

Notation

Let $n_x(q_\Lambda)=|C_x^{(\ell_-,\gamma)}\cap q_\Lambda|$, $x\in\ell_{-,\gamma}\mathbb{Z}^d\cap\Lambda$; $\underline{n}(q_\Lambda)=\{n_x(q_\Lambda), x\in\ell_{-,\gamma}\mathbb{Z}^d\cap\Lambda\}$. $\{\underline{n}(q_\Lambda)=\underline{n}\}$ denotes the collection of all q_Λ such that $\underline{n}(q_\Lambda)=\underline{n}$. By abuse of notation we define $N_0(\underline{n})$, $N_{\neq}(\underline{n})$, $N_>(\underline{n})$ respectively equal to $N_0(q_\Lambda)$, $N_{\neq}(q_\Lambda)$, $N_>(q_\Lambda)$; $q_\Lambda:\underline{n}(q_\Lambda)=\underline{n}$, and indeed $N_0(q_\Lambda)$, $N_{\neq}(q_\Lambda)$, $N_>(q_\Lambda)$ are constant on $\{q_\Lambda:\underline{n}(q_\Lambda)=\underline{n}\}$.

Recalling (11.8.1.4) we have

$$\mathcal{I}=\sum_{\underline{n}:N_0+N_>+N_{\neq}\geq 3^{-d}N_\Gamma}\int_{\underline{n}(q_\Lambda)=\underline{n}} e^{-\beta H_{\gamma,\lambda_{\beta,\gamma},\Lambda}(q_\Lambda|\chi_{\Lambda^c})}. \tag{11.8.2.1}$$

Large density cutoff

We further decompose the sum by specifying those x for which $n_x>\rho_{\max}\ell_{-,\gamma}^d$. Denote by $\mathbf{1}_{x_1,\dots,x_m}(\underline{n})$ the characteristic function of all $\underline{n}$ such that $x_1,\dots,x_m$ are the only values where $n_x > \rho_{\max}\ell^d_{-,\gamma}$; we then get

$$\mathcal{I} = \sum_{m\geq 0}\sum_{x_1,\dots,x_m}\sum_{\underline{n}:N_0+N_{\neq}\geq 3^{-d}N_\Gamma - m} \mathbf{1}_{x_1,\dots,x_m}(\underline{n}) \\ \times \int_{\underline{n}(q_\Lambda)=\underline{n}} e^{-\beta H_{\gamma,\lambda_{\beta,\gamma}}(q_\Lambda|\chi_{\Lambda^c})}, \tag{11.8.2.2}$$

Lemma 11.8.2.1 *Let* $\rho_{\max}$ *as in* (10.3.4.6) *and* γ *small enough, then for any* $\bar{q}$ *and any* $C \in \mathcal{D}^{(\ell_{-,\gamma})}$

$$\int_{|q_C|>\rho_{\max}\ell^d_{-,\gamma}} e^{-\beta H_{\gamma,\lambda_{\beta,\gamma},C}(q_C|\bar{q})} \leq e^{-4\rho_{\max}\ell^d_{-,\gamma}}. \tag{11.8.2.3}$$

Proof By (11.1.2.8)

$$\int_{|q_C|>\rho_{\max}\ell^d_{-,\gamma}} e^{-\beta H_{\gamma,\lambda_{\beta,\gamma}}(q_C|\bar{q})} \leq \sum_{n\geq\rho_{\max}\ell^d_{-,\gamma}} \frac{(\ell^d_{-,\gamma}e^{\beta b_0})^n}{n!} \leq e^{-4\rho_{\max}\ell^d_{-,\gamma}}.$$

In fact, by the Stirling formula, (A.2.1), $\sum_{n\geq\rho\ell^d_{-,\gamma}} \frac{(\ell^d_{-,\gamma}e^{\beta b_0})^n}{n!} \approx e^{\ell^d_{-,\gamma}\rho[\log(e^{\beta b_0}/\rho+1]}$ to leading orders for large ρ; hence the above bound supposing $\rho_{\max}$ as large as in the condition (10.3.4.6). □

By applying successively Lemma 11.8.2.1 in (11.8.2.2) to the cubes $C_{x_1}\cdots C_{x_m}$ appearing in (11.8.2.2) and since $1 \leq \int_{|q_C|\leq\rho_{\max}\ell^d_{-,\gamma}} e^{-\beta H_{\gamma,\lambda_{\beta,\gamma}}(q_C|\bar{q})}$ (because 1 is the contribution of the empty configuration to the above integral), we get $\mathcal{I} \leq \mathcal{I}^*_{\lambda_{\beta,\gamma}} \sum_{m\geq 0}\sum_{x_1,\dots,x_m} e^{-m\rho_{\max}\ell^d_{-,\gamma}}$ where

$$\mathcal{I}^*_\lambda := \sup_{m\geq 0} e^{-3m\rho_{\max}\ell^d_{-,\gamma}} \int_{q_\Lambda\in\mathcal{B}_m} e^{-\beta H_{\gamma,\lambda,\Lambda}(q_\Lambda|\chi_{\Lambda^c})},$$

$$\mathcal{B}_m = \{\underline{n}(q_\Lambda) : N_> = 0,\ N_0 + N_{\neq} \geq 3^{-d}N_\Gamma - m\} \tag{11.8.2.4}$$

(recall that $N_> = 0$ means that $n_x(q_\Lambda) \leq \rho_{\max}\ell^d_{-,\gamma}$ for all $x \in \ell_{-,\gamma}\mathbb{Z}^d \cap \Lambda$). Since $\sum_{m\geq 0}\sum_{x_1,\dots,x_m} e^{-m\rho_{\max}\ell^d_{-,\gamma}} = (1 + e^{-\rho_{\max}\ell^d_{-,\gamma}})^{|\Lambda|/\ell^d_{-,\gamma}}$,

$$\mathcal{I} \leq e^{(|\Lambda|/\ell^d_{-,\gamma})e^{-\rho_{\max}\ell^d_{-,\gamma}}} \mathcal{I}^*_{\lambda_{\beta,\gamma}}.$$

Recalling that in $\mathcal{B}_m$, $|q_\Lambda| \leq \rho_{\max}|\Lambda|$ and since $|\lambda(\beta) - \lambda_{\beta,\gamma}| \leq c\gamma^{1/2}$, see (11.1.0.5), there is a positive constant c_+ so that

$$\mathcal{I} \leq e^{c_+\gamma^{1/2}|\Lambda|}\mathcal{I}^*_{\lambda(\beta)}. \tag{11.8.2.5}$$

Coarse graining

In $\mathcal{B}_m$ the density is bounded and we can use the coarse graining approach of Sect. 11.1.3. Let $\tilde{\mathcal{M}}_\Lambda$ be as in Definition 11.1.3.1. N_0, $N_>$ and $N_{\neq}$ naturally extend to functions on $\tilde{\mathcal{M}}_\Lambda$ and by abuse of notation we write

$$\mathcal{B}_m = \left\{\rho_\Lambda \in \tilde{\mathcal{M}}_\Lambda : N_>(\rho_\Lambda) = 0,\ N_0(\rho_\Lambda) + N_{\neq}(\rho_\Lambda) \geq 3^{-d} N_\Gamma - m\right\}. \quad (11.8.2.6)$$

Then recalling (11.8.2.5) and using Lemma 11.1.3.4 and Theorem 11.1.3.3 (to which we also refer for the notation used below), we have

$$\log \mathcal{I} \leq -\min_{m\geq 0}\left\{3m\rho_{\max}\ell_{-,\gamma}^d + \beta \inf_{\rho_\Lambda\in\mathcal{B}_m} \tilde{\mathcal{F}}_{\gamma,\lambda(\beta),\Lambda}(\rho_\Lambda|\chi_{\Lambda^c})\right\}$$
$$+ c\gamma^{1/2}|\Lambda|. \quad (11.8.2.7)$$

$\tilde{\mathcal{F}}_{\gamma,\lambda(\beta),\Lambda}$ is defined with J_γ replaced by $\tilde{J}_\gamma$, $\tilde{J}_\gamma$ as in (11.1.3.5). We have now made a reduction to a minimization problem where densities are bounded on the scale $\ell_{-,\gamma}$.

11.8.3 Reduction to uniformly bounded densities

In (11.8.2.7) the average density is bounded on the scale $\ell_{-,\gamma}$ (by the condition $N_>(\rho_\Lambda) = 0$ in $\mathcal{B}_m$), but there still could be large density fluctuations on the finest scale $\gamma^{-1/2}$ (recall that in $\tilde{\mathcal{M}}_\Lambda$ functions are $\mathcal{D}^{(\gamma^{-1/2})}$-measurable). In Proposition 11.8.3.1 below we shall prove that fluctuations on the scale $\gamma^{-1/2}$ are "expensive" and consequently they will not occur in a minimizer.

Proposition 11.8.3.1 *Given $\rho_\Lambda \in \tilde{\mathcal{M}}_\Lambda$ ($\tilde{\mathcal{M}}_\Lambda$ as in Definition* 11.1.3.1), *denote*

$$\rho_\Lambda^{\mathrm{cf}}(r) = \min\left\{\rho_\Lambda(r), \rho_{\max}\right\}. \quad (11.8.3.1)$$

Then if $\rho_{\max}$ satisfies (10.3.4.5),

$$\tilde{\mathcal{F}}_{\gamma,\lambda(\beta),\Lambda}(\rho_\Lambda|\chi_{\Lambda^c}) \geq \tilde{\mathcal{F}}_{\gamma,\lambda(\beta),\Lambda}(\rho_\Lambda^{\mathrm{cf}}|\chi_{\Lambda^c}) + \int_\Lambda (\rho_\Lambda - \rho_\Lambda^{\mathrm{cf}}). \quad (11.8.3.2)$$

Proof Extend ρ_Λ and $\rho_\Lambda^{\mathrm{cf}}$ to $\mathbb{R}^d$ by setting them equal to 0 outside Λ. The excess $\int_\Lambda(\rho_\Lambda - \rho_\Lambda^{\mathrm{cf}})$ in (11.8.3.2) is due to the entropy term. The energy term in $\tilde{\mathcal{F}}_{\gamma,\lambda(\beta),\Lambda}(\rho_\Lambda|\chi_{\Lambda^c})$ can be written as

$$\int \{e_{\lambda(\beta)}(\tilde{J}_\gamma * [\rho_\Lambda^{\mathrm{cf}} + \chi_{\Lambda^c}]) - e_{\lambda(\beta)}(\tilde{J}_\gamma * \chi_{\Lambda^c})\}$$
$$+ \{e_{\lambda(\beta)}(\tilde{J}_\gamma * [(\rho_\Lambda - \rho_\Lambda^{\mathrm{cf}}) + \rho_\Lambda^{\mathrm{cf}} + \chi_{\Lambda^c}]) - e_{\lambda(\beta)}(\tilde{J}_\gamma * [\rho_\Lambda^{\mathrm{cf}} + \chi_{\Lambda^c}])\}.$$

By (11.1.2.1) the contribution of the last curly bracket is bounded from below by

$$\geq -b_0 \int \tilde{J}_\gamma * (\rho_\Lambda - \rho_\Lambda^{\text{cf}}) = -b_0 \int (\rho_\Lambda - \rho_\Lambda^{\text{cf}}).$$

The contribution to the free energy of the entropy term is

$$-\frac{1}{\beta}\int_\Lambda S(\rho_\Lambda) = -\frac{1}{\beta}\int_\Lambda S(\rho_\Lambda^{\text{cf}}) + \frac{1}{\beta}\int_\Lambda \{\rho_\Lambda(\log \rho_\Lambda - 1) - \rho_\Lambda^{\text{cf}}(\log \rho_\Lambda^{\text{cf}} - 1)\}.$$

The last term is bounded from below by $\geq \frac{1}{\beta}\int_\Lambda \{\rho_\Lambda - \rho_\Lambda^{\text{cf}}\} \log \rho_{\max}$, and hence (11.8.3.2) because $\rho_{\max}$: $\frac{1}{\beta}\log \rho_{\max} \geq [1 + b_0]$; see (10.3.4.5). □

There are restrictions on ρ_Λ^{cf} due to the constraint $\rho_\Lambda \in \mathcal{B}_m$. Let $\mathcal{C}'_0$ and $\mathcal{C}'_{\neq}$ be the collections defined in the beginning of Sect. 11.3 and call $\mathcal{C}'_{>}$ the collection of all cubes C of $\mathcal{D}^{(\ell_{-,\gamma})}$ in Λ' where

$$\fint_C \rho_\Lambda - \rho_\Lambda^{\text{cf}} > \frac{\zeta}{2} \tag{11.8.3.3}$$

We denote by N'_0, $N'_{\neq}$ and $N'_{>}$ the number of cubes in $\mathcal{C}'_0$, $\mathcal{C}'_{\neq}$ and $\mathcal{C}'_{>}$, all these quantities being a function of ρ_Λ.

Proposition 11.8.3.2 *Let ρ_Λ be such that $N_{>}(\rho_\Lambda) = 0$; then*

$$2\big(N'_0(\rho_\Lambda^{\text{cf}}) + N'_{>}(\rho_\Lambda)\big) + N'_{\neq}(\rho_\Lambda^{\text{cf}}) \geq N_0(\rho_\Lambda) + N_{\neq}(\rho_\Lambda). \tag{11.8.3.4}$$

Proof Assume $C \in \mathcal{C}_0(\rho_\Lambda)$; then either $C \in \mathcal{C}'_0(\rho_\Lambda^{\text{cf}})$ or $\eta^{(\zeta/2,\ell_{-,\gamma})}(\rho_\Lambda^{\text{cf}};\cdot) = s \neq 0$ on C. Then (11.8.3.3) must hold; otherwise $\eta(\rho_\Lambda;\cdot) = s$ on C against the assumption. Assume next that $C' = C_{x'}^{(\ell_{-,\gamma})}, C'' = C_{x''}^{(\ell_{-,\gamma})}$ is a pair in $\mathcal{C}_{\neq}(\rho_\Lambda)$, C' and C'' in Λ'. Call Case 1: either C or C' or both are in $\mathcal{C}'_0(\rho_\Lambda^{\text{cf}})$; Case 2: (11.8.3.3) holds somewhere in $C \cup C'$; Case 3 covers the remaining possibilities. In Cases 1 and 2 at least one among C and C' contributes to $N'_0(\rho_\Lambda^{\text{cf}}) + N'_{>}(\rho_\Lambda)$; hence the factor 2 on the l.h.s. of (11.8.3.4). In Case 3 we shall show that the pair (C', C'') is in $\mathcal{C}'_{\neq}(\rho_\Lambda^{\text{cf}})$. We argue by contradiction: suppose then that there are s' and s'' so that $\eta^{(\zeta/2,\ell_{-,\gamma})}(\rho^{\text{cf}};\cdot) = s'$ on C' and $\eta^{(\zeta/2,\ell_{-,\gamma})}(\rho^{\text{cf}};\cdot) = s''$ on C'', both $\neq 0$. Since C' and C'' are not in $\mathcal{C}'_{>}(\rho_\Lambda)$, by the previous argument $\eta^{(\zeta,\ell_{-,\gamma})}(\cdot;\rho) = s'$ on C' and $\eta^{(\zeta,\ell_{-,\gamma})}(\cdot;\rho) = s''$ on C''. Since $s' = -s'' \neq 0$, (C', C'') contributes to $N'_{\neq}(\rho^{\text{cf}})$, which yields a contradiction. □

Proof of Proposition 11.3.0.1 Let k be the value of m which minimizes the r.h.s. of (11.8.2.7), so that

$$\log \mathcal{I} \leq -\Big\{3k\rho_{\max}\ell_{-,\gamma}^d + \beta \inf_{\rho_\Lambda \in \mathcal{B}_k} \tilde{\mathcal{F}}_{\gamma,\lambda(\beta),\Lambda}(\rho_\Lambda|\chi_{\Lambda^c})\Big\} + c\gamma^{1/2}|\Lambda|. \tag{11.8.3.5}$$

Let $\rho_\Lambda \in \mathcal{B}_k$ and call $n := N'_>(\rho_\Lambda)$. By Proposition 11.8.3.1

$$\tilde{\mathcal{F}}_{\gamma,\lambda(\beta),\Lambda}(\rho_\Lambda|\chi_{\Lambda^c}) \geq n\, \ell_{-,\gamma}^d \frac{\zeta}{2} + \tilde{\mathcal{F}}_{\gamma,\lambda(\beta),\Lambda}(\rho_\Lambda^{\mathrm{cf}}|\chi_{\Lambda^c}),$$

while, by Proposition 11.8.3.2, $N'_0(\rho_\Lambda^{\mathrm{cf}}) + N'_{\neq}(\rho_\Lambda^{\mathrm{cf}}) \geq \frac{3^{-d}N_\Gamma - k}{2} - n$. Thus $\rho_\Lambda^{\mathrm{cf}} \in \mathcal{G}_{2(n+k)}$, and (11.3.0.3) then follows from (11.8.2.7), because for γ small enough $3\rho_{\max}\ell_{-,\gamma}^d > \ell_{-,\gamma}^d \zeta/2$. □

11.9 Complements to Section 11.4, cluster expansion

We shall prove here Theorem 11.4.3.1 but the analysis has a much more general validity and it is largely model independent. Its applicability follows from the fact that our weights satisfy the Peierls bounds and can be made as small as desired by taking γ correspondingly small.

The general problem of a cluster expansion is to write an expression like $\log X^+_{\gamma,\sigma_\Lambda}(\Lambda)$, with

$$X^+_{\gamma,\sigma_\Lambda}(\Lambda) = 1 + \sum_{\underline{\Gamma}\in\mathcal{B}^+_\Lambda, \underline{\Gamma}\neq\emptyset} \prod_{\Gamma\in\underline{\Gamma}} W^+_\gamma(\Gamma,\sigma), \quad \sigma \in \mathcal{X}^+, \tag{11.9.0.1}$$

as a series of polynomials of the weights. Let us start from the trivial case where Λ is so small that only one contour, say Γ, may be present. Then $X^+_{\gamma,\Lambda} = 1 + W^+_\gamma(\Gamma,\sigma)$ and $\log X^+_{\gamma,\Lambda} = \sum_{n=1}^\infty \frac{(-1)^{n+1}}{n} W^+_\gamma(\Gamma,\sigma)^n$, supposing $W^+_\gamma(\Gamma,\sigma) < 1$. We thus see a first difference with expressions like on the r.h.s. of (11.9.0.1): the same Γ appears several times and in general we shall have a sum of products of weights without the compatibility condition which is instead required in the sum in (11.9.0.1) with the condition $\{\underline{\Gamma} \in \mathcal{B}^+_\Lambda\}$. The sum on the r.h.s. of (11.9.0.1), call it S, has many terms if the region Λ is large and the idea of expanding in powers of S cannot be pursued in general because the condition $S < 1$ for its convergence fails, no matter how small we take γ, if Λ is correspondingly large. The situation, however, cannot be too catastrophic, as the following example shows. Let

$$X := \prod_{n=1}^N (1 + \epsilon_n) =: 1 + S,$$

with all $\epsilon_i \in (0,1)$ but such that $S > 1$. We cannot expand $\log X$ in a series of S; yet we can write $\log X$ as a power series of the variables ϵ_i. This suggests to look for (approximate) factorization properties of the weights $W^+_\gamma(\Gamma,\sigma)$.

A cluster expansion is in fact a power series expansion of $\log X^+_{\gamma,\Lambda,\sigma_\Lambda}$, thought of as a function of the variables $\{W^+_\gamma(\Gamma,\sigma), \Gamma \in \{\Gamma\}^+\}$, and the condition for its convergence is a condition on the weights which does not depend on Λ. As mentioned earlier we shall not prove the convergence of the expansion but only state the results. The setup for a cluster expansion is: a graph, which in our case, for the sake of definiteness, is $\{\Gamma\}^+$ (see Sect. 10.5.1), and the space $\mathcal{I}^+$ of all functions $I:\{\Gamma\}^+ \to \mathbb{N}$ such that

$$\sum_{\Gamma\in\{\Gamma\}^+} I(\Gamma) < \infty, \quad \text{the set } \{\Gamma : I(\Gamma) > 0\} \text{ is connected} \tag{11.9.0.2}$$

thus $I(\Gamma)$, $\Gamma \in \{\Gamma\}^+$, says how many "copies" we "take" of Γ. We also write

$$\mathrm{sp}(I) := \bigcup_{\Gamma:I(\Gamma)>0} \mathrm{sp}(\Gamma). \tag{11.9.0.3}$$

As we shall see the following theorem is a direct consequence of the classical cluster expansion bounds.

Theorem 11.9.0.1 *Assume that the family $\{W^+_\gamma(\Gamma;\sigma)\}$ satisfies the Peierls bounds with a constant $c > 0$. Then for any $c_{\mathrm{pol}} \in (0,c)$ there is $\gamma_{\mathrm{pol}} > 0$ and, for any $\gamma \le \gamma_{\mathrm{pol}}$, there are functions $\omega^+_\gamma(I,\sigma)$ on $\mathcal{I}^+ \times \mathcal{X}^+$ with the following properties.*

- *With $X^+_{\gamma,\Lambda,\sigma}$ as in* (11.9.0.1),

$$\log X^+_{\gamma,\Lambda,\sigma}(\{W^+_\gamma(\Gamma;\sigma)\}) = \sum_{I:\mathrm{sp}(I)\subset\Lambda} \omega^+_\gamma(I,\sigma). \tag{11.9.0.4}$$

- *For any $r \in \mathbb{R}^d$,*

$$\sum_{I:\mathrm{sp}(I)\ni r} \|\omega^+_\gamma(I,\sigma)\|_\infty \left\{\prod_\Gamma e^{\beta c_{\mathrm{pol}}(\zeta^2 \ell^d_{-,\gamma}) N_\Gamma I(\Gamma)}\right\} \le 1. \tag{11.9.0.5}$$

- *The "polymers" $\omega^+_\gamma(I,\sigma)$ have the following expression:*

$$\omega^+_\gamma(I,\sigma) = C_I \prod [W^+_\gamma(\Gamma,\sigma)]^{I(\Gamma)}, \tag{11.9.0.6}$$

where

$$C_I = \frac{1}{\prod_\Gamma I(\Gamma)!} \left\{\prod_\Gamma \frac{\partial^{I(\Gamma)}}{\partial W^+_\gamma(\Gamma,\sigma)^{I(\Gamma)}}\right\} \log X^+_{\gamma,\Lambda}(\{W^+_\gamma(\Gamma,\sigma)\})\Big|_{W^+_\gamma(\Gamma,\sigma)=0}. \tag{11.9.0.7}$$

Remarks Theorem 11.9.0.1 applies in LMP with $c = c_1/100$, see (10.5.1.9), and in the Ising case with $c = c_1/2$. The r.h.s. of (11.9.0.4) with $\omega^+_\gamma(I,\sigma)$ as in (11.9.0.6) is the [formal] power series of $\log X^+_{\gamma,\Lambda,\sigma}(\{W^+_\gamma(\Gamma;\sigma)\})$. The bound (11.9.0.5) ensures the convergence of the series, hence the proof of Theorem 11.9.0.1 is reduced to the proof of (11.9.0.5) which is a corollary of the classical statements of cluster expansion starting from the famous condition of Kotecký and Preiss [158], see also Ruelle [197], Brydges [56], and more recent work [46], [172]. Let us then see what the classical results say (we are not using the optimal bounds) and then why they yield (11.9.0.5).

Theorem CE [Cluster expansion] *Assume that there are weights $\rho(\Gamma) > 0$ on $\{\Gamma\}^+$ and $a \ge 1$ so that*

$$\sup_r \sum_{\Gamma:\mathrm{sp}(\Gamma)\ni r} \rho(\Gamma)\, e^{aN_\Gamma} \le 1. \tag{11.9.0.8}$$

Then

$$\sum_{I:I(\Gamma_0)>0} |C_I| \prod_\Gamma [\rho(\Gamma)]^{I(\Gamma)} \le e^{aN_{\Gamma_0}} \rho(\Gamma_0). \tag{11.9.0.9}$$

To prove Theorem 11.9.0.1 we set $a = 1$ and

$$\rho(\Gamma) = W_\gamma^+(\Gamma;\sigma)\, e^{\alpha N_\Gamma}, \quad \alpha = \beta c_{\mathrm{pol}} \zeta^2 \ell_{-\gamma}^d. \tag{11.9.0.10}$$

Since $c_{\mathrm{pol}} < c$, c being the constant in the Peierls bounds, for all γ small enough

$$\sup_r \sum_{\Gamma:\mathrm{sp}(\Gamma)\ni r} W(\Gamma)\, e^{\alpha N_\Gamma} e^{N_\Gamma} \le 1, \tag{11.9.0.11}$$

so that (11.9.0.8) is satisfied. Then by (11.9.0.9) and with $\omega_\gamma^+(I,\sigma)$ as in (11.9.0.6)

$$\text{l.h.s. of (11.9.0.5)} \le \sum_{\Gamma_0:\mathrm{sp}(\Gamma_0)\ni r} e^{N_{\Gamma_0}} W_\gamma^+(\Gamma;\sigma)\, e^{\alpha N_{\Gamma_0}} \le 1. \tag{11.9.0.12}$$

11.9.1 Effective hamiltonian

To interpret the r.h.s. of (11.9.0.4) as an energy we need to relate $\omega_\gamma^+(I,\sigma)$ to a potential. Let

$$\mathrm{sp}_+(\Gamma) = \mathrm{sp}(\Gamma) \cup \delta_{\mathrm{out}}^{\ell_{+,\gamma}}[c(\Gamma)], \tag{11.9.0.13}$$

and set, in analogy with (11.9.0.3),

$$\mathrm{sp}_+(I) := \bigcup_{\Gamma:I(\Gamma)>0} \mathrm{sp}_+(\Gamma). \tag{11.9.0.14}$$

Then by Theorem 10.5.1.2, when the weights $W_\gamma^+(\Gamma;\sigma)$ are the cutoff weights, or by Lemma 10.3.3.1, when the weights are the true weights, $W_\gamma^+(\Gamma;\sigma)$ depends only on the restriction of σ to $\delta_{\mathrm{out}}^{2\gamma^{-1}}[c(\Gamma)]$) so that, a fortiori, $W_\gamma^+(\Gamma;\sigma)$ depends only on the restriction to $\mathrm{sp}_+(\Gamma)$. $\omega^+_\gamma(I,\sigma)$ is then measurable on the restriction of σ to $\mathrm{sp}_+(I)$ and we can introduce [many body] potentials $\{U^+_{\gamma,\Delta}(\sigma)\}$, $\Delta\in\{D\}$ by setting

$$U^+_{\gamma,\Delta}(\sigma)\equiv U^+_{\gamma,\Delta}(\sigma_\Delta)=-\frac{1}{\beta}\sum_{I:\mathrm{sp}_+(I)=\Delta}\omega^+_\gamma(I,\sigma), \tag{11.9.0.15}$$

observing that $U^+_{\gamma,\Delta}(\sigma)$ depends only on σ_Δ (the restriction of σ to Δ).

Proof of Theorem 11.4.3.1 We have

$$\text{l.h.s. of (11.4.3.8)}\le\sum_{I:\mathrm{sp}_+(I)\ni r}e^{bN_{\mathrm{sp}_+(I)}}|\omega^+_\gamma(I,\sigma)|$$

(by definition of U_Δ) and $N_{\mathrm{sp}_+(I)}\le 3^d N_{\mathrm{sp}(I)}$. To exploit (11.9.0.5) we need to replace the condition $\mathrm{sp}_+(I)\ni r$ by $\mathrm{sp}(I)\ni r$. To this end we observe that if $\mathrm{sp}_+(I)\ni r$ and $\mathrm{sp}(I)\not\ni r$, then $\mathrm{sp}(I)\cap\delta^{\ell_{+,\gamma}}_{\mathrm{out}}[C_r]\ne\emptyset$. We then have

$$\begin{aligned}\text{l.h.s. of (11.4.3.8)}&\le\sum_{I:\mathrm{sp}(I)\cap(C_r\cup\delta^{\ell_{+,\gamma}}_{\mathrm{out}}[C_r])\ne\emptyset}e^{3^d bN_{\mathrm{sp}(I)}}|\omega^+_\gamma(I,\sigma)|\\&\le\sum_{C\subset(C_r\cup\delta^{\ell_{+,\gamma}}_{\mathrm{out}}[C_r])}e^{3^d bN_{\mathrm{sp}(I)}}\sum_{I:\mathrm{sp}(I)\supset C}|\omega^+_\gamma(I,\sigma)|.\end{aligned}$$

By using invariance under translations by $\ell_{+,\gamma}i$, $i\in\mathbb{Z}^d$ (which follows from (11.9.0.6) by the $\ell_{+,\gamma}i$-translation invariance of the weights), we have

$$\text{l.h.s. of (11.4.3.8)}\le 3^d\sum_{I:\mathrm{sp}(I)\ni r}e^{3^d bN_{\mathrm{sp}(I)}}\|\omega^+_\gamma(I,\sigma)\|_\infty.$$

We write $e^{3^d bN_{\mathrm{sp}(I)}}=e^{3^d bN_{\mathrm{sp}(I)}}e^{-\beta c_{\mathrm{pol}}(\zeta^2\ell^d_{-,\gamma})N_\Gamma I(\Gamma)}e^{\beta c_{\mathrm{pol}}(\zeta^2\ell^d_{-,\gamma})N_\Gamma I(\Gamma)}$ and bound the first two factors by $e^{-\beta[c_{\mathrm{pol}}(\zeta^2\ell^d_{-,\gamma})-3^d b]3^d}$, because it is a decreasing function of N_Γ and $N_\Gamma\ge 3^d$. Hence by (11.9.0.5) the l.h.s. of (11.4.3.8) is bounded by

$$3^d e^{-\beta[c_{\mathrm{pol}}(\zeta^2\ell^d_{-,\gamma})-3^d b]3^d}\sum_{I:\mathrm{sp}(I)\ni r}\|\omega^+_\gamma(I,\sigma)\|_\infty e^{\beta c_{\mathrm{pol}}(\zeta^2\ell^d_{-,\gamma})N_\Gamma I(\Gamma)},$$

and (11.4.3.8) follows for γ small enough. By the same argument

$$\begin{aligned}&\beta\sum_{\Delta\ni r,\Delta\cap A\ne\emptyset}|U^+_{\gamma,\Delta}(\sigma)|\\&\le\sum_{I:\mathrm{sp}(I)\cap(C_r\cup\delta^{\ell_{+,\gamma}}_{\mathrm{out}}[C_r])\ne\emptyset,\ \mathrm{sp}(I)\cap(A\cup\delta^{\ell_{+,\gamma}}_{\mathrm{out}}[A])\ne\emptyset}\|\omega^+_\gamma(I,\sigma)\|_\infty.\end{aligned}$$

We multiply $\|\omega_\gamma^+(I,\sigma)\|_\infty$ by $e^{-\beta c_{\mathrm{pol}}(\zeta^2\ell_{-,\gamma}^d)N_\Gamma I(\Gamma)}e^{\beta c_{\mathrm{pol}}(\zeta^2\ell_{-,\gamma}^d)N_\Gamma I(\Gamma)}$ and write

$$e^{-\beta c_{\mathrm{pol}}(\zeta^2\ell_{-,\gamma}^d)N_\Gamma I(\Gamma)} \le e^{-\beta c_{\mathrm{pol}}(\zeta^2\ell_{-,\gamma}^d)d_{\{D\}}(r,A)}.$$

Then

$$\beta \sum_{\Delta\ni r,\Delta\cap A\neq\emptyset} |U_{\gamma,\Delta}^+(\sigma)| \le e^{-\beta c_{\mathrm{pol}}(\zeta^2\ell_{-,\gamma}^d)N_{\ell_{+,\gamma}}(r,A)} \times \sum_{C\subset(C_r\cup\delta_{\mathrm{out}}^{\ell_{+,\gamma}}[C_r])} \sum_{I:\mathrm{sp}\supset C} \{\|\omega_\gamma^+(I,\sigma)\|_\infty e^{\beta c_{\mathrm{pol}}(\zeta^2\ell_{-,\gamma}^d)N_\Gamma I(\Gamma)}\};$$

hence (11.4.3.9). □

Chapter 12
DLR measures and the ergodic decomposition

In the previous chapter we have proved Theorem 10.3.4.2 and therefore that for γ small enough, $\gamma \le \gamma_\beta$, $\gamma_\beta > 0$; the difference between the plus and the minus diluted Gibbs measures $G^+_{\gamma,\lambda_{\beta,\gamma},\Lambda,q^+}(dq)$ and $G^-_{\gamma,\lambda_{\beta,\gamma},\Lambda,q^-}(dq)$ persists in the thermodynamic limit. We shall prove in this chapter that both $G^\pm_{\gamma,\lambda,\Lambda,q^\pm}(dq)$ converge as $\Lambda \to \mathbb{R}^d$, that the limits $\mu^\pm_\gamma \equiv \mu^\pm_{\gamma,\beta,\lambda_{\beta,\gamma}}$ are DLR measures, that they are translational invariant and that any other translational invariant DLR measure is a convex combination of the two. Proofs are given for the LMP model but they could easily be extended to the Ising model of Chap. 9, thus proving Theorem 9.1.4.2.

We shall omit instead the proof of a "local version" of the Gibbs phase rule:

Theorem 12.0.0.1 (Gibbs phase rule) *Given any $\beta' \in (\beta_c, \beta_0)$ there are δ^*, c^* and γ^* all positive so that the parameter γ_β in Theorem 10.3.4.1 is $\gamma_\beta \ge \gamma^*$ for all $|\beta - \beta'| \le \delta^*$ and the chemical potential $\lambda_{\beta,\gamma}$ of Theorem 10.3.4.1 is thus defined for all $|\beta - \beta'| \le \delta^*$. Moreover, the pair $(\beta, \lambda_{\beta,\gamma})$ defines a continuous curve in the rectangle $\{\beta \in (\beta' - \delta^*, \beta' + \delta^*), \lambda \in (\lambda(\beta') - c^*, \lambda(\beta') + c^*)\}$ of the plane (β, λ). On each point of the curve there is a phase transition with coexistence of the two distinct DLR measures $\mu^\pm_{\gamma,\beta,\lambda_{\beta,\gamma}}$; elsewhere in the rectangle there is a unique DLR measure.*

The proof of Theorem 12.0.0.1 has several subtle technical points and we thus just refer to the literature [48].

12.1 Existence of two distinct DLR measures

The theory of DLR measures for particle systems and for unbounded spins requires a technically non-trivial extension of the theory developed in Chap. 2 for the Ising model. The LMP model is an exception; the particular structure of its energy makes it almost like a bounded spin system. We leave to the reader the task of extending from the Ising case to LMP the full DLR theory. Here we shall keep the more abstract statements to a minimum and point directly to a precise formulation and then to a proof of the properties stated in the introduction to this chapter. In this section β is fixed in (β_c, β_0) and often dropped from the notation, and the chemical potential is restricted to l $[\lambda(\beta) - 1, \lambda(\beta) + 1]$ and eventually taken equal to $\lambda_{\beta,\gamma}$; see Theorem 10.3.4.1.

12.1.1 Main result

Given β and λ a probability measure μ on $\mathcal{X}$ (equipped with its Borel σ-algebra, see the next subsection) is DLR if for any bounded, measurable cylindrical function f on $\mathcal{X}$ and any bounded measurable region $\Lambda \subset \mathbb{R}^d$,

$$\mu(f) = \int_{\mathcal{X}} G_{\gamma,\lambda,\Lambda,q}(f) d\mu(q) \equiv \mu\big(G_{\gamma,\lambda,\Lambda,q}(f)\big). \tag{12.1.1.1}$$

$G_{\gamma,\lambda,\Lambda,q}$, the LMP Gibbs measure in Λ with boundary conditions q, is defined in (10.2.5.1).

Theorem 12.1.1.1 *Let $\beta \in (\beta_c, \beta_0)$, γ_β and $\lambda_{\beta,\gamma}$ as in Theorem* 10.3.4.1. *Then for any $\gamma \le \gamma_\beta$ there are two distinct DLR measures at $(\beta, \lambda_{\beta,\gamma})$ denoted by $\mu_\gamma^\pm$, and for any $r \in \mathbb{R}^d$*

$$\mu_\gamma^\pm(\{\Theta(q;r) = \pm 1\}) \ge 1 - \exp\left\{ -\beta \frac{c_1}{4} (\zeta^2 \ell_{-,\gamma}^d) \right\} \tag{12.1.1.2}$$

(c_1 *as in Theorem* 10.3.4.2). *Moreover, there are a sequence of bounded $\mathcal{D}^{(\ell_{+,\gamma})}$-measurable regions Λ_n and of configurations $q_n^\pm \in \mathcal{X}^\pm$ such that $\mu_\gamma^\pm$ is the weak limit of $G^\pm_{\gamma,\lambda_{\beta,\gamma},\Lambda_n,q_n^\pm}$ (weak limits are defined in Sect.* 12.1.2 *below).*

Theorem 12.1.1.1 will be proved in the next subsections.

12.1.2 Block spin reformulation of the LMP model

It is convenient to give $\mathcal{X}$ a product structure. Writing C_i, $i \in \mathbb{Z}^d$, for the unit cubes of $\mathcal{D}^{(1)}$, we can set-theoretically identify

$$\mathcal{X} = \prod \mathcal{X}_{C_i}, \qquad q \longleftrightarrow \{q^{(i)} := q \cap C_i, i \in \mathbb{Z}^d\}. \tag{12.1.2.1}$$

We then equip each $\mathcal{X}_{C_i}$ with its natural topology (see Sect. 10.2.3) and give $\mathcal{X}$ the corresponding product topology structure. We have thus a lattice representation for the LMP model; observe though that the "block spins" $q^{(i)}$ take values in a non-compact space and therefore the considerations of Chap. 2 cannot be directly applied.

12.1.3 Finite volume Gibbs measures

If Λ is a bounded $\mathcal{D}^{(1)}$ measurable region, then

$$G_{\gamma,\lambda,\Lambda,\bar{q}}(dq_\Lambda) = Z^{-1} e^{-\beta H_{\gamma,\lambda,\Lambda}(q_\Lambda|\bar{q}_{\Lambda^c})} \prod_{i \in \Lambda} \nu_{C_i}(dq^{(i)}), \tag{12.1.3.1}$$

where q_Λ on the l.h.s. is $q_\Lambda = (q^{(i)}, i \in \mathbb{Z}^d \cap \Lambda)$ and $\bar{q}_{\Lambda^c} = \bar{q} \cap \Lambda^c$. (12.1.3.1) follows the product structure of the free measure.

We call $\mathcal{G}_\Lambda$ the set of finite volume Gibbs measures in Λ, namely of all probabilities μ on $\mathcal{X}$ with the following property: there is a probability ν on $\mathcal{X}$ such that for any bounded, measurable function f,

$$\mu(f) = \nu(G_{\gamma,\lambda,\Lambda,\bar{q}}(f)) = \int_{\mathcal{X}} \Big(\int_{\mathcal{X}_\Lambda} f(q_\Lambda + \bar{q}_{\Lambda^c}) G_{\gamma,\lambda,\Lambda,\bar{q}}(dq_\Lambda) \Big) \nu(d\bar{q}). \tag{12.1.3.2}$$

Whenever f is independent of q_Λ, $\mu(f) = \nu(f)$ and since $\bar{q} \to G_{\gamma,\lambda,\Lambda,\bar{q}}(f)$ is independent of $\bar{q}_\Lambda$, $\mu(f) = \mu(G_{\gamma,\lambda,\Lambda,\bar{q}}(f))$ for all bounded measurable f. Thus if $\mu \in \mathcal{G}_\Lambda$ we can always take $\nu = \mu$.

In general, the definition (12.1.3.2) is more problematic as one needs to prove that $G_{\gamma,\lambda,\Lambda,\bar{q}}(f)$ is a ν-summable function of $\bar{q}$; here, by (11.1.2.8),

$$\Big| \int_{\mathcal{X}_\Lambda} f(q_\Lambda + \bar{q}_{\Lambda^c}) G_{\gamma,\lambda,\Lambda,\bar{q}}(dq_\Lambda) \Big| \le \|f\|_\infty \sum_{n=0}^{\infty} \frac{e^{\beta b_0 n}}{n!} \le c \tag{12.1.3.3}$$

uniformly in $\bar{q}$ and hence no conditions on ν are required.

12.1.4 Compactness

We shall use repeatedly in the sequel the Prohorov criterion for compactness (see for instance Parthasaraty [179], Theorem 6.7 in Chap. II), that we state in our particular context:

Theorem PC [Prohorov Criterion] *A set M of probabilities on $\mathcal{X}$ is weakly pre-compact (i.e. its weak closure is weakly compact) if for any $\epsilon > 0$ there is a compact set $K_\epsilon \subset \mathcal{X}$ such that $\mu(K_\epsilon) \ge 1 - \epsilon$ for all $\mu \in M$.*

We shall also use often in the sequel the following equivalent condition for weak convergence (see Theorem 6.1 in Chap. II of [179]):

Theorem CWC [Criterion for Weak Convergence] *Assume μ_n converges weakly to μ and let C be a closed set; then $\limsup \mu_n(C) \le \mu(C)$. Vice versa, if the inequality holds for all closed sets C, then $\mu_n \to \mu$ weakly.*

The key inequality, which has already been used in the derivation of (12.1.3.3), is behind the following lemma, which exploits the particular structure of the LMP model. Lemma 12.1.4.1 below should not be expected to hold in general. Let $C \in \mathcal{D}^{(1)}$, $\delta > 0$ and N a positive integer, write

$$A_{\delta,N,C} := \{q \in \mathcal{X}_C : |q| \le N, q \cap \{r \in C : \mathrm{dist}(r, C^c) < \delta\} = \emptyset\}. \tag{12.1.4.1}$$

$A_{\delta,N,C}$ is obviously a compact set in $\mathcal{X}_C$ and

Lemma 12.1.4.1 *For any $\bar{q} \in \mathcal{X}$ and $\lambda \in [\lambda(\beta)-1, \lambda(\beta)+1]$*

$$G_{\gamma,\lambda,C,\bar{q}}(A^c_{\delta,N,C}) \le \sum_{n=1}^{\infty} \frac{e^{\beta b_0 n}(2d\delta n)}{n!} + \sum_{n>N} \frac{e^{\beta b_0 n}}{n!}. \tag{12.1.4.2}$$

Proof Equation (12.1.4.2) follows directly from (11.1.2.8). □

Immediate corollaries are:

Corollary 12.1.4.2 *In the same context as Lemma* 12.1.4.1 *there is an integer n^* and for any $n \ge n^*$ there is $\delta_n > 0$ so that*

$$G_{\gamma,\lambda,C,\bar{q}}(A^c_{\delta_n,n,C}) \le e^{-n}, \quad \text{for all } n \ge n^*. \tag{12.1.4.3}$$

We denote by $M(\mathcal{X})$ the set of all probability measures on $\mathcal{X}$ and write

$$\mathcal{G}^0_\Lambda = \big\{\mu \in \mathcal{G}_\Lambda : \text{ there is } \nu \in M^0(\mathcal{X}) \text{ such that } \mu(f) = \nu(G_{\gamma,\lambda,\Lambda,\bar{q}}(f)) \\ \text{for all bounded cylindrical functions } f\big\}, \tag{12.1.4.4}$$

where

$$M^0(\mathcal{X}) = \big\{\mu \in M(\mathcal{X}) : \mu\big(A^c_{\delta_{k+|i|},k+|i|,C_i}\big) \le e^{-(k+|i|)},\ k \ge n^*, i \in \mathbb{Z}^d\big\}. \tag{12.1.4.5}$$

Proposition 12.1.4.3 *The sets $\mathcal{G}^0_\Lambda$ and $M^0(\mathcal{X})$ are weakly compact and $\mathcal{G}^0_\Lambda \subset M^0(\mathcal{X})$.*

Proof Let $\mu \in \mathcal{G}^0_\Lambda$ then by (12.1.4.3) for any $i \in \mathbb{Z}^d \cap \Lambda$ and $k \ge n^*$, $\mu(A^c_{\delta_{k+|i|},k+|i|,C_i}) \le e^{-(k+|i|)}$. Same bound holds for $i \in \mathbb{Z}^d \cap \Lambda^c$ and $k \ge n^*$ because $\mu(A^c_{\delta_{k+|i|},k+|i|,C_i}) = \nu(A^c_{\delta_{k+|i|},k+|i|,C_i})$, $\nu \in M^0(\mathcal{X})$, see the remark after (12.1.2.3). Thus $\mathcal{G}^0_\Lambda \subset M^0(\mathcal{X})$. For any $\mu \in M^0(\mathcal{X})$

$$\mu\bigg(\bigcap_{i\in\mathbb{Z}^d} A_{\delta_{k+|i|},k+|i|,C_i}\bigg) \ge 1 - ce^{-k}, \quad c := \sum_{i\in\mathbb{Z}^d} e^{-|i|},$$

and since $\bigcap_{i\in\mathbb{Z}^d} A_{\delta_{k+|i|},k+|i|,C_i}$ is a compact set, by the Prohorov compactness criterion, Theorem PC, it follows that $M^0(\mathcal{X})$ is weakly pre-compact as well as $\mathcal{G}^0_\Lambda$, being a subset of $M^0(\mathcal{X})$.

$\mathcal{G}^0_\Lambda$ is weakly closed: let $\mu_n \in \mathcal{G}^0_\Lambda$ converge weakly to μ. By the remark after (12.1.3.2) we have $\mu_n(f) = \mu_n(G_{\gamma,\lambda,\Lambda,\bar{q}}(f))$ for any bounded continuous cylindrical function f. Since $\bar{q} \to G_{\gamma,\lambda,\Lambda,\bar{q}}(f)$ is continuous,

$$\mu(f) = \lim_{n\to\infty} \mu_n(f) = \lim_{n\to\infty} \mu_n(G_{\gamma,\lambda,\Lambda,\bar{q}}(f)) = \mu(G_{\gamma,\lambda,\Lambda,\bar{q}}(f));$$

hence $\mu \in \mathcal{G}_\Lambda$. For $i \in \Lambda$ and $n \geq n^*$

$$\mu\big(A^c_{\delta_n,n,C_i}\big) = \mu\big(G_{\gamma,\lambda,\Lambda,\bar{q}}\big(A^c_{\delta_n,n,C_i}\big)\big) \leq e^{-n}$$

by (12.1.4.3). For $i \in \Lambda^c$ and $k \geq n^*$

$$\mu_n\big(A_{\delta_{k+|i|},k+|i|,C_i}\big) \geq 1 - e^{-(k+|i|)},$$

and since A_{δ_h,h,C_i} is closed, weak convergence to μ implies that

$$\mu\big(A_{\delta_{k+|i|},k+|i|,C_i}\big) \geq 1 - e^{-(k+|i|)}.$$

Thus $\mu \in M^0(\mathcal{X})$ and therefore $\mathcal{G}^0_\Lambda \subset M^0(\mathcal{X})$ is weakly closed and hence weakly compact, because we had already proved that $\mathcal{G}^0_\Lambda$ is weakly pre-compact. The same argument applies to $M^0(\mathcal{X})$ and Proposition 12.1.4.3 is proved. □

12.1.5 Existence of DLR measures

By the Gibbs property and Corollary 12.1.4.2,

$$\mathcal{G}^0_\Delta \supset \mathcal{G}^0_\Lambda \quad \text{for any bounded } \mathcal{D}^{(1)}\text{-measurable set } \Lambda \supset \Delta. \tag{12.1.5.1}$$

Equation (12.1.5.1) and Proposition 12.1.4.3 prove that given any increasing sequence of bounded $\mathcal{D}^{(1)}$-measurable sets Λ

$$\mathcal{G}^0 := \bigcap \mathcal{G}^0_\Lambda \quad \text{is a non-empty weakly compact set.} \tag{12.1.5.2}$$

$\mathcal{G}^0$ is evidently independent of the sequence used in its definition.

Theorem 12.1.5.1 *The set $\mathcal{G}^0$ coincides with the set of all DLR measures which is therefore a non-empty weakly compact set.*

Proof Any DLR measure is evidently in $\mathcal{G}_\Lambda$ (for any bounded $\mathcal{D}^{(1)}$-measurable Λ) and by (12.1.4.3) it is also in $M^0(\mathcal{X})$; hence it is in $\mathcal{G}^0_\Lambda$ and by the arbitrariness of Λ it is in $\mathcal{G}^0$. Vice versa if $\mu \in \mathcal{G}^0$ then μ satisfies the DLR condition for all bounded $\mathcal{D}^{(1)}$-measurable sets Λ. The statement then follows because if $\Delta \subset \Lambda$ is a measurable set, then

$$G_{\gamma,\lambda,\Lambda,\bar{q}}(f) = G_{\gamma,\lambda,\Lambda,\bar{q}}\big(G_{\gamma,\lambda,\Delta,q}(f)\big).$$ □

Theorem 12.1.5.2 *Let $\Lambda_n \nearrow \mathbb{R}^d$ be a sequence of bounded, $\mathcal{D}^{(1)}$-measurable regions and q_n a sequence in $\mathcal{X}$. Then $G_{\gamma,\lambda,\Lambda_n,q_n}$ converges weakly by subsequences and any limit point is in $\mathcal{G}^0$.*

The proof of Theorem 12.1.5.2 needs a preliminary lemma. Let Λ be a bounded $\mathcal{D}^{(\ell_{+,\gamma})}$-measurable region, $\Lambda_0 = \Lambda \setminus \delta_{\text{in}}^{\ell_{+,\gamma}}[\Lambda]$ and $\Lambda' \supset \Lambda$ a bounded, $\mathcal{D}^{(1)}$-measurable region. Given $G_{\gamma,\lambda,\Lambda',\bar{q}}$, write ν for the probability on $\mathcal{X}$ supported by configurations which have only particles in the corridor $\delta_{\text{in}}^{\ell_{+,\gamma}}[\Lambda]$, namely such that $\nu(\{q : q = q \cap \delta_{\text{in}}^{\ell_{+,\gamma}}[\Lambda]\}) = 1$, and such that their distribution, namely the marginal of ν over $\mathcal{X}_{\delta_{\text{in}}^{\ell_{+,\gamma}}[\Lambda]}$, is the same as the marginal of $G_{\gamma,\lambda,\Lambda',\bar{q}}$ over $\mathcal{X}_{\delta_{\text{in}}^{\ell_{+,\gamma}}[\Lambda]}$. Denote by $\mu_{\gamma,\lambda,\Lambda',\bar{q}}$ the measure such that for any bounded cylindrical function f,

$$\mu_{\gamma,\lambda,\Lambda',\bar{q}}(f) = \int G_{\gamma,\lambda,\Lambda_0,q}(f) d\nu(q) = \nu\big(G_{\gamma,\lambda,\Lambda_0,q}(f)\big). \tag{12.1.5.3}$$

We obviously have (the proof is omitted):

Lemma 12.1.5.3 *The measure $\mu_{\gamma,\lambda,\Lambda',\bar{q}}$ defined in* (12.1.5.3) *is in $\mathcal{G}^0_{\Lambda_0}$ and for any bounded function f cylindrical in Λ_0,*

$$\mu_{\gamma,\lambda,\Lambda',\bar{q}}(f) = G_{\gamma,\lambda,\Lambda',\bar{q}}(f). \tag{12.1.5.4}$$

Proof of Theorem 12.1.5.2 By Lemma 12.1.5.3 for any bounded cylindrical function f,

$$\mu_{\gamma,\lambda,\Lambda_n,q_n}(f) = G_{\gamma,\lambda,\Lambda_n,q_n}(f), \quad \text{for } n \text{ large enough.} \tag{12.1.5.5}$$

Given any bounded $\mathcal{D}^{(1)}$-measurable set Δ, by Lemma 12.1.5.3 the measures $\mu_{\gamma,\lambda,\Lambda_n,q_n}$ are for all n large enough in $\mathcal{G}^0_\Delta$; hence the sequence $\{\mu_{\gamma,\lambda,\Lambda_n,q_n}\}$ is weakly pre-compact and thus converges by subsequences. The limit points are in $\mathcal{G}^0_\Delta$ and by the arbitrariness of Δ they are in $\mathcal{G}^0$. Theorem 12.1.5.2 then follows using (12.1.5.5). □

12.1.6 Proof of Theorem 12.1.1.1

The proof is a corollary of Theorem 12.1.5.2 via the following lemma:

Lemma 12.1.6.1 *For any $r \in \mathbb{R}^d$ $\{q \in \mathcal{X} : \Theta(q;r) = \pm 1\}$ is a closed set.*

Proof If $q_n \to q$, then for each cube $i \in \mathbb{Z}^d$ the number of particles in $q_n^{(i)}$ is eventually equal to the number of particles in $q^{(i)}$. Thus, for all n large enough $\Theta(q;r) = \Theta(q_n;r)$, because the value of $\Theta(\cdot;r)$ is determined by the number of particles in each cube of $\mathcal{D}^{(\ell_{-,\gamma})}$ contained in $C_r^{(\ell_{+,\gamma})} \cup \delta_{\text{out}}^{\ell_{+,\gamma}}[C_r^{(\ell_{+,\gamma})}]$. □

We fix hereafter $\lambda = \lambda_{\beta,\gamma}$, a sequence Λ_n of bounded, $\mathcal{D}^{(\ell_{+,\gamma})}$-measurable regions which invades $\mathbb{R}^d$ and a sequence $q_n^+ \in \mathcal{X}^+$. The proof of Theorem 12.1.5.2

can be extended (we omit the details) to show that $G^+_{\gamma,\lambda_{\beta,\gamma},\Lambda_n,q_n^+}$ converges weakly by subsequences to a limit measure $\mu_\gamma^+ \in \mathcal{G}^0$. Let $r \in \mathbb{R}^d$, then for n large enough $r \in \Lambda_n$ and by Theorem 10.3.4.2,

$$G^+_{\gamma,\lambda_{\beta,\gamma},\Lambda_n,q_n^+}(\{\Theta(q;r)=1\}) \geq 1 - \exp\Big\{-\beta\frac{c_1}{4}(\zeta^2\ell^d_{-,\gamma})\Big\}. \tag{12.1.6.1}$$

By Lemma 12.1.6.1 $\{\Theta(q;r)=1\}$ is closed. By Theorem CWC of Sect. 12.1.4, $\mu_\gamma^+(\{\Theta(q;r)=1\}) \geq 1 - \exp\{-\beta\frac{c_1}{4}(\zeta^2\ell^d_{-,\gamma})\}$, which thus holds for any $r \in \mathbb{R}^d$. The same argument applies in the minus case and Theorem 12.1.1.1 is proved.

12.2 Properties of the plus and minus DLR measures

Throughout the sequel $\beta \in (\beta_c, \beta_0)$ and $\lambda = \lambda_{\beta,\gamma}$ are fixed and often dropped from the notation. We shall prove in this section that the DLR measures $\mu_\gamma^\pm$ of Theorem 12.1.1.1 have a trivial σ-algebra at infinity and hence they are extremal DLR measures. In the next section we shall show that they are translational invariant thus inferring that they are ergodic with respect to space translations (as a consequence of the fact that their σ-algebra at infinity is trivial). We shall finally show that $\mu_\gamma^\pm$ are the only ergodic DLR measures. The main technical tool in the proofs is the exponentially weak dependence on the boundary values of the plus [and minus] diluted Gibbs measures:

Theorem 12.2.0.1 *There are ω and c positive so that for all γ small enough, for all bounded, $\mathcal{D}^{(\ell_{+,\gamma})}$-measurable, simply connected regions Λ_1, Λ_2, for all $q_1, q_2 \in \mathcal{X}^+$ and all bounded, cylindrical functions f in a $\mathcal{D}^{(1)}$-measurable set Δ such that* $\text{dist}(\Delta, (\Lambda_1 \cap \Lambda_2)^c > 2^{10}\ell_{+,\gamma}$,

$$\big|G^+_{\gamma,\Lambda_1,q_1}(f) - G^+_{\gamma,\Lambda_2,q_2}(f)\big| \leq c\|f\|_\infty |\Delta| e^{-\omega\gamma \text{dist}(\Delta,(\Lambda_1\cap\Lambda_2)^c)}. \tag{12.2.0.1}$$

The same result holds for the minus case.

Theorem 12.2.0.1 is essentially proved by the relativized uniqueness Dobrushin theory of Sect. 11.5, which provides the most important part of the proof. Details are given in a complementary section, Sect. 12.5.

12.2.1 Uniqueness of the plus and minus measures

We start by a uniqueness result. A priori, $\mu_\gamma^\pm$ as defined in Theorem 12.1.1.1 depend on the sequence $\Lambda_n \nearrow \mathbb{R}^d$ and on the sequence of boundary conditions $q_n^\pm$. In fact, the proof of Theorem 12.1.1.1 uses compactness and thus it only gives convergence by subsequences. We shall prove here that the same limit $\mu_\gamma^\pm$ is obtained independently of $q_n^\pm$ and Λ_n.

Theorem 12.2.1.1 *Let ω and c be the positive coefficients introduced in Theorem 12.2.0.1; then for any bounded, simply connected, $\mathcal{D}^{(\ell_{+,\gamma})}$-measurable region Λ and any $q^\pm \in \mathcal{X}^\pm$*

$$\left|G^{\pm}_{\gamma,\Lambda,q^\pm}(f) - \mu^{\pm}_{\gamma}(f)\right| \le c\|f\|_\infty |\Delta| e^{-\omega\gamma \operatorname{dist}(\Delta,\Lambda^c)}, \tag{12.2.1.1}$$

for all $\Delta \subset \Lambda$ and all bounded, cylindrical functions f in Δ.

Proof Referring for simplicity to the plus case, by Theorem 12.1.1.1,

$$\left|G^{+}_{\gamma,\Lambda,q^+}(f) - \mu^{+}_{\gamma}(f)\right| = \lim_{n\to\infty} \left|G^{+}_{\gamma,\Lambda,q^+}(f) - G^{+}_{\gamma,\Lambda_n,q_n^+}(f)\right|,$$

while for n large enough

$$\left|G^{+}_{\gamma,\Lambda,q^+}(f) - G^{+}_{\gamma,\Lambda_n,q_n^+}(f)\right| \le c\|f\|_\infty |\Delta| e^{-\omega\gamma \operatorname{dist}(\Delta,\Lambda^c)}$$

by Theorem 12.2.0.1; hence (12.2.1.1). □

12.2.2 *Triviality of the tail field*

We shall prove here that $\mu^{\pm}_{\gamma}$ have a trivial σ-algebra at infinity (the tail field of the title). While referring to Chap. 2 for the general theory (in the context of Ising models) we give here a self-contained exposition focused on what is needed for the proof of the statements made in the beginning of the chapter.

Definition 12.2.2.1 Write $\mathcal{S}$ for the collection of sequences $\{\Lambda_k\}$ of the form $\{\tau_i \Delta_k\}$ where $i \in a\mathbb{Z}^d$; $a \in \{2^{-n}, n \in \mathbb{N}\}$ and $\{\Delta_k\}$ is an arbitrary but fixed increasing sequence of $\mathcal{D}^{(\ell_{+,\gamma})}$-measurable cubes of sides $2^k \ell_{+,\gamma}$ which invades the whole space.

The definition is given in view of the applications in the next section where we need to consider also translates of the sequence $\{\Delta_k\}$. Observe that sequences in $\mathcal{S}$ are not necessarily $\mathcal{D}^{(\ell_{+,\gamma})}$-measurable.

Theorem 12.2.2.2 *For all γ small enough*

$$\mu^{\pm}_{\gamma}(\mathcal{X}^{\pm}_{\text{tail}}) = 1, \tag{12.2.2.1}$$

where $\mathcal{X}^{\pm}_{\text{tail}} = \{q \in \mathcal{X} : \lim_{k\to\infty} G_{\gamma,\Lambda_k,q}(f) = \mu^{\pm}_{\gamma}(f)$, for any $\{\Lambda_k\} \in \mathcal{S}$ and for any bounded, measurable cylindrical function $f\}$.

Proof Since the number of sequences $\{\Lambda_k\} \in \mathcal{S}$ is countable, it suffices to prove (12.2.2.1) for only one sequence $\{\Lambda_k\}$. Moreover, after taking countably many intersections, it suffices to prove that for any $\mathcal{D}^{(\ell_{+,\gamma})}$-measurable cube Δ

$$\mu^{+}_{\gamma}\left(\left\{q \in \mathcal{X} : \lim_{k\to\infty} G_{\gamma,\Lambda_k,q}(f) = \mu^{+}_{\gamma}(f)\right\}\right) = 1 \tag{12.2.2.2}$$

uniformly over all bounded functions f cylindrical in Δ; we are referring for simplicity only to the plus case.

Scheme of the proof

The convergence to $\mu_\gamma^+(f)$ in (12.2.2.2) would follow from (12.2.1.1) if instead of the conditional measure $G_{\gamma,\Lambda_k,q}$ we had the plus, diluted Gibbs measure $G^+_{\gamma,\Lambda_k,q}$ and the heart of the proof is to reduce to such a case. For almost all q (with respect to μ_γ^+) there are infinitely many "circuits" where $\Theta = 1$, and if we could replace the sequence $\{\Lambda_k\}$ by another sequence where the boundaries of the regions are the above circuits, we would be in business. Such a new sequence of regions, however, depends on q while in the definition of the σ-algebra at infinity we need "deterministic" regions. On the other hand, there is no hope to find a "deterministic" sequence of regions such that in a set of configurations which has probability 1, $\Theta \equiv 1$ on the boundaries of such regions, because the probability of a fluctuation has a small yet positive probability.

Thus we cannot avoid to deal with boundary conditions which do not have Θ constantly equal to 1. However, as a consequence of the Peierls bounds, the cubes at the boundaries where $\Theta < 1$ are few (a small fraction of the total number) and it is plausible that the analysis with plus boundary conditions extends to such a case. The conditional here is necessary because a rigorous proof does not seem straightforward and we shall proceed with a much easier argument (which, however, is not "constructive"). The idea in fact is to prove that the boundary conditions for which we do not find a plus circuit "close to the boundaries" have such a small probability that by a Borel–Cantelli argument they can be neglected. In this way we shall not characterize explicitly the "good" boundary condition, but we only prove that they occur with probability 1.

The plus circuit

We fix arbitrarily a bounded set Δ and consider functions $f = f(q_\Delta)$. We call $\Lambda_{k;0}$ the union of all cubes of $\mathcal{D}^{(\ell_{+,\gamma})}$ contained in Λ_k (recall that Λ_k may not be $\mathcal{D}^{(\ell_{+,\gamma})}$-measurable) and define the random set $\hat{N}_{k,\text{out}}$ as follows. $\hat{N}_{k,\text{out}}(q)$ is the union of $\Lambda^c_{k;0}$ with all the maximal connected components A of the set $\{r \in \Lambda_{k;0} : \Theta(q;r) < 1\}$ such that $\delta_{\text{out}}^{\ell_{+,\gamma}}[A] \cap \Lambda^c_{k;0} \neq \emptyset$. We call $\hat{N}_{k,\text{in}}$ the complement of $\hat{N}_{k,\text{out}}$ and observe that, by construction, $\Theta(q;r) = 1$ for all $r \in \delta_{\text{in}}^{\ell_{+,\gamma}}[\hat{N}_{k,\text{in}}]$. We can separate in the set $\{\hat{N}_{k,\text{out}} = N_{\text{out}}\} := \{q : \hat{N}_{k,\text{out}}(q) = N_{\text{out}}\}$ the constraints localized in N_{out} from those in N_{in}, as we have:

$$\{\hat{N}_{k,\text{out}} = N_{\text{out}}\} = X_{N_{\text{out}}} \cap X_{N_{\text{in}}}, \tag{12.2.2.3}$$

where $X_{N_{\text{out}}}$ is the set of all q such that $\Theta(q;r) = 0$, $r \in \Lambda_{k;0} \cap \{N_{\text{out}} \setminus \delta_{\text{in}}^{\ell_{+,\gamma}}[N_{\text{out}}]\}$ and $\theta(q;r) = 1$ on $\delta_{\text{in}}^{\ell_{+,\gamma}}[N_{\text{out}}]$. $X_{N_{\text{in}}}$ is the set of all q such that $\theta(q;r) = 1$ on

$\delta_{\rm in}^{\ell_{+,\gamma}}[N_{\rm in}]$ as well as on $\delta_{\rm in}^{\ell_{+,\gamma}}[B]$, $B = N_{\rm in} \setminus \delta_{\rm in}^{\ell_{+,\gamma}}[N_{\rm in}]$. Observe that the sets $X_{N_{\rm out}}$ and $X_{N_{\rm in}}$ are cylindrical in $N_{\rm out}$ and $N_{\rm in}$, respectively, and this will be essential in the proof of the following.

Proposition 12.2.2.3 (Strong DLR property) *With the above notation, for any k*

$$\mu_\gamma^+(f) = \sum_{N_{\rm out}} \mu_\gamma^+\big(\mathbf{1}_{\{\hat{N}_{\rm out}=N_{\rm out}\}} G^+_{\gamma,N_{\rm in},q}(f)\big). \tag{12.2.2.4}$$

Proof We obviously have

$$\mu_\gamma^+(f) = \sum_{N_{\rm out}} \mu_\gamma^+\big(\mathbf{1}_{\{\hat{N}_{k,\rm out}=N_{\rm out}\}} f\big).$$

We introduce the shorthand $\mathbf{1}_{N_{\rm in}} = \mathbf{1}_{N_{\rm in}}$, $\mathbf{1}_{N_{\rm out}} = \mathbf{1}_{N_{\rm out}}$, use first (12.2.2.3) and then the DLR property to get $\mu_\gamma^+(\mathbf{1}_{\{\hat{N}_{k,\rm out}=N_{\rm out}\}} f) = \mu_\gamma^+(\mathbf{1}_{N_{\rm out}} G_{\gamma,N_{\rm in},q}(\mathbf{1}_{N_{\rm in}} f))$, and hence

$$\mu_\gamma^+(f) = \sum_{N_{\rm out}} \mu_\gamma^+\big(\mathbf{1}_{\{N_{\rm out}\}} G_{\gamma,N_{\rm in},q}(\{N_{\rm in}\}) G^+_{\gamma,N_{\rm in},q}(f)\big). \tag{12.2.2.5}$$

Write $g = \mathbf{1}_{\{N_{\rm out}\}} G^+_{\gamma,N_{\rm in},q}(f)$, observe that g is measurable on $\mathcal{X}_{N_{\rm out}}$; then using backwards the DLR property and (12.2.2.3), we get (12.2.2.4). □

Our next goal is to prove that with large probability Λ is well inside $\hat{N}_{k,\rm out}$. Given r, call $A(q;r)$ the maximal connected component of $\{\Theta(q;\cdot) < 1\}$ which contains r ($A(q;r)$ may be empty). Writing $\mathrm{diam}(A)$ for the diameter of the set A, we define

$$\mathcal{B}_k := \big\{q : \mathrm{diam}(A(q;r)) < 2^{k-1}\ell_{+,\gamma}, \ \text{for all } r \in \delta_{\rm in}^{\ell_{+,\gamma}}[\Lambda_{k;0}]\big\}. \tag{12.2.2.6}$$

Notice that

$$\hat{N}_{k,\rm out}(q) \cap \Lambda_{k-2} = \emptyset, \quad \text{for all } q \in \mathcal{B}_k. \tag{12.2.2.7}$$

Proposition 12.2.2.4 *There are $\epsilon_k > 0$ so that for all γ small enough*

$$\mu_\gamma^+(\mathcal{B}_k) \ge 1 - \epsilon_k, \quad \sum_{k=1}^{\infty} \sqrt{\epsilon_k} < \infty. \tag{12.2.2.8}$$

Proof We have

$$\mu_\gamma^+(\mathcal{B}_k) \ge 1 - \frac{|\delta_{\rm in}^{\ell_{+,\gamma}}[\Lambda_k]|}{\ell_{+,\gamma}^d} \sup_{r \in \delta_{\rm in}^{\ell_{+,\gamma}}[\Lambda_k]} \mu_\gamma^+\big(\mathrm{diam}(A(\cdot;r)) \ge 2^{k-1}\ell_{+,\gamma}\big). \tag{12.2.2.9}$$

As the event $\{\text{diam}(A(\cdot;r)) \geq 2^{k-1}\ell_{+,\gamma}\}$ is cylindrical, the probability on the r.h.s. is

$$\lim_{n\to\infty} G^+_{\gamma,\Lambda_n,q^+}\left(\text{diam}(A(\cdot;r)) \geq 2^{k-1}\ell_{+,\gamma}\right). \tag{12.2.2.10}$$

Since $\{\text{diam}(A(\cdot;r)) \geq 2^{k-1}\ell_{+,\gamma}\} \subset \{$there is Γ such that $c(\Gamma) \ni r$ and $N_\Gamma \geq 2^{k-1}\}$; then by (9.2.8.3) (once the Peierls bounds are established, estimates based on counting the contours are the same in the Ising case and LMP)

$$\mu^+_\gamma\left(\text{diam}(A(\cdot;r)) \geq 2^{k-1}\ell_{+,\gamma}\right) \leq \sum_{\Gamma:c(\Gamma)\ni r, N_\Gamma \geq 2^{k-1}} e^{-\beta(c_1/2)\,(\zeta^2\ell^d_{-,\gamma})N_\Gamma}, \tag{12.2.2.11}$$

and using (9.2.8.4),

$$\mu^+_\gamma\left(\text{diam}(A(\cdot;r)) \geq 2^{k-1}\ell_{+,\gamma}\right) \leq e^{-\beta(c_1/4)\,(\zeta^2\ell^d_{-,\gamma})2^{k-1}}, \tag{12.2.2.12}$$

which, with (12.2.2.9), proves (12.2.2.8), as $|\delta^{\ell_{+,\gamma}}_{\text{in}}[\Lambda_k]| \leq 2d[2^k\ell_{+,\gamma}]^{d-1}\ell_{+,\gamma}$. □

Theorem 12.2.2.5 *Define*

$$F_k = \left\{q : G_{\gamma,\Lambda_k,q}(\mathcal{B}_k) \geq 1 - \sqrt{\epsilon_k}\right\}. \tag{12.2.2.13}$$

Then

$$\mu^+_\gamma(F_k) \geq 1 - \sqrt{\epsilon_k}, \tag{12.2.2.14}$$

with ϵ_k *as in* (12.2.2.8) *and for all* $q \in F_k$ *and all measurable, bounded functions* f *cylindrical in* Δ,

$$\left|G_{\gamma,\Lambda_k,q}(f) - \mu^+_\gamma(f)\right| \leq 2\|f\|_\infty\sqrt{\epsilon_k} + c\|f\|_\infty|\Delta|\,e^{-\omega\gamma\,\text{dist}(\Delta,\Lambda^c_{k-2})}, \tag{12.2.2.15}$$

with c *and* ω *as in* (12.2.1.1).

Proof By (12.2.2.8) and using the DLR property,

$$\begin{aligned}1 - \epsilon_k \leq \mu^+_\gamma(\mathcal{B}_k) &= \mu^+_\gamma\left(G_{\gamma,\Lambda_k,\sigma}(\mathcal{B}_k)\right)\\ &\leq \mu^+_\gamma(F_k) + (1-\sqrt{\epsilon_k})\left(1 - \mu^+_\gamma(F_k)\right);\end{aligned}$$

hence (12.2.2.14). Let $q \in F_k$; then

$$\begin{aligned}&\left|G_{\gamma,\Lambda_k,q}(f) - \sum_{B\subset\Lambda^c_{k-2}} G_{\gamma,\Lambda_k,q}\left(\mathbf{1}_{\hat{N}_{k,\text{out}}=B}\,G^+_{\gamma,B^c,q}(f)\right)\right|\\ &\quad\leq \|f\|_\infty G_{\gamma,\Lambda_k,q}\left(\mathcal{B}^c_k\right) \leq \|f\|_\infty\sqrt{\epsilon_k}.\end{aligned}$$

Hence

$$\begin{aligned} &\left|G_{\gamma,\Lambda_k,q}(f)-\mu_\gamma^+(f)\right| \\ &\le 2\|f\|_\infty\sqrt{\epsilon_k}+\sup_{B\subset\Lambda_{k-1}^c}\ \sup_{q^+\in\mathcal{X}^+}\left|G^+_{\gamma,B^c,q^+}(f)-\mu_\gamma^+(f)\right|, \end{aligned}$$

so that (12.2.2.17) follows from (12.2.1.1). The bound is uniform in all f cylindrical in Δ and Theorem 12.2.2.5 is proved. □

Proof of Theorem 12.2.2.2 We use Proposition 12.2.2.4 and Theorem 12.2.2.5 to implement a Borel–Cantelli argument from which Theorem 12.2.2.2 will follow. Let F_k be as in (12.2.2.13). By (12.2.2.14) and (12.2.2.8),

$$\mu_\gamma^+(F)=1,\quad F:=\bigcup_n\bigcap_{k\ge n}F_k. \tag{12.2.2.16}$$

By (12.2.2.15)

$$\lim_{k\to\infty}G_{\gamma,\Lambda_k,q}(f)=\mu_\gamma^+(f),\quad \text{for all } q\in F, \tag{12.2.2.17}$$

and (12.2.2.2) follows from (12.2.2.16) and (12.2.2.17). □

12.3 Decomposition of translational invariant DLR measures

In the sequel we shall study the translational invariant DLR measures. We shall prove that $\mu_\gamma^\pm$ are translational invariant and that any other translational invariant DLR measure is a convex combination of μ_γ^+ and of μ_γ^-. Paradoxically we shall first prove the latter statement and only afterwards show that $\mu_\gamma^\pm$ are translational invariant.

12.3.1 Translational invariant DLR measures

We begin by recalling some notation.

Notation and definitions

Let $a>0$ and $\tau_i^{(a)}: a\mathbb{Z}^d\to a\mathbb{Z}^d$, $i\in a\mathbb{Z}^d$, the translation by i, $\tau_i^{(a)}(r)=r+i$. By abuse of notation we denote by the same symbol also its dual actions: $\tau_i^{(a)}(q)$, $\tau_i^{(a)}(f)$ and $\tau_i^{(a)}(\mu)$, respectively on particle configurations, on functions of the particle configurations and on probabilities on the space of particle configurations.

A probability ν is translational invariant if for any mesh $a > 0$, $\tau_i^{(a)}(\nu) = \nu$ for all $i \in a\mathbb{Z}^d$ (see Sect. A.1 for the elementary properties of translations in Ising spaces, as they obviously extend to the present case).

Theorem 12.3.1.1 *For all γ small enough, if m is any translational invariant DLR measure, then there is a unique $u \in [0, 1]$ so that*

$$m = u\mu_\gamma^+ + (1-u)\mu_\gamma^-, \quad u \in [0, 1]. \tag{12.3.1.1}$$

The proof presented in Sects. 12.3.3–12.3.4 follows a classical argument by Gallavotti and Miracle Sole [129], introduced in the proof of the analogous statement for the Ising model at low temperatures. The analogue of Theorem 12.3.1.1 for the Ising model of Chap. 9 has been proved in [61].

12.3.2 Reduction to intensive variables

The intuition developed so far that contours are small and rare does not apply here because we do not control the boundary conditions: the measures we have to deal with may be produced by weird boundary conditions which force the presence of many contours. The key ingredient in the proof will be that using translational invariance the expectation of a cylindrical function f can be replaced by the expectation of its spatial average. We shall then have reduced th case to estimating objects which scale as the volume, so that we shall be able to replace the boundary conditions which scale as the surface with negligible error, thus reducing to plus boundary conditions.

We start by observing that while the proof that $\mu_\gamma^\pm$ are translational invariant is far from trivial, invariance by multiples of $\ell_{+,\gamma}$ is almost immediate:

Lemma 12.3.2.1 *$\mu_\gamma^\pm$ are invariant under $\{\tau_i^{(\ell_{+,\gamma})}\}$, $i \in \ell_{+,\gamma}\mathbb{Z}^d$.*

Proof We shall only prove that $\tau_i(\mu_\gamma^+) = \mu_\gamma^+$, $i \in \ell_{+,\gamma}\mathbb{Z}^d$; the minus case is similar and we omit it. For any bounded, measurable cylindrical function f, $[\tau_i(\mu_\gamma^+)](f) = \lim_{n\to\infty}[\tau_i(G^+_{\gamma,\Lambda_n,q})](f)$, where $\Lambda_n \to \mathbb{R}^d$ is an increasing sequence of $\mathcal{D}^{(\ell_{+,\gamma})}$-measurable cubes and $q \in \mathcal{X}^+$. On the other hand by, (A.1.9) $\tau_i(G^+_{\gamma,\Lambda_n,q}) = G^+_{\gamma,\tau_i(\Lambda_n),\tau_i(q)}$ and by (12.2.1.1) $\lim_{n\to\infty} G^+_{\gamma,\tau_i(\Lambda_n),\tau_i(q)}(f) = \mu_\gamma^+(f)$. □

We fix arbitrarily a bounded, measurable cylindrical function f. Let Λ be a $\mathcal{D}^{(\ell_{+,\gamma})}$-measurable cube in $\mathbb{R}^d$ and N_Λ the cardinality of $\Lambda \cap \ell_{+,\gamma}\mathbb{Z}^d$, namely the number of $\mathcal{D}^{(\ell_{+,\gamma})}$-cubes contained in Λ. The intensive variable in the title of the subsection is then

$$\mathcal{A}_\Lambda(f) = \frac{1}{N_\Lambda}\sum_{i\in\Lambda\cap\ell_{+,\gamma}\mathbb{Z}^d} f_i, \quad f_i := \tau_i(f) \tag{12.3.2.1}$$

($\mathcal{A}_\Lambda$ stands for "average in Λ"). Since m in Theorem 12.3.1.1 is translational invariant, $m(f) = m(\mathcal{A}_\Lambda(f))$. Moreover, by Lemma 12.3.2.1

$$\mu^{\pm}_{\gamma}(f_i) = \mu^{\pm}_{\gamma}(f), \quad i \in \Lambda, \ f_i = \tau_i(f). \tag{12.3.2.2}$$

The random sets $\hat{N}_{\text{out}}$ and the random integers $\hat{M}_{\text{out}}$

$\hat{N}_{\text{out}}$ is the union of Λ^c with all the maximal connected components A of the set $\{r \in \Lambda : \Theta(\cdot; r) = 0\}$ such that $\delta^{\ell_{+,\gamma}}_{\text{out}}[A] \cap \Lambda^c \neq \emptyset$. $\hat{M}_{\text{out}} = \frac{|\hat{N}_{\text{out}} \cap \Lambda|}{\ell^d_{+,\gamma}}$ is the number of cubes of $\mathcal{D}^{(\ell_{+,\gamma})}$ in $\hat{N}_{\text{out}}$. Finally, we call $\hat{N}_{\text{in}}$ the complement of $\hat{N}_{\text{out}}$.

Lemma 12.3.2.2 *If Δ is a maximal connected component of $\hat{N}_{\text{in}}$, then $\Theta(\cdot; r)$, $r \in \delta^{\ell_{+,\gamma}}_{\text{in}}[\Delta]$, is a non-zero constant.*

Proof Since Δ is maximal $\Theta(\cdot; r) \neq 0$ for all $r \in \delta^{\ell_{+,\gamma}}_{\text{in}}[\Delta]$. By Theorem 6.4.3.1 Δ is simply connected $\delta^{\ell_{+,\gamma}}_{\text{in}}[\Delta]$ is connected and $\Theta(\cdot; r)$ is constant, because $\{\Theta = 1\}$ and $\{\Theta = -1\}$ are disconnected. □

12.3.3 Estimates with few contours

Denoting by $\ell_{+,\gamma} L$ the side of Λ so that $N_\Lambda = L^d$ we say that $\hat{N}_{\text{out}}$ is "small" if $\hat{M}_{\text{out}} \leq L^{d-1+1/4}$ and "large" in the opposite case; the choice 1/4 above is not critical. We then write

$$m(f) = m\big(\mathcal{A}_\Lambda(f)\mathbf{1}_{\hat{M}_{\text{out}} > L^{d-1+1/4}}\big) + m\big(\mathcal{A}_\Lambda(f)\mathbf{1}_{\hat{M}_{\text{out}} \leq L^{d-1+1/4}}\big). \tag{12.3.3.1}$$

We postpone to the next subsection the proof that

$$\lim_{L \to \infty} m(\hat{M}_{\text{out}} > L^{d-1+1/4}) = 0, \tag{12.3.3.2}$$

and study here the second term on the r.h.s. of (12.3.3.1), where $\hat{M}_{\text{out}}$ is "small" and consequently there are only "few contours" attached to the boundaries of Λ; hence the title of the subsection.

If $\hat{M}_{\text{out}} \leq L^{d-1+1/4}$, then $\hat{N}_{\text{in}} \neq \emptyset$ and by Lemma 12.3.2.2 it splits into

$$\hat{N}_{\text{in}} = \hat{N}^+_{\text{in}} \cup \hat{N}^-_{\text{in}}, \quad \text{where } \Theta = \pm 1 \text{ on } \delta_{\text{in}}[\hat{N}^\pm_{\text{in}}]. \tag{12.3.3.3}$$

By the strong DLR property, see Proposition 12.2.2.3 where an analogous statement is proved, for any bounded measurable function g,

$$m\big(g\mathbf{1}_{\hat{M}_{\text{out}} \leq L^{d-1+1/4}}\big) = \sum_{\Delta_\pm} m\big(\mathbf{1}_{\hat{N}^\pm_{\text{in}} = \Delta_\pm, \hat{M}_{\text{out}} \leq L^{d-1+1/4}}[G^+_{\gamma,\Delta_+,q} \times G^-_{\gamma,\Delta_-,q}](g)\big). \tag{12.3.3.4}$$

Define next

$$\Delta_\pm^0 = \{i \in \Delta_\pm \cap \ell_{+,\gamma}\mathbb{Z}^d : \mathrm{dist}(i, \Delta_\pm^c) > \ell_{+,\gamma} L^{1/(4d)}\}. \tag{12.3.3.5}$$

Then there is $c > 0$ so that if $\hat{M}_{\mathrm{out}} \le L^{d-1+1/4}$, then $\mathrm{card}(\Delta_+^0 \cup \Delta_-^0)$ is bounded from below by

$$L^d - \sum_{i \in \{(\hat{N}_{\mathrm{out}} \cap \Lambda) \cup \delta_{\mathrm{out}}^{\ell_{+,\gamma}}[\Lambda]\} \cap \ell_{+,\gamma}\mathbb{Z}^d} \mathrm{card}\left(j \in \ell_{+,\gamma}\mathbb{Z}^d : |i-j| \le \ell_{+,\gamma} L^{1/(4d)}\right)$$
$$\ge L^d - cL^{d-1+1/4}(L^{1/(4d)})^d = L^d - cL^{d-1/2}. \tag{12.3.3.6}$$

If L is large enough and $i \in \Delta_\pm^0$, then $\tau_i f$ is measurable on $\mathcal{X}_{\Delta_\pm}$ and we get from (12.3.3.4)

$$\Big| m\big(\mathcal{A}_\Lambda(f)\mathbf{1}_{\hat{M}_{\mathrm{out}} \le L^{d-1+1/4}}\big) - \frac{1}{L^d} \sum_{\{\Delta_\pm\}} m\Big(\mathbf{1}_{\hat{N}_{\mathrm{in}}^\pm = \Delta_\pm, \hat{M}_{\mathrm{out}} \le L^{d-1+1/4}}$$
$$\times \Big\{ \sum_{i \in \Delta_+^0} G_{\gamma,\Delta_+,q}^+(f_i) + \sum_{i \in \Delta_-^0} G_{\gamma,\Delta_-,q}^-(f_i) \Big\}\Big)\Big|$$
$$\le \frac{cL^{d-1/2}}{L^d} \|f\|_\infty. \tag{12.3.3.7}$$

By Theorem 12.2.1.1, for all $i \in \Delta_\pm^0$,

$$\sup_{q \in \mathcal{X}^\pm} |G_{\gamma,\Delta_\pm,q}^\pm(f_i) - \mu_\gamma^\pm(f_i)| \le \|f\|_\infty o(L^{1/(4d)}), \tag{12.3.3.8}$$

$o(L^{1/(4d)})$ denoting a function which vanishes when its argument diverges.

By (12.3.2.2), $\mu_\gamma^\pm(f_i) = \mu_\gamma^\pm(f)$, so that

$$\Big| m\big(\mathcal{A}_\Lambda(f)\mathbf{1}_{\hat{M}_{\mathrm{out}} \le L^{d-1+1/4}}\big) - \frac{1}{L^d} \sum_{\{\Delta_\pm\}} m\big(\{\hat{N}_{\mathrm{in}}^\pm = \Delta_\pm, \hat{M}_{\mathrm{out}} \le L^{d-1+1/4}\}\big)$$
$$\times \{\mathrm{card}(\Delta_+^0)\mu_\gamma^+(f) + \mathrm{card}(\Delta_-^0)\mu_\gamma^-(f)\}\Big|$$
$$\le \{cL^{-1/2} + o(L^{1/(4d)})\}\|f\|_\infty,$$

Let

$$u_{\pm,L} = \sum_{\{\Delta_\pm\}} m\big(\{\hat{N}_{\mathrm{in}}^\pm = \Delta_\pm, \hat{M}_{\mathrm{out}} \le L^{d-1+1/4}\}\big) \frac{\mathrm{card}(\Delta_\pm^0)}{L^d}. \tag{12.3.3.9}$$

Since $m(\mathcal{A}_\Lambda(f)) = m(f)$ because m is translational invariant,

$$\big| m(f) - \{u_{+,L}\mu_\gamma^+(f) + u_{-,L}\mu_\gamma^-(f)\}\big|$$

$$\le \big[o(L^{1/(4d)}) + cL^{-1/2} + m\big(\hat{M}_{\rm out} > L^{d-1+1/4}\big)\big]\|f\|_\infty. \tag{12.3.3.10}$$

Using (12.3.3.2), we deduce that the r.h.s. of (12.3.3.10) vanishes as $L \to \infty$; hence in the same limit $\{u_{+,L}\mu_\gamma^+(f)+u_{-,L}\mu_\gamma^-(f)\}$ converges to $m(f)$. On the other hand, by compactness, there is a subsequence L_n such that

$$\lim_{n\to\infty} u_{\pm,L_n} = u_\pm; \tag{12.3.3.11}$$

hence

$$m(f) = u_+\mu_\gamma^+(f) + u_-\mu_\gamma^-(f). \tag{12.3.3.12}$$

By taking $f = 1$, we then deduce that $u_+ + u_- = 1$, thus concluding the proof of (12.3.1.1), pending the validity of (12.3.3.2).

12.3.4 Estimates with many contours

Here we shall prove (12.3.3.2) and thus conclude the proof of (12.3.1.1). We start by observing that $\hat{M}_{\rm out} \le \tilde{M}_{\rm out}$, where the latter is defined by setting $\tilde{M}_{\rm out}(q) = \hat{M}_{\rm out}(q')$ where q' is obtained from q by erasing the particles of q which are in $A := \delta_{\rm in}^{\ell_+,\gamma}[\Lambda_0]$, $\Lambda_0 = \Lambda \setminus \delta_{\rm in}^{\ell_+,\gamma}[\Lambda]$; see Fig. 12.1. Thus $q' \cap A = \emptyset$ while $q' \cap A^c = q \cap A^c$. By the definition of q', $\{r : \Theta(q';r) = 0\} \supset \{r : \Theta(q;r) = 0\}$; hence $\hat{M}_{\rm out} \le \tilde{M}_{\rm out}$. Thus

$$m\big(\hat{M}_{\rm out} > L^{d-1+1/4}\big) \le m\big(\tilde{M}_{\rm out} > L^{d-1+1/4}\big). \tag{12.3.4.1}$$

With B the union of A, $\delta_{\rm in}^{\ell_+,\gamma}[\Lambda]$, $\delta_{\rm out}^{\ell_+,\gamma}[\Lambda]$ and $\Lambda_1 = \Lambda \cup \delta_{\rm out}^{\ell_+,\gamma}[\Lambda]$, we claim:

$$m\big(\tilde{M}_{\rm out} > L^{d-1+1/4}\big) \le e^{(\beta b_0+1)|B|} G_{\gamma,\Lambda\setminus B,\emptyset}\big(\tilde{M}_{\rm out} > L^{d-1+1/4}\big), \tag{12.3.4.2}$$

$G_{\gamma,\Lambda\setminus B,\emptyset}$ being the Gibbs measure with the empty configuration as boundary conditions. Proof: by the DLR property the l.h.s. of (12.3.4.2) is bounded by the integral over $q_1 \in \mathcal{X}$, $q_2 \in \mathcal{X}_{\Lambda\setminus B}$ and $q_3 \in \mathcal{X}_{\Lambda\cap B}$ of

$$\mathbf{1}_{\tilde{M}_{\rm out} > L^{d-1+1/4}} \frac{e^{-\beta H_{\gamma,\Lambda\cap B}(q_3|q_2+(q_1)\cap\Lambda^c)}}{Z_{\gamma,\Lambda\cap B,q_3+(q_1)\cap\Lambda^c}} \nu_{\Lambda\cap B}(dq_2) G_{\gamma,\Lambda\setminus B,\emptyset}(dq_3) m(dq_1);$$

(12.3.4.2) then follows, because $\tilde{M}_{\rm out}$ depends only on q_3, the energy is bounded using (11.1.2.8), and the partition function is ≥ 1.

Claim: there is a constant c so that for any $q^+_{\Lambda_1^c} \in \mathcal{X}^+_{\Lambda_1^c}$:

$$1 \le e^{c|B|} G_{\gamma,B,q^+_{\Lambda_1^c}}\big(\{q_B \cap A = \emptyset\} | \Theta(q_B + q^+_{\Lambda_1^c}; r) = 1, r \in \delta_{\rm out}^{\ell_+,\gamma}[\Lambda]\big). \tag{12.3.4.3}$$

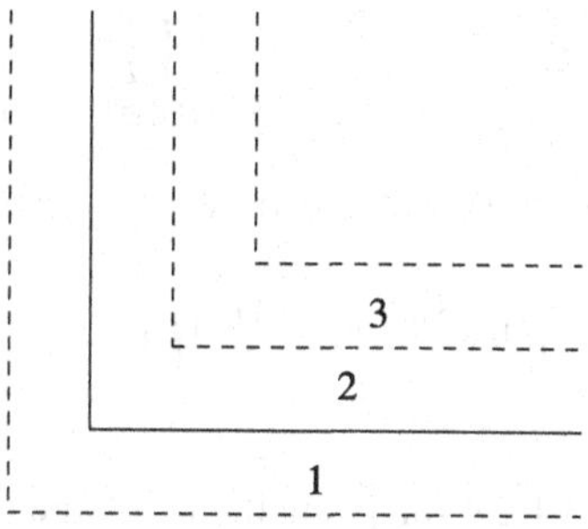

Fig. 12.1 The *thick line* is a portion of the boundary of Λ. The distance between the lines is $\ell_{+,\gamma}$. The region 1 is $\delta_{\text{out}}^{\ell_{+,\gamma}}[\Lambda]$, 2 is $\delta_{\text{in}}^{\ell_{+,\gamma}}[\Lambda]$ and 3 is A. B is the union of the three corridors, Λ_0 is Λ minus the corridor 2. Λ_1 is the union of Λ and the set 1

Proof: write the probability as a ratio of two partition functions. By (11.1.2.6) the one in the numerator is bounded from below,

$$Z_{\gamma,\Lambda_0\setminus A,\emptyset}\int_{\mathcal{X}^+_{\delta_{\text{in}}^{\ell_{+,\gamma}}[\Lambda]}} e^{-c'|B|}d\nu_{\delta_{\text{in}}^{\ell_{+,\gamma}}[\Lambda]} \geq Z_{\gamma,\Lambda_0\setminus A,\emptyset}e^{-(c'+c)|B|},$$

the last inequality holding by Lemma 11.7.1.1. Analogously, the one in the denominator is bounded from above by $Z_{\gamma,\Lambda_0\setminus A,\emptyset}e^{(c'+c)|B|}$; hence (12.3.4.3).

Thus $m(\tilde{M}_{\text{out}} > L^{d-1+1/4}) \leq$ r.h.s. of (12.3.4.2) $\times$ r.h.s. of (12.3.4.3), which is equal to $e^{(\beta b_0+1+c)|B|}$ times

$$G_{\gamma,\Lambda_1,q^+_{\Lambda_1^c}}\big(\{q_B\cap A=\emptyset\}\cap \tilde{M}_{\text{out}} > L^{d-1+1/4}\,|\,\Theta(q_B+q^+_{\Lambda_1^c};r)=1, r\in\delta_{\text{out}}^{\ell_{+,\gamma}}[\Lambda]\big),$$

because in the "corridor" A there are no particles. For the same reason we can replace $\tilde{M}_{\text{out}}$ by $\hat{M}_{\text{out}}$ on the r.h.s. and after having done that we bound by dropping the condition $\{q_B\cap A=\emptyset\}$ so that

$$m\big(\hat{M}_{\text{out}} > L^{d-1+1/4}\big) \leq e^{(\beta b_0+c+1)|B|}\sup_{q^+\in\mathcal{X}^+} G^+_{\gamma,\Lambda_1,q^+}\big(\hat{M}_{\text{out}} > L^{d-1+1/4}\big), \tag{12.3.4.4}$$

observing that $(\beta b_0+c+1)|B| \leq c'\ell^d_{+,\gamma}L^{d-1}$. We write more explicitly $\hat{M}_{\text{out}}$, call $\underline{\Gamma}$ the contours of q_{Λ_1}; then, since $\Theta=1$ on $\delta_{\text{in}}^{\ell_{+,\gamma}}[\Lambda]$,

$$\hat{M}_{\text{out}} = \sum_{\Gamma\in\underline{\Gamma}:\text{sp}(\Gamma)\cap\delta_{\text{in}}^{\ell_{+,\gamma}}[\Lambda]\neq\emptyset} N_\Gamma .$$

We label by $i\in\{1,\ldots,K\}$ the cubes of $\delta_{\text{in}}^{\ell_{+,\gamma}}[\Lambda]$ and recall that the side of Λ is $\ell_{+,\gamma}L$, $K\leq 2d(\ell_{+,\gamma}L)^{d-1}\ell_{+,\gamma}/\ell^d_{+,\gamma}=2dL^{d-1}$. If $\underline{\Gamma}$ is the collection of contours we call $n_i=N_\Gamma$ if $\Gamma\in\underline{\Gamma}$ and $\text{sp}(\Gamma)\ni i$, $i=1,\ldots,K$. Then $\sum_{i=1}^K n_i\geq \hat{M}_{\text{out}}$, and using the Peierls bounds,

$$G^+_{\gamma,\Lambda_1,q^+}\big(\hat{M}_{\text{out}} > L^{d-1+1/4}\big)$$

$$\le \sum_{n_1+\cdots+n_K>L^{d-1+1/4}} \prod_{i:n_i\ge 1} \sum_{\Gamma:\mathrm{sp}(\Gamma)\ni 0, N_\Gamma=n_i} e^{-\beta(c_1/2)(\zeta^2\ell^d_{-,\gamma})N_\Gamma},$$

which by (9.2.8.4) is bounded by

$$\le e^{-\beta(c_1/4)(\zeta^2\ell^d_{-,\gamma})[L^{d-1+1/4}]}\Big(1+\sum_{\Gamma:\mathrm{sp}(\Gamma)\ni 0, N_\Gamma=n_i} e^{-\beta(c_1/4)(\zeta^2\ell^d_{-,\gamma})N_\Gamma}\Big)^K.$$

By Lemma 3.1.2.4 for γ small enough the bracket is bounded by 2^K; thus

$$\le \exp\{-\beta(c_1/4)(\zeta^2\ell^d_{-,\gamma})[L^{d-1+1/4}]+(\log 2)L^{d-1}\},$$

which concludes the proof of (12.3.3.2).

12.4 The plus and minus DLR measures are ergodic

In this section we shall prove that $\mu^{\pm}_\gamma$ are translational invariant and ergodic so that (12.3.1.1) is the ergodic decomposition of m.

12.4.1 Preliminary lemmas

By (A.1.9) we have

$$\tau_i(G_{\gamma,\Delta,q}) = G_{\gamma,\tau_i(\Delta),\tau_i(q)} \tag{12.4.1.1}$$

(a similar property was used in the proof of Lemma 12.3.2.1). It follows from (12.4.1.1) that $\tau_i(\mu^{\pm}_\gamma)$ are DLR measures (it is true in general that translates of DLR measures are DLR). With $\mathcal{S}$ as in Definition 12.2.2.1 and with $i \in a\mathbb{Z}^d$, $a \in \{2^{-n}, n \in \mathbb{N}\}$, define

$$\mathcal{X}^{\pm}_{\text{tail};\, i} = \Big\{q \in \mathcal{X} : \lim_{k\to\infty} G_{\gamma,\Lambda_k,q}(f) = [\tau_i(\mu^{\pm}_\gamma)](f), \text{ for any } \{\Lambda_k\} \in \mathcal{S} \text{ and any bounded, measurable cylindrical function } f\Big\}, \tag{12.4.1.2}$$

so that $\mathcal{X}^{\pm}_{\text{tail; i=0}} = \mathcal{X}^{\pm}_{\text{tail}}$, the tail set of Theorem 12.2.2.2 relative to $\mu^{\pm}_\gamma$.

Lemma 12.4.1.1 *For any $i \in a\mathbb{Z}^d$, $a \in \{2^{-n}, n \in \mathbb{N}\}$,*

$$\mathcal{X}^{\pm}_{\text{tail; i}} = \tau_{-i}(\mathcal{X}^{\pm}_{\text{tail}}), \quad [\tau_i(\mu^{\pm}_\gamma)](\mathcal{X}^{\pm}_{\text{tail; i}}) = 1. \tag{12.4.1.3}$$

Moreover, $\tau_i(\mu^{\pm}_\gamma) \neq \mu^{\pm}_\gamma$ if and only if $\mathcal{X}^{\pm}_{\text{tail; i}} \cap \mathcal{X}^{\pm}_{\text{tail}} = \emptyset$.

Proof We claim that $\mathcal{X}^+_{\text{tail}} \subseteq \tau_{-i}(\mathcal{X}^{\pm}_{\text{tail; i}})$. Proof. Fix i and $q \in \mathcal{X}^+_{\text{tail}}$. Then for any $\{\Lambda_k\} \in \mathcal{S}$ and any bounded, measurable, cylindrical function g

$$\mu^+_\gamma(g) = \lim_{k\to\infty} G_{\gamma,\tau_{-i}(\Lambda_k),q}(g),$$

because $\tau_{-i}(\Lambda_k) \in \mathcal{S}$ by Definition 12.2.2.1. Then we write $f = \tau_i(g)$, $g = \tau_{-i}(f)$ and $[\tau_i(\mu^+_\gamma)](f) = \mu^+_\gamma(g) = \lim_{k\to\infty} G_{\gamma,\tau_{-i}\Lambda_k,q}(g)$ so that by (12.4.1.1)

$$[\tau_i(\mu^+_\gamma)](f) = \lim_{k\to\infty} [\tau_i(G_{\gamma,\tau_{-i}(\Lambda_k),q})](f) = \lim_{k\to\infty} G_{\gamma,\Lambda_k,\tau_i(q)})(f),$$

which shows that $\tau_i(q) \in \mathcal{X}^{\pm}_{\text{tail; i}}$. Hence $q \in \tau_{-i}(\mathcal{X}^{\pm}_{\text{tail; i}})$, and the claim is proved.

By the claim, $1 = \mu^+_\gamma(\mathcal{X}^+_{\text{tail}}) \le \mu^+_\gamma(\tau_{-i}(\mathcal{X}^+_{\text{tail; i}})) = [\tau_i(\mu^+_\gamma)](\mathcal{X}^+_{\text{tail};i})$, which is therefore equal to 1. Repeating the whole argument with μ^+_γ and $\tau_i(\mu^+_\gamma)$ inverted we have $\tau_{-i}(\mathcal{X}^{\pm}_{\text{tail; i}}) \subseteq \mathcal{X}^+_{\text{tail}}$ and hence (12.4.1.3).

Finally, if $\tau_i(\mu^{\pm}_\gamma) \ne \mu^{\pm}_\gamma$ then $\mathcal{X}^{\pm}_{\text{tail; i}} \cap \mathcal{X}^{\pm}_{\text{tail}} = \emptyset$, because if there was $q \in \mathcal{X}^{\pm}_{\text{tail; i}} \cap \mathcal{X}^{\pm}_{\text{tail}}$, then

$$\mu^+_\gamma(f) = \lim_{k\to\infty} G_{\gamma,\Lambda_k,q}(f) = [\tau_i(\mu^{\pm}_\gamma)](f),$$

for all bounded, measurable cylindrical functions f, which contradicts the assumption $\tau_i(\mu^{\pm}_\gamma) \ne \mu^{\pm}_\gamma$. If vice versa $\mathcal{X}^{\pm}_{\text{tail; i}} \cap \mathcal{X}^{\pm}_{\text{tail}} = \emptyset$ then $\tau_i(\mu^{\pm}_\gamma)$ and $\mu^{\pm}_\gamma$ are mutually singular and hence $\tau_i(\mu^{\pm}_\gamma) \ne \mu^{\pm}_\gamma$. Lemma 12.4.1.1 is proved. □

Lemma 12.4.1.2 *For any $i \in a\mathbb{Z}^d$, $a \in \{2^{-n}, n \in \mathbb{N}\}$, $\tau_i(\mu^{\pm}_\gamma)$ is invariant under translations in $\ell_{+,\gamma}\mathbb{Z}^d$.*

Proof Let $j \in \ell_{+,\gamma}\mathbb{Z}^d$; then $\tau_j(\tau_i(\mu^{\pm}_\gamma)) = \tau_i(\tau_j(\mu^{\pm}_\gamma)) = \tau_i(\mu^{\pm}_\gamma)$, the last equality holding by Lemma 12.3.2.1. □

Lemma 12.4.1.3 *For all γ small enough the following holds: for any $i \in a\mathbb{Z}^d$, $a \in \{2^{-n}, n \in \mathbb{N}\}$, μ^-_γ and $\tau_i(\mu^+_\gamma)$ are mutually singular and $\mathcal{X}^+_{\text{tail; i}} \cap \mathcal{X}^-_{\text{tail}} = \emptyset$.*

Proof By the same argument as used at the end of the proof of Lemma 12.4.1.1, it suffices to show that $\mu^-_\gamma \ne \tau_i(\mu^+_\gamma)$ and, by Lemma 12.4.1.2, to consider τ_i with $i \in C_0^{(\ell_{+,\gamma})} \cap a\mathbb{Z}^d$. Let Λ be a $\mathcal{D}^{(\ell_{+,\gamma})}$-measurable cube, $|q|$ the number of particles in a configuration q; then $\mu^-_\gamma(\frac{|q_\Lambda|}{|\Lambda|}) = \mu^-_\gamma(\frac{|q_{C_0^{(\ell_{+,\gamma})}}|}{|C_0^{(\ell_{+,\gamma})}|})$ because μ^-_γ is invariant under translations in $\ell_{+,\gamma}\mathbb{Z}^d$. $\mu^-_\gamma(\frac{|q_\Lambda|}{|\Lambda|})$ is then bounded by

$$(\rho_{\beta,-} + \zeta) + \mu^-_\gamma\left(\mathbf{1}_{\Theta(\cdot;0)>-1}\frac{|q_{C_0^{(\ell_{+,\gamma})}}|}{\ell^d_{+,\gamma}}\right) \le \rho_{\beta,-} + \zeta + c[e^{-\beta(c_1/4)\zeta^2\ell^d_{-,\gamma}}]^{1/2},$$

$$c = \ell_{+,\gamma}^{-d} \mu_\gamma^-(|q_{C_0^{(\ell_{+,\gamma})}}|^2)^{1/2} \le \ell_{+,\gamma}^{-d/2} \Big(\sum_{C_i^{(1)} \subset C_0^{(\ell_{+,\gamma})}} \mu_\gamma^-(|q_{C_i^{(1)}}|^2) \Big)^{1/2},$$

having used Cauchy–Schwartz and (10.3.4.2). By (11.1.2.8)

$$\mu_\gamma^-(|q_{C_i^{(1)}}|^2) \le \sum_{n\ge 1} \frac{e^{\beta b_0 n} n^2}{n!} < \infty.$$

Thus, in conclusion

$$\mu_\gamma^-\Big(\frac{|q_\Lambda|}{|\Lambda|}\Big) \le \rho_{\beta,-} + (\zeta + c' e^{-\beta(c_1/8)\zeta^2 \ell_{-,\gamma}^d}). \tag{12.4.1.4}$$

For any $j \in C_0^{(\ell_{+,\gamma})} \cap a\mathbb{Z}^d$, $\tau_j(\Lambda) \supset \Lambda_0$, $\Lambda_0 = \Lambda \setminus \delta_{\text{in}}^{\ell_{+,\gamma}}[\Lambda]$. Let N_Λ and N_{Λ_0} the number of $\mathcal{D}^{(\ell_{+,\gamma})}$-cubes in Λ and Λ_0; then for any $i \in C_0^{(\ell_{+,\gamma})} \cap a\mathbb{Z}^d$

$$[\tau_i(\mu_\gamma^+)]\Big(\frac{|q_\Lambda|}{|\Lambda|}\Big) \ge \mu_\gamma^+\Big(\frac{|q_{\Lambda_0}|}{|\Lambda|}\Big) = \frac{N_{\Lambda_0}}{N_\Lambda} \mu_\gamma^+\Big(\frac{|q_{C_0^{(\ell_{+,\gamma})}}|}{\ell_{+,\gamma}^d}\Big).$$

We then write $\mu_\gamma^+(\frac{|q_{C_0^{(\ell_{+,\gamma})}}|}{\ell_{+,\gamma}^d}) \ge \mu_\gamma^+(\frac{|q_{C_0^{(\ell_{+,\gamma})}}|}{\ell_{+,\gamma}^d} \mathbf{1}_{\Theta(\cdot;0)=1})$ and get

$$\begin{aligned}[\tau_i(\mu_\gamma^+)]\Big(\frac{|q_\Lambda|}{|\Lambda|}\Big) &\ge \frac{N_{\Lambda_0}}{N_\Lambda}(\rho_{\beta,+} - \zeta)(1 - e^{-\beta(c_1/4)\zeta^2 \ell_{-,\gamma}^d}) \\ &\ge \rho_{\beta,+} - \Big(\zeta + \rho_{\beta,+}\Big\{ e^{-\beta(c_1/4)\zeta^2 \ell_{-,\gamma}^d} + \frac{N_\Lambda - N_{\Lambda_0}}{N_\Lambda}\Big\}\Big),\end{aligned}$$

which is strictly larger than the r.h.s. of (12.4.1.4) for Λ large and γ small. Thus $\tau_i(\mu_\gamma^+) \ne \mu_\gamma^-$. □

12.4.2 Translational invariance

We shall first prove translational invariance for special values of the mesh; the general case will follow by a density argument.

Proposition 12.4.2.1 *For all γ small enough, the measures $\mu_\gamma^\pm$ are invariant under the action of the translation group $\tau_i^{(a)}$, for any $a \in \{2^{-n}, n \in \mathbb{N}\}$.*

Proof We consider explicitly only the statement about μ_γ^+. By Lemma 12.3.2.1, the measure

$$\nu := \frac{1}{|C_0^{(\ell_{+,\gamma})}|} \sum_{i \in C_0^{(\ell_{+,\gamma})} \cap a\mathbb{Z}^d} \tau_i^{(a)}(\mu_\gamma^+) \tag{12.4.2.1}$$

is translational invariant under $\tau_i^{(a)}$. Then by (12.3.1.1) there is $u \in [0,1]$, so that

$$\nu = u\mu_\gamma^+ + (1-u)\mu_\gamma^-. \tag{12.4.2.2}$$

By Lemma 12.4.1.1 $\mu_\gamma^-(\mathcal{X}_{\text{tail}}^-) = 1$, while by Lemma 12.4.1.3, for all i, $\tau_i^{(a)}(\mu_\gamma^+)(\mathcal{X}_{\text{tail}}^-) = 0$; hence $\nu(\mathcal{X}_{\text{tail}}^-) = 0$ by (12.4.2.1), and by (12.4.2.2) $(1-u) \times \mu_\gamma^-(\mathcal{X}_{\text{tail}}^-) = 0$, which yields $u = 1$, namely $\nu = \mu_\gamma^+$. Now, if there were i so that $\tau_i^{(a)}(\mu_\gamma^+) \neq \mu_\gamma^+$, then by Lemma 12.4.1.1, $\mu_\gamma^+(\mathcal{X}_{\text{tail; i}}^+) = 0$ and $[\tau_i^{(a)}(\mu_\gamma^+)](\mathcal{X}_{\text{tail; i}}^+) = 1$ which contradicts (12.4.2.1) (as we have proved that $\nu = \mu_\gamma^+$). □

We shall prove that μ_γ^+ is invariant under $\tau_i^{(a)}$, $i \in a\mathbb{Z}^d$, $a > 0$ (the minus case is similar and we omit it) by showing that for any bounded, measurable cylindrical function f, $\mu_\gamma^+(f) = \mu_\gamma^+(\tau_i^{(a)}(f))$. By Proposition 12.4.2.1 for any $n \in \mathbb{N}$ and $i_n \in \mathbb{Z}^d$, $\mu_\gamma^+(g) = \mu_\gamma^+(\tau_{2^{-n}i_n}(g))$. Take $g = \tau_i^{(a)}(f)$; then $\mu_\gamma^+(\tau_i^{(a)}(f)) = \mu_\gamma^+(\tau_{2^{-n}i_n+i}(f))$. We choose i_n so that $\lim_{n\to\infty} 2^{-n}i_n = -i$ and we thus need to show that for any $\delta > 0$,

$$\lim_{\epsilon\to 0} |\mu_\gamma^+(\tau_{-\epsilon}(f)) - \mu_\gamma^+(f)| \le \delta, \quad \epsilon \in \mathbb{R}^d. \tag{12.4.2.3}$$

Let f be cylindrical in a cube Δ and let Λ be a $\mathcal{D}^{(\ell_{+,\gamma})}$-measurable cube which contains Δ such that the distance of Δ from Λ^c is ≥ 1. Write $\Lambda^+ = \Lambda \cup \delta_{\text{out}}^{\ell_{+,\gamma}}[\Lambda]$. Denote by $\mathbf{1}_\rho$ the characteristic function of $q \in \mathcal{X} : |q_{\Lambda^+}| \le \rho|\Lambda^+|$ ($|q_{\Lambda^+}|$ is the number of particles in q_{Λ^+}). By (11.1.2.8) the probability (for a DLR measure) that this does not happen vanishes as $\rho \to \infty$. We then write

$$\mu_\gamma^+(f) = \mu_\gamma^+(\mathbf{1}_\rho f) + o(\rho); \tag{12.4.2.4}$$

$o(\rho)$ is a function which vanishes as $\rho \to \infty$. By the DLR property

$$\mu_\gamma^+(\mathbf{1}_\rho f) = \int G_{\gamma,\Lambda,q'}(\mathbf{1}_\rho f) d\mu_\gamma^+(q'). \tag{12.4.2.5}$$

We claim that $G_{\gamma,\Lambda,q}(q_\Lambda \neq \emptyset) \le |\Lambda|(e^{\beta b_0 + e^{\beta b_0}|\Lambda|})$ for all bounded measurable sets Λ and all configurations q. Indeed by (11.1.2.8) the energy is bounded by $e^{\beta b_0 n}$ if there are n particles in Λ, so that the probability is bounded by $\sum_{n\ge 1} \frac{(e^{\beta b_0}|\Lambda|)^n}{n!}$; hence the claim. Then if $\Lambda' := \{r \in \Lambda : r' := r - \epsilon \in \Lambda\}$,

$$G_{\gamma,\Lambda,q'}(\mathbf{1}_\rho f) = \int_{q_\Lambda \in \mathcal{X}_{\Lambda'}} \mathbf{1}_\rho(q_\Lambda + q'_{\Lambda^c}) f(q_\Lambda) e^{-\beta H_{\gamma,\Lambda}(q_\Lambda|q'_{\Lambda^c})} d\nu_\Lambda(q_\Lambda) + o(\epsilon). \tag{12.4.2.6}$$

After the change of variables $q_\Lambda \to q_\Lambda - \epsilon$ by which each particle position r in q goes into $r - \epsilon$, $G_{\gamma,\Lambda,q'}(\mathbf{1}_\rho f)$ becomes equal to

$$\int_{(q_\Lambda+\epsilon)\in\mathcal{X}_{\Lambda'}} \mathbf{1}_\rho((q_\Lambda+\epsilon) + q'_{\Lambda^c}) f((q_\Lambda+\epsilon)) e^{-\beta H_{\gamma,\Lambda}((q_\Lambda+\epsilon)|q'_{\Lambda^c})} d\nu_\Lambda(q_\Lambda) + o(\epsilon). \tag{12.4.2.7}$$

Observe that $f((q_\Lambda+\epsilon)) = [\tau_{-\epsilon}(f)](q_\Lambda)$; $|H_{\gamma,\Lambda}((q_\Lambda+\epsilon)|q'_{\Lambda^c}) - H_{\gamma,\Lambda}((q_\Lambda)|q'_{\Lambda^c})|$ vanishes as $\epsilon \to 0$ for each fixed ρ uniformly in the set $\mathbf{1}_\rho((q_\Lambda+\epsilon)+q'_{\Lambda^c}) = 1$. Thus, $G_{\gamma,\Lambda,q'}(\mathbf{1}_\rho f)$ becomes

$$\int_{(q_\Lambda+\epsilon)\in\mathcal{X}_{\Lambda'}} \mathbf{1}_\rho((q_\Lambda+\epsilon)+q'_{\Lambda^c})\tau_{-\epsilon}(f)e^{-\beta H_{\gamma,\Lambda}(q_\Lambda|q'_{\Lambda^c})}d\nu_\Lambda(q_\Lambda)+o(\epsilon). \quad (12.4.2.8)$$

We then extend the integral to $q_\Lambda \in \mathcal{X}_\Lambda$ with an error $o(\epsilon)$. Moreover, with an error $o(\rho)$ we drop the characteristic function $\mathbf{1}_\rho((q_\Lambda+\epsilon)+q'_{\Lambda^c})$, and thus

$$G_{\gamma,\Lambda,q'}(\mathbf{1}_\rho f) = G_{\gamma,\Lambda,q'}(\tau_{-\epsilon}(f)) + o(\rho) + o(\epsilon), \quad (12.4.2.9)$$

and, going back to (12.4.2.4)–(12.4.2.5),

$$\mu^+_\gamma(f) = \int G_{\gamma,\Lambda,q'}(\tau_{-\epsilon}(f))d\mu^+_\gamma(q') + o(\rho) + o(\epsilon)$$

which proves (12.4.2.3) setting ρ large enough and letting $\epsilon \to 0$.

12.4.3 Ergodicity

The measures $\mu^\pm_\gamma$ are ergodic with respect to the group $\tau^{(a)}_i$, $i \in a\mathbb{Z}^d$, for any mesh $a > 0$; the statement follows from the general theory because the tail field is trivial. For self-consistency we give here a proof in the L^2 context. We shall prove that for any bounded, measurable cylindrical function f,

$$\lim_{n\to\infty} \mu^+_\gamma\left(|\mathcal{A}^{(a)}_{\Lambda_n}(f - \mu^+_\gamma(f))|^2\right) = 0 \quad (12.4.3.1)$$

where $\Lambda_n \nearrow \mathbb{R}^d$ is an increasing sequence of $\mathcal{D}^{(a)}$-measurable cubes and

$$\mathcal{A}^{(a)}_{\Lambda_n}(g) = N^{-1}_{a,\Lambda_n} \sum_{i\in\Lambda_n\cap a\mathbb{Z}^d} \tau^{(a)}_i(g), \quad N_{a,\Lambda_n} := \mathrm{card}(\Lambda_n \cap a\mathbb{Z}^d). \quad (12.4.3.2)$$

By translation invariance we need to prove that for any $\epsilon > 0$,

$$\left|\mu^+_\gamma(\{f - \mu^+_\gamma(f)\}\tau^{(a)}_i(f))\right| \le \epsilon, \quad (12.4.3.3)$$

as soon as i is large enough. Let $\{\Delta_k\} \in \mathcal{S}$, see Theorem 12.2.2.2; then by Theorem 12.2.2.5 given any $\epsilon' > 0$, for all for k large enough

$$\mu^+_\gamma(q \in \mathcal{X} : |G_{\gamma,\Delta_k,q}(f - \mu^+_\gamma(f))| < \epsilon') \ge 1 - \sqrt{\epsilon_k}, \quad (12.4.3.4)$$

where $\epsilon_k \to 0$ as $k \to \infty$. Let i in (12.4.3.3) be so large that $i \notin \Delta_k$. Then using the DLR property,

$$\left|\mu^+_\gamma(\{f - \mu^+_\gamma(f)\}\tau^{(a)}_i(f))\right| \le \|f\|_\infty\epsilon' + \|f\|^2_\infty\sqrt{\epsilon_k}; \quad (12.4.3.5)$$

hence (12.4.3.3) after choosing ϵ' small enough and k large enough.

12.5 Complements: Decay from the boundaries

In analogy with Theorem 10.5.1.1, we shall prove that besides partition functions we can also express expectations as expectations on $\mathcal{X}^+$. We shall then use the results of Sect. 11.4.2 to reduce the system to a lattice gas and exploit the couplings in Theorem 11.5.4.1 to prove Theorem 12.2.0.1.

12.5.1 A representation formula for Gibbs measures

Our aim here is to rewrite in an advantageous way the expectations $G^+_{\gamma,\Lambda,q^+}(f)$ (Λ a bounded $\mathcal{D}^{(\ell_+,\gamma)}$-measurable set, here and in the sequel) in order to have a formalism which allows us to exploit the couplings introduced in the previous chapter.

Notation

We fix a bounded set Δ and a bounded, measurable cylindrical function f on Δ, $f(q) = f(q \cap \Delta)$; let $\Lambda \supset \Delta$. Given Λ and $\underline{\Gamma} \in \mathcal{B}^+_\Lambda$, we set

$$c(\underline{\Gamma}) = \bigcup_{\Gamma \in \underline{\Gamma}} c(\Gamma), \qquad \text{ext}(\underline{\Gamma}) = \Lambda \setminus c(\underline{\Gamma}), \qquad W^+_\gamma(\underline{\Gamma}; q) = \prod_{\Gamma \in \underline{\Gamma}} W^+_\gamma(\Gamma; q). \tag{12.5.1.1}$$

Recall from Sect. 10.5.1 that $\mathcal{B}^{+,\text{ext}}_\Lambda$ is the subset of $\mathcal{B}^+_\Lambda$ of collections $\underline{\Gamma} = (\Gamma_1, \dots, \Gamma_n)$ made exclusively of external contours, namely such that all $c(\Gamma_i)$ are mutually disconnected. Let $\underline{\Gamma} \in \mathcal{B}^{+,\text{ext}}_\Lambda$ and write (the dependence on f, Λ and Δ is not made explicit):

$$\hat{K}(\underline{\Gamma}) = \{\Gamma \in \underline{\Gamma} : c(\Gamma) \cap \Delta \neq \emptyset\}, \tag{12.5.1.2}$$

$$\hat{D}(\underline{\Gamma}) = \{\Delta \cap \text{ext}(\underline{\Gamma})\} \cup \Big\{ \bigcup_{\Gamma \in \hat{K}(\underline{\Gamma})} \delta^{\ell_+\gamma}_{\text{out}}[c(\Gamma)] \Big\}, \tag{12.5.1.3}$$

$$F(q, \underline{\Gamma}) = \frac{N^+_\gamma(\underline{\Gamma}; q; f)}{N^+_\gamma(\underline{\Gamma}; q; 1)}, \qquad q : q_{\text{ext}(\underline{\Gamma})} \in \mathcal{X}^+_{\text{ext}(\underline{\Gamma})}, \tag{12.5.1.4}$$

where, putting $\mathcal{X}^0(\Gamma) = \{q_{c(\Gamma)} : \eta(q_{c(\Gamma)}; r) = \eta_\Gamma(r), r \in \text{sp}(\Gamma), \Theta(q_{c(\Gamma)}; r) = \pm 1, r \in \delta^{\ell_+,\gamma}_{\text{in}}[\text{int}^\pm(\Gamma)]\}$ and $\mathcal{X}^0(\underline{\Gamma}) = \bigcap_{\Gamma \in \underline{\Gamma}} \mathcal{X}^0(\Gamma)$, and $N^+_\gamma(\underline{\Gamma}; q; f)$ is equal to

$$\int_{\mathcal{X}^0(\underline{\Gamma})} e^{-\beta H_{\gamma, c(\underline{\Gamma})}(q'_{c(\underline{\Gamma})} | q_{\text{ext}(\underline{\Gamma})})} f(q'_{c(\underline{\Gamma})}, q_{\text{ext}(\underline{\Gamma})}) d\nu_{c(\underline{\Gamma})}(q'_{c(\underline{\Gamma})}). \tag{12.5.1.5}$$

Theorem 12.5.1.1 *With the above notation*

$$\|F\|_\infty \le \|f\|_\infty, \quad F(q,\underline{\Gamma}) = F(q_{\hat{D}(\underline{\Gamma})}, \hat{K}(\underline{\Gamma})), \tag{12.5.1.6}$$

$$G^+_{\gamma,\Lambda,q^+}(f) = \sum_{\underline{\Gamma}\in\mathcal{B}^+_\Lambda} \int_{\mathcal{X}^+_\Lambda} F(q_\Lambda; \phi_{\text{ext}}(\underline{\Gamma}))d\mu^+_{\gamma,\Lambda,q^+}(q_\Lambda,\underline{\Gamma}), \tag{12.5.1.7}$$

where $\phi_{\text{ext}}(\underline{\Gamma})$ *is the subset of external contours in* $\underline{\Gamma}$ *(obtained by deleting from* $\underline{\Gamma}$ *all* Γ' *with* $c(\Gamma')\subset c(\Gamma)$ *for some other* $\Gamma\in\underline{\Gamma}$*);* $d\mu^+_{\gamma,\Lambda,q^+}(q_\Lambda,\underline{\Gamma})$ *is equal to*

$$(Z^+_{\gamma,\Lambda,q^+})^{-1}e^{-\beta H_{\gamma,\Lambda}(q_\Lambda|q^+_{\Lambda^c})}W^+_\gamma(\underline{\Gamma};q_\Lambda)\mathbf{1}_{q_\Lambda\in\mathcal{X}^+_\Lambda}d\nu_\Lambda(q_\Lambda). \tag{12.5.1.8}$$

Z^+_{γ,Λ,q^+} *is the plus diluted partition function defined in* (10.3.2.2).

Proof The first statement in (12.5.1.6) is a direct consequence of the definition; the second one follows because the contribution to the r.h.s. of (12.5.1.4) of the contours $\Gamma_i : c(\Gamma_i)\cap\Delta=\emptyset$ cancels between numerator and denominator. By Theorem 10.5.1.1, Z^+_{γ,Λ,q^+} is the normalization factor in (12.5.1.8).

By (12.5.1.5) and (12.5.1.4) $G^+_{\gamma,\Lambda,q^+}(f)$ is equal to

$$\sum_{\underline{\Gamma}\in\mathcal{B}^{+,\text{ext}}_\Lambda} \int_{\mathcal{X}^+_{\text{ext}(\underline{\Gamma})}} e^{-\beta H_{\gamma,\text{ext}(\underline{\Gamma})}(q_{\text{ext}(\underline{\Gamma})}|q^+_{\Lambda^c})} N^+_\gamma(\underline{\Gamma};q;f)d\nu_{\text{ext}(\underline{\Gamma})}(q_{\text{ext}(\underline{\Gamma})})$$

$$= \sum_{\underline{\Gamma}\in\mathcal{B}^{+,\text{ext}}_\Lambda} \int_{\mathcal{X}^+_{\text{ext}(\underline{\Gamma})}} e^{-\beta H_{\gamma,\text{ext}(\underline{\Gamma})}(q_{\text{ext}(\underline{\Gamma})}|q^+_{\Lambda^c})}$$
$$\times F(q_{\hat{D}(\underline{\Gamma})},\underline{\Gamma})N^+_\gamma(\underline{\Gamma};q;1)d\nu_{\text{ext}(\underline{\Gamma})}(q_{\text{ext}(\underline{\Gamma})}).$$

We then write $N^+_\gamma(\underline{\Gamma};q;1)$ in terms of $W^+_\gamma(\underline{\Gamma};q)$, and (12.5.1.8) follows using the same iterative scheme as in the proof of Theorem 10.5.1.1. Details are omitted. □

12.5.2 Reduction to a lattice gas

To apply the theory of Sect. 11.5 we preliminarily need to reduce the system to a lattice gas. Using the notation in Sect. 11.4.1 we approximate the r.h.s. of (12.5.1.7) by

$$\mu^+_{\gamma,a,\Lambda,q^+}(F^{(a)}) := \sum_{\underline{\Gamma}\in\mathcal{B}^+_\Lambda}\sum_{\sigma_\Lambda\in\mathcal{X}^+_{\Lambda,a}} F^{(a)}(\sigma_\Lambda;\underline{\Gamma})\mu^+_{\gamma,a,\Lambda,q^+}(\sigma_\Lambda,\underline{\Gamma}), \tag{12.5.2.1}$$

where $\mu^+_{\gamma,a,\Lambda,q^+}$ is the probability on $\mathcal{X}^+_{\Lambda,a}\times\mathcal{B}^+_\Lambda$ defined in (11.4.1.6), while $F^{(a)}$ is as in (11.4.2.1). By Theorem 11.4.2.1 the above expression converges as $a\to 0$

to the r.h.s. of (12.5.1.7) and hence to $G^+_{\gamma,\Lambda,q^+}(f)$, so that (12.2.0.1) will be proved once we show that there are c and $\omega > 0$ so that for all γ small enough and all $\mathcal{D}^{(\ell_+,\gamma)}$-measurable bounded regions Λ_1 and Λ_2 containing Δ

$$|\mu^+_{\gamma,a,\Lambda_1,q_1}(F^{(a)}) - \mu^+_{\gamma,a,\Lambda_2,q_2}(F^{(a)})| \le c\|f\|_\infty |\Delta| e^{-\omega\gamma \operatorname{dist}(\Delta,(\Lambda_1\cap\Lambda_2)^c)} \tag{12.5.2.2}$$

for all a small enough. We rewrite

$$\mu^+_{\gamma,a,\Lambda,q^+}(F^{(a)}) = m^+_{\gamma,a,\Lambda,q^+}(p^+_{\gamma,\Lambda,\sigma_\Lambda}(F^{(a)})), \tag{12.5.2.3}$$

$m^+_{\gamma,a,\Lambda,q^+}$ being the marginal distribution of $\mu^+_{\gamma,a,\Lambda,q^+}$ on $\mathcal{X}^+_{\Lambda,a}$ (obtained by summing out the contours) and $p^+_{\gamma,\Lambda,\sigma_\Lambda}$ the conditional probability given σ_Λ:

$$p^+_{\gamma,\Lambda,\sigma_\Lambda}(\underline{\Gamma}) = \frac{W^+_\gamma(\underline{\Gamma},\sigma_\Lambda)}{X^+_{\gamma,\sigma_\Lambda}(\Lambda)}, \qquad X^+_{\gamma,\sigma_\Lambda}(\Lambda) \text{ as in (11.1.1.6).} \tag{12.5.2.4}$$

We claim that $m^+_{\gamma,a,\Lambda,q^+}$ is the same as the measure in (11.4.4.4) with the interpolating parameter $u = 1$. Indeed, by summing (11.4.1.6) over $\underline{\Gamma}$ we get

$$m^+_{\gamma,a,\Lambda,q^+}(\sigma_\Lambda) = (Z^+_{\gamma,a,\Lambda,q^+})^{-1}(a^d)^{|\sigma_\Lambda|} e^{-\beta H_{\gamma,\Lambda}(\sigma_\Lambda|q^+_{\Lambda^c})} X^+_{\gamma,\sigma_\Lambda}(\Lambda), \tag{12.5.2.5}$$

and then write $X^+_{\gamma,\sigma_\Lambda}(\Lambda) = e^{-\beta K^+_{\gamma,\Lambda}(\sigma_\Lambda)}$, see (11.4.3.1), obtaining the expression (11.4.4.4).

It may seem that with (12.5.2.3) we are in business because we have from the previous chapter good decay properties of $m^+_{\gamma,a,\Lambda,q^+}$; the problem is that unlike the original f, the function $p^+_{\gamma,\Lambda,\sigma_\Lambda}(F^{(a)})$ is not cylindrical and this requires some extra work. Recalling that $\hat{D}$ and $\hat{K}$ are defined in (12.5.1.3) and (12.5.1.2), we have for any $\sigma_\Lambda \in \mathcal{X}^+_{\Lambda,a}$

$$p^+_{\gamma,\Lambda,\sigma_\Lambda}(F^{(a)}) = \sum_{D,K} F(\sigma_D, K) p^+_{\gamma,\Lambda,\sigma_\Lambda}(\{\hat{D} = D, \hat{K} = K\}), \tag{12.5.2.6}$$

recalling that the ranges of values of D and K depend on the set Δ associated to f.

12.5.3 Couplings

By using the analysis of the previous chapter we shall prove that there is a coupling of $m^+_{\gamma,a,\Lambda_1,q_1}$ and $m^+_{\gamma,a,\Lambda_2,q_2}$ which gives agreement with large probability on a large set Λ. Given Δ, Λ_1 and Λ_2 as in Theorem 12.2.0.1, let

$$\Lambda = \bigcup_{x\in\Delta} C_x, \tag{12.5.3.1}$$

where for each $x \in \Delta$ we choose a $\mathcal{D}^{(\ell_{+,\gamma})}$-measurable cube C_x such that

$$|C_x| \le R^d, \qquad \mathrm{dist}(x, C_x^c) > \frac{R}{100}, \qquad \mathrm{dist}(C_x, (\Lambda_1 \cap \Lambda_2)^c) > \frac{R}{2}, \quad (12.5.3.2)$$

with $R := \mathrm{dist}(\Delta, (\Lambda_1 \cap \Lambda_2)^c)$ (which by assumption is large, i.e. $\ge 2^{10}\ell_{+,\gamma}$). By Proposition 11.6.2.1 there is a coupling Q of $m^+_{\gamma,a,\Lambda_1,q_1}$ and $m^+_{\gamma,a,\Lambda_2,q_2}$ so that

$$Q(\sigma_\Lambda = \sigma'_\Lambda) \ge 1 - c|\Delta| R^d e^{-\gamma\omega R/4}. \qquad (12.5.3.3)$$

Going back to (12.5.2.3), we see that $|\mu^+_{\gamma,a,\Lambda_1,q_1}(F^{(a)}) - \mu^+_{\gamma,\Lambda_2,q_2}(F^{(a)})|$ is bounded uniformly in a by

$$\left|E_Q\left(\mathbf{1}_{\sigma'_\Lambda=\sigma''_\Lambda}[p^+_{\gamma,\Lambda_1,\sigma'_{\Lambda_1}}(F^{(a)}) - p^+_{\gamma,\Lambda_2,\sigma''_{\Lambda_2}}(F^{(a)})]\right)\right| + c|\Delta|R^d e^{-\gamma\omega R/4}.$$

The set A

Write A for the collection of all values (D, K) of $\hat{D}$ and $\hat{K}$ so that $D \subset \Lambda$ and all Γ in $\hat{K}$ are such that $c(\Gamma) \cup \delta^{\ell_{+,\gamma}}_{\mathrm{out}}[c(\Gamma)] \subset \Lambda$.

If $\sigma'_\Lambda = \sigma''_\Lambda$, by (12.5.2.6) $|p^+_{\gamma,\Lambda_1,\sigma'_{\Lambda_1}}(F_1^{(a)}) - p^+_{\gamma,\Lambda_2,\sigma''_{\Lambda_2}}(F_2^{(a)})|$ is bounded by

$$\|f\|_\infty \Bigg(p^+_{\gamma,\Lambda_1,\sigma'_{\Lambda_1}}((\hat{D},\hat{K}) \notin A) + p^+_{\gamma,\Lambda_1,\sigma''_{\Lambda_2}}((\hat{D},\hat{K}) \notin A)$$
$$+ \sum_{(D,K)\in A} \left|p^+_{\gamma,\Lambda_1,\sigma'_{\Lambda_1}}((\hat{D},\hat{K}) = (D,K)) - p^+_{\gamma,\Lambda_2,\sigma'_{\Lambda_2}}((\hat{D},\hat{K}) = (D,K))\right|\Bigg),$$

and (12.5.2.2) follows from the following theorem proved in Sect. 12.5.6.

Theorem 12.5.3.1 *There are $\omega > 0$ and c so that for all a and γ small enough and all $\sigma'_{\Lambda_1}, \sigma''_{\Lambda_2}$,*

$$p^+_{\gamma,\Lambda_1,\sigma'_{\Lambda_1}}((\hat{D},\hat{K}) \notin A) + p^+_{\gamma,\Lambda_1,\sigma''_{\Lambda_2}}((\hat{D},\hat{K}) \notin A) \le c|\Delta| e^{-\gamma\omega R}. \qquad (12.5.3.4)$$

Moreover, if $\sigma'_{\Lambda_1}, \sigma''_{\Lambda_2}$ are such that $\sigma'_\Lambda = \sigma''_\Lambda$ then

$$\sum_{(D,K)\in A} \left|p^+_{\gamma,\Lambda_1,\sigma'_{\Lambda_1}}((\hat{D},\hat{K}) = (D,K)) - p^+_{\gamma,\Lambda_2,\sigma'_{\Lambda_2}}((\hat{D},\hat{K}) = (D,K))\right|$$
$$\le c|\Delta| e^{-\gamma\omega R}. \qquad (12.5.3.5)$$

12.5.4 Percolation disagreement

The proof of (12.5.3.5) is based on the disagreement percolation scheme introduced by van der Berg and Maes [206]. The analogy here is with the classic coupling of two independent random walks: they run independently till they meet, afterwards the coupling matches their jumps so that they remain always together.

We fix hereafter σ'_{Λ_1} and σ''_{Λ_2} assuming $\sigma'_{\Lambda} = \sigma''_{\Lambda}$ and write

$$P_{\Lambda_1,\Lambda_2}(\underline{\Gamma}', \underline{\Gamma}'') = p^+_{\gamma,\Lambda_1,\sigma'_{\Lambda_1}}(\underline{\Gamma}')p^+_{\gamma,\Lambda_2,\sigma''_{\Lambda_2}}(\underline{\Gamma}''), \tag{12.5.4.1}$$

which defines a (product) probability on $\mathcal{B}^+_{\Lambda_1} \times \mathcal{B}^+_{\Lambda_2}$. We denote by $E_{P_{\Lambda_1,\Lambda_2}}$ its expectation and we shall prove (12.5.3.5) using the formula

$$\begin{aligned} &p^+_{\gamma,\Lambda_1,\sigma'_{\Lambda_1}}\big((\hat{D}, \hat{K}) = (D, K)\big) - p^+_{\gamma,\Lambda_2,\sigma''_{\Lambda_2}}\big((\hat{D}, \hat{K}) = (D, K)\big) \\ &\quad = E_{P_{\Lambda_1,\Lambda_2}}\big(\mathbf{1}_{(\hat{D},\hat{K})(\underline{\Gamma}')=(D,K)} - \mathbf{1}_{(\hat{D},\hat{K})(\underline{\Gamma}'')=(D,K)}\big). \end{aligned} \tag{12.5.4.2}$$

We shall estimate the r.h.s. of (12.5.4.2) by conditioning on a random set which depends on $\underline{\Gamma}', \underline{\Gamma}''$.

The random sets $\hat{N}_{\rm out}$ and $\hat{N}_{\rm in}$

- The random set $\hat{N}_{\rm out}(\underline{\Gamma}', \underline{\Gamma}'')$ is the union of Λ^c with all the maximal connected components of $c(\underline{\Gamma}') \cup c(\underline{\Gamma}'')$ which are connected to Λ^c. $\hat{N}_{\rm in}(\underline{\Gamma}', \underline{\Gamma}'')$ denotes the complement of $\hat{N}_{\rm out}(\underline{\Gamma}', \underline{\Gamma}'')$.
- The random configuration $(\underline{\Gamma}')_{\rm in}$ (which depends on $(\underline{\Gamma}', \underline{\Gamma}'')$) is defined by $(\underline{\Gamma}')_{\rm in} = \{\Gamma \in \underline{\Gamma}' : c(\Gamma) \in \hat{N}_{\rm in}(\underline{\Gamma}', \underline{\Gamma}'')\}$; we also define $(\underline{\Gamma}')_{\rm out} = \{\Gamma \in \underline{\Gamma} : c(\Gamma) \in \hat{N}_{\rm out}(\underline{\Gamma}', \underline{\Gamma}'')\}$. $(\underline{\Gamma}'')_{\rm in}$ and $(\underline{\Gamma}'')_{\rm out}$ are defined analogously.
- $N_{\rm out}$ (i.e. without the hat) will denote a possible value of $\hat{N}_{\rm out}$; $N_{\rm in}$ the complement of $N_{\rm out}$; $N^+_{\rm out} = N_{\rm out} \cup \delta^{\ell_+,\gamma}_{\rm out}[N_{\rm out}]$.

Lemma

- *Given $\underline{\Gamma}', \underline{\Gamma}''$ write $N_{\rm out} := \hat{N}_{\rm out}(\underline{\Gamma}', \underline{\Gamma}'')$; then $(\underline{\Gamma}')_{\rm in} \in \mathcal{B}^+_{N_{\rm in}}$ and $(\underline{\Gamma}')_{\rm out} \in \mathcal{B}^+_{N^+_{\rm out}\cap\Lambda_1}$; the same for $\underline{\Gamma}''$ with Λ_2 instead of Λ_1. Moreover,*

$$\begin{aligned} \underline{\Gamma}' = (\underline{\Gamma}')_{\rm in} \cup (\underline{\Gamma}')_{\rm out}, \qquad \underline{\Gamma}'' = (\underline{\Gamma}'')_{\rm in} \cup (\underline{\Gamma}'')_{\rm out}, \\ \hat{N}_{\rm out}((\underline{\Gamma}')_{\rm out}, (\underline{\Gamma}'')_{\rm out}) = N_{\rm out}. \end{aligned} \tag{12.5.4.3}$$

- *Vice versa, suppose we are given $N_{\rm out}$, $\underline{\Gamma}'_2 \in \mathcal{B}^+_{N^+_{\rm out}\cap\Lambda_1}$, $\underline{\Gamma}''_2 \in \mathcal{B}^+_{N^+_{\rm out}\cap\Lambda_2}$, $\underline{\Gamma}'_1 \in \mathcal{B}^+_{N_{\rm in}}$, $\underline{\Gamma}''_1 \in \mathcal{B}^+_{N_{\rm in}}$ such that $\hat{N}_{\rm out}(\underline{\Gamma}'_2, \underline{\Gamma}''_2) = N_{\rm out}$. Then denoting $\underline{\Gamma}' = \underline{\Gamma}'_1 \cup \underline{\Gamma}'_2$,*

$\underline{\Gamma}'' = \underline{\Gamma}_1'' \cup \underline{\Gamma}_2''$

$$\hat{N}_{\text{out}}(\underline{\Gamma}', \underline{\Gamma}'') = N_{\text{out}}, \qquad \underline{\Gamma}_1' = (\underline{\Gamma}')_{\text{in}}, \qquad (\underline{\Gamma}')_{\text{out}} = \underline{\Gamma}_2', \\ (\underline{\Gamma}'')_{\text{in}} = \underline{\Gamma}_1'', \qquad (\underline{\Gamma}'')_{\text{out}} = \underline{\Gamma}_2''. \tag{12.5.4.4}$$

Proof Let $\Gamma \in \underline{\Gamma}'$; then either $c(\Gamma) \subset N_{\text{out}}$ or $c(\Gamma)$ is not connected to N_{out}, otherwise by definition it would contribute to N_{out} and hence it would be contained in N_{out}. As the analogous property holds for $\underline{\Gamma}''$, we have (12.5.4.3) and we see that $(\underline{\Gamma}')_{\text{in}} \in \mathcal{B}^+_{N_{\text{in}}}$, $(\underline{\Gamma}')_{\text{out}} \in \mathcal{B}^+_{N^+_{\text{out}} \cap \Lambda_1}$ (with the analogous property for $\underline{\Gamma}''$). The same arguments prove (12.5.4.4). □

Proposition 12.5.4.1 *Denoting by $\{\hat{N}_{\text{out}} = N_{\text{out}}\}$ the set of all $(\underline{\Gamma}', \underline{\Gamma}'')$ such that $\hat{N}_{\text{out}}(\underline{\Gamma}', \underline{\Gamma}'') = N_{\text{out}}$, $P_{\Lambda_1,\Lambda_2}(\underline{\Gamma}', \underline{\Gamma}''|\{\hat{N}_{\text{out}} = N_{\text{out}}\})$ is equal to*

$$P_{N^+_{\text{out}} \cap \Lambda_1, N^+_{\text{out}} \cap \Lambda_2}((\underline{\Gamma}')_{\text{out}}, (\underline{\Gamma}'')_{\text{out}})|\{\hat{N}_{\text{out}} = N_{\text{out}}\}) P_{N_{\text{in}}, N_{\text{in}}}((\underline{\Gamma}')_{\text{in}}, (\underline{\Gamma}'')_{\text{in}}). \tag{12.5.4.5}$$

Proof We claim that given N_{out} there is a number X such that for any $(\underline{\Gamma}', \underline{\Gamma}'')$ in $\{\hat{N}_{\text{out}} = N_{\text{out}}\}$

$$P_{\Lambda_1,\Lambda_2}(\underline{\Gamma}', \underline{\Gamma}'') = X P_{N^+_{\text{out}} \cap \Lambda_1, N^+_{\text{out}} \cap \Lambda_2}(\underline{\Gamma}_2', \underline{\Gamma}_2'') \times P_{N_{\text{in}}, N_{\text{in}}}(\underline{\Gamma}_1', \underline{\Gamma}_1'') \tag{12.5.4.6}$$

having written $\underline{\Gamma}_i', \underline{\Gamma}_i''$, $i = 1, 2$, for the above decomposition of $\underline{\Gamma}'$ and $\underline{\Gamma}''$. Indeed, $P_{\Lambda_1,\Lambda_2}(\underline{\Gamma}', \underline{\Gamma}'')$ is equal to (omitting the dependence on σ'_{Λ_1} and σ''_{Λ_2})

$$\frac{W^+_\gamma(\underline{\Gamma}_1') W^+_\gamma(\underline{\Gamma}_2')}{X_\gamma(\Lambda_1)} \frac{W^+_\gamma(\underline{\Gamma}_1'') W^+_\gamma(\underline{\Gamma}_2'')}{X_\gamma(\Lambda_2)},$$

which yields (12.5.4.6) with

$$X = \frac{X_\gamma(N_{\text{in}})^2 X_\gamma(N_{\text{out}} \cap \Lambda_1) X_\gamma(N_{\text{out}} \cap \Lambda_2)}{X_\gamma(\Lambda_1) X_\gamma(\Lambda_2)}.$$

We sum both sides of (12.5.4.6) over $(\underline{\Gamma}', \underline{\Gamma}'') \in \{\hat{N}_{\text{out}} = N_{\text{out}}\}$ and rewrite the sum on the r.h.s. by a sum over $\underline{\Gamma}_1' \in \mathcal{B}^+_{N_{\text{in}}}$, $\underline{\Gamma}_1'' \in \mathcal{B}^+_{N_{\text{in}}}$, $\underline{\Gamma}_2' \in \mathcal{B}^+_{N^+_{\text{out}} \cap \Lambda_1}$, $\underline{\Gamma}_2'' \in \mathcal{B}^+_{N^+_{\text{out}} \cap \Lambda_2}$ with the constraint $\hat{N}_{\text{out}}(\underline{\Gamma}_2', \underline{\Gamma}_2'') = N_{\text{out}}$ (having used (12.5.4.3)). We then have $P_{\Lambda_1,\Lambda_2}(\hat{N}_{\text{out}} = N_{\text{out}}) = X P_{N^+_{\text{out}} \cap \Lambda_1, N^+_{\text{out}} \cap \Lambda_2}(\hat{N}_{\text{out}} = N_{\text{out}})$ which is used to identify X. Inserting such a value in (12.5.4.6) we get (12.5.4.5). □

12.5.5 Counting contours

Claim *There are constants c and $\omega > 0$ so that*

$$P_{\Lambda_1,\Lambda_2}\big(\hat{N}_{\text{in}}(\underline{\Gamma}', \underline{\Gamma}'') \not\supset \Delta\big) \le c|\Delta| e^{-\gamma \omega R}. \tag{12.5.5.1}$$

Proof $\hat{N}_{\rm in}(\underline{\Gamma}',\underline{\Gamma}'')\not\supset\Delta$ occurs in Case 1: among all contours $\Gamma\in\underline{\Gamma}'\cup\underline{\Gamma}''$ there is at least one, call it Γ, such that $c(\Gamma)\supset\Lambda$; then $\hat{N}_{\rm int}=\emptyset$. Case 2: if not in Case 1, each maximal connected component of $c(\underline{\Gamma}')\cup c(\underline{\Gamma}'')$ in $\hat{N}_{\rm out}(\underline{\Gamma}',\underline{\Gamma}'')$ which is not contained in Λ^c must intersect $\delta_{\rm in}^{\ell_{+,\gamma}}[\Lambda]$. Then there must be a cube $C\in\delta_{\rm in}^{\ell_{+,\gamma}}[\Lambda]$ and a connected, $\mathcal{D}^{(\ell_{+,\gamma})}$-measurable set D such that $D\supset C$, $D\subset\{{\rm sp}(\underline{\Gamma}')\cup{\rm sp}(\underline{\Gamma}'')\}$ and, recalling (12.5.3.1)–(12.5.3.2), the diameter of D must exceed $R/100$.

Case 1. By (12.5.3.1)–(12.5.3.2) there must be $x\in\Delta\cap\mathbb{Z}^d$ and a contour $\Gamma\in\underline{\Gamma}'\cup\underline{\Gamma}''$ such that $c(\Gamma)\ni x$ with the diameter of ${\rm sp}(\Gamma)$ exceeding $R/100$. The probability for this to happen is bounded by

$$2\sum_{\Gamma:c(\Gamma)\ni 0,|{\rm sp}(\Gamma)|\ge R/100} e^{-\beta(c_1/2)(\zeta^2\ell^d_{-,\gamma})N_\Gamma}$$

(the factor 2 because Γ may either belong to $\underline{\Gamma}'$ or to $\underline{\Gamma}''$); hence

$$\le 2e^{-\beta(c_1/4)(\zeta^2(\ell_{-,\gamma}/\ell_{+,\gamma})^d)R/100}\sum_{\Gamma:c(\Gamma)\ni 0} e^{-\beta(c_1/4)(\zeta^2\ell^d_{-,\gamma})N_\Gamma}$$

$$\le 2e^{-\beta(c_1/4)(\zeta^2(\ell_{-,\gamma}/\ell_{+,\gamma})^d)R/100},$$

having used (9.2.8.4). By summing over all $x\in\Delta\cap\mathbb{Z}^d$, we get (12.5.5.1).

Case 2. Let $C\in\delta_{\rm in}^{\ell_{+,\gamma}}[\Lambda]$; then the probability that there is a set D as specified in Case 2 is bounded by

$$\le\sum_{D\ni 0,|D|\ge R/100} 2^{N_D}\big(3^{(\ell_{+,\gamma}/\ell_{-,\gamma})^d}\big)^{N_D} e^{-\beta(c_1/2)(\zeta^2\ell^d_{-,\gamma})N_D},$$

where N_D is the number of $\mathcal{D}^{(\ell_{+,\gamma})}$-cubes in D. Indeed each cube of D may belong to a contour of $\underline{\Gamma}'$ or of $\underline{\Gamma}''$, and hence the factor 2^{N_D}; the other combinatorial factor counts how many contours there are with given spatial support. By the same argument used in the proof of (9.2.8.4) the above is $\le e^{-\beta(c_1/4)(\zeta^2(\ell_{-,\gamma}/\ell_{+,\gamma})^d)}$. Then summing over all $C\in\delta_{\rm in}^{\ell_{+,\gamma}}[\Lambda]$ we get $\le|\Delta|\frac{|\delta_{\rm out}^{\ell_{+,\gamma}}[C_x]|}{\ell^d_{+,\gamma}}e^{-\beta(c_1/4)(\zeta^2(\ell_{-,\gamma}/\ell_{+,\gamma})^d)}$; hence (12.5.5.1). □

12.5.6 Conclusions

We shall use (12.5.5.1) to prove (12.5.3.4) and (12.5.3.5) and thus Theorem 12.5.3.1; Theorem 12.2.0.1 then follows for what we said just before Theorem 12.5.3.1. It follows from Proposition 12.5.4.1 that if $\hat{N}_{\rm in}\supset\Delta$, then all contours Γ (if any) such that $c(\Gamma)\cap\Delta\neq\emptyset$ are contained in $\hat{N}_{\rm in}$, and therefore $c(\Gamma)$ is disconnected from Λ^c. Thus if $\hat{N}_{\rm in}(\underline{\Gamma}',\underline{\Gamma}'')\supset\Delta$ then $(\underline{\Gamma}',\underline{\Gamma}'')\in A$, A being the random set defined before

Theorem 12.5.3.1, and (12.5.3.4) follows from (12.5.5.1). To prove (12.5.3.5) we write the probability on the r.h.s. of (12.5.4.2) by conditioning on $\{\hat{N}_{\text{out}} = N_{\text{out}}\}$. If $\hat{N}_{\text{out}}(\underline{\Gamma}', \underline{\Gamma}'') \cap \Delta = \emptyset$ the conditional probability vanishes, by Proposition 12.5.4.1; hence (12.5.3.5) follows from (12.5.5.1).

12.6 Notes and reference to Part III

In the sixties Kac introduced the notion of Kac potentials to derive the van der Waals theory of liquid–vapor phase transitions in a statistical mechanics context. The program was carried through in several papers [147–149]. Lebowitz and Penrose extended the results to a more general class of systems introducing and systematically using coarse graining techniques [160] (which have been used extensively in this book). Non-attractive long range forces are considered in [131–133, 138]. All these works deal with the actual limit $\gamma \to 0$, and next order corrections in γ have been studied in [161, 162] and [169]. [69] and [45] are the first works with $\gamma > 0$ kept fixed; they prove phase transitions at small γ for the same Ising system of Chap. 9; in [40] the case of an Ising system with Kac potentials and a reference hamiltonian is considered. The structure of DLR measures is studied in [61].

The scaling properties of the Kac potentials indicate a simple way to make an interaction sufficiently "small and long range" to have mean field behavior: just take γ small enough. The question is completely open if we do not have Kac potentials but just a given interaction. The issue is discussed in [47].

The behavior at the critical point (for Kac potentials) is not covered in the above references and neither in the text. In particular it is not known, I believe, if the Ising model of Chap. 9 at the critical point has a first or a second order phase transition, I understand that the general belief is in the latter. In mean field theory it is of second order, which would seem to indicate that also with Kac potentials it is of second order; in $d = 1$ with interactions which decay as r^{-2} the transition is, however, of first order [3], while it is of second order in the mean field approximation. In this matter the interplay between critical temperature and fluctuations is considered as a very relevant issue; see for instance [176] for experimental evidence. Fluctuations have been proved to be responsible for the shift at finite γ of the critical temperature from the mean field value: in [71] it is proved that in $d = 2$, $\beta_c(\gamma) \geq 1 + c\gamma^2 \log \gamma^{-1}$, 1 the mean field critical value, $c > 0$ being related to the covariance of the magnetization density fluctuations.

A more complex phase diagram appears in the Potts model, i.e. a generalization of Ising with the Potts spin $\sigma(x)$ which can take $Q \geq 3$ values [35]. In its simplest version the spin–spin interaction is $J(x, y)\mathbf{1}_{\sigma(x) \neq \sigma(y)}$, i.e. each pair of distinct spins pays a penalty $J(x, y)$. The case where J is a Kac potential has been studied in [129] where it is proved that at the critical temperature the phase transition is of first order with coexistence of Q ordered states (where the density of a spin is larger than the others) and a disordered phase where instead all spins have the same density. The result holds in $d \geq 2$ dimensions for γ small enough. Since the n.n. Potts model with $Q = 3$ is supposed to have a second order phase transition in $d = 2$, this would mean that there is a phase transition triggered by the range of the interaction. It may thus happen in such a scenario that when the range changes from say 100 to 101 we go from a second to a first order phase transition! The mean field behavior of the Potts model has also been observed at large spatial dimensions and for long range reflection positive interactions [34, 36]. The general question of the mean field behavior in a system with long range interactions is discussed in [77]. Phase transitions for an Ising system with Kac potentials and a periodic magnetic field is proved in [33].

While in the limit $\gamma \to 0$ phase transitions occur at all dimensions, when γ is fixed the general theory tells us that the Ising ferromagnetic Kac potential of Chap. 9 has no phase transitions. The typical configurations at $\beta > 1$, however, have a non-trivial spatial structure, where the coarse grained spin magnetization has long strings where (modulo fluctuations) the $+m_\beta$ and $-m_\beta$ values alternate, the typical length of the strings scales like $e^{c\gamma^{-1}}$, $c > 0$, the instanton free energy of the L–P functional [70]. The typical configurations in $d = 1$ when a random external magnetic field acts on the system have been characterized in [72, 73]. It is proved that the length of the strings reduces drastically from exponential to order of γ^{-2} and the lengths of the strings are to leading order deterministic (determined by the realizations of the magnetic field), while in the absence of the field, they are random (and, in a suitable limit, independent of each other). Typical configurations of rotators in $d = 1$ have been studied in [60].

The first proof of phase transitions in the continuum is due to Ruelle [198], with his analysis of the Widom–Rowlinson model, where he shows that there are two distinct DLR measures when the chemical potential is large enough. In its simplest version the Widom–Rowlinson model consists of two species of particles with a hard core repulsion among particles of different species, while particles of the same species do not interact with each other. Ruelle proved that at large densities the Peierls argument in the Ising model can be extended to such a case. As in the Ising case, the proof exploits the symmetry between the two species. An extension to systems where the symmetry is absent has been considered in [54, 55], where a generalization of the Pirogov–Sinai theory has been proposed. A more probabilistic-framed version of Ruelle's proof is in [78], and extensions to the quantum case (with Boltzmann statistics) are in [68] and [145].

The first result, I believe, on the occurrence of a phase transition in $d \geq 2$ in a one component system is due to Lebowitz, Mazel and Presutti in [164]. The analysis is based on an extension of the Pirogov–Sinai theory. The classical reference for the Pirogov–Sinai approach is Sinai's book on phase transitions [201]. See also Zahradnik [209, 210], and the already mentioned works by Bricmont, Kuroda, Lebowitz for an extension of the theory to models in the continuum [54, 55]. In the present book, in Chaps. 10–11, I have presented a more detailed analysis than in the original paper [164]; it is an elaboration of unpublished notes of the author for a course he gave at the Institute H. Poincaré in Paris (June–July 1998). The analysis of the tail field and the results on the ergodic decomposition in Chap. 12 are original; the Gibbs phase rule, Theorem 12.0.0.1, is proved in [48]. The quantum version of the LMP model has been studied in [12, 13]; the two papers are currently under revision.

The two species Widom–Rowlinson model is a continuum version of the lattice Ising model; in the same way, once extended to $Q \geq 3$ species, it becomes a continuum version of the lattice Potts model. The existence of Q distinct (ordered) states at large density has been proved in [136]. With respect to its lattice analogue, the Potts model has a much richer structure, because, besides the densities of each spin, in the continuum the total density is also an order parameter (on the lattice the total density is 1). A lattice gas version in which also vacant sites are allowed has been studied in [79] using reflection positivity techniques. The mean field behavior of the continuum Potts model has been characterized in [137]. The analysis has been extended to Kac potentials in [102, 103] with a proof that at the critical temperatures there are (as in the lattice case) $Q+1$ DLR measures, with Q ordered states (each one with a dominant spin density) and the disordered phase, where all spins have the same density. The new feature of the continuum model is that the total density has a discontinuity going through the critical point as the magnetization is larger in the ordered than in the disordered state. The effect is known in physics as magnetostriction. It states that the lattice spacing shrinks when the system gets magnetized, which in the continuum Potts model translates into an increase of density when ordered states appear. The analysis requires a remarkable technical improvement due to the fact that the relativized Dobrushin uniqueness condition, which holds in LMP (when $\beta \in (\beta_c, \beta_0)$), is no longer satisfied. The proof in [102] involves ideas in the spirit of the Dobrushin–Shlosman theory [110]. It seems plausible that a similar approach may be used in LMP as well for $\beta \geq \beta_0$.

Phase transitions occur also in $d = 1$ if the interaction decays at least as slow as r^{-2}. A proof in the continuum that the thermodynamics of the system has a phase transition has been obtained by Johansson [146]. The analysis of DLR measures with Kac potentials which decay as $r \to \infty$ with $r^{-2+\alpha}$, $\alpha \in [0, 1)$, is open, also in the lattice. It would be interesting to examine the nature of the critical point which Aizenman, the Chayes's and Newman [3], have proved to be of first order in the lattice with an interaction which decays as r^{-2}. In mean field theory instead it is of second order, and it would therefore be interesting to determine what happens with Kac potentials when γ is finite. Another interesting issue concerns the behavior in the presence of random magnetic fields. It is believed that with the parameter α one can simulate the spatial dimensions $\alpha \leq 1/2$ corresponding to $d \geq 3$. In the Ising model with random magnetic fields, the critical dimension is $d = 3$, which is the minimal dimension for the occurrence of a phase transition; see Aizenman and Wehr [2], Bricmont and Kupianen [53], and the recent book by Bovier [44]. In a recent paper, Cassandro, Orlandi and Picco [74], have proved, in agreement with such a belief, that indeed there is a phase transition in the lattice with a random magnetic field at some $\alpha > 1/2$.

Part IV
Appendices

Appendix A
Ising model, topology and measure theory

In Sect. A.1, we collect notation and definitions on translations in Ising spaces; in Sect. A.2 we recall the Stirling formula; in Sect. A.3 topological properties of Ising spaces; in Sects. A.4 and A.5 some basic notions of probability which are then used, from Sect. A.6, on to study conditional probabilities. In this part I follow Rohlin [194]. Since any probability in a Polish space can be represented as a probability on $\mathcal{X}$, the theory extends to such spaces, "abstract Lebesgue measures" in Rohlin's terminology.

A.1 The group of translations

We denote by $\tau_i : \mathbb{Z}^d \to \mathbb{Z}^d$, $i \in \mathbb{Z}^d$, the map which translates by i:

$$\tau_i(x) = x + i, \quad x \in \mathbb{Z}^d \tag{A.1.1}$$

and write $\tau_i(\Delta) = \Delta + i = \{y \in \mathbb{Z}^d : y = x + i, x \in \Delta\}$, $\Delta \subset \mathbb{Z}^d$. τ_i is invertible and $\tau_i^{-1} = \tau_{-i}$; more generally $\tau_{i+j} = \tau_i \circ \tau_j$, where $\circ$ denotes composition.

As a general rule, if ϕ is an invertible map from a space X onto a space Y, then the functions f on X are transformed into functions $\phi(f)$ on Y by the relation

$$\phi(f)(\phi(x)) = f(x), \qquad \phi(f)(y) = f(\phi^{-1}(y)). \tag{A.1.2}$$

By applying (A.1.2) to τ_i and considering the configurations σ as functions on $\mathbb{Z}^d$,

$$\tau_i(\sigma)(\tau_i(x)) = \sigma(x), \qquad \tau_i(\sigma)(x) = \sigma(x - i). \tag{A.1.3}$$

By applying again (A.1.2) to a next level, if f is a function on $\mathcal{X}$,

$$\tau_i(f)(\tau_i(\sigma)) = f(\sigma), \qquad \tau_i(f)(\sigma) = f(\tau_i^{-1}(\sigma)); \tag{A.1.4}$$

while, if μ is a probability and f a bounded measurable function,

$$\tau_i(\mu)(\tau_i(f)) = \mu(f), \qquad \tau_i(\mu)(f) = \mu(\tau_i^{-1}(f)). \tag{A.1.5}$$

Let f be a cylindrical function on Δ; then $\tau_i(f)$ is cylindrical on $\tau_i(\Delta)$. To prove it, we need to show that $\tau_i(f)(\sigma) = \tau_i(f)(\sigma')$ whenever $\sigma = \sigma'$ on $\tau_i(\Delta)$. For any such σ and σ' we have for any $x \in \Delta$

$$\tau_i^{-1}(\sigma)(x) = \sigma(x + i) = \sigma'(x + i) = \tau_i^{-1}(\sigma')(x).$$

Thus, we have $\tau_i^{-1}(\sigma) = \tau_i^{-1}(\sigma')$ on Δ, and hence

$$\tau_i(f)(\sigma) = f(\tau_i^{-1}(\sigma)) = f(\tau_i^{-1}(\sigma')) = \tau_i(f)(\sigma').$$

The above applies also to $H_\Delta(\sigma_\Delta|\sigma'_{\Delta^c})$ regarded as a function of the two variables σ and σ' (and cylindrical respectively on Δ and Δ^c). Then

$$\tau_i(H_\Delta)(\sigma|\sigma') = H_\Delta\big([\tau_i^{-1}(\sigma)]_\Delta|[\tau_i^{-1}(\sigma')]_{\Delta^c}\big) = H_{\tau_i(\Delta)}(\sigma|\sigma'), \tag{A.1.6}$$

having used, in the last equality, that the interaction is translational invariant.

Denoting by δ_{σ^*} the probability supported by the single configuration σ^*,

$$\tau_i(\delta_{\sigma^*})(f) = \delta_{\sigma^*}(\tau_i^{-1}(f)) = f(\tau_i(\sigma^*)) = \delta_{\tau_i(\sigma^*)}(f). \tag{A.1.7}$$

Writing

$$G_\Delta(f|\sigma_{\Delta^c}) = \sum_{\sigma'} c_{\Delta,\sigma}(\sigma')\delta_{\sigma'}(f), \quad c_{\Delta,\sigma}(\sigma') = \frac{e^{-\beta H_\Delta(\sigma'|\sigma)}}{Z_\Delta(\sigma)}, \tag{A.1.8}$$

we have

$$\tau_i(G_\Delta(f|\sigma_{\Delta^c})) = \sum_{\sigma'} c_{\Delta,\sigma}(\sigma')\delta_{\tau_i(\sigma')}(f) = \sum_{\sigma''} c_{\Delta,\sigma}(\tau_i^{-1}(\sigma''))\delta_{\sigma''}(f);$$

and, using (A.1.6) and translational invariance of the interaction,

$$\begin{aligned} c_{\Delta,\sigma}(\tau_i^{-1}(\sigma'')) &= \frac{e^{-\beta H_\Delta(\tau_i^{-1}(\sigma'')|\tau_i^{-1}(\tau_i(\sigma))}}{Z_\Delta(\sigma)} \\ &= \frac{e^{-\beta H_{\tau_i(\Delta)}(\sigma''|\tau_i(\sigma))}}{Z_\Delta(\sigma)} = c_{\tau_i(\Delta),\tau_i(\sigma)}(\sigma''), \end{aligned}$$

so that

$$\tau_i(G_\Delta(\cdot|\sigma_{\Delta^c})) = G_{\tau_i(\Delta)}(\cdot|(\tau_i(\sigma)_{\tau_i(\Delta)^c}). \tag{A.1.9}$$

More generally,

$$\tau_i(\mathcal{G}_\Delta) = \mathcal{G}_{\tau_i(\Delta)}. \tag{A.1.10}$$

Indeed, if $\mu \in \mathcal{G}_\Delta$,

$$\begin{aligned} \tau_i(\mu)(f) = \mu(\tau_i^{-1}(f)) &= \int G_\Delta(\tau_i^{-1}(f)|\sigma_{\Delta^c}))\mu(d\sigma) \\ &= \int (\tau_i G_\Delta(\cdot|\sigma_{\Delta^c}))(f)\mu(d\sigma) = \int G_{\tau_i(\Delta)}(f|\tau_i(\sigma)_{\tau_i(\Delta)^c})\mu(d\sigma) \\ &= \int G_{\tau_i(\Delta)}(f|\sigma_{\tau_i(\Delta)^c})\tau_i(\mu)(d\sigma), \end{aligned}$$

where we have used that $\int h(\tau_i(\sigma))\mu(d\sigma) = \mu(\tau^{-1}(h)) = \tau_i(\mu)(h)$.

A.2 Stirling formula

The Stirling formula, extensively used in the text, is

$$n! = n^{n+1/2}e^{-n}\sqrt{2\pi}\left(1+0\left(\frac{1}{\sqrt{n}}\right)\right). \tag{A.2.1}$$

We will use (A.2.1) to estimate

$$\binom{n}{k} =: e^{nS_n(k/n)} \tag{A.2.2}$$

as n and k diverge. Here, $nS_n(k/n)$ is an entropy and $S_n(k/n)$ the corresponding entropy density. By (A.2.1)

$$\binom{n}{k} = \frac{e^{nS(k/n)}}{[2\pi n\,\frac{k}{n}(1-\frac{k}{n})]^{1/2}}\left(1+0\left(n^{-1/2},k^{-1/2},(n-k)^{-1/2}\right)\right), \tag{A.2.3}$$

where

$$S(x) = -x\log x - (1-x)\log(1-x), \quad x \in [0,1]. \tag{A.2.4}$$

To compare $S(k/n)$ and $S_n(k/n)$ we first observe that $S_n(x) = S(x)$ for $x = 0, 1$. For the other values of k/n, (A.2.3) gives

$$\left|nS_n(k/n) - nS(k/n)\right| \le c\log n, \tag{A.2.5}$$

with c a constant. For $m \in M_\Lambda = \{-1, -1+2/|\Lambda|, \dots, 1-2/|\Lambda|, 1\}$, we have

$$e^{|\Lambda|I_{|\Lambda|}(m)} = \text{card}\left\{\sigma_\Lambda \in \mathcal{X}_\Lambda : \sum_{x\in\Lambda}\sigma_\Lambda(x) = m|\Lambda|\right\}. \tag{A.2.6}$$

Then

$$I_{|\Lambda|}(m) = S_{|\Lambda|}\left(\frac{m+1}{2}\right), \tag{A.2.7}$$

and setting

$$I(m) = -\frac{1-m}{2}\log\frac{1-m}{2} - \frac{1+m}{2}\log\frac{1+m}{2}, \tag{A.2.8}$$

$$\sup_{m\in M_\Lambda}\left||\Lambda|I_{|\Lambda|}(m) - |\Lambda|I(m)\right| \le c\log|\Lambda|. \tag{A.2.9}$$

A.3 Ising model. Topological properties

We regard $\mathcal{X} = \{-1,1\}^{\mathbb{Z}^d}$ as a topological space with the product topology, which means that a sequence $\sigma^{(n)} \to \sigma$ if and only if for any $x \in \mathbb{Z}^d$ there is n_x, so that

$\sigma^{(n)}(x) = \sigma(x)$ for all $n \geq n_x$. By the Tykonoff theorem, Theorem I.8.5 in [114], $\mathcal{X}$ is a compact space. We denote by $C(\mathcal{X})$ the Banach space of continuous functions on $\mathcal{X}$ with sup norm, and we call the functions on $\mathcal{X}$ cylindrical which depend on finitely many spins, i.e. such that $f(\sigma_\Delta, \sigma_{\Delta^c}) = f(\sigma_\Delta, \sigma'_{\Delta^c})$ for all $\sigma_\Delta \in \mathcal{X}_\Delta$, $\sigma_{\Delta^c}, \sigma'_{\Delta^c} \in \mathcal{X}_{\Delta^c}$, f is then called cylindrical in Δ. Cylindrical sets are sets whose characteristic functions are cylindrical. Elementary cylinders are the cylindrical sets whose characteristic functions have the form $\mathbf{1}_{\sigma_\Delta}$, Δ bounded, $\sigma_\Delta \in \mathcal{X}_\Delta$. $\mathcal{C}$ denotes the collection of all the elementary cylindrical sets.

Theorem A.1 *All cylindrical functions are continuous; cylindrical sets are both open and closed. The set of all cylindrical functions is dense in $C(\mathcal{X})$, and the collection of elementary cylinders is a basis of open sets in $\mathcal{X}$.*

Proof Let f be a cylindrical function in Δ and $\sigma^{(n)} \to \sigma$. Then there is n_Δ so that $\sigma^{(n)}(x) = \sigma(x)$ for all $n \geq n_\Delta$; hence $f(\sigma^{(n)}) = f(\sigma)$ for all $n \geq n_\Delta$. Thus $f \in C(\mathcal{X})$. The inverse images of the values of f are cylindrical sets, and since f is continuous they are both open and closed. Hence any cylindrical set is both open and closed.

It directly follows from the definition of convergence of sequences $\sigma^{(n)}$ that the collection of elementary cylinders is a basis of open sets in $\mathcal{X}$. The collection of all cylindrical functions is a closed sub-algebra in $C(\mathcal{X})$ which contains the identity and separates the points; then by Stone's theorem, IV.6.16 in [114], it is dense in $C(\mathcal{X})$. □

For any σ and σ' in $\mathcal{X}$ and $R > 0$, define

$$\begin{aligned} d(\sigma, \sigma') &:= \sum_{x \in \mathbb{Z}^d} e^{-|x|} |\sigma(x) - \sigma'(x)|, \\ B_R(\sigma) &= \{\sigma' \in \mathcal{X} : d(\sigma, \sigma') < R\}. \end{aligned} \tag{A.3.1}$$

Theorem A.2 *$d(\sigma, \sigma')$ is a metric in $\mathcal{X}$, and the topology induced by $d(\sigma, \sigma')$ is equivalent to the product topology of $\mathcal{X}$; namely, given any $\sigma \in \mathcal{X}$, for any $\epsilon > 0$, there is a cylinder $C \in \mathcal{C}$ such that $\sigma \in C$ and $C \subset B_\epsilon(\sigma)$; vice versa, for any cylinder $C \in \mathcal{C}$ such that $\sigma \in C$, there is $\epsilon > 0$ so that $B_\epsilon(\sigma) \subset C$.*

Proof It directly follows from (A.3.1) that $d(\sigma, \sigma') = d(\sigma', \sigma)$, $d(\sigma, \sigma'') \leq d(\sigma, \sigma') + d(\sigma', \sigma'')$ and that $d(\sigma, \sigma') = 0$ iff $\sigma = \sigma'$; thus d is a metric and the sets $B_{n^{-1}}(\sigma)$ are a countable basis of open sets for the topology induced by the metric $d(\cdot, \cdot)$. Given σ and $\epsilon > 0$, let Λ be a bounded set in $\mathbb{Z}^d$ such that $\sum_{x \notin \Lambda} 2e^{-|x|} < \epsilon$. Call $C_{\sigma_\Lambda} = \{\sigma' : \sigma'_\Lambda = \sigma_\Lambda\}$ then $C_{\sigma_\Lambda} \subset B_\epsilon(\sigma)$. Vice versa, given any Λ and C_{σ_Λ}, let $\epsilon = \min\{e^{-|x|}, x \in \Lambda\}$; then $B_\epsilon(\sigma) \subset C_{\sigma_\Lambda}$. □

Definition

- *σ-algebras.* A σ-algebra Σ is a collection of subsets of $\mathcal{X}$ such that $\mathcal{X} \in \Sigma$, Σ is closed under countable unions and intersections and the complement A^c of

any set $A \in \Sigma$ is also in Σ. $\mathcal{B}$ denotes the Borel σ-algebra of $\mathcal{X}$, namely the σ-algebra generated by $\mathcal{C}$ which means the minimal σ-algebra which contains all the elementary cylinders. $\mathcal{B}_\Delta$ denotes the σ-algebra generated by the cylinders with basis in Δ, Δ possibly unbounded.
- *Monotone classes.* A family $\mathcal{S}$ of Borel sets is "a monotone class" if (i) $\mathcal{X} \in \mathcal{S}$; (ii) if A_1 and A_2 are in $\mathcal{S}$ and $A_1 \subset A_2$, then also $A_2 \setminus A_1$ is in $\mathcal{S}$; (iii) if $\{A_n\}$ is an increasing sequence of elements of $\mathcal{S}$, then $\bigcup A_n \in \mathcal{S}$.

A proof of the following theorem can be found in Chap. 0, Sect. 2 of the book by Revuz–Yor [189].

Theorem A.3 (Monotone class theorem) *If a collection of Borel sets $\mathcal{R}$ is closed under finite intersections, then the smallest monotone class which contains $\mathcal{R}$ contains the minimal σ-algebra which contains $\mathcal{R}$.*

A.4 Space of probabilities

In the sequel, by default a probability always means a σ-additive, positive measure μ on the Borel σ-algebra $\mathcal{B}$ of $\mathcal{X}$, namely $\mu : \mathcal{B} \to [0, 1]$, $\mu(\mathcal{X}) = 1$, $\mu(\emptyset) = 0$ and for any sequence $\{A_i\}$, $A_i \in \mathcal{B}$, $A_i \cap A_j = \emptyset$ whenever $i \neq j$,

$$\mu\Big(\bigcup_i A_i\Big) = \sum_i \mu(A_i). \tag{A.4.1}$$

By abuse of notation we will write $\mu(f)$, $f \in L^1(\mathcal{X}, \mu)$, for the integral of f with respect to μ with the usual convention of writing $\mu(f) = \int f \, \mu(d\sigma)$, even when f is defined on a subset of $\mathcal{X}$ which has μ-measure 1. We will say that $\mu_n \to \mu$ weakly, if $\mu_n(f) \to \mu(f)$, for all $f \in C(\mathcal{X})$, $C(\mathcal{X})$ being the space of all continuous functions on $\mathcal{X}$. We will denote by $\mu(f)$ the integral of f with respect to μ.

Theorem A.4 *The following two statements hold.*

- *$\mu_n \to \mu$ weakly if and only if $\mu_n(C) \to \mu(C)$ for all $C \in \mathcal{C}$.*
- *The space of all probabilities, denoted by $M(\mathcal{X})$, is weakly compact.*

Proof If $\mu_n \to \mu$ weakly, then $\mu_n(C) = \mu_n(\mathbf{1}_C) \to \mu(\mathbf{1}_C) = \mu(C)$, because $\mathbf{1}_C$, being a cylindrical function, is continuous. Vice versa, by Theorem A.1 any continuous function can be approximated in sup norm by a cylindrical function; hence by a linear combination of characteristic functions of elementary cylinders. Then, if $\mu_n(C) \to \mu(C)$, $C \in \mathcal{C}$, also $\mu_n(f) \to \mu(f)$ for any $f \in C(\mathcal{X})$, so that $\mu_n \to \mu$ weakly. The statement that $M(\mathcal{X})$, is weakly compact is Theorem 6.4 in [179]; recall that, by Theorem A.2, $\mathcal{X}$ can be thought of as a compact metric space. □

A.5 Extensions theorems

We call a function $m_0 : \mathcal{C} \to [0,1]$ a pre-probability, if, for any bounded set $\Lambda \subset \mathbb{Z}^d$,

$$\sum_{\sigma_\Lambda} m_0(\{\sigma' : \sigma'_\Lambda = \sigma_\Lambda\}) = 1, \tag{A.5.1}$$

and, for any $\Delta \subset \Lambda$ and any σ_Δ

$$\sum_{\sigma_{\Lambda\setminus\Delta}} m_0(\{\sigma' : \sigma'_{\Lambda\setminus\Delta} = \sigma_{\Lambda\setminus\Delta}; \sigma'_\Delta = \sigma_\Delta\}) = m_0(\{\sigma' : \sigma'_\Delta = \sigma_\Delta\}), \tag{A.5.2}$$

$m_0 : \mathcal{A} \to [0,1]$, where $\mathcal{A}$ is the algebra of cylinders (whose elements are finite unions of elementary cylinders), is an additive probability, if $m_0(\mathcal{X}) = 1$, $m_0(\emptyset) = 0$, and

$$m_0\left(\bigcup_{i=1}^n A_i\right) = \sum_{i=1}^n \mu(A_i) \tag{A.5.3}$$

for any finite sequence of mutually disjoint sets $A_i \in \mathcal{A}$.

Theorem A.5 *Any pre-probability extends uniquely to an additive probability on $\mathcal{A}$, which, in turn, extends uniquely to a probability.*

Proof The compatibility conditions (A.5.1)–(A.5.2) allow one to extend uniquely a pre-probability to an additive probability on $\mathcal{A}$. The further extension to a probability is granted by the Caratheodory theorem. This states that an additive, normalized measure m on an algebra $\mathcal{U}$ extends uniquely to the minimal σ-algebra which contains $\mathcal{U}$, if, for any sequence B_n of mutually disjoint sets in $\mathcal{U}$ whose union B is in also in $\mathcal{U}$, it happens that

$$m(B) = \sum_{n=1}^{\infty} m(B_n). \tag{A.5.4}$$

Equation (A.5.4) is automatically satisfied in our case, because any set in $B \in \mathcal{A}$ can only be written as a finite union of disjoint elements of $\mathcal{A}$. In fact, $\mathcal{X}$ is compact and any cylinder (i.e. any set in $\mathcal{A}$) is simultaneously closed and open. □

A.6 Existence of conditional probabilities

Let μ be a probability and Σ a σ-algebra contained in $\mathcal{B}$. We call a pair $(\mathcal{X}', \{\mu(\cdot|\Sigma)(\sigma), \sigma \in \mathcal{X}'\})$ a "version of the conditional probability of μ given Σ," where $\mathcal{X}' \subset \mathcal{X}$ and for any $\sigma \in \mathcal{X}'$, $\mu(\cdot|\Sigma)(\sigma)$ is a probability on $\mathcal{X}$. Moreover, the pair must be such that the following holds.

- $\mathcal{X}' \in \Sigma$ and $\mu(\mathcal{X}') = 1$.

- For any $A \in \mathcal{B}$, $\mu(A|\Sigma)(\sigma)$ is $\Sigma \cap \mathcal{X}'$-measurable and, for any $B \in \Sigma$,

$$\mu(A \cap B) = \int_{B \cap \mathcal{X}'} \mu(A|\Sigma)(\sigma)\,\mu(d\sigma). \tag{A.6.1}$$

By a density argument, which is omitted, it then follows that for any bounded measurable function f, the integral $\mu(f|\Sigma)(\cdot)$ is $\Sigma \cap \mathcal{X}'$-measurable and, for any $B \in \Sigma$,

$$\mu(f\mathbf{1}_B) = \int_{B \cap \mathcal{X}'} \mu(f|\Sigma)(\sigma)\,\mu(d\sigma). \tag{A.6.2}$$

The basic example Let Σ have finitely many elements, say $A_1, \ldots, A_k$. Using the notation, $A_n^1 = A_n$ and $A_n^0 = A_n^c$, we consider the family of sets

$$\left\{ C(u) := \bigcap_{n=1}^{k} A_n^{u_n},\ u = (u_1, \ldots, u_k) \in \{0, 1\}^k \right\}. \tag{A.6.3}$$

After excluding those which are empty, (A.6.3) defines a partition π of $\mathcal{X}$ and any $A_i \in \Sigma$ is a union of some of the atoms of π. For any probability μ call $C_1, \ldots, C_n$ the atoms of π (ordered arbitrarily) which have positive measure. Call $\pi(\sigma) = C_i$ if $\sigma \in C_i$, and define

$$\mu(A|\Sigma)(\sigma) := \frac{\mu(A \cap \pi(\sigma))}{\mu(\pi(\sigma))} \equiv \mu\big(A|\pi(\sigma)\big); \tag{A.6.4}$$

$(\mathcal{X}, \{\mu(\cdot|\Sigma)(\sigma), \sigma \in \mathcal{X}\})$ is obviously a version (the canonical one) of the conditional probability of μ given Σ.

As we shall see, this example is typical in the sense that the conditional probability is μ-modulo 0 made by measures each one supported by an atom of a partition of $\mathcal{X}$. However, the partition has in general uncountably many atoms, but they are "countably generated."

Definition (Measurable partitions) π is a measurable partition if there is a sequence $\{A_i, i \in \mathbb{N}\}$ of elements of $\mathcal{B}$, such that the atom $\pi(\sigma)$ of π which contains σ is

$$\pi(\sigma) := \bigcap_{n=1}^{\infty} A_n^{u_n}, \quad u = (u_1, \ldots, u_k, \ldots) \in \{0, 1\}^{\mathbb{N}}.$$

We then say that π is generated by $\{A_i, i \in \mathbb{N}\}$.

Theorem A.6 *For any probability μ and any σ-algebra $\Sigma \subset \mathcal{B}$, there exists a version $(\mathcal{X}', \mu(\cdot|\Sigma)(\cdot))$ of the conditional probability of μ given Σ. Moreover, for any $B \in \Sigma$,*

$$\mu(B|\Sigma)(\sigma) = \mathbf{1}_{\sigma \in B}, \quad \textit{for } \mu\textit{-almost all } \sigma \in \mathcal{X}'. \tag{A.6.5}$$

Proof For any $A \in \mathcal{B}$, consider $B \to \mu(A \cap B)$, $B \in \Sigma$, as a measure on the σ-algebra Σ. This measure is absolutely continuous with respect to the marginal of μ on Σ, i.e. the restriction of μ to Σ. Then, by the Radon–Nikodym theorem [114], it has a density; namely, there exists $\mathcal{X}_A \in \Sigma$, $\mu(\mathcal{X}_A) = 1$, and a non-negative function $D_A(\sigma)$, $\sigma \in \mathcal{X}_A$, which is Σ-measurable; and it is such that

$$\mu(A \cap B) = \int_{B \cap \mathcal{X}_A} D_A(\sigma)\, \mu(d\sigma). \tag{A.6.6}$$

$D_A(\sigma)$ is called the Radon–Nikodym derivative of $\mu(A \cap \cdot)$ with respect to $\mu(\cdot)$.

Claim *There is a set $\mathcal{X}' \in \Sigma$ such that $\mu(\mathcal{X}') = 1$; $\mathcal{X}' \subset \mathcal{X}_C$ for any $C \in \mathcal{C}$; for any $\sigma \in \mathcal{X}'$, $\{D_C(\sigma), C \in \mathcal{C}\}$ is a pre-probability.*

Proof of the claim We start by proving that if A_1 and A_2 are disjoint, then, calling $\mathcal{X}_{A_1,A_2} = \mathcal{X}_{A_1} \cap \mathcal{X}_{A_2} \cap \mathcal{X}_{A_1 \cup A_2}$,

$$\mu\big(\{\sigma \in \mathcal{X}_{A_1,A_2} : D_{A_1}(\sigma) + D_{A_2}(\sigma) = D_{A_1 \cup A_2}(\sigma)\}\big) = 1. \tag{A.6.7}$$

Call $B_+ = \{\sigma \in \mathcal{X}_{A_1,A_2} : D_{A_1 \cup A_2}(\sigma) > D_{A_1}(\sigma) + D_{A_2}(\sigma)\}$ and B_- the set with the reverse inequality. Then $B_\pm \in \Sigma$, and

$$\begin{aligned} 0 &= \mu\big((A_1 \cup A_2) \cap B_\pm\big) - \mu(A_1 \cap B_\pm) - \mu(A_2 \cap B_\pm) \\ &= \int_{B_\pm \cap \mathcal{X}_{A_1,A_2}} \big(D_{A_1 \cup A_2} - D_{A_1} - D_{A_2}\big)\, \mu(d\sigma); \end{aligned}$$

hence $\mu(B_\pm) = 0$, and (A.6.7) is proved.

We define $\mathcal{X}'$ by taking intersections of all sets $\mathcal{X}_{A_1,A_2}$ as above with A_1 and A_2 in $\mathcal{C}$. Since there are countably many intersections, $\mathcal{X}' \in \Sigma$ and $\mu(\mathcal{X}') = 1$. Equations (A.5.1)–(A.5.2) then follow directly from (A.6.7), and we have completed the proof of the claim. □

By the claim and the extension theorem from pre-probabilities to probabilities, we conclude that there is a probability $\mu(\cdot|\Sigma)(\sigma)$, $\sigma \in \mathcal{X}'$, which extends $D_C(\sigma)$, namely such that

$$\mu(C|\Sigma)(\sigma) = D_C(\sigma), \quad \text{for all } C \in \mathcal{C}. \tag{A.6.8}$$

We have thus proved that $\mathcal{X}' \in \Sigma$, $\mu(\mathcal{X}') = 1$, $\sigma \to \mu(C|\Sigma)(\sigma)$ is Σ-measurable for any $C \in \mathcal{C}$ and, for all $\sigma \in \mathcal{X}'$, (A.6.8) holds.

By (A.6.6) and (A.6.8)

$$\mu(C \cap B) = \int_{B \cap \mathcal{X}'} \mu(C|\Sigma)(\sigma)\, \mu(d\sigma), \quad \text{for all } C \in \mathcal{C} \text{ and all } B \in \Sigma. \tag{A.6.9}$$

Let $\mathcal{S}$ be the collection of all sets A such that $\sigma \in \mathcal{X}' \to \mu(A|\Sigma)(\sigma)$ is Σ-measurable and (A.6.1) holds. By (A.6.9), $\mathcal{S} \supset \mathcal{C}$. Moreover, if $\mathcal{S}$ contains A and $A' \subset A$, it contains $A \setminus A'$ as well, as we are going to show. We have

$$\mu(A|\Sigma)(\sigma) = \mu(A'|\Sigma)(\sigma) + \mu(A \setminus A'|\Sigma)(\sigma), \quad \sigma \in \mathcal{X}',$$

because $\mu(\cdot|\Sigma)(\sigma)$ is a probability. Then $\mu(A \setminus A'|\Sigma)(\sigma)$ is $\Sigma \cap \mathcal{X}'$-measurable, as a difference of two such functions, and

$$\begin{aligned}\mu\big((A \setminus A') \cap B\big) &= \mu(A \cap B) - \mu(A' \cap B) \\ &= \int_{B \cap \mathcal{X}'} \mu(A|\Sigma)(\sigma) - \mu(A'|\Sigma)(\sigma)\, \mu(d\sigma) \\ &= \int_{B \cap \mathcal{X}'} \mu(A \setminus A'|\Sigma)(\sigma)\, \mu(d\sigma).\end{aligned}$$

By a similar argument, if $\mathcal{S}$ contains an increasing sequence A_n, it contains $\bigcup_n A_n$ as well; hence $\mathcal{S}$ is a monotone class which contains $\mathcal{C}$ and, by Theorem A.3, $\mathcal{S} = \mathcal{B}$. We have thus proved (A.6.1) and hence that $\{\mu(\cdot|\Sigma)(\sigma), \mathcal{X}'\}$ is a version of the conditional probability of μ given Σ.

Finally, to prove (A.6.5), we take B and $B' \in \Sigma$, and since $\mu(B \cap B') = \mu(B \cap B' \cap \mathcal{X}')$,

$$\mu(B \cap B') = \int_{B' \cap \mathcal{X}'} \mathbf{1}_B(\sigma)\, \mu(d\sigma).$$

Now use (A.6.1) to rewrite it as

$$\mu(B \cap B') = \int_{B' \cap \mathcal{X}'} \mu(B|\Sigma)(\sigma)\, \mu(d\sigma),$$

so that

$$0 = \int_{B' \cap \mathcal{X}'} \{\mathbf{1}_B(\sigma) - \mu(B|\Sigma)(\sigma)\}\, \mu(d\sigma).$$

By taking $B' = B$, $\mu(B|\Sigma)(\sigma) = 1$ for μ almost all $\sigma \in B$; by taking $B' = B^c$, we have

$$0 = \int_{B^c \cap \mathcal{X}'} \mu(B|\Sigma)\, \mu(d\sigma),$$

so that $\mu(B|\Sigma)(\sigma) = 0$ for μ almost all $\sigma \in B^c$. Hence we have (A.6.5). □

A.7 Uniqueness

We will prove here that the conditional probabilities are modulo zero uniquely defined.

Theorem A.7 *If $(\mathcal{X}', \mu'(\cdot|\Sigma)(\sigma))$ and $(\mathcal{X}'', \mu''(\cdot|\Sigma)(\sigma))$ are two versions of the conditional probability of μ, given Σ, then*

$$\mu(\mathcal{X}_0) = 1, \quad \mathcal{X}_0 = \{\sigma \in \mathcal{X}' \cap \mathcal{X}'' : \mu'(\cdot|\Sigma)(\sigma) = \mu''(\cdot|\Sigma)(\sigma)\}, \tag{A.7.1}$$

and $(\mathcal{X}^0, \mu'(\cdot|\Sigma)(\sigma))$ is also a version of the conditional probability of μ given Σ.

Proof For any $C \in \mathcal{C}$ and $B \in \Sigma$,

$$\int_{B\cap\mathcal{X}'\cap\mathcal{X}''} \big(\mu'(C|\Sigma) - \mu''(C|\Sigma)\big)\,\mu(d\sigma) = 0.$$

We can take for B the sets B^+ and B^-, where

$$B^+ = \big\{\sigma \in \mathcal{X}' \cap \mathcal{X}'' : \mu'(C|\Sigma) > \mu''(C|\Sigma)\big\},$$

while B^- is defined by the reverse inequality. Then the same argument as used to prove the claim in the proof of Theorem A.6 shows that $\mu'(C|\Sigma) = \mu''(C|\Sigma)$ with μ-probability 1. Thus,

$$\mu(\mathcal{X}_0^*) = 1, \quad \mathcal{X}_0^* = \{\sigma \in \mathcal{X}' \cap \mathcal{X}'' : \mu'(C|\Sigma)(\sigma) = \mu''(C|\Sigma)(\sigma), \text{ for all } C \in \mathcal{C}\}.$$

If two probabilities agree on $\mathcal{C}$, they are equal, by Theorem A.5, hence $\mathcal{X}_0^* = \mathcal{X}_0$, and (A.7.1) is proved. Since $\mu(\mathcal{X}_0) = 1$, by (A.6.1), $\mu'(\mathcal{X}_0|\Sigma)(\sigma) = 1$ for μ almost all $\sigma \in \mathcal{X}'$. Moreover, writing $\mu(A \cap B) = \mu(A \cap B \cap \mathcal{X}_0)$, by (A.6.1),

$$\mu(A \cap B) = \int_{B\cap\mathcal{X}'} \mu'(A \cap \mathcal{X}_0|\Sigma)(\sigma)\,\mu(d\sigma) = \int_{B\cap\mathcal{X}_0} \mu'(A|\Sigma)(\sigma)\,\mu(d\sigma),$$

where in the last equality we have used that $\mu'(\mathcal{X}_0|\Sigma)(\sigma) = 1$ for μ almost all $\sigma \in \mathcal{X}'$ and that $\mathcal{X}' \supset \mathcal{X}_0$. Thus, $\{\mu'(\cdot|\Sigma)(\sigma), \mathcal{X}_0\}$ is a version of the conditional probability of μ given Σ. □

A.8 Support properties

In this section, we will generalize the example of finite σ-algebras considered in Sect. A.6, characterizing the support of the conditional probabilities, establishing a correspondence between σ-algebras and "measurable partitions," as defined in Sect. A.6. The result has an essential role in the sequel and in the applications to the theory of DLR states; the whole analysis is based on Rohlin's theory of conditional probabilities in abstract Lebesgue spaces [194].

Let μ be a probability, Σ a σ-algebra, $(\mathcal{X}', \mu(\cdot|\Sigma)(\sigma))$ a version of the conditional probability. For any $\sigma \in \mathcal{X}'$, call

$$\begin{aligned} \Omega_\sigma &= \big\{\sigma' \in \mathcal{X}' : \mu(\cdot\,|\Sigma)(\sigma') = \mu(\cdot\,|\Sigma)(\sigma)\big\}, \\ \mathcal{X}'' &= \big\{\sigma \in \mathcal{X}' : \mu(\Omega_\sigma\,|\Sigma)(\sigma) = 1\big\}. \end{aligned} \tag{A.8.1}$$

The collection $\{\Omega\}$ of all $\Omega_\sigma, \sigma \in \mathcal{X}'$ is the partition of $\mathcal{X}'$ we have been referring to. We will call Ω-measurable a set $A \subset \mathcal{X}'$ which is union of sets Ω_σ.

Theorem A.8 *The set $\mathcal{X}''$ in* (A.8.1) *is in $\mathcal{B}$; it is Ω-measurable, and $\mu(\mathcal{X}'') = 1$. $(\mathcal{X}'', \mu(\cdot|\Sigma)(\sigma))$ is a version of the conditional probability of μ given Σ.*

Proof If $\sigma \in \mathcal{X}''$, i.e. $\mu(\Omega_\sigma\,|\Sigma)(\sigma) = 1$, then for all $\sigma' \in \Omega_\sigma$:

$$\mu(\Omega_\sigma\,|\Sigma)(\sigma') = \mu(\Omega_\sigma\,|\Sigma)(\sigma) = 1$$

so that $\Omega_\sigma \in \mathcal{X}''$. By the same argument, if $\sigma \in \mathcal{X}' \setminus \mathcal{X}''$, then $\Omega_\sigma \cap \mathcal{X}'' = \emptyset$; hence $\mathcal{X}''$ is Ω-measurable. The partition Ω of $\mathcal{X}'$ is measurable (measurable partitions are defined in Sect. A.6) and generated by the countable family of sets

$$S(C, a, b) = \{\sigma \in \mathcal{X}' : a \le \mu(C|\Sigma)(\sigma) \le b\}, \tag{A.8.2}$$

where $C \in \mathcal{C}$, $0 \le a \le b \le 1$ and a, b are rational numbers. In fact, $S(C, a, b)$ and $S^0(C, a, b) := \mathcal{X}' \setminus S(C, a, b)$ are both Ω-measurable and because the measures $\mu(\cdot|\Sigma)(\sigma)$ are uniquely determined by the values $\mu(C|\Sigma)(\sigma)$, $C \in \mathcal{C}$. Thus,

$$\bigcap_{C,a,b} S(C, a, b)^{1_{S(C,a,b)}(\sigma)} = \Omega_\sigma, \quad \sigma \in \mathcal{X}'. \tag{A.8.3}$$

Define next

$$\mathcal{X}_{C,a,b} = \{\sigma \in \mathcal{X}' : \mu(S(C, a, b)|\Sigma)(\sigma) = \mathbf{1}_{S(C,a,b)}(\sigma)\}, \tag{A.8.4}$$

which is also Ω-measurable, in $\Sigma \cap \mathcal{X}'$, and hence in $\mathcal{B}$. Then, by (A.6.5), $\mu(\mathcal{X}_{C,a,b}) = 1$, so that

$$\mu(\mathcal{X}^*) = 1, \quad \mathcal{X}^* = \bigcap_{C,a,b} \mathcal{X}_{C,a,b}. \tag{A.8.5}$$

$\mathcal{X}^*$ is Ω-measurable, in $\Sigma \cap \mathcal{X}'$, and in $\mathcal{B}$. We are going to prove that $\mathcal{X}^* = \mathcal{X}''$, thus completing the proof of the theorem. If $\sigma \in \mathcal{X}''$, then $\mu(\Omega_\sigma\,|\Sigma)(\sigma) = 1$, and by (A.8.3), $\mu(S(C, a, b)^{1_{S(C,a,b)}(\sigma)}|\Sigma)(\sigma) = 1_{S(C,a,b)}(\sigma)$, so that $\sigma \in \mathcal{X}_{C,a,b}$ and since this holds for all C, a, b, we have $\sigma \in \mathcal{X}^*$, which proves that $\mathcal{X}'' \subset \mathcal{X}^*$. Vice versa, if $\sigma \in \mathcal{X}^*$, then for any choice of the $\{0, 1\}$ valued function $\epsilon_{C,a,b}$,

$$\mu\Big(\bigcap_{C,a,b} S(C, a, b)^{\epsilon_{C,a,b}}\Big|\Sigma\Big)(\sigma) = \prod_{C,a,b} \mathbf{1}_{S(C,a,b)^{\epsilon_{C,a,b}}}(\sigma); \tag{A.8.6}$$

hence by (A.8.3), $\mu(\Omega_\sigma|\Sigma)(\sigma) = 1$. □

A.9 Space translations and the ergodic decomposition

As an example of applications of the previous theory, we study here the problem of determining the conditional probabilities of a translational invariant measure μ given the σ-algebra Σ of all translational invariant Borel sets. The question is non-empty, because a Bernoulli measure on $\mathcal{X}$ is translational invariant and there exist translational invariant Borel sets, like $\mathcal{X}$ itself and $\emptyset$.

We begin by constructing a family of measures $\mathcal{A}_\sigma$. We will prove later that they are a version of the conditional probability given Σ of any translational invariant measure μ. Let $\Delta_n \to \mathbb{Z}^d$ be an increasing sequence of cubes, and define, for any Borel set C, the frequency of visits to C by

$$p_\sigma(C) = \lim_{n\to\infty} \frac{1}{|\Delta_n|} \sum_{i\in\Delta_n} \mathbf{1}_{\tau_i(\sigma)\in C}, \tag{A.9.1}$$

wherever the limit exists. Since there are translational invariant measures, we can apply the Birkhoff theorem to conclude that there exists a set of σ where the frequency of visits to C is well defined, the measure of such a set being 1 for any translational invariant measure. We then call

$$\mathcal{X}' = \{\sigma : p_\sigma(C) \text{ exists for all } C \in \mathcal{C}\}. \tag{A.9.2}$$

Again, $\mathcal{X}'$ is non-empty, being a countable intersection of sets of full measure, for any translational invariant measure.

Claim *$\mathcal{X}'$ is in Σ; $\{p_\sigma(C), C \in \mathcal{C}\}$ is a translational invariant pre-probability and, for any $C \in \mathcal{C}$, $\sigma \to p_\sigma(C)$ is $\Sigma \cap \mathcal{X}'$-measurable.*

Proof As $p_\sigma(C)$ is a Cesaro average, $\sigma \in \mathcal{X}'$ if and only if its translates are in $\mathcal{X}'$. For the same reason $p_\sigma(C) = p_\sigma(\tau_i(C))$, $i \in \mathbb{Z}^d$; hence $p_\sigma(\cdot)$ is translational invariant. The same argument also shows that $\sigma \to p_\sigma(C)$ is translational invariant. Moreover, $\sigma \to p_\sigma(C)$, $\sigma \in \mathcal{X}'$, is measurable being a limit of measurable functions. Thus, the claim is proved. □

Since, for any $\sigma \in \mathcal{X}'$, $\{p_\sigma(C), C \in \mathcal{C}\}$ is a pre-probability, there is a unique probability $\mathcal{A}_\sigma$ which extends $\{p_\sigma(C), C \in \mathcal{C}\}$. We next claim that

$$\text{for any } \sigma \in \mathcal{X}', \quad \mathcal{A}_\sigma \text{ is translational invariant.} \tag{A.9.3}$$

In fact, the translate $\tau_i(\mathcal{A}_\sigma)$ of $\mathcal{A}_\sigma$ agrees with $\mathcal{A}_\sigma$ on $\mathcal{C}$, because by the above claim p_σ is translational invariant. Then $\tau_i(\mathcal{A}_\sigma) = \mathcal{A}_\sigma$ and (A.9.3) is proved.

Since $\mathcal{A}_\sigma = p_\sigma$ on $\mathcal{C}$, it inherits the measurability properties of the latter and, as a function on $\mathcal{X}'$, $\sigma \to \mathcal{A}_\sigma(C)$ is $\Sigma \cap \mathcal{X}'$-measurable, for any $C \in \mathcal{C}$. Call

$$\mathcal{S} = \{A \in \mathcal{B} : \sigma \to \mathcal{A}_\sigma(A) \text{ is } \Sigma \cap \mathcal{X}'\text{-measurable}\}. \tag{A.9.4}$$

It is easy to see that $\mathcal{S}$ is a monotone class and, since it contains $\mathcal{C}$, $\mathcal{S} = \mathcal{B}$. Summarizing, we have proved that $\{\mathcal{A}_\sigma, \sigma \in \mathcal{X}'\}$ is a family of translational invariant probabilities; $\sigma \to \mathcal{A}_\sigma(A)$ is $\Sigma \cap \mathcal{X}'$-measurable, for any Borel set A, $\mathcal{A}_\sigma(C) = p_\sigma(C)$, for any $\sigma \in \mathcal{X}'$ and any $C \in \mathcal{C}$. Finally, $\mu(\mathcal{X}') = 1$ for any translational invariant measure μ.

Theorem A.9 *For any pair (μ, Σ), μ translational invariant, Σ the σ-algebra of translational invariant Borel sets, $\mathcal{A}_\sigma$, with $\sigma \in \mathcal{X}'$, is a version of the conditional probability of μ given Σ and, for any Borel set A,*

$$\mathcal{A}_\sigma(A) = p_\sigma(A) \quad \textit{for } \mu \textit{ a.a. } \sigma. \tag{A.9.5}$$

Proof We have checked already the measurability condition; we thus only need to prove (A.6.1).

Claim *For any Borel set A and any $B \in \Sigma$,*

$$\mu(A \cap B) = \int_B p_\sigma(A)\, \mu(d\sigma). \tag{A.9.6}$$

Proof By (A.9.1), which, by the Birkhoff theorem, holds μ-a.s. (μ is invariant)

$$\begin{aligned}\int_B p_\sigma(A)\, \mu(d\sigma) &= \int_B \left\{ \lim_{n\to\infty} \frac{1}{|\Delta_n|} \sum_{i\in\Delta_n} \mathbf{1}_{\tau_i(\sigma)\in A} \right\} \mu(d\sigma) \\ &= \lim_{n\to\infty} \int_B \left\{ \frac{1}{|\Delta_n|} \sum_{i\in\Delta_n} \mathbf{1}_{\tau_i(\sigma)\in A} \right\} \mu(d\sigma) \\ &= \lim_{n\to\infty} \frac{1}{|\Delta_n|} \sum_{i\in\Delta_n} \mu(B \cap \tau_i A).\end{aligned}$$

Since B is invariant, $\mu(B \cap \tau_i A) = \mu(\tau_i(B \cap A))$, and since μ is invariant, we have $\mu(\tau_i(B \cap A)) = \mu(B \cap A)$; hence (A.9.6) holds and the claim is proved. □

Since $p_\sigma(C) = \mathcal{A}_\sigma(C)$ for $\sigma \in \mathcal{X}'$ and $C \in \mathcal{C}$, (A.9.6) shows that for any $C \in \mathcal{C}$ and $B \in \Sigma$,

$$\mu(C \cap B) = \int_B \mathcal{A}_\sigma(C)\, \mu(d\sigma)$$

which proves (A.6.1) limited to the cylinders. But at this point we can use again the monotone class argument given after (A.6.9), and prove that the family $\mathcal{S}$ of Borel sets A for which (A.6.2) holds is a monotone class. Since $\mathcal{S} \supset \mathcal{C}$, $\mathcal{S} = \mathcal{B}$; hence $\mathcal{A}_\sigma$, $\sigma \in \mathcal{X}'$ is a version of the conditional probability of μ given Σ.

It remains to prove (A.9.5). We have already proved that $\sigma \to \mathcal{A}_\sigma(A)$ is $\Sigma \cap \mathcal{X}'$-measurable; also $\sigma \to p_\sigma(A)$ is μ-almost surely measurable, being μ-almost sure limit of measurable functions, by (A.9.1). Obviously it is translational invariant, too. Thus, (A.9.5) will follow from proving that for all $B \in \Sigma$,

$$\int_B \mathcal{A}_\sigma(A)\,\mu(d\sigma) = \int_B p_\sigma(A)\,\mu(d\sigma). \tag{A.9.7}$$

The l.h.s. of (A.9.7) is equal to $\mu(A \cap B)$, because $\mathcal{A}_\sigma$ is the conditional probability of μ given Σ, but the r.h.s. is also equal to $\mu(A \cap B)$ by (A.9.6). □

Let

$$\mathcal{X}'' = \{\sigma \in \mathcal{X}' : \mathcal{A}_\sigma(\Omega^0_\sigma) = 1\}, \quad \Omega^0_\sigma = \{\sigma' \in \mathcal{X}' : \mathcal{A}_\sigma = \mathcal{A}_{\sigma'}\}. \tag{A.9.8}$$

As a consequence of the support properties of Sect. A.8, if μ is invariant, then

$$\mu(\mathcal{X}'') = 1. \tag{A.9.9}$$

For any $\sigma \in \mathcal{X}''$ and any Borel set A,

$$\begin{aligned} 1 = \mathcal{A}_\sigma(\Omega^0_\sigma) &\le \mathcal{A}_\sigma\big(\{\sigma' \in \mathcal{X}' : \mathcal{A}_{\sigma'}(A) = \mathcal{A}_\sigma(A)\}\big) \\ &= \mathcal{A}_\sigma\big(\{\sigma' \in \mathcal{X}' : p_{\sigma'}(A) = \mathcal{A}_\sigma(A)\}\big), \end{aligned} \tag{A.9.10}$$

having used, in the last equality, (A.9.5), with μ replaced by $\mathcal{A}_\sigma$. Equation (A.9.10) shows that $\mathcal{A}_\sigma$, $\sigma \in \mathcal{X}''$, satisfies one of the several equivalent conditions for ergodicity, and it is therefore an ergodic measure. We have thus proved the following theorem.

Theorem A.10 *For any pair (μ, Σ) as in Theorem* A.9, *$\{\mathcal{X}'', \mathcal{A}_\sigma\}$, is a version of the conditional probability of μ given Σ. For any $\sigma \in \mathcal{X}''$, $\mathcal{A}_\sigma$ is an invariant, ergodic measure. The decomposition of μ into the conditional probability $\mathcal{A}_\sigma$, $\sigma \in \mathcal{X}''$, gives, for any Borel set A,*

$$\mu(A) = \int_{\mathcal{X}''} \mathcal{A}_\sigma(A)\,\mu(d\sigma)$$

which is the ergodic decomposition of μ into its ergodic components $\mathcal{A}_\sigma$, $\sigma \in \mathcal{X}''$.

A.10 Limits of decreasing σ-algebras

In this and in the next section we will consider monotone sequences of σ-algebras, starting here with the decreasing ones. A sequence Σ_n of σ-algebras is decreasing, if, for all n, $\Sigma_n \supseteq \Sigma_{n+1}$. The results of this section have been used in Chap. 2 to study the decomposition at infinity of the DLR measures. In such a case $\Sigma_n = \mathcal{B}_{\Delta_n^c}$, the σ-algebra generated by the spins $\sigma(x)$, $x \in \Delta_n^c$, Δ_n an increasing sequence of bounded regions in $\mathbb{Z}^d$ which invades the whole space.

We fix below a probability μ, a decreasing sequence Σ_n, and, for each n, a version $\{\mathcal{X}_n, \mu(\cdot|\Sigma_n)(\sigma)\}$ of the conditional probability of μ given Σ_n. We call Σ the σ-algebra of sets B which are contained in all Σ_n (recall that Σ_n is a decreasing sequence) and we define

$$\mathcal{X}' = \left\{ \sigma \in \bigcap_{n\geq 1} \mathcal{X}_n : \lim_{n\to\infty} \mu(C|\Sigma_n)(\sigma) =: p(C;\sigma) \text{ for all } C \in \mathcal{C} \right\}. \quad \text{(A.10.1)}$$

Theorem A.11 *The set $\mathcal{X}'$ is in Σ and $\mu(\mathcal{X}') = 1$. For any $\sigma \in \mathcal{X}'$, $\{p(C;\sigma), C \in \mathcal{C}\}$ is a pre-probability (see Sect.* A.5) *so that it extends uniquely to a probability, denoted by $\mu(\cdot|\Sigma)(\sigma)$. $\{\mathcal{X}', \mu(\cdot|\Sigma)(\sigma)\}$ is a version of the conditional probability of μ given Σ and, for any Borel set A,*

$$\lim_{n\to\infty} \mu(A|\Sigma_n)(\sigma) = \mu(A|\Sigma)(\sigma), \quad \mu\text{-a.s.} \quad \text{(A.10.2)}$$

Proof The main ingredient in the proof is the Doob's martingale convergence theorem, for which we refer to the literature, see Doob [113]. Applied to our case, it states that for any bounded measurable function f,

$$\lim_{n\to\infty} \mu(f|\Sigma_n)(\sigma) \quad \text{exists } \mu\text{-a.s.} \quad \text{(A.10.3)}$$

The statement holds both for decreasing and increasing sequences of σ-algebras; here, we apply it to the former, in the next section to the latter.

By (A.10.3), and since $\mathcal{C}$ is countable, we have $\mu(\mathcal{X}') = 1$. We will next prove that $\mathcal{X}' \in \Sigma$. $\mathcal{X}^* := \bigcap_{n\geq 1} \mathcal{X}_n$ is obviously in Σ. $\limsup_{n\to\infty} \mu(C|\Sigma_n)(\sigma)$ and $\liminf_{n\to\infty} \mu(C|\Sigma_n)(\sigma)$ are $\Sigma \cap \mathcal{X}^*$-measurable as well as the set where they are equal. Hence $\mathcal{X}' \in \Sigma$.

$p(C;\sigma), C \in \mathcal{C}$, is a pre-probability, because it is a limit of probabilities. Thus, it extends uniquely to a probability $\mu(\cdot|\Sigma)(\sigma)$, $\sigma \in \mathcal{X}'$. We still need to prove, however, that the notation is correct, namely that $\mu(\cdot|\Sigma)(\sigma)$ is indeed a version of the conditional probability of μ given Σ. We proceed by essentially the same argument as after (A.6.9). We thus denote by $\mathcal{S}$ the collection of all Borel sets A such that $\sigma \to \mu(A|\Sigma)(\sigma)$, $\sigma \in \mathcal{X}'$, is $\Sigma \cap \mathcal{X}'$-measurable and (A.6.1) holds. We have already proved that $\mathcal{S} \supset \mathcal{C}$.

Let us now prove (A.6.1) with $A = C \in \mathcal{C}$. If $B \in \Sigma$, then $B \in \Sigma_n$, for any n; hence

$$\mu(C \cap B) = \int_{B\cap\mathcal{X}_n} \mu(C|\Sigma_n)(\sigma)\, \mu(d\sigma) = \int_{B\cap\mathcal{X}'} \mu(C|\Sigma_n)(\sigma)\, \mu(d\sigma)$$

which, by the Lebesgue dominated convergence theorem, yields (A.6.1) in the limit $n \to \infty$. This completes the proof that $\mathcal{S} \supset \mathcal{C}$. The proof that $\mathcal{S}$ is a monotone class is just like the proof after (A.6.9) and it is omitted; then, by the monotone class theorem, we conclude that $\mathcal{S} = \mathcal{B}$, thus, establishing that $\mu(\cdot|\Sigma)(\sigma)$ is a version of the conditional probability of μ given Σ.

It remains to prove (A.10.2). We preliminarily notice that for any Borel set A,

$$\lim_{n\to\infty} \mu(A|\Sigma_n)(\sigma) \quad \text{exists and it is } \Sigma\text{-measurable } \mu \text{ modulo } 0. \tag{A.10.4}$$

The proof of (A.10.4) when $A = C \in \mathcal{C}$ applies in the same way to Borel sets (it uses (A.10.3) and the fact that limits of measurable functions are measurable).

We next fix a Borel set A and, given any $B \in \Sigma$, we have

$$\mu(A \cap B) = \int_{B\cap\mathcal{X}'} \mu(A|\Sigma)(\sigma)\, \mu(d\sigma) = \int_{B\cap\mathcal{X}'} \mu(A|\Sigma_n)(\sigma)\, \mu(d\sigma).$$

By letting $n \to \infty$,

$$\int_{B\cap\mathcal{X}'} \mu(A|\Sigma)(\sigma) - \left\{\lim_{n\to\infty} \mu(A|\Sigma_n)(\sigma)\right\} \mu(d\sigma) = 0. \tag{A.10.5}$$

Since the integrand is Σ-measurable modulo 0, by the arbitrariness of $B \in \Sigma$, it must be 0 almost everywhere. □

A.11 Limits of increasing σ-algebras

In this section we complement the analysis of the previous one by studying an increasing sequence Σ_n of σ-algebras, namely such that for all n, $\Sigma_n \subseteq \Sigma_{n+1}$. We call

$$\mathcal{U} = \{B : \text{ there is } n \text{ so that } B \in \Sigma_n\}; \tag{A.11.1}$$

$\mathcal{U}$ is an algebra of sets, and we denote by Σ, called the limit of Σ_n, the minimal σ-algebra which contains $\mathcal{U}$.

Given a probability μ and an increasing sequence Σ_n, we denote by $\mu(\cdot|\Sigma_n)(\sigma)$, $\sigma \in \mathcal{X}_n$, a version of the conditional probability of μ given Σ_n. The exact analogue of the case of decreasing σ-algebras holds.

Theorem A.12 *In the above setup, there exists $\mathcal{X}'$, $\mu(\mathcal{X}') = 1$, such that $\mathcal{X}' \subset \mathcal{X}_n$, for all n, and*

$$\text{for any } \sigma \in \mathcal{X}' \quad \lim_{n\to\infty} \mu(C|\Sigma_n)(\sigma) =: p(C;\sigma) \quad \text{for all } C \in \mathcal{C}. \tag{A.11.2}$$

$p(C;\sigma)$ is a pre-probability, and it extends uniquely to a probability denoted by $\mu(\cdot|\Sigma)(\sigma)$, $\sigma \in \mathcal{X}'$. Now, $\mu(\cdot|\Sigma)(\sigma)$ is a version of the conditional probability of μ given Σ and for any set $A \in \mathcal{B}$,

$$\lim_{n\to\infty} \mu(A|\Sigma_n)(\sigma) = \mu(A|\Sigma)(\sigma), \quad \mu\text{-a.s.} \tag{A.11.3}$$

Proof We use again Doob's convergence result (A.10.3) which holds for increasing sequences as well. Exactly as for decreasing sequences, we then conclude that

(A.11.2) holds with $\mu(\mathcal{X}') = 1$; moreover, $p(C;\sigma), C \in \mathcal{C}$, is a pre-probability because it is a limit of probabilities; thus, it extends uniquely to a probability $\mu(\cdot|\Sigma)(\sigma)$, $\sigma \in \mathcal{X}'$, but we still need to prove that the notation is correct; namely, that it is a version of the conditional probability given Σ.

Let $\mathcal{U}$ be as in (A.11.1). If $B \in \mathcal{U}$, $B \in \Sigma_n$ for all n large enough and for such values of n,

$$\mu(C \cap B) = \int_B \mu(C|\Sigma_n)(\sigma)\, \mu(d\sigma), \quad \text{for any } C \in \mathcal{C}.$$

By letting $n \to \infty$, we have, for any $C \in \mathcal{C}$,

$$\mu(C \cap B) = \int_B \mu(C|\Sigma)(\sigma)\, \mu(d\sigma). \tag{A.11.4}$$

We next claim that (A.11.4) extends to all $B \in \Sigma$. Suppose that $\mu(C) > 0$ and divide both sides by $\mu(C)$, so that they are normalized, additive measures on $B \in \mathcal{U}$. Moreover, if $B \in \mathcal{U}$ is countable union of disjoint sets $B_i \in \mathcal{U}$, then

$$\mu(C \cap B) = \sum_i \mu(C \cap B_i),$$

so that, by (A.11.4),

$$\begin{aligned}\int_B \mu(C|\Sigma)(\sigma)\, \mu(d\sigma) &= \sum_i \int \mathbf{1}_{B_i} \mu(C|\Sigma)(\sigma)\, \mu(d\sigma)\\ &= \int \Big\{\sum_i \mathbf{1}_{B_i}\Big\} \mu(C|\Sigma)(\sigma)\, \mu(d\sigma),\end{aligned}$$

where the last equality uses the Lebesgue dominated convergence theorem.

Hence (A.11.4) holds for $B = \bigcup_i B_i$, and, by the Caratheodory theorem, (A.11.4) holds for all $B \in \Sigma$. By the same argument used below (A.6.9) (based on the "monotone class theorem"), (A.11.4) extends also to all $C \in \mathcal{B}$.

To conclude that $\mu(\cdot|\Sigma)(\sigma)$ is a version of the conditional probability of μ given Σ, we need to prove that for any $A \in \mathcal{B}$, $\mu(A|\Sigma)(\sigma)$ is $\Sigma \cap \mathcal{X}'$-measurable ($\mathcal{X}'$ is the set where $\mu(\cdot|\Sigma)(\sigma)$ has been defined). Call $\mathcal{K}$ the collection of Borel sets A such that $\mu(A|\Sigma)(\sigma)$ is $\Sigma \cap \mathcal{X}'$-measurable. $\mathcal{K}$ is easily seen to be a monotone class and we need to show that it contains $\mathcal{C}$. Let $C \in \mathcal{C}$. $\mu(C|\Sigma)(\sigma)$, $\sigma \in \mathcal{X}'$, is the limit of $\mu(C|\Sigma_n)(\sigma)$, which is $\Sigma_n \cap \mathcal{X}'$-measurable, hence $\Sigma \cap \mathcal{X}'$-measurable; hence $\mu(C|\Sigma)(\sigma)$ is $\Sigma \cap \mathcal{X}'$-measurable. We have thus proved that $\mu(\cdot|\Sigma)(\sigma)$ is a version of the conditional probability of μ given Σ.

It remains to prove (A.11.3). Recalling the definition (A.11.1) of $\mathcal{U}$, for any $B \in \mathcal{U}$ and any Borel set A, for all n large enough

$$\mu(A \cap B) = \int_B \mu(A|\Sigma)(\sigma)\, \mu(d\sigma) = \int_B \mu(A|\Sigma_n)(\sigma)\, \mu(d\sigma). \tag{A.11.5}$$

By (A.10.3),

$$\int_B \mu(A|\Sigma)(\sigma)\,\mu(d\sigma) = \int_B \left\{\lim_{n\to\infty} \mu(A|\Sigma_n)(\sigma)\right\} \mu(d\sigma). \tag{A.11.6}$$

The same argument used earlier, following (A.11.4), shows that (A.11.6) remains valid for all $B \in \Sigma$. Moreover, the function $\sigma \to \{\lim_{n\to\infty} \mu(A|\Sigma_n)(\sigma)\}$, defined on a set $\mathcal{X}''$ of full measure, is $\Sigma \cap \mathcal{X}''$-measurable, by the same argument used for $\mu(\cdot|\Sigma)(\sigma)$. (A.11.3) is thus a consequence of (A.11.6). □

Appendix B
Geometric measure theory

B.1 Basic notions

We state here without proofs some basic results from geometric measure theory, with the book of Evans and Gariepy [116] as a general reference.

In view of the applications in Chap. 7, we take the unit torus $\mathcal{T}$ as the spatial domain of the theory and say that a function f on $\mathcal{T}$ has a bounded variation, $f \in BV(\mathcal{T})$, if its gradient Df (in the sense of distributions) is a vector valued Radon measure whose total variation measure $d\mu$ has finite mass $\|\mu\|$ given by the variational formula

$$\|\mu\| = \mu(\mathcal{T}) = \sup_{\phi \in C^1(\mathcal{T},\mathbb{R}^d),\ \|\phi\|_\infty \le 1} \left| \int_{\mathcal{T}} dr\, f \mathrm{div}\phi \right|. \tag{B.1.1}$$

If E is a C^1 set, then its characteristic function $\mathbf{1}_E$ is in $BV(\mathcal{T})$, the total variation $d\mu$ of $D\mathbf{1}_E$ is the usual Hausdorff (or area) measure $dH^{d-1}(r)$ on ∂E, and for any $\phi \in C(\mathcal{T}, \mathbb{R}^d)$ (denoting by $\langle \cdot, \cdot \rangle$ the scalar product in $\mathbb{R}^d$)

$$\int_{\mathcal{T}} \langle D\mathbf{1}_E, \phi \rangle = -\int_{\mathcal{T}} dr\, \mathbf{1}_E \mathrm{div}\phi = -\int_{\partial E} dH^{d-1}(r) \langle \nu(r), \phi \rangle, \tag{B.1.2}$$

where $\nu(r)$ is the outward unit normal to ∂E at r. Thus $D\mathbf{1}_E = -dH^{d-1}(r)\nu(r)$ and $d\mu = dH^{d-1}$.

If E is a general BV set, i.e. $\mathbf{1}_E \in BV(\mathcal{T})$, and $d\mu$ is the total variation measure of $D\mathbf{1}_E$, then there is a set $\partial^* E \subset \partial E$, called the "reduced boundary" of E, and a unit vector valued function $\nu(r)$ on $\partial^* E$ so that, for any $\phi \in C(\mathcal{T}, \mathbb{R}^d)$,

$$\int_{\mathcal{T}} \langle D\mathbf{1}_E, \phi \rangle = -\int_{\partial^* E} d\mu(r) \langle \nu(r), \phi \rangle; \tag{B.1.3}$$

namely,

$$\|\mu\| = \int_{\partial^* E} d\mu, \qquad D\mathbf{1}_E(r) = -d\mu(r)\nu(r), \quad r \in \partial^* E. \tag{B.1.4}$$

By comparing (B.1.3) and (B.1.2), we can interpret $\nu(r)$ as a generalized unit normal to the boundary of E. The interpretation is strengthened by the following theorem which states that BV sets can be regarded, measure theoretically, as C^1 sets.

Theorem *Let $E \in BV(\mathcal{T})$ and $D\mathbf{1}_E(r) = -d\mu\nu(r)$. Then for any $\epsilon > 0$ there are C^1 hyper-surfaces $S_1, \ldots, S_m$, whose closures are disjoint from each other, and compact sets $K_1, \ldots, K_m$ with $K_i \subset S_i \cap \partial^\star E$, so that*

$$d\mu|_{K_i} = dH^{d-1}|_{K_i}$$

and

$$\int_{\mathcal{T}} d\mu - \sum_{i=1}^{m} \int_{K_i} d\mu \leq \epsilon. \tag{B.1.5}$$

Moreover, the normal to S_i at $r \in K_i$ is the same as the unit vector $\nu(r)$ in (B.1.4), *and we have*

$$\max_{i=1,\ldots,m} \max_{r,r' \in S_i} \left|\nu(r) - \nu(r')\right| \leq \epsilon. \tag{B.1.6}$$

B.2 Polygonal approximations

We state here three approximation theorems of BV sets, which have been used in Chap. 7 in the proofs of the upper and lower bounds of Gamma convergence. The first one (used in the upper bound) states that a BV set can be approximated in variation by poliedrical sets: a sequence u_n in $BV(\mathcal{T})$ converges to $u \in BV(\mathcal{T})$ in variation, if, denoting by $d\mu_n$ and $d\mu$ the corresponding total variation measures,

$$\lim_{n\to\infty} |u_n - u|_{L^1(\mathcal{T})} = 0, \qquad \lim_{n\to\infty} \|\mu_n\| = \|\mu\|. \tag{B.2.1}$$

In [4] the following is proved.

Theorem B.1 *If $E \in BV(\mathcal{T})$, there are poliedrical sets E_n such that $\mathbf{1}_{E_n}$ converges in variation to $\mathbf{1}_E$.*

As we are going to see, convergence in variation of $\mathbf{1}_{E_n}$ implies trivially weak convergence of the normals. We first show that $D\mathbf{1}_{E_n}$ converges in the weak* topology of measures to $D\mathbf{1}_E$; namely,

$$\lim_{n\to\infty} \int_{\mathcal{T}} \langle D\mathbf{1}_{E_n}, \phi\rangle = \int_{\mathcal{T}} \langle D\mathbf{1}_E, \phi\rangle, \quad \text{for all } \phi \in C(\mathcal{T}, \mathbb{R}^d). \tag{B.2.2}$$

The limit in (B.2.2) clearly holds for any $\phi \in C^1(\mathcal{T}, \mathbb{R}^d)$, by integration by parts and the assumption that $\mathbf{1}_{E_n} \to \mathbf{1}_E$ in $L^1(\mathcal{T})$. The result extends to $\phi \in C(\mathcal{T}, \mathbb{R}^d)$, because $C^1(\mathcal{T}, \mathbb{R}^d)$ is dense in $C(\mathcal{T}, \mathbb{R}^d)$ in sup norm and, if $u \in BV$ and $d\mu$ is the corresponding total variation measure,

$$\left|\int_{\mathcal{T}} \langle Du, \psi\rangle\right| \leq \|\psi\|_\infty \int_{\mathcal{T}} d\mu_u. \tag{B.2.3}$$

Convergence of the normals is actually true in a stronger sense.

Theorem B.2 *In the same context as in Theorem* B.1 *for any* $g \in C(\mathcal{T} \times B_1(\mathbb{R}^d))$, $B_1(\mathbb{R}^d)$ *being the unit ball in* $\mathbb{R}^d$,

$$\lim_{n\to\infty} \int_{\partial^* E_n} d\mu_n(r) g(r, \nu(r)) = \int_{\partial^* E} d\mu(r) g(r, \nu(r)), \tag{B.2.4}$$

where $d\mu_n$ *and* $d\mu$ *are the total variation measures corresponding to* E_n *and* E.

For the lower bound of Gamma convergence in Chap. 7, we have used Theorem B.3 below, which states that a BV set is with good approximation made of essentially flat parts plus a small remainder. We use the following notation: E denotes a BV set, μ the associated total variation measure and for any $\epsilon > 0$, $S_j^{(\epsilon)}$, $K_j^{(\epsilon)}$, $j = 1, \dots, m_\epsilon$, are the sets introduced in Theorem B.1. In particular, therefore, each $K_j^{(\epsilon)} \sqsubset S_j^{(\epsilon)}$, with $S_j^{(\epsilon)}$ a C^1 surface which has a well defined unit normal $\nu(r)$. In view of the applications in Chap. 7, we also set

$$u = m_\beta(\mathbf{1}_{E^c} - \mathbf{1}_E). \tag{B.2.5}$$

Theorem B.3 *For any* $\epsilon > 0$, *there are* $n \ge 1$ *disjoint measurable sets* Σ_i, *each one contained in some* $K_j^{(\epsilon)}$, n *cubes* R_i, *all of side* h, *and* n *unit vectors* ν_i, ν_i *orthogonal to a face of* R_i, *with the following properties*:

$$\begin{gathered} \sup_{r\in\Sigma_i} |\nu(r) - \nu_i| < \epsilon, \\ \left| h^{d-1} - \int_{\Sigma_i} d\mu \right| < \epsilon h^{d-1}, \qquad \left| n h^{d-1} - \int_{\mathcal{T}} d\mu \right| < \epsilon. \end{gathered} \tag{B.2.6}$$

Moreover, calling $\chi_{R_i} := m_\beta(\mathbf{1}_{R_i^+} - \mathbf{1}_{R_i^-})$, *with* $R_i^\pm$ *the upper and lower halves of* R_i *with respect to the direction* ν_i,

$$\int_{R_i} dr |\chi_{R_i} - u| < \epsilon h^d, \quad i = 1, \dots, n. \tag{B.2.7}$$

References

1. Ahlers, G.: Over two decades of pattern formation, a personal perspective. In: Brey, J.J., Marro, J., Rubi, J.M., San Miguel, M. (eds.) 25 Years of Non Equilibrium Statistical Mechanics. Lecture Notes in Physics, vol. 1995, pp. 91–124. Springer, Berlin (1995)
2. Aizenman, M., Wehr, J.: Rounding of first order phase transitions in systems with quenched disorder. Commun. Math. Phys. **130**, 489–528 (1990)
3. Aizenman, M., Chayes, J., Chayes, L., Newman, C.: Discontinuity of the magnetization in one dimensional $1/(x-y)^2$ percolation, Ising and Potts models. J. Stat. Phys. **50**, 1–40 (1988)
4. Alberti, G., Bellettini, G.: A non local anisotropic model for phase transition. Part I: the optimal profile problem. Math. Ann. **310**, 527–560 (1998)
5. Alberti, G., Bellettini, G.: A non local anisotropic model for phase transition: asymptotic behaviour of rescaled energies. Eur. J. Appl. Math. **9**, 261–284 (1998)
6. Alberti, G., Bellettini, G., Cassandro, M., Presutti, E.: Surface tension in Ising systems with Kac potentials. J. Stat. Phys. **82**, 743–796 (1996)
7. Alikakos, N.D., Bates, P.W., Chen, X.: Periodic traveling waves and locating oscillating patterns in multidimensional domains. Trans. Am. Math. Soc. **351**, 2777–2805 (1999)
8. Alikakos, N.D., Fife, P.C., Fusco, G., Sourdis, C.: Analysis of the heteroclinic connection in a singularly perturbed system arising from the study of crystalline grain boundaries. Interfaces Free Bound. **8**, 159–183 (2006)
9. Ambrosio, L., Fusco, N., Pallara, D.: Functions of Bounded Variations and Free Discontinuity Problems. Oxford Science Publications, New York (2000)
10. Andersson, H., Britton, T.: Stochastic Epidemic Models and Their Statistical Analysis. Lecture Notes in Statistics. Springer, New York (2000)
11. Asselah, A., Giacomin, G.: Metastability for the exclusion process with mean-field interaction. J. Stat. Phys. **93**, 1051–1110 (1998)
12. Baffioni, F., Kuna, T., Merola, I., Presutti, E.: A liquid vapor phase transition in quantum statistical mechanics. Preprint (2004)
13. Baffioni, F., Kuna, T., Merola, I., Presutti, E.: A relativized Dobrushin uniqueness condition and applications to Pirogov–Sinai models. Preprint (2004)
14. Barles, G., Souganidis, P.E.: A new approach to front propagation problems: theory and applications. Arch. Ration. Mech. Anal. **141**, 237–296 (1998)
15. Bates, P., Chen, F.: Periodic traveling waves for a nonlocal integro-differential model. Electr. J. Differ. Equ. **26**, 19 (1999)
16. Bates, P.W., Chmaj, A.: An integro-differentiable model for phase transitions: stationary solutions in higher dimensions. J. Stat. Phys. **95**, 1119–1139 (1998)
17. Bates, P.W., Chmaj, A.: On a discrete convolution model for phase transitions. Arch. Ration. Mech. Anal. **150**, 281–305 (1999)
18. Bates, P.W., Fife, P.C., Ren, X., Wang, X.: Traveling waves in a convolution model for phase transitions. Arch. Ration. Mech. Anal. **138**, 105–136 (1997)
19. Bellettini, G., Cassandro, M., Presutti, E.: Constrained minima of non local free energy functionals. J. Stat. Phys. **84**, 1337–1349 (1996)
20. Bellettini, G., Buttà, P., Presutti, E.: Sharp interface limits for non local anisotropic interactions. Arch. Ration. Mech. Anal. **159**, 109–135 (2001)
21. Bellettini, G., De Masi, A., Dirr, N., Presutti, E.: Stability of invariant manifolds in one and two dimensions. Nonlinearity **20**, 537–582 (2007)
22. Bellettini, G., De Masi, A., Dirr, N., Presutti, E.: Tunnelling in two dimensions. Commun. Math. Phys. **269**, 715–763 (2007)
23. Benois, O., Bodineau, T., Presutti, E.: Large deviations in the van der Waals limit. Stoch. Process. Their Appl. **75**, 89–104 (1998)

24. Bertini, L., Cancrini, N.: The stochastic heat equation: Feynman–Kac formula and intermittency. J. Stat. Phys. **783**, 1131–1138 (1995)
25. Bertini, L., Giacomin, G.: Stochastic Burgers and KPZ equations from particle systems. Commun. Math. Phys. **183**, 571–607 (1997)
26. Bertini, L., Presutti, E., Rüdiger, B., Saada, E.: Dynamical critical fluctuations and convergence to a stochastic non linear PDE in one dimension. Theory Probab. Appl. **38**, 689–741 (1993)
27. Bertini, L., Buttà, P., Rüdiger, B.: Interface dynamics and Stefan problem from a microscopic conservative model. Rend. Mat. Appl. **19**, 547–581 (2000)
28. Bertini, L., Brassesco, S., Buttà, P., Presutti, E.: Front fluctuations in one dimensional stochastic phase field equations. Ann. Inst. Henri Poincaré **3**, 29–86 (2002)
29. Bertini, L., Brassesco, S., Buttà, P., Presutti, E.: Stochastic phase field equations: existence and uniqueness. Ann. Inst. Henri Poincaré **3**, 87–98 (2002)
30. Bertini, L., Buttà, P., Presutti, E., Saada, E.: Interface fluctuations in a conserved system: derivation and long time behavior. Markov Processes Relat. Fields **9**, 1–34 (2003)
31. Billingsley, P.: Convergence of Probability Measures. Wiley, New York (1968)
32. Billingsley, P.: Probability and Measure, 2nd edn. Wiley, New York (1986)
33. Bisceglie, C.: Phase coexistence: variational and probabilistic techniques. Ph.D. thesis, Università L'Aquila (2006)
34. Biskup, M., Chayes, L.: Rigorous analysis of discontinous phase transitions via mean field bounds. Commun. Math. Phys. **238**, 53–93 (2003)
35. Biskup, M., Chayes, L., Kotecký, R.: Coexistence of partially disordered/ordered phases in an extended Potts model. J. Stat. Phys. **99**, 1169–1206 (2000)
36. Biskup, M., Chayes, L., Crawford, N.: Mean-field driven first order phase transitions in systems with long-range interactions. J. Stat. Phys. **122**, 1139–1193 (2006)
37. Blömker, D., Maier-Paape, S., Wanner, T.: Spinodal decomposition for the Cahn–Hilliard–Cook equation. Commun. Math. Phys. **223**, 553–582 (2001)
38. Blömker, D., Maier-Paape, S., Wanner, T.: Second phase spinodal decomposition for the Cahn–Hilliard–Cook equation. Trans. Am. Math. Soc. **360**, 449–489 (2008)
39. Bodineau, T.: The Wulff construction in three and more dimensions. Commun. Math. Phys. **207**, 197–229 (1999)
40. Bodineau, T., Presutti, E.: Phase diagram of Ising systems with additional long range forces. Commun. Math. Phys. **189**, 287–298 (1997)
41. Bodineau, T., Presutti, E.: Surface tension and Wulff shape for a lattice model without spin flip symmetry. Ann. Inst. Henri Poincaré **4**, 847–896 (2003)
42. Bonaventura, L.: Interface dynamics in an interacting spin system. Nonlinear Anal. **25**, 799–819 (1995)
43. Bonetto, F., Gallavotti, G., Gentile, G.: Aspects of the Ergodic, Qualitative, Statistical Aspects of Motion. Springer, Berlin (2004)
44. Bovier, A.: Statistical Mechanics of Disordered Systems. A Mathematical Perspective. Cambridge Series in Statistical and Probabilistic Mathematics. Cambridge University Press, Cambridge (2006)
45. Bovier, A., Zahradník, M.: The low-temperature phase of Kac–Ising models. J. Stat. Phys. **87**, 311–332 (1997)
46. Bovier, A., Zahradník, M.: A simple inductive approach to the problem of convergence of cluster expansions of polymer models. J. Stat. Phys. **100**, 765–778 (2000)
47. Bovier, A., Zahradník, M.: Cluster expansions and Pirogov–Sinai theory for long range spin systems. Markov Processses Relat. Fields **8**, 443–478 (2002)
48. Bovier, A., Merola, I., Presutti, E., Zahradnik, M.: On the Gibbs phase rule in the Pirogov–Sinai regime. J. Stat. Phys. **114**, 1235–1267 (2004)
49. Braides, A.: Gamma-Convergence for Beginners. Oxford Lecture Series in Mathematics and Its Applications, vol. 22. Oxford University Press, Oxford (2002)
50. Brassesco, S., Buttà, P.: Interface fluctuations for the $d = 1$ stochastic Ginzburg–Landau equation with nonsymmetric reaction term. J. Stat. Phys. **93**, 1111–1142 (1998)

51. Brassesco, S., De Masi, A., Presutti, E.: Brownian fluctuations of the interface in the $d = 1$ Ginzburg–Landau equation with noise. Annal. Inst. Henri Poincaré **31**, 81–118 (1995)
52. Brassesco, S., Buttà, P., De Masi, A., Presutti, E.: Interface fluctuations and couplings in the $d = 1$ Ginzburg–Landau equation with noise. J. Theor. Probab. **11**, 25–80 (1998)
53. Bricmont, J., Kupiainen, A.: Phase transition in the three-dimensional random field Ising model. Commun. Math. Phys. **116**, 539–572 (1988)
54. Bricmont, J., Kuroda, K., Lebowitz, J.L.: The structure of Gibbs states and phase coexistence for nonsymmetric continuum Widom–Rowlinson models. Z. Wahrscheinlichkeitstheor. Verw. Geb. **67**, 121–138 (1984)
55. Bricmont, J., Kuroda, K., Lebowitz, J.L.: First order phase transitions in lattice and continuous systems: extension of Pirogov–Sinai theory. Commun. Math. Phys. **101**, 501–538 (1985)
56. Brydges, D.: A short course on cluster expansion. In: Osterwalder, K., Stora, R. (eds.) Les Houches 1984. North Holland, Amsterdam (1986)
57. Buttà, P.: On the validity of an Einstein relation in models of interface dynamics. J. Stat. Phys. **72**, 1401–1406 (1993)
58. Buttà, P.: Motion by mean curvature by scaling a non local equation: convergence at all times in 2d-case. Lett. Math. Phys. **31**, 41–55 (1994)
59. Buttà, P., De Masi, A.: Fine structure of the interface motion. Diff. Integral Equ. **12**, 207–259 (1999)
60. Buttà, P., Picco, P.: Large deviation principle for one dimensional vector spin models with Kac potentials. J. Stat. Phys. **92**, 101–150 (1998)
61. Buttà, P., Merola, I., Presutti, E.: On the validity of the van der Waals theory in Ising systems with long range interactions. Markov Processes Relat. Fields **3**, 63–88 (1997)
62. Buttà, P., De Masi, A., Rosatelli, E.: Slow motion and metastability for a non local evolution equation. J. Stat. Phys. **112**, 709–764 (2003)
63. Cahn, J.W.: Phase separation by spinodal decomposition in isotropic systems. J. Chem. Phys. **42**, 93–99 (1965)
64. Calderoni, P., Pellegrinotti, A., Presutti, E., Vares, M.E.: Transient bimodality in interacting particle systems. J. Stat. Phys. **55**, 523–578 (1989)
65. Carlen, E., Carvalho, M., Esposito, R., Lebowitz, J.L., Marra, R.: Free energy minimizers for a two-species model with segregation and liquid–vapour transition. Nonlinearity **16**, 1075–1105 (2003)
66. Carlen, E.A., Carvalho, M.C., Esposito, R., Lebowitz, J.L., Marra, R.: Droplet minimizers for a non-local free functional. Work in progress
67. Carvalho, M.C., Carlen, E., Esposito, R., Lebowitz, J.L., Marra, R.: Displacement convexity and minimal fronts at phase boundaries. Preprint ArXiv 0706.0133v1 [math-FA] (2007)
68. Cassandro, M., Picco, P.: Existence of a phase transition in a continuous quantum system. J. Stat. Phys. **103**, 841–856 (2001)
69. Cassandro, M., Presutti, E.: Phase transitions in Ising systems with long but finite range interactions. Markov Processes Relat. Fields **2**, 241–262 (1996)
70. Cassandro, M., Orlandi, E., Presutti, E.: Interfaces and typical Gibbs configurations for one dimensional Kac potentials. Probab. Theory Relat. Fields **96**, 57–96 (1993)
71. Cassandro, M., Marra, R., Presutti, E.: Upper bounds on the critical temperature for Kac potentials. J. Stat. Phys. **88**, 537–566 (1997)
72. Cassandro, M., Orlandi, E., Picco, P.: Typical configurations for one-dimensional random field Kac model. Ann. Probab. **27**, 1414–1467 (1999). (Correction Ann. Probab. **34**, 1641–1643 (2006))
73. Cassandro, M., Orlandi, E., Picco, P., Vares, M.E.: One-dimensional random field Kac's model: localization of the phases. Electron. J. Probab. **10**, 786–864 (2005)
74. Cassandro, M., Orlandi, E., Picco, P.: Phase transition in the 1d random field Ising model with long range interactions. To be published (2008)
75. Cercignani, C., Illner, R., Pulvirenti, M.: The Mathematical Theory of Dilute Gases. Applied Mathematical Sciences, vol. 106. Springer, New York (1994)
76. Cerf, R., Pisztora, A.: On the Wulff crystal in the Ising model. Ann. Probab. **28**, 947–1017 (2000)

77. Chayes, L.: Mean field analysis of low dimensional systems. Preprint
78. Chayes, J.T., Chayes, L., Kotecký, R.: The analysis of the Widom–Rowlinson model by stochastic geometric methods. Commun. Math. Phys. **172**, 551–569 (1995)
79. Chayes, L., Kotecký, R., Shlosman, S.: Aggregation and intermediate phases in dilute spin systems. Commun. Math. Phys. **171**, 203–232 (1995)
80. Chayes, L., Kotecký, R., Shlosman, S.B.: Staggered phases in diluted systems with continuous spins. Commun. Math. Phys. **189**, 631–640 (1997)
81. Chen, X.: Existence uniqueness and asymptotic stability of traveling waves in nonlocal evolution equations. Adv. Differ. Equ. **2**, 125–160 (1997)
82. Chen, X., Oshita, Y.: Periodicity and uniqueness of global minimizers of an energy functional containing a long range interaction. SIAM J. Math. Anal. **37**, 1299–1332 (2006)
83. Chmaj, A., Ren, X.: Homoclinic solutions of an integral equation: existence and stability. J. Diff. Eq. **155**, 17–43 (1999)
84. Choksi, R.: Scaling laws in microphases separation of diblock copolymers. J. Noninear Sci. **11**, 223–236 (2001)
85. Comets, F.: Nucleation for a long range magnetic model. Ann. Inst. Henri Poincaré, B Probab. Stat. **23**, 135–178 (1987)
86. Comets, F., Eisele, Th., Schatzmann, M.: On secondary bifurcations for some nonlinear convolution equations. Trans. Am. Math. Soc. **296**, 661–702 (1986)
87. Dal Maso, G.: An Introduction to Gamma Convergence. Birkhäuser, Boston (1993)
88. De Masi, A., Gobron, T.: Liquid–vapor interfaces and surface tension in a mesoscopic model of fluid with nonlocal interactions. J. Stat. Phys. **115**, 643–679 (2004)
89. De Masi, A., Presutti, E.: Mathematical Methods for Hydrodynamic Limits. Lecture Notes in Mathematics, vol. 1501. Springer, Berlin (1991)
90. De Masi, A., Kipnis, C., Presutti, E., Saada, E.: Microscopic structure at the shock in the asymmetric simplee exclusion. Stochastics **27**, 151–165 (1989)
91. De Masi, A., Orlandi, E., Presutti, E., Triolo, L.: Motion by curvature by scaling non local evolution equations. J. Stat. Phys. **73**, 543–570 (1993)
92. De Masi, A., Orlandi, E., Presutti, E., Triolo, L.: Uniqueness of the instanton profile and global stability in non local evolution equations. Rend. Mat. **14**, 693–723 (1993)
93. De Masi, A., Orlandi, E., Presutti, E., Triolo, L.: Glauber evolution with Kac potentials I. Mesoscopic and macroscopic limits, interface dynamics. Nonlinearity **7**, 1–67 (1994)
94. De Masi, A., Orlandi, E., Presutti, E., Triolo, L.: Stability of the interface in a model of phase separation. Proc. R. Soc. Edinb. **124**, 1013–1022 (1994)
95. De Masi, A., Pellegrinotti, S., Presutti, E., Vares, M.E.: Spatial patterns when phases separate in an interacting particle system. Ann. Probab. **22**, 334–371 (1994)
96. De Masi, A., Gobron, T., Presutti, E.: Travelling fronts in non local evolution equations. Arch. Ration Mech. Anal. **132**, 143–205 (1995)
97. De Masi, A., Orlandi, E., Presutti, E., Triolo, L.: Glauber evolution with Kac potentials II. Fluctuations. Nonlinearity **9**, 27–51 (1996)
98. De Masi, A., Orlandi, E., Presutti, E., Triolo, L.: Glauber evolution for Kac potentials III. Spinodal decomposition. Nonlinearity **9**, 53–114 (1996)
99. De Masi, A., Olivieri, E., Presutti, E.: Spectral properties of integral operators in problems of interface dynamics and metastability. Markov Processes Relat. Fields **4**, 27–112 (1998)
100. De Masi, A., Olivieri, E., Presutti, E.: Critical droplet for a non-local mean field equation. Markov Processes Relat. Fields **6**, 439–471 (2000)
101. De Masi, A., Dirr, N., Presutti, E.: Instability of interface under forced displacements. Ann. Inst. Henri Poincaré **7**, 471–511 (2006)
102. De Masi, A., Merola, I., Presutti, E., Vignaud, Y.: Potts models in the continuum. Uniqueness and exponential decay in the restricted ensembles. J. Stat. Phys. (2008, to appear)
103. De Masi, A., Merola, I., Presutti, E., Vignaud, Y.: Coexistence of ordered and disordered phases in Potts models in the continuum. Preprint (2008)
104. Dellacherie, C., Meyer, P.-A.: Probabilités et potentiel. Publications de l'Institut de Mathematique de Strasbourg, vol. XV. Hermann, Paris (1975)

105. Dirr, N., Luckhaus, S.: A Stefan problem with surface tension as the sharp interface limit of a non lcoal system of phase field type. J. Stat. Phys. **114**, 1085–1113 (2004)
106. Dobrushin, R.L.: Existence of phase transition in two and three dimensional Ising models. Theory Probab. Appl. **10**, 193–313 (1965)
107. Dobrushin, R.L.: Prescribing a system of random variables by conditional distributions. Theory Probab. Appl. **15**, 458–456 (1970)
108. Dobrushin, R.L.: Investigation of Gibbsian states for three dimensional lattice systems. Theory Probab. Appl. **18**, 253–271 (1973)
109. Dobrushin, R.L., Pechersky, E.A.: A criterion of the uniqueness of Gibbsian fields in the non compact space. Lect. Notes Math. **1021**, 97–110 (1982)
110. Dobrushin, R.L., Shlosman, S.: Completely analitical interactions. J. Stat. Phys. **46**, 983–1014 (1987)
111. Dobrushin, R.L., Zahradnik, M.: Phase diagrams for continuous spin models. Extensions of Pirogov–Sinai theory. In: Dobrushin, R.L. (ed.) Mathematical Problems of Statistical Mechanics and Dynamics. pp. 1–123. Kluwer, Dordrecht (1987).
112. Dobrushin, R.L., Kotecký, R., Shlosman, S.: The Wulff Construction: A Global Shape for Local Interactions. Amer. Math. Soc., Providence (1992)
113. Doob, J.L.: Measure Theory. Graduate Texts in Mathematics, vol. 143. Springer, New York (1994)
114. Dunford, N., Schwartz, J.T.: Linear Operators. Part I. Interscience Publishers, New York (1957)
115. Ermentrout, G.B., McLeod, J.B.: Existence and uniqueness of travelling waves for a neural network. Proc. R. Soc. Edinb. **23A**, 461–478 (1993)
116. Evans, L.C., Gariepy, R.: Measure Theory and Fine Properties of Functions. Studies in Advanced Math. CRC Press, Boca Raton (1992)
117. Evans, L.C., Soner, H.M., Souganidis, P.E.: Phase transitions and generalized motion by mean curvature. Commun. Pure Appl. Math. **XLV**, 1097–1123 (1992)
118. Feller, W.: An Introduction to Probability Theory and Its Applications, vols. 1 and 2. Wiley, New York (1966)
119. Ferrari, P.: Shocks in one dimensional processes with drift. In: Grimmett, G. (ed.) Probability and Phase Transitions. NATO Adv. Sci. Ins. Ser. C, Math. Phys. Sci. (Cambridge 1993), vol. 420, pp. 35–48. Kluwer, Dordrecht (1994)
120. Ferrari, P., Kipnis, C., Saada, E.: Microscopic structure of traveling waves. Ann. Prob. **19**, 226–244 (1991)
121. Fife, P., McLeod, J.B.: A phase plane discussion of convergence to travelling front for non linear diffusion. Arch. Ration Mech. Anal. **75**, 281–314 (1981)
122. Foguel, S.R.: The Ergodic Theory of Markov Processes. Van Nostrand, New York (1969)
123. Fritz, J.: An Introduction to the Theory of Hydrodynamic Limits. Lectures in Mathematical Sciences. The University of Tokyo, Tokyo (2001)
124. Fritz, J., Rüdiger, B.: Time dependent critical fluctuations of a one dimensional local mean field model. Probab. Theory Relat. Fields **103**, 381–407 (1996)
125. Fröhlich, J., Pfister, C.: On the absence of spontaneous symmetry breaking and of crystalline ordering in two-dimensional systems. Commun. Math. Phys. **81**, 277–298 (1981)
126. Funaki, T.: The scaling limit for a stochastic PDE and the separation of phases. Probab. Theory Relat. Fields **102**, 221–288 (1995)
127. Funaki, T., Spohn, H.: Motion by mean curvature from the Ginzburg–Landau $\nabla\phi$ interface model. Commun. Math. Phys. **185**, 1–36 (1997)
128. Gallavotti, G.: Statistical Mechanics, a Short Treatise. Text and Monographs in Physics. Springer, Berlin (1999)
129. Gallavotti, G., Miracle-Sole, S.: Equilibrium states of the Ising model in the two phase region. Phys. Rev. **3B**, 2555–2559 (1972)
130. Gärtner, J., Presutti, E.: Shock fluctuations in a particle system. Ann. Inst. Henri Poincaré **53**, 1–14 (1990)
131. Gates, D.J., Penrose, O.: The van der Waals limit for classical systems. I. A variational principle. Commun. Math. Phys. **15**, 255–276 (1969)

132. Gates, D.J., Penrose, O.: The van der Waals limit for classical systems. II. Existence and continuity of the canonical pressure. Commun. Math. Phys. **16**, 231–237 (1970)
133. Gates, D.J., Penrose, O.: The van der Waals limit for classical systems. III. Deviation from the van der Waals–Maxwell theory. Commun. Math. Phys. **17**, 194–209 (1970)
134. Gayrard, V., Presutti, E., Triolo, L.: Density oscillations at the interface between vapour and liquid. J. Stat. Phys. **108**, 863–884 (2002)
135. Georgii, H.O.: Gibbs Measures and Phase Transitions. de Gruyter Studies in Mathematics, vol. 9. Walter de Gruyter & G., Berlin (1988)
136. Georgii, H.O., Haggstrom, O.: Phase transition in continuum Potts models. Commun. Math. Phys. **181**, 507–528 (1996)
137. Georgii, H.O., Miracle-Sole, S., Ruiz, J., Zagrebnov, V.: Mean field theory of the Potts gas. J. Phys. A **39**, 9045–9053 (2006)
138. Gerardi, A., Marchioro, C., Oliveiri, E., Presutti, E.: van der Waals–Maxwell theory, Lebowitz–Penrose limit and superstable interactions. Commun. Math. Phys. **29**, 219–231 (1973)
139. Giacomin, G.: Phase separation and random domain patterns in a stochastic particle model. Stoch. Process. Their Appl. **51**, 25–624 (1994)
140. Giacomin, G., Lebowitz, J.L., Presutti, E.: Deterministic and stochastic hydrodynamic equations arising from simple microscopic model systems. In: Carmona, R.A., Rozovskii, B. (eds.) Stochastic Partial Differential Equations: Six Perspectives, Mathematical Surveys and Monographs of AMS, vol. 64, pp. 107–149. Amer. Math. Soc., Providence (1999).
141. Giuliani, A., Lebowitz, J.L., Lieb, E.H.: Ising models with long range antiferromagnetic and short range ferromagnetic interactions. Preprint (2006)
142. Gobron, T., Merola, I.: First-order phase transition in Potts models with finite-range interactions. J. Stat. Phys. **126**, 507–583 (2007)
143. Grant, C.P.: Spinodal decomposition for the Cahn–Hilliard equation. Commun. Partial Differ. Equ. **18**, 453–490 (1993)
144. Higuchi, Y.: On the absence of non translationally invariant Gibbs states for the two dimensional Ising system. In: Fritz, J., Lebowitz, J.L., Szasz, D. (eds.) Random Fields. North Holland, Amsterdam (1981)
145. Ioffe, D.: A note on the quantum Widom–Rowlison model. J. Stat. Phys. **106**, 375–384 (2002)
146. Johansson, K.: On separation of phases in one-dimensional gases. Commun. Math. Phys **169**, 521–561 (1995)
147. Kac, M., Uhlenbeck, G., Hemmer, P.C.: On the van der Waals theory of vapor–liquid equilibrium. I. Discussion of a one dimensional model. J. Math. Phys. **4**, 216–228 (1963)
148. Kac, M., Uhlenbeck, G., Hemmer, P.C.: On the van der Waals theory of vapor–liquid equilibrium. II. Discussion of the distribution functions. J. Math. Phys. **4**, 229–247 (1963)
149. Kac, M., Uhlenbeck, G., Hemmer, P.C.: On the van der Waals theory of vapor–liquid equilibrium. III. Discussion of the critical region. J. Math. Phys. **5**, 60–74 (1964)
150. Kato, T.: Perturbation Theory for Linear Operatots. Die Grundleheren der Mathematischen Wissenschaften in Einzeldarstellungen Band, vol. 132. Springer, New York (1966)
151. Katsoulakis, M., Souganidis, P.E.: Interacting particle systems and generalized evolution of fronts. Arch. Ration. Mech. Anal. **127**, 133–157 (1994)
152. Katsoulakis, M.A., Souganidis, E.: Generalized motion by mean curvature as a macroscopic limit of stochastic Ising models with long range interactions and Glauber dynamics. Commun. Math. Phys. **169**, 61–97 (1995)
153. Katsoulakis, M.A., Souganidis, P.E.: Stochastic Ising models and anisotropic front propagation. J. Stat. Phys. **87**, 63–89 (1997)
154. Khinchin, A.I.: Mathematical Foundations of Statistical Mechanics. Dover, New York (1949)
155. Kipnis, C., Landim, C.: Scaling Limits of Interacting Particle Systems. Grund. math. Wissenschaften, vol. 320. Springer, Berlin (1999)
156. Kohn, R.V., Otto, F., Reznikoff, M.G., Vanden-Eijnden, E.: Action minimization and sharp interface limits for the stochastic Allen–Cahn Equation. Commun. Pure Appl. Math. **60**, 393–438 (2007)

157. Kohn, R.V., Reznikoff, M.G., Tonegawa, Y.: The sharp interface limit of the action functional for Allen Cahn in one space dimension. Calc. Var. Partial Diff. Equ. **25**, 503–534 (2006)
158. Kotecký, R., Preiss, D.: Cluster expansion for abstract polymer models. Commun. Math. Phys. **103**, 491–498 (1986)
159. Laanait, L., Messager, A., Miracle-Sole, S., Ruiz, J., Shlosman, S.: Interfaces in Potts model I: Pirogov–Sinai theory of the Fortuin–Kasteleyn representation. Commun. Math. Phys. **140**, 81–91 (1991)
160. Lebowitz, J.L., Penrose, O.: Rigorous treatment of the Van der Waals–Maxwell theory of the liquid vapour transition. J. Math. Phys. **7**, 98–113 (1966)
161. Lebowitz, J., Stell, G., Baer, S., Theumann, W.: Separation of the interaction potential into two parts in Statistical Mechanics—Part I. J. Math. Phys. **6**, 1282 (1965)
162. Lebowitz, J., Stell, G., Baer, S., Theumann, W.: Separation of the interaction potential into two parts in Statistical Mechanics—Part II. J. Math. Phys. **7**, 1532 (1966)
163. Lebowitz, J.L., Orlandi, E., Presutti, E.: A particle model for spinodal decomposition. J. Stat. Phys. **63**, 933–974 (1991)
164. Lebowitz, J.L., Mazel, A., Presutti, E.: Liquid–vapor phase transitions for systems with finite-range interactions. J. Stat. Phys. **94**, 955–1025 (1999)
165. Mac Isaac, A.B., Whitehead, J.P., Robinson, M.C., De'Bell, K.: Striped phases in two dimensional dipolar ferromagnets. Phys. Rev. B. **51**, 16033–16045 (1995)
166. Maier-Paape, S., Wanner, T.: Spinodal decomposition for the Cahn–Hilliard equation in higher dimensions. I. Probability and wavelength estimate. Commun. Math. Phys. **195**, 435–464 (1998)
167. Maier-Paape, S., Wanner, T.: Spinodal decomposition for the Cahn–Hilliard equation in higher dimensions: nonlinear dynamics. Arch. Ration. Mech. Anal. **151**, 187–219 (2000)
168. Maier-Paape, S., Stoth, B., Wanner, T.: Spinodal decomposition for multicomponent Cahn–Hilliard systems. J. Stat. Phys. **98**, 871–896 (2000)
169. Merola, I.: Asymptotic expansion of the pressure in the inverse interaction range. J. Stat. Phys. **95**, 745–758 (1999)
170. Messager, A., Miracle-Sole, S., Ruiz, J., Shlosman, S.: Interfaces in the Potts model II: Antonov's rule and rigidity of the order disorder interface. Commun. Math. Phys. **140**, 275–290 (1991)
171. Meyn, S.P., Tweedie, R.L.: Generalized resolvents and Harris recurrence of Markov processes. In: Doeblin and Modern Probability. Blaubeuren, 1991. Contemp. Math., vol. 149, pp. 227–250. Amer. Math. Soc., Providence (1993).
172. Miracle-Sole, S.: On the convergence of cluster expansions. Physica A **279**, 244–249 (2000)
173. Modica, L., Mortola, S.: Un esempio di Gamma convergenza. Boll. Unione Mat. Ital B(5) **14**, 285–299 (1977)
174. Müller, S.: Singular perturbations as a selection criterion for periodic minimizing sequences. Calc. Var. Partial Differ. Equ. **1**, 169–204 (1993)
175. Naimark, M.A.: Normed Rings. Noordhoff, Groningen (1964)
176. Oh, J., Ahlers, G.: Thermal-noise effect on the transition to Rayleigh–Benard convection. Phys. Rev. Lett. **91**, 094501.1-4 (2003)
177. Olivieri, E., Vares, M.E.: Large Deviations and Metastability. Encyclopedia of Mathematics and its Applications, vol. 100. Cambridge University Press, Cambridge (2005)
178. Orlandi, E., Triolo, L.: Travelling fronts in non local models for phase separation in external field. Proc. R. Soc. Edinb. **127A**, 823–835 (1997)
179. Parthasarathy, K.R.: Probability Meaures on Metric Spaces. Academic, New York (1967)
180. Peierls, R.: On Ising's model of ferromagnetism. Proc. Camb. Philol. Soc. **32**, 477–481 (1936)
181. Penrose, O.: Foundations of Statistical Mechanics. A Deductive Treatment. Dover, New York (2005). (Work originally published by Pergamon Press, Oxford, 1970)
182. Penrose, O., Lebowitz, J.L.: Rigorous treatment of metastable states in the van der Waals–Maxwell theory. J. Stat. Phys. **3**, 211–236 (1971)

183. Penrose, O., Lebowitz, J.L.: Toward a rigorous molecular theory of metastability. In: Montroll, E.W., Lebowitz, J.L. (eds.) Fluctuation Phenomena, 2nd edn. North-Holland, Amstersam (1987).
184. Pfister, C.: Thermodynamical aspects of classical lattice systems. In: Lectures at the IV Brasilian school of Mathematics, Mambucaba, Brasil (2000)
185. Pirogov, S.A., Sinai, Ya.: Phase diagrams of classical lattice systems. Theor. Math. Phys. **25**, 1185–1192 (1975). (Theor. Math. Phys. **25**, 358–369, in Russian)
186. Pirogov, S.A., Sinai, Ya.: Phase diagrams of classical lattice systems: continuation. Theor. Math. Phys. **26**, 39–49 (1976)
187. Presutti, E., Wick, W.D.: Macroscopic fluctuations in a one dimensional mechanical system. J. Stat. Phys. **52**, 497–502 (1988)
188. Revuz, D.: Markov Chains. North Holland, Amsterdam (1975)
189. Revuz, D., Yor, M.: Continuous martingales and Brownian motion. Grundlehner der mathematischen Wiessenschaften, vol. 293. Springer, Berlin (1991)
190. Rezakhanlou, F.: Hydrodynamic limit for a system with finite range interactions. Commun. Math. Phys. **129**, 445–480 (1990)
191. Rogers, R.C.: Nonlocal variational problems in nonlinear electro-magneto elasto-statics. SIAM J. Math. Anal. **19**, 1329–1347 (1988)
192. Rogers, R.C.: A nonlocal model for the exchange energy in ferromagnetic materials. J. Integral Equ. Appl. **3**, 85–127 (1991)
193. Rogers, R.C.: Some remarks on nonlocal interactions and hysterisis in phase transitions. Contin. Mech. Thermodyn. **8**, 65–73 (1996)
194. Rohlin, V.A.: On the fundamental ideas of measure theory. Am. Math. Soc. Transl. Series 1 **10**, 1–52 (1962)
195. Rohlin, V.A.: Lectures on ergodic theory. Am. Math. Soc. Transl. **22**, 1–54 (1962)
196. Rothman, D.H., Zaleski, S.: Lattice-Gas Cellular Automata. Simple Models of Complex Hydrodynamics. Collection Aléa-Saclay: Monographs and Texts in Statistical Physics, vol. 5. Cambridge University Press, Cambridge (1997)
197. Ruelle, D.: Statistical Mechanics: Rigorous Results. W.A. Benjamin, New York (1969)
198. Ruelle, D.: Existence of a phase transition in a continuous classical system. Phys. Rev. Lett. **27**, 1040–1041 (1971)
199. Sewell, G.L.: Quantum Mechanics and Its Emergent Macrophysics. Princeton University Press, Princeton (2002)
200. Simon, B.: The Statistical Mechanics of Lattice Gases, vol. 1. Princeton University Press, Princeton (1993)
201. Sinai, Ya.G.: Theory of Phase Transitions: Rigorous Results. Akademiai Kiadó, Budapest (1982)
202. Spohn, H.: Large Scale Dynamics of Interacting Particle System. Texts and Monographs in Physics. Springer, Berlin (1991)
203. Spohn, H.: Interface motion in models with stochastic dynamics. J. Stat. Phys. **71**, 1081–1133 (1993)
204. Theil, F.: A proof of crystallization in two dimensions. Commun. Math. Phys. **262**, 209–236 (2006)
205. Triolo, L.: Space structures and different scales for many-component biosystems. Markov Processes Relat. Fields **11**, 389–404 (2005)
206. van den Berg, J.: A uniqueness condition for Gibbs measures with applications to the two dimensional Ising antiferromagnet. Commun. Math. Phys. **152**, 161–166 (1993)
207. Varhadan, S.R.S., Yau, H.T.: Diffusive limit of lattice gases with mixing conditions. Asian J. Math. **1**, 623–678 (1997)
208. Villani, C.: Optimal transportation, dissipative PDE's and functional inequalities. Lecture Notes in Math., vol. 1813. Springer, Berlin (2003)
209. Zahradnik, M.: An alternate version of Pirogov–Sinai theory. Commun. Math. Phys. **93**, 559–581 (1984)
210. Zahradnik, M.: A short course on the Pirogov–Sinai theory. Rend. Mat. Appl. **18**, 411–486 (1998)

Index

Theoretical and Mathematical Physics

Scaling Limits in Statistical Mechanics and Microstructures in Continuum Mechanics
By E. Presutti

Statistical Methods in Quantum Optics 2
Non-Classical Fields
By H. J. Carmichael

An Introduction to Riemann Surfaces, Algebraic Curves and Moduli Spaces
2nd Edition
By M. Schlichenmaier

Concepts and Results in Chaotic Dynamics: A Short Course
By P. Collet and J.-P. Eckmann

The Theory of Quark and Gluon Interactions
4th Edition
By F. J. Ynduráin

From Microphysics to Macrophysics
Methods and Applications of Statistical Physics
Volume I
By R. Balian

From Microphysics to Macrophysics
Methods and Applications of Statistical Physics
Volume II
By R. Balian

Titles published before 2006 in *Texts and Monographs in Physics*

The Statistical Mechanics of Financial Markets
3rd Edition
By J. Voit

Magnetic Monopoles
By Y. Shnir

Coherent Dynamics of Complex Quantum Systems
By V. M. Akulin

Geometric Optics on Phase Space
By K. B. Wolf

General Relativity
By N. Straumann

Quantum Entropy and Its Use
By M. Ohya and D. Petz

Statistical Methods in Quantum Optics 1
By H. J. Carmichael

Operator Algebras and Quantum Statistical Mechanics 1
By O. Bratteli and D. W. Robinson

Operator Algebras and Quantum Statistical Mechanics 2
By O. Bratteli and D. W. Robinson

Aspects of Ergodic, Qualitative and Statistical Theory of Motion
By G. Gallavotti, F. Bonetto and G. Gentile

The Frenkel-Kontorova Model
Concepts, Methods, and Applications
By O. M. Braun and Y. S. Kivshar

The Atomic Nucleus as a Relativistic System
By L. N. Savushkin and H. Toki

The Geometric Phase in Quantum Systems
Foundations, Mathematical Concepts, and Applications in Molecular and Condensed Matter Physics
By A. Bohm, A. Mostafazadeh, H. Koizumi, Q. Niu and J. Zwanziger

Relativistic Quantum Mechanics
2nd Edition
By H. M. Pilkuhn

Physics of Neutrinos and Applications to Astrophysics
By M. Fukugita and T. Yanagida

High-Energy Particle Diffraction
By E. Barone and V. Predazzi

Foundations of Fluid Dynamics
By G. Gallavotti

Many-Body Problems and Quantum Field Theory An Introduction
2nd Edition
By Ph. A. Martin, F. Rothen, S. Goldfarb and S. Leach

Statistical Physics of Fluids
Basic Concepts and Applications
By V. I. Kalikmanov

Statistical Mechanics A Short Treatise
By G. Gallavotti

Quantum Non-linear Sigma Models
From Quantum Field Theory
to Supersymmetry, Conformal Field Theory,
Black Holes and Strings
By S. V. Ketov

Perturbative Quantum Electrodynamics and Axiomatic Field Theory
By O. Steinmann

The Nuclear Many-Body Problem
By P. Ring and P. Schuck

Magnetism and Superconductivity
By L.-P. Lévy

Information Theory and Quantum Physics
Physical Foundations for Understanding
the Conscious Process
By H. S. Green

Quantum Field Theory in Strongly Correlated Electronic Systems
By N. Nagaosa

Quantum Field Theory in Condensed Matter Physics
By N. Nagaosa

Conformal Invariance and Critical Phenomena
By M. Henkel

Statistical Mechanics of Lattice Systems
Volume 1: Closed-Form and Exact Solutions
2nd Edition
By D. A. Lavis and G. M. Bell

Statistical Mechanics of Lattice Systems
Volume 2: Exact, Series
and Renormalization Group Methods
By D. A. Lavis and G. M. Bell

Fields, Symmetries, and Quarks
2nd Edition
By U. Mosel

Renormalization An Introduction
By M. Salmhofer

Multi-Hamiltonian Theory of Dynamical Systems
By M. Błaszak

Quantum Groups and Their Representations
By A. Klimyk and K. Schmüdgen

Quantum The Quantum Theory of Particles, Fields, and Cosmology
By E. Elbaz

Effective Lagrangians for the Standard Model
By A. Dobado, A. Gómez-Nicola,
A. L. Maroto and J. R. Peláez

Scattering Theory of Classical and Quantum N-Particle Systems
By J. Derezinski and C. Gérard

Quantum Relativity A Synthesis
of the Ideas of Einstein and Heisenberg
By D. R. Finkelstein

The Mechanics and Thermodynamics of Continuous Media
By M. Šilhavý

Local Quantum Physics Fields, Particles, Algebras
2nd Edition
By R. Haag

Relativistic Quantum Mechanics and Introduction to Field Theory
By F. J. Ynduráin

Supersymmetric Methods in Quantum and Statistical Physics
By G. Junker

Path Integral Approach to Quantum Physics An Introduction
2nd printing
By G. Roepstorff

Finite Quantum Electrodynamics
The Causal Approach
2nd Edition
By G. Scharf

From Electrostatics to Optics
A Concise Electrodynamics Course
By G. Scharf

Geometry of the Standard Model of Elementary Particles
By A. Derdzinski

Quantum Mechanics II
By A. Galindo and P. Pascual

Generalized Coherent States and Their Applications
By A. Perelomov

The Elements of Mechanics
By G. Gallavotti